Cellular Radio and Personal Communications

Selected Readings

Edited by

Dr. Theodore S. Rappaport
Mobile and Portable Radio Research Group
Bradley Department of Electrical Engineering
Virginia Polytechnic Institute and State University
Blacksburg, Virginia

TABLE OF CONTENTS

Introduction 1

Chapter 1 Introduction to Mobile Radio Systems 3

Mobile Radio—an Overview
Andy D. Kucar 5

Trends in Cellular and Cordless Communications
David J. Goodman 19

The Wireless Revolution
Theodore S. Rappaport 31

Chapter 2 Cellular Concept—System Design Fundamentals 43

Cellular Mobile Radio—An Emerging Technology
John Oeting 45

Elements of Cellular Mobile Radio Systems
William C. Y. Lee 51

Handover and Channel Assignment in Mobile Cellular Networks
Sirin Tekinay and Bijan Jabbari 61

Smaller Cells for Greater Performance
William C. Y. Lee 67

Chapter 3 Mobile Radio Propagation 73

Propagation Models and Evaluation of Current Models
IEEE Vehicular Technology Committee on Radio Propagation 75

Indoor Radio Communication for Factories of the Future
Theodore S. Rappaport 117

914 MHz Path Loss Prediction Models for Indoor Wireless
Communications in Multifloored Buildings
Scott Y. Seidel and Theodore S. Rappaport 127

Properties of Mobile Radio Propagation above 400 MHz
Doug O. Reudink 139

Chapter 4 Modulation Techniques 157

The Choice of a Digital Modulation Scheme in a Mobile Radio System
Pritpal S. Mundra, T.L. Singhal, Rakesh Kapur 159

Modulation Methods for PCNs
William T. Webb 163

Defining, Designing, and Evaluating Digital Communication Systems
Bernard Sklar 169

GMSK Modulation for Digital Mobile Radio Telephony
Kazuaki Murota, Kenkichi Hirade 179

Modems for Emerging Digital Cellular Mobile Radio Systems
Kamilo Feher **187**

Spread Spectrum for Mobile Communications
Raymond L. Pickholtz, Laurence B. Milstein, David L. Schilling **199**

Bit Error Simulation for π/4 DQPSK Mobile Radio Communications
Using Two-Ray and Measurement-Based Impulse Response Models
Victor Fung, Theodore S. Rappaport, Berthold Thoma **209**

Chapter 5   Channel Coding, Equalization and Diversity **223**

The Application of Error Control to Communications
Elwyn R. Berlekemp, Robert E. Peile, Stephen P. Pope **225**

Adaptive Equalization
Shahid U.H. Qureshi **239**

Adaptive Equalization for TDMA Digital Mobile Radio
John G. Proakis **279**

Diversity Techniques for Mobile Radio Reception
John D. Parsons, Miguel Henze, P.A. Ratliff,
Michael J. Withers **289**

A Comparison of Specific Space Diversity Techniques for Reduction
of Fast Fading in UHF Mobile Radio Systems
William C. Jakes, Jr. **301**

Chapter 6   Speech Coding **313**

Speech Codecs for Personal Communications
Raymond Steele **315**

Coding Speech at Low Bit-rates
Nuggehally S. Jayant **323**

Speech Coding
James L. Flanagan, Manfred R. Schroeder, Bishnu S. Atal,
Ronald E. Crochiere, Nuggehally S. Jayant, Jose M. Tribolet **329**

Signal Compression: Technology, Targets, and Research Directions
Nuggehally S. Jayant **357**

Chapter 7   Multiple Access Techniques **381**

Spectrum Efficiency in Cellular
William C.Y. Lee **383**

Capacity of Digital Cellular TDMA Systems
Krister Raith and Jan Uddenfeldt **391**

On the Capacity of a Cellular CDMA System
Klein S. Gilhousen, Irwin M. Jacobs, Roberto Padovani,
Andrew J. Viterbi, Lindsay A. Weaver, Charles E. Wheatley III **401**

Advantages of CDMA and Spread Spectrum Techniques over
FDMA and TDMA in Cellular Mobile Radio Applications
Peter Jung, Paul W. Baier, Andreas Steil **411**

A Comparison of CDMA and TDMA Systems
Bjorn Gudmundson, Johan Skold, Jon K. Ugland **419**

Chapter 8 Cellular Networking **423**

An Overview of Signaling System No. 7
Abdi R. Modarressi and Ronald A. Skoog **425**

The European (R)evolution of Wireless Digital Networks
Jens C. Arnbak **443**

Cellular Packet Communications
David J. Goodman **453**

Chapter 9 Cellular Systems and Standards **463**

Wireless Network Access for Personal Communications
Donald C. Cox **465**

What are GSM and DCS?
C. Dechaux, R. Scheller **485**

DECT, a universal cordless access system
R.J. Mulder **495**

The CDMA Standard
David P. Whipple **501**

Overview of Cellular CDMA
William C. Y. Lee **511**

INTRODUCTION

The cellular and personal communications field is undergoing enormous growth. Just in the past few years, cellular telecommunications and paging companies have seen the number of subscribers increase at rates of 40% or more. All indications are that this growth will continue at an even more rapid pace as wireless communications becomes ubiquitous throughout the world. Following on the heels of the personal computer revolution of the 1980s, wireless communications appears to be a technology that will shape our society. Many believe that the merging of wireless communications and personal computers will redefine the way in which mankind works and recreates in the 21st century.

Although most cellular radio systems have only been in existence since the early 1980s, they enjoy population penetration rates that are approaching 10%. The demand for engineers and scientists with knowledge or experience in this area is great, and job opportunities abound in all aspects of the wireless communications industry. New technologies and communications services being deployed throughout the world are now promising a new era of wireless telecommunications known as *Personal Communications*. Personal communications uses many of the same ideas developed for cellular radio systems, but provides greater capacity, functionality, and flexibility through the advances of digital modulation, speech coding, multiple access techniques, system design, and networking. All of these areas, incidentally, are covered in this book.

Advances in the field of wireless communications are happening daily. It is difficult for the active researcher, let alone the practicing engineer, to keep up with the latest in theory and techniques that are propelling the age of wireless communications. It *is* possible, however, to identify some key technical papers which provide fundamental insight into the principles and practices of today's wireless personal communication systems, and this is what we have attempted to do. These papers are an excellent collection of references for those interested in wireless telecommunications, whether as a student, a practicing engineer, or researcher.

Many of these papers have been used to teach a first year graduate course called Cellular Radio and Personal Communications on the campus of Virginia Tech. With the assistance of Rias Muhamed and Varun Kapoor, two graduate students who have worked in industry and who appreciate the difficulties of learning a subject in the absence of a text, we have carefully selected papers which are particularly well written and may serve as stand-alone tutorials. All articles are being reprinted by permission of the publishers, and we thank them for their cooperation.

The suggestions, advice, and experience of Mr. Muhamed and Mr. Kapoor, along with input from IEEE reviewers and over 100 students who have completed the cellular course at Virginia Tech have been incorporated to form this book. It is our hope that you will find it useful as a single source of reference for independent learning as well as for classroom use, as you pursue the exciting field of wireless personal communications.

THEODORE S. RAPPAPORT
MOBILE & PORTABLE RADIO RESEARCH GROUP
BRADLEY DEPARTMENT OF ELECTRICAL ENGINEERING
VIRGINIA POLYTECHNIC INSTITUTE AND STATE UNIVERSITY
BLACKSBURG, VIRGINIA

Chapter 1

INTRODUCTION TO MOBILE RADIO SYSTEMS

Mobile Radio: An Overview

Technology is giving us impressive communications, information, and navigation systems. The social and economic impact will be substantial.

Andy D. Kucar

Reprinted from *IEEE Communications Magazine*, pp. 72-85, Nov., 1991.

T he focus of this presentation is on terrestrial and satellite mobile radio communications. This includes cellular radio systems such as existing North American AMPS, Japanese MCS, Scandinavian NMT and British TACS, and the proposed GSM, Digital AMPS, and spread spectrum CDMA; cordless telephony systems such as existing CT1, CT2 and the proposed CT2Plus, CT3, and DECT; mobile radio data systems such as ARDIS and RAM; projects known as PCN, PCS, and FPLMTS; satellite mobile radio systems such as existing INMARSAT and OmniTRACS and the proposed INMARSAT X, MSAT, Iridium, Globalstar, and ORBCOMM.

Following a brief prologue and historical overview, the paper discusses such technical issues as the repertoire of systems and services, management of the airwaves, the operating environment, service quality, network issues and cell size, channel coding and modulation, speech coding, diversity, multiplex and multiple access (FDMA, TDMA, CDMA).

Also addressed are the potential economic and sociological impacts of mobile radio communications in the wake of the redistribution of airwaves at the World Administrative Radio Conference WARC '92. Most existing mobile radio communications systems collect the information on network behavior, users' positions, etc., with the purpose of enhancing the performance of communications, improving handover procedures and increasing the system capacity. Coarse positioning usually is achieved inherently, while more precise navigation can be achieved by employing LORAN-C and/or GPS signals, or some other means, at the marginal expense in cost and complexity.

The traffic load peaks of many mobile radio communications systems coincide in time and space with vehicular traffic congestion and traffic accidents. It might be expected that improved traffic management provided by vehicle information systems would improve the traffic safety, relieve the thirst for airwaves, and enhance the performance of mobile radio communications systems. Recent developments related to these topics are briefly described.

Andy Kucar is President of 4U Communications Research.

Prologue

Mobile radio systems provide their users with opportunities to travel freely within the service area and simultaneously communicate with any telephone, fax, data modem, and electronic mail subscriber anywhere in the world. These systems allow users to determine their own positions; to track the precious cargo; to improve the management of fleets of vehicles and the distribution of goods; to improve traffic safety; to provide vital communication links during emergencies, search and rescue operations, etc. These tieless (wireless, cordless) communications, the exchange of information, determination of position, course and distance traveled are made possible by the unique property of the radio to employ an aerial (antenna) for radiating and receiving electromagnetic waves.

When the user's radio antenna is stationary over a prolonged period of time, the term fixed radio is used. A radio transceiver capable of being carried or moved around, but stationary when in operation, is called a portable radio. A radio transceiver capable of being carried and used by a vehicle or by a person on the move is called mobile radio.

Individual radio users may communicate directly or via one or more intermediaries, which may be passive radio repeater(s), base station(s) or switch(es). When all intermediaries are located on the Earth, the terms terrestrial radio system and radio system have been used. When at least one intermediary is satellite-borne, the terms satellite radio system and satellite system have been used. According to the location of a user, the terms land, maritime, aeronautical, space, and deep-space radio systems have been used.

The second unique property of all terrestrial and satellite radio systems is that they all share the same natural resource: the airwaves (frequency bands and the space).

Recent developments in microwave monolithic integrated circuit (MMIC), application specific integrated circuit (ASIC), analog/digital signal processing (A/DSP) and battery technology, supported by computer aided design (CAD) and robotic manufacturing, allow a viable implemen-

 0163-6804/91/$01.00 1991© IEEE

5

1908	Public radio telephone between ships and the land in Japan was established.
1921	Police car radio dispatch service was introduced in Detroit USA Police Department.
1945	During WW II, significant progresses in design and widespread use of mobile radio were made.
1958	LORAN-C commercial operation started. The initial development was started during WW II.
1964	Railway Telephone service on Japanese Tokaido bullet train was introduced.
1968	Carterphone Decision. FCC allows non-Bell equipment to be connected to (Bell) network.
1971	Fully automatic radiotelephone system, the B network, was introduced in West Germany. Later extended to the corresponding networks in Austria, Luxemburg, and the Netherlands.
1974	US Federal Communications Commission allocated 40 MHz frequency band, paving the way for estabilishing what is now known as advanced Mobile Phone Service (AMPS).
1976	MARISAT consortium initiated commercial service for mobile maritime users, providing full duplex voice, data, and teleprinter services, worldwide.
1979	Mobile communications systems MCS-L1 introduced by NTT Japan.
1982	The Conference of European Postal and Telecommunications Administrations (CEPT) established Groupe Special Mobile (GSM) with the mandate to define future Pan-European cellular radio standard.
1982	INMARSAT began providing similar services as MARISAT.
1982	Cospas - 1 inclined orbit satellite was launched, with a search and rescue package compatible with Future Global Maritime Distress and Safety Sysem (FGMDSS) on board.
1983	SARSAT search and rescue instrument package was placed on board of U.S. National Oceanic and Atmospheric Administration satellite NOAA-8 and launched.
1984	January 01, divestiture (breakup) of AT&T.
1984	The first interagency tests of Global Positioning System (GPS) receivers conducted in California.
1985	Total Access Communications System (TACS) was introduced in UK.
1985	CD900 cellular mobile radio system was introduced in West Germany.
1987	Japan launched its own experimental satellite *ETS - V* supported by extensive study and experimental work.
1988	Geostar introduced its Link One radiodetermination services. The radiodetermination information is obtained from a LORAN-C receiver and sent over a L-band satellite payload toward the ground central station.
1988	Qualcomm/Omninet started its two way data communication and radiodetermination (using a LORAN-C receiver) OmniTRACS services.
1988	The second high capacity land mobile communications system (MCS-L2) was introduced in Japan.
1990	Pegasus rocket has been launched from the wing of a B-52; the rocket injected its 423 lb payload into a 273 x 370 nm 94° inclined orbit.
	• After almost two decades of studies and experiments, sponsored by Canadian and U.S. tax payers, North American mobile satellite systems MSAT is entering its realization stage,
	• The European community is sponsoring studies and experiments for Pan-European mobile satellite systems PROSAT and PRODAT.
	• Future Australian AUSSAT satellites will include, among other services, L band payloads for mobile communicaitons and receivers for radiodetermination services.
	• Experimental field trials of CT2, CT3, DECT, GSM, CDMA, TDMA, FDMA mobile radio communications systems in progress, worldwide.

Table 1 A summary of events related to the mobile radio communications

6

Today, cordless (wireless, fiberless) radio systems offer telepoint services similar in scope to those provided by the public telephone.

tation of miniature radio transceivers. The continuous flux of market forces (excited by the possibilities of a myriad of new services and great profits), international and domestic standard forces (who manage common natural resource: the airwaves), and technology forces (capable to create viable products) acted harmoniously and created a broad choice of communications (voice and data), information, and navigation systems, which propelled an explosive growth of mobile radio services for travelers.

Is space the limit?

A Glimpse of History

Many things have an epoch, in which they are found at the same time in several places, just as the violets appear on every side in spring.
Farkas Wolfgang Bolyai, in 1823.

Late in the nineteenth century Heinrich Rudolf Hertz, Nikola Tesla, Guglielmo Marconi, and other scientists experimented with the transmission and reception of electromagnetic waves. The birth of mobile radio generally is accepted to have occurred in 1897, when Marconi was credited with the patent for wireless telegraph. Since that time, mobile radio communications have provided safe navigation for ships and airplanes, saved many lives, dispatched diverse fleets of vehicles, won battles, generated new businesses, etc. A summary of some of the key historical developments related to commercial mobile radio communications is provided in Table 1.

Satellite mobile radio systems launched in the 1970s and early 1980s use UHF frequency bands of approximately 400 MHz, with bands of approximately 1.5 GHz used for communications and navigation services. In the 1950s and 1960s, numerous private mobile radio networks, citizen band (CB) mobile radio, ham operator mobile radio, and portable home radio telephones used diverse types and brands of radio equipment and chunks of airwaves located anywhere in the frequency band from near 30 MHz to 3 GHz.

In the 1970s, Ericsson introduced the Nordic Mobile Telephone (NMT) system, and AT&T Bell Laboratories introduced Advanced Mobile Phone Service (AMPS). The impact of these two public land mobile telecommunication systems on the standardization and prospects of mobile radio communications may be compared with the impact of Apple and IBM on the personal computer industry. In Europe systems resembling AMPS competed with NMT systems; in the rest of the world, AMPS, backed by Bell Laboratories' reputation for technical excellence and the clout of AT&T, became de facto and de jure the technical standard. (The British TACS and the Japanese MCS-L1 are based on AMPS.)

In 1982, the Conference of European Postal and Telecommunications Administrations (CEPT) established Groupe Special Mobile (GSM) with the mandate to define future Pan-European cellular radio standards. On January 1, 1984, during the phase of explosive growth of AMPS and similar cellular mobile radio communications systems and services, the divestiture (breakup) of AT&T occurred.

Le roi est mort, vive le roi.

Panta Rhei

Based on the solid foundation established in the BD (before divestiture) era, the buildup of mobile radio systems and services in the AD (after divestiture) era is continuing at a 20 percent to 50 percent rate per year worldwide. Terrestrial mobile radio systems offer analog voice and low-to medium-rate data services compatible with existing public switching telephone networks in scope, but with poorer voice quality and lower data throughput. Satellite mobile radio systems currently offer analog voice, low- to medium-rate data, radiodetermination and global distress safety services for travelers. By the end of 1988 (1990) there were approximately 2 (4.5) million cellular telephones in North America, and an additional 2 (4.5) million in the rest of the world. There are approximately 20 million cordless phones and about nine million pagers in North America, and about the same number in the rest of the world. Considerable progress has been made in recent years.

Equipment miniaturization and price are important constraints on the systems providing these services. In the early 1950s mobile radio equipment used a considerable amount of a car's trunk space and challenged the capacity of a car's alternator/battery source while in transmit mode; today, the pocket-sized (7.7 ounces or 218 grams) handheld cellular radio telephone provides 45 minutes of talk capacity. The average cost of the least expensive models of battery-powered cellular mobile radio telephones has dropped proportionally, and has broken the $500 barrier.

There is a rapidly expanding market for portable communications, primarily devoted to the indoor (in building, around building) environment. Today, these cordless (wireless, fiberless) radio systems offer telepoint services similar in scope to those provided by the public telephone booths. Their objectives are to provide a broad range of services similar to those currently offered by the Public Switched Telephone Network (PSTN) and the planned Integrated Services Digital Network (ISDN).

Mobile satellite systems are expanding in many directions: large and powerful single unit geostationary systems; medium-sized, low orbit multisatellite systems; and small-sized, low orbit multisatellite systems, launched from a plane. The growth and profit potentials of the mobile radio communications market attracted the major manufacturers in the areas of networks, systems, and switching. This caused profound changes in research and development, standardization, and the decision-making processes in the mobile radio communications industry. In the search for El Dorado the mobile radio communications industry is following two main paths: terrestrial and satellite. The terrestrial mobile radio pioneers, now accompanied by large marketing teams, favor existing cellular radio systems concepts. The newcomers with telephony, switching, and software backgrounds promote cordless telephony (CT), personal communications networks (PCN), and personal communications systems (PCS). Those individuals with a background in administration promote future public land mobile telecommunications systems (FPLMTS) concepts. The satellite mobile radio pioneers build on existing and new geostationary satellite systems, while the newcomers promote inclined orbit concepts. The promoters of each concept may further be subdivided into analog and digital; frequency division multiple access (FDMA), time division multiple access (TDMA), and spread spectrum code division multiple access (CDMA), etc.

Omnia mutantur, nos et mutamur in illis.

Repertoire of Systems and Services

The variety of services offered to travelers essentially consists of information in analog and/or digital form. Although most of today's traffic consists of analog voice transmitted by analog frequency codulation FM (or phase modulation PM), digital signaling and a combination of analog and digital traffic might provide superior frequency re-use capacity, processing, and network interconnectivity. By using a powerful and affordable microprocessor and digital signal processing chips, a myriad of different services particularly well-suited to the needs of people on the move could be realized economically. A brief description of a few elementary systems/services currently available to travelers will follow. Some of these elementary services can be combined within the mobile radio units for a marginal increase in cost and complexity compared with the cost of a single service system. For example, a mobile radio communications system can include a positioning receiver, digital map, etc.

Terrestrial Systems

In a terrestrial mobile radio network, a repeater usually was located at the nearest summit offering maximum service area coverage. As the number of users increased, the available frequency spectrum became unable to handle the increased traffic, and a need for frequency re-use arose. The service area was split into many small sub-areas called cells, and the term cellular radio was born. The frequency re-use offers an increased system capacity, while the smaller cell size can offer an increased service quality, but at the expense of increased complexity of the user's terminal and network infrastructure. The trade-offs between real estate availability (base stations) and cost, the price of equipment (base and mobile), network complexity and implementation dynamics dictate the shape and size of cellular network.

Satellite systems employ one or more satellites to serve as base station(s) and/or repeater(s) in a mobile radio network. The position of satellites relative to the service area is of crucial importance for the coverage, service quality, price, and complexity of the overall network.

When a satellite circles the Earth in 24-hour periods, the term geosynchronous orbit has been used. An orbit that is inclined with respect to the equatorial plane is called an inclined orbit. An orbit with a 90° inclination is called a polar orbit. A circular geosynchronous (24-hour) orbit over the equatorial plane (0° inclination) is known as geostationary orbit (since from any point at the surface of the Earth the satellite appears to be stationary). This orbit is particularly suitable for the land mobile services at low latitudes, and for maritime and aeronautical services at latitudes less than 80°.

Systems that use geostationary satellites include INMARSAT, MSAT, and AUSSAT. An elliptical geosynchronous orbit with the inclination angle of 63.4° is known as Tundra orbit. An elliptical 12-hour orbit with the inclination angle of 63.4° is known as Molniya orbit.

Both Tundra and Molniya orbits have been selected for the coverage of the northern latitudes and the area around the North Pole. For users at those latitudes the satellites appear to wander around the zenith for a prolonged period of time. The coverage of a particular region (regional coverage) and the entire globe (global coverage) can be provided by different constellations of satellites, including those in inclined and polar orbits. For example, inclined circular orbit constellations have been proposed for GPS (18-24 satellites, 55°-63° inclination), Globalstar (48 satellites, 47° inclination), and Iridium (77 satellites, 90° inclination, polar orbits) systems. All three systems will provide the global coverage. ORBCOM system employs Pegasus launchable low-orbit satellites to provide uninterrupted coverage of the Earth below +60° latitudes, and an intermittent but frequent coverage over the polar regions.

Satellite antenna systems can have one (single-beam global system) or more beams (multibeam spot system). The multibeam satellite system, similar to the terrestrial cellular system, employs antenna directivity to achieve better frequency re-use, at the expense of system complexity.

Radio paging is a non-speech, one-way (from base station toward travelers) personal selective calling system with alert, without message or with defined message such as numeric or alphanumeric. A person wishing to send a message contacts a system operator by public switched telephone network (PSTN), and delivers his message. After an acceptable time (queueing delay), a system operator forwards the message to the traveler by radio repeater (FM broadcasting transmitter, VHF or UHF dedicated transmitter, satellite, cellular radio system). After receiving the message, a traveler's small (roughly the size of a cigarette pack) receiver (pager) stores the message into its memory, and on demand either emits alerting tones or displays the message. Examples include the Swedish system, which uses a 57 kHz subcarrier on FM broadcasting transmitters; United States systems that employ 150 MHz, 450 MHz, and 800 MHz mobile radio frequencies; the RPCl system used in the United Kingdom, United States, Australia, New Zealand, the People's Republic of China, and Finland, which employs 150 MHz mobile radio frequencies; and the Japanese system that operates at approximately 250 MHz.

Global Distress Safety System (GDSS) geostationary and inclined orbit satellites transfer emergency calls sent by vehicles to the central earth station. Examples are: COSPAS, Search And Rescue Satellite Aided Tracking system (SARSAT), Geostationary Operational Environmental Satellites (GOES), and SEarch and REscue Satellite (SERES). The recommended frequency for this transmission is 406.025 MHz.

Global Positioning System (GPS)

United States Department of Defense Navstar GPS 18-24 planned satellites in inclined orbits emit L band (L1 = 1575.42 MHz, L2 = 1227.6 MHz)

By using a powerful and affordable micro-processor and digital signal processing chips, a variety of services can be offered economically to people on the move.

8

Parameter	US	Sweden	Japan	Australia
TX freq. band, MHZ				
base	935-940 851-866	76.0-77.5	850-860	865.00-870.00 415.55-418.05
mobile	896-901 806-821	81.0-82.5	905-915	820.00-825.00 406.10-408.60
Duplexing method	FDD/semi, full	FDD/semi	FDD/semi	FDD/semi, full
RF channel bw, kHz	12.5 25.0	25.0	12.5	25.0 12.5
RF channel rate, kb/s	≤9.6	1.2	1.2	≤9.6
Number of traffic ch.	400 600	60 ?	799	200
Modulation type: voice	FM	FM	FM	FM
data	FSK	MSK-FM	MSK-FM	FSK

Table 2. Comparison of dispatch systems

spread spectrum signals from which an intelligent microprocessor-based receiver will be able to extract extremely precise time and frequency information, and accurately determine its own three-dimensional position, velocity, and acceleration worldwide. The coarse accuracy of <100 m available to commercial users has been demonstrated by using a hand-held receiver. An accuracy of meters or centimeters is possible by using the precise (military) codes and/or differential GPS (additional reference) principles. Glonass is the USSR's counterpart of the United State's GPS. Similar systems have been studied by the European Space Agency (Navsat) and by West Germany (Granas, Popsat, and Navcom).

Loran C is the 100 kHz frequency navigation system that provides a positional accuracy between 10 m and 150 m. A user's receiver measures the time difference between the master station transmitter and secondary stations signals, and defines his hyperbolic line of position. North American Loran C coverage includes the Great Lakes and the Atlantic and Pacific coasts, with decreasing signal strength and accuracy as the user approaches the Rocky Mountains from the East. Similar radio-navigation systems are the 100 kHz Decca and 10 kHz Omega.

RadioDetermination Satellite Service (RDSS) uses L (1610.0 MHz to 1626.5 MHz) and S (2483.5 MHz to 2500.0 MHz) band spread spectrum code division multiple access (CDMA) signals translated by dislocated satellites to provide radiodetermination services on a primary basis and any associated nonvoice data services on an ancillary basis. Both the United States GEOSTAR (which covers North America) and the future French CNES LOC-STAR system (which covers Europe, Africa, and the Middle-East) are technically capable of providing RDSS, digital voice, and data.

The Inmarsat communications system consists of three operational geostationary payloads located at 26° W (Atlantic Ocean), 63° E (Indian Ocean), and 180° W (Pacific Ocean). The Standard A L-band system, by employing a 0.79 m to 1.95 m diameter pointing antenna and approximately 200 kg of above-/below-deck equipment, can provide analog voice telephony, telex, facsimile, up to 56 kb/s data, group call broadcasting, and emergency calls to maritime users. The Standard B system will provide digital voice (about 9.6 kb/s), data and telex services by employing smaller equipment then Standard A. The Standard C system, which employs a small antenna (about the size of a half liter can) and a small transceiver (roughly the size of a

Parameter	AMPS	MCS-L1 MCS-L2	NMT	C450	TACS	GSM	PCN	IS-54
TX freq., MHz:								
base	869-894	870-885	935-960	461-466	935-960	890-915	1710-1785	869-894
mobile	824-849	925-940	890-915	451-456	890-915	935-960	1805-1880	824-849
Multiple access	FDMA	FDMA	FDMA	FDMA	FDMA	TDMA	TDMA	TDMA
Duplexing method	FDD	FDD	FDD	FDD	FDD	FDD	FDD	FDD
Channel bw, kHZ	30.0	25.0 12.5	12.5	20.0 10.0	25.0	200.0	200.0	30.0
Traffic channels per RF channel	1	1	1	1		8	16	3
Total traffic ch.	832	600 1200	1999	222 444	1000	125 x 8	375 x 16	832 x 3
Voice: syllabic comp.	analog 2:1	analog 2:1	analog 2:1	analog 2:1	analog 2:1	RELP	RELP	VSELP
speech rate, kb/s						13.0	6.7	8.0
modulation	PM	PM	PM	PM	PM	GMSK	GMSK	π/4
peak dev., kHz	±12	±5	±5	±4	±9.5			
ch. rate, kb/s	-	-	-	-	-	270.8	270.8	48.6
Control: modulation	digital FSK	digital FSK	digital FFSK	digital FSK	digital FSK	digital GMSK	digital GMSK	digital π/4
bb waveform	Manch.	Manch.	NRZ	NRZ	Manch.	NRZ	NRZ	NRZ
peak dev., kHZ	±8	±4.5	±3.5	±2.5	±6.4			
ch. rate, kb/s	10.0	0.3	1.2	5.3	8.0	270.8	270.8	48.6
Channel coding base→mobile	BCH (40,28)	BCH (43,31)	B1 burst	BCH (15,7)	BCH (40,28)	RS (12,8)	RS (12,8)	Conv. 1/2
mobile→base	(48,36)	a.(43,31) p.(11,07)	burst	(15,7)	(48,36)	(12,8)	(12,8)	1/2

Table 3. Comparison of cellular mobile radio systems

9

Parameter	CT2Plus	CT3	DECT	CDMA
Multiple access method	(F/T)DMA	TDMA	TDMA	CDMA
Duplexing method	TDD	TDD	TDD	FDD
RF channel bw, MHz	0.10	1.00	1.73	2x1.25
RF channel rate, kb/s	72	640	1152	1228.80
Number of traffic ch. per one RF channel	1	8	12	32
Burst/frame length, ms	1/2	1/16	1/10	n/a
Modulation Type	GFSK	GMSK	GMSK	?QPSK
Coding	Cyclic, RS	CRC 16	CRC 16	Conv 1/2, 1/3
Transmit power, mW	≤10	≤ 80	≤100	≤10
Transmit power steps	2	1	1	many
TX power range, dB	16	0	0	≥ 80
Vocoder type / Vocoder rate, kb/s	ADPCM fixed 32	ADPCM fixed 32	ADPCM fixed 32	CELP up to 8
Max data rate, kb/s	32	ISDN 144	ISDN 144	9.6
Processing delay, ms	2	16	16	80
Re-use efficiency Minimum / Average / Maximum	1/25 / 1/15 / (Note 1)1/02	1/15 / 1/07 / (Note 1)1/02	1/15 / 1/07 / (Note 1)1/02	1/4 / 2/3 / 3/4
Theor. number of vc. per cell and 10 MHz	100x1	10x8	6x12	4x32
Practical per 10 MHz Minimum / Average / Maximum	4 / 7 / (Note 1) 50	5-6 / 11-12 / (Note 1) 40	5-6 / 11-12 / (Note 1) 40	(Note 2) 32 (08) / 85 (21) / 96 (24)

Table 4 Comparison of Digital Cordless Telephone Systems
Note 1: The capacity (in the number of voice channels) for a single isolated cell.
Note 2: The capacity in parentheses may correspond to a 32 kb/s vocoder. Re-use efficiency and associate capacities reflect our own estimates.
Source: 4U Communications Research, 1991.09.15.

The first generation of the UK's cordless telephones (coded CT1) was developed as the answer to the large quantities of imported mobile radio equipment.

telephone book directory) can offer up to 600 b/s data. Standard M system is planned for land mobile and maritime mobile users, while aeronautical systems will provide data and voice services to air travelers.

Volna is a Soviet system of satellites which, in conjunction with the satellite More, and with L-band transponders on the Raduga and Gorizont satellites, will provide worldwide voice and data services to a fleet of ships and aircraft.

Airphone is a public, fully automatic air-to-ground telephone system that operates in the 900 MHz band using 6 kHz SSB transmission. Each ground transceiver, by emitting an effective isotropic radiated power of 3 dBW, serves a cell with a radius of about 400 km. An aircraft uses two blade antennas, four transceivers (each radiating 7 dBW), a telephone set and an airborne computer that directs all call logistics.

Dispatch two-way radio land mobile or satellite system, with or without connection to the PSTN, consists of an operating center controlling the operation of a fleet of vehicles such as aircraft, taxis, police cars, tracks, rail cars, etc. The summary of some existing and planned terrestrial systems, including MOBITEX RAM and ARDIS, is given in Table 2. OmniTRACS dispatch system employs Ku-band geostationary satellite located at 103° W to provide two-way digital messaging and position reporting (derived from incorporated satellite-aided LORAN C receiver), throughout the contiguous United States (CONUS).

Cellular radio or public land mobile telephone system offers a full range of services to the traveler that are equivalent to those provided by PSTN. Some of the operating cellular radio systems are: the North American Advanced Mobile Phone Service (AMPS), the Japanese land mobile communications systems MCS-L1 and MCS-L2, the Nordic Mobile Telephone systems NMT-450 and NMT-900, the German C450, the Italian public land mobile radio communication system at 450 MHz, the French radiotelephone multiservice network at 200, 400 MHz RADIOCOM 2000, and the United Kingdom Total Access Communication System (TACS). The technical characteristics of some existing and planned systems are summarized in Table 3.

Cordless Telephony

The first generation of the United Kingdom's cordless telephones (coded CT1) was developed as the answer to the large quantities of imported, technically superior yet unlicensed mobile radio equipment. The simplicity and cost-effectiveness of CT1 analog radio and base station products using eight RF channels and FDMA scheme stem from their applications limited to incoming calls from a limited number of mobile users to the isolated telepoints. As the number of users grew, so did the co-channel interference levels, while the quality of the service deteriorated. Anticipating this situation, the second generation digital cordless telecommunications radio equipment and common air interface standards (CT2/CAI), incompatible with the CT1 equipment, have been developed. CT2

Vehicle information system is a synonym for the variety of systems and services aimed toward traffic safety and location.

schemes employ digital voice, but the same FDMA principles as CTI schemes. Network and frequency re-use issues necessary to accommodate anticipated residential, business, and telepoint traffic growth have not been addressed adequately. Recognizing these limitations and anticipating the market requirements, different frequency division multiple access (FDMA), time division multiple access (TDMA), code division multiple access (CDMA), and hybrid schemes aimed at cellular mobile and DCT services have been developed. The technical characteristics of some schemes are given in Table 4.

Future Public Land Mobile Telecommunications Systems (FPLMTS) is a huge international administrative project, for which tasks and objectives are presented in CCIR Report 1153. It discusses different terrestrial and satellite mobile radio communications and broadcasting systems, the transmission of data, voice, and images, at rates between 8 kb/s and 1,920 kb/s, and a very broad range of services and technical and administrative issues.

Amateur satellite services started in 1965, when the OSCAR 3 satellite was launched. Successive OSCAR/AMSAT satellites used 144 MHz, 432 MHz, 1270 MHz, and 2400 MHz carrier frequencies. The USSR's Iskra satellites use 21/29 MHz and RS-3 satellites use 145/29 MHz carrier frequencies.

Vehicle information system is a synonym for the variety of systems and services aimed toward traffic safety and location. This includes: traffic management, vehicle identification, digitized map information and navigation, radio-navigation, speed sensing, and adaptive cruise control, collision warning and prevention, etc. Some of the vehicle information systems can easily be incorporated in mobile radio communications transceivers to enhance the service quality and capacity of respective communications systems.

Embarrass du choix.

Airwaves Management

The airwaves (the frequency spectrum and the space surrounding us) are a limited natural resource shared among several different radio users (military, government, commercial, public, and amateur). Its sharing (e.g., among different users, services described in the previous section, television and sound broadcasting, etc.), coordination, and administration is an ongoing process exercised on national and international levels. National administrations (e.g., the United States Federal Communications Commission (FCC), the Canadian Department of Communications (DOC) in Canada, etc.), in cooperation with users and industry, set the rules and procedures for planning and utilization of scarce frequency bands. These plans and utilizations must be further coordinated internationally.

The International Telecommunications Union (ITU) is a specialized agency of the United Nations, stationed in Geneva, Switzerland. The ITU has more than 150 government members, responsible for all policies related to radio, telegraph, and telephone. According to the ITU, the world is divided into three regions: Region 1—Europe, including the Soviet Union, Outer Mongolia, Africa,

and the Middle East west of Iran; Region 2—the Americas and Greenland; and Region 3—Asia (excluding parts west of Iran and Outer Mongolia), Australia, and Oceania. Historically, these three regions have developed, more or less independently, their own frequency plans, which best suit local purposes.

With the advent of satellite services and globalization trends, the coordination between different regions becomes more urgent. Frequency spectrum planning and coordination is performed through such ITU bodies as: the Comite Consultatif de International Radio (CCIR), the International Frequency Registration Board (IFRB), the World Administrative Radio Conference (WARC), and the Regional Administrative Radio Conference (RARC).

Through its study groups, CCIR deals with technical and operational aspects of radio communications. Results of these activities have been summarized in the form of reports and recommendations published every four years.

The IFRB serves as a custodian of a common and scarce natural resource, namely, the airwaves. In its capacity, the IFRB records radio frequencies, advises the members on technical issues, and contributes on other technical matters. Based on the work of CCIR, IFRB, and the national administrations, ITU members convene at appropriate RARC and WARC meetings, where documents on frequency planning and utilization (the Radio Regulations) are updated. Actions on a national level follow.

The managing of airwaves becomes even more interesting. CCIR's big brother, CCITT, became involved with the mobile radio communications. Restructuring of ITU (including CCIR and IFRB) also has been proposed.

The far-reaching impact of mobile radio communications on economies and the well-being of the three main trading blocks, other developing and third world countries, potential manufacturers, and users makes the airways (frequency spectrum) even more important. While the battle for the mobile radio market, the airwaves and global competitiveness intensifies, the World Administrative Radio Conference, WARC'92, is scheduled to be held in Granada, Spain. At that conference, the airwaves map of the world is expected to be redrawn.

Si vis pacem, para bellum.

Operating Environment

While traveling, a customer of cellular mobile radio systems may experience sudden changes in signal quality caused by his movements relative to the corresponding base station and surroundings, multipath propagation, and unintentional jamming (e.g., manmade noise, adjacent channel interference, and co-channel interference inherent to the cellular systems). Such an environment belongs to the class of nonstationary random fields, of which experimental data is difficult to obtain. Their behavior is hard to predict and model satisfactorily. When reflected signal components become comparable in level to the attenuated direct component, and their delays are comparable to the inverse of the channel bandwidth, frequency selective fading occurs. The reception is further degraded by movements of a user, relative to reflection points and relay station, causing the Doppler fre-

11

quency shifts. The simplified model of this environment is known as the Doppler affected multipath Rayleigh channel.

The existing and planned cellular mobile radio systems employ sophisticated narrowband and wideband filtering, interleaving, coding, modulation, equalization, decoding, carrier and timing recovery, and multiple access schemes. The cellular mobile radio channel involves a dynamic interaction of signal arrived via different paths, adjacent and co-channel interference, and noise. Most channels exhibit some degree of memory, which description requires higher order statistics of spatial and temporal multidimensional random vectors (amplitude, phase, multipath delay, Doppler frequency, etc.) to be employed. This may require the evaluation of usefulness of existing radio channel models and eventual development of more accurate ones.

Cell engineering, prediction of service area, and service quality, in an ever-changing mobile radio channel environment, is a difficult task. The average path loss depends on terrain microstructure within a cell, with considerable variation between different types of cells (i.e., urban, suburban, and rural environments). A variety of models based on experimental and theoretic work have been developed to predict path radio propagation losses in a mobile channel. Unfortunately, none of them are universally applicable. In almost all cases, an excessive transmitting power is necessary to provide an adequate system performance. The curves of path propagation loss L in decibels (dB) versus the distance d in kilometers (km), based on some of these models, are summarized in Fig. 1.

The first generation mobile satellite systems employ geostationary satellites (or payloads piggybacked on a host satellite) with small 18 dBi antennas covering the entire globe. When the satellite is positioned directly above the traveler (at zenith), a near constant signal environment is experienced—the Gaussian channel. The traveler's movement relative to the satellite is negligible (i.e., Doppler frequency is practically equal to zero). As the traveler moves (north, south, east or west) the satellite appears lower on the horizon. In addition to the direct path, many significant strength-reflected components are present, resulting in a degraded performance. Frequencies of these components fluctuate due to movement of the traveler relative to the reflection points and the satellite. This environment is known as the Doppler affected Ricean channel. An inclined orbit satellite located for a prolonged period of time above 45° latitude north and 106° longitude west could provide travelers all over the United States and Canada, including the far North, a service quality unsurpassed by either geostationary satellite or terrestrial cellular radio. Similarly, a satellite located at 45° latitude north and 15° longitude east could provide travelers in Europe with improved service quality.

A typical return link budget (from a traveler to the central station via a satellite) of the second generation geostationary satellite system for medium rate data services (and digitized voice) follows. The traveler's mobile radio employs a 10 W transmitter and a small 3 dBi gain antenna. On its way toward a satellite (uplink), the 1.6 GHz signal experiences a line-of-sight (LOS) loss of 188.9

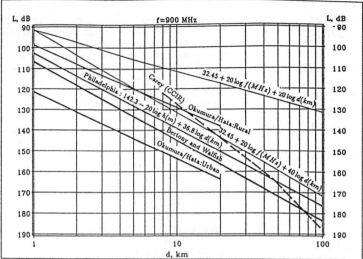

Figure 1. The curves of path propagation loss L in decibels (db) vs. the distance d in kilometers (km)

dB. In a satellite, this signal is amplified, converted to 12 GHz, then once again amplified, and retransmitted toward the earth control station suffering a LOS loss of 205.8 dB in downlink, and a delay of 2 x 0.125 s.

Huge LOS losses put an enormous burden on satellite energy resources, while long delays limit voice transmission quality, and the selection of efficient data protocols. Inclined orbit satellite systems can offer a low startup cost, a near Gaussian channel environment and improved service quality. Low orbit satellites, positioned closer to the service area, can provide high signal levels and short (a few milliseconds long) delays, and offer compatibility with the cellular terrestrial systems. These advantages need to be weighed against network complexity, intersatellite links, tracking facilities, etc.

Service Quality

The primary and the most important measure of service quality should be customer satisfaction. The customer's needs, both current and future, should provide guidance to a service provider and an equipment manufacturer for both the system concept and product design stages. Acknowledging the importance of each step of the complex service process and architecture, attention is limited here to a few technical merits of quality.

Guaranteed quality level usually is related to a percentage of the service area coverage for an adequate percentage of time.

Data service quality can be described by the average bit error rate (e.g., BER < 10^{-5}), packet BER (PBER < 10^{-2}), control PBER (CPER < 10^{-3}), signal processing delay (1-10 ms), multiple access collision probability (< 20 percent), the probability of a false call (false alarm), the probability of a missed call (miss), the probability of a lost call (synchronization loss), etc.

Voice quality usually is expressed in terms of the mean opinion score (MOS) of subjective evaluations by service users. MOS marks are: bad = 0, poor = 1, fair = 2, good = 3, and excellent = 4. The MOS for PSTN voice service, pooled by leading service providers, relates the poor MOS mark to a signal-to-noise ratio (S/N) in a voice channel

12

of S/N≈ 35 dB, while an excellent score corresponds to S/N > 45 dB. Currently, users of mobile radio services are giving poor marks to the voice quality associated with a S/N ≈15 dB and an excellent mark for S/N > 25 dB. It is evident that there is a significant difference (20 dB) between the PSTN and mobile services. If digital speech is employed, both the speech and the speaker recognition have to be assessed. For more objective evaluation of speech quality under real conditions (with no impairments, in the presence of burst errors during fading, random bit errors at BER = 10^{-2}, Doppler frequency offsets, truck acoustic background noise, ignition noise, etc.), additional tests such as

the diagnostic acceptability measure (DAM), diagnostic rhyme test (DRT), Youden square rank ordering, and Sino-Graeco-Latin square tests, can be performed.

Network Issues and Cell Size

To understand ideas and technical solutions offered in existing schemes, and in proposals such as cordless telephony (CT), digital cordless telecommunications (DCT), personal communications service (PCS), personal communications network (PCN), etc., one also needs to analyze the reasons for their introduction and success. Cellular mobile services are flourishing at an annual rate of 20 percent to 40 percent worldwide. These systems (e.g., AMPS, NMT, TACS, and MCS), use frequency division multiple access (FDMA) and digital modulation schemes for access, command and control purposes and analog phase/frequency modulation schemes for the transmission of an analog voice. Most of the network intelligence is concentrated at fix elements of the network, including base stations, which seem to be well-suited to networks with a modest number of medium- to large-sized cells.

To satisfy the growing number of potential customers, more cells and base stations were created by the cell splitting and frequency re-use process. Technically, the shape and size of a particular cell is dictated by the base station antenna pattern and the topography of the service area. Current terrestrial cellular radio systems employ cells with a radius of 0.5 km to 50 km. The maximum cell size usually is dictated by the link budget, particularly the gain of a mobile antenna and available output power. This situation arises in a rural environment, where the demand on capacity is very low and cell splitting is not economical. The minimum cell size usually is dictated by the need for an increase in capacity, particularly in downtown cores. Practical constraints (e.g., real estate availability and price, and construction dynamics) limit the minimum cell size to 0.5 km to 2 km. In such types of networks, however, the complexity of the network and the cost of service grow exponentially with the number of base stations, while the efficiency of present handover procedures becomes inadequate.

Antennas with an omnidirectional pattern in a horizontal direction, but with about 10 dBi gain in vertical direction, provide the frequency re-use efficiency of $N_{FDMA} = 1/12$. Base station antennas with similar directivity in vertical direction and 60° directivity in horizontal direction (a cell is divided into six sectors) can provide the re-use efficiency $N_{FDMA} = 1/4$. This results in a three-fold increase in the system capacity; if CDMA is employed instead of FDMA, an increase in re-use efficiency $N_{FDMA} = 1/4 \rightarrow N_{CDMA} = 2/3$ may be expected.

Recognizing some of the limitations of existing schemes and anticipating the market requirements, the research in time division multiple access (TDMA) schemes aimed at cellular mobile and DCT services, and in code division multiple access (CDMA) schemes aimed toward mobile satellite systems, cellular and personal mobile applications have been initiated. Although employing different access schemes, TDMA (CDMA) network

ACSSB	Amplitude Companded Single Side Band
AM	Amplitude Modulation
APK	Amplitude Phase Keying Modulation
BLQAM	Blackman Quadrature Amplitude Modulation
BPSK	Binary Phase Shift Keying
CPFSK	Continuous Phase Frequency Shift Keying
CPM	Continuous Phase Modulation
DEPSK	Differentially Encoded PSK (with carrier recovery)
DPM	Digital Phase Modulation
DPSK	Differential Phase Shift Keying (no carrier recovery-
DSB-AM	Double SideBand Amplitude Modulation
DSB-SC-AM	Double SideBand Suppressed Carrier AM
FFSK	Fast Frequency Shift Keying (MSK)
FM	Frequency Modulation
FSK	Frequency Shift Keying
FSOQ	Frequency Shift Offset Quadrature modulation
GMSK	Gaussian Minimum Shift Keying
GTFM	Generalized Tamed Frequency Modulation
HMQAM	Hamming Quadrature Amplitude Modulation
IJF	Intersymbol Jitter Free (SQORC)
LPAM	L-ary Pulse Amplitude Modulation
LRC	LT symbols long Raised Cosine pulse shape
LREC	LT symbols long Rectangularly EnCoded pulse shape
LSRC	LT symbols long Spectrally Raised Cosine scheme
MMSK	Modified Minimum Shift Keying
MPSK	M-ary Phase Shift Keying
MQAM	M-ary Quadrature Amplitude Modulation
MQPR	M-ary Quadrature Partial Response
MQPRS	M-ary Quadrature Partial Response System
MSK	Minimum Shift Keying
m-h	multi-h CPM
OQPSK	Offset (staggered) Quadrature Phase Shift Keying
PM	Phase Modulation
PSK	Phase Shift Keying
QAM	Quadrature Amplitude Modulation
QAPSK	Quadrature Amplitude Phase Shift Keying
QPSK	Quadrature Phase Shift Keying
QORC	Quadrature Overlapped Raised Cosine
SQAM	Staggered Quadrature Amplitude Modulation
SQPSK	Staggered Quadrature Phase Shift Keying
SQORC	Staggered Quadrature Overlapped Raised Cosine
SSB	Single Side Band
S3MQAM	Staggered class 3 Quadrature Amplitude Modulation
TFM	Tamed Frequency Modulation
TSI QPSK	Two-Symbol-Interval QPSK
WQAM	Weighted Quadrature Amplitude Modulation
XPSK	Cross-correlated PSK
π/4 QPSK	π/4 shift QPSK
3MQAM	Class 3 Quadrature Amplitude Modulation
4MQAM	Class 4 Quadrature Amplitude Modulation
12PM3	12 state PM with 3 bit correlation

Table 5. Modulation schemes; glossary of terms
(Source: 4U Communications Research, 1991.09.15)

13

concepts rely on a smart mobile/portable unit that scans time slots (codes) to gain information on network behavior, free slots (codes), etc., improving frequency re-use and handover efficiency while, hopefully, keeping the complexity and cost of the overall network at reasonable levels.

Some of the proposed system concepts depend on low gain (0 dBi) base station antennas deployed in a license-free, uncoordinated fashion; small size cells (10 m to 1000 m in radius) and an emitted isotropic radiated power of about 10 mW (+10 dBm) per 100 kHz have been anticipated. A frequency re-use efficiency of $N = 1/9$ to $N = 1/36$ has been projected for DCT systems. $N = 1/9$ corresponds to the highest user capacity with the lowest transmission quality, while $N = 1/36$ has the lowest user capacity with the highest transmission quality. This significantly reduced frequency re-use capability of proposed system concepts will result in significantly reduced system capacity, which needs to be compensated for by other means, including new spectra.

In practical networks, the need for a capacity (and frequency spectrum) is distributed unevenly in space and time. In such an environment, the capacity and frequency re-use efficiency of the network may be improved by dynamic channel allocation, where an increase in the capacity at a particular hot spot may be traded for the decrease in the capacity in cells surrounding the hot spot, the quality of the transmission, and network instability.

To cover the same area (space) with increasingly smaller cells, one must employ more and more base stations. A linear increase in the number of base stations in a network usually requires an exponential increase in the number of connections between base stations, switches and network centers. These connections can be realized by fixed radio systems (providing more frequency spectra will be available for this purpose), or, more likely, by a cord (wire, cable, fiber, etc.). Therefore, the following postulate can apply:

There will be a plenty of cord in cordless telephony. (PI)

The first generation geostationary satellite system antenna beam covers the entire earth (i.e., the cell radius equals ≈ 6500 km). The second generation geostationary satellites will use larger multibeam antennas providing 10 to 20 beams (cells) with a radius of 800 km to 1,600 km. Low-orbit satellites (e.g., Iridium) will use up to 37 beams (cells) with a radius of 670 km. The third generation geostationary satellite systems will be able to use very large reflector antennas (roughly the size of a baseball stadium), and provide 80 to 100 beams (cells) with a cell radius of ≈ 200 km. If such a satellite is tethered to a position 400 km above the earth, the cell size will decrease to ≈ 2 km in radius, which is comparable in size with today's small size cell in terrestrial systems. Yet, such a satellite system might have the potential to offer an improved service quality due to its near optimal location with respect to the service area. Similar to the terrestrial concepts, an increase in the number of satellites in a network will require an increase in the number of connections between satellites and/or earth network management and satellite tracking centers, etc. Additional factors that need to be taken into consideration include price, availability, reliability, and

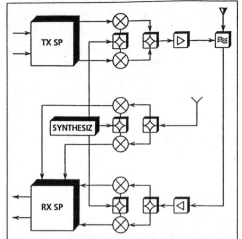

Figure 2. *The conceptual transmitter (TX) and receiver (RX) of a mobile system*

timeliness of the launch procedures, a few large versus many small satellites, tracking stations, etc.

Coding and Modulation

The conceptual transmitter (TX) and receiver (RX) of a mobile system are shown in Fig. 2. The transmitter signal processor (TX SP) accepts analog voice and/or data and transforms (by analog and/or digital means) these signals into a form suitable for a double-sided suppressed carrier amplitude modulator, also called quadrature amplitude modulator (QAM). Both analog and digital input signals may be supported, and either analog or digital modulation may result at the transmitter output. Coding and interleaving also can be included. Often, the processes of coding and modulation are performed jointly; we will call this joint process modulation. A list of typical modulation schemes suitable for transmission of voice and/or data over Doppler-affected Ricean channel, which can be generated by this transmitter, is given in Table 5. These particular modulations, however, also can be generated by means different than that suggested in Fig. 2.

Existing cellular radio systems such as AMPS, TACS, MCS, and NMT employ hybrid (analog and digital) schemes. For example, in access mode AMPS uses a digital modulation scheme (BCH coding and FSK modulation). While in information exchange mode, the frequency modulated analog voice is merged with discrete SAT and/or ST signals and occasionally blanked to send a digital message. These hybrid modulation schemes exhibit a constant envelope and as such allow the use of d.c. power-efficient nonlinear amplifiers. On the receiver side, these schemes can be demodulated by an inexpensive but efficient limiter/discriminator device. They require modest to high $C/N = 10$ dB to 20 dB, are very robust in adjacent (a spectrum is concentrated near the carrier) and co-channel interference (up to $C/I = 0$ dB, due to capture effect) cellular radio environment, and react quickly to the signal fade outages (no carrier, code or frame synchronization). Frequency-selective and Doppler-affected mobile radio channels will cause modest to significant degradations, known as the random phase/frequency modulation.

Tightly filtered modulation schemes, such as

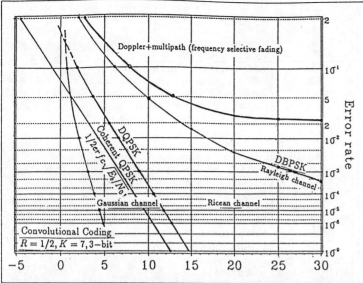

Figure 3. Error rate vs. E_b/N_o curves for digital 4QAM codulation schemes

π/4 QPSK additionally filtered by a square root raised-cosine filter, exhibit a nonconstant envelope, which demands (quasi) linear, less d.c. power-efficient amplifiers to be employed. On the receiver side, these schemes require complex demodulation receivers, a linear path for signal detection, and a nonlinear one for reference detection, differential detection, or carrier recovery. When such a transceiver operates in a selective fading multipath channel environment, additional countermeasures (inherently sluggish equalizers, etc.) are necessary to improve the performance by reducing the bit error rate foor. These modulation schemes require modest C/N = 8 dB to 16 dB and perform modestly in adjacent and/or co-channel (up to C/I = 8 dB) interference environment. Codulation schemes employed in spread spectrum systems use low-rate coding schemes and mildly filtered modulation schemes. When equipped with sophisticated amplitude gain control on the transmit and receive side, and robust rake receiver, these schemes can provide superior C/N = 4 dB to 10 dB and C/I < O dB performance.

The error rate performance of some digital codulation schemes operating in three different mobile channel environments is summarized in Fig. 3. It might be expected that a codulation scheme (in conjunction with an appropriate access scheme) tailored to a particular local environment can provide a significant improvement in performance and/or capacity over an ad hoc selected combination. Therefore, the following can apply:

> *For every mobile radio channel, there is an optimal codulation scheme. (P2)*

> *For every codulation scheme, there is an optimal mobile radio channel. (P3)*

Speech Coding

Human vocal tract and voice receptors, in conjunction with language redundancy (coding), are well-suited for face-to-face conversation. As the channel changes (e.g., from telephone channel to mobile radio channel), different coding strategies are necessary to protect the loss of information. In (analog) companded PM/FM mobile radio sys-

tems, speech is limited to 4 kHz, compressed in amplitude (2:1), pre-emphasized, and phase/frequency modulated. At a receiver, inverse operations are performed. Degradation caused by these conversions and channel impairments results in lower voice quality. Finally, the human ear and brain have to perform the estimation and decision processes on the received signal.

In digital schemes sampling and digitizing of an analog speech (source) are performed first. Then, by using knowledge of properties of the human vocal tract and the language itself, a spectrally efficient source coding is performed. A high rate 64 kb/s, 56 kb/s and ADPCM 32 kb/s digitized voice complies with CCITT recommendations for toll quality, but may be less practical for the mobile environment. One is primarily interested in 8 kb/s to 16 kb/s rate speech coders, which might offer satisfactory quality, spectral efficiency, robustness, and acceptable processing delays in a mobile radio environment. A summary of the major speech coding schemes is provided in Table 6.

At this point, a partial comparison between analog and digital voice should be made. The quality of 64 kb/s digital voice, transmitted over a telephone line, is essentially the same as the original analog voice (they receive nearly equal MOS). What does this near equal MOS mean in a radio environment?

A mobile radio conversation consists of one (mobile to home) or a maximum of two (mobile to mobile) mobile radio paths, which dictate the quality of the overall connection. The results of a comparison between analog and digital voice schemes in different artificial mobile radio environments have been widely published. Generally, systems that employ digital voice and digital codulation schemes appear to perform well under modest conditions, while analog voice and analog codulation systems outperform their digital counterparts in fair and difficult (near threshold, in the presence of strong co-channel interference) conditions. Fortunately, present technology can offer a viable implementation of both analog and digital systems within the same mobile/portable radio telephone unit. This would give every individual a choice of either an analog or digital scheme, better service quality and higher customer satisfaction. Tradeoffs between the quality of digital speech, the complexity of speech and channel coding, as well as d.c. power consumption, must be assessed carefully and compared with analog voice systems.

Macro and Micro Diversity

Macro Diversity

In a cellular system, the base station is usually located in the barocenter of the service area (center of the cell), as illustrated in Fig. 4a. Typically, the base antenna is omnidirectional in azimuth, but with about 6 dBi to 10 dBi gain in elevation, and serves most of the cell area (e.g., > 95 percent). Some parts within the cell may experience a lower quality of service because the direct path signal may be attenuated due to obstruction losses caused by buildings, hills, trees, etc.

The closest neighboring base stations (the first tier) serve corresponding neighboring area cells by using different sets of frequencies, eventually caus-

15

ing adjacent channel interference. The second closest neighboring base stations (the second tier) might use the same frequencies (frequency re-use) causing co-channel interference. If the same real estate (base stations) is used in conjunction with 120° directional (in azimuth) antennas, the designated area may be served by three base stations, as illustrated in Fig. 4b. In this configuration one base station serves three cells by using three 120° directional antennas. Therefore, the same number of existing base stations equipped with new directional antennas and additional combining circuitry is required to serve the same number of cells, yet in a different fashion.

The mode of operation in which two or more base stations serve the same area is called the macro diversity. Statistically, three base stations are able to provide a better coverage of an area similar in size to the system with a centrally located base station. The directivity of a base station antenna (120° or even 60°) provides additional discrimination against signals from neighboring cells, therefore, reducing adjacent and co-channel interference (i.e., improving re-use efficiency and capacity). Effective improvement depends on the terrain configuration and the combining strategy and efficiency. However, it requires more complex antenna systems and combining devices.

Micro Diversity

Micro diversity refers to the condition in which two or more signals are received at one site (base or mobile).

Space diversity systems employ two or more antennas spaced a certain distance apart from one another. A separation of only $\lambda/2 \approx 15$ cm, which is suitable for implementation on the mobile side, can provide a notable improvement in some mobile radio channel environments. Micro space diversity is routinely used on cellular base sites. Macro diversity is also a form of space diversity.

Field-component diversity systems employ different types of antennas receiving either the electric or the magnetic component of an electromagnetic signal.

Frequency diversity systems employ two or more different carrier frequencies to transmit the same information. Statistically, the same information signal may or may not fade at the same time at different carrier frequencies. Frequency hopping and very wide band signaling can be viewed as frequency diversity techniques.

Time diversity systems are used primarily for the transmission of data. The same data is sent through the channel as many times as necessary, until the required quality of transmission is achieved (auto-

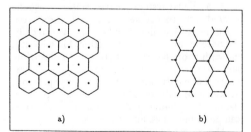

Figure 4 The cellular system concepts: a) Barocentric base stations, center excited cellular system; b) Corner base stations, corner excited cellular system.

ADM	Adaptive Delta Modulation
ADPCM	Adaptive Differential Pulse Code Modulation
ACIT	Adaptive Code sub-band excited Transform (GTE)
APC	Adaptive Predictive Coding
APC-AB	APC with Adaptive Bit Allocation
APC-HQ	APC with Hybrid Quantization
APC-MQL	APC with Maximum Likelihood Quantization
AQ	Adaptive Quantization
ATC	Adaptive Transform Coding
BAR	Backward Adaptive Reencoding
CELP	Code Excited Linear Prediction
CVSDM	Continuous Variable Slope Delta Modulation
DAM	Diagnostic Acceptability Measure
DM	Delta Modulation
DPCM	Differential Pulse Code Modulation
DRT	Diagnostic Rhyme Test
DSI	Digital Speech Interpolation
DSP	Digital Signal Processing
HCDM	Hybrid Companding Delta Modulation
LDM	Linear Delta Modulation
LPC	Linear Predictive Coding
MPLPC	Multi Pulse LPC
MSQ	Multipath Search Coding
NIC	Nearly Instantaneous Companding
PVXC	Pulse Vector eXcitation Coding
PWA	Predicted Wordlength Assignment
QMF	Quadrature Mirror Filter
RELP	Residual Excited Linear Prediction
RPE	Regular Pulse Excitation
SBC	Sub Band Coding
TASI	Time Assigned Speech Interpolation
TDHS	Time Domain Harmonic Scaling
VAPC	Vector Adaptive Predictive Coding
VCELP	Vector Code Excited Linear Prediction
VEPC	Voice Excited Predictive Coding
VQ	Vector Quantization
VQL	Variable Quantum Level Coding
VSELP	Vector-Sum Excited Linear Prediction
VXC	Vector eXcitation Coding

Table 6. Digitized voice; glossary of terms

matic repeat request ARQ). "Would you please repeat your last sentence" is a form of time diversity used in a speech transmission.

The improvement of any diversity scheme is strongly dependent on the combining techniques employed, i.e., the selective (switched) combining, the maximal ratio combining, the equal gain combining, the feedforward combining, the feedback (Granlund) combining, majority vote, etc.

Multiplex and Multiple Access

Communications networks for travelers have two distinct directions: the forward link (from the base station via satellite to the traveler) and the return link (from a traveler via satellite to the base station). In the forward direction, a base station distributes information to travelers according to the previously established protocol, i.e., no multiple access is involved. In the reverse direction, many travelers make attempts to access one of the base stations. This occurs in so-called control channels, in a particular time slot, at particular frequency, or by using a particular code. If collisions occur, customers must wait in a queue and try again until success is achieved. If successful (i.e., if no collision occurred), a particular customer will automatically exchange the necessary information for call setup. The network management center (NMC)

16

The strengths of FDMA schemes seem to be fully exploited in narrowband channel environments.

will verify the customer's status, his credit rating, etc. Then the NMC may assign a channel frequency, time slot, or code on which the customer will be able to exchange information with his correspondent. The optimization of the forward and reverse links may require different coding, modulation schemes, and bandwidths in each direction.

In forward link, there are three basic distribution (multiplex) schemes: one uses discrimination in frequency between different users and is called frequency division multiplex (FDM); a second discriminates in time and is called time division multiplex (TDM); and the third has different codes based on spread spectrum signaling, which is known as code division multiplex (CDM). It should be noted that hybrid schemes using a combination of basic schemes also can be developed.

In reverse link, there are three basic access schemes: one uses discrimination in frequency between different users and is called frequency division multiple access (FDMA); a second discriminates in time and is called time division multiple access (TDMA); and a third which has different codes based on spread spectrum signaling is known as code division multiple access (CDMA). It should be noted that hybrid schemes using combination of basic schemes also can be developed.

A performance comparison of multiple access schemes is a difficult task. The strengths of FDMA schemes seem to be fully exploited in narrowband channel environments. To avoid the use of equalizers, channel bandwidths as narrow as possible should be employed. Yet, in such narrowband channels the quality of service is limited by the maximal expected Doppler frequency and practical stability of frequency sources. Current practical limits are approximately 5 kHz.

The strengths of both TDMA and CDMA schemes seem to be fully exploited in wideband channel environments. TDMA schemes need many slots (and bandwidth) to collect information on network behavior. Once the equalization is necessary (at bandwidths > 20 kHz), the data rate should be made as high as possible to increase frame efficiency and freeze the frame to ease equalization. High data rates, however, require high RF peak powers and a large amount of signal processing power, which may be difficult to achieve in handheld units. Current practical bandwidths are approximately 0.1 MHz to 1.0 MHz.

CDMA schemes need large spreading (processing) gains (and bandwidth) to realize spread spectrum potentials, yet high data rates also require a large amount of signal processing power, which may be difficult to achieve in handheld units. Current practical bandwidths are approximately 1.2 MHz. Narrow frequency bands seem to favor FDMA schemes, since both TDMA and CDMA schemes require more spectra to fully develop their potentials. Once the adequate power spectrum is available, however, the latter two schemes may be better suited for a complex (micro)cellular network environment. Multiple access schemes also are message sensitive. The length and type of message, and the kind of service will influence the choice of multiple access, ARQ, frame, and coding. Therefore, the following can apply:

For every mobile radio channel, there is an optimal access scheme. (P3)

For every access scheme, there is an optimal mobile radio channel. (P4)

For every type of service, there is an optimal access scheme. (P5)

For every access scheme, there is an optimal type of service. (P6)

Hybrid Schemes

For various reasons a number of existing and proposed multiple access schemes are hybrid. For example, the GSM system employs a TDMA scheme with eight slots per 200 kHz wide RF channel; channels are further distributed in an FDM fashion. There is an optional frequency-hopping pattern (code division) available. The transmitter and receiver are separated by 45 MHz, and frequency division duplex (FDD) mode of operation is assumed. Therefore, this rather complex hybrid scheme is denoted as FDM/FDD/TDMA/CDM.

A slightly less complex scheme is the newly proposed North American digital cellular system (denoted IS-54 in Table 3). This system employs a TDMA scheme with three slots per 30 kHz wide RF channel. Channels are further distributed in an FDM fashion, the transmitter and receiver are separated by 45 MHz, and a frequency division duplex (FDD) mode of operation is assumed. Therefore, this hybrid scheme is denoted as FDM/FDD/TDMA.

The CT2Plus system proposes a time division duplex (TDD) mode of operation in 100 kHz wide RF channel. Channels are distributed in an FDM fashion, access channels use TDMA, and information channels use FDM mode of operation. Therefore, this rather complex hybrid scheme is denoted as FDM/TDD/TDMA/FDM.

The design and evaluation process of such hybrid schemes requires careful balancing between complexity and technical advantages and disadvantages of each of the elementary schemes.

System Capacity

The recent surge in the popularity of cellular radio, and mobile service in general, has resulted in an overall increase in traffic and a shortage of available system capacity in large metropolitan areas. Current cellular systems exhibit a wide range of traffic densities, from low in rural areas to overloading in downtown areas, with large daily variations between peak hours and quiet night hours. It is a great system engineering challenge to design a system that will make optimal use of the available frequency spectrum, offering a maximal traffic throughput (e.g., Erlangs/MHz/service area) at an acceptable service quality, constrained by the price and size of the mobile equipment.

In a cellular environment, the overall system capacity in a given service area (space) is a product of many factors (with complex interrelationships), including the available frequency spectra, service quality, traffic statistics, type of traffic, type of protocol, shape and size of service area, selected antennas, diversity, frequency re-use capability, spectral efficiency of coding and modulation schemes, efficiency of multiple access, etc.

In the 1970s so-called analog cellular systems employed omnidirectional antennas and simple or no diversity schemes offering modest capacity, which satisfied a relatively low number of customers. Analog cellular systems of the 1990s employ up to 60° sectorial antennas and improved diversity schemes. This latter combination resulted in a three-fold to fivefold increase in capacity. A further (twofold) increase in capacity can be expected from narrowband analog systems (25 kHz → 12.5 kHz), which was practically demonstrated in Japan (Ll → L2 system). However, slight degradation in service quality might be expected. These improvements spurred the current growth in capacity, the overall success and prolonged life of analog cellular radio.

There also are numerous marketing results, where a tenfold to twentyfold increase in capacity has been claimed (watch for small print!). In this kind of campaign new digital systems of the twenty-firtst century, operating under ideal conditions, usually are compared with the old systems of the 1970s, operating under the worst conditions.

There are numerous ways of increasing the capacity of cellular radio; acquiring new frequency spectra is perhaps the easiest one.

Conclusion

In this contribution, a broad repertoire of terrestrial and satellite systems and services for travelers is briefly described. The technical characteristics of the dispatch, cellular, and cordless telephony systems are tabulated for ease of comparison. Issues such as operating environment, service quality, network complexity, cell size, channel coding and modulation (codulation), speech coding, macro and micro diversity, multiplex and multiple access, and the mobile radio communications system capacity, are discussed. Presented data reveals significant differences between existing and planned terrestrial cellular mobile radio communications systems, and between terrestrial and satellite systems. These systems use different frequency bands, bandwidths, codulation schemes, protocols, etc., meaning they are not compatible.

What are the technical reasons for this incompatibility? In this paper, performance dependence on multipath delay (related to the cell size and terrain configuration), Doppler frequency (related to the carrier frequency, data rate and the speed of vehicles), and message length (may dictate the choice of multiple access) are briefly discussed. A system optimized to serve the travelers in the Great Plains may not perform well in mountainous Switzerland. A system optimized for downtown cores may not be well-suited to a rural environment. A system employing geostationary (above equator) satellites may not be able to serve travelers at high latitudes adequately. A system appropriate for slow-moving vehicles may fail to function properly in a high Doppler shift environment. Additionally, a system optimized for voice transmission may not be good for data transmission. A system designed to provide a broad range of services to everyone, everywhere, may not be as good as a system designed to provide a particular service in a particular local environment, just as a world champion decathlete may not be as successful in competitions with specialists in particular disciplines.

However, there are many opportunities where compatibility between systems, their integration, and frequency sharing may offer improvements in service quality, efficiency, cost, and capacity (and therefore availability). Terrestrial systems offer a low startup cost and a modest cost-per-user in densely populated areas. Satellite systems may offer a high quality of service and may be the most viable solution to serve travelers in scarcely populated areas, on oceans, and in the air. Terrestrial systems are confined to two dimensions, and radio propagation occurs in the near horizontal sectors. Barostationary satellite systems use the narrow sectors in the user's zenith nearly perpendicular to the Earth's surface having the potential for frequency re-use and an increase in the capacity in downtown areas during peak hours. A call setup in a forward direction (from the PSTN via base station to the traveler) may be a cumbersome process in a terrestrial system when a traveler to whom a call is intended is roaming within an unknown cell. However, this is easily realized in a global beam satellite system.

England expects this day that every man will do his duty.
Horatio Nelson, 21,10,1805.

Terrestrial systems offer a low startup cost and a modest cost-per-user in densely populated areas.

Biography

Andy D. Kucar (M '80, SM '90) received a Dipl. Tech degree (summa cum laude) in electronics from the Technical College Rijeka in La Guardia's Fiume; in 1974 a Dipl. Ing. degree (financial reward) in electrical engineering, and in 1980 an M.S. in electrical engineering, both from Zagreb University; and in 1987 the Ph.D. degree in electrical engineering from University of Ottawa, Canada. Between 1971 and 1973, Dr. Kucar served as a research and teaching assistant (microwave radio and radar systems) in the Department of RF and Microwaves at the faculty of electrical engineering at Zagreb University. In 1974 he was an engineer for Radioindustrija Zagreb. Between September 1974 and September 1982 he worked for Iskra Ljubljana. From September 1982 to December 1988, he worked for University of Ottawa Digital Communications Group. Dr. Kucar worked for Telesat Canada from 1985 to 1987. From 1987 to 1990 he worked for Bell Northern Research radio division. In November 1990 he established 4U Communications Research, a company specializing in excellence in terrestrial and satellite radio communications, where he serves as a president. He has organized and chaired numerous sessions on international conferences including ICC, GLOBECOM, and VTC.

18

Reprinted from *IEEE Communications Magazine*, pp. 31-39, June, 1991.

Trends in Cellular and Cordless Communications

David J. Goodman

AS PRODUCTS THAT ALLOW PEOPLE TO USE THE worldwide public telecommunications network in a natural, convenient manner, cellular telephones and residential cordless telephones enjoy widespread public approval. Both are relatively new, introduced in the 1980s in most places, and the demand for both is increasing rapidly. To meet this demand, increase quality, and enlarge the range of applications, new second generation systems are emerging. The second generation will be characterized by digital speech transmission and enhanced capabilities of wireless terminals to exchange signaling and control information with the remainder of the network. Meanwhile, researchers and planners are devoting increasing attention to third generation technologies to meet the needs of the next century.

Looking at today's cellular products, we find at least six different, incompatible standards employed in different parts of the world [1]. All rely on analog frequency modulation for speech transmission and in-band signaling to move control information between terminals and the rest of the network during a call. Second generation cellular will conform to at least three different standards: one for Western Europe—Groupe Special Mobile (GSM) [2], one for North America—an Electronic Industry Association Interim Standard (IS-54) [3], and a third for Japan.

While the European and Japanese standards will be applied to completely new cellular systems operating in dedicated frequency bands, the North American standard specifies dual-mode operation. It incorporates the first generation standard, Advanced Mobile Phone System (AMPS) [4], and adds a digital voice transmission capability to new subscriber equipment. Thus, IS-54 is an enhancement to, rather than a replacement for, present cellular technology. The situation in North America is enriched, not to mention complicated, by at least three other dual-mode AMPS enhancements in various stages of development. Thus, in cellular, North America and Europe seem to be moving in opposite directions. Europe will migrate from a collection of incompatible analog systems to a single digital system, while North America, which has enjoyed from the outset continental compatibility of all cellular equipment, will see the deployment of various dual-mode technologies, all sharing a common analog standard, but with radically different advanced technologies.

Cordless telephones, which today are stand-alone consumer products, do not require any interoperability specifications at all. Each cordless telephone comes with its own base station and needs only to be compatible with that base station. In fact,

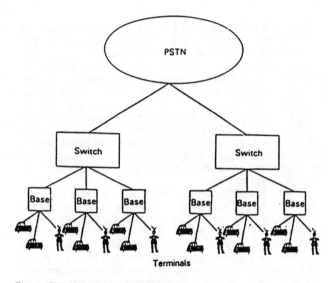

Fig. 1. Cellular network architecture.

billing security and privacy are achieved if the cordless phone is barred from operating with any other base station. Second generation cordless telephones, on the other hand, will be designed as components of a larger network. Products that conform to a British Standard, Cordless Telecommunications, second generation (CT2) [5], or a Pan-European standard, Digital European Cordless Telecommunications (DECT) [6], will have access to public base stations at telepoints, and to base stations installed as components of private branch exchanges in business environments. Security and privacy will be established through registration and handshaking procedures.

As industry plans and implements second generation networks in the early 1990s, researchers and network planning experts turn their attention to the more distant future. In various organizations, a vision of third generation wireless information networks is coming into focus. The idea is to create a single network infrastructure that will make it possible for all people to transfer economically any kind of information between any desired locations. The new network will merge the separate first and second generation cellular and cordless services, and also encompass other means of wireless access such as paging, dispatch, public safety, and wireless local area networks.

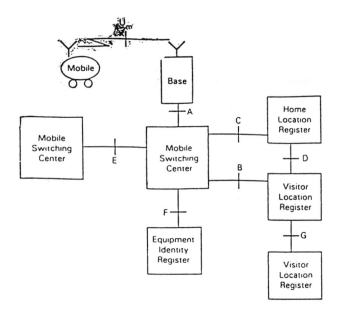

Fig. 2. Pan-European cellular network architecture.

Current work on advanced wireless systems is focused on problems that arise when information terminals are not connected physically to the rest of the network. For example, the location of a terminal is unknown prior to the start of a communication and subject to change as the communication progresses. In addition, the transmission channel linking the terminal and the fixed part of the network is unpredictable, subject to change, and often impaired by poor propagation conditions and by noise and interference. A third constraint on wireless access systems is limited communications bandwidth. While in fixed systems, transmission capacity is limited only by economics, government policies restrict the capacity of wireless access channels.

At least four different groups are currently exploring network architectures and enabling technologies for realizing the ambitious goals of the third generation in the face of formidable obstacles. Task Group 8/1 of the International Consultative Committee on Radio (CCIR) [7] provides an international forum for the exchange of ideas and the formulation of guidelines on service requirements, spectrum allocations, and technologies. The hope is that third generation systems will conform to a mutually compatible set of standards accepted throughout the world instead of the five or more incompatible standards of the second generation. Within the framework of the European Commission's Research on Advanced Communications in Europe (RACE) program [8], 25 European organizations are working together to create enabling technologies for third generation services. A principal aim is to provide users of wireless terminals access to the projected Broadband Integrated Services Digital Network (BISDN).

In the U.S., Bell Communications Research (Bellcore), owned by the regional Bell operating companies, has conducted an extensive set of studies on universal portable radio communications [9]. In addition to facilitating user mobility, the Bellcore approach would eliminate the need for the wires that connect residences to the facilities of the local telephone company. Another ambitious research program focused on third generation networks is found in the Rutgers University Wireless Information Network Laboratory (WINLAB) under the sponsorship of the Federal Communications Commission and fifteen wireless networking companies in the U.S., Europe, and Japan.

The purpose of this article is to trace the evolution of wireless information networks from the present first generation

systems to next century's third generation. In the following sections we examine four important issues: the goal a network is designed to achieve, network architecture, radio transmission technology, and the control channels that make it possible for wireless terminals and the remainder of the network to coordinate their operations. In each section, we describe present networks, refer to the properties of four second generation networks, GSM, IS-54, CT2, and DECT, and then describe early work in progress at the Rutgers WINLAB [10].

Goals

The original purpose of cellular mobile radio was to provide telephone service to a large population of users in their automobiles [11]. Prior mobile services, based on wide area radio transmission, were severely limited in capacity. The cellular approach promised virtually unlimited capacity through cell splitting. As the popularity of cellular radio escalated in the 1980s, however, industry encountered practical limits. With cells becoming smaller and smaller, it is increasingly difficult to place base stations at the locations that offer the necessary radio coverage. The problem of establishing a new cell site at the optimum location is particularly acute in the large congested cities where capacity requirements are most pressing. In addition to capacity bottlenecks, the utility of first generation cellular phones in Europe is diminished by the proliferation of incompatible standards, which makes it impossible for a person to use the same cellular phone in different countries.

These limitations of first generation cellular services motivate the principal goals of the second generation: higher capacity and, in Europe, a continental system with full international roaming and handoff. In Europe, these goals are served by new spectrum allocations and by the formulation of a Pan-European cellular standard, GSM. In North America, where no new cellular frequency bands are available, higher capacity is derived from intensive application of advanced transmission techniques including efficient speech coding, error correcting channel codes, and bandwidth efficient modulation.

In the case of cordless telephones, the main restrictions of first generation products are small coverage areas and vulnerability to interference. Second generation cordless telephones are designed not only for residential use, the main application of the first generation, but also for multicell business environments and for public pay telephone services—telepoint. In contrast to present products, a second generation cordless telephone will have access to a network of base stations, rather than a single base station purchased with the telephone. The cordless telephone will automatically choose a transmission channel that offers interference-free communication.

While they differ in many ways, cellular and cordless phones serve the same basic aim: to provide wireless access to the worldwide Public Switched Telecommunications Network (PSTN). The PSTN is evolving rapidly from a means of connecting any pair of voice telephones in the world to a versatile, intelligent, multimedia ISDN. It follows that an important purpose of the new cellular and cordless systems is to grant their users access to as many features as possible of the emerging ISDN. Just as every basic rate ISDN connection includes a logical channel—D channel—for network control information, second generation wireless information networks will incorporate dedicated control channels linking base stations and wireless terminals.

The vision of the third generation is unified wireless access replacing the diverse and incompatible second generation networks with a single means of wireless access to advanced information services. Third generation subscribers will be able to move information of all kinds between any desired locations. They will use the same terminals indoors and outdoors, in suburban and rural areas, and in crowded cities. The wireless terminal will function well in a fixed position or moving at hundreds of kilometers per hour on a high-speed train. Third

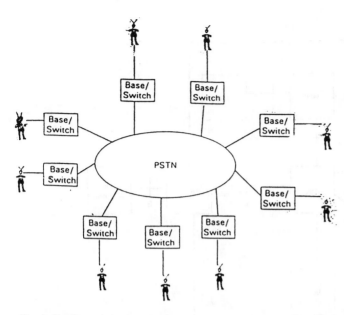

Fig. 3. Cordless telephone architecture.

generation networks will accommodate a truly mass market of users in an era when wireless access will serve as the most common interface to information services.

Network Architecture

Relative to present systems, second generation cellular systems will have an increased number of open, rather than proprietary, interfaces between network elements. Network control will be more dispersed than at present. Functions now performed by the system controller at the mobile switching office will be distributed among base stations and wireless terminals. Second generation cordless telephones will have access to more than one base station and one base station will serve many cordless telephones simultaneously. In establishing a link with a specific base station, the cordless phone will automatically choose the best available channel and change channels in response to changing signal strength and interference conditions during a call.

One view of the third generation is a network with radically distributed control of information routing and of access to radio channels. Information in packet format will find its way to the correct destination with the aid of addresses carried in packet headers. Wireless terminals will autonomously decide which base station to use and move from one base station to another without the intervention of a central controller.

Cellular

Three network elements, shown in Figure 1, are essential for providing wireless access to a fixed information network: wireless terminals, base stations, and switches. A base station exchanges radio signals with wireless terminals, and a switch connects the wireless information network with the fixed network, the PSTN. Cellular systems are organized in the hierarchical fashion of Figure 1. Each base station serves dozens of wireless terminals simultaneously, and a switch connects the user signals from up to 100 base stations to trunks of the public telephone network. In addition to making these connections, the cellular switch controls the assignment of radio channels to wireless terminals.

To serve roamers (subscribers in locations remote from their home service areas), a number of proprietary communication links between cellular switches have been established.

The trend in second generation cellular systems is to manage roaming and intersystem handoff with standardized signaling systems linking cellular switches and databases. The architecture of the Pan-European digital cellular system, GSM, shown in Figure 2, includes, in addition to the three network elements of Figure 1, three databases. Two of them, the home location register and the visitor location register, contain information that enables the system to provide call delivery to roaming subscribers. Because all of the labeled interfaces in Figure 2 are GSM standards, cellular service providers will have great flexibility in configuring their networks and obtaining equipment from a variety of manufacturers. In the second generation North American cellular system, IS-54, the base station to switch interface will remain proprietary. However, industry is working toward a standard interface between cellular switching offices to facilitate roaming and intersystem handoff [12].

Another trend in second generation cellular is to distribute network control functions more evenly between switches, base stations, and mobile terminals. An important feature of both North American and Pan-European digital cellular is mobile assisted handoff. In first generation cellular, the system controller in the mobile switching office assembles information from several base stations about the strength of the signal path from each active mobile terminal. On the basis of these measurements, the controller decides when the terminal should tune to a channel at a different base station. With the number of subscribers and the number of cells increasing rapidly, the volume of handoffs is approaching the processing capacity of cellular controllers.

Second generation cellular telephones will measure during a call the strength of signals received from neighboring base stations. These measurements will be reported to the serving base station, which will determine the need for a handoff and communicate to the system controller at the switching office the preferred new cell. In this way, a large fraction of the information processing and control requirements associated with handoff is moved from the system controller to the base stations and to the mobile terminals. This will avoid overloading the controller and facilitate faster handoff.

Cordless

Viewed as a wireless information network, a first generation cordless phone is as simple as can be. A single base station communicates with one and only one wireless terminal, and the "switch" is merely a make or break connection between the base station and a two-wire subscriber line. Channel selection is solely the task of each cordless telephone. The simplest phones have access to only one radio channel installed in the factory. More expensive products have multichannel capability, usually with the user choosing the channel on the basis of perceived communication quality. If a neighbor uses the same channel, the user will change to a different channel, hunting to find a clear one. All this occurs with no coordination at all between telephones. Figure 3 shows the equivalent network architecture.

With the arrival of second generation technology, cordless telephones will change from stand-alone consumer items to elements of a geographically dispersed network. Network control will remain distributed, with wireless terminals competing with little or no central coordination for available channels. However, their owners will subscribe to public telepoint services and use the cordless phones at thousands of locations with base stations connected to subscriber lines of the public telephone network. In addition, the same instruments will operate in business environments where cordless telephone technology is incorporated in wireless private branch exchanges. In large business premises, cordless telephones will have access to several base stations that provide intercell handoff as a user moves from one location to another during the course of a call.

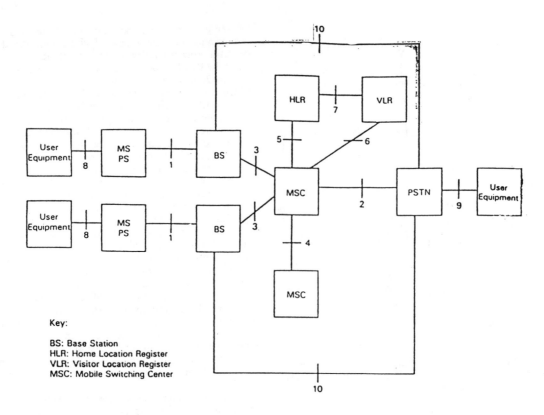

Fig. 4. *Future public land mobile telecommunications system reference model.*

Billing security and privacy will be achieved through the use of handshaking protocols that require base stations and cordless telephones periodically (on the order of once per second) to exchange identification codes while a call is in progress.

Third Generation

CCIR Task Group 8/1, in its work on "Future Land Public Mobile Telecommunications Systems," has developed the functional reference model [13] of Figure 4, which is a slightly modified version of the GSM model in Figure 3. In Figure 4, the wireless terminal is designated Mobile Station (MS) and Portable Station (PS), and a standard interface (number 8) links it to subscriber equipment such as a telephone, a computer, or a facsimile machine.

Anticipating a mass market demand for access to future wireless networks, researchers at the Rutgers WINLAB are studying a cellular packet switch to implement the functions indicated in Figure 4. Packet communications in this application will serve to:

• Reduce substantially the burden on the central controller of radio resource management
• Efficiently multiplex speech, network control information, and a wide variety of user data transmissions
• Harmonize with the anticipated BISDN which will be based on packetized asynchronous transfer mode communications [14]

In the cellular packet switch of Figure 5, all of the base stations in a service area are connected to a metropolitan area network which is also connected to trunks of the public switched telephone network and facilities of a BISDN. Network interface units associated with each Wireless terminal Interface Unit (WIU), central office Trunk Interface Unit (TIU), and Base station Interface Unit (BIU) contain simple routing tables. Based on the information in these routing tables, which

change in a very dynamic manner, the interface units direct each information packet to the correct destination even as users move frequently between base station service areas. All of this is possible because each packet contains, in addition to user information or network control information, address fields indicating the origin and destination of the packet. The routing is based on a combination of virtual circuit packet switching and datagram switching. The principal call-processing task of the cellular controller in Figure 5 is to assign a virtual circuit identifier to a wireless terminal and a PSTN or BISDN trunk at the beginning of each call.

Other interfaces to the metropolitan area network include a Visitor location register Interface Unit (VIU), a Home location register Interface Unit (HIU), and a Gateway Interface Unit (GIU) for connection to another cellular packet switch. Together, they implement in a nonhierarchical manner all of the CCIR interfaces in Figure 4 and a new interface, 11, that links base stations directly. This interface is valuable in providing seamless, mobile-controlled handoff with no loss of information as terminals change cells [15].

Transmission Technologies

In all first generation cellular and cordless products, the means of speech transmission is analog frequency modulation over radio channels that occupy frequency bands ranging in width from 12.5–30 kHz. Frequency division multiple access separates the signals of different terminals and simultaneous two-way communications are based on frequency division duplex. The second generation networks, all using digital speech transmission, will employ a variety of access technologies including narrowband time division with eight channels per carrier (GSM), narrowband time division with three channels per carrier (IS-54), frequency division (CT2), and time division with twelve channels per carrier (DECT). Continuous two-way

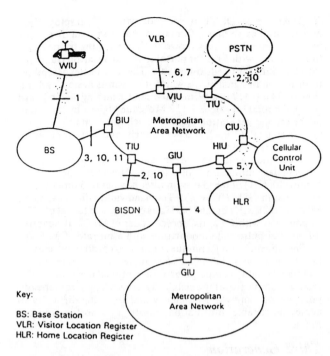

Key:

BS: Base Station
VLR: Visitor Location Register
HLR: Home Location Register

Fig. 5. Cellular packet switch.

communications take place in the two second generation cellular systems by means of frequency division duplex with 45 MHz spacing between the upstream and downstream carriers. Both cordless systems rely on time division duplex for two-way communications.

GSM: Pan-European Digital Cellular

All four systems use burst mode mobile-to-base transmission based on slot and frame structures with varying degrees of complexity. The timing patterns of GSM are by far the most intricate, with time intervals defined as short as 920 ns (one quarter a bit) and as long as 3 h 28 min 53.76 s (encryption hyperframe). The GSM multiple access technique is a combination of frequency division with carriers spaced at 200 kHz intervals and time division with eight logical channels per carrier. During a call, each terminal has access to a two-way digital traffic channel and a separate two-way control channel.

Figures 6 and 7 show how these channels are organized. Figure 6 shows a GSM multiframe of 120 ms duration divided into 26 frames, each containing eight time slots. In the multiframe, 24 frames are for user information carried in logical traffic channels. The other two frames convey system control information in associated control channels. Each call in progress is assigned one of these logical control channels which provide GSM with an out-of-band signaling capability that does not exist in first generation cellular systems. Figure 7 shows the composition of a GSM time slot. Two bursts of 58 b account for most of the transmission time in a slot. Fifty seven data bits carry user information, while the other bit is a flag used to distinguish speech from other transmissions. There is an equalizer training sequence consisting of 26 b in the middle of the time slot, and the slot begins and ends with three "tail bits," all logical zeros. The guard time of 8.25 b intervals prevents overlap at a base station of signal bursts arriving from different terminals. With 156.25 b/577 μs, the transmission rate is 270.833 kb/s. At this rate, an adaptive equalizer is essential in the multipath transmission environment of a cellular system. The bit interval is 3.69 μs and GSM mandates that receivers operate accurately when multiple signal paths have

delay differences up to 16 μs, which is more than four bit intervals.

The GSM modulation is Gaussian minimum shift keying, in which the modulator bandpass filter has a 3 dB cutoff frequency of 81.25 kHz (0.3 times the bit rate). The modulation efficiency of 271 kb/s signals operating with a 200 kHz carrier spacing is 1.35 b/s/Hz. The GSM speech coder is referred to as linear predictive coding with regular pulse excitation [16]. The source rate is 13 kb/s and the transmission rate, including error detecting and correcting codes, is 22.8 kb/s.

In early GSM implementations, each full rate traffic channel will occupy all 24 frames in the multiframe of Figure 6. The associated control channel will be carried in one other frame. Later implementations will add spectrum efficiency by reducing the allocation to 12 frames per half rate channel per multiframe. The speech transmission rate will be 11.4 kb/s and a carrier will have a capacity of 16 user channels. Then both associated control channel frames in Figure 6 will be occupied.

IS-54 North American Digital Cellular

The carrier spacing of IS-54 is 30 kHz, as in the first generation AMPS system [4]. Operating companies will selectively convert analog channels to digital operation in order to relieve traffic congestion at cellular base stations. Each digital channel operates at 48.6 kb/s carrying three user signals. The IS-54 timing structure, shown in Figure 8, is far simpler than that of GSM. The frame duration is 40 ms divided into six 6.67 ms time slots. One time slot carries 324 b, including 260 b of user information ("data" in Figure 8) and 12 b of system control information (Slow Associated Control Channel—SACCH). The remaining 52 b carry a time synchronization signal (28 b), a digital verification color code (12 b), and in the mobile-to-base direction a 6 b guard time interval, when no energy is transmitted, followed by a 6 b ramp interval to allow the transmitter to reach its full output power level. Like GSM, IS-54 defines full rate and half rate logical traffic channels. Initial IS-54 implementations will include only full rate channels, which occupy two slots per 40 ms frame. With 260 b of user information per slot, full rate channels operate at 13 kb/s. Eventually, IS-54 will introduce 6.5 kb/s half rate channels that occupy one slot per frame.

The 28 b synchronization field contains a known bit pattern that allows the receiver to establish bit synchronism and to train an adaptive equalizer. The system specifies six different synchronization patterns, one for each slot in the 40 ms frame. This allows the receiver to lock onto its assigned time slots. The digital verification color code plays the role of the supervisory audio tone in the analog AMPS system. There are 256 8 b color codes, protected by a (12;8;3) Hamming code. Each base station is assigned one of these codes and the verification procedure prevents a receiver from locking onto an interfering signal from a distant cell.

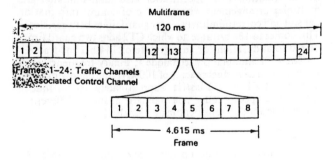

Fig. 6. GSM multiframe and frame structure.

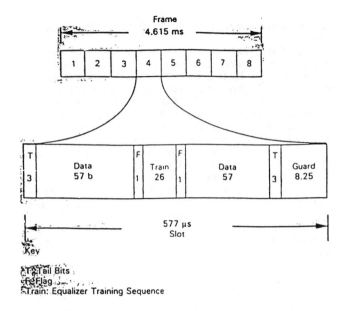

Fig. 7. GSM time slot structure

In a departure from the constant envelope modulations of first generation wireless access systems and other second generation systems, IS-54 adopts a linear modulation technique, Differential Quaternary Phase Shift Keying (DQPSK). The principal advantage of this modulation method is bandwidth efficiency. The transmission rate is 48.6 kb/s with a channel spacing of 30 kHz, which comes to 1.62 b/s/Hz, a 20% improvement over GSM. The main penalty of linear modulation is power efficiency, which affects the weight of hand held portable terminals and time between battery chargings. The specific method of modulation is π/4 shifted DQPSK with root cosine rolloff filtering at the transmitter and receiver. The rolloff factor is 0.35, which places a zero at 16.4 kHz in the baseband spectrum of the transmitted signal. The IS-54 speech coder is a type of codebook excited linear predictive coding referred to as Vector Sum Excited Linear Prediction (VSELP) [17]. The source rate is 7.95 kb/s and the transmission rate is 13 kb/s.

CT2

Of the four second generation wireless information networks covered in this article, CT2 is the only one that uses frequency division multiple access. Carriers spaced at 100 kHz each convey one conversation, using time division duplex for two-way transmission. The bit rate of CT2 is 72 kb/s and its timing structure, shown in Figure 9, is the simplest of the systems we describe. A frame of 2 ms duration contains two time slots, one for portable station to fixed station (mobile-to-base) transmission and the other for fixed-to-portable. Within each slot, there are 64 b for user information and 4 b for system control information. (The specification also admits an alternate multiplex arrangement with only 2 b of control information per slot and correspondingly longer guard times between slots.) This comes to 136 b/frame, to which CT2 adds two guard times of 8 b total duration for an aggregate of 144 b/2 ms frame.

The CT2 modulation technique is binary frequency shift keying. With a channel spacing of 100 kHz, the bandwidth efficiency of CT2 is 0.72 b/s/Hz, approximately half of that of GSM. The speech coder is a standard adaptive differential pulse code modulator operating at 32 kb/s [18].

DECT

Ten carriers, each occupying a bandwidth of 1.728 MHz, will each carry 12 channels in a Time Division Multiple Access

(TDMA) format. DECT, in common with CT2, employs time division duplex so that information moves in both directions over the same carrier. As shown in Figure 10, the frame duration is 10 ms, with 5 ms for portable-to-fixed station transmission and 5 ms for fixed-to-portable. The transmitter assembles information in signal bursts which it transmits in slots of duration 10/24 = 0.417 ms. With 480 b/slot (including a 64 b guard time), the total bit rate is 1.152 Mb/s, the highest by far of any of the second generation systems. Each slot contains 64 b for system control (C, P, Q, and M channels) and 320 b for user information (I channel).

In common with GSM, the modulation technique of DECT is Gaussian minimum shift keying. However, the relative bandwidth (BT) of the Gaussian filter is wider (0.5 times the bit rate) than in GSM (BT = 0.3). The bandwidth efficiency, 1.152 Mb/s in 1.728 MHz or 0.67 b/s/Hz, is comparable to that of CT2. In common with CT2, the speech coder of DECT is adaptive differential pulse code modulation with a bit rate of 32 kb/s.

The transmission techniques of the two cordless systems reflect the emphasis of designers on simple, low-power terminals and signal quality comparable to conventional telephones. The relative complexity of the cellular systems derives from the importance of bandwidth efficiency and from the difficulty of achieving reliable information transfer in a hostile environment.

Third Generation

The multiple access scheme under investigation at Rutgers WINLAB is Packet Reservation Multiple Access (PRMA) [19]. PRMA, a close relative of reservation ALOHA [20] [21], can be viewed as a combination of TDMA and slotted ALOHA.

In common with TDMA, the channel bit stream is organized in slots and frames. Within a frame, terminals recognize each slot as either available or reserved on the basis of a feedback packet broadcast in the previous frame from the base station to all of the terminals. As in slotted ALOHA, terminals with new information to transmit contend for access to available slots. At the end of each slot, the base station broadcasts the feedback packet that reports the result of the transmission. A terminal that succeeds in sending a packet to the base station obtains a reservation for exclusive use of the corresponding time slot in subsequent frames.

The frame duration is chosen so that a speech terminal generates one packet per frame. Upon gaining a reservation, the terminal continues to use its reserved slot until it no longer has an information packet to transmit. The base station, on receiving no packet in the reserved slot, informs all terminals that the slot is once again available. In subsequent frames, terminals with new information contend for that slot, and for other available slots. The contention mechanism is based on a permission probability, p, a design constant for all terminals in a given PRMA system. A contending terminal waits for an available slot and, with probability p, transmits a packet to the base in the available slot. In our studies to date, we have assumed that the transmission succeeds if and only if one terminal transmits in a slot.

Table I. PRMA Variables

Channel Rate	720 kb/s
Source Rate	32 kb/s
Frame Duration	16 ms
Overhead	64 b/packet
Maximum Delay	32 ms

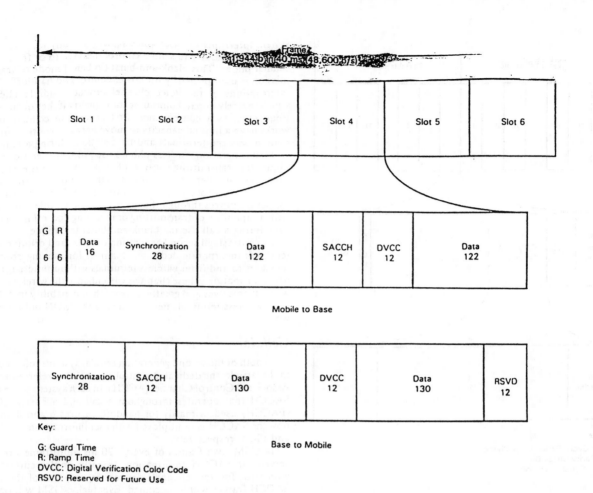

Key:

G: Guard Time
R: Ramp Time
DVCC: Digital Verification Color Code
RSVD: Reserved for Future Use

Mobile to Base

Base to Mobile

Fig. 8. IS-54 slots and frame.

Because each terminal with a reservation has exclusive use of one slot in each frame, the terminals with reservations share the channel as in TDMA. With terminals accessing the channel autonomously, PRMA lends itself well to operation with speech activity detectors. A speech activity detector classifies a speech signal at each instant as either talking or silent so that the voice terminal generates bursts of packets corresponding to talkspurts. No packets are generated during the silent gaps that punctuate the talkspurts.

Figure 11 illustrates the operation of PRMA. There are eight time slots per frame and the base station feedback packets for frame k − 1 have established that in frame k, six slots are reserved by terminals 11, 5, 3, 1, 8, and 2. Two slots, 3 and 7, are available in frame k. At the beginning of the frame, terminals 6 and 4 are contending for access to the channel. Both of these terminals obtain permission to transmit in slot 3 and because their packets collide, neither obtains a reservation. In slot 7, both terminals fail to obtain permission to transmit and thus both remain in the contending state at the beginning of frame k + 1. Meanwhile, in frame k − 1, terminal 3 transmitted the final packet in its talkspurt. Therefore, in frame k (slot 4) it does not use its reservation. The base station feedback packet for slot 4 of frame k indicates that slot 4 is available in frame k + 1.

In frame k + 1, neither terminal 6 nor terminal 4 has permission to transmit in slot 3. In slot 4, terminal 4 has permission, but terminal 6 does not. Thus, terminal 4 gains a reservation for slot 4. Terminal 6 obtains permission to transmit in slot 7 and reserves that slot in frame k + 2. In frame k + 1, terminal 8

gives up its reservation of slot 6, and a talkspurt begins at terminal 12 which enters the contending state. In frame k + 2, terminal 12 gains a reservation (slot 3) and terminal 1 releases its reservation (slot 5).

At any time the number of available slots depends on the number of terminals in the talking state, a quantity that is subject to statistical fluctuation. When there are few available slots, terminals with new talkspurts are subject to relatively long delays in obtaining reservations. Unlike packet data systems, which respond to congestion by storing packets until they can be transmitted, packet speech systems must deliver pack-

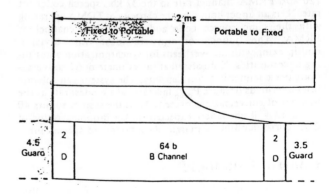

Fig. 9. CT2 slot and frame.

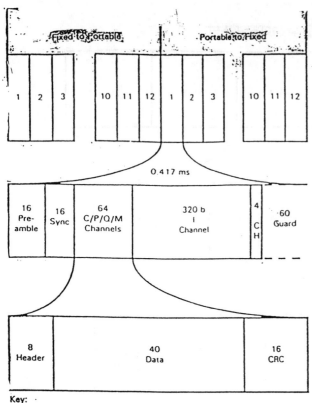

0.417 ms

| 16 Preamble | 16 Sync | 64 C/P/Q/M Channels | 320 b I Channel | 4 C H | 60 Guard |

| 8 Header | 40 Data | 16 CRC |

Key:

CRC: Cyclic Redundancy Check
Ch: Parity Check Bit
Sync: Synchronization

Fig. 10. DECT slot and frame.

ets promptly. In PRMA, any packet held beyond a certain specified limit is discarded by its terminal. Thus, a measure of PRMA performance is P_{drop}, the probability of packet dropping. To estimate system capacity, we set the limit $P_{drop} \leq 0.01$, and find the maximum number of terminals that can operate within this limit.

Table I shows the design values of a trial PRMA configuration that would be appropriate for indoor operation. For the parameters of Table I, Figure 12 displays the relationship of P_{drop} to the number of simultaneous conversations. The maximum number of simultaneous conversations with $P_{drop} \leq 0.01$ is 37, which we define as system capacity. This number can be compared with $N_c \doteq 22.5$ channels per carrier, the ratio of the 720 kb/s PRMA channel rate to the 32 kb/s speech coder bit rate. N_c is an upper bound on the capacity of TDMA operating with the same source code and transmission channel as PRMA. The bound conforms to ideal time division multiplexing with no overheads for synchronization and timing uncertainties. Note also that the estimate of 37 conversations per channel is a soft capacity. The system can tolerate brief overloads because P_{drop} increases at a modest rate as the number of conversations exceeds 37. If the system admits 40 users, all will encounter expected packet dropping ratios of 0.02, corresponding to marginally substandard speech quality.

Control Channels

For call setup, cellular systems rely on broadcast control channels (base-to-mobile) and shared random access control

system interrupts the user voice signal (typically for 100 ms) when it is necessary to send a control message to or from the cellular phone. This blank-and-burst (in band) mode of transmission is used primarily for power control and handoff. The interruptions of the voice channel produce audible clicks, which severely impair communication quality if they occur too frequently. As a consequence, first generation cellular networks have a limited capacity to move network control information between terminals and base stations during a call.

First generation cordless phones use product-specific techniques for transmitting control information. With the exception of Japan, where the generous spectrum allocation includes two shared control channels, all cordless telephone control information moves through the voice channel. Cordless phones with a capability of transmitting or receiving control information during a call use the blank-and-burst technique.

By contrast, all second generation cordless and cellular systems will incorporate dedicated channels for moving control messages to and from wireless terminals without interrupting user information. These channels will enhance the wireless networks themselves and greatly increase their capability to deliver to wireless terminals new features of the ISDN public network.

Cellular

In both of the second generation cellular systems referred to in this article, the dedicated control channels are referred to as Associated Control Channel (ACCH). In each system there is a SACCH that operates throughout a call and a Fast ACCH (FACCH) used primarily for handoff. Figures 6 and 8 show how the SACCH is multiplexed with user information in GSM and IS-54, respectively.

In GSM, two frames of every 120 ms multiframe are reserved for SACCH information. In early implementations, with eight full rate channels per carrier, only one of the two SACCH frames will be occupied. Eventually, GSM will operate with 16 half rate channels per carrier, and they will use all 16 slots in the two SACCH frames in Figure 6. Thus, each active terminal owns a slot in one SACCH frame. A SACCH message contains 184 information bits. Error control coders process this information to produce 456 channel bits which are

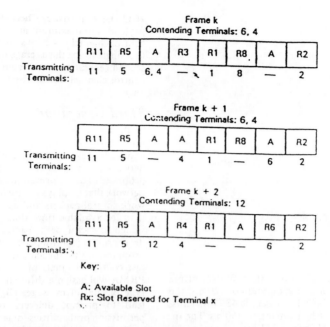

Fig. 11. PRMA protocol operation example.

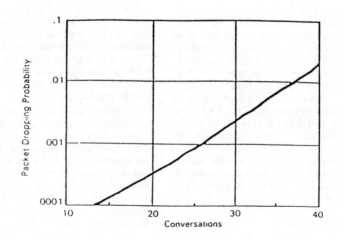

Fig. 12. PRMA packet dropping probability.

16 Cyclic Redundancy Check (CRC) bits in addition to the 48 b payload. The maximum information throughput of the DECT control channel is 4.8 kb/s. With each control word contained in a single frame the average delay is 5 ms. Clearly, among the second generation systems surveyed in this article, the DECT control channel has the best throughput and delay performance.

Third Generation

Packet communication networks perform statistical multiplexing of many kinds of information. Because the header of each packet identifies the type of information carried, the network can process and route each packet in the manner appropriate to the information carried. In a wireless information network that employs the PRMA transmission technique, network control packets and user information packets both contend for available time slots. The permission probability assigned to each type of packet ensures that the system meets delay criteria (control packets or user data packets) and packet dropping limits (speech packets). Corresponding to slow and fast control channels of second generation cellular systems, PRMA will assign a different permission probability to each type of control message. High-permission probabilities will lead to prompt delivery of important messages. Low-permission probabilities for less urgent messages will limit network congestion and maintain channel throughput.

Preliminary studies are underway to determine design principles and combined user-information, control-information capacity [23]. Given the bursty, intermittent nature of control message transmission, we anticipate that the statistical multiplexing provided by a packet network will lead to greater overall efficiency than second generation techniques which assign a constant proportion of network resources to control channels.

System Implementation

Second generation wireless information networks, conforming to the CT2 cordless telephone standard, are presently in operation in the United Kingdom, where telepoint services are available at a growing number of public locations. While initial networks use proprietary transmission techniques, future base stations and terminals will conform to a common air interface that will insure compatibility of cordless phones with the base stations of all network operators. The two second generation cellular systems, GSM and IS-54, are due for introduction in 1991, the year in which the DECT standard will be established in final form. Frequency allocations for DECT and initial products are scheduled to be in place in 1992. Meanwhile, at least one business cordless system with a transmission technique similar to that of the DECT standard has appeared on the market [24].

In North America, a government policy that allows cellular operating companies considerable flexibility in adopting transmission technologies has simulated lively activity outside of the official standards process. The most prominent effort is by Qualcomm which, in cooperation with several operating companies and equipment vendors, is developing a dual mode system with digital transmission based on spread spectrum Code Division Multiple Access (CDMA) [25]. Initial equipment demonstrations of CDMA and enthusiastic capacity estimates by Qualcomm have led many companies to defer decisions on deployment of dual mode systems pending the outcome of further trials. In response to this uncertainty, Motorola has developed analog dual mode cellular equipment that divides each 30 kHz AMPS channel into three 10 kHz bands, each carrying a frequency modulated voice signal and control information on a subcarrier. It appears likely that some companies will adopt this analog approach to capacity enhancement while others

interleaved and distributed over four time slots. With one time slot per multiframe and a multiframe duration of 120 ms, the total transmission time of a SACCH message is 480 ms. Thus, the transmission bit rate is 456 b in 480 ms, or 950 b/s. The information rate is 184 b in 480 ms, or 383 b/s.

In IS-54, a control channel message has a length of 66 b, which in the SACCH are protected by a rate 1/2 forward error correcting coder that generates 132 b per message. This information is interleaved over 12 time slots and transmitted in the SACCH fields as shown in Figure 8. In a system operating with full rate digital traffic channels, the total delay (12 slots) of the SACCH is six IS-54 frames, or 240 ms, and the information bit rate is 300 b/s. With half rate channels, the SACCH rate information rate will be 150 b/s and the message delay will be 480 ms.

In both GSM and IS-54, the FACCH preempts user information frames to achieve rapid handoff. While this mode of operation is a return to the blank-and-burst technique of the first generation, both systems use signal processing to estimate the speech that was deleted to make room for a FACCH message.

Cordless

Figures 9 and 10 indicate that as in IS-54, each burst of user information in CT2 and DECT is accompanied by system control information. The CT2 control channels are referred to as D channels. As indicated in Figure 9, each time slot contains four D-channel bits for a total bit rate of 2 kb/s in each direction (or 1 kb/s with terminals that operate with two, rather than four, D-channel bits per time slot). Control messages consist of one to six 64 b code words. A code word is distributed over 16 frames and, therefore, has a delivery time of 32 ms. Each code word is protected by 16 error-detecting parity bits. The receiver acknowledges code words that satisfy all parity checks. Upon detecting transmission errors, the receiver requests retransmission of the affected code word. With 48 information bits per code word, the maximum information throughput (in the absence of retransmission) of the CT2 control channel is 3 kb/s, far in excess of the signaling capability of the second generation cellular systems.

DECT control information is carried by 64 b in every time slot of an established communication. These bits are assigned to one of four logical channels (either the C, P, Q, or M channel) depending on the nature of the control information. Thus the raw data rate is 6.4 kb/s. As in CT2, DECT relies on error detection and retransmission for accurate delivery of control information, rather than the forward error correction of the cellular systems. Thus, every 64 b control word contains

gain experience with one or both of the digital approaches. Meanwhile, Hughes Network Systems has demonstrated an extended TDMA technique that will provide capacity enhancements to IS-54 systems. All in all, therefore, North America is moving from a single continental analog cellular system to at least three dual mode systems, just as Europe is converging from five incompatible analog systems to the GSM standard.

GSM will gain even more momentum in Europe with the introduction of Personal Communication Networks (PCN) in the United Kingdom. Adding to two existing cellular service operators, British regulatory authorities have recently licensed three new companies to introduce PCN in 1992. All three will operate with the GSM transmission standard, but in newly allocated frequency bands around 1.8 GHz. In the U.S., the Federal Communications Commission has authorized one company, PCN America, to operate PCN on an experimental basis in two cities. Like the United Kingdom's PCN, these networks will be aimed primarily at people moving about in cities at pedestrian speeds with hand held telephones. In a departure from other second generation networks, the American PCN will use a spread spectrum multiple access technique. One purpose of the field trial will be to determine whether spread spectrum PCN can share frequency bands assigned to other applications, such as point-to-point microwave [26].

It is certain that as the second generation systems mature, efforts to transform the third generation from vision to reality will accelerate. In contrast to other areas of information technology, such as high-speed transmission, where technical capability precedes demand, there is already widespread public perception of the benefits of wireless access to information networks. The latent demand is held in check by high prices and limited quality, while everyone eagerly awaits the technology that will make it possible to use advanced information services in the natural and convenient manner afforded by wireless information networks.

References

[1] J. Walker, ed., *Mobile Information Systems*, Artech House, p. 75, 1990.

[2] B. J. T. Mallinder, "An Overview of the GSM System," *Digital Cellular Radio Conf.*, Hagen, Germany, pp. 1a/1–1a/13, Oct. 1988.

[3] Electronic Industries Association, "Dual-Mode Subscriber Equipment-Network Equipment Compatibility Specification," *Interim Standard 54*, Dec. 1989.

[4] Electronic Industries Association, "Mobile-Standard Land Station Compatibility Specification," *American National Standard EIA/TIA 553–1989*, Apr. 1989.

[5] R. S. Swain and D. W. J. Holmes, "The Digital Cordless Telecommunications Common Air Interface," *British Telecom Tech. J.*, vol. 8, no. 1, pp. 12–18, Jan. 1990.

[6] H. Ochsner, "DECT—Digital European Cordless Telecommunications," *39th IEEE Vehicular Tech. Conf.*, San Francisco, pp. 718–721, May 1989.

[7] M. H. Callendar, "International Standards for Personal Communications," *39th IEEE Vehicular Tech. Conf.*, San Francisco, pp. 722–728, May 1989.

[8] D. Grillo and G. MacNamee, "European Perspectives on Third Generation Personal Communication Systems," *40th IEEE Int'l Vehicular Tech. Conf.*, Orlando, pp. 135–140, May 1990.

[9] D. C. Cox, "Universal Portable Radio Communications," *Proc. IEEE*, vol. 75, no. 4, pp. 436–477, Apr. 1987.

[10] D. J. Goodman, "Cellular Packet Communications," *IEEE Trans. on Commun.*, vol. 38, no. 8, pp. 1,272–1,280, Aug. 1990.

[11] W. R. Young, "Advanced Mobile Phone Service: Introduction Background and Objectives," *Bell Sys. Tech. J.*, vol. 58, no. 1, part 3, pp. 1–14, Jan. 1979.

[12] Electronic Industries Association, "Cellular Radiotelecommunications Intersystem Operations," *Interim Standard 41*, Feb. 1988.

[13] International Consultative Committee on Radio (CCIR), "Future Public Land Mobile Telecommunication Systems," Report M/8, XVII Plenary Assembly, Dusseldorf, 1990.

[14] R. Gallassi, G. Rigolio, and L. Verri, "Resource Management and Dimensioning in ATM Networks," *IEEE Network Mag.*, vol. 4, no. 3, pp. 8–17, May 1990.

[15] K. S. Meier-Hellstern, G. P. Pollini, D. J. Goodman, "A Wireless Service for the IEEE 802.6 Metropolitan Area Network," Rutgers University, WINLAB Technical Report.

[16] P. Vary et al., "Speech Codec for the European Mobile Radio System," *Int'l Conf. on Acoustics, Speech, and Signal Processing*, New York, pp. 227–230, Apr. 1988.

[17] I. A. Gerson and M. A. Jasiuk, "Vector Sum Excited Linear Prediction (VSELP) Speech Coding at 8 kb/s," *Int'l Conf. on Acoustics, Speech, and Signal Processing*, Albuquerque, pp. 461–464, Apr. 1990.

[18] CCITT, "32-kb/s Adaptive Differential Pulse Code Modulation (ADPCM)," Recommendation G.721, fascicle III.3, Malaga-Torremolinos, 1984.

[19] D. J. Goodman and S. X. Wei, "Efficiency of Packet Reservation Multiple Access," *IEEE Trans. on Vehicular Tech.*, vol. 40, no. 1, pp. 170–176, Feb. 1991.

[20] S. Tasaka, "Stability and Performance of the R-ALOHA Protocol," *IEEE Trans. on Comp.*, vol. C-32, no. 8, pp. 717–726, Aug. 1983.

[21] S. S. Lam, "Packet Broadcast Networks—A Performance Analysis of the R-ALOHA Protocol," *IEEE Trans. on Comp.*, vol. C-29, no. 7, pp. 596–603, July 1980.

[22] W. C. Y. Lee, *Mobile Cellular Telecommunications Systems*, McGraw Hill, pp. 78–91, 1988.

[23] S. Nanda, "Analysis of Packet Reservation Multiple Access: Voice Data Integration for Wireless Networks," *IEEE Globecom Conf.*, San Diego, pp. 1,984–1,988, Dec. 1990.

[24] C. Buckingham, G. K. Wolterink, and D. Åkerberg, "A Business Cordless PABX Telephone System on 800 MHz Based on the DECT Technology," *IEEE Commun. Mag.*, vol. 29, no. 1, pp. 105–110, Jan. 1991.

[25] K. S. Gilhousen et al., "On the Capacity of a Cellular CDMA System," *IEEE Trans. on Vehicular Tech.*, to be published May 1991.

[26] D. L. Schilling et al., "Shared Spectrum Capability of CDMA Personal Communications Networks," *IEEE Soutcon/91 Conf. Record*, Atlanta, pp. 129–138, Mar. 1991.

Biography

David J. Goodman received a B.S. degree from Rensselaer Polytechnic Institute, a M.S. degree from New York University, and a Ph. D. degree from Imperial College, University of London. He is a Fellow of the Institute of Electrical and Electronic Engineers and the Institution of Electrical Engineers (London).

He is Director of the Wireless Information Network Laboratory (WINLAB) an industry/university/government cooperative research center at Rutgers, the State University of New Jersey. He is also Professor and Chairperson of Electrical and Computer Engineering at Rutgers. In 1988, prior to joining Rutgers, he enjoyed a twenty year research career at AT&T Bell Laboratories where he was a Department Head in the Communication Systems Research Laboratory. Mr. Goodman's research innovations in digital signal processing, speech coding, and wireless information networks are embodied in several international standards and commercial products.

Reprinted from *IEEE Communications Magazine*, pp.52-71, Nov., 1991.

The Wireless Revolution

Rapid multinational progress will soon make global wireless communication a ubiquitous reality.

Theodore S. Rappaport

O ver the past three years, the interest in wireless communications has been nothing less than spectacular. Cellular radio systems around the world have been enjoying 33 percent to 50 percent growth rates. Many paging services have been gaining customers at a rate of 30 percent to 70 percent or more per year, and within the last two years there has been intense corporate research and development aimed at commercializing new wireless communication services called PCS (personal communications services). Meanwhile, new digital cellular technologies have been installed in Europe, and developing nations are beginning to install cellular infrastructure.

The first wide-scale adoption of a wireless personal communications system was in citizens band (CB) radio during the late 1960s and early 1970s. Although it was a victim of its own success due to a rapid and uncontrolled saturation of the radio spectrum, and suffered severely from lack of traffic management and services, the staggering acceptance of CB radio was a clear indication that consumers wanted an inexpensive, portable means of communicating. A more modern wireless device that has enjoyed widespread popularity is the cordless telephone. By the time you read this article, it is likely that more than 65 million cordless telephones will have been sold in the United States alone (although it is estimated that almost half of the cordless phones sold have been discarded or are not used). The convenience of portable telecommunications offered by a device as simple as the cordless phone is clearly in demand. Although the design solutions for a ubiquitous wireless communications network will be tremendously complex, all market indications show that consumers wish to have a small device, similar in size to a cordless phone, that would allow them to make and receive phone calls wherever they are. The idea behind wireless personal communications networks (PCNs) is to make communications truly personal, so that anyone can place a call to anyone else. Much like the stereo Walkman™ or the laptop computer, PCN will permit truly ubiquitous access regardless of the location of the user at the time of access. Today, the terms PCN and PCS often are used interchangeably. PCN refers to a concept by which a person can use a single communicator anywhere in a large geographic area. PCS refers to a service that may not embody all of the PCN concepts, but is more personalized (i.e., lightweight terminal, better performance, more flexibility, user options, etc.) than a present-day cellular telephone.

Current Demand

The premise that wireless personal communications is emerging as a key, wide-sweeping technology that will dramatically impact our society is supported in numerous sources, including worldwide trade journals and government agency reports. As an example, in the United States there were more than 6.3 million cellular telephone users as of September 1991 [18]. This compares with 25,000 users in 1984, and 2.5 million U.S. users in late 1989 [1]. It is clear to most industry experts that the Cellular Telephone Industry Association's (CTIA) 1989 projections of 10 million United States cellular users by 1995 will be exceeded in late 1992, and cellular radio carriers are enjoying exponential increases in service subscriptions. This demand for mobile/portable telecommunications is a worldwide trend and is particularly acute in Europe. For example, cellular telephone in Sweden already enjoys a 6.6 percent adult market penetration [18], and this figure has been increasing by more than 0.1 percent per month. Finland, Norway, and France have been experiencing similar growth rates. In Hong Kong, more than 50 percent of the adult population own or have operated cellular telephones. Although no spectrum has yet been allocated to emerging PCS in the United States, industry analysts already are projecting annual U.S. revenues to be between $33 billion and $55 billion by the year 2000 [17].

In the United Kingdom, viewed by many as the leading country for PCS initiatives, three major companies are investing hundreds of millions of dollars to install an infrastructure that may eventually allow citizens to use small 10 mW portable terminals to place and receive calls in populated areas throughout the country. At worst, PCS will provide relief for users who operate in congested 900 MHz United Kingdom cellular markets. At best PCS will offer customers the ability to use a single wireless communications unit for home, office, or automobile, thereby obviating the need for a traditional wired phone to the home. Two of the three companies involved, Microtel and Unitel, are actually consortia consisting of many leading telecommunications companies. The third United Kingdom PCN service provider, Mercury, is the leading non-wireline service provider of England's

Theodore Rappaport is a faculty member of the Virginia Polytechnic Institute, the State University in Blacksburg, Virginia, and at the Mobile and Portable Radio Research Group (MPRG).

0163-6804 91/S01.00 1991© IEEE

present analog cellular system. These PCN companies are expected to become some of the biggest advertisers in the United Kingdom throughout this decade, as they strive to pioneer and market the revolutionary PCN service [19].

The pervasiveness of PCN will be made available by an immense infrastructure of low-power, suitcase-sized base stations which will provide portable subscribers with wireless access to the local telephone loop in populated areas. PCN hopes to be able to offer wireline communications quality using radio as a transmission medium. Base stations will be placed on lamp posts, roofs, and ceilings of buildings and concourses, and in other locations where people congregate. Backbone links that connect PCN base stations to other PCN base stations and to the public switched telephone network (PSTN) will be supplied by one of two methods: either via the existing telephone wire or fiber plant, or via line-of-sight microwave links. Of course, the wide-scale deployment of such an extensive high-grade wireless telephone system will require engineering tools and techniques and antenna designs that allow rapid and accurate propagation prediction and system design. The radio coverage of each base station will intentionally be limited by low transmitter power, so that the same frequencies can be re-used many times within a few city blocks. Depending on regulatory decisions by the British Post Office, PCN could compete directly with wired residential telephone service in the United Kingdom. Thus, it is conceivable that the customers could bypass the telephone company and use a single PCN terminal for communications at home or in the office.

In the United States, more than 65 experimental licenses have been issued within the last year to regional bell operating companies (RBOCs), non-wireline cellular service providers, manufacturers, cable companies, and new start-up companies hoping to pioneer PCN service. Experimental FCC licenses that allow limited use of radio transmissions for portable FAX service, wireless spread spectrum PABX systems, and microcellular and personal communication services have been granted and trials are in various phases.

To increase the competition and development time of new wireless technologies and services, the FCC created a special incentive, called the pioneer's preference, that offers the exclusive use of reallocated radio spectrum to companies that first demonstrate new technologies or concepts for PCN or other new wireless services. Many of the United States PCN experimenters have filed petitions for rulemaking to the FCC requesting new dedicated or shared spectrum for PCN services, and hope to secure an advantageous position through pioneer's preference. Industry and government experts view the demand for wireless personal communications to be so great that the FCC has indicated that underutilized portions of existing radio spectrum could be subject to reallocation in order to accommodate consumer demand.

Several good tutorial articles that describe the impetus and technological challenges behind cellular radio and PCN can be found in the August 1990 and September 1990 issues of the *IEEE Communications Magazine*. The different second generation standards that have emerged around the world are described in several invited papers in the May 1991 Issue of the *IEEE Transactions on Vehicular Technology (VT)* which features digital cellular technologies. As indicated in the *VT* special issue, it is conceivable that 50 million to 75 million subscribers could be using wireless systems for various types of personal communications by the mid-1990s. This is corroborated by a recent Morgan Stanley report that predicts cellular and PCN systems will achieve at least 12 percent market penetration in many developed countries by the end of this century.

Recent Events in Wireless

In large United States markets such as Los Angeles and New York, where hundreds of thousands of users can access the cellular radio spectrum, the 832 cellular analog-FM voice channels are unable to accommodate the number of users, and methods to improve capacity and cellular system design are desperately needed. For those not familiar with cellular radio, in each market (i.e., city) there are two cellular service providers: the radio common carrier (non-wireline provider), called the A channel provider, and the wireline, or B channel, provider. Each of the two service providers are allocated 416 duplex voice channels in a 25 MHz spectrum allocation. Each voice channel is comprised of a 30 kHz base-to-mobile link and a 30 kHz mobile-to-base link.

Over the past four years, numerous standards have been proposed for digital cellular radio communication interfaces throughout the world. Digital modulation offers improved spectral efficiency and simultaneously offers better speech intelligibility for a given carrier-to-interference ratio (C/I). More importantly, digital modulation accommodates powerful digital speech coding techniques that further reduce the spectrum occupancy of voice users. Using digital transmission formats, service providers will be able to offer customers additional features, such as dynamically allocated data services, encryption, etc.

In early 1990, the CTIA and the Telecommunications Industry Association (TIA) approved Interim Standard 54, which specifies a dual-mode cellular radio transceiver that uses both the analog FM (the present-day United States Advanced Mobile Phone System, or AMPS, standard) and a linearized $\pi/4$ Differential Quadrature Phase Shift Keying modulation format with a Code Excited Linear Predictive (CELP) speech coder (called the U.S. Digital Cellular, or USDC standard) [20].

The USDC standard offers roughly three times the capacity improvement over AMPS by providing three voice channels that are time-division multiplexed (TDMA) on a single AMPS 30 kHz FM voice link. With further speech coding improvements, six times capacity is likely to be achieved by 1994. The dual mode equipment will allow a graceful transition from analog FM to digital cellular, since cellular operators will be able to change out analog channels to digital channels depending on their geographic capacity demands. In this manner, customers with analog phones will be assured service in any market until some announced time in the future.

A rural cellular carrier that does not suffer great capacity demand would likely stay with AMPS for as long as possible, probably several years (until the mid-1990s), while a metropolitan operator would likely change out AMPS for USDC more

In large United States markets, the 832 cellular analog-FM channels are unable to accommodate the number of users.

In contrast with USDC, the European digital cellular system (Group Special Mobile, or GSM) was developed for a new spectrum allocation in the 900 MHz band.

rapidly (within two years). It is interesting (and also a bit troubling to those who invested a great deal of time and money into developing the IS-54 standard and who plan to abide by it) that since the introduction of the USDC standard IS-54, numerous vendors have introduced their own competing standards that are incompatible with IS-54[20]. Now virtually all major cellular radio service providers are conducting field trials to evaluate new, competing standards, so it is unclear if U.S. digital cellular radio systems will be completely compatible throughout the country, even though this was the major goal of USDC.

In contrast to USDC, the European digital cellular system (called Group Special Mobile, or GSM) was developed for a brand new spectrum allocation in the 900 MHz band. That is, GSM was developed to ensure that a single access and equipment standard would be used throughout the European continent.

Unlike USDC, which hoped to make a seamless transition of America's analog system to digital, GSM was developed from scratch and made pan-European compatibility its primary objective. Today, in most European countries, the cellular systems and standards are unique to the individual countries. In fact, at the time of this writing, a cellular phone that works in France cannot be used in England. GSM operates in new spectrum at 900 MHz dedicated for use throughout Europe. Details of the GSM specification are given in [21, 25] and equipment will be available to the pan-European community by the time this article is published.

GSM is the world's first TDMA cellular system standard, and uses a constant envelope modulation format to gain power efficiency (constant envelope modulation allows more efficient class-C power amplifiers to be used) over spectral efficiency (constant envelope modulation has a larger bits per hertz of RF occupancy than does linear modulation). GSM uses an equalizer and slow frequency hopping to overcome multipath effects that cause intersymbol interference and, thus, irreducible bit error rates. As the world's first digital cellular radio standard that has been adopted by a large market, GSM is viewed as a front-runner for early implementation of PCN throughout the world. Depending on the success of United Kingdom 1800 MHz PCN initiatives, GSM equipment could be implemented on a worldwide scale if spectrum allocations are made available in other countries.

In February 1990, just after CTIA adopted IS-54, QUALCOMM, Inc. proposed to CTIA the use of spread spectrum and sophisticated base station signal processing to offer capacity improvements ten times greater than AMPS (see the paper by Gilhousen, et. al. in the May 1991 *IEEE Trans. VT* special issue on Digital Cellular Technologies). Major cellular radio service providers and manufacturers have steadily supported QUALCOMM as they develop prototype spread spectrum cellular phones that will be ready by mid-1992 [22]. Work conducted two decades ago [26], and more recently [29], confirms that spread-spectrum holds great promise for accommodating huge capacity with simple frequency management, although the capacity is highly dependent on radio path loss within the service area. Of course, higher capacity means higher revenues and less churn (loss of customers) for cellular service providers, so the QUAL-

COMM proposal received immediate attention. Today, many United States companies (e.g., American Personal Communications, PCN America, and Omnipoint Data) are looking at the viability of overlaying TDMA, FDMA, and spread spectrum mobile radio systems on existing point-to-point microwave links, used by utility companies, banks, and public safety offices, in the 1850 MHz -1990 MHz band. There are many others who are aggressively conducting research and experimentation to determine overlay possibilities, and their experimental results may be obtained at the FCC office in Washington, D.C. Recent FCC experiment reports filed by American Personal Communications [36] and Telesis Technology Laboratories [37] indicate that even in the largest urban United States markets, it may be possible to break the existing 1850 MHz to 1990 MHz band into 5 MHz segments which could be used simultaneously by line-of-sight microwave users and PCS providers.

The concept behind overlay is that low-power PCN service could be offered directly on top of the existing terrestrial microwave band. The reasoning is that since existing microwave links use directional antennas, and thus do not radiate over a large geographical area, it could be possible to intertwine PCS frequencies over a geographical area such that existing microwave links are not interfered with by overlaid PCS. Recent experiments and analysis seem to suggest that existing microwave links have not been engineered throughout large markets to take full advantage of the spectrum.

References [23] and [24] discuss the concept behind overlay, and some techniques that could be used to minimize interference between existing and new users. Recent propagation measurements to test the levels of interference caused by PCN subscriber units to existing fixed microwave users were presented in [24] and more recently have been detailed in FCC reports [37, 38] as part of experimental PCN licenses.

Presently, major telecommunication companies are researching the effectiveness of overlay systems from a capacity standpoint. If a sufficient grade of service could be offered through spectrum sharing between new wireless service providers and existing line-of-sight microwave licensees, then the FCC would be able to instantly accommodate market demand for wireless cellular/PCN without a major new spectrum allocation. The fear, of course, is that there could be an unsatisfactory degradation of service to both the existing microwave users and the overlayed PCN users. The overlay concept is being tested by numerous companies in the United States under the FCC experimental license program. At present the FCC has not made a decision on their position on the matter.

The cable industry also has been watching the sudden growth in cellular/PCN throughout the world. Cable companies have a massive RF network installed throughout populated regions, and it is obvious that telecommunication services could be offered over the existing cable plant. In fact, PCN base stations could be installed easily in residential areas by splicing existing cable runs and tapping on base stations mounted on lamp posts or inside buildings. Indeed, several United States cable companies are conducting experiments to determine the feasibility of PCS using the cable plant. When one realizes that United States cable operators already have spectrum allocated for microwave feeds in the 2 GHz and 13 GHz bands, the possibility of utilizing the cable spectrum for PCN services instead of point-to-point feeders presents a lucrative new business opportunity for the cable industry.

For local loop, or premises applications, which merge voice and data, Bellcore [15] and the European Technological Standards Institute (ETSI) have proposed digital TDMA standards that offer between 400 kilobits per second (kb/s) and 1100 kb/s data rates in office and residential environments. Although the widespread deployment of such standards likely will rely on new dedicated spectrum for the services, significant engineering manpower has been devoted to develop the standards, and a great deal can be learned from the research.

Bellcore's system exploits the slow time-varying nature of indoor channels and uses antenna polarization diversity to improve link performance between a base station (port) and mobile terminal (portable). Bellcore's proposal limits the data rate to 450 kb/s based on an extensive measurement program that determined worst case multipath channels in a large number of buildings, houses, and cities. Bellcore's approach minimizes the complexity and battery drain of the portable, since power-hungry adaptive equalizers can be avoided. Also, Bellcore's system uses a novel oversampling demodulation technique that allows a receiver to lock coherently onto the incoming modulation with only a couple of bits of overhead. Reference [15] provides additional details about the Bellcore system, and [21, 25, 34, and 35] provide additional information about ETSI and the DECT standard.

Funding in Wireless

The burgeoning wireless communications industry has created an interesting problem for wireless manufacturers and service providers throughout the world. Because the field of cellular radio and personal communications is changing so rapidly, and since the field involves system concepts seldom taught at universities, wireless companies are having difficulty finding entry-level graduates with sufficient education to make an immediate contribution in research or design. In particular, a large number of universities presently do not offer undergraduate or graduate courses on the topics of mobile radio propagation or wireless communication system design. Consequently, recruiters are forced to raid competing companies for more senior personnel, and resign themselves to spending six months to two years to train new graduates in the art and science of mobile and portable radio.

In an informal survey of some of the largest cellular radio companies in the United States, the author has learned that engineers with knowledge of mobile radio communications change jobs often, and are in extremely high demand. Particularly in the past two years, since cellular radio has enjoyed a 50 percent annual growth rate and new digital systems have been proposed and tested, engineers with experience in cell-site design or computer simulation expertise in mobile radio propagation, traffic modeling, antenna design, and digital signal processing have been highly sought after by the wireless industry.

The National Aeronautics and Space Admin-

istration (NASA), through its research programs in space communications and mobile satellite communications experiments (MSAT-X), and more recently through its Advanced Communications Technology Satellite (ACTS) program, has provided the bulk of United States federal research funding in the mobile communications area. Although mobile satellite systems do not promise nearly the same capacity (users per MHz) as do land-based microcellular and PCS systems, there is remarkable commonality between the technologies used in satellite and cellular mobile systems. Factors such as modulation, coding, multiple access, antenna design, network control, and propagation are fundamental to both types of wireless communication systems, as are the techniques used to design and analyze such systems.

Through the NASA MSAT program alone, at least 11 United States universities and numerous companies have been able to conduct multiple-year research projects that have produced both knowledgeable graduates and innovative technologies. Today, many graduates of NASA-funded communications research programs are working in the satellite or cellular industries. Examples of two different satellite systems for personal communications are described in [38] and [39].

Until recently the National Science Foundation has shown little interest in funding experimental or theoretical work applied to wireless communications. It is ironic that an NSF-sponsored project at Purdue University in the mid-1970s provided the first study of a spread spectrum cellular radio system [26], a study that created intense interest in spread spectrum access approaches for cellular radio communications which now (17 years later) are being extensively commercialized by numerous companies. This summer, however, NSF awarded Rutgers University with a cooperative University/Industry Center. The Rutgers Wireless Information Network (WINLAB), founded by Professor David Goodman in 1988-1989, is working on network and access solutions for third generation wireless personal communication systems, and was the first United States academic laboratory formed for wireless personal communications education and research. Steadily, more universities are recruiting faculty interested in the wireless communications field.

The Defense Advanced Research Projects Agency (DARPA) has funded projects to develop small, low-powered, wireless devices, and is supporting research that will lead to technology development for rapidly deployable local area communication systems. New design and fabrication technologies range from advanced silicon integrated circuits, to enhanced speech coding algorithms and circuit fabrication, to software tools for propagation prediction and installation. The personnel development, knowledge base, resulting technologies, and system design tools from these projects will not only improve the United States military capability, but also has relevance to the U.S. consumer wireless personal communications industry.

Virginia Tech's Mobile and Portable Radio Research Group (MPRG) is a new group within the university's Bradley Department of Electrical Engineering. Founded in the spring of 1990, it is conducting basic and applied research in the areas of radio propagation measurement and prediction, communication system design using measure-

NASA has provided the bulk of United States federal research funding in the mobile communications area

In Canada, numerous initiatives are under way to encourage academic participation in research for wireless personal communications.

ment-based propagation models, and simulation of various digital modulation, diversity, and radio equalizer techniques. The group's mission is to help the United States wireless industry through the development of analysis tools and computing techniques for emerging wireless personal communication systems, while providing quality research opportunities for graduate and undergraduate students.

MPRG also is providing opportunities for technical interchange. An EE graduate student lecture series in the fall of 1989 featured key researchers from the cellular radio industry, and the First Virginia Tech Symposium on Wireless Personal Communications was held June 1991 in Blacksburg, Virginia. The symposium featured 20 invited speakers and panel discussions by industry experts over a three-day period. It was attended by 175 people from 22 states and nine countries. The symposium, now scheduled as an annual June event on the Virginia Tech campus, provided an opportunity for students and engineers to learn both fundamentals and trends in wireless communications. The necessity for academic research groups like MPRG becomes clear when one realizes the enormous, yet sudden, activity in the wireless field, and the growing demand for young graduates who can make immediate contributions to a dynamic field.

It is worth noting that in contrast with the United States government agencies, the European, Canadian, and Japanese governments have made substantial funding commitments to research laboratories and university programs focusing on wireless personal communications and related technologies. For example, in Europe, the RACE program (Research and Development in Advanced Communications in Europe) has committed over $100 million per year to the European community during the period 1987-1992 [27].

A significant portion of those funds has been spent on collaborative industry/university research in wireless communications, with the goal of concurrently expanding the knowledge base and pool of technical experts. RACE appears to be yielding large dividends. The European community is widely recognized as the world leader in creating and accepting new digital cellular radio system techniques (many European countries presently enjoy more than a 5 percent cellular market penetration, compared with approximately 3 percent in the United States), and it is where the concept of personal communications was first put into wide scale practice.

Casual conversations with researchers across the world indicate that more than 50 European Ph.D. students are tackling dissertations dealing with antennas and propagation for emerging wireless personal communication systems. This does not include other areas of PCN, such as modulation, coding, diversity, and system design.

In Canada, numerous initiatives are under way to encourage academic participation in research for wireless personal communications. The Telecommunications Research Institute of Ontario (TRIO) program is enhancing the technological competitiveness of Canadian industry through university/industry partnerships in telecommunications research. TRIO has grown steadily since it was founded in 1987.

This year, the Ontario province and Canadian industry will provide more than $6 million for university communications research in Ontario. A significant portion of those funds is being used for mobile and satellite systems research for wireless personal communications. The mobile and satellite systems program supports more than 50 graduate students at five universities, and involves approximately 20 Canadian communications companies. Other Canadian provinces also are providing research support for regional universities active in wireless personal communications.

On a federal level, The Canadian Institute for Telecommunications Research (CITR) was established in 1989, and is providing research grants to universities throughout Canada. CITR's budget is more than $4 million annually (overhead is waived on CITR funding) and is focusing on two major areas: broadband and wireless communications. Examples of CITR projects in the antennas and propagation area include propagation and diversity techniques for indoor wireless communications at millimeter waves, fading issues for 20 GHz to 30 GHz personal satellite communications systems, and new cellular system design techniques. Universities involved with CITR wireless research include Carleton University, Concordia, Ecole Polytechnic, Laval University, McGill, Queens, Simon Frazier University, the University of British Columbia, the University of Calgary, the University of Ottawa, the University of Toronto, the University of Victoria, the University of Waterloo, and the University of Western Ontario.

Japanese research programs in wireless personal communications abound, and there are numerous universities active in the area that are making fundamental contributions. As an example, Kyoto University has been a major contributor in the wireless communications field, and has an active graduate program in propagation and communication system design. Federal funding for academic research in Japan often involves interaction with industrial laboratories [27]. It seems clear that most federal governments view the tremendous impact that wireless communications will make on the world's economy as a golden opportunity, and are hoping their investments will result in new technologies and a body of experts who can engineer and develop new wireless personal communication systems.

Early efforts to obtain government funding from the National Science Foundation failed, but industry response has been excellent and after one year, 15 major wireless companies (including many regional Bell Operating Companies, major radio manufacturers, and computer manufacturers), DARPA, and the FBI have provided a funding base of over $1 million. Collaborative research projects with Purdue University, Northeastern University, and the University of California at San Diego have provided cross-fertilization of ideas and knowledge, and are helping to make the United States approach to wireless communications research more synergistic.

At MPRG, students receive an educational experience that provides them with a solid understanding of the theory and practice of mobile radio communications and emerging personal communication systems. At the same time, they are tasked with developing and validating analyses, models, and research tools for the wireless industry. These tools help transfer knowledge from the MPRG laboratory into industry and academia.

In 1990, the first year of MPRG, five M.S.E.E. students graduated with expertise in RF filter design, indoor radio propagation measurement and prediction, adaptive noise-cancellation techniques, and urban radio propagation prediction. Presently, there are 18 graduate and six undergraduate MPRG students pursuing degrees with an emphasis on wireless communications. As a result of student theses, several analysis and simulation software tools have been developed for internal use, and are also being used for research and development by a number of companies and universities. An electrical engineering graduate course dedicated to the topics of cellular radio and personal communications was offered in the spring of 1991, and enjoyed an enrollment of 34 students, making it the most popular graduate course in the EE curriculum at Virginia Tech that term. Senior elective courses on radio wave propagation and satellite communications also are popular EE courses at Virginia Tech.

Although the research and educational mission of MPRG concerns itself with more than just propagation, the group received its first research contracts in the area of propagation measurement and prediction, and continues to maintain an active program in this area. There is extreme interest in and demand for propagation measurements and models for the proper design of emerging wireless services such as PCN, and more powerful site-specific channel modeling techniques and tools are needed.

The United States wireless industry, however, often finds conducting their own measurements and propagation research to be a time-consuming and expensive task, and many industrial players view it as an expensive luxury that involves high-priced consultants. MPRG has emerged as a sensible and cost-effective alternative, since it has an established equipment arsenal and provides research expertise in the area of propagation measurement and prediction. By pooling resources in an industrial affiliates program, MPRG is generating useful tools and basic propagation models that can be shared by all affiliate members, and the resulting value of the research is much greater than the cost to an individual member. All results are published so the entire research community benefits from the knowledge base.

Here, we briefly present some of the basic parameters used to describe multipath channel characteristics, and show results from propagation measurements in urban, microcellular, and indoor channels. These measurements involve wideband characterizations, where a very short RF burst is sent and the echoes are received, as well as narrowband (CW) characterizations that measured signal fading over large temporal and spatial spans. As is subsequently shown, antenna experiments have revealed that polarization can dramatically reduce the multipath time delay spread and the fluctuation of delay spread in mobile channels. Results presented in this article are not meant to be interpreted as definitive work in mobile radio propagation. On the contrary, it is hoped that this article will spark interest and ongoing discussions in propagation modeling and prediction for personal communications.

Propagation Measurement

A large part of MPRG research has dealt with measuring, and then statistically modeling, the path loss and time dispersion of multipath radio channels. Measurements and models have been made in many different environments: traditional urban cellular radio channels with base station antenna heights exceeding many tens of meters [1, 2]; urban cellular radio channels with lower antenna heights, on the order of 15 to 20 meters [2]; and in-building channels within sports arenas, factories, and office buildings [3, 4], and open plan office buildings [5]. Also, impulsive noise measurements have been made inside many types of buildings at three bands between 900 MHz and 4.0 GHz [28], including two of the license-free Industrial, Scientific, and Medical (ISM) bands.

For typical urban mobile radio channels, it has been reported that the coherence bandwidth (the bandwidth over which the received signal strength will likely be within 90 percent of any other frequency from the same source) is between 10 kHz to 500 kHz [6]. Consequently, to sufficiently resolve multipath components (in the time domain) that cause frequency selective fading, a channel sounder for urban channels should possess an RF bandwidth several times larger than the maximum coherence bandwidth. This thinking led us to use a 500 ns probing pulse (4 MHz RF bandwidth) in [1, 2]. For indoor measurements, we have used probes that have durations on the order of 5 ns, thereby providing measurements that span over 400 MHz and which resolve individual multipath echoes to about 1.6 m separation distance. (The baseband pulse is DSB modulated so there is a bandwidth expansion factor of two at RF. Thus, the baseband complex envelope response has a 200 MHz resolution bandwidth). Time domain techniques rather than swept frequency techniques are used since it is easy to identify the location and intensity of reflecting objects in the channel.

This information is vital for successful development of site-specific propagation models that can recreate the channel impulse response. Our measurement systems are easy to assemble, test, and deploy, and provide instantaneous channel measurements with excellent time delay resolution. Post-detection integration is employed to improve the signal to noise ratio of the measured power delay profiles. Our approach, though, requires more peak power than the direct sequence systems used in [7, 8, and 9] for a specified coverage distance. Broadband discone, yagi, and helical antennas have been used to ensure no pulse spreading is attributed to impedance mismatch.

Important parameters that indicate multipath dispersion in mobile radio channels are illustrated in Fig. 1. The rms delay spread (σ_τ) measures how spread of the channel power delay profile about the centroid, and the excess delay spread (X dB) indicates the maximum excess delay at which multipath energy falls to X dB below the peak received level. These parameters are useful measures for comparing different multipath channels and have been used to determine approximate bit error rates for digital modulation schemes without equalization.

Historically, time dispersion parameters (e.g., for ionospheric channels) were computed by using a

A large part of MPRG research has dealt with measuring, and then statistically modeling, the path loss and time dispersion of multipath radio channels.

Particularly in indoor channels, individual multipath components fade very little between two fixed terminals or terminals moved along a small area.

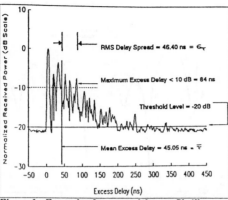

Figure 1. Example of a power delay profile illustrating important channel parameters

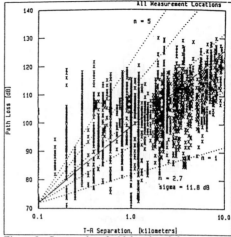

Figure 2. Scatter plot of wideband path loss measured in cellular and microcellular channels in Europe[2]

temporal average of the channel impulse response during a time period over which the channel appeared wide-sense stationary. In mobile systems, however, the time variation of the impulse response is due primarily to motion, so the parameters may be computed over a spatial average during which the channel appears wide-sense stationary [9, 30].

Particularly in indoor channels, individual multipath components fade very little between two fixed terminals or terminals moved along a small area [31, 32]. Statistical processing on an extensive indoor propagation data base showed that individual multipath components fade in a lognormal sense over small temporal and spatial intervals, with a standard deviation of only a couple of dB. Simultaneous CW measurements showed that the narrowband fading between two fixed terminals is Ricean, but the CW fading can be either Ricean, log-normal, or Rayleigh when the receiver is moved over a small area. Deep fades of individual multipath components are primarily due to shadowing as a terminal is moved, or results from the phasor sum of unresolvable multipath components within a resolution cell.

Knowledge of the channel time dispersion to temporal resolutions much greater (smaller duration) than the bit durations of a communication signal, and how the time dispersiveness changes over space, is important because these factors determine the instantaneous bit error floor that occurs because of data smearing. By performing the time convolution of transmitted data bits with accurate spatially (time) varying impulse response models, it becomes possible to predict burst errors and conduct real-time system design exper-

iments using computer simulation instead of prototype hardware [33].

Historically, path loss has been found to be closely linked to the separation distance between transmitter and receiver, so a simple model for the path loss at some distance $r > r_o$ from a transmitter can be expressed as

$$P(r) = P(r_o) (r_o /r)^n \qquad (1)$$

The exponent n in Equation (1) represents the best-fit (in a mean square sense) average power law at which signal power decays with respect to a free space measurement at r_o, the close-in reference distance. Measurements have shown that field measurements are generally log-normally distributed about the average distance-dependent power law given in (1), independent of r.

Different time dispersion and path loss results for urban microcellular measurements reported in [2] are shown in Fig. 2 and Table 1. The measurements were made throughout several existing cellular markets in Germany using a 500 ns probing pulse and existing cellular base station antenna heights that ranged from 20 m to 93 m in height [2]. Data in Fig. 2 and Table 1 assume r_o is 100 m. Using the assumption that path loss is log-normally distributed about the mean power law, the standard deviation σ in dB completely specifies the distribution of received power as a function of distance. For a best-fit path loss exponent, σ will be minimized over the entire scatter plot.

Reference [29] shows that the selection of the

	Antenna Height (m)	n	σ(dB)	Maximum T-R Separation (km)	Maximum Rms Delay Spread (μs)	Maximum Excess Delay Spread (μs) (10 dB)
Hamburg	40	2.5	8.3	8.5	2.7	7.0
Stuttgart	23	2.8	9.6	6.5	5.4	5.8
Düsseldorf	88	2.1	10.8	8.5	4.0	15.9
Frankfurt (PA Bldg.)	20	3.8	7.1	1.3	2.9	12.0
Frankfurt (Bank Bldg.)	93	2.4	13.1	6.5	8.3	18.4
Kronberg	50	2.4	8.5	10.0	19.6	51.3
All (100m)		2.7	11.8	10.0	19.6	51.3
All (1 km)		3.0	8.9	10.0	19.6	51.3

Table 1. Important channel statistics for the various base locations and the entire data set

IEEE Communications Magazine • November 1991

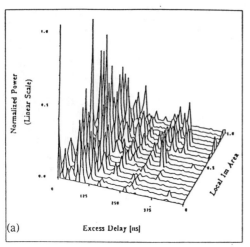

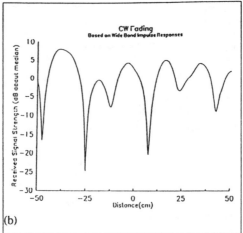

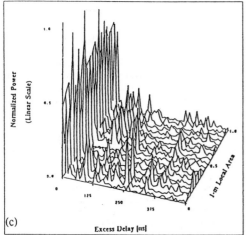

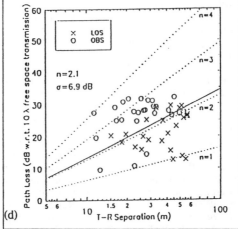

MPRG research has focused on measuring indoor and microcellular channels and developing models for such channels.

Figure 3 (These waveforms are examples produced by SIRCIM. SIRCIM provides the user with data files that contain amplitudes, phases, and time delays of the channel impulse response, and computes fading statistics, large-scale path loss, best fit exponent and standard dev. User can specify building type, topography (LOS or OBS), and T-R separation distance. SIRCIM is based on extensive measurements made in more than 10 different buildings.

close-in reference distance is important for system design considerations and capacity analysis. The close-in reference distance is a leverage point from which a best-fit mean path loss exponent extends. Historically, a reference distance of 1 km was used as early land mobile systems strived to provide coverage over tens of kilometers. Emerging microcellular systems, however, will cover deliberately smaller regions so that spectrum re-use can be employed more extensively in a specific market. By diminishing the size of the radio cells, more users can be accommodated within a given spectrum allocation; consequently capacity can be increased at the expense of more base stations and infrastructure. Thus, path loss models must use a close-in reference distance that is several times smaller than the distance of the most distant user within a coverage zone.

Using field measurements, it has been shown that the value of the path loss exponent can change depending on the free space reference distance chosen, which indicates that simple d^n path loss models are subject to interpretation of the close-in leverage point selected. For microcellular and PCN systems, reference distances on the order of 1 m to 100 m are appropriate.

MPRG research has focused on measuring indoor and microcellular channels and developing models for such channels, since it is our belief that indoor environments and street-level systems will serve the largest number of wireless users in the coming decades. Extensive indoor propagation measurements have been and continue to be made with the goal of deriving site-specific modeling approaches and installation tools based on physical descriptions of building interiors [5, 13].

Along the way, we have used statistical modeling procedures [10] to reproduce, on a personal computer, extensive propagation measurements given in [13] so that research can focus on indoor radio communication system design using realistic computer-generated impulse responses. Also, more recent measurements [3, 4] and measurements reported in the literature [14] have been used to generate (on a computer) impulse response and path loss measurements in traditional-partitioned and soft-partitioned (Herman-Miller office partitions) office buildings. The statistical channel models are useful for determining, through simulation, irreducible bit error rates, modulation performance, diversity implementations, and robust equalization methods.

The propagation simulator, called SIRCIM (Simulation of Indoor Radio Channel Impulse response

ly predict signal strength contours.

Figures 5, 6, and 7 illustrate how the simple site-specific attenuation model can accurately predict coverage throughout the work space. A schematic of a typical 59 m x 59 m open plan office building with movable cloth partitions and concrete walls is shown in Fig. 6. Dark lines in Fig. 6 denote concrete walls that extended from floor to ceiling. Lighter lines represent 2.0 m tall soft partitions. Fig. 5 and 7 can be overlayed on Figure 6, and show measured and predicted signal strength contours based on the site-specific model in [5]. A simple distant-dependent path loss model would provide circular contours of constant radius about the transmitting antenna.

Although this modeling work is preliminary, it demonstrates that simple descriptions about the building topography could be used to predict coverage areas and interference zones with much better accuracy than models used today. MPRG researchers are continuing measurements that will help to develop accurate site specific models. These models will then be incorporated into an automated system design tool that will optimally locate base stations for minimum interference and consequently maximum capacity. MPRG is working to exploit knowledge of the propagation environment to improve and automate system installation without measurements.

Work in [4] has shown that antenna polarization can play a big part in reducing the delay spread (i.e., improving the bit error performance). In [15], Cox describes the Bellcore UDPC system as using polarization diversity to open the eye in digital modulation techniques. Our work [4] shows that, indeed, polarization diversity can be used to select the best channel at a particular location. Our work also shows that circularly polarized (C-P) directional antennas, when used in line-of-sight channels, can provide a much lower delay spread than linear polarized antennas with similar directionality [4].

Figure 8 shows how the instantaneous rms delay spread changes as a mobile receiver is moved over a 2.5λ track. The identical track was traversed with a receiver using omnidirectional and directional linear polarized antennas, and directional circular polarized antennas. Note that the

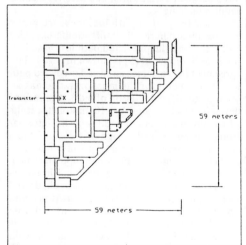

Figure 6. Blueprint of open plan building measured in Fig. 5

ITEM	LOSS (dB)
In Office Buildings	
Concrete Block Wall	13
Loss From One Floor	20-30
Loss From One Floor and one Wall	40-50
Fade observed when transmitter turned a right angle corner in a corridor	10-15
Light Textile inventory	3-5
Chain link fenced in area 20 ft. high which contains tools, inventory, and people	5-12
Metal Blanker-12 square feet	4-7
Metallic Hoppers which hold scrpa metal for recycling-10 square feet	3-6
Small Metal Pole-6 in. diameter	3
Metal Pulley System used to hoist metal inventory-4 sq. ft.	6
Light Machinery < 10 sq. ft.	1-4
General Machinery - 10-20 square feet	5-10
Heavy Machinery > 20 sq. ft.	10-12
Metal catwalk/stairs	5
Light Textile	3-5
Heavy Textile inventory	8-11
Arerea where workers inspect metal finished products for defects.	3-12
Metallic inventory	4-7
Large I-beam - 16-20 in	8-10
Metallic inventory racks - 8 square feet	4-9
Empty Cardboard inventory boxes	3-6
Concrete Block Wall	13-20
Ceiling Duct	1-8

Table 3 Measured signal loss due to common obstructions in buildings*

* These data were collected by comparing signal strengths on either side of the obstruction

Narrowband measurements have shown how the path loss exponent can be affected by building type, or location within a building.

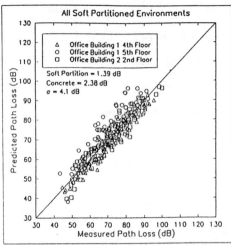

Figure 4. Best fit line for a simple two-parameter model used to predict signal loss due to obstructions

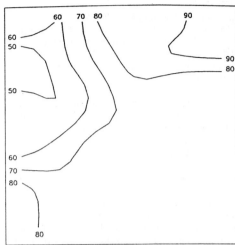

Figure 5. Measured and predicted signal strength contours based on site-specific model [5]

Measurements), is a valuable tool for MPRG communications research, and also is being used by 40 companies and universities. The models used in SIRCIM are detailed in [10]. Although similar in nature to the SURP simulation program for urban radio propagation [11], SIRCIM is based on measurements made over small scale distances, and thus allows synthesis of the phase of individual multipath components based on the Doppler shift and a random scattering model. By computing the spatially varying phasor sum of multipath components, SIRCIM recreates CW fading envelopes identical in nature to those measured in the field. MPRG plans to update the software as more measurements become available, and as user experiences dictate.

A useful result from [3] is that propagation characteristics are very similar at both 1.3 GHz and 4.0 GHz, which means the SIRCIM models (based on measurements at 1.3 GHz) will hold up to at least 4.0 GHz, and probably at somewhat higher frequencies. Typical examples of the data produced by SIRCIM are shown in Fig. 3. Data files that contain the amplitudes, phases, time delays, and path loss for individual multipath components are produced by SIRCIM and written to disk for later use in bit error simulation. These files may then be used to test bit error rates and system designs without prototype hardware.

Narrowband measurements have shown how the path loss exponent, and the deviation about the best-fit average path loss model, can be affected by building type, or location within a building. Table 2, extracted from [5], indicates that in different buildings, the floors can offer different values of attenuation.

Measured attenuation factors for various obstacles in indoor environments are presented in Tables 3 and 4 [4, 5]. In [5], a simple two-parameter statistical model was developed to model the loss due to each partition or each wall encountered between a transmitter and receiver located within a building.

A scatter plot of path loss measurements made on three different floors of two different office buildings, and the predicted path loss based on the simple model in [5] is shown in Fig. 4. The agreement between measured and predicted path loss is very good for the most part, and has an overall standard deviation of 4 dB. A standard deviation much greater than 10 dB usually results when only distance (and no site-specific information) is used to predict signal strength from a data base of several different types of buildings.

For measurements in Fig. 4, the transmitter antenna was mounted 1.8 m above ground, and the receiver was located at desk height, slightly shadowed by movable office cubicles (soft partitions) and obstructed by concrete walls. Using a simple model that assumes 1.4 dB loss for each cubicle wall and 2.4 dB for each concrete wall (the walls did not span the entire floor), and free space propagation everywhere else, it was possible to close-

	FAF(dB)	σ(dB)	# Locations
Office Building 1:			
Through 1 floor	12.9	7.0	104
Through 2 floors	18.7	2.8	18
Through 3 floors	24.4	1.7	18
Through 4 floors	27.0	1.5	18
Office Building 2:			
Through 1 floor	16.2	2.9	40
Through 2 floors	27.5	5.4	42
Through 3 floors	31.6	7.2	40

	n	σ(dB)	# Locations
All Buildings			
All Locations	3.14	16.3	646
Same Floor	2.76	12.9	501
Through 1 Floor	4.19	5.1	144
Through 2 Floors	5.04	6.5	60
Through 3 Floors	5.22	6.7	58
Grocery Store	1.81	5.2	89
Retail Store	2.18	8.7	137
Office Building 1:			
Entire Building	3.54	12.8	320
Same Floor	3.27	11.2	238
W. Wing 5th Floor	2.68	8.1	104
Central Wing 5th	4.01	4.3	118
W. Wing 4th Floor	3.18	4.4	120
Office Building 2:			
Entire Building	4.33	13.3	100
Same Floor	3.25	5.2	37

Table 2. Attenuation values for different buildings and floors

40

Site-specific propagation models that predict, with good accuracy, the shadowing losses and the diffraction effects in urban canyons are needed for system design.

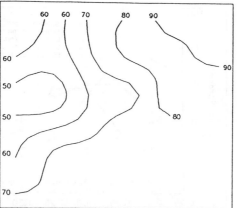

Figure 7. *Predicted contour plot of signal strength using best fit model of Fig. 3. Agreement with measured data is better than simply using T-R separation alone [5]*

C-P helical antenna offers much less delay spread, and smaller delay spread variability, than do the L-P omni or yagi antennas. Also, directional antennas reduce the delay spread when compared with omni antennas. We also have seen this on cross-campus links that illuminate several buildings at a time. In outdoor links especially, it appears that when aligned off-axis, directional C-P antennas offer much more multipath resistance than linear polarized antennas. The multipath reduction is likely due to cancellation of odd-bounce multipath, and offers a significant performance gain since it reduces the time dispersion of the channel. Further, this finding indicates that an accurate propagation prediction tool must consider polarization effects.

Obstacle Description	Attenuation (dB)
2.5 m storage rack with small metal parts (loosely packed)	4.6
4 m metal box storage	10-12
5 m storage rack with paper products (loosely packed)	2-4
5 m storage rack with paper products (tightly packed)	6
5 m storage rack with large metal parts (tightly packed)	20
Typical N/C machine	8-10
Semi-automated Assembly Line	5-7
0.6 m square reinforced conrete pillar	12-14
Stainless Steel Piping for Cook-Cool Process	15
Concrete wall	8-15
Concrete floor	10

Table 4 *Shadowing effects of common factory equipment*

MPRG is conducting additional measurements up to 30 GHz around campus to provide data for building penetration loss, floor-to-floor loss for different-shaped buildings, the correlation of signal strengths over small distances, and the importance of antenna pattern and polarization on system design.

Referring back to Fig. 2, there is a large amount of scatter about the best fit, indicating that surrounding buildings have a large impact on the measured path loss between a transmitter and receiver. Site-specific propagation models that predict, with good accuracy, the shadowing losses and the diffraction effects in urban canyons are needed for system design, and good progress is being made by numerous researchers throughout the world. A recent paper shows the viability of ray tracing and shadowing for accurate propagation prediction for microcellular systems [16]. Additionally, unpublished work at MPRG shows that in fact only a few rays and simple diffraction methods can be used most of the time to get surprisingly good prediction (within 3 dB) of measured signals in microcellular environments.

Conclusion

The wireless personal communications age is coming soon. This paper has attempted to put in perspective some recent trends and events that are shaping the wireless personal communications revolution. The paper also gave insight into the research and educational activities at Virginia Tech, and pointed out some tools and techniques used to characterize radio propagation. New models for propagation prediction will lead to appropriate spectrum allocations and robust system designs.

One area of research that is of great importance but was not mentioned is environmental safety. As important as research aimed at developing wireless personal communications, good, objective in-depth experiments must be performed jointly and immediately by engineers and medical scientists to determine the health risks associated with continuous and pulsed microwave radiation. Although worldwide standards exist, it is unclear to many if those standards actually represent safe levels for humans. If a cause and effect relationship exists between cancers and microwave radiation, this must be made known immediately and the physical mechanisms must be learned to combat radiation effects in the future. A universal, responsible approach to this potential problem, and the public's perception of the problem, should be undertaken immediately by government, industry, and academia. Within a decade, low-level RF radiation at microwave frequencies will be near our bodies and all around us. In 20 to 30 years, wireless communications probably will be as pervasive as utility lines and house wiring are today. We must be certain that the wireless personal communications age will be an environmentally safe age, as well.

References

[1] T. S. Rappaport, S. Y. Seidel, and R. Singh, "900 MHz Multipath Propagation Measurements for U.S. Digital Cellular Radiotelephone," *IEEE Trans. Veh. Technol.*, pp. 132-39, May 1990.
[2] S. Y. Seidel, et.al., "Path Loss, Scattering, and Multipath Delay Statistics in four European Cities for Digital Cellular and Microcellular Radiotelephones," *IEEE Trans. Veh. Technol.*, vol. 40, no. 4, Nov. 1991.

[3] D. A. Hawbaker and T. S. Rappaport, "Indoor Wideband Radiowave Propagation Measurements at 1.3 GHz and 4.0 GHz," *IEEE Electronics Letters,* vol. 26, no. 21, pp. 1800-02, October 10, 1990.

[4] D. A. Hawbaker, "Indoor Radio Propagation Measurements, Models, and Communication System Design Issues," Masters Thesis in Electrical Engineering, Virginia Polytechnic Institute and State University, Blacksburg, VA, May 1991. Also see paper in Globecom '91

[5] S. Y. Seidel and T. S. Rappaport, "900 MHz Path Loss Measurements and Prediction Techniques for In-Building Communication System Design," 1991 IEEE Vehicular Technology Conference, St. Louis, MO, May 21, 1991. To be published in *IEEE Trans. Antennas Propag.* early 1992

[6] R. W. Lorenz "Impact of Frequency-Selective Fading on Digital Land Mobile Radio Communications at Transmission Rates of Several Hundred kbit/s," *IEEE Trans. Veh. Technol.,* vol. VT-35, no. 3, pp. 122-28, Aug. 1987.

[7] R. J. C. Bultitude, "Propagation Characteristics on Microcellular Urban Mobile Radio Channels at 910 MHz," *IEEE J. in Sel. Areas Commun.,* vol. 7, no. 1, pp. 31-39, Jan. 1989.

[8] D. M. J. Devasirvatham, "Multipath Time Delay Spread in the Digital Portable Radio Environment," *IEEE Commun. Magazine,* vol. 25, no. 6, pp. 13-21, June 1987.

[9] D. C. Cox, "Time and Frequency-domain Characterizations of Multipath Propagation at 910 MHz in a Suburban Mobile-Radio Environment," *Radio Sci.,* vol. 7, no. 12, pp. 1069-81, Dec. 1972.

[10] T. S. Rappaport, S. Y. Seidel, and K. Takamizawa, "Statistical Channel Impulse Response Models for Factory and Open Plan Building Radio Communication System Design " *IEEE Trans. on Commun.* vol. 39, no. 5, pp. 794-807, May 1991.

[11] H. Hashemi, "Simulation of the Urban Radio Propagation Channel," *IEEE Trans. Veh. Technol.,* vol. VT-28, pp. 213-25, Aug. 1979.

[12] J. I. Smith, "A Computer Generated Multipath Fading Simulation for Mobile Radio," *IEEE Trans. Veh. Technol.,* vol. 24, no. 3, pp. 39-40, Aug. 1975.

[13] T. S. Rappaport, "Characterization of UHF Multipath Radio Channels in Factory Buildings," *IEEE Trans. Antennas and Propag.,* vol. 37, no. 8, pp. 1058-69, Aug. 1989.

[14] D. Molkdar, "Review on Radio Propagation Into and Within Buildings," *IEEE Proc.-H,* vol. 138, no. 1, pp. 61-73, Feb. 1991.

[15] D. C. Cox, "A Radio System Proposal for Widespread Low-Power Tetherless Communications," *IEEE Trans. Comm.,* vol. 39, no. 2, pp. 324-35, Feb. 1991.

[16] F. Ikegami, T. Takeuchi, and S. Yoshida, "Theoretical Prediction of Mean Field Strength for Urban Mobile Radio," *IEEE Trans. Antennas Propag.,* vol. 39, no. 3, pp. 299-302, March 1991.

[17] D. Bishop, "Personal Communications Service," *Mobile Radio Technol.,* p. 4, July 1991.

[18] "RSA Report and International Cellular Penetration," *Cellular Bus.,* pp. 79-81, October, 1991.

[19] J. Loeber, "Personal Communications in the UK," 1991 Eastern Communications Forum, Engineering Communications Forum, Washington, D.C., April 30, 1991.

[20] Electronics Industry Association/Telecommunications Industry Association Interim Standard IS-54 "Cellular System Dual-Mode Mobile Station-Base Station Compatibility Specification," May 1990.

[21] A. Maloberti, "Radio Transmission Interface of the Digital Paneuropean Mobile System," 1989 IEEE Vehic. Technol. Conf. Proc., Orlando, FL, pp. 712-17, May 1989.

[22] A. Salmasi, "Cellular and Personal Communications Networks Based on the Application of Code Division Multiple Access (CDMA)," Proc of First Virginia Tech. Symp. on Wireless Personal Commun., Blacksburg, VA, pp. 10.1-.9, June 5, 1991.

[23] R. L. Pickholtz, L. B. Milstein, and D. L. Schilling, "Spread Spectrum for Mobile Communications," *IEEE Trans. Veh. Technol.,* vol. 40, no. 2, pp. 313-22, May 1991.

[24] L. B. Milstein, R. L. Pickholtz, and D. L. Schilling, "Shared Spectrum Consideration in the Design of a Direct Sequence PCN," Proc. of First Virginia Tech. Symp. on Wireless Personal Commun., Blacksburg, VA, June 5, 1991, pp. 12.1-.2, June 5, 1991.

[25] K. Raith and J. Uddenfeldt, "Capacity of Digital Cellular TDMA Systems," *IEEE Trans. Veh. Tech.,* vol. 40, no. 2, May 1991, pp. 323-32, May 1991.

[26] G. R. Cooper and R. W. Nettleton, "A Spread-Spectrum Technique for High-Capacity Mobile Communications," *IEEE Trans. Veh. Tech.,* vol. 27, no. 4, pp. 264-75, Nov. 1978.

[27] IEEE Spectrum Special Issue on Research and Development, October 1990.

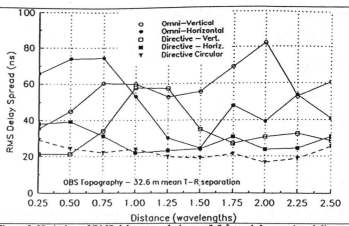

Figure 8 Variation of RMS delay spread along a 2.5 λ track for omni and directional antennas. Gain of omni antenna was 1.5 dBi, gains of linear directional and CP directional antennas were approximately 10 dBi. Frequency = 1.3 GHz.

[28] K. L. Blackard, T. S. Rappaport, and C. W. Bostian, "Radio Frequency Noise Measurements and Models for Indoor Wireless Communications at 918 MHz, 2.44 GHz, and 4.0 GHz," Proc. of 1991 IEEE Int'l Commun. Conf., Denver, CO, pp. 19.3-.5, June 23, 1991.

[29] T. S. Rappaport and L. B. Milstein, "Effects of Path Loss and Fringe User Distribution in CDMA Cellular Frequency Reuse," Proc. of 1990 IEEE Global Commun. Conf., San Diego, CA, pp. 500-06, Dec. 2, 1990.

[30] T. S. Rappaport, "Indoor Radio Communications for Factories of the Future," *IEEE Commun. Magazine,* vol. 27, no. 5, pp. 15-24, May 1989.

[31] T. S. Rappaport and C. D. McGillem, "UHF Fading in Factories," *IEEE J. Sel. Areas Commun.,* vol. 7, no. 1, pp. 40-48, Jan. 1989.

[32] R. Ganesh and K. Pahlavan "Statistics of Short Time Variations of Indoor Radio Propagation," Proc. 1991 IEEE Int'l Commun. Conf., Denver, CO, pp. 1-6, June 22, 1991.

[33] V. Fung and T. S. Rappaport, "Bit-Error Simulation of π/4 DQPSK in Flat and Frequency-Selective Fading Mobile Radio Channels with Real Time Applications," Proc. 1991 IEEE Int'l Commun. Conf., Denver, CO, pp. 19.3.1-3.5, June 23, 1991.

[34] H. Ochsner, "DECT—Digital European Cordless Telecommunications," 1989 IEEE Veh. Technol. Conf. Proc., Orlando, FL, pp. 718-721, May 1989.

[35] D. J. Goodman, "Trends in Cellular and Cordless Communications," *IEEE Commun. Magazine,* vol. 29, no. 6, pp. 31-40, June 1991.

[36] "Report on Spectrum Sharing in the 1850-1990 MHz band Between PCS and Private Operational Fixed Microwave Service," American Personal Communications, vols. I-II, filed at FCC, Washington, D.C., July 1991.

[37] "Experimental License Progress Report," Telesis Technologies Laboratory, filed at FCC, Washington, D.C., August 22, 1991.

[38] P. Estabrook, *et al.,* "A 20/30 GHz Personal Access Satellite System Design," IEEE ICC 89, Boston, MA, pp. 7.4.1-4.5, June 12, 1989.

[39] B. Bertiger, "The Iridium System," Proc. of First Virginia Tech. Symp. on Wireless Personal Commun., Blacksburg, VA, pp. 4.1-.9, June 5, 1991.

Biography

Theodore Rappaport [SM '91] received a Ph.D. in EE from Purdue University in 1987. He is an assistant professor in the Bradley Department of Electrical Engineering at Virginia Polytechnic Institute and State University in Blacksburg, Virginia. Dr. Rappaport is a faculty member of the Mobile and Portable Radio Research Group (MPRG).

Chapter 2

CELLULAR CONCEPT -- SYSTEM DESIGN FUNDAMENTALS

Reprinted from *IEEE Communications Magazine*, pp. 10-15, Nov., 1983

Cellular Mobile Radio— An Emerging Technology

JOHN OETTING

Frequency reuse and cell splitting.

CELLULAR MOBILE RADIO holds the promise of meeting the increasing demand for mobile telephone channels. In this paper, we describe the cellular mobile radio concept, with emphasis on the two key features of a cellular system: frequency reuse and cell splitting. We also briefly discuss the history of mobile radio and recent FCC decisions resulting in the present situation. Finally, we describe two methods for achieving improved coverage of remote locations: a satellite-augmented terrestrial radio system and an HF-augmented system.

Mobile telephone service has traditionally been expensive and cumbersome, with a grade-of-service substantially lower than that provided by fixed landlines. Despite these problems, the demand for service has far outstripped the available capacity in many major cities, resulting in long waiting lists for service. Modern technology has made possible a significant increase in the efficiency of mobile radio through frequency reuse and sophisticated system control. In this paper, we discuss recent FCC decisions that have changed the future of mobile radio. We then describe the cellular concept in some detail and discuss some potential future applications.

Background

Although mobile radio dates back as far as 1921, progress has been exceedingly slow until recent years. Since the mid-50's, the capacity of mobile radio systems has been limited to a few dozen channels in the 150-MHz and 450-MHz bands. In 1974, the FCC decided to alleviate this problem by allocating 115 MHz of bandwidth in the 806-to-947 MHz band for cellular mobile radio applications as follows:

- 40 MHz for wireline common carrier use,
- 30 MHz for private services, and
- 45 MHz as a reserve for future use.

Under pressure from the Radio Common Carriers

(RCC's[1]), the FCC later decided to split the 40-MHz band such that the wireline common carriers and the RCC's would each get 20 MHz. With 30-kHz channel spacing, each 20-MHz band can accommodate 333 full-duplex voice channels. In a typical system, 312 channels are used for voice, and the remaining channels are used for signaling and supervision. By means of frequency reuse, such a system could potentially support up to 10 000 customers with a grade-of-service comparable to the landline telephone system.

The FCC is currently awarding licenses to wireline and radio common carriers for the top 30 market areas in the United States. In markets where more than one carrier has applied for licenses, litigation may delay implementation of cellular mobile radio for several years. In many markets, applicants are reaching agreements among themselves and forming joint ventures so that the FCC can award licenses in these cities almost immediately.[2] Applications for the second group of 30 cities (markets 31–60) were due on November 8, 1982 and applications for markets 61–90 were due on January 7, 1983. Although a large number of firms will likely be awarded licenses in various parts of the United States, FCC regulations require that the systems be designed so that all mobile (mounted in automobiles or trucks) and portable (hand-held) radio telephones are interoperable with every cellular system. This feature will enable users to travel from city to city with the same radio-telephone equipment.

Currently, there are two cellular mobile radio experiments in progress in the United States. AT&T is testing their Advanced Mobile Phone Service (AMPS) in the Chicago area, where they have set up a 10-cell experiment serving 2000 customers. In October 1982, the FCC authorized AT&T to build a commercial system in Chicago. AT&T expects to field a 17-cell operational system by November

---

[1] RCC's are competitors to the large wireline common carriers for mobile telephone service. Since 1949, the FCC has allocated separate channels to the RCC's to provide mobile telephone service.

[2] These cities may have cellular mobile radio service prior to the end of the 1983 calendar year.

See Dr. Harold Ware's article about the FCC and cellular mobile radio on page 16.

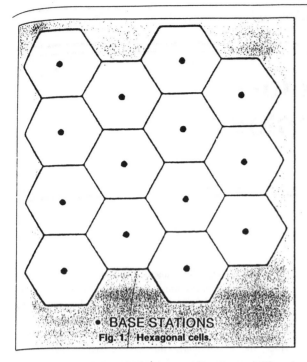

• BASE STATIONS
Fig. 1. Hexagonal cells.

1983. The other United States experiment is being conducted by the American Radio Telephone Service (ARTS) in the Baltimore/Washington and Northern Virginia area. This system, which uses equipment designed by Motorola, has been tested successfully with both mobile and portable equipment (while the AMPS experiment is limited to mobile equipment). Currently, the Phase II system [3] consists of four cells, with a fifth under construction. The Phase III system will implement a commercially viable system with 7 cells and 48 voice channels.

Because of litigation among the FCC, the wireline common carriers, and the RCC's, the United States has fallen behind some foreign countries in implementing cellular mobile radio. In Japan, a 900-MHz cellular radio system consisting of 13 cells went into operation in the Tokyo metropolitan area in December 1979 [4]. NEC, the supplier for the Japanese system, has also delivered systems in Australia and Mexico City. L.M. Ericsson, a Swedish cellular radio equipment manufacturer, has more than 5000 mobile units in service in Sweden and about 10 000 in the other Scandinavian countries. The Phase I program, which has been operational for almost a year, will eventually consist of 100 cells (50 have been in operation since May 1982). The Canadian Department of Communications will begin awarding licenses early in 1983. Unlike the other foreign systems, the Canadian system should be compatible with United States cellular mobile radios. Finally, the United Kingdom has announced plans to provide the world's first nationwide cellular mobile radio system, with the goal of providing wide national coverage by 1985.

The Cellular Concept

Cellular mobile radio differs from previous mobile radio designs in two critical areas: *frequency reuse* and *cell*

splitting. With conventional mobile radio systems, the objective is to have each fixed base station cover as large an area as possible by using antennas mounted in high towers and the maximum affordable power. Each base station is assigned a group of disjoint channels and the system configuration does not change for the lifetime of the system.

With cellular systems, the power radiated by the base stations is kept to a minimum and the antennas are located just high enough to achieve the desired coverage. These procedures enable nonadjacent cells to use the same set of frequencies, which is the frequency reuse feature mentioned above.

As the demand for service increases, the number of channels assigned to a cell will become insufficient to provide the required grade-of-service. At this point, cell splitting can be used to increase the number of customers that can be served in a given area *without increasing the number of available channels.* This process works by subdividing the congested cell into smaller cells (each with its own base station), reducing the antenna height and transmitted power of the new base stations, and reusing the same frequencies in some efficient pattern.

To illustrate the concept, consider the idealized case of an array of hexagonal cells shown in Fig. 1. The base stations

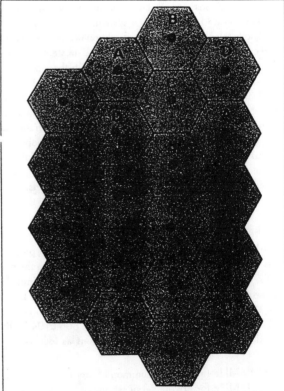

A = CELL USING FIRST SET OF FREQUENCIES
B = CELL USING SECOND SET OF FREQUENCIES
C = CELL USING THIRD SET OF FREQUENCIES
D = CELL USING FOURTH SET OF FREQUENCIES

Fig. 2. A four-frequency-set plan.

11

are indicated by the dots near the center of each cell. To minimize co-channel interference, it is desirable to avoid adjacent cells using the same set of frequencies. Figure 2 shows a possible 4-frequency set plan that has this desirable property. Figure 3 shows a 7-frequency set plan that exhibits somewhat larger spacing between cells using the same frequency. Finally, if it is desirable to avoid adjacent cells using adjacent frequencies, the minimum number of frequency sets is twelve. Figure 4 shows a twelve-frequency set plan that, in conjunction with suitable selection of the frequencies in each set, can accomplish the goals of avoiding adjacent frequencies in adjacent cells while maintaining the maximum frequency spacing within each cell.[*]

Base stations act as the interface between the mobile telephone subscriber and the cellular system. In current cellular designs, each base station is connected to a Central Control Station (CCS)[1] by dedicated wirelines (see Fig. 5). The CCS is involved in setting up calls, coordinating the activities of the base stations, and providing billing and built-in test functions. It acts as the interface between the cellular radio system and the landline telephone network. In the Chicago test, the CCS occupies a position in the switching

---

[*]To minimize interference within a cell, all of the plans discussed above are designed so that the nth set of frequencies includes frequency n, $n + N$, $n - 2N$, . . . , where N is the number of sets required ($N = 4$, 7, or 12). This procedure maximizes the spacing between frequencies used in a given cell.

[1]AT&T refers to the Central Control Station as the Mobile Telephone Switching Office (MTSO), while ARTS calls their central controller an Area Radio Controller (ARC). For generality, we will use the CCS terminology.

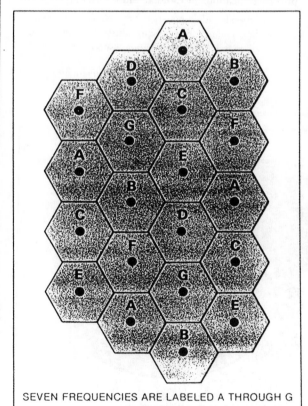

SEVEN FREQUENCIES ARE LABELED A THROUGH G

Fig. 3.   A seven-frequency-set plan.

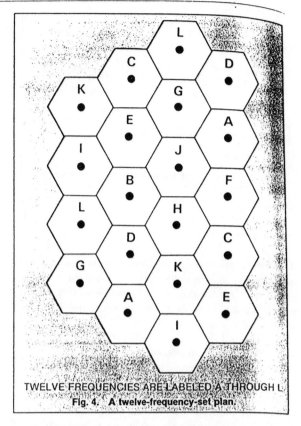

TWELVE FREQUENCIES ARE LABELED A THROUGH L

Fig. 4.   A twelve-frequency-set plan.

hierarchy below a class 5 (local) office and uses a number 1/1A Electronic Switching System (ESS). AT&T plans to treat the RCC cellular systems as PBX's connected to a class 5 central office.

While a mobile call is in progress, the CCS periodically monitors the SNR measured at the base station to insure that a high-quality circuit is maintained. If the SNR falls below some preassigned threshold, the CCS can make the decision to switch the mobile user to another base station. The system then signals the mobile to switch to another channel. This process, called *handoff*, is transparent to the mobile radio user but enables him to move from cell to cell during a call without interruption of service.

Propagation

Cellular radio systems in the United States will transmit on frequencies ranging from 825–890 MHz. (The mobile units transmit in the 825–845 MHz band and receive in the 870–890 MHz band, so that each full-duplex channel has a T/R separation of 45 MHz.) The advantage of these relatively high frequencies is that they penetrate buildings well and thus provide good portable coverage. The disadvantage is that shadowing by obstacles in the transmission path can be a problem. Even leafy trees are opaque at these frequencies, while valleys, hills, and mountains can cause significant gaps in coverage.

Intensive measurements have been made in the upper UHF band, and a number of propagation models have been developed with varying degrees of success. Because of multipath effects, the signal levels suffer short-term variations

12

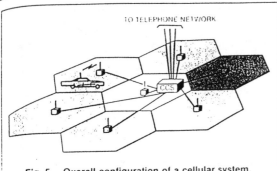

Fig. 5.   Overall configuration of a cellular system.

that roughly obey a Gaussian distribution (the field strength is lognormal and the loss in dB is Gaussian). The mean path loss is approximated by:

$$L = k - 10n \log D$$

where k is a constant that depends on the characteristics of the transmitter/receiver, n is the path loss exponent (typically $n = 4$), and D is the distance between the transmitter and receiver. Typically, the short-term variations about the mean path loss have a standard deviation of 6-10 dB.

Superimposed on these rapid variations are slower variations caused by shadowing from local terrain features. This fading is slow compared to the Rayleigh fading discussed above, but nevertheless, a 5-dB change in signal level over 100 feet of vehicle movement is typical.

Phase distortion is another effect of multipath propagation. This distortion limits the maximum signal bandwidth to a value between 40 and 250 kHz, depending on whether the receiver is in an urban or a rural area. If the signal bandwidth exceeds this so-called "coherence bandwidth," frequency selective fading becomes a problem.

Special Services

Cellular mobile radio systems are being designed to achieve a grade-of-service comparable to that on a landline telephone channel. However, since the types of impairments found on mobile radio links are not always the same as those found on landline links, it is likely that conventional wireline modems will not work well over mobile links. On the other hand, the channel spacing for mobile links is 30 kHz, so that special purpose digital modems operating at data rates of 10 kb/s and higher are feasible.

As an example, the AMPS system transmits 10 kb/s over the 30 kHz signaling and supervision channels, which are identical to the voice channels from the standpoint of propagation effects. This data rate is achieved with simple binary FSK modulation. Using fifth-order time diversity with majority vote decisions and a high-rate BCH code, the actual throughput is only about 1200 b/s.

Another special service that may be provided by cellular systems is roaming. A roaming capability would enable mobile units to move to other metropolitan areas and the highways in between without losing connectivity. It is very likely that cellular mobile radio systems will allow roaming vehicles to place calls from wherever they happen to be. However, a problem arises if a land telephone places a call to a mobile unit that has roamed outside its normal service area. The roaming vehicle may not want his whereabouts disclosed to an unknown caller and the caller may not wish to incur long distance charges for what he thinks is a local call.

Other special services furnished by ESS offices will probably be available with mobile phones. These include voice privacy, three-way calling, call waiting, and speed calling (automatic dialing of frequently called numbers).

Future Applications of Cellular Mobile Radio

Cellular mobile radio appears to have significant potential for providing enough channels to satisfy the current and near-term demand for mobile telephone service. In addition,

---

The primary difference is that the channel characteristics are never fixed, but vary with movement of the vehicle and changes in its environment. The channel characteristics may vary abruptly when the vehicle is handed off to another base station.

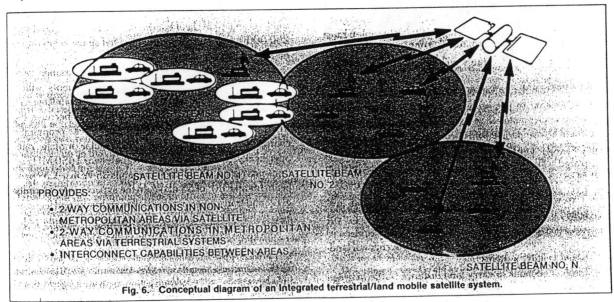

PROVIDES:
• 2-WAY COMMUNICATIONS IN NON-METROPOLITAN AREAS VIA SATELLITE
• 2-WAY COMMUNICATIONS IN METROPOLITAN AREAS VIA TERRESTRIAL SYSTEMS
• INTERCONNECT CAPABILITIES BETWEEN AREAS

SATELLITE BEAM NO. 1   SATELLITE BEAM NO. 2   SATELLITE BEAM NO. N

Fig. 6.   Conceptual diagram of an integrated terrestrial/land mobile satellite system.

13

ACTUAL BEAM CROSS SECTIONS ARE CIRCULAR. CIRCLES INSCRIBED IN HEX-AGONS WOULD REPRESNT 0.5 BEAMWIDTH. BEAM POWER DENSITY AT CORNERS OF HEXAGONS IS-4 dB RELATIVE TO BEAM CENTERS. NUMBERS IN HEXAGONS REFER TO CHANNEL SETS OF FREQUENCY REUSE PATTERN.

Fig. 7. Idealized hexagonal cell pattern for a 69-beam satellite covering the United States.

mobile radio has excellent potential as an aid to reestablishing communications after a natural disaster. Some of this potential was demonstrated after the recent eruption of Mt. St. Helens.

In this section, we discuss some of the potential future applications of cellular mobile radio. In particular, we discuss two methods of extending the range of cellular systems:

- Access through compatible satellite systems
- Access via HF Radio

Access Through Compatible Satellite Systems

The use of cellular mobile radio in the conventional manner described above has several disadvantages. The most important are that the mobile vehicle must be within line-of-sight range of a base station antenna, and coverage of cellular mobile radio may be limited to fairly large urban areas (areas with populations in excess of perhaps 200 000 people). Even optimistic estimates project that only 10% of the land area in the contiguous United States will be covered by cellular mobile radio systems in the year 2000, leaving more than 20% of the population without service.

As a solution to this problem, many technologists have suggested implementation of a satellite system for rural areas that would be compatible with the urban cellular mobile radio systems [7-9]. The proposed satellite augmented system, the Integrated Terrestrial/Land Mobile Satellite System (T/LMSS), would provide the following services:

- Two-way satellite communication of voice and low-speed data between base stations and vehicles in areas not served by terrestrial systems.
- Similar services in urban areas using conventional land-based cellular systems.
- Interconnections to carry long distance traffic between areas.

The concept is illustrated in Fig. 6.

The 1979 WARC authorized the use of satellite mobile radio in the cellular mobile radio band (806–890 MHz). Other countries, particularly Canada (with its huge amount of sparsely populated land), are extremely interested in the T/LMSS concept. The Canadian government has held numerous discussions with NASA about the possibility of a multibeam satellite to serve both Canada and the United States. However, considerable controversy exists over the economic viability of such a system.

Because of the ability to control satellite sidelobes much better than terrestrial coverage areas, it may be possible to use a three-frequency plan for the overlaid satellite system. Figure 7 shows how such a plan could be implemented in the United States using a 69-beam satellite with 0.5° beamwidth. Using 10 MHz of bandwidth,* this system could serve 200 000 subscribers, with an estimated space segment cost of $80 million per year.

To demonstrate the feasibility of satellite mobile radio, NASA has conducted a number of experiments with its ATS-1, ATS-3, and ATS-6 geostationary satellites. The ATS-6 experiments are particularly relevant because they were conducted at frequencies that have propagation characteristics similar to those of cellular mobile radio. Specifically, the ATS-6 receives signals near 1650 MHz and transmits near 1550 MHz. It transmits 35 W in a 1.5° pencil beam for an EIRP of 53 dBW. In the experiments, a number of L-band transceivers were installed on trucks and jeeps and field tested across the United States and in Panama [7].

The jeeps were used to provide voice communications to

_____
* To conserve precious bandwidth in the cellular mobile band, signals transmitted between the mobiles and the satellite would use these links between the satellite and base stations would use a higher frequency band, perhaps the K-band (12–14 GHz).

remote areas under actual and simulated emergency conditions. Voice intelligibility was reasonably good and the rapid fading characteristic of terrestrial systems was noticeably absent. The principal investigators reported that satellite communications were more reliable than any of their terrestrial systems, especially in rugged terrain. After the eruption of Mt. St. Helens in May of 1980, the ATS-3 satellite was used to provide voice communications for coordinating the search and rescue operations. In this case, the satellite link overcame many of the telephone communications problems that normally occur in such operations.

The use of satellites to augment the terrestrial cellular radio network depends upon a number of political and economic factors as well as technical considerations. NASA-Lewis is currently conducting studies to define an operational system and to design and demonstrate a system that includes a multibeam satellite.

Access Through HF Radio

An alternative scheme has been proposed to provide telephone access to remote areas without the use of satellites. Some work has been done in developing HF radiotelephone systems capable of interfacing with existing telephone systems without the need for operators. The best example of such an HF system is the Radiotelephone with Automatic Channel Evaluation (RACE) being developed by the Canadian Department of Communications [10]. The goals of this R&D effort are to develop a system with the following characteristics:

- Circuit availability greater than 90%,
- automated operation,
- direct dial access to and from conventional telephone systems,
- good speech quality, and
- low cost.

The basic idea of the RACE system is to have a master station (similar to the CCS of cellular mobile systems) which provides access to the telephone network. The master station cycles through each of the assigned frequencies and transmits short digital messages that are used by the remote terminal to evaluate channel quality. At the beginning of a call, the remote terminal goes through a handshaking procedure with the master station to determine the best available channel and the call is transmitted on that channel. (In the case of a call between two remote units, a direct HF link between the two units is set up so that the master station is not involved beyond the handshaking stage.)

An experimental system was developed by the Canadian Department of Communications in order to test the concept. The system consisted of two (non-mobile) remote stations and a master station. It used 8 frequencies covering the HF band from 3.2–20.5 MHz so that a broadband antenna was required. During the three month test period (April, May, and June of 1980), communication was possible on the 60 km link 99.9% of the time and on the 1000 km link 98.3% of the

time. Acceptable error rates were achieved 84.2% of the time on the short link and 77.7% of the time on the long link.

Summary

Cellular mobile radio is an important new technology that will provide greatly expanded mobile telephone services in the near future. These services will be available in major metropolitan areas through Wireline Common Carriers such as AT&T, as well as through independent Radio Common Carriers. In addition to providing more people with mobile telephone service, cellular mobile radio will also make possible a variety of new services.

Since cellular service will be confined to major urban areas, especially in the near term, a large segment of the population and large land areas will lack connectivity with the network. As potential solutions to this problem, we described two concepts: a satellite-augmented system, and an HF radiotelephone system. Neither of these systems is currently planned for United States operation, and their future will depend to a large extent on political and economic factors rather than technical considerations.

References

[1] *Bell Syst. Tech. J.*, special issue devoted to the Advanced Mobile Phone Service (AMPS), Jan. 1979.
[2] F. H. Blecher, "Advanced Mobile Phone Service," *IEEE Trans. Veh. Technol.*, pp. 238–244, May 1980.
[3] "Phase 2 Report—Developmental Cellular Mobile and Portable Radiotelephone System in the Washington-Baltimore-Northern Virginia Area," American Radio Telephone Service, Inc.
[4] K. Izumi et al., "Advanced technology for mobile radio telelphone systems," *Proc. NTC '80*, pp. 19.3.1–5, December 1980.
[5] "EIA Interim Standard CIS-3," Cellular System Mobile Station-Land Station Compatibility Specification, Electronic Industries Association, July 1981.
[6] Federal Communications Commission, "Cellular system mobile station—land station compatibility specification," *OST Bulletin*, no. 53, April 1981.
[7] R. E. Anderson, "Satellite augmentation of terrestrial cellular mobile radio telephone systems," *Proc. 1981 IEEE Conf. Veh. Technol.*, pp. 191–196.
[8] G. H. Knouse and P. A. Castruccio, "The concept of an Integrated Terrestrial/Land Mobile Satellite System," *IEEE Trans. Veh. Technol.*, pp. 97–101, August 1981.
[9] R. E. Anderson et al., "Satellite-aided mobile communications: experiments, applications, and prospects," *IEEE Trans. Veh. Technol.* pp. 54–61, May 1981.
[10] S. M. Chow, G. W. Irvine, and B. McLarnon, "RACE—an automatic HF radio telephone system for communications in remote areas," *Communications Research Center Report no. 1338-E*, Canadian Dept. Commun., December 1980.

John D. Oetting was born in Merchantville, NJ on July 31, 1947. He received his B.S. (summa cum laude), M.S., and Ph.D. degrees from the University of Pennsylvania, Philadelphia, in 1969, 1971, and 1973, respectively.

For the last nine years, he has worked at Booz, Allen & Hamilton, Bethesda, Md. performing systems engineering studies in the areas of coding and modulation, low probability of intercept (LPI)/antijam (AJ) communications, and most recently, cellular mobile radio. Prior to this experience, he learned some valuable lessons working for two hardware contractors. In his spare time, he is course director of a three-day short course on AJ communications for the Continuing Education Institute.

Dr. Oetting is a member of Tau Beta Pi and Eta Kappa Nu, and a Member of the IEEE. ∎

15

Elements of Cellular Mobile Radio Systems

WILLIAM C. Y. LEE, FELLOW, IEEE

Abstract—A major concern in a cellular mobile radio system is the co-channel interference. Therefore, the reduction of co-channel interference becomes a main thrust for the system design engineers. We use the co-channel interference reduction factor as a design criterion and predict the signal-to-interference (S/I) ratios in different system configurations. The handoff mechanism and algorithmic considerations, the traffic capacity and procedure for splitting cells, and the near-end-to-far-end ratio interference and reduction are the elements described.

I. INTRODUCTION

THE PURPOSE OF THIS paper is to introduce a simple methodology which will enable us to understand better how each element affects a cellular mobile radio system. The first recognition is that hexagonal shaped cells are artificial and such a shape cannot be generated in the real world. Engineers draw hexagonal-shaped cells on a layout to simplify the planning and design of a cellular system. This paper also illustrates a simple mechanism which makes the cellular system implementable based on hexagonal cells. Otherwise a statistical approach will be used in dealing with a real-world situation. Fortunately the outcomes resulting from these two approaches are very close, yet the latter does not provide a clear physical picture. Besides, today these hexagonal-shaped cells have already become a widely promoted symbol. for cellular mobile systems. Use of hexagonal cells here can be easily adapted by the readers for this reason.

A major problem facing today's radio communication industry is the limitation of available radio frequency spectrum. A conventional mobile system is usually designed by selecting one or more channels from an allocation for use in autonomous geographical zones shown in Fig. 1. The communication coverage area of each zone is normally planned to be as large as possible. The user who starts a call in one zone has to reinitiate the call as he moves into a new zone (see Fig. 1).

In this kind of system, the number of active users is limited to the number of channels assigned to that zone. Also, this system is not a desirable radio telephone system since there is no guarantee that every call can be a complete call without a handoff capability. The handoff will be addressed later.

In setting allocation policy, the Federal Communications Commission (FCC) seeks systems which need minimal bandwidth but provide high usage and consumer satisfaction. One

Manuscript received March 10, 1985; revised January 25, 1986. This work was presented at Seminar on Cellular Radio for Voice and Data Communications, New York, NY, November 8, 1984.

The author was with ITT Defense Communications Division, 492 River Road, Nutley, NJ. He is now with Pactel Mobile Companies, 3355 Main Street, Irvine, CA 92714. Telephone (714) 553-6042.

IEEE Log Number 8609783.

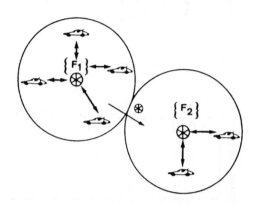

{F1} AND {F2} ARE TWO SETS OF FREQUENCY CHANNELS

⊕ REINITIATING CALLS

⊗ CELL SITE

IN A CONVENTIONAL MOBILE SYSTEM

- HIGH POWER
- LARGER CELL
- NO HANDOFFS

Fig. 1. Conventional mobile system.

system which fits this requirement is the recently developed cellular mobile radio system. Recent articles in the *Bell System Technical Journal* [1], and by Blecher [2] and Oetting [3] provide an overall structural view of a cellular mobile system. To make the cellular system they describe work properly, more information will be needed, and will be introduced in this paper. Since the general and more realistic models are very technical, this paper will try to offer a simple explanation of a complicated system. Therefore, certain assumptions will be simplified, and the results are more easily understood.

II. BRIEF DESCRIPTION OF MOBILE RADIO ENVIRONMENT

A mobile radio signal $r(t)$ illustrated in Fig. 2 can be characterized by two components $m(t)$ and $r_0(t)$ based on natural physical occurrences.

$$r(t) = m(t) \cdot r_0(t). \qquad (1)$$

$m(t)$ is called local mean, long-term fading, or log-normal fading which is due to the terrain contour between the base station and the mobile unit. r_0 is called multipath fading, short-term fading, or Rayleigh fading which is due to the waves

Reprinted from *IEEE Transactions on Vehicular Technology*, Vol. VT-35, No. 2, May, 1986.

51

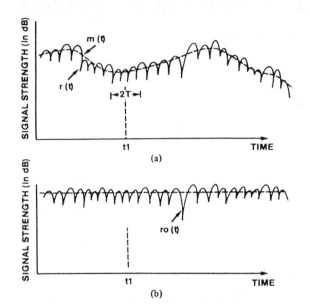

Fig. 2. A mobile radio signal fading representation. (a) A mobile radio signal fading. (b) A short-term signal fading.

reflected from the surrounding buildings and man-made structures. The long-term fading $m(t)$ can be obtained from the following:

$$m(t_1) = \int_{t_1-T}^{t_1+T} r(t)\ dt \tag{2}$$

where $2T$ is the time interval for averaging $r(t)$. T can be determined based on the fading rate of $r(t)$; usually 20 to 40 fades [4]. Therefore $m(t)$ is the envelope of $r(t)$ as shown in Fig. 2(a). $m(t)$ is also found to be a log-normal distribution based on its characteristics caused by the terrain contour.

The short-term fading r_0 is obtained by

$$r_0 \ (\text{dB}) = r(t) - m(t) \quad (\text{dB}) \tag{3}$$

as shown in Fig. 2(b). $r_0(t)$ follows a Rayleigh distribution assuming that only reflected waves from local surroundings are the ones received (a normal situation for the mobile radio environment). Therefore the term Rayleigh fading is often used.

III. Concept of Reuse Frequency Channels

A frequency channel consists of a pair of frequencies, one for each direction of transmission, that is used for a full duplex system. A particular frequency channel, say F_1, used by a user in one geographical zone called a cell, say C_1, with a coverage radius R, can be used by another user in another cell with the same coverage radius at a distance D away.

This frequency reuse concept is the core of the cellular mobile radio system. In this frequency reuse system, users in different geographical locations (different cells) may simultaneously use the same frequency channel. If the system is not properly planned this arrangement can cause interference to occur. Interference due to the common use of the same channel is called co-channel interference, and co-channel interference is our major concern. Schemes to reduce co-

channel interference are developed and described below. The impact on the handoff mechanism, cell splitting, and frequency management are also addressed.

IV. Reduction of Co-Channel Interference

Reusing an identical frequency channel in different cells is limited by co-channel interference to become a major problem. In order to reduce this co-channel interference, several effective schemes are used.

A. Co-channel Interference Reduction Factor

Assume that the size of all cells is roughly the same. The cell size is determined by the coverage area of the signal strength in each cell. As long as the cell size is fixed, co-channel interference is independent of the transmitted power of each cell. Actually, co-channel interference is a function of a parameter Q defined as

$$Q = D/R \tag{4}$$

where R is the radius of cells and D is the separation between two co-channel cells.

The parameter Q is the co-channel interference reduction factor. When the ratio Q increases, the co-channel interference decreases. Furthermore, the separation D in (4) is a function of K_0 and S/I where K_0 is the number of co-channel interfering cells in the first tier as shown in Fig. 3, and S/I is the received signal-to-interference ratio at the desired mobile receiver:

$$\frac{S}{I} = \frac{S}{\sum_{k=1}^{K_0} I_k}. \tag{5}$$

In a fully equipped hexagonal-shaped cellular system, there are always six co-channel interfering cells in the first tier as shown in Fig. 3, i.e., $K_0 = 6$. Co-channel interference can be experienced both at the cell site and at mobile units in the center cell. If the interference is much greater than the noise from all other sources, then the signal-to-interference ratio at the mobile units caused by the six interfering sites is (on the average) the same as the S/I received at the center cell site caused by interfering mobile units in the six cells, according to the reciprocity theorem of radio propagation. The S/I can be expressed as

$$\frac{S}{I} = \frac{R^{-\gamma}}{\sum_{k=1}^{K_0} D_k^{-\gamma}} \tag{6}$$

where γ is a propagation path-loss slope [5] determined by the actual terrain environment; γ usually lies between two and five. K_0 is the number of cochannel interfering cells, and is equal to six in a fully developed system, as shown in Fig. 3. The co-channel interfering cells in the second tier cause weaker interference than those in the first tier. Therefore, the co-channel interference from the second-tier interfering cells can be negligible.

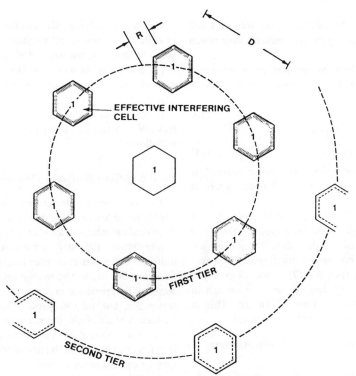

Fig. 3. Six effective interfering cells of cell #1.

B. The Normal Case in an Omnidirectional Antenna System

Assume that all D_k are the same for simplicity as shown in Fig. 4, then

$$\frac{S}{I} = \frac{R^{-\gamma}}{6 \cdot D^{-\gamma}} = \frac{Q^{\gamma}}{6}. \qquad (7)$$

Thus

$$Q^{\gamma} = 6 \cdot \frac{S}{I}$$

and

$$Q = (6 \cdot S/I)^{1/\gamma}. \qquad (8)$$

In (8), the value of S/I is based on the required system performance, and the specified value of γ based on the terrain environment. With given values of S/I and γ, the co-channel interference reduction factor Q can be determined. Normal cellular practice is to specify S/I to be 18 dB or higher based on subjective tests and the criterion that 75 percent of the users say voice quality is "good" or "excellent" in 90 percent of the total covered area on a flat terrain [6]. Since the S/I of 18 dB is measured from the acceptance of voice quality from present cellular mobile receivers, this acceptance implies that both the mobile radio multipath fading and the co-channel interference becomes ineffective at that level.

The path-loss slope γ is equal to about four in a mobile radio environment [5]. Then (8) becomes

$$Q = (6 \times 63.1)^{1/4} = 4.41. \qquad (9)$$

which is very close to $Q = 4.6$ from a simulation [6].

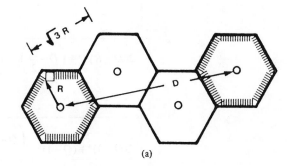

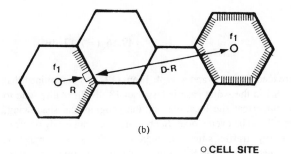

CELL SITE

MOBILE UNIT

Fig. 4. Determination of S/I in an omnidirectional antenna system. (a) Normal case in an omnidirectional antenna system. (b) Worst case in an omnidirectional antenna.

The ninetieth percentile of the total coverage area in a cell would be achieved by increasing the transmitted power which does not affect (9).

The factor Q can be related to the number of frequency reuse patterns (a set of cells) in a hexagonal-shape cellular system by

$$Q = \sqrt{3N}. \tag{10}$$

Substituting Q from (9) into (10) yields

$$N = 7. \tag{11}$$

Equation (11) indicates that a seven-cell reuse pattern[1] is needed for a S/I of 18 dB. The seven-cell reuse pattern is shown in Fig. 5.

Based on $Q = D/R$, the determination of D can be reached by choosing a radius R in (9). Usually, a greater value of Q than that shown in (9) should be desirable. The greater the value of Q, the lower the co-channel interference. In a real environment, (6) is always true, but (7) is not. Since (9) is derived from (7), the value Q may not be large enough to maintain a signal-to-interference ratio of 18 dB. This is particularly true in the worst case as shown below.

C. Worst-Case Design in an Omnidirectional Antenna System

In the worst case, the mobile unit is at the cell boundary R, and its distances from two co-channel interfering sites are $D - R$ (see Fig. 4). The others can be approximately $D + R/2$, $D - R/2$, D, and $D + R$ as figured out from Fig. 3.

Then the signal-to-interference ratio is

One way of solving this problem is to increase the number of cells N in a seven-cell frequency reuse pattern. However, when N increases, the number of frequency channels assigned in a cell must become smaller (assuming a fixed total allocation), and the efficiency of applying the frequency reuse scheme decreases.

Instead of increasing the number N in a set of cells, let us leave $N = 7$ and introduce a directional-antenna arrangement as follows.

D. Worst-Case Design in Directional Antenna System

The co-channel interference can be reduced by using directional antennas. This means that each cell is divided into three or six sectors and uses three or six directional antennas at a base station; the three-sector case is shown in Fig. 5. To illustrate the worst-case situation, two co-channel cells are shown in Fig. 6(a). The mobile unit in the right shaded cell-sector will experience interference from the left shaded cell-sector site, but not vice versa. This is because the front-to-back ratio of a cell-site directional antenna is at least 10 dB or more in a mobile radio environment. The worst co-channel interference case in the interfered directional-antenna sectors may be calculated as follows.

The S/I at the mobile unit in the worst case is when the unit is at the position B, at which the distance between the mobile unit and the interfering antenna is $D + R/2$. Because of the use of directional antennas, the number of principal interferers is reduced from six to two as shown in Fig. 5. The value of S/I

$$\frac{S}{I} = \frac{R^{-4}}{2(D-R)^{-4} + \left(D - \frac{R}{2}\right)^{-4} + D^{-4} + \left(D + \frac{R}{2}\right)^{-4} + (D+R)^{-4}}$$

$$= \frac{1}{\dfrac{2(Q+1)^4 + (Q-1)^4}{(Q^2-1)^4} + \dfrac{\left(Q+\frac{1}{2}\right)^4 + \left(Q-\frac{1}{2}\right)^4}{\left(Q^2-\frac{1}{4}\right)^4} + \dfrac{1}{Q^4}}$$

$$= 49.56 \; (=) \; 17 \; \text{dB} \tag{12}$$

provided $Q = 4.6$ and $\gamma = 4$. From (12) it is determined that the S/I in the worst case is less than 18 dB. In reality, due to the imperfect site locations and the geographical shadowing, the S/I can be even worse. Such an instance can easily occur in a heavy traffic situation; therefore, the system must be designed around the S/I in the worst case. In that case, a co-channel interference reduction factor of $Q = 4.6$ is insufficient.

can be obtained by the following expression:

$$(S/I)_{\text{worst case}} = \frac{R^{-4}}{K_0 \left(D + \frac{R}{2}\right)^{-4}}$$

$$= \frac{(Q+0.5)^4}{K_0} \tag{13}$$

where $Q = 4.6$ and $K_0 = 2$. Then (13) becomes

$$(S/I)_{\text{worst case}} = 25.3 \; \text{dB}. \tag{14}$$

---

[1] In this seven-cell reuse pattern, the total allocated frequency band is divided into seven subsets. Each particular subset of frequency channels is assigned to one of seven cells.

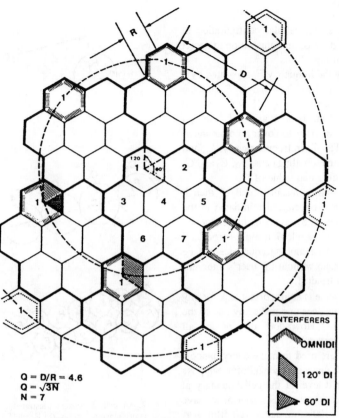

Q = D/R = 4.6
Q = √3N
N = 7

INTERFERERS
OMNIDI
120° DI
60° DI

Fig. 5.   The interfering cells shown in a seven-cell system.

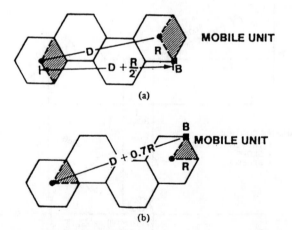

(a)

(b)

Fig. 6.   Determination of S/I in a directional antenna system. (a) Worst case in a 120° directional antenna system ($n = 7$). (b) Worst case in a 60° directional antenna system ($n = 7$).

The S/I received by a mobile unit from the 120° directional antenna sector system expressed in (14) well exceeds 18 dB in a worst case. Equation (14) shows that using directional antenna sectors can improve the signal-to-interference ratio, i.e., reduce the co-channel interference.

We may also divide a cell into six sectors by using six 60° beam directional antennas as shown in Fig. 6(b). In this case, only one instance of interference can occur in each sector as shown in Fig. 5.

$$\frac{S}{I} = \frac{(Q+0.7)^4}{1} \ (=) \ 29 \ \text{dB (in a seven-cell system)}$$

$$(15)$$

which shows a further reduction of co-channel interference.

The price we have to pay is more antennas mounted on an antenna mast and more frequent handoffs due to the greater chance of the mobile units traveling across the six sectors in a cell. Furthermore, the assignment of the proper frequency channel to the mobile unit in each sector is more difficult because the locating device at the cell site cannot accurately determine the location of a mobile unit in each sector. In addition, the trunking efficiency becomes poor also.

V. HANDOFF MECHANISM

Once the value of Q is determined, the cellular mobile system is ready to be implemented in its service area. For a given value of radius R, a specified value of D is associated because of the value Q.

$$D = Q \cdot R. \qquad (16)$$

For a startup system, the radius R of a cell can be very large. That means a large coverage area and light traffic density. The size of the cell is determined only by the required signal-to-noise ratio received at the cell boundary rather than by the signal-to-interference ratio. Therefore, the size of the cell can be enlarged by increasing the antenna-height, the antenna gain, or the transmitted power.

55

Although we have shown a real cellular configuration in Fig. 5, it is easy to use a one-dimensional illustration to depict the handoff concept shown in Fig. 7.

Two co-channel cells using the frequency F_1 separated by a distance D are shown in Fig. 7(a). The radius R and the distance D are governed by the value of Q. Now we have to fill in with other frequency channels such as F_2, F_3, and F_4 between two co-channel cells in order to cover a communication system in a whole area. The fill-in frequencies F_2, F_3, and F_4 are also assigned to their corresponding cells, C_2, C_3 and C_4 (see Fig. 7(b)) according to the same value of Q.

Suppose a mobile unit is starting a call in cell C_1 then it moves to C_2. The call can be dropped and reinitiated in the new cell C_2, or it can be carried on by changing the frequency channel from F_1 to F_2 while the mobile unit moves from cell C_1 to cell C_2. This process of changing frequencies can be done automatically by the system without the user's intervention. This process is called a handoff.

The handoff processing scheme is an important task for any successful mobile system. How does one make any one of the necessary handoffs successful? How does one reduce all unnecessary handoffs in the system?

The following approaches are used to make every handoff successful and to eliminate all unnecessary handoffs. Suppose that -100 dBm is a threshold level at the cell boundary at which a handoff would be taken. Given this scenario, we have to set up a level higher than -100 dBm, say -100 dBm $+ \Delta$ dB, and when the received signal reaches this level, a handoff request is initiated. If the value of Δ is fixed and large, then the time it takes to lower -100 dBm $+ \Delta$ to -100 dBm is longer. During this time, many situations, such as turning back toward the cell site or stopping can happen due to the direction and speed of the moving vehicles. Then the signals will never drop below -100 dBm. It means that many unnecessary handoffs may occur just because we have taken the action too early. If Δ is small, then there is not enough time for the call to handoff at the cell site, and many calls can be lost while they are handed off. Therefore, Δ should be varied according to the path-loss slope of the received signal strength and the level crossing rate of the signal strength as shown in Fig. 8.

Let the value of Δ be 10 dB in the above example, it means that the level of requesting a handoff is not 90 dBm. Then we can calculate the velocity V of the mobile unit based on the predicted level crossing rate [7] at a -10 dB level with respect to the root mean square (rms) level which is at -90 dBm.

$$V = \frac{n \cdot \lambda}{\sqrt{2\pi} \cdot (0.27)} \text{ ft/s}$$

$$= n \cdot \lambda \text{ mi/h} \tag{17}$$

where n is the level crossing rate (crossings/s) counting positive slopes and λ is the wavelength in feet.

Here, two pieces of information, the velocity of vehicle V, and the path-loss slope γ, can be used to determine the value of Δ so that the number of unnecessary handoffs can be reduced and the required handoffs can be completed successfully.

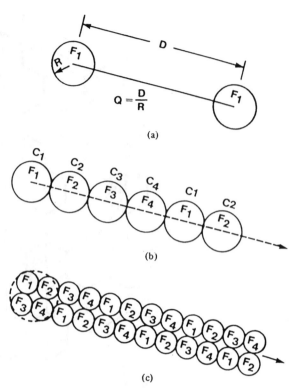

(a)

(b)

(c)

Fig. 7. A cellular system illustrated in one dimension. (a) Co-channel interference reduction. (b) Handoff occurrence. (c) Cell splitting.

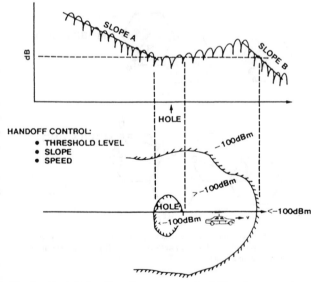

Fig. 8. Handoff algorithms for unnecessary handoffs and necessary handoffs.

There are circumstances when handoffs are necessary but cannot be made. One such circumstance occurs when the mobile unit is located at a signal strength hole within a cell but not at the boundary (see Fig. 8). Another happens when the mobile unit approaches a cell boundary, but no channels in the new cell are available.

In the first circumstance, the call must be kept in the old frequency channel until the call is dropped due to an unacceptable signal level. In the second case, the new cell must rearrange its frequency assignment or the call will be dropped.

The system switching office usually controls the frequency assignment in each cell and can rearrange channel assignments or split cells when this occurs. Cell splitting is described in the following section.

VI. CELL SPLITTING

A. Traffic Load After Splitting

When traffic density starts to build up and the frequency channels F_i in each cell C_i cannot provide enough mobile calls, the original cell can be split into smaller cells. Usually the new radius is one-half of the original radius (see Fig. 7(c)):

$$\text{a new cell radius} = 1/2 \text{ the old cell radius.} \quad (18)$$

Then based on (18), the following equation is true:

$$\text{a new cell area} = 1/4 \text{ the old cell area.} \quad (19)$$

Let each new cell carry the same maximum traffic load of the old cell; then

$$\text{a new traffic load per unit area} = 4X \text{ (the traffic load per unit area of the old cell).} \quad (20)$$

B. Transmitted Power After Splitting

The transmitted power P_{t_2} for a new cell due to its reduced size can be determined from the transmitted power P_{t_1} of the old cell.

If we assume that the received power at the cell boundary is P_r, then the following equations can be deduced from (4):

$$P_r = \alpha \cdot P_{t_1} \cdot R^{-\gamma} \quad \text{(the received power at the boundary of the old cell)} \quad (21)$$

$$P_r = \alpha \cdot P_{t_2} \cdot \left(\frac{R}{2}\right)^{-\gamma} \quad \text{(the received power at the boundary of the new cell)} \quad (22)$$

where α is a constant. To set up an identical received power P_r at the boundaries of two different sized cells, dropping the parameter P_r by combining (21) and (22), we find

$$P_{t_2} = P_{t_1}(1/2)^{-\gamma}. \quad (23)$$

For a typical mobile radio environment, $\gamma = 4$, then (23) becomes

$$P_{t_2} = \frac{P_{t_1}}{16} \quad (12 \text{ dB less than } P_{t_1}). \quad (24)$$

The new transmitted power must be 12 dB less than the old transmitted power. The new value of Q_2 after cell splitting is equal to the value of Q since both D and R were split in half.

When cell splitting occurs, the value of Q is always kept constant. The traffic load can increase four times in the same area after splitting the original cell into four subcells. Each subcell can again be split into four subcells, which would

allow traffic to increase 16 times. As the cell splitting goes on, the general formula can be expressed as

new traffic load

$$= (4)^m \text{ (the traffic load of startup cell)} \quad (25)$$

where m is the number of splittings. For $m = 4$ it means that an original startup cell has split four times. The traffic load is 256 times larger than the traffic load of the startup cell.

C. Cell Splitting Technique

The technique of cell splitting in real time is described below.

Cell splitting should proceed gradually over a system to prevent dropped calls. Suppose that the area, just halfway between two old $2A$ sectors, needs more traffic capacity as indicated in Fig. 9. We can take the middle point between two old $2A$ sectors and name it new $2A$. The new $1A$ sector can be found by rotating the old $1A$-$2A$ line (shown in Fig. 9) clockwise $120°$ [6]. Then the orientation of the new set of seven cells is determined. To maintain service for ongoing calls while doing the cell splitting we let the channels assigned in the old $2A$ sector separate into two groups:

$$2A = (2A)' + (2A)'' \quad (26)$$

where $(2A)'$ is the frequency channels used in both new and old cells, but in the small sectors and $(2A)''$ is the frequency channels only used in the old cells.

At the early splitting stage, only a few channels are in $(2A)'$. Gradually, more channels will be transferred from $(2A)''$ to $(2A)'$. When all the channels are transferred in $(2A)'$ and no channels are left in $(2A)''$, the cell splitting procedure is completed. The software algorithm program should have no problem handling the cell splitting procedure.

VII. ADJACENT CHANNEL INTERFERENCE [8]

Usually, a given set of serving channels are assigned to each cell site. If all the channels are simultaneously transmitted at one cell-site antenna, enough band isolation between channels is required for the multichannel combiner to reduce intermodulation products. This requirement is no different than other nonmobile radio systems. Assume that band separation requirements can be resolved, for example say using multiple antennas instead of one antenna at the cell site. There is another adjacent channel interference that is unique to the mobile radio system.

In the mobile radio system, most mobile units are in motion. Their relative positions change from time to time. There is no fixed frequency plan that can be carried out to avoid mutual interference.

The following situation is shown in Fig. 10. Distance d_0 between a calling mobile transmitter and a base-station receiver is much larger than the distance d_1 between interfered mobile transmitter and the same base-station receiver with the interfered mobile transmitter close enough to override the desired base-station signal. This interference, which is based

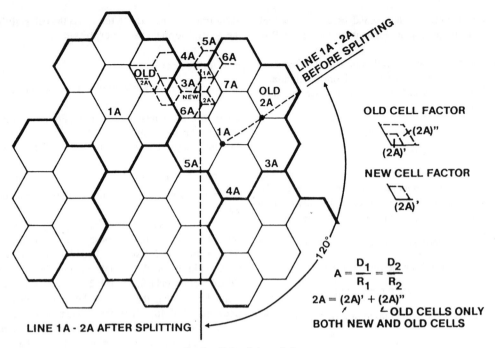

Fig. 9. Cell splitting techniques.

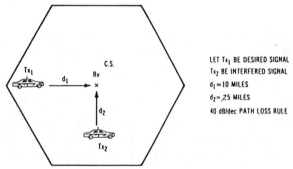

Fig. 10. Near-end-to-far-end ratio interference.

on the distance ratio, can be expressed as

$$S/I = \left(\frac{d_0}{d_I}\right)^{-\gamma} \qquad (27)$$

The ratio d_I/d_0 is the near-end-to-far-end ratio. From (27) the near-end-to-far-end ratio affects the signal-to-adjacent channel interference ratio according to the relative positions due to the motion of the mobile units.

For example, if the calling mobile unit is 10 mi away from the base station receiver and the interfering mobile unit is 0.25 mi away from the base station receiver, then the signal-to-interference received at the base station receiver with $\gamma = 4$ is

$$\frac{S}{I} = \left(\frac{d_0}{d_I}\right)^{-4} = (40)^{-4} \; (=) \; -64 \quad \text{dB}. \qquad (28)$$

This kind of interference can be reduced only by frequency separation with a specified filter characteristic. Assume that a filter of channel B has a 24 dB/oct. slope [1], then a 24 dB loss

occurs at the edge of the channel, $B/2$. Then from $B/2$ to B results in another 24 dB loss, from B to $2B$ results another 24 dB loss and so forth. In order to achieve a loss of 64 dB, we may have to double the frequency band more than two times as

$$G = \frac{64}{L} = \frac{64}{24} = 2.67$$

where L is the filter characteristic. The frequency band separation for 64 dB isolation is

$$2^{2.67} \left(\frac{B}{2}\right) = 3.18B. \qquad (29)$$

Therefore, a minimum separation of four channels is needed to satisfy an isolation criterion of 64 dB. The general formula for requiring a channel separation based on the filter characteristic L is expressed as follows:

$$\text{frequency band separation} = 2^G \cdot B/2 \qquad (30)$$

where G is indicated as[2]

$$G = \frac{\gamma \cdot \log_{10}\left(\frac{d_0}{d_I}\right)}{L}.$$

If a separation of 4B is needed for two adjacent channels in a cell in order to avoid the near-end-to-far-end ratio interference, it is then implied that a minimum separation of 4 B is required among all adjacent channels in one cell.

Since we distribute the total frequency channels in a set of N

---

[2] For greater accuracy, 3 dB should be added to G in order to take care of both sides of adjacent channel interference.

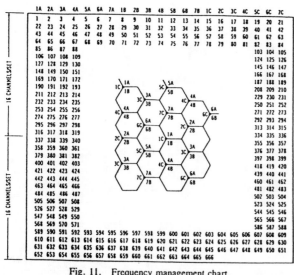

Fig. 11.　Frequency management chart.

cells, each cell only has $1/N$ of the total frequency channels. We denote, $\{F_1\}$, $\{F_2\}$, $\{F_3\}$, $\{F_4\}$, $\cdots$ for the sets of frequency channels assigned in their corresponding cells, C_1, C_2, C_3, C_4.

Now the question is how to make a good frequency management chart to properly assign the N sets of frequency channels and avoid the problems indicated above. The following section addresses how cellular system engineers solve this problem.

VIII. FREQUENCY MANAGEMENT

A frequency management plan is very crucial in a cellular system. The problems indicated in the frequency reuse, the handoff, cell splitting, and the near-end-to-far-end ratio interference, depend upon a good frequency management plan. An existing commercial frequency management chart can be used to illustrate its advantages. The chart is shown in Fig. 11.

There are 21 frequency channel sets. Those 21 sets can be grouped in different ways as long as the adjacent-channel separation meets the requirement. For a seven-cell reuse pattern, three subsets can be used in each cell and the adjacent-channel separation is $6\,B$ which is adequate. In each set, say $1A$, the adjacent channel separation is $20\,B$ which is more than required to reduce the adjacent-channel (near-end-to-far-end ratio) interference. The distance D for a seven-cell reuse pattern is $4.6\,R$. For the sake of further reducing the co-channel interference, the insert in the center of the chart discloses a system using three 120° or six 60° directional antennas at each cell site in the system. The directivity is a means to help reduce the co-channel interference.

IX. SUMMARY AND CONCLUSION

The cellular mobile system is a high capacity system. It can be designed to use spectrum efficiency while providing the highest quality telephone services. The techniques we have discussed include frequency reuse, handoff, and cell splitting. The major problems which this system faces are co-channel interference and adjacent channel (near-end to far-end) interference. Several techniques can be used to solve those problems including reduction of transmitted power for small cells, the requirement of co-channel cell separation for co-channel interference reduction, and frequency channel separation for adjacent channel isolation.

ACKNOWLEDGMENT

The author thanks Dr. Robert Powers, FCC Chief Scientist, for his stimulating discussion of this paper.

REFERENCES

[1] "Advanced mobile phone services," Special Issue, Bell Syst. Tech. J., vol. 58, Jan. 1979.
[2] F. H. Blecher, "Advanced mobile phone service," IEEE Trans. Veh. Technol., vol. VT-29, pp. 238–244, May 1980.
[3] J. Oetting, "Cellular mobile radio—An emerging technology," IEEE Commun. Magazine, vol. 21, no. 8, pp. 10–15, Nov. 1983.
[4] W. C. Y. Lee, "Estimate of local average power of a mobile radio signal," IEEE Trans. Veh. Technol., vol. VT-34, pp. 22–27, Feb. 1985.
[5] ——, Mobile Communications Engineering. New York: McGraw-Hill, 1982, pp. 101, 374.
[6] V. H. MacDonald, "The cellular concept," Bell System Tech. J., vol. 58, pp. 15–43, Jan. 1979.
[7] W. C. Y. Lee, "Statistical analysis of the level crossings and duration of fades of the signal from an energy density mobile radio antenna," Bell Syst. Tech. J., vol. 46, pp. 417–448, Feb. 1967.

William C. Y. Lee (M'64–SM'80–F'82) received the B.S. degree from the Chinese Naval Academy, Taiwan, China, in 1953, and the M.S. and Ph.D. degrees from The Ohio State University, Columbus, both in electrical engineering, in 1960 and 1963, respectively.

From 1964 to 1979 he worked for Bell Telephone Laboratories. He created both fading and propagation path loss models, searched effective schemes in reducing multipath fading, studied mobile antennas and polarization effects, and participated in designing and analyzing the Bell System Advanced Mobile Phone Service (AMPS) system. He developed a software program based on his UHF propagation model used for the mobile telephone system design. He joined ITT Defense Communications Division, in 1979 where he worked on military communication systems and adaptive antenna array theory. He is now Vice President of Corporate Technology for PacTel Mobile Communications, Irvine, CA.

Dr. Lee is an Associate Editor of IEEE TRANSACTIONS ON VEHICULAR TECHNOLOGY. He is the author of two books: Mobile Communications Engineering, published by McGraw-Hill in 1982, and Mobile Communication Design Fundamentals, published by Howard W. Sams Book Co. in 1986.

Handover and Channel Assignment in Mobile Cellular Networks

Quick and timely handover has a crucial effect on how users perceive quality of service, however, handover strategies should not be too complicated.

Sirin Tekinay and Bijan Jabbari

Reprinted from *IEEE Communications Magazine,* pp. 42-46, Nov., 1991.

Bijan Jabbari is associate professor of Electrical and Computer Engineering at George Mason University's School of Information Technology and Engineering in Fairfax, Virginia

Sirin Tekinay (Student Member, IEEE) is currently completing requirements for the Ph.D. degree at George Mason University's School of Information Technology and Engineering in Fairfax, Virginia.

T he rapid growth in the demand for mobile communications has led the industry into intense research and development efforts towards a new generation of cellular systems. One of the important objectives in the development of the new generation is improving the quality of cellular service, with handovers nearly invisible to the Mobile Subscriber (MS). In general, the handover function is a most frequently encountered network function and has direct impact on the perceived quality of service. It provides continuation of calls as the MS travels across cell boundaries, where new channels are assigned by the new Base Station (BS) and the Mobile Switching Center (MSC).

The system performance characteristics include probability of blocking of new traffic, probability of forced termination of ongoing calls, delay in channel assignment, and total carried traffic. There is a tradeoff between the quality of service and implementation complexity of the channel allocation algorithms, number of database lookups and spectrum utilization. In selecting a channel assignment strategy, the objective is to achieve a high degree of spectrum utilization for a given quality of service with the least possible number of database lookups and simplest possible algorithms employed at the BS and/or the MSC. Handover prioritization schemes are channel assignment strategies that allocate channels to handover requests more readily than originating calls. Prioritization schemes provide improved performance at the expense of reduction in the total admitted traffic.

In this article, we provide a taxonomy of the channel assignment strategies along with the complexity in each cellular component. Next, we consider various handover scenarios and the roles of the BS and MSC. We then discuss the prioritization schemes and define the required intelligence distribution among the network components.

Strategies and Functionality

Efficient utilization of the scarce spectrum allocated for cellular communications is certainly one of the major challenges in cellular system design. All of the proposed strategies suggest the reusage of the same radio frequencies in noninterfering cells. Channel assignment strategies can be classified into fixed [1], flexible [2] and dynamic [3] (see Fig. 1). Table I provides a summary of these strategies, along with the role assumed by the MSC with each of them. The MSC function common to all channel assignment strategies is the storage and update of information on which MS is being served on which channel. This information is essential for network-directed criteria (involved in other network functions as well) such as location information of MSs, control traffic loads and overall traffic loads. In the descriptions of various channel assignment strategies that follow, we focus on the case where all cells under consideration belong to the same MSC.

Fixed Channel Assignment Strategies

The common underlying theme in all fixed assignment strategies is the permanent assignment of a set of channels to each cell. The same set of radio frequencies is reused by another cell at some distance away. The minimum distance at which radio frequencies can be reused with no interference is called the "cochannel reuse distance," which is accepted to be three cell units in the seven-cell cluster model.

The basic fixed assignment strategy (see Fig. 2) implies that a call attempt at a cell site can only be served by the unoccupied channels of the predetermined set of channels at that cell site; otherwise, the call is blocked. Here, the only role of the MSC is to inform the new BS, and receive a confirmation or rejection message from the new BS, about the handover. The MSC keeps track of serving channels for the purpose of updating stored information regarding the location of the MS.

Other fixed assignment methods are variations of the basic strategy described above, with various channel-borrowing methods (see Fig. 3). We will demonstrate the role of the MSC with the simple borrowing, hybrid assignment, and borrowing-with-channel-ordering strategies.

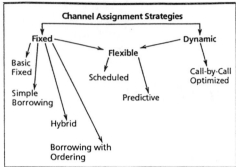

Figure 1. Classification of channel allocation strategies

In the simple borrowing strategy, if all permanent channels of a cell are busy, a channel can be borrowed from a neighboring cell, provided that this channel does not interfere with the existing calls. When a channel is borrowed, additional cells are prohibited from using it. The MSC supervises the borrowing procedure, following an algorithm that favors channels of cells with the most unoccupied channels to be borrowed. The algorithm "locks" the borrowed channel toward the cells that are one or two cell units away from the borrower cells. The MSC keeps record of free, serving and borrowed (therefore, locked) channels and informs all involved BSs about locked channels. The reward of increased storage requirement at the MSC and the need for database lookups is a lower call blocking probability up to a certain traffic level. In heavy traffic, since borrowed channels are locked for at least five additional cells, channel utilization efficiency is degraded.

This trend is improved by the hybrid channel assignment strategy proposed in [4]. In this strategy, permanent channels of a cell are divided into two groups: one group can be used only locally, i.e., within the cell; the other can be borrowed. The ratio of the numbers of channels in the two groups is determined *a priori*, depending on an estimation of the traffic conditions. In addition to its duties in the simple borrowing strategy, in the hybrid channel assignment strategy, the MSC has to label all channels with respect to the group to which they belong.

The borrowing-with-channel-ordering strategy suggested in [5] introduces a further improvement on the channel-borrowing concept. It elaborates on the idea of hybrid assignment by dynamically varying the local-to-borrowable channel ratio according to the changing traffic conditions. Each channel has a different adjustable

Channel Assignment Strategy	MSC Functionalities
Fixed Assignment	Inform new BS Keep track of serving channels
Simple Borrowing	Keep track of free/serving/locked channels Inform all involved BSs
Hybrid Assignment	Assign a set of fixed and borrowable channels to each cell at an optimum ratio depending on estimated traffic load Keep track of free/serving/locked channels Inform all involved BSs
Borrowing with Channel Ordering	Adjust fixed/borrowable channels ratio according to traffic load to each channel Assign a probability of being either used for a local call or borrowed Keep track of free/serving/locked channels Inform all involved BSs
Flexible (Scheduled)	Assign flexible channels at scheduled time according to stored estimation pattern Keep track of free/serving/flexible channels
Flexible (predictive)	Assign flexible channels according to changing traffic Keep track of free/serving flexible channels
Dynamic (Call-by-Call)	Assign channels upon request by evaluating channel reuse distance and future call blocking probability Keep track of free/serving channels

Table 1. MSC roles with different channel assignment strategies

probability of being borrowed and is ranked with respect to this probability, so that channels toward the bottom of the list are more likely to be borrowed, and vice versa. Each time a call is attempted, an algorithm at either the MSC or BS is run to choose the most "appropriate" channel among all free channels, looking at their associated probabilities. If this is part of the BS functionality, the MSC must be informed of the resulting assignment. The MSC determines and updates each channel's probability of being borrowed, based on the traffic conditions, by using an adaptive algorithm. The channel assignment strategy can be made more complex by allowing intracellular handover, i.e., immediate reallocation of a released higher-rank channel to a call existing on a lower-rank channel. The aim of such reallocation is to minimize the number of calls on the relatively more "borrowable" channels in order to reduce the locking effect of borrowed channels in additional cells. Reallocation is achieved by a comparison algorithm accommodated at either the BS or MSC, which is invoked each time a channel is freed.

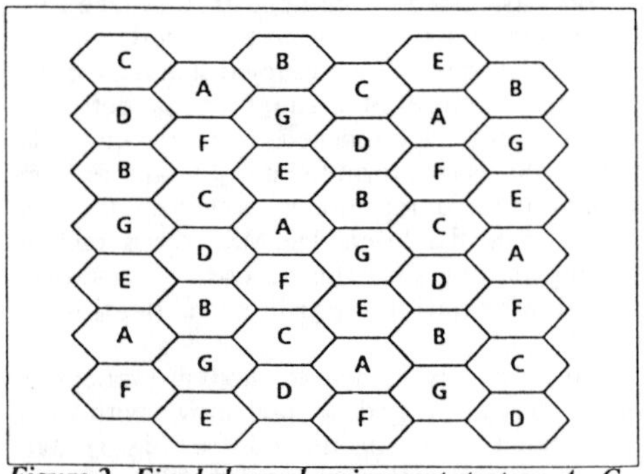

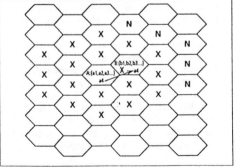

Figure 3. Borrowing strategies. Channel a4 is borrowed and now locked to cells marked "N." Cells marked "X" were already prohibited from using a4.

Figure 2. Fixed channel assignment strategy. A - G denote different sorts of channels permanently assigned to cells

Dynamic Channel Assignment Strategies

In contrast to fixed assignment, in dynamic assignment strategies cells have no channels to themselves but refer all call attempts to the MSC, which manages all channel assignment in its region. Each time a call attempt arrives, the BS asks the MSC for the channel with the minimum cost to be assigned. The cost function depends on the future blocking probability, usage frequency of the candidate channel, the reuse distance of the channel, and so on. The MSC decides, on a call-by-call basis, which channel to assign to which call attempt by searching for the available channel for which the cost function is minimum. It needs to have information regarding channel occupancy distributions under current traffic conditions and other network-directed criteria, as well as radio channel measurements of individual MSs.

Flexible Channel Assignment Strategies

Flexible channel assignment strategies combine aspects of both the fixed and dynamic strategies in the sense that each cell is assigned a set of permanent channels that typically will suffice under light traffic loads. The MSC holds a set of flexible channels and assigns these to cells whose permanent channels have become inadequate under increasing traffic loads. The distribution of these emergency channels among the cells in need of them is carried out by the MSC in either a scheduled or predictive manner [2].

If the flexible channels are reassigned on a scheduled basis, it is assumed that future changes in traffic distribution are pinpointed in time and space. The change in assignment of flexible channels is then made at the predetermined peaks of traffic change.

In the predictive assignment strategy, the traffic intensity or, equivalently, the blocking probability, is constantly measured at every cell site so that the reallocation of the flexible channels can be carried out by the MSC at any point in time.

Flexible assignment strategies, like call-by-call dynamic strategies, require the MSC to have up-to-date information about the traffic pattern in its area and other network-directed criteria in order to manage its set of flexible channels efficiently.

Possible Handover Scenarios

The channel assignment strategies described above are used whenever a new call or handover request is received by the BS or MSC. Some assignment strategies prioritize handover requests in order to protect ongoing calls from forced termination. Before describing the handover prioritization schemes, we review the handover process (see Table II).

The decision that a handover shall take place can be made by both the MS and the BS by monitoring the channel quality. If the decision is made by the MS alone, a handover request is provided to the BS. The new BS is determined by the MS or MSC. If it is determined by the MS, the candidate BS is provided to the MSC. We note that the decisions made by the MS are based on radio channel measurements only, whereas the MSC is in a position to judge according to a collection of criteria, including network-directed ones such as the traffic distribution in the area.

Radio Channel Measurements

From the viewpoint of the network, the detection of the need for handover and its timely execution are challenging tasks. Momentary fadings in the communication channels between the MS and BS may occur due to geographical and environmental factors well within the cell. This means that the decrease in the power level of these channels should be observed for a certain amount of time before it can be concluded that the MS is actually moving away from the BS. On the other hand, if there actually is a need for handover, it must be responded to as soon as possible in order to minimize the risk of forced termination of the call. In order to detect the need for handover, the MS needs to take measurements on the channel it is currently using as well as the broadcast channels of the neighboring cells. Different standards for cellular operations specify different procedures for these

Task	MS	BS	MSC
Radio Channel Measurements	Make periodic measurements on current and neighboring broadcast channels. Send results to BS		
	Start measurements. Send results to BS	Monitor backwards channels. Give measurment order to MS	
Issue Handover Request		Send measurement results to MSC. Request handover	
			Evaluate handover request. Inform new BS
		Evaluate handover requests. Request handover	Inform new BS

Task	MS	New BS	MSC
Confirm/ Disconfirm Handover		Accept/block/ delay (queue) handover request	Permit/drop/ delay handover

Table 2 Intelligence distribution among MS, BS and MSC in handover procedures

measurements (see bibliography).

According to the Telecommunications Industry Association (TIA) standards, the BS monitors the backward channels of all MSs with which it is communicating. When it detects a significant drop in the power level, it sends the MS a measurement order. Upon receiving the measurement order, the MS starts taking measurements. The measurement results are reported to the BS with the frequency prescribed in the measurement order. The Pan-European GSM standards suggest that the MS should take measurements all the time and report the results periodically to the BS. This eliminates the need for the BS to constantly monitor all backward channels. A promising method of radio channel measurements would be interactively varying the intervals between the taking and/or reporting of measurements.

Roles of the BS and MSC in Handover Procedures

The BS receives either measurement results only, which it has to evaluate to decide whether a handover is necessary, or the measurement results together with the next BS selected by the MS. In the first case, the BS issues the handover request, if necessary, and sends it to the MSC. Then the MSC picks the best BS to serve the continuation of the call. In the second case, the BS merely sends the MSC the request for handover to the candidate BS specified by the MS. In both cases, the MSC informs the new BS of the handover request.

The new BS, depending on the channel assignment strategy (and possibly the handover prioritization scheme), may accept, block, or queue the handover request. It informs the MSC regarding the status of the handover request. Depending on the response of the new BS, the MSC may permit, delay, or drop the handover request.

Handover Policies

In some channel assignment strategies, the BS handles handover requests in exactly the same manner as it handles originating calls. Obviously, such schemes suggest that the probability of forced termination of an ongoing call due to unsuccessful handover equals the probability of blocking an originating call. From the MS's point of view, however, forced termination of an ongoing call is significantly less desirable than blocking a new call attempt. Therefore, methods for decreasing the probability of forced termination by prioritizing handovers at the expense of a tolerable increase in

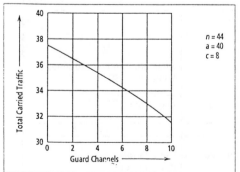

Figure 4. Decrease in total traffic as a function of the number of guard channels [6]

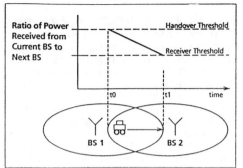

Figure 5. Handover and receiver thresholds. Linear motion from BS 1 to BS 2 is assumed; handover must occur in [t0,t1]

call blocking probability have been devised in order to increase the quality of cellular service. We now present two prioritization schemes.

The Guard Channel Concept

The "guard channel" concept was introduced in the mid-1980s [6, 7]. It offers a generic means of improving the probability of successful handover by simply reserving a number of channels exclusively for handovers. The remaining channels can be shared equally between handovers and originating calls.

The penalty is the reduction of total carried traffic (see Fig. 4) due to the fact that fewer channels are granted to originating calls, and it is the originating calls and not the ongoing calls that really add to the total traffic. This disadvantage can be bypassed by allowing the queuing of originating calls. Intuitively, we can say that the latter method is feasible because originating calls are considerably less sensitive to delay than handover requests.

Another shortcoming of the employment of guard channels, especially with fixed channel assignment strategies, is the risk of inefficient spectrum utilization. Careful estimation of channel occupancy time distributions[1] is essential in order to minimize this risk by determining the optimum number of guard channels.

With flexible or dynamic channel assignment strategies, the guard channel concept is revisited in a modified manner. Cells do not keep guard channels in their possession. The MSC can keep a collection of channels only for handover requests, or it can have a number of flexible channels with associated probabilities of being allocated for handover requests.

Queuing of Handover Requests

The queuing of handover requests, with or without the employment of guard channels, is another generic prioritization scheme offering reduced probability of forced termination. There is again a tradeoff between the increase in service quality and the corresponding decrease in total carried traffic. Before we discuss its consequences, we briefly describe this scheme.

Handover can occur in the time interval during which the ratio of the power levels received from the current and next BSs is between the "handover threshold" and the "receiver threshold" (see Fig. 5). The handover threshold is set at the point where the power received from the BS of a neighboring cell site has started to exceed the power received from the current BS by a certain amount

With flexible or dynamic channel assignment strategies, the guard channel concept is revisited in a modified manner.

[1]*Unencumbered call duration is generally accepted to have an exponential distribution. Due to the memoriless property of the exponential distribution, the duration of the remaining portion of the call after a handover is also exponentially distributed. For a complete analysis of channel occupancy time distributions, the reader is referred to [8].*

One of the aims of our current research is to improve the quality of cellular service by modifying the queue discipline in queuing handovers.

and/or for a certain time. The receiver threshold is the point at which the received power from the BS is at the minimum acceptable level. At this point, since communicating with the current BS is no longer possible, the call will be terminated unless a successful handover to an eligible cell has already occurred. Queuing handover requests is made possible by the existence of the time interval that the MS spends between these two thresholds. The maximum possible waiting time in the queue is given by this interval. The allowable queue size needs to be determined. Computation of the queue size requires knowledge of the traffic pattern of the area, the major factor of which is the expected number of handover requests. In the case of high demand for handovers, the assumption of infinite queue size introduces an undesirably large decrease in total carried traffic [6]. Furthermore, the probability of forced termination is still strictly greater than zero, because the handover request can only wait until the receiver threshold is reached. This is why handover requests are much more sensitive to delay in service than originating calls. Indeed, queuing handovers has been widely discussed; some are in favor of it because of the decrease in the probability of forced termination it offers, while others argue that the delay insensitivity of originating calls makes it more feasible to queue new call attempts rather than handover requests.

One of the aims of our current research is to improve the quality of cellular service by modifying the queue discipline in queuing handovers [9]. The queuing system is not viewed as "first come first serve." A handover request is ranked according to how close the MS stands to (and, possibly, how fast it is approaching) the receiver level. The necessary radio channel measurements are already made; therefore, the only additional complexity in implementing the modification is a fairly simple comparison algorithm to be run continuously on the stored handover requests.

Summary

In this article, we have reviewed various handover scenarios and suggested several ways of distributing intelligence between the MS, BS, and MSC, focusing on their respective roles in these scenarios. We have described the effect of different channel assignment strategies and handover prioritization schemes on BS and MSC functions.

The main criteria used to compare the performance of a cellular system model under different assumptions are probability of call blocking, probability of forced termination, total carried traffic, delay in channel assignment, and number of database lookups. These criteria together define the cost function, the minimization of which, along with quality of service improvement,

is the objective. We have proposed a method of prioritizing handover requests by queuing them in such a way that the one with the maximum probability of forced termination is served first.

References

[1] M. Zhang and T. P. Yum, "Comparisons of Channel-Assignment Strategies in Cellular Mobile Telephone Systems," *IEEE Trans. on Vehicular Tech.*, vol. 38, Nov. 1989.
[2] J. Tajima and K. Imamura, "A Strategy for Flexible Channel Assignment in Mobile Communication Systems," *IEEE Trans. on Vehicular Tech.*, vol. 37, May 1988.
[3] D. C. Cox and D. O. Reudnik, "Increasing Channel Occupancy in Large-Scale Mobile Radio Systems: Dynamic Channel Assignments," *IEEE Trans. on Vehicular Tech.*, vol. 22, Nov. 1973.
[4] T. J. Kahwa and N. D. Georganas, "A Hybrid Channel Assignment Scheme in Large-Scale, Cellular-Structured Mobile Communication Systems," *IEEE Trans. on Commun.*, vol. 26, Apr. 1978.
[5] S. M. Elnoubi, R. Singh, and S. C. Gupta, "A New Frequency Channel Assignment Algorithm in High Capacity Mobile Communication Systems," *IEEE Trans. on Vehicular Tech.*, vol. 31, Aug. 1982.
[6] R. A. Guerin, Ph.D. thesis, Dept. of Electrical Engineering, CA Inst. of Tech., Pasadena, CA, 1986.
[7] D. Hong and S. S. Rappaport, "Traffic Model and Performance Analysis for Cellular Mobile Radio Telephone Systems with Prioritized and Nonprioritized Handoff Procedures," *IEEE Trans. on Vehicular Tech.*, vol. 35, Aug. 1986.
[8] R. A. Guerin, "Channel Occupancy Time Distribution in a Cellular Radio System," *IEEE Trans. on Vehicular Tech.*, vol. 34, Aug. 1987.
[9] S. Tekinay and B. Jabbari, "A Measurement-Based Queueing Scheme for Protection of Handovers in Mobile Cellular Systems," Tech. Rep., George Mason Univ., 1991.

Bibliography

International Consultative Committee for Telephone and Telegraph (CCITT) Blue Book Recs. Q.1001, Q.1002, Q.1003, Q.1004, and Q.1005, vol. VI, fascicle VI.12.
Electronics Industry Association (EIA)/TIA Proj. No. 2078, "Cellular Radiotelecommunications Intersystem Operations: Intersystem Handoff," IS-41.2, Feb. 1991.
European Telephone Standards Institute (ETSI) TC GSM, Rec. GSM 03.09, "Handover Procedures," Feb. 1990.
ETSI TC GSM, Rec. GSM 08.06, "Signalling Transport Mechanisms between the BSS and MSC Procedures," June 1990.

Biography

Bijan Jabbari (Senior Member, IEEE) received the B.S. degree from Arya-Mehr University, Tehran, Iran, in 1974, the M.S. and Ph.D. degrees from Stanford University, Stanford, CA, in 1977 and 1981, respectively, all in electrical engineering, and the M.S. degree in engineering-economic systems from Stanford University in 1979. From 1979 to 1981 he was with Hewlett Packard. After graduation, from 1981 to 1983, he was an Assistant Professor at Southern Illinois University, Carbondale, IL. From 1983 to 1985 he was with Satellite Business Systems (now MCI Telecommunications), McLean, VA, where he managed programs on systems requirements definition, system specification, and architecture of the SBS next generation communications system. In 1985, he became Director at M/A-COM Telecommunications for development of Advanced Data Communications Networks. In 1988, he joined the School of Information Technology and Engineering at George Mason University, Fairfax, VA, where he is currently associate professor of Electrical and Computer Engineering. His current research activities include architecture and protocols and performance modeling of broadband telecommunications, intelligent networks, control and signalling for fixed and mobile telecommunications networks. He is a member of Eta Kappa Nu and the Association for Computing Machinery.

Sirin Tekinay (Student Member, IEEE) received the B.S. and M.S. degrees from Bogazici University, Istanbul, Turkey, in 1989 and 1991 respectively, in electrical engineering. She is currently completing requirements for the Ph.D. degree at the School of Information Technology and Engineering, George Mason University, Fairfax, VA. Her area of research is mobile cellular communication networks with concentration on call control, traffic analysis, and performance evaluation. Ms. Tekinay is a member of Eta Kappa Nu and the IEEE Communications Society.

Smaller Cells for Greater Performance

A new microcell architecture that reduces interference, increases system capacity, improves voice quality, and demands fewer handoffs is ideally suited for PCS systems

Dr. W. C. Y. Lee

Reprinted from *IEEE Communications Magazine*, pp. 19-23, Nov., 1991.

T he Advanced Mobile Phone Service (AMPS) cellular system at 850 MHz, as used in North America, is a high-capacity system. Its spectrum utilization is based on the frequency re-use concept, in which a frequency can be re-used repeatedly in different geographical locations. Various locations using the same frequencies are called cochannel cells.

The minimal separation (D_S) required between two nearby cochannel cells is based on specifying a tolerable cochannel interference, which is measured by a required carrier-to-interference ratio $(C/I)_S$. The $(C/I)_S$ ratio also is a function of the minimum acceptable voice quality of the system. In an AMPS system, $(C/I)_S$ is equal to about 18 dB (the point at which 75 percent of the users call the system "good" or "excellent"), and the minimal required separation D_S, based on $(C/I)_S = 18$ dB, is about 4.6R, where R is the radius of the cell. In a cellular system, the number of cells K in a cell re-use pattern is a function of the cochannel separation D_S. For $D_S = 4.6R$, then K = 7. This means that a cluster of seven cells can share the entire allocated spectrum. In each of the two bands allocated for cellular, there are 395 voice channels; each cell can have 57 channels on average.

In 1991, the conventional cellular systems in use since 1984 began to reach their capacity in the larger markets. In order to increase system capacity, we may take approaches based on what will be called the cochannel interference reduction factor (CIRF), q_S, which is defined as [1]

$$q_S = D_S / R \overset{\Delta}{=} \sqrt{3K} \qquad (1)$$

D_S is the minimum required distance between any two cochannel cells in a cellular system (see Fig. 1a) where D_S is corresponding to the required carrier-to-interference ratio (C/I) received at both the cell site and the mobile unit in a cell. R is the cell radius, K is the number of cells in a cell re-use pattern. K is also called cell re-use factor. The three approaches for increasing capacity are stated as follows. The first two are conventional approaches; the third one is the new approach. Equation

(1) is derived from an idealized hexagonal cell layout and is commonly used.

• Split cells. In this approach, capacity can be increased by reducing R but keeping q_S unchanged as in Equation (1) (shown in Fig. 1b), i.e., rescaling the system. When R is smaller than one mile or one kilometer, the cell is commonly called a microcell. As a first-order approximation, every time R is reduced by one half, capacity increases by four. The measurement of capacity in this approach is the number of channels per square kilometer. The cell splitting approach leads to an increase in radio capacity [2]. Splitting cells is system scale independent, i.e., the value of q_S remains unchanged. This approach can be used in any analog or digital system.

• Reduce the cell re-use factor (also called "a reduction of the required D/R" approach). In this approach, we seek to increase capacity by determining methods by which D can be reduced, i.e., forming a new configuration, but keeping R unchanged in Equation (1) (shown in Fig. 1c). Therefore, q_S can be reduced, and thus is the cell re-use factor K, as shown in Equation (1). The value of D_S, however, is a function of the required $(C/I)_S$. For example, if a new cellular system can achieve a frequency re-use factor of K = 3, then the capacity of the new system can be obtained by comparing it with the AMPS system of K = 7. Since K is reduced from K = 7 to K = 3, the capacity is increased by 7/3 = 2.33 times. The measurement of radio capacity in this approach is the number of channels per cell.

$$m = \frac{\text{total voice channels}}{K} \qquad (2)$$

The reduction of the cell re-use factor approach would increase radio capacity m, as shown in Equation (2).

In the past, sectorization was used to reduce the value of K in an analog system. When the cochannel interference in a cell increases, either a three-sector or six-sector cell configuration should be used in order not to expand the required cochannel cell separation D_S. In other words, with a given interference, the sectorization seems to be able to reduce

William C. Y. Lee is vice president of research and technology at PacTel Corporation.

 0163-6804/91/$01.00 1991© IEEE **19**

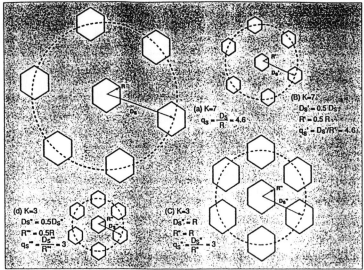
Figure 1 Four cases of Expression of CIRF

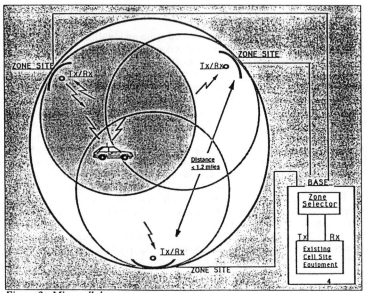
Figure 2 Microcellular concept

the value of D_s. However, when we take a good look at sectorization, the method of assigning a set of frequency channels in each sector is the same as if assigned in a cell. The handoffs occur as the vehicle passes among the sectors, the same as among cells. Therefore, if in a K = 7 system of three-sector cells, the number of channels per sector, assuming a total of 395 voice channels, is

$$\frac{395}{7 \times 3} = 19 \text{ channels/sectors}$$

In a K = 4 system of six-sector cells, the number of channels per sector is

$$\frac{395}{4 \times 6} = 16 \text{ channels/sectors}$$

From the above two values, there is not much difference in radio capacity between the two cellular configurations. In order to gain further capacity by using the splitting cells concept as stated in the split cells approach, the size of each sector can be reduced. The problem for the sectorization is that

the trunking efficiency of the utilized channels decreases. The cell usage with the same number of channels in an omni cell is much higher than that of a sector cell. Therefore, sectorization is not an effective method of reducing q_s.

• Reduce the required D/R by a new microcell approach. The concept of the new microcell system is shown in Fig. 1d. In this case, not only the cell radius reduces, but the CIRF also reduces. Furthermore, there is no degradation in trunking efficiency; it is a true K = 3 system. The advantages of this system include both reduction of cochannel interference and confining the cochannel interference relative to the signal to a small area. It will be described in the following section.

Description of New Microcell System Design

Generally this new microcell consists of three zones, as shown in Fig. 2[4]. (It can be more than three zones when needed.) Each zone has a zone site, and one of the three zone sites usually is colocated with a base site. All radio transmitters and the receivers that serve a microcell are installed at the base site. Every zone site physically shares the same radio equipment installed at the base. To serve a vehicle from a zone site, an 800 MHz cellular signal can be up-converted to a microwave or optical signal at the base and then down-converted back to an 800 MHz signal at the zone site to serve the vehicle at that zone as if the vehicle is located at the base.

Conversely, the received cellular signal, after boasting with a low-noise amplifier at a zone site, can be up-converted to either a microwave or optical signal, then down-converted to 800 MHz at the base. In this case, the zone site only requires an up/down converter, power amplifier and a low-noise broadband pre-amplifier, which is easy to install because of the small size and the light weight of the zone-site apparatus.

Signal coming from mobile unit. A mobile unit driving in a microcell sends a signal. Each zone site receives the signal and passes it through its up/down converter, up-converting the signal and sending it through either the microwave or optical signal medium, then down-converting the signal at the base site. Thus, the mobile signals received from all zones are sent back to the base site. A zone selector located at the base site is used to select a proper zone to serve the mobile unit by choosing the zone having the strongest signal strength. Then the base site delivers the cellular signal to the proper zone site through its up-converting processing.

Signal coming from base site. The proper zone site receives the cellular signal from the base site through a down-converting process and transmits to the mobile unit after amplification. Therefore, although the receivers at three zones are all active, only the transmitter of one zone is active in that particular frequency to serve that particular mobile unit. When the mobile unit is moved from zone to zone, the assigned channel frequency remains unchanged. The zone selector at the base site simply shifts the transmitting signal (base-to-mobile) from one zone to another zone according to the mobile unit's location. Only one active base-

IEEE Communications Magazine • November 1991

transmitting zone at one time is serving a vehicle (one assigned frequency) in a cell. Therefore, no hand-off action is needed when the mobile unit is entering a new active zone.

Analysis of Capacity and Voice Quality

In order to prove the increase of capacity and the improvement of voice quality in this new microcell system as shown in Fig. 1d, we may calculate the cochannel interference reduction factor (CIRF), q_S, which is a key element in designing the cellular system. In the conventional macrocell system, q_S is used for taking care of both the voice quality and the capacity since they are related. In this microcell system, which is different from the macrocell systems, there are two CIRFs to be considered. One CIRF q_{S1} is used to measure the voice quality and the other, CIRF q_{S2}, is used to measure the radio capacity, because in this microcell system the voice quality and the capacity are measured differently. The microcell system is shown in Fig. 3.

The CIRF between two active base transmitting cochannel zones. This is a new CIRF q_{S1} defined as $q_{S1} = D_1/R_1$, where D_1 is the distance between one active zone in one microcell and the corresponding active zone in the other microcell, as shown in Fig. 3. R_1 is the radius of each zone. Although the antenna is mounted at the edge of each zone, the real coverage area of each zone is used for estimating interference. Therefore, the radius R_1 of a real coverage area is used to confine the zone area.

There are many values of q_{S1}, depending on which two active cochannel zones are considered. Among them, the two closest cochannel active zones are the worst case to be used for measuring CIRF, q_{S1}. As we know, in an AMPS system, C/I has to be 18 dB, thus implying that q_S must be 4.6 in order to maintain an acceptable voice quality when using 30kHz analog FM radios. In the AMPS system, the earlier simulation shows that $q_{S1} = 4.6$ was adequate for omni-directional cells [5].

When the cell site antenna height is normally 100 ft. to 150 ft. high and the ground is not flat, however, the cochannel interference received on the reverse link (mobile-to-base) is larger than expected. Therefore, a sectorization architecture was introduced for the macrocells. In a microcell system, the microcell antenna height is always lower than 100 ft., normally 40 ft. to 50 ft., and generally the ground in a small area around the antenna is flat. Under this condition the cochannel interference on the reverse link is reduced, and the sectorization arrangement becomes unnecessary for a K = 7 system configuration. This is supported by measured data. Since the same radios are used in the microcell system, q_{S1} has to be at least the same as 4.6 in order to be back on the K = 7 configuration.

By construction, it is shown that the q_{S1} of the two closest cochannel active zones in their corresponding microcells is 4.6, as shown in Fig. 3. In this microcell, normally, the q_{S1} between any two active zones in two corresponding cochannel microcells is always equal or greater than 4.6, as shown in Fig. 3. It is proven that the voice quality in this microcell system based on $q_{S1} \geq 4.6$ is equal or

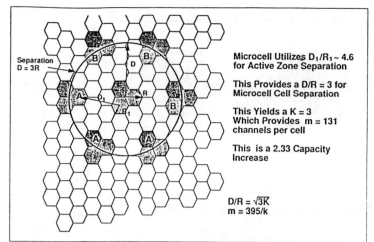

Microcell Utilizes $D_1/R_1 \sim 4.6$ for Active Zone Separation

This Provides a D/R = 3 for Microcell Cell Separation

This Yields a K = 3 Which Provides m = 131 channels per cell

This is a 2.33 Capacity Increase

$D/R = \sqrt{3K}$
$m = 395/k$

Figure 3. Microcell application

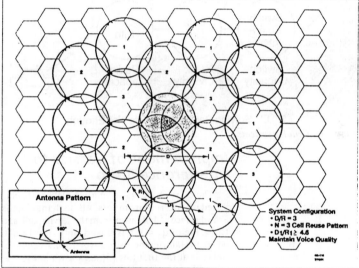

Antenna Pattern

System Configuration
• D/R = 3
• N = 3 Cell Reuse Pattern
• $D_1/R_1 \geq 4.6$
Maintain Voice Quality

Figure 4. A K-3 microcell system

better than the voice quality of AMPS system. This q_{S1} is used to measure the voice quality only in the new microcell system.

The CIRF between two cochannel microcells. The radio capacity is based on the separation of two neighboring cochannel cells. In the microcell system, the CIRF, q_{S2} is defined as $q_{S2} = D/R$ where D is the distance between two cochannel microcells and R is the microcell radius as shown in Fig. 3. In this case $q_{S2} = D/R = 3$, equivalent to K = 3 shown in Equation (1). The three zones per microcell and the K = 3 system is illustrated in Fig. 4. The antenna pattern for each zone coverage is 160°, as depicted in Fig. 4. Thus, the entire cell is covered. Since K is reduced from K = 7 of the AMPS system to K = 3, the microcell system has increased 7/3 = 2.33 times, as is shown by Equation (2). Therefore, q_{S2} is used to measure the capacity.

The frequency assignment in a K = 3 system is shown in Fig. 4. The total allocated 395 channels can be divided into three groups. The first group consists of channels 1, 4, 7, 10, etc. The second group consists of channels 2, 5, 8, 11, etc. The channels of the third group are 3, 6, 9, 12, etc. Each group will be assigned to each cell according to the cell number shown in Fig. 4.

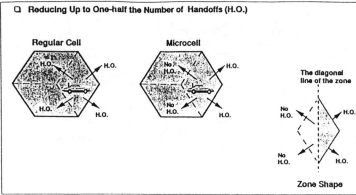

Figure 5   Reduction of handoffs in microcell system

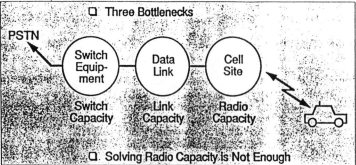

Figure 6   Cellular systems

Improvement of Carrier-to-Interference Ratio

The $q_S = D_S/R = 4.6$ in an AMPS system is based on two requirements: all cochannel cells are a distance of 4.6R away from the serving cell, and the value of $q_S = 4.6$ is based on C/I = 18 dB, where the interference is received from six cochannel cells at the first tier [3], as shown in Fig. 1a. The C/I of 18 dB at RF averaged over Rayleigh fading, provides a good or excellent signal to the user in an analog system.

In a microcell system, the separation D_1 of any two nearest cochannel zones between two active zones in two corresponding microcells is $4.6R_1$, as shown in Fig. 3. All other cochannel zones in their microcells are separated farther than $4.6R_1$. For consideration of the worst case scenario, we can choose between an active zone in the center cell and individual cochannel zones in six corresponding microcells, and calculate the C/I received at that zone in the center cell of the microcell system, as shown in the following (in Fig. 4).

$$\frac{C}{I} = \frac{R_1^{-4}}{\sum_{i=1}^{6} D_i^{-4}} = \frac{R_1^{-4}}{3(4.6R_1)^{-4} \cdot 3(5.75R_1)^{-4}} = 105(=)20dB \quad (3)$$

3 A-zones   3 B-zones

Equation (3) shows that C/I in a microcell system according to the worst case is 2 dB better than that of an AMPS system. Since Equation (3) shows the worst case, where all the cochannel zones are in either A or B zones, the C/I can be even better if all of the active cochannel zones are other than A or B zones. The C/I in the normal case then is always greater than 20 dB. The

C/I ≥ 20 dB is at least 2 dB better than the required $(C/I)_S$ of the AMPS system. This proves that the voice quality of the microcell system is always better than that of the AMPS system. One remark is that this calculation is based on the signal coverage in each zone regardless of the type of antenna. This is because the shape of the coverage takes care of the antenna pattern.

Reduction of Handoffs

The definition of handoffs is to hand off a frequency to another frequency while the vehicle enters a new cell or a new sector. Within each microcell, no handoffs from zone-to-zone are needed; zone-to-zone switchings are handled by a zone selector. The active zone follows the mobile unit as it moves from one zone to another. The channel frequency assigned to the mobile unit remains unchanged.

In this section we may roughly estimate how many handoffs can be eliminated relative to a microcell plan in which three zones are used. In a regular cell, there are three sectors. The car can move in any one of three directions, as shown in Fig. 5. When a car moves through the other two sectors, it needs handoffs. When it enters, as well as when moving out of a cell, a handoff occurs. In a microcell, a car moving to either of the other two zones does not need handoffs. A handoff only occurs when the car moves in or out of the cell. Since the shape of a zone is based on the hexagonal cell, it is diamond shaped. Symmetry to its diagonal line can be observed from the left side of the diagonal line where no handoffs are required, and on the right side handoffs are needed. Therefore, we may estimate that only one half of the handoffs required in a regular cell configuration will occur in a microcell configuration. This reduction in handoffs in the microcell system makes a great contribution to the system capacity.

System Capacity

In any cellular system, system capacity is the overall capacity of each system, and can differ from system to system. System capacity may be capped by three limiting elements: radio capacity, control link capacity, and switch capacity, as shown in Fig. 6.

• Radio capacity is the element most often addressed in the public domain and in published literature [3].

• Link capacity/switch capacity are the two elements often overlooked in measuring system capacity. Control link capacity measures the capacity of the fast control link between the cell site and the switches. If the number of microwave links or T1 carrier lines are not sufficient, a bottleneck will result. Switch capacity measures the capacity of traffic at the switching office. Again, if the switch is not big enough to handle the radio capacity, a bottleneck problem occurs.

Among these three elements (capacities), the weakest element must be used to gauge the system's capacity. Therefore, improving the radio capacity in the system is not enough. Improving system capacity requires upgrading the lowest capacity of the three. With this in mind, every system operator should be aware that radio capacity is not the entire problem nor the entire solution.

In a microcell system design, because fewer inter-cell handoffs are needed as compared to a regular system, both the switching load and the control link load are cut roughly by half, leaving two times the load to be handled by the present capacity. The easing of half the load means twice the load can be added back onto the system. This roughly two times (2.33 times to be exact) radio capacity is exactly what the microcell system will offer without changing the present switching equipment.

Attributes of Microcell

The new microcell design contains many attributes:

• Increased system capacity. Based on the cell reuse pattern (reduced from K = 7 to K = 3), it is 2.33 times the AMPS system capacity.

• Voice quality improved. The voice quality of the microcell system always is better than the quality of AMPS.

• Interference reduced. (a) Since the antennas of all zone sites in one cell are facing toward each other, the interference signal has to cross the cell before interfering with the neighboring cell. Furthermore, the coverage is only served in one active zone; therefore, the interference is very weak compared with the interference from a transmitter from the center of a regular cell. (b) The three zone sites receiving the mobile signal simultaneously from three zones form a three-branch different-site diversity that is suitable for low-power portable units. It is increasing the probability of signal reception at the base due to diversity schemes. (c) The microcell system is the best arrangement to control interference. The active zone follows the vehicles.

• Adaptability. This microcell design can be added to any vendors' system without modifying the

vendor's hardware or software.

• Size of the zone apparatus. The zone up/down converters are small, and they can be mounted on the side of a building or on poles. Therefore, it is a PCS (Personal Communications Service) type system because of the tight control of interference. Also, it is easy to remount from pole to pole when the signal coverage requirement is changed.

Conclusion

This new microcell system is easy to implement, and poses a very low risk investment. Capacity, based on the K = 3 system, is 2.33 times higher than the existing analog system of K = 7. Furthermore, this microcell system can provide better voice quality than the AMPS system. It also can be used with digital cellular with slight modification. A microcell can serve in a small area; therefore, it is suitable for inbuilding communications. This microcell system is currently being implemented in west L.A.

References

[1] W. C. Y. Lee, "Mobile Cellular Telecommunications Systems," McGraw-Hill, p. 57, 1989.
[2] V. H. MacDonald, "The Cellular Concept" *BSTJ*, vol. 58, p. 15, Jan. 1979.
[3] W. C. Y. Lee, "Spectrum Efficiency in Cellular," *IEEE Trans. on Vehic. Tech.*, 69-75, May 1989.
[4] W. C. Y. Lee, "Cellular Telephone System," U.S. Patent 4,932,049, June 5, 1990.
[5] Gary D. Ott, "Vehicle Location in Cellular Mobile Radio Systems," *IEEE Trans. on Vehic. Tech.*, VT-26, pp. 43-6, Feb. 1977.

Biography

William C. Y. Lee, a Fellow of IEEE, received his B.S. degree from the Chinese Naval Academy, Taiwan, and his M.S. and Ph.D. degrees from Ohio State University at Columbus in 1954, 1960, and 1963, respectively. From 1959 to 1963 he was a research assistant at the Electroscience Laboratory at Ohio State University. He was with AT&T Bell Laboratories from 1964 to 1979. Mr. Lee worked for the ITT Defense Communications Division from 1979 to 1985. In 1985 he joined PacTel Mobile Companies. Currently, he is the vice president of research and technology at PacTel Corporation. Mr. Lee has written more than 150 technical papers and three textbooks.

In a microcell system design both the switching load and the control link load are cut roughly by half.

Chapter 3

MOBILE RADIO PROPAGATION

Coverage Prediction for Mobile Radio Systems Operating in the 800/900 MHz Frequency Range

IEEE VEHICULAR TECHNOLOGY SOCIETY COMMITTEE ON RADIO PROPAGATION

Reprinted from *IEEE Transactions on Vehicular Technology*, Vol. 37, No. 1, Feb. 1988.

Preface

Samuel R. McConoughey, Board member and former President of the IEEE Vehicular Technology Society, recognized the need for industry-accepted propagation models for predicting reliable service areas and co-channel interference in the 806–947 MHz segment of the radio spectrum. To address the growing need for these models in private and common carrier services with conventional, trunked, cellular and radio paging systems, in 1982 Mr. McConoughey presented a resolution to the Board of Directors of the Vehicular Technology Society. This resolution proposed the formation of an Ad Hoc Committee to study existing radio frequency propagation models and to make appropriate recommendations.

Neal H. Shepherd served as Chairman of the committee since early 1983. The Secretary for the first eight meetings was Floyd Shipley, followed by Robert P. Eckert for the remaining 28 meetings.

Special credit is due Henry Bertoni, who assumed the enormous task of editing this publication to provide continuity. Members of the committee submitted documents in accordance with an outline, which were then reviewed before acceptance by the Ad Hoc Committee. Following is a list of contributors.

Contributors	Affiliation
Nadia S. Adawi	Vision Systems, Inc. 850 N. Burlington Street Arlington, VA 22203
Henry L. Bertoni	Center for Advanced Technology in Telecommunications Polytechnic University 333 Jay Street Brooklyn, NY 11201
Joseph R. Child	Computer Sciences Corp. 3328 Glenmore Drive Falls Church, VA 22041
William A. Daniel[†]	Federal Communications Commission 1919 M Street, N.W. Washington, DC 20554
John E. Dettra	Dettra Communications Inc. 2021 K Street, N.W., Suite 309 Washington, DC 20006
Robert P. Eckert[†]	Federal Communications Commission 1919 M Street, N.W. Washington, DC 20554
Earl H. Flath, Jr.	Consultant 13634 Braemar Circle Dallas, TX 75234
Robert T. Forrest	Communications Technology Assoc., Inc. P.O. Box 4579 Lynchburg, VA 24502

[†] The views expressed here are those of the authors and do not necessarily reflect the views of the Federal Communications Commission.

William C. Y. Lee   PacTel Mobile Companies
                             2355 Main Street, P.O. Box 19707
                             Irvine, CA 92714

Samuel R. McConoughey†   Federal Communications Commission
                             1919 M Street, N.W.
                             Washington, DC 20554

John P. Murray   John Murray Associates
                             1823 Folsom Street
                             Boulder, CO 80302

Herbert Sachs   Sachs/Freeman Associates, Inc.
                             14300 Gallant Fox Lane
                             Bowie, MD 20715

George L. Schrenk   Comp Comm, Inc.
                             Station House, Suite 412, 900 Haddon Ave.
                             Collingswood, NJ 08108

Neal H. Shepherd   Consultant
                             1914 McGuffey Lane
                             Lynchburg, VA 24503

Floyd D. Shipley   Consultant
                             1201 Ridgewood Terrace
                             Arlington, TX 76012

The Committee's major objective was to recommend modeling approaches which would be suitable for predicting service areas for mobile systems operating in the 806–947 MHz band. Although each model studied had salient features, which met some of the needs, none were accurate enough to warrant a recommendation.

The following list provides deficiencies noted in various models:

- none of the models studied provided an adjustment for reflections from buildings or hills;
- none of the models studied made an allowance for transmission loss due to foliage;
- the Longley–Rice model underestimated the transmission loss for most typical mobile situations;
- the Okumura urban model generally overestimated the transmission loss for most United States cities;
- both TIREM and Longley–Rice models provided serious discontinuities in the transmission loss values when changing from one mode of propagation to another;
- none of the models studied provided data on sector transmission loss distributions due to multipath reception, except to assume a loss deviation of 8.2 dB.

Definitions have been included in Section II to provide a basis for comparing propagation models and field measurements.

Other members of the Committee attending one or more of the meetings were: Tom Aitchison, Virgil R. Arens, Robert Bultitude, Rick Burke, Darnyl DeLawder, George Dewire, William E. Frazier, Ernest Freeman, Simon Goldman, Vick Graziano, George Hagn, Ed Hanley, David Hodgin, Keith W. Kaczmarek, William H. Keller, Ken Kelly, William K. Kokorelis, David Land, John B. Lomax, R. Singh Lunayach, Don Mazak, James Mikulaski, John A. Moffet, Joseph Moffitt, Phil Rice, J. T. Roussos, Fred Schaefer, L. G. Schimpf, Victor Tawil, Mark Whitty, William Wickline and Don Yost.

The Propagation Committee has been made a permanent technical committee of the Vehicular Technology Society, whose purpose will be to

- provide a propagation model for predicting reliable service areas and co-channel interference for all types of mobile radio systems;
- prepare standard methods of measurement of radio frequency propagation in land mobile bands.

Neal H. Shepherd

I. Introduction

DURING THE PAST DECADE, broad applications of land mobile services have been authorized by regulatory agencies in the 800 and 900 MHz portions of the radio spectrum. These decisions opened the door for trunked conventional land mobile; for cellular radio telephone; for 800 MHz conventional and nationwide paging; and for other applications.

It is generally recognized that there is no clear consensus of methods to evaluate coverage of mobile systems in these bands. There are many reasons for this, but perhaps the most significant factor is the limited experience with application of radio propagation to land mobile systems in the 800/900 MHz portion of the spectrum. This constraint applies to the entire land mobile community; there are no procedures within the mobile engineering community that are as yet generally acknowledged or accepted as providing accurate and reliable 800/900 MHz propagation information.

This constraint on engineering analysis, particularly in the frequency range of 800/900 MHz, has been identified by the Institute of Electrical and Electronics Engineers as an area requiring immediate attention. As a result, early in 1983 the IEEE Vehicular Technology Society organized an Ad Hoc Committee on Radio Propagation Models for land mobile applications in that band. The committee was supported by a broad group of technical experts from government, industry, and universities whose common goal was the improvement of UHF propagation loss calculations. This is the Final Report of the committee, which became a standing committee in 1986.

The Committee's task was to recommend radio propagation models for land mobile radio services operating in the 800/900 MHz radio spectrum. These propagation models were to provide the industry and governments with statistical methods

Manuscript received February 4, 1988.
IEEE Log Number 8821063.

of predicting reliable service areas and interference. The radio services to be addressed included conventional and trunked private land mobile, common carrier cellular land mobile, and private and common carrier radio paging systems. Common carrier air-ground and satellite mobile systems were to be addressed at a later time.

Based on the above task statement, the objectives of the Committee on Radio Propagation Models were as follows:

1) to identify current capabilities to estimate propagation in the 800/900 MHz band;
2) to define propagation modeling requirements for 806/947 MHz land mobile applications;
3) to recommend propagation modeling approaches to meet the defined requirements.

These objectives will be explored in depth in this report, through a comprehensive evaluation of today's needs and capabilities in 800/900 MHz propagation loss determinations. More specifically, the report will:

1) identify the technical factors that affect 800/900 MHz propagation;
2) summarize current knowledge in making 800/900 MHz propagation estimates;
3) explore the relationship among the propagation path, the environment, and the equipment involved;
4) address such issues as prediction accuracy, automated versus manual calculations, the relationship of reliability criteria to system compliance testing;
5) indicate current deficiencies in prediction methods and the collection of propagation data, and outline a program that addresses these limitations;
6) recommend modeling approaches that are most appropriate to employ when evaluating particular land mobile systems.

II. Definitions

To assist the reader, the following list of definitions is included. The reader is encouraged to proceed to subsequent material, and return to these definitions as needed rather than mastering the definitions now.

Analytical: Using, subject to, or capable of being subjected to a methodology involving algebra and calculus.

"Area" propagation model: A propagation model in which median transmission loss calculations are based on generalized characteristics of the areas surrounding the transmitter and receiver, as well as the intervening area. Features such as local environment, terrain roughness, building density, etc. are used to modify a median transmission loss equation, in order to adapt it to the service area.

Base station noise: A composite of individual noise sources assumed to be uniformly distributed around the base station antenna.

Bit error rate (BER): Ratio of the number of bits of a message incorrectly received to the number of bits transmitted [1].

Calling probability: 1) The calling probability is the ratio of expected successful selective calls relative to the total number of trials [2]; 2) the standard calling probability is defined as an 80 percent probability of successful calling [3].

Cornu spiral: A plot in the complex plane of the Fresnel integral. The Fresnel zone clearance value "v" is the running parameter along the curve.

Coverage area: A collection of sectors for which there is a stated probability of receiving a signal exceeding a given level. It may sometimes be convenient to describe the coverage area in terms of a bounding contour within which lower signal level subareas ("holes") are identified. Note: the stated probability does not always guarantee a grade of service. See Service Area.

Coverage probability: The probability, applying to a designated collection of sectors, that a specified signal level will be available for reception during (at least) a specific percent of the time. For example, $F(50, 90)$ means that in 50 percent of the sectors designated, the specified signal level will be exceeded 90 percent of the time. For most systems, time variability is less significant than location variability and loss deviation, and may be neglected by assuming 100 percent for the time fraction.

Ducting: Confinement of electromagnetic wave propagation to a restricted atmospheric layer by steep gradients in the index of refraction with altitude [1].

Effective antenna elevation: The height of the radiation center of the base station antenna above the average elevation of the ground between distances of 3 and 15 km from the base station in the direction of the mobile. This is equivalent to height above average terrain, as defined by the Federal Communications Commission. The mobile antenna height is defined as the height of its radiation center above ground level (adapted from [4]).

Error burst: A group of bits in which two successive

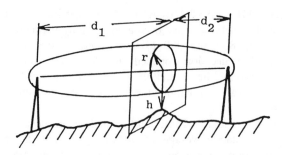

Fig. 1. Fresnel zone about the line connecting two antennas.

erroneous bits are always separated by less than a given number (X) of correct bits. The last erroneous bit in the burst and the first erroneous bit in the following burst are accordingly separated by (X) correct bits or more. Number (X) should be specified when describing an error burst [1].

Fresnel ellipse: The locus of points formed by the intersection of the surface of the first Fresnel zone surrounding the line connecting transmitting and receiving antennas with: 1) any plane containing the two antennas; or 2) a plane such as the surface of the earth, that intersects the first Fresnel zone.

Fresnel integral: An integral function used for numerical calculations of diffraction results.

Fresnel radius: The radius r of the circle formed by the intersection of the first Fresnel zone with a plane perpendicular to the line between transmitting and receiving antennas. Referring to Fig. 1, for transmitter and receiver at distances d_1 and d_2 from the plane, and assuming that they are large compared to wavelength, the value of r in meters is given by

$$r \simeq 548 \sqrt{\frac{d_1 d_2}{f(d_1 + d_2)}}, \qquad d_1, d_2 \text{ in km and } f \text{ in MHz.}$$

The radius of the nth Fresnel zone is found by multiplying r by $\sqrt{n}$. If d_2 is about $0.1 d_1$, or less, and if $f = 900$, then $r \simeq 18.3 \sqrt{d_2}$.

Fresnel zones: 1) The region in space between successive ellipsoids of revolution whose foci are the transmitting and receiving antennas. The surfaces of the ellipsoids are defined by the condition that the combined distances from any point on the surface to the receiving antenna and to the transmitting antenna be greater than the direct distance between antennas by an integer multiple of one-half wavelength. The first Fresnel zone is the region within the innermost ellipsoid. 2) The area formed by the intersection of the regions of definition 1 with any plane perpendicular to the straight line between the transmitting and receiving antennas. The first such zone is a circular disk, and higher zones are circular rings.

Fresnel zone clearance ratio: The effect of diffraction is evaluated in terms of the Fresnel zone clearance. The Fresnel zone clearance ratio is the ratio of the distance from the direct line between antennas to the path obstacle (distance h in Fig.

1) and the Fresnel radius r. It is expressed in terms of the parameter "v" used with the Cornu spiral, where $v = \sqrt{2}h/r$.

Height above average terrain (height AAT): See Effective antenna elevation.

Location variability: The standard deviation of a sector transmission loss distribution. The sector transmission loss distribution is typically found to be log normal.

Log normal distribution function: The integral of the log normal probability density function.

Log normal probability density function: A probability density function describing some random variable, given by [1]

$$f(X) = \frac{1}{X\sigma\sqrt{2\pi}} \exp\left[-(\ln X)^2/(2\sigma^2)\right].$$

Loss deviation (LD): The decibel difference between the 50 and 90 percent values of the transmission loss distribution. When the transmission loss distribution is described by the Weibull variation, loss deviation can be found from any two points $n = 1, 2$ on the distribution using the formula [2]:

$$LD = \frac{1.884(X_1 - X_2)}{\ln[\ln(1/P_1)] - \ln[\ln(1/P_2)]}$$

where

P_n percent/100

X_n transmission loss at P_n.

Median transmission loss: The median value of the sector transmission loss distribution for a collection of sectors located in similar environments and at equal distances from the base station.

Message acceptance rate (MAR): The number of correctly received messages of specified length, divided by the total number of messages sent through a fully loaded system.

Message error rate: 1−(message acceptance rate).

Open environment: See Rural environment.

"Point-to-point" propagation model: A propagation model in which median transmission loss calculations are based on specific characteristics of the path along the great circle between the transmitter and the receiver.

Point-to-random multipoint: A mode of operation in which one fixed station is in a defined location, and the other fixed stations are randomly located. This mode combines the effects of mobile variability, where the dominant variations are primarily affected by location, and point-to-point variability, where the dominant variations are primarily affected by time.

Propagation loss: The total reduction in radiant power density. The propagation loss for any path traversed by a point on a wave front is the sum of the spreading loss and the attenuation loss for that path [1].

Propagation model: An empirical or mathematical expression used to compute propagation path loss [1].

Radio propagation path: For a radio wave propagating from one point to another, the great-circle route between the transmitter and receiver antenna sites [1].

Rayleigh density function: A probability density function, describing the behavior of some variable, given by [5]

$$f(X; \sigma) = \frac{X}{\sigma^2} \exp\left[-\frac{1}{2}\left(\frac{X}{\sigma}\right)^2\right].$$

Rayleigh fading: Radio fading due to multipath which follows the Rayleigh probability curve stating the natural distribution of random variables [6].

Rural environment: A radio environment comprised of an area where there are few obstacles like tall trees or buildings in the propagation path, and with cleared areas approaching 300 to 400 m across (as for instance, farm land, open fields, etc.) [7], [8].

Sector: A region over which the variations of transmission loss can be used to construct a transmission loss distribution. A sector can have any shape; typically the minimum and maximum dimensions are 4 and 20 wavelengths, respectively [2].

Sector transmission loss (STL): The transmission loss corresponding to the 50 percent location or median value of the distribution of transmission loss within a sector. When the transmission loss distribution for a sector is described by the Weibull variation, sector transmission loss can be found from any two points $n = 1, 2$ on the distribution using the formula [2]:

$$STL = X_1 + (X_1 - X_2)\frac{0.367 + \ln[\ln(1/P_1)]}{\ln[\ln(1/P_1)] - \ln[\ln(1/P_2)]}$$

where

P_n percent/100

X_n transmission loss at P_n.

Sector transmission loss distribution: The probability distribution of the sector transmission loss for sectors located in similar environments and at equal distance from the base station. The 50 percent value of the sector transmission loss distribution is the value used for propagation models for the specified environmental parameters [2], and is called the median transmission loss.

Service area: 1) The area within which a radio system provides either generally satisfactory service or a specific quality of service [1]; 2) the area within which radio service is required.

Service probability: The probability that a designated collection of sectors will receive a specified grade of service for a specific percent of the time. For example, $S(50, 90)$ means that in 50 percent of the sectors designated, the specified grade of service will be exceeded 90 percent of the time.

Signal-to-interference ratio (S/I): The ratio of the magnitude of the signal to that of the interference or noise. Note: The ratio may be in terms of peak values or root mean square (rms) values, and is often expressed in decibels. The ratio may be a function of the bandwidth of the system [1].

Signal-to-noise ratio (S/N): The ratio of the value of the signal to that of the noise. Notes: 1) This ratio is usually in terms of peak values in the case of impulse noise and in terms of the rms values in the case of random noise. 2) Where there is a possibility of ambiguity, suitable definitions of the signal and noise should be associated with the term, as for example, peak signal to peak noise ratio, root mean square signal to root mean square noise ratio, peak-to-peak signal to peak-to-peak noise ratio, etc. 3) The ratio may be often expressed in decibels. 4) This ratio may be a function of the bandwidth of the transmission system. 5) In mobile systems, the S/N may be defined as the ratio of a specified speech energy spectrum to the energy of the noise in the same spectrum [1].

Signal strength: The level of radio signal presented to the antenna terminals of a receiver, expressed in dBW.

SINAD ratio: A measure expressed in decibels of the ratio of 1) the signal-plus-noise-plus-distortion to 2) noise-plus-distortion produced at the output of a receiver that is the result of a modulated-signal input [1].

SINAD sensitivity: The minimum standard modulated carrier-signal input required to produce a specified SINAD ratio at the receiver output [1].

Spectrum amplitude: The vector sum of the voltages produced by an impulse in a given bandwidth divided by that bandwidth. Note: Spectrum amplitude is usually expressed in volts/Hz or in dBμV/MHz (across a resistance of 50 Ω) see Section 1 of [9].

Strip noise: A composite impulsive noise source formed by a long narrow boundary. The most common source results from automobiles on a multilane highway.

Suburban environment: A radio environment comprised of a village or highway with scattered houses, small buildings, and trees, often near the mobile unit [7], [8].

Transmission loss: In a system consisting of a transmitting antenna, receiving antenna, and the intervening propagation medium, the ratio of the power radiated from the transmitting antenna to the resultant power that would be available from an equivalent loss-free receiving antenna. For this report, both antennas are assumed to be vertical half-wave dipoles and the ratio is expressed in decibels [1].

Transmission loss distribution: The distribution of the percentage of locations within a sector where the transmission loss is less than the indicated value. It is best described by a Weibull distribution [2].

Transmission quality: For mobile communications, the measure of the minimum usable speech-to-noise ratio, with reference to the number of correctly received words in a specified speech sequence [1].

Urban environment: A radio environment comprised of an area which is heavily built up, crowded with large buildings and multistory residences, or a large village closely interspersed with multistory houses and thickly grown trees [7], [8].

Vegetation effects: The effects of biomass on transmission loss. These effects may be due to reflection, attenuation, or diffraction by local vegetation, or by vegetation along the path.

Weibull distribution: A class of distributions, the cumulative distribution functions of which have the form:

$$f(X; \sigma', \lambda) = 1 - \exp\left[-(X/\sigma')\lambda\right]$$

where σ' and λ are the scale and shape factors, respectively. The Rayleigh distribution function is a special case of the Weibull function, with $\lambda = 2$ and $\sigma' = \sqrt{2}\sigma$ [5].

Weibull fading: The variation of transmission loss at different locations within a sector due to multipath reception.

REFERENCES

[1] F. Jay, Ed., *IEEE Standard Dictionary of Electrical and Electronic Terms*, 3rd ed. IEEE Std. 100-1984, New York, 1984.
[2] N. H. Shepherd, "Radio wave loss deviation and shadow loss at 900 MHz," *IEEE Trans. Veh. Technol.*, vol. VT-26, pp. 309-313, 1977.
[3] Pub. 489-6, Int. Electrotech. Commission (IEC), Geneva, 1974.
[4] "Propagation," CCIR Xth Plenary Assembly, Geneva, Rep. 239, vol. II, 1963.
[5] A. M. Wood, F. A. Graybill, and D. C. Boes, *Introduction to the Theory of Statistics*, 3rd Ed. New York: McGraw-Hill, 1974.
[6] E. E. Smith, *Glossary of Communications*. Chicago, IL: Telephony, 1971.
[7] W. C. Jakes, Jr., *Microwave Mobile Communications*. New York: Wiley, 1974.
[8] Y. Okumura *et al.*, "Field strength and its variability in VHF and UHF land-mobile radio service," *Rev. Elec. Commun. Lab.*, vol. 16, pp. 825-873, 1968.
[9] Draft Pub. 489, Subcommittee 12F, pt. 9, Int. Electrotech. Commission (IEC), Geneva, 1983.

Some of these references are reprinted in *Land-Mobile Communications Engineering*, D. Bodson, G. F. McClure, and S. R. McConoughey, Eds. New York: IEEE Press, 1984. The table of contents of this book is listed in Appendix I.

III. Background

A. Types of Systems

A variety of communication systems employ radio propagation in the 800/900 MHz frequency band. A listing and brief description of a number of these systems is given below.

1) Trunked and Conventional: A trunk is a one- or two-way channel provided as a common artery between switching equipment. A trunked radio system is a method of operation in which a number of radio frequency channel pairs are assigned to mobile and base stations in the system for use as a trunk group.

Conventional land mobile systems operate on one or more radio frequency channels, but are not employed as a trunked group. In general, conventional systems are manually switched between channels, while trunked systems are automatically switched.

Propagation prediction requirements for trunked and conventional land mobile systems are similar, since both normally operate with a single base station and require the greatest practical communication range.

2) Cellular Mobile Radio: An advanced land mobile system characterized by the ability to accommodate large numbers of subscribers through efficient frequency reuse [1]–[3]. The area to be served by a cellular system is subdivided into smaller service areas called "cells." A cell is the area reliably served by one transmitter location. The cells are configured in such a way as to permit "handoff," that is, the smooth transfer of a call from a channel in one cell to a different channel in an adjacent cell as the subscriber moves through the service area.

Each cell is assigned a number of discrete frequencies which may be reused in other cell sites that are sufficiently far away to avoid interference. Because many channels are available for use by subscribers in each cell, a significantly higher number of subscribers per channel can be achieved than in conventional land mobile systems.

This increased trunking efficiency coupled with frequency reuse makes it possible to serve a great many subscribers and still maintain a grade of service comparable to that of a land line network.

A cellular system can grow internally by cell splitting or sectoring. Cell splitting often involves the introduction of smaller corner-excited (directional) cells into prior clusters of circular cells. This involves additions of cell sites. Cell splitting can be accomplished by adding many smaller cells to the existing larger cells serving the same area. This enables a system to accommodate a higher traffic density because a cell site's entire complement of channels, which previously served a larger cell area, can now be devoted to a smaller area. Cell sectoring is the use of multiple directional antennas at a single cell site to decrease the distance required between cells for frequency reuse without interference. The original cell is partitioned into a number of wedge-shaped sectors, each with its own set of channels. These directional cells and the sectorized cells require less separation for frequency reuse than omnidirectional cells would because the radiation in each sector is suppressed in all directions except for the narrow arc defining that sector.

3) Air/ground: Air/ground land mobile systems are similar to cellular systems, except the mobile station is aboard an aircraft. The cells are much larger in size than ground mobile cellular systems.

4) Paging: A land mobile system that provides one-way signaling between one or more base stations and personal receivers. It may provide tone-only, tone-with-optical-read or tone-plus-voice. For tone-only service, oral, visual or tactile signaling devices are received. For tone-plus-voice service, the audio circuit in the addressed receiver is activated after the tone signal is received. Transmission loss to paging receivers is greatly affected by the environment near the receiver.

5) Mobile-to-Mobile: Communication between two mobile stations without the use of a base station repeater in a mobile-to-mobile system. Due to a relatively large transmission loss even for short distances between mobile stations, such a system has limited use.

6) Point-to-Point (e.g., Control Links): Point-to-point transmissions between base stations are used to control the operation of remote base stations. The prediction of transmission loss between such stations can usually be done with few or no errors.

7) Satellite: A satellite mobile system provides communication between earth stations and active satellites for communication between mobile and base stations [4], [5]. The earth station may be fixed or mobile, and is intended to be used while either in motion or at unspecified fixed locations.

8) Radiating Cable: Radiating cable, also referred to as leaky cable, can be used in systems requiring communication inside buildings, mines, and tunnels where shielding by the environment restricts mobile radio coverage [6]. Such radiating cables are usually connected to a remotely operated base station, but can also be used as part of a passive or active repeater using outside antennas. In the 900 MHz frequency range, it is important to compare the transmission loss in the environment with the attenuation of a radiating cable, in order that the mode with the least attenuation is selected.

9) Simulcast (Paging and Mobile): Simulcast base stations can be used to extend the coverage area of a paging or mobile system without the requirement of additional channel frequencies being assigned. The operating frequency of each base station must be adjusted and maintained to within 3 parts in 10^9 while the audio group delay is held within 100 μs.

B. Modes of Propagation

Propagation in the 800/900 MHz band, for the most part, takes place via space waves. At this frequency groundwaves are attenuated very rapidly with distance, and skywaves pass readily through the ionosphere with little energy being reflected back to earth. Space waves are subject to absorption, reflection, refraction, and scattering by the troposhere and by the surface of the earth and obstacles in their paths. For the relatively short service ranges of land mobile systems using this band, propagation is usually by direct or reflected wave with some diffraction over obstacles. Interference, however, may occur at distances beyond the normal horizon via diffraction over the earth or by refraction or scattering by the troposphere.

Because of the multiplicity of factors involved, most coverage and interference predictions depend on a statistical approach. However, deterministic analyses are still required to account for isolated and gross terrain features. Following is a brief discussion of the various propagation mechanisms and factors pertinent to the 800/900 MHz band. More detailed equations and qualifications are discussed in connection with specific propagation models described in a later section.

1) Free Space Propagation: In free space, electromagnetic energy spreads out uniformly in all directions from a source. The amount of energy available to an antenna of a given effective area is inversely proportional to the square of the distance from the source [7]. Generally, free space conditions can be deemed to prevail if there is clearance for the first Fresnel zone for the path between transmitter and receiver and no reflections are present from the surface of the earth or from hills, trees, buildings, etc. In land mobile operations, free space conditions are rarely encountered at the mobile unit, unless it is located on an elevated roadway or near the crest hill.

2) Reflections from the Surface of the Earth: Signals which arrive at the receiving antenna after reflection from the surface of the earth may interfere constructively or destructively with the direct wave depending on the relative phase of the two waves [7]. The relative phase and amplitude of the reflected wave is determined by the difference in path length and reflection coefficient of the surface.

When the surface is smooth, specular reflections occur and the angle of incidence is equal to the angle of reflection. For most land mobile systems, reflections occur at small grazing angles so that the magnitude of the reflection coefficient is approximately unity, indicating no loss or attenuation, and its phase is 180°. For path lengths where the curvature of the earth has to be considered, the divergence of the reflected rays causes a reduction of the specularly reflected signal.

When surface roughness is comparable to a wavelength, the reflected signal has a specular component which is coherent with the incident wave and a diffuse component whose amplitude is Rayleigh distributed. The reflected signal can also be blocked or partially blocked by terrain or structures along the path.

For paths a few miles in length over water or smooth terrain and with a first Fresnel zone clearance of 0.577 or more, the resultant signal decreases monotonically with distance and the theoretical value can be found from nomograms developed by Bullington [7]. At 900 MHz for normal mobile antenna heights, measured values of transmission loss are usually 10 to 20 dB greater than the theoretical value.

In air-to-ground systems, multiple reflections may take place from surface features such as mountains, flat terrain, and buildings. Generally, the signal will exhibit a series of predictable nulls as the separation distance between the ground station and aircraft increases (see Appendix IV).

3) Diffraction Over a Smooth Earth: Diffraction of radio waves around the curvature of the earth is affected by the frequency and polarization of the waves, the properties of the ground, and the geometry of the path. The path geometry is affected by the properties of the atmosphere through which the waves pass. At 900 MHz the path loss increases rapidly with distance beyond the horizon. A series of papers beginning with one by Sommerfeld in 1909 and one by Watson in 1918 developed analytical expressions for computing the field in the diffraction zone. The history of these developments was reviewed in a paper by Ekhert in 1986 [8]. The mathematics are complex and require a computer for any degree of precision.

4) Refraction: The speed of electromagnetic waves through the earth's atmosphere is less than the speed through a vacuum. The refractive index of the atmosphere is the ratio of the speed in a vacuum to that in the atmosphere. This ratio at the surface of the earth in temperate climates is approximately 1.000340. It varies with the temperature, pressure, and moisture content of the air and can be calculated from measured values of these quantities or can be measured directly with a refractometer [9].

For computational convenience, radio refractivity N is defined in terms of the refractive index of the atmosphere n through the following [9]:

$$N = (n - 1) \times 10^6.$$

In terms of the physical properties of the atmosphere, the radio refractivity is given by

$$N = 77.6(P/T) + 3.73(e/T)$$

where

P atmospheric pressure (in millibars)
e water vapor pressure (in millibars)
T absolute temperature (°K).

Radio refractivity decreases with height above the earth being an average of about $N = 301$ at the surface and about 260 at 1 km above the surface in temperate climates. This gradient causes radio waves passing through the atmosphere to be bent back toward the earth, so that signals radiated at angles above the horizon may be bent around the curvature of the earth and received at distances beyond the normal horizon. In designing communications systems, this bending effect is taken into account by drawing path profiles assuming the radius of the earth to be 4/3 of its actual value [10]. With this assumption, ray paths can be represented as straight lines, greatly simplifying ray tracing procedures. If more precise calculations are needed, an effective "earth's radius" is calculated based on the refractive gradient for a specific location and season of the year. Contour maps showing surface refractivity and the refractivity gradient for all parts of the world can be found in a number of references [11].

Relatively low levels of energy are propagated great distances beyond the horizon via scattering from local irregularities in the refractive index [12]. For this mode of propagation, called troposcatter, energy is scattered from parts of the atmosphere visible to both the transmitter and receiving antennas with an attenuation proportional to the inverse distance between antennas.

5) Superrefraction and Ducting: In coastal areas, the movement of large masses of air due to advection and subsidence result in layers of air with refractive indexes significantly different from those of layers above and below. These steep gradients in refractivity can cause radio waves to bend sharply back toward earth and can produce greatly enhanced fields and cause interference at distances well beyond the normal radio horizon. Superrefraction occurs where the decrease in N per kilometer exceeds 100. Ducting, wherein a radio wave is trapped between elevated layers of air or between an elevated layer and the surface of the earth, can occur when the rate of decrease of N exceeds 157 per kilometer. This phenomenon can produce signals at or above the free space value at distances of hundreds of miles beyond the radio horizon. Superrefraction and ducting conditions are present for significant percentages of the time in the Southern California and Gulf Coast areas of the U.S.

C. Effects of Buildings and Trees

In urban areas propagation is generally dominated by shadow loss and reflections caused by buildings and trees. In the case of land mobile systems these are usually in the environment surrounding the mobile. Signals arrive from all directions with random amplitude and phase, and with a spread in arrival times of several microseconds. It is impossible to describe the received signal using deterministic models. The signal at a vehicle moving through the resulting standing wave pattern exhibits a Weibull distribution. The statistical treatment is discussed in detail in Appendix III.

1) Reflections from Buildings and Trees: In a given situation a reflection from an individual building may be dominant. This may happen because of the building's height, size, orientation, or specific path configurations. Reflections may make a high signal available in areas deeply shadowed by buildings or terrain. The gain produced by a large reflecting surface can even result in signal levels in excess of free space values, as observed by Shepherd [13]. The strength of the reflected signal is determined by the height and width of the portion of a building visible to both terminals and is affected by obstacles within the first Fresnel zone of the incident or reflected ray and by the nature of the reflecting surface—type of material, size, shape, and orientation of sheathing elements. The vertical and horizontal width of the reflecting surface will determine the effective beamwidth of the reflected signal in the vertical and horizontal directions, respectively.

In the case of hills, the factors that influence the amplitude and beamwidth of the reflected fields include the size, shape, and orientation of the reflecting surface and its coefficient of reflection which may be affected by trees or other vegetation.

Composite signals reflected from wind-disturbed foliage or moving automobiles may exhibit a short-term variability of several decibels when measured between fixed terminals or a base station and a stationary vehicle.

2) Attenuation Due to Trees: At 900 MHz shadowing, scattering, and absorption by vegetation can introduce substantial path losses. Over the past 45 years a number of studies have been carried out to characterize and model attenuation from vegetation. Fifty published works covering reports of measurement data and empirical and theoretical models were reviewed by Weissberger in 1982 [14]. Results of recent measurements can be found in [15].

Weissberger concluded that a traditional exponential decay model was appropriate for those situations where propagation is likely to occur through a grove of trees rather than by diffraction over the top. After reviewing several exponential decay models, which are based on specific attenuation in terms of dB per meter of path length, and comparing them with several sets of available data at frequencies from 230 MHz to 95 GHz, he developed a modified exponential decay model (MED) which is applicable where a ray path is blocked by dense, dry, in-leaf trees found in temperate climates.

$$L = 1.33 F^{0.284} d_f^{0.588}, \qquad \text{for } 14 \le d_f \le 400$$
$$= 0.45 F^{0.284} d_f, \qquad \text{for } 0 \le d_f < 14$$

where L is loss in dB, F is frequency in GHz, and d_f is depth of trees in meters.

The difference in path loss for trees with and without leaves has been found to be about 3 to 5 dB. For a frequency of 900 MHz, the foregoing expressions reduce to

$$L = 1.291 d_f, \qquad \text{for } 14 \le d_f \le 400$$
$$= 0.437 d_f, \qquad \text{for } 0 \le d_f < 14.$$

An empirical–theoretical model by Kinase [16] was found to be applicable to situations in which one antenna is located well above the foliage and the area in the vicinity of the second antenna is covered by a given percentage of clutter, which may be either buildings or trees. A recent extension of this model can be found in CCIR vol. 5 [17].

When both antennas are clear of trees in their immediate vicinity, the principal mode of propagation occurs by diffraction over the trees. A criterion for determining sufficient clearance is based on the takeoff angle to the tops of the trees. Estimates of this angle range from 8° to 26°. There is considerable uncertainty among investigators as to when a knife-edge versus a rounded obstacle model is appropriate.

3) Building Penetration: Attenuation of signals penetrating a building is measured by taking the difference between the median signal level of several sectors located on the streets in front and beside the building in question, and the median signal level of sectors located inside the building. Building attenuation is dependent on the type of construction and the materials used, as well as the size of the building [17]–[25]. Table I lists construction elements and corresponding attenuations that are typical [25].

Given these element attenuations, the attenuation of a multifloor office building can be calculated by applying the method illustrated in the following example. A corridor is located on the first floor and 15 m from the nearest exterior

TABLE I
ATTENUATION OF CONSTRUCTION MATERIALS

Construction Element	Attenuation (dB)	Standard Deviation (dB)
8 in. concrete block wall	7	1
Wood and brick siding	3	0.5
Aluminum siding	2	0.5
Metal walls	12	4
Attenuation past office furnishings (dB/m)	1	0.3

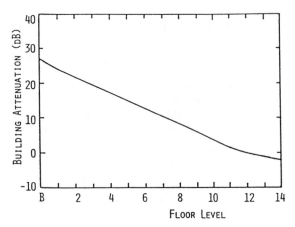

Fig. 1. Variation with height of the signal reduction inside tall buildings (average over seven buildings in downtown Washington, DC).

wall of a 50 m square concrete block building. A single wall board partition separates the corridor from the exterior wall.

Element	Attenuation (dB)
Exterior wall	7
Office furnishings	15
Interior partition	2
Total Attenuation	24

On higher floors the signal exhibits a height gain relative to the signal at street level. This height gain can be estimated from Fig. 1, which is based on attenuation measurements made on different floors of seven buildings in downtown Washington, DC.

D. Statistical Variability of the Desired Signal

Measurements of transmission loss between base station and mobile antennas, made as the mobile position varies, form a statistical distribution. The median of this distribution for a given sector is called the sector transmission loss, and can have long-term variations, such as season-to-season and year-to-year. However, within a given sector the statistical variation about the median is relatively constant. Values of transmission loss greater than the median have more significant effect on signal quality than do values less than the median. Therefore, the statistical model which best describes those values of transmission loss greater than the median value will be used.

1) Within a Sector: The variation of the signal in a sector is frequently assumed to be Rayleigh distributed. Such an assumption is not valid under all physical conditions [26]. The Nakagami distribution (also called the m distribution), however, assumes that the received signal is a sum of vectors with both random amplitude and random phase, which leads to a better fit with experimental data. The Weibull distribution affords the same flexibility as the Nakagami distribution and closely matches experimental data for those transmission loss measurements within a sector that are greater than the median value. It further provides the advantage of giving straight line plots on Weibull graph paper, which can be identified as a single slope parameter called loss deviation.

In the 800/900 MHz bands, loss deviation varies from about 3 to 30 dB. This variation is considerable compared to the Rayleigh distribution value of 8.2 dB. In general, loss deviation can be related to the following propagation conditions [13]. Loss deviation:

- increases as the size of either antenna decreases;
- increases as the median loss for a given sector increases above the median value for all sectors at a given distance from the base station;
- increases in locations where the absolute area of a reflecting surface is greater than one-half the distance from the reflector to either the transmitter or receiver antenna;
- is greatest when several (2–4) signals of nearly equal amplitude are present;
- decreases in locations where a strong single source signal is present;
- decreases in locations where large numbers of signal sources are present.

A further discussion of the statistics of path loss is given in Appendix III.

2) Between Sectors: Although a sector is defined as being within a relatively small area, i.e., up to 20 wavelengths, the signal variation in adjacent sectors and sometimes in a large number of those nearby, may have about the same median value of signal amplitude and loss deviation. In other cases, however, both the median value and the loss deviation from one sector to the next can change rapidly due to such environmental factors as reflectors, obstructions, and foliage.

Reflectors, such as tall buildings and hills, reradiate signals by passive reflection which is similar to radiation or reception by very high-gain antennas. This radiation, due to high gain (narrow beamwidth), frequently causes large variations in signal amplitude from one sector to the next.

Further variations in a sector's median signal value occur behind a reflector since it also acts as an obstruction. The sector transmission loss for a large number of sectors located at a specified distance from the base station antenna can be combined to obtain a sector transmission loss distribution. In general, such a distribution is log normal, and its standard deviation gives the location variability for the specified environmental parameters. Both foliage and trees without foliage will cause the sector transmission loss to increase for each sector located in or behind the trees.

E. Interference

Various sources of interference are found in the 800/900 MHz band. These include noise, both impulsive and nonimpulsive, other communications systems, and intermodulation within the transmitter and receiver. These sources are discussed below.

1) Impulsive Noise: The rms amplitude of impulsive noise decreases at the rate of 28 dB per decade of frequency in the frequency bands above 200 MHz, which results in a noise level of about 8 dB lower at 800 MHz, as compared to 450 MHz. The lower impulsive noise level at 800 MHz permits the effective use of lower noise figure receivers than is possible at lower frequencies.

a) Typical receiver noise figure of mobile and base receivers is about 10 dB. The effective noise figure of a receiver and its antenna is degraded by loss in the transmission line. It is the effective noise figure at a receiver antenna which must be used to determine degradation of performance due to impulsive noise or consideration for attenuation of the transmission line must be used for determining the spectrum amplitude of impulsive noise at the receiver input.

Transmission lines for base receivers normally have greater attenuation than for the case of a mobile receiver. This greater attenuation can usually be offset by RF amplifiers with a noise figure as low as 3 dB. The best available receivers, transmission lines and RF amplifiers will provide effective noise figures of 11 dB. A noise figure of 11 dB should provide a receiver sensitivity of 0.25 μV for narrow-band FM receivers. For wider band width cellular receivers the effective sensitivity will be slightly greater.

b) Characterization of impulsive noise, for the purpose of determining degradation of the performance of receiving systems, utilizes two basic parameters: spectrum amplitude; and impulse rate. These two parameters are effectively displayed by a graph of the noise amplitude distribution (NAD).

c) Impulsive noise sources of sufficient amplitude to be considered a potential for creating degradation of performance of receiver systems are spherics, electrical power lines, and ignition-type engines. Impulsive noise radiated by spherics can exceed received spectrum amplitudes of 60 dBμV/MHz. Its impulse rate is very low, and as a result does not create significant interference except for digital transmission where very low error rates are required.

Impulsive noise radiated by electrical power lines can produce received spectrum amplitudes near a line exceeding 40 dBμV/MHz at impulse rates of 200 i/s. However, the

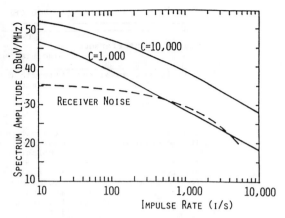

Fig. 2. Noise amplitude distribution (NAD) for strip and receiver noise. Reference sensitivity is 0.25 μV, and signal and noise frequencies are 850 MHz.

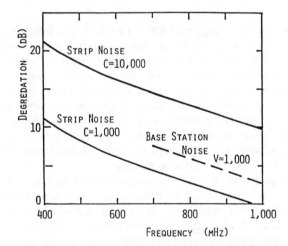

Fig. 3. Degradation of reference sensitivity due to ignition noise. Reference sensitivity is 0.25 μV, and signal frequency is 850 MHz.

occurrence of such noise is not often enough to be further considered. Impulsive noise radiated by ignition-type engines is the only source that will be considered of significant spectrum amplitude and impulse rate to cause degradation of receiver performance.

d) Ignition noise must be considered both for vehicles near the base station and for vehicles in the vicinity of the mobile (strip noise). Two examples of the NAD of strip noise for 10 000 and 1000 vehicles per hour are illustrated by Fig. 2.

Also shown in Fig. 2 are the spectrum amplitude and impulse rates for receiver noise, which have an effective sensitivity of 0.25 μV. Comparison of the receiver noise with the strip noise will indicate degradation for any spectrum amplitude of the strip noise exceeding the receiver noise.

The NAD of base station noise for high vehicle densities of 1000 vehicles/km² is about equivalent in interference to the NAD for strip noise of $C = 1000$ vehicles/h.

e) Degradation of receiver performance by ignition noise is shown in Fig. 3. This figure gives the probable degradation of receiver performance for both base station and two types of strip noise as a function of operating frequency.

TABLE II
SOURCES OF INTERFERENCE

Source of interference	Degree of Interference to	
	Base station receiver	Mobile station receiver
Television transmitter sideband	Strength depends frequency separation	Strength depends on related location of station
Transmitter noise	Strong on frequencies near band edges	Strength depends on relative location of base station
Transmitter intermodulation	Strength depends on design of transmitter combiner	Strength depends on design of transmitter combiner
Receiver intermodulation	Not a serious problem except when a common antenna is used with two or more transmitters	Strong if the design of the transmitter combiner has reduced transmitter inter-modulation
Receiver desensitization	Not important except on frequencies near band edges	Not important unless transmitter noise has been filtered out

The frequency range from 400 to 1000 MHz was included to illustrate the decreased degradation as a function of frequency.

2) Nonimpulsive Interference: Interference to both the base and mobile station receivers can come from a number of different sources. In the case of a base station receiver the interference can be co-site, from a distance, or some combination of both. Interference to mobile receivers is subject to wider variation, which is dependent on receiver location and time of day. The more important types of interference are listed in Table II and discussed below.

In general the potential for interference has been reduced by two approaches. The first of these was accomplished by the design of equipment and systems with state-of-the-art performance. Transmitter characteristics, such as sideband noise and intermodulation-produced radiation, along with receiver characteristics, such as selectivity, intermodulation rejection, and desensitization immunity, have been controlled. The second approach was to assign transmitter and receivers with 45 MHz frequency separation.

a) Television transmitters operating on UHF channels 67–69, adjacent to the land mobile base receiver frequencies starting at 806 MHz, radiate sideband noise covering a wide frequency range. The resulting interference can be significant for a base station receiver located within six miles of a channel 68 transmitter and within 100 miles from a channel 69 transmitter. Mobile receivers are less susceptible to TV transmitter noise due to a wider frequency separation.

b) Transmitter noise in typical systems decreases by about 65 dBc/10 kHz close to the operating frequency. It decreases in amplitude beyond 1 MHz and is no longer a problem at 20 MHz. The use of 45 MHz frequency spacing between the base station transmitter and receivers of a duplex channel should prevent noise from being a problem at a base station for the channel. However, the frequency spacing between the transmitter of one channel and the receiver for another channel can be much less than 45 MHz, leading to the possibility of interference due to transmitter noise.

Transmitter noise interference to mobile receivers is much more serious, since the frequency spacing can be as close as 25 kHz. At close frequency spacing, less than 1 MHz, the interference range can be as great as ten miles.

c) Transmitter nonlinearities in the output stage can generate third-order products when an external carrier is received through the transmitting antenna. The carrier frequency of the IM product is given by $2f_B - f_A$ where f_A is the frequency of the transmitter of the external carrier and f_B is the frequency of the transmitter that generates the IM product. The amplitude of the product varies from about 0 to 6 dB below the amplitude of the external carrier. A similar product can also be generated by transmitter A, in which case its frequency will be $2f_A - f_B$. The amplitude of either product is a direct function of the attenuation between the output of one transmitter and that of the other transmitter.

d) Receiver nonlinearities can cause interference on the same channel frequencies as those mentioned above for transmitter generated intermodulation. Although difficult, it is usually possible to locate the source of the intermodulation products through the use of attenuators, isolators, or filters. After a source is located, the effect of interference can be reduced by adding isolators, filters, or relocating antennas.

Intermodulation products generated in receivers are more numerous than those generated in transmitters. Higher order products, as well as low-order products, can cause serious interference. The potential for interference increases rapidly as the number of radio systems at a given site increases.

The intermodulation products generated in base station receivers, due to the 45 MHz transmitter/receiver spacing, is not usually a serious problem. Base station receivers, when operated on a common antenna with two or more transmitters, can receive high-order generated product interference from any of a number of possible generators in the antenna system.

3) Interference due to Multipath Fading Within a Sector: Rapid fading of a received signal due to motion of a mobile station or any reflective objects in the transmission path causes a degradation of the receiver sensitivity at both the base and mobile station. The degree of sensitivity degradation is a

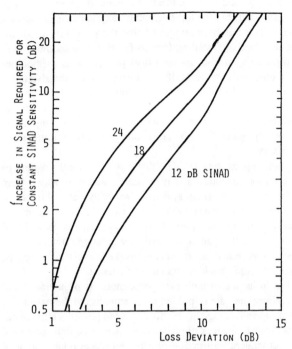

Fig. 4. Degradation of SINAD sensitivity due to multipath fading.

function of the loss deviation of the received signal. In FM receiver, the improvement ratio also decreases from a typical 3:1 to 1:1 ratio, when the loss deviation increases from 0 to about 10 dB. Fig. 4 shows typical values of degradation of SINAD sensitivity for loss deviations between 1 and 14 dB. Each curve gives the required increase in the median value of a desired signal level to maintain the indicated SINAD sensitivity.

F. Antenna Characteristics

The design of an 800/900 MHz mobile radio communication system must include considerations for the characteristics of antennas used in fixed base stations, as well as those used on the mobiles. Specific characteristics of the antennas chosen for use in a system have different effects upon system operation. These must be well understood if optimum system design is to be achieved. The following paragraphs highlight considerations related to the choice of antennas for use in 800/900 MHz mobile radio systems. For completeness, references are given where more detailed information can be obtained from the literature.

1) Antennas in Mobile Radio: One of the primary characteristics of an antenna is its radiation pattern. This is a graphical representation [27] of the radiation properties of the antenna as a function of spatial coordinates. Normally, radiation patterns are referred to as isotropic (equal radiation in all directions), omnidirectional (equal radiation in all directions in a given plane), or directional. Although an isotropic radiator is not physically realizable, its theoretical radiation pattern is often used as a reference for comparison of different antennas.

Unless received power measurements are being made in the field, it is unlikely that a systems engineer will encounter actual radiation patterns, as relative power gain (often just gain) patterns are more often employed in systems design. These show the ratio of the power gain of an antenna as a function of spatial coordinates to the power gain of a reference antenna in its reference direction. Here, power gain refers to 4π times the ratio of radiation intensity [27] in a given direction to the net power accepted by the antenna from a connected transmitter. Although an isotropic pattern is sometimes used as a reference, it is more common in land mobile work to reference power gain patterns to that of the omnidirectional H-plane pattern of a free space half-wave dipole with equivalent excitation. Relative power gain is then expressed in dB_d, where the subscript indicates the dipole reference. Manufacturers' specifications often quote antenna power gain to the nearest 0.01 dB. It is well known, however, that the best accuracy that can be expected from simulated free space antenna measurements [28] is 0.5 dB. Specification of antenna gain to greater accuracies in system design documents is therefore not a recommended practice.

The number of different shapes of relative gain patterns that can be radiated by different antennas is limitless. Often, however, the patterns of directive antennas exhibit one prominent radiation lobe. This lobe is visualized as the main "beam" of the antenna, leading to a subsequent classification of directive antennas by the width of this main beam (beamwidth) in a specified plane at its half-power points.

One immediately obvious use for a knowledge or specification of the relative gain pattern of an antenna is to protect a particular radio system against unwanted interference. More involved requirements for specific pattern shapes are discussed in later paragraphs.

Antenna polarization is perhaps the next important characteristic of an antenna. An antenna is said to be polarized identically to the electromagnetic waves radiated by it. Wave polarization can be linear or elliptical and is defined by the locus formed by the tip of the electric field vector as the wave propagates away from the antenna. Linearly polarized waves are classified by the direction of their electric field vector with reference to some reference plane. The land mobile industry takes this reference plane to be the earth's surface. A wave having an electric field directed horizontally is therefore classified as horizontally polarized. This classification is extended to the antenna from which it is radiated and similarly for vertically polarized antennas.

The sense of polarization most often employed at mobile system base stations is vertical. While polarization is unchanged by reflection or transmission at plane homogeneous dielectric or conducting interfaces [29], measurements indicate that both vertical and horizontal components of electric fields are present at the receiver. The rotated component of field arises through scattering from nonplaner and rough surfaces [30]. It has been found that the statistics of the orthogonal components of field are uncorrelated [31]. Polarization diversity has therefore been suggested as a means for the reduction of fading. An associated phenomenon, though not a polarization change, is a reversal of the direction of rotation for elliptical polarization. This can occur on reflection from a plane conducting interface, because the tangential component of an electric field undergoes a phase reversal, whereas the normal component remains unchanged.

Another very important antenna characteristic is bandwidth. This is the range of frequencies over which the performance of the antenna, with respect to some characteristic, conforms to a specified standard [27]. The specifications for each characteristic vary, and are set in order to meet the requirements of each particular application. Pattern bandwidth and impedance bandwidth are often distinguished from each other. Associated with pattern bandwidth are gain, sidelobe level, beamwidth, polarization, and beam direction. Impedance bandwidth influences feed point impedance and radiation efficiency, and is of particular importance in the new 800/900 MHz band where there is a 45 MHz separation between allocated transmit and receive frequencies. This requirement for large antenna bandwidths complicates antenna designs and has been the subject of recent investigations [32]. Detailed information regarding this subject can be found in the reference. It should be noted, however, that manufacturer specified voltage standing-wave ratio (VSWR)/bandwidth curves are often smoothed and the detailed shape of such characteristics, although extremely important in some digital applications, is often obscured. The impedance characteristic of an antenna as seen from the transmitter or receiver is also greatly influenced by the connecting transmission line, and the operating environment of the antenna. Possible mismatches in feed lines and connectors are important considerations. Bandwidths specified by manufacturers are typically the frequency range over which the maximum VSWR measured at discrete frequencies at the antenna feed point is less than 2:1.

2) Base Station Antenna Considerations: Factors which must be considered in the specification and siting of a base station antenna include: relative power gain as a function of azimuth, polarization, transmission line losses, height, impedance matching, proximity to support towers, buildings and other antennas, environmental noise, and coverage ability. While some of these and the methods by which they can be dealt with are obvious, others require further considerations. These will be discussed in the following paragraphs.

Unless polarization diversity is planned [31], nearly all base station antennas utilize vertical polarization. This is reflected by typical commercially available designs which include vertical dipole fed corner reflectors and vertically spaced arrays of vertically polarized folded dipole elements.

Base station antenna gain in any particular direction is dependent upon the proximity of the antenna to other potential radiators, the multipath environment, and the surrounding topography (including man-made obstacles). The antenna gain specified by manufacturers is measured on an antenna range in a simulated "free space" environment. The best accuracy that can be expected from such measurements is 0.5 dB. However, 3 dB variations in the measured azimuthal pattern of land mobile system "omnidirectional" antennas are not uncommon. If the antenna is side mounted on a conducting support tower [33] its azimuthal gain pattern may be altered significantly in a manner which is difficult to predict. Patterns may also be altered by the presence of transmission lines running up the tower behind the antenna. It is obvious that top

mounting the antenna is a means by which these problems can be eliminated, but the top position on any tower is not always available. It has been suggested that the proximity problems can be solved by installing four co-fed directive antennas so as to produce a quasi-omnidirectional pattern. This method has been employed successfully in Europe. One disadvantage, however, is that the element spacings required for such a system are greater than that which can be achieved using tower mounted antennas. For this reason, antenna mounting at the top of orthogonal faces on tall buildings has been found to be necessary.

Another proximity problem that can influence the pattern of base station antennas is that of nearby antennas which form parts of other co-sited systems. Normally, at a horizontal spacing of four wavelengths, antenna isolations of approximately 25 dB can be attained. More isolation can be achieved through vertical spacing of vertically polarized antennas, but precautions must be exercised to keep vertically spaced antennas within each others cone of silence.

The influence of multipath propagation is pronounced and can reduce the gain of phased array antennas to the extent that any design advantage in effective radiated power is eliminated. This is due to the lack of phase coherence in the multipath field and subsequent destruction of required phase relationships at the antenna elements. To avoid this problem base station antennas should be mounted high above surrounding obstacles, which reduces multipath, provides higher effective gain, and also allows advantage to be taken of a reported [34] 6 dB gain advantage for every doubling of the antenna height. Although this gain can probably not be realized in an urban environment, coverage to any parts of the system that are in a suburban or rural area separated from the base station by flat or smooth rolling terrain should be improved.

3) Mobile Antenna Considerations: Some of the constraints on antennas for mobile operation such as size, ruggedness, appearance, and cost are obvious and require little discussion. There are, however, a number of more subtle constraints and characteristics, particularly in the 800/900 MHz band, some of which can be used to gain unexpected advantages in mobile system operation.

In a previous section, it was explained that if vertical polarization is used at the base station, most of the waves impinging on the mobile antenna can be considered to be essentially vertically polarized. For this reason most commercially available mobile antennas are vertically polarized, and this sense of polarization is assumed for the antenna characteristics discussed in the following paragraphs.

The quarterwave rooftop antenna [34], which is a good mobile antenna for 450 and 150 MHz, exhibits a significantly poorer performance in the 800/900 MHz band. This is because the effective area of such an antenna in the band of interest is less by a factor of about thirty. To achieve higher theoretical gain, commercially available antennas for this band are usually of the collinear array type, consisting of 5/8 or 1/2 wavelength top sections, a phasing coil, and a 1/4 wavelength lower section. When measured in a multipath environment, however, the same antennas exhibit substantially less gain [32]. As for the base station antenna, this is a result of the lack

TABLE III
ATTENUATION OF SEMIFLEXIBLE CORRUGATED COPPER LINES

	Foam Filled		Air Filled		
	1/2"	7/8"	1-5/8"	7/8"	1-5/8"
Loss/100 ft	2.2	1.2	0.8	1.2	0.6
Loss/100 m	6.8	4.1	2.5	3.8	2.1

of phase coherence in the multipath fields. It is therefore recommended that systems designers anticipate vehicular antenna performance that falls somewhat short of published manufacturers range measurement results. In addition, it can be expected that the *in situ* azimuthal gain pattern of a vehicular mounted antenna will be dependent upon mounting position and will invariably be different from that measured on a range.

Although changes in the azimuthal pattern of a vehicular antenna as a result of the proximity of the vehicle body are reported [34] to have little effect on average received field strength, azimuthal pattern shape has been found [35] to influence the observed received signal fading rate. The elevation angles from which signals arrive at a mobile unit in an urban environment have been measured [34] to be limited to about 30°. Advantage can therefore be taken of extra antenna gain offered by antennas with vertical directivity without a significant reduction in received signal power due to the exclusion of waves impinging from high elevation angles.

One final point that might be considered with regard to mobile antennas is the possibility of obtaining some power gain by increasing height. For a large range of frequencies [36] and for several base station antenna heights, it has been observed that 3 dB gain advantage can be obtained for a 3 m high antenna as compared with the gain of a 1.5 m high antenna. Due to obvious constraints on mobile antenna height, however, it is considered impractical to attempt to achieve more than about 1 dB height gain advantage.

G. System Characteristics for Multisource Signals

Most systems are designed to operate at virtually the maximum effective radiated power permitted in that service, and from an antenna that is located as high above average terrain as possible for that geographical area. Therefore, in order to expand the area of coverage of the base station, additional antenna locations may be used. Paging systems and signaling transmissions (on two-way systems) may be simulcast, all transmitters broadcasting the same signal at the same time. For operation of such simulcast systems, consideration should be given to: the frequency stability of the transmitter 3 parts in 10^9 or better); use of a radio control link from a central control point to avoid appreciable audio delay; identical modulation characteristics in all base transmitters; audio group delay within 100 μs at each base transmitter; and the frequency of each transmitter adjusted to within one or two Hertz of the reference transmitter.

Mobile (two-way) systems have additional considerations because of the mobiles. If the power output of the mobile unit is considerably less than that of the base station, remote receiving locations may be used if the received signals are processed by a voting receiver that electronically selects the best signal. In cellular radio, the frequency pair is automatically switched to the frequency of the next cell to expand the area of coverage.

H. Fixed System Losses

Far more attention must be given to the passive components between the transmitter RF power output stage and the input to the antenna in the 800/900 MHz bands than is required in the VHF band. Transmission line losses are greater in the 800/900 MHz band, and the quality of mechanical installation, as well as aging of the components (especially connectors), can easily lead to a high VSWR, thus reducing the RF power radiated from the antenna. The number of connections should be kept to a minimum. Air filled transmission lines have less loss than solid or foam dielectric filled lines, per unit of size and length; however, air lines will require more maintenance after installation to keep the lines pressurized with dry air or gas to prevent condensation.

1) Cable/Connector Characteristics: All transmission lines, even so called "jumper-cables" used in combiner cabinets, should be of solid copper outer conductor rather than braided to reduce leakage. Typical losses (at 900 MHz) of lines are listed in Table III.

Although not having any effect on the electrical characteristics of the transmission line, all exterior transmission lines should be jacketed with a polyethylene sheath. Not only will the jacket protect the copper from the elements, but it will also provide a bond to the tape wrapped around the connectors to make the connections waterproof. All connectors in the RF portion of the system should be of the N type.

Not only is it "good engineering," but it may be required at many antenna sites to insert one bandpass cavity with each transmitter. The cavity filter reduces the RF pollution due to out-of-channel emissions and prevents channel signals from mixing in the output stage of the transmitter. The typical insertion loss of a cavity is 0.5 dB; however, many filters have settings of 1, 2, or 3 dB which will give a greater selectivity to the response curve of the cavity.

If the transmitter is to be used in an environment which has other transmitters operating in the same band, an isolator should be inserted in the system. An isolator has an insertion loss of approximately 0.5 dB at the desired frequency but offers an attenuation of more than 25 dB to an unwanted signal external to the system. An isolator is a circulator with an appropriate load termination attached. A harmonic filter should be used with all isolators to reject the second harmonic. Normal loss for a harmonic filter is 0.2 dB. Duplexers, used to isolate the transmitter and receiver when using a common transmission line and broad-band antenna, are either of a "bandpass" or "band-reject" type. For a single transmit frequency and a separate receive frequency system, a duplexer with a combination of bandpass and band-reject sections is preferred. The bandpass portion gives the advantages of a cavity and the reject portion offers a higher rejection of the unwanted signal than a cavity. In multichannel systems, as in trunked and cellular systems which require a very large bandwidth, only the cavity type is used.

2) Duplexers and Diplexers: The isolation between the transmitter and receiver is a function of the number of cavities used in either branch of the duplexer. A diplexer is essentially a duplexer used as a transmitter combiner when the frequency separation between the two transmitters is approximately 2 MHz or more.

3) Combiners (Transmitters): There are two basic types of transmitter-combiners depending on the frequency separation between the transmitting frequencies. If the transmitting frequencies are closely spaced, less than 0.25 MHz–0.30 MHz, a hybrid ferrite type combiner must be used. However, the hybrid device has an insertion loss of 3.2 dB.

If the transmitting frequencies are widely spaced, greater than 0.30 MHz, a junction combiner (multiport connector) can be used. The junction itself has no real significant loss; however, the isolating components necessary can make the loss for a two-transmitter combiner comparable to a hybrid type combiner.

a) Hybrid-ferrite type transmitter combiners can be used with frequency separations as close as 0.025 MHz and are always used with at least one isolator and harmonic filter. Minimum isolation is 70 dB. If additional isolation is needed, another isolator may be inserted in series with the first. If three or four transmitters are to be combined, the outputs of two hybrid couplers can be fed into a third hybrid coupler.

The insertion loss of the hybrid, 3.2 dB, is additive, thus making the loss of a four-transmitter hybrid combiner 7.1 dB. The hybrid combiner can be further expanded for eight transmitters and 16 transmitters by adding hybrids to the pyramid.

b) Junction type transmitter combiners can be used if the frequency separation is 0.30 MHz or greater. The junction is a multiport connector with one port for the antenna and a port for each transmitter. Isolation is accomplished by the number of isolators and cavities in each leg of the combiner.

In a multifrequency system, rather than using a separate duplexer in each paired channel, all receivers should be connected to one multicoupler which is fed by one antenna. Since receive systems are plagued by intermodulation and receiver desensitization, it is usually more manageable to deal with the location of one antenna and a band-pass filter for all the receiving frequencies. Most multicouplers have built-in preamplifiers which give an overall gain to the system.

REFERENCES

[1] *Special Issue on Advanced Mobile Phone Service, Bell Syst. Tech. J.*, vol. 58, no. 1, 1979.

[2] W. C. Y. Lee, *Mobile Communications Engineering*. New York: McGraw-Hill, 1982.

[3] J. F. Whitehead, "Cellular system design: An emerging engineering discipline," *IEEE Commun. Mag.*, vol. 24, pp. 8–15, 1986.

[4] R. E. Anderson, R. L. Frey, J. R. Lewis, and R. T. Milton, "Satellite-aided mobile communications: Experiments, applications and prospects," *IEEE Trans. Veh. Technol.*, vol. VT-30, pp. 54–61, 1981.

[5] G. H. Knouse and P. A. Castruccio, "The concept of an integrated terrestrial/land mobile satellite system," *IEEE Trans. Veh. Technol.*, vol. VT-30, pp. 97–101, 1981.

[6] Q. V. Davis, Guest Ed., Special Issue on Leaky Feeder Radio Communication Systems, *The Radio and Electronic Engineer*, vol. 45, no. 5, 1975.

[7] K. Bullington, "Radio propagation variations at VHF and UHF," *Proc. IRE*, vol. 38, pp. 27–32, 1950.

[8] R. P. Eckert, "Modern methods for calculating ground-wave field strength over a smooth spherical earth," FCC/OET R86-1, Feb. 1986.

[9] "The formula for the radio refractive index," CCIR XV Plenary Assembly, Geneva, Recommendation 453, vol. 5, 1982.

[10] "Effects of large-scale tropospheric refraction on radiowave propagation," CCIR XV Plenary Assembly, Geneva, Rep. 718-1, vol. 5, 1982.

[11] "Radiometeorological data," CCIR XV Plenary Assembly, Geneva, Rep. 563-2, vol. 5, 1982.

[12] "Effects of Small-Scale Spatial or Temporal Variations of Refraction on Radiowave Propagation," CCIR XV Plenary Assembly, Geneva, Rep. 881, vol. 5, 1982.

[13] N. H. Shepherd, "UHF radio wave propagation in Dallas, Texas base to mobile stations for vertical polarization," Gen. Elec. Tech. Inform. Series, R75-MRD-1, Mar. 1975.

[14] M. A. Weissberger, "An initial critical summary of models for predicting the attenuation of radio waves by trees," ESD-TR-81-101, Electromagn. Compat. Analysis Center, Annapolis, MD, July 1982.

[15] W. J. Vogel and J. Goldhirsh, "Tree attenuation at 869 MHz derived from remotely piloted aircraft measurements," *IEEE Trans. Antennas Propagat.*, vol. AP-34, pp. 1460–1464, 1986.

[16] A. Kinase, "Influences of terrain irregularities and environment clutter surroundings on the propagation of broadcasting waves in the UHF and VHF bands," Japan Broadcasting Corp., Tokyo, Japan, NHK Tech. Monograph 14, Mar. 1969.

[17] "Methods and statistics for estimating field strength values in the land mobile services using the frequency range 30 MHz to 1 GHz," CCIR XV Plenary Assembly, Geneva, Rep. 567, vol. 5, 1982.

[18] P. I. Wells and P. V. Tryon, "The attenuation of UHF radio signals by houses," *IEEE Trans. Veh. Technol.*, vol. VT-26, pp. 358–362, 1977.

[19] H. H. Hoffman and D. C. Cox, "Attenuation of 900 MHz radio waves propagating into a metal building," *IEEE Trans. Antennas Propagat.*, vol. AP-30, pp. 808–811, 1982.

[20] D. C. Cox, R. R. Murray, and A. W. Norris, "Measurements of 800-MHz radio transmission into buildings with metallic walls," *Bell Syst. Tech. J.*, vol. 62, pp. 2695–2718, 1983.

[21] E. H. Walker, "Penetration of radio signals into buildings in the cellular radio environment," *Bell Syst. Tech. J.*, vol. 62, no. 9, pp. 2719–2734, 1984.

[22] D. C. Cox, R. R. Murray, and A. W. Norris, "800 MHz attenuation measured in and around suburban houses," *Bell Labs. Tech. J.*, vol. 63, pp. 921–954, 1986.

[23] D. C. Cox, R. R. Murray, and A. W. Norris, "Antenna height dependence of 800 MHz attenuation measured in houses," *IEEE Trans. Veh. Technol.*, vol. VT-34, pp. 108–115, 1985.

[24] J. Horikoshi, K. Tanaka, and T. Morinaga, "1.2 GHz band wave propagation measurements in concrete buildings for indoor radio communications," *IEEE Trans. Veh. Technol.*, vol. VT-35, pp. 146–152, 1986.

[25] *General Electric Systems Application Manual*, sec. 80-A1, Table IV-3, General Electric Corp., Lynchburg, VA, Dec. 1972.

[26] N. H. Shepherd, "Radio wave loss deviation and shadow loss at 900 MHz," *IEEE Trans. Veh. Technol.*, vol. VT-26, pp. 309–313, 1977.

[27] C. A. Balanis, *Antenna Theory*. New York: Harper & Row, 1982, ch. 2.

[28] Unpublished report to the EIA committee on the calibration of the gain reference standard antenna for EIA standard RS-329 by Nat. Bur. Stand., Boulder, CO.

[29] S. R. Seshadri, *Fundamentals of Transmission Lines and Electromagnetic Fields*. Reading, MA, Addison Wesley, 1971, ch. 6.

[30] F. G. Bass *et al.*, "Very high frequency radio wave scattering by a

disturbed sea surface," *IEEE Trans. Antennas Propagat.*, vol. AP-16, pp. 554–559, 1968.

[31] W. C. Y. Lee, *Mobile Communications Engineering.* New York: McGraw-Hill, 1982, ch. 5.

[32] J. S. Belrose, "Vehicular antennas for 800 MHz mobile radio," in *Conf. Rec., IEEE Veh. Technol. Conf.*, Toronto, May 1983.

[33] W. A. Wickline, "Cellular demands superior antenna radiation control," *Mobile Radio Technol.*, vol. 2, Feb. 1984.

[34] W. C. Jakes, *Microwave Mobile Communications.* New York: Wiley, 1974, ch. 3.

[35] J. R. Stidham, "Experimental study of UHF mobile radio transmission using a directive antenna," *IEEE Trans. Veh. Commun.*, vol. VC-15, pp. 16–24, 1966.

[36] Y. Okumura *et al.*, "Field strength and its variability in VHF and UHF land mobile service," *Rev. Elec. Comm. Lab.*, vol. 16, pp. 825–873, 1968.

Some of these references are reprinted in *Land-Mobile Communications Engineering*, D. Bodson, G. F. McClure, and S. R. McConoughey, Eds. New York: IEEE Press, 1984. The table of contents for this book is listed in Appendix I.

IV. Propagation Models

A variety of experimentally or theoretically based models have been developed to predict radio propagation in land mobile systems. Many of the models that have been reported in the literature are summarized and compared in this section. We start by listing and summarizing the major features of the models. The models are then compared in terms of the environmental and propagation factors they account for, and in terms of the output information they provide. Finally, the median transmission loss predictions of various models are compared for a given base antenna site. A detailed evaluation of some models is given in Section V.

A. Summary of Current Models

In listing the various models, a common notation and set of units have been adopted for all. The distance D between base and mobile antennas is in kilometers, while the height of h of the base antenna is in meters. L is used for median transmission loss in dB between half-wave dipoles for 900 MHz operating frequency.

1) Free Space: This fundamental theoretical model predicts the losses due to radial spreading for propagation in an ideal region without boundaries [1], [2]. Transmission loss predicted by this model is given by

$$L = 87 + 20 \log D \quad \text{(dB)}.$$

This model is useful as a reference for comparing transmission loss given by other models. When the Fresnel clearance is 0.6 or greater, this model can be used to estimate the loss associated with the fields that propagate directly (without reflection off the surface of the earth) from transmitting to receiving antenna.

2) Plane Earth: The plane earth model has been derived theoretically taking into account the presence of an idealize plane earth with finite conductivity [3]. For 900 MHz, median transmission loss to a 1.5 m high mobile antenna is given by

$$L = 111.6 - 20 \log h + 40 \log D \quad \text{(dB)}.$$

This expression is equal to the free space loss when $D = h/17$ (km).

This model is also useful as a reference for comparing other models. Calculated results for land mobile service at 900 MHz are not sensitive to the values chosen for the ground constants. The model is an oversimplification in that it does not include important factors such as terrain profile, vegetation and buildings.

3) Bullington Nomograms: In addition to free space and plane earth effects, the Bullington nomograms give field strength and median transmission loss as a function of antenna heights for certain idealized types of intervening terrain, specifically smooth curved earth with: 1) no obstructions; 2) a single sharp ridge; and 3) hills [1], [2]. The smooth earth and single ridge results come directly from physical theory. The treatment of hills is based on an approximate solution to the theoretical problem of diffraction by two knife edges located at the respective horizons of the transmitting and receiving antennas.

Computer versions of these nomograms are in common use. The model is an oversimplification since vegetation, buildings, and other terrain factor are not accounted for.

4) Egli: This model consists of empirical formulas that provide a "terrain factor" to be applied to the theoretical plane-earth field strength [4]. The terrain factor for 900 MHz has median value of 27.5 dB. The variation of field strength with antenna heights and distance is that of the underlying plane earth model. Median transmission loss for a 1.5 m high mobile antenna is equal to

$$L = 139.1 - 20 \log h + 40 \log D \quad \text{(dB)}.$$

The Egli model is a systematic interpretation of measurements covering a wide span of frequencies (90 to 1000 MHz) with some sacrifice of the accuracy that might be obtained by examining narrower bands individually.

5) Carey: Curves give the $F(50, 50)$ and $F(50, 10)$ field strength versus distance for propagation under average terrain conditions with: 1) mobile antenna height of 1.8 m; 2) base station antenna heights ranging from 30 to 1500 m above average terrain; and 3) distances up to 130 km for the $F(50, 50)$ curves, and up to 240 km for the $F(50, 10)$ curves [5]. The frequency bands explicitly covered are those used for common carrier mobile service at the time. Although 900 MHz frequencies are not mentioned, the Carey curves for 450–460 MHz were based on a CCIR recommendation (Geneva, 1963) covering the entire band from 450 to 1000 MHz. Median transmission loss for this model can be found for engineering purposes from the following expressions that have been fitted to the published curve:

$$L = 110.7 - 19.1 \log h + 55 \log D \quad \text{(dB)}, \quad \text{for } 8 \leqslant D < 48$$

$$L = 91.8 - 18 \log h + 66 \log D \quad \text{(dB)}, \quad \text{for } 48 \leqslant D < 96.$$

The Carey curves themselves are included in FCC Rules and Regulations and are being used for cellular radio licensing applications (the analytic fit to the Carey curves given above should not be used for application purposes). The curves are derived from CCIR curves for television broadcasting. The latter were adjusted downward by 9 dB to account for the 1.8 m height of mobile station antennas. Note that $F(L, T)$ denotes field strength exceed at L percent of locations during T percent of the time. At distances involved in land mobile service, only the location variability is significant, and the time variability can be considered as $T = 100$.

6) Tech. Note 101: This model consists of curves, theoretical equations, and empirical formulas for predicting cumulative distributions of sector transmission loss for a wide range of frequencies over almost any type of terrain and in several climatic regions [6].

Application of this model usually requires a succession of fairly detailed calculations, and at each step one must take care to find the curves or formulas that are exactly appropriate.

Many specific ideas presented in Tech. Note 101 are utilized in other models, most notably Longley–Rice and TIREM. See Section V for further discussion.

7) R-6602-LM: Calculation procedures for $F(50, 50)$ and $F(50, 10)$ field strength as a function of distance and base station antenna height are given in this report [7]. Mobile antenna height is 1.8 m, but an equation is given for adjusting heights within the range 1.8 to 9 m.

This method was developed to assist rulemaking activities in Docket 18261 having to do with land mobile sharing with TV broadcast curves appearing in the FCC Rules. At distances involved in land mobile service, only the location variability is significant and the time variability can be considered as $T = 100$.

8) Longley–Rice Point-to-Point Prediction: This model is in the form of a computer code for predicting long-term median transmission loss over irregular terrain [8]. The method is applicable for radio frequencies above 20 MHz. The point-to-point prediction procedure requires detailed terrain profiles. From the profiles, one must determine the distances to the respective radio horizon, the horizon elevation angles, and effective antenna heights. These distances, angles, and heights are then supplied as input to the computer program.

The report [8] covers two distinct situations. In one of these, both ends of the radio path are known and fixed so that a terrain profile can be prepared. For such cases of point-to-point prediction, the computer program is basically an automated version of the procedures of Tech. Note 101 (see above). The computer programs have been revised several times. This model is discussed further in Section V.

9) Longley–Rice Area Prediction: For random paths in an area where the variations in terrain elevation are characterized statistically, this model predicts long-term median transmission loss [8]. Estimates of variability are provided as program output. These estimates are of the variability with respect to location and time, and an estimate is also given of the standard error of prediction.

This report [8] introduces methods for estimating effective antenna heights, horizon distances and angles. The estimates are made in terms of a terrain roughness factor, the heights of the antennas above ground, and the type of antenna siting (random or preferred).

10) Okumura: This model gives a method for predicting field strength and service area for a given terrain for frequencies in the range 150–2000 MHz, distance of 1 to 100 km, base station effective antenna heights of 30 to 1000 m, and receiver antenna heights typical of land mobile applications [9]. The basic median field strength curve for 900 MHz applies to an urban area. Correction factors are given for suburban, open and isolated mountain areas, rolling hills, sloping terrain and mixed land-sea paths.

The model is based on measurements made in Tokyo, Japan, and surrounding suburbs at 200, 453, 922, 1310, 1430, and 1920 MHz. Statistical analysis of measurements was used to determine distance and frequency dependence of median field strength, location variability and antenna height gain factors. The urban curves with suburban correction factors seem to be most suited for cities in the U.S.

11) Terrain Integrated Rough Earth Model (TIREM): In the form of a computer program, this model was developed by the Electromagnetic Compatibility Analysis Center (ECAC) [10]. It predicts propagation loss between two points taking into account the frequency, atmospheric and ground constants, and the characteristics of the terrain profile between the two points. A digitized data base of terrain elevations gives necessary information on the profile, and the model then selects the appropriate propagation algorithm and calculates loss.

TIREM is one of seven point-to-point propagation models included in [10]. The other six, however, are not appropriate for 900 MHz land mobile propagation. The inputs to the program consist of the frequency (40 MHz to 20 GHz), polarization, ground permittivity and conductivity, atmospheric refractivity modulus, absolute humidity, and transmitter and receiver antenna structural heights, site elevations, latitudes and longitudes. The program gives as outputs the path loss and fading statistics. It may be difficult to adapt the computer program to operate at a particular facility because of the requirement to read a digitized data base. This model is discussed in greater detail in Section V.

12) General Electric Slide Rule: This model is in the form of a slide rule for calculating the maximum range for mobile, portable, and point-to-point communications in the land mobile bands; alternatively, it may be used to calculate required transmitter power for a given range [11]. Actual mobile antenna height is used unless effective height (read from a table) is higher. Percentage of locations to be covered may be specified from 50 to 99.9.

Received signals (microvolts across a 50 Ω input impedance) are related by the slide rule to transmitter output power after the device has been set for the desired range, percentage of locations, antenna heights, and antenna system gain/loss adjustment.

13) EPM-73, High Antenna/Low Antenna: Step-by-step procedures are given for estimating median transmission loss with an associated standard deviation [12]. Input parameters are antenna heights, frequency, distance, polarization and type of soil. The procedure differs according to whether the ratio of antenna height to wavelength is large or small, and for land mobile applications it may be necessary to make both calculations and use the larger of the resulting loss estimates. EPM-73 was developed with the intention of providing a model that would give, with minimum input information, reasonably accurate estimates of expected loss together with associated uncertainties.

14) Lee: The Lee model predicts point-to-point transmission by using two components [13]. The first is an area-to-area path-loss prediction, while the second component gives point-to-point prediction for each distance. The constants provided are for 30 m base antenna height and 3 m mobile antenna height. Transmission loss expressions are different for different cities. A typical example is for Philadelphia where median transmission loss is given by

$$L = 142.3 - 20 \log h + 36.8 \log D \quad \text{(dB)}.$$

Corrections for sloping terrain and a path obstruction are also given.

TABLE I
SUMMARY OF MODEL FEATURES

Model	Above average terrain	Above street level	Effective height	Mobile (1-3 m)	Terrain data	Building data	Foliage data	Hill shape	Distance	Free space	Diffraction by smooth earth	Reflection from earth surface	Reflection from hills	Diffraction by hills	Atmospheric refractivity	Building penetration	Loss deviation	Location variability	Time fading	Median trans- mission loss
	Antenna Height																			
	Input Parameters Treated									Propagation Factors Included							Output Parameters			
Bullington	N	E	L	E	L	N	N	L	E	E	E	L	N	E	N	N	N	L	L	E
Egli	E	N	L	E	N	N	N	N	E	N	N	L	N	N	N	N	N	E	N	E
Carey	E	N	N	N	N	N	N	N	E	N	N	N	N	N	N	N	N	L	L	E
Tech. Note 101	N	E	E	E	E	N	L	L	E	E	E	E	N	E	E	L	N	E	E	E
R-6602-LM	E	L	N	L	N	N	N	N	E	N	N	N	N	N	N	N	N	L	L	E
Longley-Rice Pt.-to-Pt.	N	E	E	E	E	N	N	L	E	E	E	E	N	L	E	N	N	E	E	E
Longley-Rice Area	N	E	L	E	N	N	N	N	E	E	E	E	N	N	E	N	E	N	E	E
Okumura	N	E	E	E	L	N	L	L	E	E	N	N	N	L	N	N	L	E	N	E
TIREM	E	E	E	E	E	N	N	L	E	E	E	E	N	E	E	N	L	E	E	E
Gen. Electric Slide Rule	N	E	L	E	N	N	N	N	L	L	L	L	N	N	N	N	N	E	N	E
Lee	N	E	L	N	L	N	N	L	L	N	N	L	N	L	N	N	N	N	N	E
CCIR Report 567	E	L	N	N	N	N	N	N	E	E	E	N	N	N	N	E	N	E	E	E
Bertoni & Walfisch	N	E	L	E	L	E	N	N	E	E	N	N	N	N	N	N	N	N	N	E

*Rating Scale: N = not treated; L = limited treatment; E = extensive treatment

It is difficult to know how to apply a given expression to an actual case, other than the three cities discussed. A mobile antenna height of 3 m is twice the value usually given to this parameter and may result in errors up to 6 dB. This model is described in detail in Appendix VI.

15) CCIR: Curves give field strength at 900 MHz frequencies for 50 percent of locations and 50 percent of the time in urban areas (to be used with caution in other areas) for mobile antenna heights of 1.5 m and base antenna heights between 30 and 1000 m [14]. To adjust for mobile antenna heights of 3 m instead of 1.5 m, a height-gain factor of 3 dB is suggested. Standard deviations are given as a function of distance and terrain irregularity, and this information is to be used in conjunction with the assumption that propagation variations with location and time are characterized in decibel quantities by the Gaussian distribution.

Treatment of the median transmission loss for frequencies near 900 MHz in this report is based on the work of Okumura *et al. [9]* (also see Okumura model above).

16) Bertoni and Walfish: This theoretical model was developed specifically to predict the effect of buildings on the median transmission loss [15], [16]. It applies to those urban and suburban environments where the buildings are of fairly uniform height and are built in rows with small separation between neighboring buildings. For level terrain and well within the radio horizon, the median transmission loss at 900 MHz between half-wave dipoles is predicted to be

$$L = 147.2 + A - 18 \log (h - h_b) + 38 \log D \quad \text{(dB)}$$

where

$$A = 5 \log \left[\left(\frac{d}{2} \right)^2 + (h_b - h_m)^2 \right] - 9 \log d + 20 \log \{\tan^{-1} [2(h_b - 2h_m)/d]\}.$$

Here h_m is the height of the mobile antenna, h_b is average building height, and d is the center-to-center spacing of the rows of buildings (typically one-half the narrow dimension of the blocks). For example if $d = 50$ m, $h_b = 12$ m and $h_m = 1.5$ m, then $A = -9$ dB.

In an alternate form to that given above, the model accounts for local terrain slope in the vicinity of the mobile. It does not, however, incorporate terrain roughness factors or treat obstructing terrain features, such as hills. The model is discussed in detail in Appendix V.

B. Comparison of Factors Treated

The models listed above were developed from various perspectives and with different original intentions. As a result, the models have for their inputs different environmental and path parameters, and account for different propagation factors. While all give median transmission loss as an output, some also give location variability and time fading information, but none give loss deviation. A comparison of the extent to which the models make use of various input parameters and propagation factors, and the output information they give, is shown in Table I. The letter N is used to indicate that a

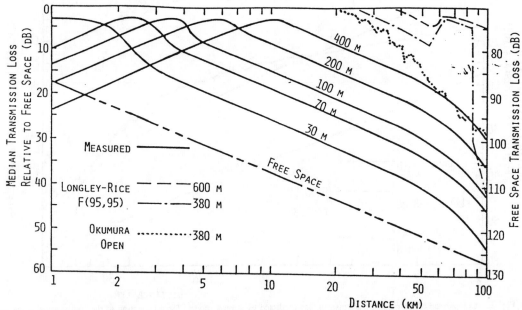

Fig. 1. Comparison of measured median transmission loss for level terrain and open areas with predictions of the Okumura and Longley-Rice models.

parameter or factor is not treated, while limited or extensive treatments are indicated by L or E, respectively.

Certain input parameters have significant influence on the accuracy of predicting median transmission loss by a given model. In particular, terrain data, including hill shape, building data and foilage data are individually and jointly indispensible for making accurate predictions. It should be noted that all of the models evaluated have failed to give extensive treatment to all three parameters. Even though two models give limited treatment of foliage data, such data is seldom available, nor in the proper format to use as input to a computer prediction program. While foliage always increases transmission loss, large isolated buildings and hills may either increase or decrease the transmission loss, depending on whether they block or provide an additional scattering path between the base and mobile antennas. When any one of the three types of input data cited above is neglected, the resulting error in the prediction of median transmission loss is usually unacceptable.

Both the accuracy of predicting the output parameters, and the type of parameters given, determine the usefulness of the model for predicting service probability in a given area. The most important output parameters for determining service probability are median transmission loss, location variability and loss deviation. While all models give median transmission loss, and some give the location variability, none give an extensive treatment of the loss deviation. As a result, errors can be expected in determining the service probability, no matter which model is used.

C. Comparison of Median Transmission Loss Predictions

The values of median transmission loss predicted by several of the models described above have been compared with each other, and with measurements, for various radials away from the Sears Tower in Chicago. This base station site offers

different terrain conditions ranging from the smooth lake surface to the highly irregular terrain South and West of the Sears Tower. Figs. 1-3 compare Longley-Rice, Okumura, slide rule, Carey, and TIREM models with measured median transmission loss data. Each value of median transmission loss shown is relative to free space transmission loss. The total median transmission loss for each distance can be found by adding the free space loss values to those given for each model.

Fig. 1 shows five median transmission loss curves measured for different antenna heights. These curves are typical of open areas with smooth terrain. Starting from a point close to the base station, the median transmission loss initially decreases at a rate of about 6 dB per doubling of distance out to a distance of about 25 times the base antenna height h. This decrease is due to ground reflections and to the patterns of the vertically oriented dipole antennas. Beyond 25 times the base antenna height, there is an initial sharp increase of median transmission loss as the mode of propagation switches from that of free space to propagation over a plane earth. In the plane earth mode, median transmission loss increases at a constant 6 dB per doubling of distance out to about 60 km, where diffraction loss due to the earth's curvature increases the slope to 25 dB per doubling of distance. These five median transmission loss curves, which have been smoothed to show average conditions, are useful as standards of comparison with predicted losses for the models shown on the three figures.

The predictions of the Okumura open model for $h = 380$ m are also shown in Fig. 1. It is seen to give lower values of median transmission loss than measured for distances out to 60 km. Predictions obtained from the Longley-Rice model using terrain data for propagation over Lake Michigan are also shown in Fig. 1 for $h = 380$ m and 600 m. The resulting curves illustrate three serious problems in the Longley-Rice model. First, for distances less than about 55 km, the predicted median transmission loss is considerably less than would be

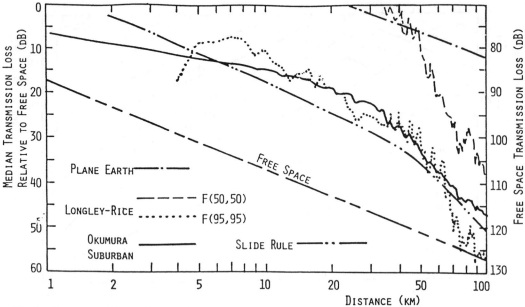

Fig. 2.   Comparison of the median transmission loss predicted by several models for propagation to the south of the Sears Tower for base station antenna height of 380 m.

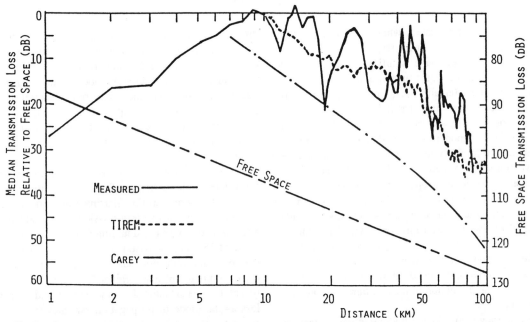

Fig. 3.   Comparison of measured median transmission loss south of the Sears Tower to predictions given by the Carey and TIREM models for base station antenna height of 380 m.

measured. Secondly, discontinuities occur at distances beyond 55 km, which are probably caused by changes in the mode of propagation assumed by the model. Finally, for large distances the slope of the Longley–Rice predictions increases to 55 dB per doubling of distance, which is 25 dB greater than that measured. It should be noted that the curves have been plotted for the $F(95, 95)$ location/time probability levels since the predictions for median probability $F(50, 50)$ give even smaller transmission loss than the $F(95, 95)$ predictions.

Fig. 2 shows a comparison of median transmission loss relative to free space as predicted by five models for $h = 380$ m. The two Longley–Rice curves were obtained using terrain data for the Chicago area and averaging the results computed along seven radials located in a southerly direction from the

Sears Tower. The Okumura suburban model has been used to obtain the predictions shown in Fig. 2, rather than the urban model, which gives values of median transmission loss that are as much as 20 dB higher. The suburban model gives values closer to those measured in most U.S. cities. Predictions made by the G.E. slide rule closely follow the Okumura suburban curve. For distances between 5 and 65 km, the $F(95, 95)$ Longley–Rice predictions are also close to the Okumura suburban model. On the other hand, the plane earth and $F(50, 50)$ Longley–Rice models give predictions that differ by more than 25 dB from those of the Okumura suburban model.

A comparison between TIREM, Carey, and measured data is shown in Fig. 3. The measurements were taken for propagation between the Sears Tower and a mobile traveling southward in a sector centered on a bearing angle of 200°. The TIREM results were obtained by averaging computations made using terrain data along seven radicals lying between the bearing angles 168.75° and 236.25°. The TIREM model predictions of median path loss are seen to be in close agreement with the measured data, while the Carey median transmission loss values are on the average about 15 dB greater. These differences are caused in part by the gradual rise of the terrain south of Chicago.

REFERENCES

[1] K. Bullington, "Radio propagation above 30 megacycles," *Proc. IRE*, vol. 35, pp. 1122–1136, 1947.

[2] ——, "Radio propagation for vehicular communications," *IEEE Trans. Veh. Technol.*, VT-26, pp. 295–308, 1977.

[3] C. Burrows, "Radio propagation over plane earth - field strength curves," *Bell. Syst. Tech. J.*, 16, Jan. 1937.

[4] J. Egli, "Radio propagation above 40 Mc over irregular terrain," *Proc. IRE*, vol. 45, pp. 1383–1391, 1957.

[5] R. Carey, "Technical factors affecting the assignment of facilities in the domestic public land mobile radio service," FCC, Washington, DC, Rep. R-6406, 1964.

[6] P. L. Rice, A. G. Longley, K. A. Norton, and A. P. Barsis, "Transmission loss predictions for tropospheric communication circuits," U.S. Government Printing Office, Washington, DC, NBS Tech. Note 101, issued May 1965; revised May 1966 and Jan. 1967.

[7] J. Damelin, W. A. Daniel, H. Fine and G. Waldo, "Development of UHF propagation curves for TV and FM broadcasting," FCC, Washington, DC, Rep. R-6602, 1966.

[8] A. G. Longley and P. L. Rice, "Prediction of tropospheric radio transmission over irregular terrain. A. Computer method—1968," ESSA Tech. Rep. ERL 79-ITS 67, U.S. Government Printing Office, Washington, DC, July 1968.

[9] Y. Okumura *et al.*, "Field strength and its variability in VHF and UHF land-mobile radio service," *Rev. Elec. Commun. Lab.*, vol. 9, pp. 825–873, 1968.

[10] "Master Propagation System (MPS 11) User's Manual," U.S. Dep. Commerce, Nat. Telecommun. Inform. Serv., NTIS Accession no. PB 83-178624. (Computer program tape available from NTIS as PB-173971.)

[11] "Range and transmitter power calculator," General Electric. Mobile Radio Dep., Lynchburg, VA, 1977.

[12] M. Lustarten and J. Madison, "An empirical propagation model (EPM-73)," *IEEE Trans. Electromagn. Commun.*, vol. EMC-19, pp. 301–309, 1977.

[13] W. C. Y. Lee, *Mobile Communications Engineering.* New York: McGraw-Hill, 1982, ch. 3, 4.

[14] "Methods and statistics for estimating field strength values in the land mobile services using the frequency range 30 MHz," CCIR Plenary Assembly, Geneva, Rep. 567-2, 1982.

[15] H. L. Bertoni and J. Walfisch, "A diffraction based theoretical model for prediction of UHF path loss in cities," in *Proc. AGARD Conf. Terrestrial Propagat. Characteristics in Modern Syst.*, Ottawa, 1986, pp. 8-1–8-9.

[16] J. Walfisch and H. L. Bertoni, "A theoretical model of UHF propagation in urban environments," *IEEE Trans. Antennas Propagat.*, to be published.

Some of these references are reprinted in *Land-Mobile Communications Engineering*, D. Bodson, G. F. McClure, and S. R. McCononghey, Eds. New York: IEEE Press, 1984. The table of contents for this book is listed in Appendix I.

V. Evaluation of Current Models

In this section we evaluate in detail several of the propagation models as they apply to 800/900 MHz land mobile communications. In order to evaluate how various propagation models perform, we will discuss predictions obtained in the following basic situations:

1) smooth terrain between base station and mobile;
2) smooth terrain with a knife-edge obstacle located close to the mobile; and
3) smooth terrain with a knife-edge obstacle at the midpoint of the path.

The case of smooth terrain between base station and mobile gives insight into how models handle the various propagation modes, such as line-of-sight, diffraction, and scatter, and how the models make transitions between the modes.

Of particular interest are those models that employ detailed terrain path profiles and are sufficiently defined that they have been implemented as computer programs so that all "engineering judgements" surrounding the interpretation of terrain path profiles have been reduced to computer algorithms. Accordingly, these models can be interfaced to digital terrain data bases for predicting coverage on a point-to-point basis at various receive locations surrounding the base station.

In studying the predictions of different models it is convenient to utilize the median transmission loss relative to the free-space transmission loss. The separation out of the free-space transmission loss removes the basic minimum transmission loss that is common to all models and leaves the specific loss that is predicted by each model. Thus, in order to obtain the total median transmission loss, the appropriate free-space attenuation must always be added to the results presented here.

Two basic antenna heights will be considered, namely 500 ft (152 m) and 200 ft (61 m). In all cases the mobile antenna is assumed to be 6 ft (1.8 m) above street level. Both configurations will be evaluated over smooth-earth paths from 1 to 100 km. In addition, as appropriate, all results in this section are for:

- frequency of 900 MHz;
- vertical polarization;
- surface refractivity of 301 (N-units);
- a 4/3 effective earth radius;
- continental temperature climate;
- relative ground dielectric constant of 15;
- ground conductivity of 0.005 s/m.

As appropriate, the prediction of signal variabilities will also be discussed, in which case the predictions deal with "corrections" to be made to the basic median transmission loss predictions.

In order to act as a reference point for the various models, results for the two basic antenna heights are presented in Figs. 1 and 2 for the Bullington smooth-earth model, and for the

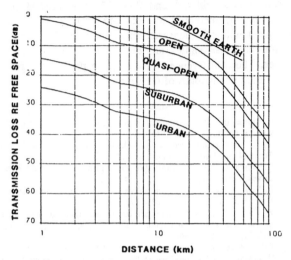

Fig. 1. Median transmission loss relative to the free-space transmission loss for Bullington smooth earth and Okumura urban, suburban, quasi-open, and open at 900 MHz; H_T = 500 ft (152 m); H_R = 6 ft (1.8 m).

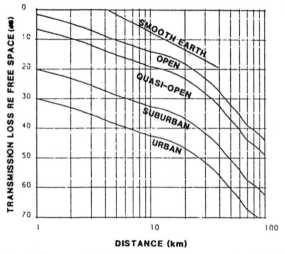

Fig. 2. Median transmission loss relative to the free-space transmission loss for Bullington smooth earth and Okumura urban, suburban, quasi-open, and open at 900 MHz; H_T = 200 ft (61 m); H_R = 6 ft (1.8 m).

Okumura model with all four environments (urban, suburban, quasi-open, and open).

A. National Bureau of Standards (NBS) Technical Note 101

NBS Tech. Note 101 [1] was first published in 1965, then revised in 1966 and again in 1967. The two-volume report describes detailed prediction methods for radio systems using propagation through the troposphere. It is intended for point-to-point transmission paths with frequencies from 40 MHz to 100 GHz over all kinds of terrain and long time intervals. Distances range from a few kilometers to hundreds of kilometers, with an emphasis on long radio links.

The detailed prediction methods described in Tech. Note 101 utilize the path geometry of the terrain profile and the atmospheric refractivity of the troposphere to predict median transmission loss. For transmission paths within the radio horizon, the transmission loss is computed using geometric optics (ray) theory. Fresnel–Kirchhoff knife-edge diffraction is used for common horizon paths and is extended to predict the diffraction loss over isolated round obstacles. A modification of the Van der Pol–Bremmer method is utilized to compute the loss in the far diffraction region of double horizon paths whose distance is only slightly longer than the radio horizon. For longer distances, forward scatter theory is used to make troposcatter predictions of the median transmission loss.

Tech. Note 101 describes these techniques in detail with many equations and figures. The report, however, does *not* contain any listing of a computer implementation of these techniques. Because there is no publicly available computer program that specifically implements Tech. Note 101, no results are presented here for comparison. Tech. Note 101 is included here because of the extent to which it is referenced in radio propagation literature. The significant contribution of Tech. Note 101 for 900 MHz propagation is the in-depth discussion of propagation fundamentals, particularly as they relate to radio transmission through the troposphere. Many of the specific ideas presented in Tech. Note 101 are utilized in other prediction methods. Most notable, the Longley–Rice computer method is a direct application of the Tech. Note 101 information. Also, various parts of TIREM are based on Tech. Note 101 concepts.

B. Longley–Rice Methods

In 1968, Longley and Rice published an ESSA Technical Report [2] presenting both the theory and an associated computer program for predicting long-term median transmission loss over irregular terrain for frequencies between 20 MHz and 10 GHz. Many of the equations used were derived from Tech. Note 101. This report presents methods for calculating the median transmission loss relative to the free-space transmission loss (often referred to as the reference attenuation, A_{cr}(dB)) from characteristics of the propagation path. The Longley–Rice model is sometimes referred to as the "ITS irregular terrain model" where ITS stands for the Institute for Telecommunication Sciences.

For a particular transmission path, a series of parameters must be defined, namely:

- frequency;
- polarization;
- path length;
- antenna heights above ground;
- surface refractivity;
- effectuve earth's radius;
- climate;
- ground conductivity;
- ground dielectric constant.

In addition, a series of path-specific parameters are required, namely:

- effective antenna heights;
- horizon distances of the antennas;
- horizon elevation angles;
- the angular distance for a transhorizon path;
- terrain irregularity of the path.

If a detailed terrain path profile is available, these path-specific parameters can be determined from the specific terrain path profile. The resulting prediction is referred to as a "point-to-point" mode prediction. If a detailed terrain path profile is not available, this report gives techniques for estimating these path-specific parameters. The resulting prediction using these estimated parameters is referred to as an "area" mode prediction.

The 1968 Longley–Rice Report [2] recognizes the importance of procedures for determining the effective heights of the transmitting and receiving antennas. On [2, p. 12] they state:

> Further study of definitions of effective antenna height appears to be the most urgent requirement for improving these predictions for low antenna heights over irregular terrain. Different definitions of $h_{e1,2}$ may be found appropriate for line-of-sight and diffraction formulas.

On pages 3-1 and 3-2 of the report they define the effective antenna heights to be the elevations of the antennas above the "dominant reflecting plane," and go on to discuss how the "dominant reflecting plane" might be determined. It is important to note, however, that this 1968 report does *not* give a computer program that implements the concepts they discuss for determining either the effective antenna heights or the terrain irregularity. The computer program they give in this report assumes that the effective antenna heights have already been determined as they are to be specified as input to the program. It is also important to note that this report does *not* describe how to compute the various types of signal variabilities, nor does it provide a computer program for calculating signal variabilities.

No attempt will be made here to describe the basic Longley–Rice procedure. The reader is referred to the original 1968 report [2] and to a more recent 1982 NTIA Report [3] that lists computer programs that determine effective antenna heights and the terrain irregularity parameter from terrain profiles.

There are a series of computer program versions of the Longley–Rice model, which are described in the following quote taken from the 1982 report [3]:

> The original "Longley–Rice" model was published in 1968. Shortly afterward a new version was developed which improved the formulation for the forward scatter prediction, and later the computer implementation was changed to improve its efficiency and increase the speed of operation. Since then, minor but important modifications have been made in the line-of-sight calculations.
>
> To keep track of the various versions, most of which are presently being used at some facility, we have recently begun numbering them in serial fashion. Following the original (which might be called version 0), here is a list of the more important versions, together with approximate dates when they

were first distributed:

1.0   January 1969
1.1   August 1971
1.2   March 1977
1.2.1   April 1979
2.0   May 1970
2.1   February 1972
2.2   September 1972

Version 1.2.1 corrects an error in version 1.2; it is the currently recommended version and is the one whose implementation is listed in Appendix A. The second series, beginning with version 2.0, used considerably modified diffraction calculations and tried to incorporate a groundwave at low frequencies. It is not now recommended and is no longer maintained by its developers.

Subsequent to the 1982 report, a 1985 Memorandum [4] containing several modifications was issued. This memo refers to the newly altered model as version 1.2.2 and gives it the date of September 1984. All results presented here for the Longley–Rice model are based upon version 1.2.2.

1) Definition of Path Parameters: Because of the importance of effective antenna heights and terrain irregularity, we shall now discuss in detail exactly how they are determined in version 1.2.2 of the Longley–Rice model. Program QLRPFL first defines a range of interest for determining the effective antenna heights and terrain irregularity. The distance from each antenna $X_{L1,2}$ is computed as follows:

$$X_{L1,2} = \min (15 h_{G1,2}, 0.1 d_{L1,2}) \qquad (1)$$

where $h_{G1,2}$ is the elevation of each antenna above ground level and $d_{L1,2}$ is the horizon distance of each antenna.

The range of interest starts at X_{L1} from the transmitter and goes to X_{L2} from the receiver. For a 6 ft (1.8 m) mobile antenna, $15 h_{G2} = 27.45$ m, so that $X_{L2} = 27.45$ m unless d_{L2} is less than 274.5 m, which will not often be the case. For a 500 ft (152 m) transmitting antenna, $15 h_{G1} = 2,286$ m, and $X_{L1} = 2286$ m, unless d_{L1} is less than 22.86 km, which, if there is an obstacle, will often be the case. Once the range of interest is determined, the determination of effective antenna heights falls into one of the following situations.

Line-Of-Sight: A least squares fit to the terrain profile is made over the range of interest X_{L1} to X_{L2} and the least squares elevation values h_{XL1} and h_{XL2} are determined. Using these values, the effective antenna heights are then computed from

$$H_{e1,2} = h_{G1,2} + \text{DMIN (ground level at ant}_{1,2}, h_{XL1,2}) \quad (2)$$

where DMIN is the Fortran positive difference function that returns the first argument minus the minimum of the two arguments. Using these effective elevations, new line-of-sight distances, D_L are computed. If the sum of the new D_L is less than the total path distance, then H_e and D_L have a "correction" applied to them.

Transhorizon: A least squares fit of the terrain profile is made over the range X_{L1} to $0.9 d_{L1}$ for the transmitter, and another least squares fit of the terrain is made over the range

$$(\text{path distance} - 0.9 d_{L2}) \text{ to } X_{L2}$$

and the least square elevation values h_{XL1} and h_{XL2} are determined. With these values, the effective antenna heights are then computed from

$$H_{e1,2} = h_{G1,2} + \text{DMIN (ground level at ant}_{1,2}, h_{XL1,L2}). \quad (3)$$

To determine terrain irregularity, program DLTHX selects up to 25 points using linear interpolation on the path profile between X_{L1} and X_{L2} as a basis for fitting a straight line. The routine QTILE is then used to determine the interdecile range $\Delta h(d)$ of terrain heights above and below this straight line. The terrain irregularity parameter is then computed by applying the following correction:

$$\Delta h = \Delta h(d)/[1 - 0.8 \exp(-\min(20, d/5 \times 10^4))] \quad (4)$$

where

$$d = (\text{path distance} - X_{L2}) - X_{L1}. \qquad (5)$$

The 1982 report [2] also contains a computer program called AVAR for computing the various types of signal variabilities.

2) Median Transmission Loss Predictions: The Longley–Rice model and associated computer programs have two distinct parts that will be considered separately. The first part is the prediction of median transmission loss. The second part is the prediction of signal variabilities. The basic median transmission loss prediction techniques are described in the original 1968 Longley–Rice Report [2]. Between the original 1968 report and the 1982 report there have also been several publications proposing modifications to the basic Longley–Rice procedure to resolve difficulties that have been observed when comparing median transmission loss predictions to actual measurements.

A 1970 ESSA Tech. Report [5] proposes an equation that is obtained empirically from a series of measurements for calculating the reference attenuation for line-of-sight paths and a revised procedure utilizing Fresnel–Kirchhoff knife-edge diffraction along with an allowance for ground reflections for calculating the reference attenuation for single-horizon paths.

In an appendix to her 1978 OT Report [6], Longley states that "The Longley–Rice model was originally developed for use with rather low antennas in irregular terrain, and was tested against data from measurements made with low antennas at frequencies from 20 to 100 MHz. Very few of these original test paths had terminals within radio line-of-sight. Later tests against data obtained with higher antennas and at higher frequencies showed rather poor agreement with the line-of-sight area predictions." This section describes in detail a series of changes to the original Longley–Rice model to overcome these deficiencies.

The 1978 OT Report [6] deals with radio propagation in urban areas and contains a correction factor to account for the urban environment. As described in this report [6, p. 31], most of the data used in the formulation of the original Longley–Rice model came from open areas, towns, and small cities. This report developed an "allowance for the additional attenuation due to urban clutter near the receiving antenna." This allowance was developed by comparing results from the

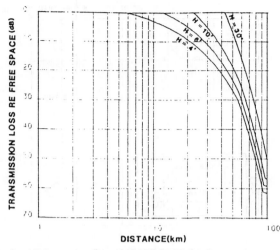

Fig. 3. Median transmission loss relative to the free-space transmission loss for Longley–Rice over smooth earth at 900 MHz; H_T = 500 ft (152 m); H_R = 4, 6, 10, and 30 ft.

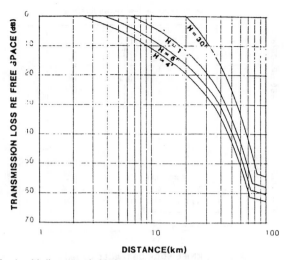

Fig. 4. Median transmission loss relative to the free-space transmission loss for Longley–Rice over smooth earth at 900 MHz; H_T = 200 ft (61 m); H_R = 4, 6, 10, and 30 ft.

1977 modified version of the Longley–Rice computer method with Okumura's curves. All comparisons utilized smooth earth. The Okumura predictions consistently showed additional power loss in urban areas compared to the Longley–Rice predictions. Accordingly, an empirical equation was developed that gives the difference between the two models; this difference is referred to as the "urban factor (UF)." Longley states [6, pp. 32–33] that the Longley–Rice computer model, "with the urban factor added, should adequately predict the median attenuation for moderately large cities in rather smooth terrain."

In order to understand better how the Longley–Rice model functions in predicting median signal levels for 900 MHz mobile communication systems, a series of smooth-earth test cases have been run using version 1.2.2 of the Longley–Rice program in the point-to-point prediction mode. All test cases predict the median transmission loss relative to the free-space transmission loss. It should be noted that version 1.2.2 does *not* utilize either the corrections proposed in [5] or the urban factor proposed in [6].

The first set of test cases we will consider are predictions of the median transmission loss relative to the free-space transmission loss over smooth-earth as a function of path length. Results obtained for a transmitting elevation of 500 ft (152 m) and receiver elevations of 4, 6, 10, and 30 ft are shown in Fig. 3. Similar results obtained for a transmitting elevation of 200 ft (61 m) and receiver elevations of 4, 6, 10, and 30 ft are given in Fig. 4. Examination of these results shows that the Longley–Rice point-to-point prediction method gives a smooth monotonically increasing median transmission loss relative to the free-space transmission loss as the path distance over smooth earth is increased from 1 to 100 km. The transition to the scattering region for large path distances is evident in both figures—particularly in Fig. 4 for the 200 ft transmitting elevation.

Comparison of Figs. 3 and 4 with the Okumura and Bullington results presented in Figs. 1 and 2 shows that Longley–Rice 6 ft predictions essentially agree with Bul-

lington smooth-earth predictions for distances where the median transmission loss relative to the free-space transmission loss is small. The Longley–Rice predictions start to increase over the Bullington smooth-earth predictions as distance increases in order to provide a continuous transition into the diffraction region. We have also compared Longley–Rice diffraction predictions with Bullington diffraction predictions and find agreement within the accuracy to which the Bullington nomograms can be read. Because of the difficulties in accurately reading the Bullington diffraction nomograms, results obtained from them are not shown in Figs. 3 and 4. Thus, Longley–Rice predictions and Bullington predictions essentially agree when the path loss mechanism is either totally smooth earth or totally diffraction. The Longley–Rice procedures provide a continuous transition between these two modes whereas Bullington procedures do not. Longley–Rice 6 ft predictions disagree with all four environmental modes of the Okumura smooth-earth predictions. The shape of the curves, as well as the values, is significantly different.

The second set of test cases we will consider are point-to-point mode predictions of median transmission loss relative to the free-space transmission loss over smooth earth as single knife-edge obstacles are introduced into specific paths. These results give insight into how the Longley–Rice point-to-point prediction mode handles the introduction of obstacles into the Fresnel zone. Results obtained for a transmitting elevation of 500 ft (152 m), a receiver elevation of 6 ft (1.8 m), and a path distance of 12 km are shown in Fig. 5. A knife-edge obstacle is introduced at a distance of 1 km from the receiver and the height of the knife-edge obstacle is increased from 0 to 20 m. This figure shows that there is a 3.3 dB discontinuity in the median transmission loss at the precise point where the Longley–Rice model changes from the line-of-sight mode to the transhorizon mode. This occurs at the point where the knife-edge obstacle intersects the central ray connecting the transmit and receive antennas.

Fig. 6 shows results obtained for the same case as above, except that the knife-edge obstacle is introduced at a distance

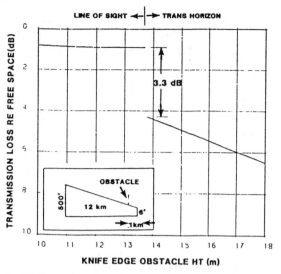

Fig. 5. Median transmission loss relative to the free-space transmission loss for Longley–Rice over a 12 km smooth earth path as a function of the height of a knife-edge obstacle located 1 km from a 6 ft (1.8 m) receiver at 900 MHz; $H_T = 500$ ft (152 m).

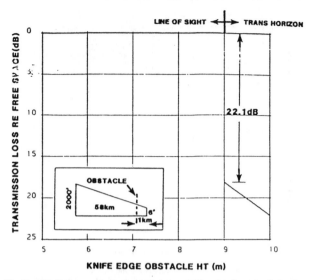

Fig. 7. Median transmission loss relative to the free-space transmission loss for Longley–Rice over a 58 km smooth earth path as a function of the height of knife-edge obstacle located at 1 km from a 6 ft (1.8 m) receiver at 900 MHz; $H_T = 2000$ ft (610 m).

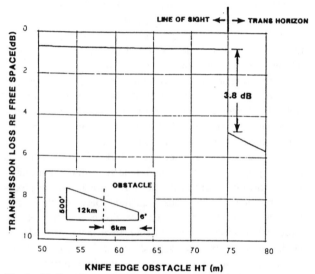

Fig. 6. Median transmission loss relative to the free-space transmission loss for Longley–Rice over a 12 km smooth earth path as a function of the height of a knife-edge obstacle at 6 km (the midpoint of the path) from a 6 ft (1.8 m) receiver at 900 MHz; $H_T = 500$ ft (152 m).

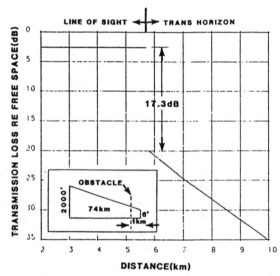

Fig. 8. Median transmission loss relative to the free-space transmission loss for Longley–Rice over a 74 km smooth earth path as a function of the height of a knife-edge obstacle located at 1 km from a 6 ft (1.8 m) receiver at 900 MHz; $H_T = 2000$ ft (610 m).

of 6 km from the receiver (the midpoint of the path) and the height of the knife-edge obstacle is increased from 0 to 80 m. This figure shows that there is a 3.8 dB discontinuity in the median transmission loss at the precise point where the Longley–Rice model changes from the line-of-sight mode to the transhorizon mode. As before, this occurs at the point where the knife-edge obstacle intersects the central ray connecting the transmit and receive antennas.

In Fig. 7 we show results obtained for a transmitting elevation of 2000 ft (610 m), a receiver elevation of 6 ft (1.8 m), and a path distance of 58 km. A knife-edge obstacle is introduced at a distance of 1 km from the receiver and the height of the knife-edge obstacle is increased from 0 to 10 m. This figure shows that there is a 22.1 dB discontinuity at the

point where the knife-edge obstacle intersects the central ray connecting the transmit and receive antennas.

Similar results are shown in Fig. 8 when the path distance is changed to 74 km. In this case there is a 17.3 dB discontinuity at the point where the knife-edge obstacle intersects the central ray connecting the transmit and receive antennas.

The 2000 ft transmitter, 6 ft receiver, 58 km path case of Fig. 7 was also studied with midpath knife-edge obstacles. At the point where the knife-edge intersected the central ray connecting transmitting and receive antennas, the Longley–Rice program returned an error flag indicating that "internal calculations show parameters out of range." The program, however, went on to produce results and returned a discontinuity significantly larger than that shown in Fig. 7. Investigation of the range criteria in the Longley–Rice LPROP subroutine show that the distances between transmitter/receiver and the nearest obstacle that defines the actual horizon of each antenna must be: 1) not less than 0.1 times the smooth-earth line-of-sight transmitter/receiver distances; and 2) not greater than three times the smooth-earth line-of-sight transmitter/receiver distances.

Some consequences of the foregoing conditions are seen for a 6 ft mobile antenna and a 500 ft base station antenna. The 6 ft antenna has a 5.6 km horizon over a smooth 4/3 radius earth. Thus, a horizon-defining obstacle closer than 0.56 km to the receiver or further than 16.8 km will violate the restrictions in the Longley–Rice program and return an error code. For the 500 ft antenna, the horizon is 50.9 km over a smooth 4/3 radius earth so that a horizon-defining obstacle closer than 5.1 km or further than 152.7 km will violate the restrictions in the Longley–Rice program and return an error code. Thus for the geometries of Figs. 5 and 6, any single knife-edge obstacle that defines the horizon of both the transmitter and receiver must occur in the region that is between 5.1 km from the transmitter and 0.56 km from the receiver. The knife-edge obstacles used in Figs. 5 and 6 satisfy these conditions.

If the transmitting antenna is raised to 2000 ft, its horizon is 101.8 km for a smooth 4/3 radius earth. Thus, a horizon-defining obstacle closer than 10.2 km or further than 305.4 km will violate the restrictions in the Longley–Rice program and return an error code. For a 6 ft receiver, any single knife-edge obstacle that defines the horizon of both the transmitter and receiver must occur in the region that is *both* between 10.2 and 305.4 km from the transmitter *and* also between 0.56 and 16.8 km from the receiver. While the knife-edge obstacle used in Fig. 7 satisfies these conditions, a knife-edge obstacle at the midpoint of a 58 km path does not. Similarly for the 74 km path of Fig. 8, the knife-edge obstacle satisfies these conditions. A knife-edge obstacle at the midpoint of a 74 km path does not satisfy both of these conditions.

A third test case we will consider are point-to-point mode predictions of median transmission loss relative to the free-space transmission loss over smooth-earth as a function of distance behind a knife-edge obstacle. These results give further insight into how the Longley–Rice point-to-point prediction model handles single knife-edge obstacles. Fig. 9 shows results obtained for a transmitting elevation of 500 ft (152 m) and a receiver elevation of 6 ft (1.8 m). Here a knife-edge obstacle of height 80 m is introduced at a distance of 6 km from the transmitter. Median transmission loss relative to the free-space transmission loss is computed for the receiver located at distances from the transmitter ranging from 8 to 100 km. The median transmission loss relative to the free-space transmission loss without the single knife-edge obstacle is also plotted. As readily apparent from Fig. 9, there is a 4.1 dB discontinuity at the point where the central ray between the transmit and receive antennas is no longer "blocked" by the knife-edge obstruction. The differences between the predictions at distances greater than 20 km are due to the way the computer program computes the effective antenna heights and horizon angles.

As is readily apparent from the preceding results, the

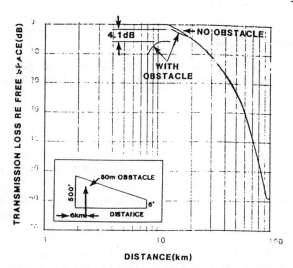

Fig. 9. Median transmission loss relative to the free-space transmission loss for Longley–Rice over a smooth earth path as a function of the path distance for a knife-edge obstacle located at 6 km from a 500 ft (152 m) transmitter at 900 MHz; $H_R = 6$ ft (1.8 m).

Longley–Rice point-to-point prediction model gives a discontinuous transition from the line-of-sight mode to the transhorizon mode when the transition is caused by single knife-edge obstacles. The Longley–Rice point-to-point prediction model does not provide for the increase in attenuation that would be expected as an obstacle penetrates the first Fresnel zone prior to the obstacle penetrating the central ray between the transmit and receive antennas. This failure to utilize first Fresnel zone penetration is significant as 900 MHz mobile propagation almost always has first Fresnel zone penetration near the mobile receiver.

3) Signal Variability Predictions: The original 1968 Longley–Rice Report [2] and associated computer programs do not describe how to compute signal variabilities. Several subsequent publications have treated location variability. A 1976 OT Report [7] develops equations for predicting location variability as a function of wavelength and the terrain irregularity parameter. A 1978 report [6] compared location variability results based on the 1976 report with results reported by Okumura and Egli and concluded that the 1976 equations predict "more variability than that observed by Okumura in Japan" but agree "with the relationship shown by Egli in 1957." Finally, the 1982 report [3] contains a detailed discussion on "Statistics and Variability" and also includes computer code (as part of the Longley–Rice model, version 1.2.1) that implements the procedures discussed. The report (on page 28) specifically *excludes* the "short-term or small displacement variability that is usually attributed to multipath propagation." Three basic types of variability are defined:

1) time variability—variations of local hourly medians on a specific path with time;
2) location variability—variations in long-term statistics that occur from path to path;
3) situation variability—variations in location variability that occur from situation to situation.

The report and associated computer program define four

different variability modes for combining these three basic types of variability, namely:

1) single message mode—time, location, and situation variability are combined together to give a confidence level;
2) individual mode—reliability is given by time availability, while confidence is a combination of location and situation variability;
3) mobile mode—reliability is a combination of time and location variability, and confidence is given by the situation variability;
4) broadcast mode—reliability is given by the two-fold statement of *at least* q_T of the time in q_L of the locations, with confidence given by the situation variability.

In addition, they provide an option whereby location variability is eliminated, as it should be when a well-engineered path is being treated in the point-to-point mode. A second option is also provided for eliminating direct situation variability, as it should be when considering interference problems. With this last option, it must be noted that there may still be a small residual situation variability.

The procedures for predicting variability in the program AVAR depend significantly upon the type of climate specified:

- equatorial;
- continental subtropical;
- maritime subtropical;
- desert;
- continental temperate;
- maritime temperate over land;
- maritime temperate over sea.

These prediction procedures also depend significantly upon the terrain irregularity parameter, whose determination is discussed in detail in Section 5.2 of [3]. On page 15, they give an equation that is used by their computer program to convert from the terrain irregularity parameter determined over a specified path distance to the asymptotic terrain irregularity parameter. It should be noted that version 1.2.2 of the Longley–Rice model does *not* utilize the location variability results published by Longley in 1976 [7].

In order to understand better how the Longley–Rice model functions in predicting signal variabilities for 900 MHz mobile communication systems, a series of test cases have been run using version 1.2.2 in the point-to-point prediction mode. All test cases predict the signal variability relative to the median transmission loss for a 500 ft (152 m) base antenna and a 6 ft (1.8 m) mobile antenna for path distances from 1 to 100 km. Local terrain irregularity parameter values of 0, 1, 5, 10, 50, and 100 m were utilized; unless otherwise noted, these path-specific terrain irregularity values were corrected by the program to their asymptotic values.

The first set of signal variability test cases consider predictions of the single message mode variability (also referred to as the confidence level variability) as a function of terrain irregularity and path length. Fig. 10 shows the standard deviation in decibels of the single message mode variability (as determined by setting $Q_c = 0.84$). As readily apparent from

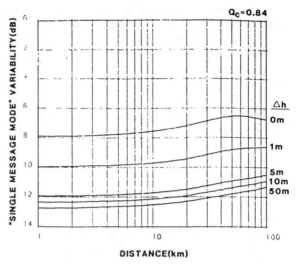

Fig. 10.   Longley–Rice single message mode variability as a function of path length at 900 MHz; $H_T = 500$ ft (152 m); $H_R = 6$ ft (1.8 m); $\Delta h = 0, 1, 5, 10,$ and 50 m with asymptotic correction; $\Delta h = 100$ m curve virtually indistinguishable from 50 m curve.

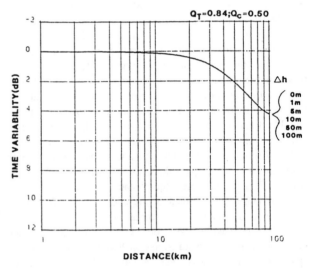

Fig. 11.   Longley–Rice time variability as a function of path length at 900 MHz; $H_T = 500$ ft (152 m); $H_R = 6$ ft (1.8 m); 50 percent confidence level; $\Delta h = 0, 1, 5, 10, 50,$ and 100 m with asymptotic correction.

this figure, the standard deviation of the single message mode variability (confidence level) varies significantly (approximately 4 dB) as the terrain irregularity goes from 0 to 5 m, whereas the difference from 50 to 100 m is so small that the 100 m curve is not plotted as it is virtually indistinguishable from the 50 m curve.

The second set of signal variability test cases consider predictions of the time variability as a function of terrain irregularity and path length. Fig. 11 shows the standard deviation in decibels of the time variability for a confidence level of 50 percent (as determined by setting $Q_t = 0.84$; $Q_l = 0.5$; $Q_c = 0.5$ in the broadcast mode). As readily apparent from this figure, the standard deviation of the time variability does *not* vary as a function of terrain irregularity.

Third, location variability predictions are considered as a function of terrain irregularity and path length. The standard

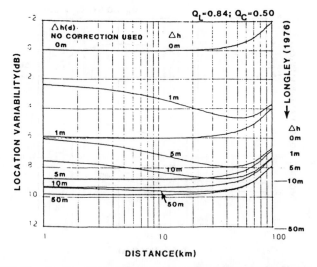

Fig. 12. Longley–Rice location variability as a function of path length at 900 MHz; H_T = 500 ft (152 m); H_R = 6 ft (1.8 m); 50 percent confidence level; Δh = 0, 1, 5, 10, and 50 m, both with and without asymptotic correction; Δh = 100 m curve virtually indistinguishable from 50 m curve. Right margin gives the values of location variability determined from equations published by Longley in 1976.

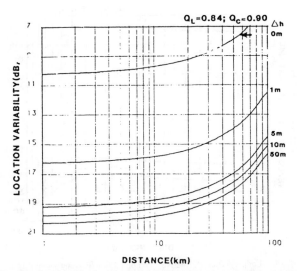

Fig. 13. Longley–Rice location variability as a function of path length at 900 MHz; H_T = 500 ft (152 m); H_R = 6 ft (1.8 m); 90 percent confidence level; Δh = 0, 1, 5, 10, and 50 m with asymptotic correction; Δh = 100 m curve virtually indistinguishable from 50 m curve.

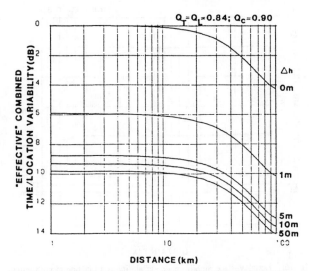

Fig. 14. Longley–Rice "effective" combined time/location variability as a function of path length at 900 MHz; H_T = 500 ft (152 m); H_R = 6 ft (1.8 m); 50 percent confidence level; Δh = 0, 1, 5, 10, and 50 m with asymptotic correction; Δh = 100 m curve virtually indistinguishable from 50 m curve.

deviation in decibels of the location variability for a confidence level of 50 percent (as determined by setting Q_1 = 0.84; Q_t = 0.5; Q_c = 0.5 in the broadcast mode) are shown in Fig. 12. In order to understand the significance of the difference between asymptotic terrain irregularity values and path-specific terrain irregularity values, results were obtained for both types of terrain-irregularity values. Fig. 12 shows how location variability is impacted by terrain irregularity. The margin in Fig. 12 also shows the values of location variability determined from the equations published by Longley in 1976. As readily apparent from this figure, the standard deviation of the location variability depends significantly upon whether the terrain irregularity parameter is the path-specific value or the asymptotic value. Also, the location variability for a confidence level of 50 percent varies significantly (almost 9 dB) as the terrain irregularity goes from 0 to 5 m, whereas the difference from 50 to 100 m is too small to plot. Fig. 13 shows that the standard deviation of the location variability for a confidence level of 90 percent (as determined by setting Q_1 = 0.84, Q_t = 0.5; Q_c = 0.90 in the broadcast mode) also varies significantly (approximately 9 dB) as the terrain irregularity goes from 0 to 5 m, whereas the difference from 50 to 100 m is too small to plot.

The fourth set of signal variability test cases consider the "effective" combined time/location variability as a function of terrain irregularity and path length. Fig. 14 shows that the standard deviation in dB of the "effective" combined time/location variability for a confidence level of 50 percent (as determined by setting Q_1 = 0.84; Q_t = 0.84; Q_c = 0.5 in the broadcast mode) varies significantly (almost 9 dB) as the terrain irregularity goes from 0 to 5 m, whereas the difference from 50 to 100 m is too small to plot. Fig. 15 shows that the "effective" combined time/location variability for a confidence level of 90 percent (as determined by setting Q_1 = 0.84; Q_t = 0.84; Q_c = 0.90 in the broadcast mode) also varies

significantly (approximately 9 dB) as the terrain irregularity goes from 0 to 5 m, whereas the difference from 50 to 100 m is too small to plot.

The fifth and final set of signal variability test cases we consider are predictions of the mobile mode variability as a function of terrain irregularity and path length. Fig. 16 shows that the standard deviation in decibels of the mobile mode variability for a confidence level of 50 percent (as determined by setting Q_r = 0.84; Q_c = 0.5 in the mobile mode) varies significantly (almost 9 dB) as the terrain irregularity goes from 0 to 5 m, whereas the difference from 50 to 100 m is too small to plot. Fig. 17 shows that the mobile mode variability for a confidence level of 90 percent (as determined by setting Q_r = 0.84; Q_c = 0.90 in the mobile mode) also varies significantly

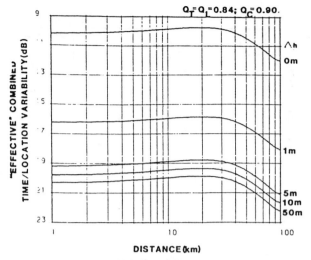

Fig. 15. Longley–Rice "effective" combined time/location variability as a function of path length at 900 MHz; H_T = 500 ft (152 m); H_R = 6 ft (1.8 m); 90 percent confidence level; Δh = 0, 1, 5, 10, and 50 m with asymptotic correction; Δh = 100 m curve virtually indistinguishable from 50 m curve.

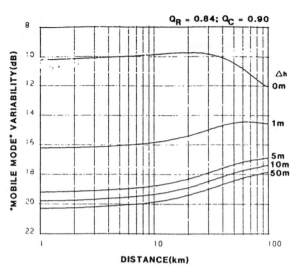

Fig. 17. Longley–Rice mobile mode variability as a function of path length at 900 MHz; H_T = 500 ft (152 m); H_R = 6 ft (1.8 m); 90 percent confidence level; Δh = 0, 1, 5, 10, and 50 m with asymptotic correction; Δh = 100 m curve virtually indistinguishable from 50 m curve.

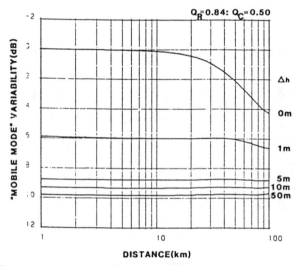

Fig. 16. Longley–Rice mobile mode variability as a function of path length at 900 MHz; H_T = 500 ft (152 m); H_R = 6 ft (1.8 m); 50 percent confidence level; Δh = 0, 1, 5, 10, and 50 m with asymptotic correction; Δh = 100 m curve virtually indistinguishable from 50 m curve.

(approximately 9 dB) as the terrain irregularity goes from 0 to 5 m, whereas the difference from 50 to 100 m is too small to plot.

The preceding results show that the terrain irregularity value, if less than 10 m, can significantly impact signal variability predictions. It is important to note that signal variabilities also depend upon the effective transmitting and receiving elevations.

4) Digital Terrain Data: Several important matters must be noted concerning the use of the Longley–Rice point-to-point prediction model with digital terrain data bases. The Longley–Rice point-to-point prediction model expects both the transmitter and receiver antenna elevations to be input relative to ground level. While this is appropriate for a mobile, specifica-

tion of the base station antenna relative to ground level will usually lead to erroneous results. Digital terrain data bases do not provide accurate elevations at all points; rather they provide accurate elevations only at discrete data points and use interpolation procedures to provide elevations at other points. Thus ground elevations at other than the actual discrete data points are subject to error. If an antenna elevation is specified relative to ground level with the ground level being determined by interpolation from the digital terrain data base, the resulting antenna elevation will usually be in error. To ensure that the antenna have the proper elevation, either the transmitter height must always be specified in AMSL, or the actual ground elevation of the transmitter antenna must be input and utilized.

The Longley–Rice point-to-point prediction model utilizes effective antenna heights as part of its computation of transmission loss. These predictions are quite sensitive to the actual value determined/used for the effective height of the receiver. In mobile system, the receive antenna is low so that any small change in the surrounding terrain can result in a rather large change in the effective height of the receive antenna and a corresponding sizeable change in the predicted median transmission loss. Depending upon the accuracy of the digital terrain information utilized, sizeable variations in the effective height of the receiver can often be due to digital data base factors, such as whether the elevation information is rounded to the nearest 10 ft or nearest 10 m rather than due to actual physical situations. It should be noted that there are two versions of the National Geophysical and Solar-Terrestrial Data Center (NGSDC) 30-s data base in use; the original version had its elevations rounded to the nearest 10 m whereas the revised version has its elevations rounded to the nearest 10 ft. It is our experience that significantly different results can often be obtained for both median transmission loss and signal variability predictions as a result of the difference between these two digital terrain data bases.

In addition to effective antenna height, small changes in

terrain elevations near the mobile receiver can intercept the central ray between the base transmitter and the mobile receiver and thus introduce a sizeable change in median transmission loss relative to the free-space transmission loss. Moreover, if the "terrain obstacle" is too close to the receiver, an error flag will indicate that program internal parameters are out of range. Depending upon the accuracy of the digital terrain information utilized, the presence or absence of an obstacle can often be due to digital data base factors rather than due to any actual physical situation.

When a single horizon-defining obstacle is present, the elevation of the horizon-defining obstacle is excluded from the determination of effective antenna heights. In order for this to function properly in version 1.2.2, the path profile must utilize an increment that is somewhat finer than 0.1 times the shortest actual horizon distance. Otherwise, the elevation of the horizon-defining obstacle may impact the determination of the effective antenna heights and thus the resultant computation of the median transmission loss relative to the free-space transmission loss. The Longley–Rice model is also relatively sensitive to where horizon-defining obstacles are located. Accordingly, it is imperative that the error-flags built into the program be utilized. The program will return an error flag that indicates that internal parameters are out of range and still go on to return the median transmission loss. When computing mobile system coverage, i.e., coverage over a number of receive points, a sizeable number of situations will occur that are beyond the stated range of applicability of the model and thus give rise to error flags. Dependence upon median transmission loss values without monitoring the value of the built-in error-flag can easily be fraught with peril.

The signal variability predictions are extremely sensitive to the terrain irregularity parameter when terrain irregularity is less than approximately 10 m. Depending upon the accuracy of the digital terrain information utilized, if the terrain is relatively smooth, sizeable variations in signal variability predictions can be expected—variations that are due to digital data base factors, such as whether the elevation information is rounded to the nearest 10 ft or nearest 10 m, rather than due to actual physical situations. Proper use of the model requires that median signal level predictions be compared against field strength measurements of median signal levels. Signal variability corrections should *not* be used as means for modifying model predictions before comparison with measured median signal level data. To use such a method for modifying the model's predictions of median signal level totally defeats the usefulness/applicability of the model.

5) Significant Factors Not Considered: The Longley–Rice prediction model does not contain any provisions for determining corrections due to local environmental factors near the mobile receiver or for predicting the effects of either buildings or foliage. Moreover, it does not contain any provisions for predicting the "short-term" variability (loss deviation) that is due to multipath propagation.

C. NTIA Propagation Models

The National Telecommunications and Information Administration (NTIA) provides propagation analysis techniques in the form of time-sharing computer programs to interested agencies and organizations on a cost-reimbursable basis. As described in NTIA's Telecommunications Analysis Services Reference Guide [8], there are three different propagation analysis programs:

- CSPM—communication system performance model
- RAPIT—radio propagation over irregular terrain
- COVERAGE.

The communication system performance model program is described in Section 10 of this reference as follows:

The CSPM program creates detailed contour plots of basic transmission loss, field strength, power density, etc. from one or more transmitters in a given geographical area. It uses the ITS irregular terrain model (ITM) in the point-to-point mode for determining path loss along radials of about one degree azimuthal intervals around each transmitter. The ITM propagation model is applicable to analyze mobile, broadcast, or radar coverage and interference problems in the 20 MHz to 20 GHz band. ITM is the same propagation model that is used in both COVERAGE (section 6) and RAPIT (section 5) where it is discussed in more detail.

The RAPIT program is described in Section 5 of this reference. There are two models in program RAPIT, the area model and the point-to-point model.

The area predictions model is described in Section 5.3 as follows:

Program RAPIT contains two different models (or algorithms) which are to be used under different circumstances. The first of these is the "Area Prediction" model designed by Longley and Rice (1968), and meant to be used when only moderate information about the propagation paths is known .... A recent ITS report (Hufford, *et al.*, 1982) gives additional guidance on the use of the Area Prediction model.

The point-to-point predictions model is described in Section 5.4 as follows:

The second propagation model is designed to be used on a fixed path where certain gross features of the intervening terrain profile are known. The model is known as the point-to-point mode of the irregular terrain model and is based upon the ESSA '70 model (Longley, *et al.*, 1971).

The COVERAGE program is described in Section 6 of this reference as "a Telecommunications Analysis Services program which predicts the field strength coverage contours and populations within these contours for broadcast or base stations." Section 6.4 states:

Three propagation models are available in COVERAGE. The first model is based on the FCC propagation curves for TV and FM broadcast stations. . . .

The second and third models are based on the "ITS irregular terrain (ITM) model," successor to the Longley-Rice model (Longley and Rice, 1968; Hufford, Longley and Kissick, 1982). The two models are applicable to ground-based systems from 20 MHz to 20 GHz. The second model is commonly referred to as the ITM in the *area* prediction mode and the third as the ITM in the *point-to-point* mode. The difference between the later two models is that the point-to-point model requires detailed terrain information about the particular propagation path involved while the area prediction model requires less information.

The foregoing quotes indicate that NTIA's model is a ersion of the Longley–Rice model. Regrettably, however, NTIA's documentation does not relate their version of the Longley–Rice model to the version numbers defined in [3]. Absent this information, and absent copies of the actual Fortran programs used by NTIA, it is not possible to comment in depth on NTIA's specific implementation of the Longley–Rice model. Longley–Rice methods have been covered extensively in this Section, and no further discussion will be presented. All concerns and findings previously presented in this section about Longley–Rice methods should also apply to the NTIA implementation of this model.

D. The TIREM Model

TIREM is an acronym for terrain integrated rough earth model. It is one of a series of point-to-point propagation models in the Master Propagation System (MPS11) developed by the office of United States Department of Commerce, National Telecommunications and Information Administration, Annapolis, MD [9]. As part of this system, the model is referred to as TIREM11; the computer program is available from NTIS [9]. For propagation problems in the 40 MHz to 20 GHz band where the latitude and longitude of the transmitter and receiver are known, TIREM11 is the preferred model in the master propagation system.

The TIREM11 model is described in chapter 2 of the MPS11 documentation [9]. Unfortunately, this documentation is rather sketchy and does not always agree with the computer program. According to comments in the computer program, the TIREM11 program is based on the propagation model TIREM used at the Electromagnetic Compatibility Analysis Center (ECAC). The computer program is still evolving, but until early in 1987, there were no version numbers output on runs; accordingly, it is very difficult to track the evolution of this program. In January 1987 version number 1.3 appears on TIREM output from a revised version wherein significant changes were made to beyond-the-horizon prediction techniques. All discussions here that refer to the computer code utilize a version obtained from NTIA during the fall of 1986.

1) TIREM11 Propagation Modes: As described in chapter 2 of the MPS11 documentation, TIREM11 computes the median basic transmission loss in a series of steps. First, the terrain profile is examined and a series of basic parameters are determined, e.g., radio horizon distances, effective antenna heights, path angular distances, etc. Using the radio horizon distances, a determination is made whether the given path is within the horizon or beyond it. The ultimate selection of a propagation mode is based upon a number of additional parameters such as Fresnel zone clearance, etc. Chapter 2 of the M. 311 documentation discusses 12 different TIREM11 modes.

Line-of-sight propagation modes: For line-of-sight paths, there are three TIREM11 propagation modes. The selection of which mode to use is based on the minimum ratio along the entire path of the ray clearance above the terrain h to the width

of the first Fresnel zone at that point r as depicted in Section II, fig. 1. The three line-of-sight modes are:

$h/r \geqslant 1.5$ Line-of-sight, free-space mode (mode 8). The ray is well above the terrain and free-space loss is used.

$h/r \leqslant 0.5$ Line-of-sight, rough-earth mode (mode 7). The ray is very near the earth's surface, and an empirical, rough-earth formulation is used to compute the loss.

$0.5 < h/r < 1.5$ Line-of-sight, transition mode (mode 5). A weighted combination of the free-space and rough-earth losses is used.

Beyond line-of-sight propagation modes: For beyond line-of-sight paths, chapter 2 of the MPS11 documentation discusses nine different TIREM11 modes, namely:

- knife-edge diffraction, beyond line-of-sight (mode 2);
- rough-earth diffraction (mode 6);
- effective knife-edge diffraction (mode 1);
- effective knife-edge/rough-earth diffraction (mode 4). This is a weighted combination of effective knife-edge diffraction and rough-earth diffraction;
- tropospheric scatter (mode 9);
- effective double knife-edge (mode 12);
- diffraction-scatter 1 (mode 10). This mode combines diffraction and troposcatter losses;
- diffraction-scatter 2 (mode 11). This mode combines diffraction and troposcatter losses with effective double knife-edge losses;
- diffraction-scatter 3 (mode 13). This mode combines troposcatter and effective double knife-edge losses.

2) TIREM11 Median Transmission Loss Predictions: Since line-of-sight propagation modes are of considerable importance, we will examine in detail exactly how TIREM11 computes the median transmission loss for line-of-sight paths. As previously described, there are three different modes used for line-of-sight paths. For each of these modes, there are two parts to the median transmission loss—the free-space transmission loss and the median transmission loss relative to the free-space transmission loss.

Let us first consider the line-of-sight, rough-earth case (mode 7). Study of the TIREM11 computer code, in conjunction with the TIREM11 documentation and references, shows that in this case TIREM11 computes the median transmission loss relative to the free-space transmission loss utilizing either a version of the Longley–Rice method or an empirical equation described in [5]. Longley and Reasoner developed this expression to fit experimental communication link data and report that it gives excellent agreement with measured values for the 84 line-of-sight paths they studied. This empirical expression for the attenuation in decibels below free space, A_{cs}, is

$$A_{cs} = 9[1 + \exp(-0.01 \Delta h)]$$

$$- 3.5 \log_{10}(\min h_{e1,2}/\lambda) + 0.07d \quad (6)$$

where d is the path length in kilometers, Δh is the terrain roughness in meters, and λ is the wavelength in meters. The terms $h_{el,2}$ are the effective antenna heights in meters of each antenna along the great circle path between the antennas. As detailed in this report, page 51, all of these parameters are to be computed using methods described by Longley and Rice [2] and Rice *et al.* [1].

Study of the computer program shows that TIREM11 does not compute $h_{el,2}$ as defined by these reports, but rather computes each effective antenna height as the maximum of the structural height above terrain and the corrected height above the local average terrain, where the local average is computed over the first ten miles from each antenna (if $d < 10$ mi, the entire distance between the antennas is used). In contrast, Longley and Rice [2] define the effective antenna heights to be elevations above the dominant reflecting plane (or the structural heights, should they be greater), as determined by a least squares fit to terrain elevations that are visible to both antennas over the entire path.

Study of the computer program shows that TIREM11 computes the terrain roughness parameter by determining the interdecile range of terrain heights, above and below a straight line fitted by least squares to elevations above sea level, over the entire path between the transmitter and receiver. It also uses the asymptotic correction for terrain roughness that was developed by Longley and Rice. Thus, TIREM11 uses the same techniques for determining terrain roughness as described by Longley and Rice [2]. It should be noted, however, that in version 1.2.2 of the Longley-Rice program, the determination of effective antenna heights and the terrain roughness parameter are somewhat different; they define a concept they call "range of interest" and use this concept to determine the range over which effective antenna heights and terrain roughness are determined.

The logical branches of the TIREM11 computer program show that the TIREM11 program computes the median transmission loss relative to the free-space transmission loss as follows. If the path distance is greater than 1 km, the effective heights of both the transmitter and the receiver are less than 3000 m, and the frequency greater than 200 MHz, then the median transmission loss relative to the free-space transmission loss is computed from the Longley-Reasoner empirical equation. If the frequency is less than 150 MHz, then the median transmission loss relative to the free-space transmission loss is computed using a version of the Longley-Rice method. Finally, if the frequency is between 150 MHz and 200 MHz, then the median transmission loss relative to the free-space transmission loss is computed as a weighted average of the value determined from the Longley-Reasoner empirical equation and the value determined from the Longley-Rice method.

For line-of-sight paths where $h/r \leq 0.5$ (mode 7), the TIREM11 median transmission loss is the sum of the free-space transmission loss and the value determined above for the transmission loss relative to the free-space transmission loss. For 900 MHz land mobile communication systems, the TIREM11 median transmission loss is *always* determined by the Longley-Reasoner empirical equation. For line-of-sight paths where $0.5 < h/r < 1.5$ (mode 5), a weighted combination of the free-space transmission loss and the rough-earth transmission loss determined above is used. For line-of-sight paths where $h/r \geq 1.5$ (mode 8), only the free-space transmission loss is used.

3) Physical Consequences for 900 MHz Land Mobile Communication Systems: For 900 MHz land mobile communication systems, it is important to understand the consequences of TIREM11's use of the Longley-Reasoner empirical equation and the above separation into three modes based on the parameter h/r. Since 900 MHz land mobile communication systems almost always have $h_{e1} > h_{e2}$, for line-of-sight conditions the base station antenna height is only used to determine h/r and, from this value, the regions where the various modes apply. Within mode 7, where $h/r \leq 0.5$, the median transmission loss is determined solely by the effective height of the mobile receiver, the terrain roughness, and the path length. Changing the base station height, so long as the h/r remains less than 0.5, has *no* effect on the median transmission loss. As a result TIREM11 has a sizeable line-of-sight region wherein changes to the transmitter height have absolutely no effect on the transmission loss, a result that is contrary to basic propagation theory. Mobile propagation measurements, e.g., Okumura [10], show a very definite dependence on the height of the base station antenna.

The Longley-Reasoner equation (6) has very little dependence on path length. Study of this equation shows that the path dependence is contained in only the last term, $0.07 d$, where d is expressed in kilometers. Thus a change in path length from 10 to 20 km, assuming transmitter and receiver heights such that both paths have line-of-sight and $h/r \leq 0.5$, only changes the median transmission loss relative to the free-space transmission loss by an additional 0.7 dB. For the same path change from 10 to 20 km, the free-space transmission loss increases by 6 dB; thus the total median transmission loss increase is 6.7 dB for this path change. This is considerably less than the change predicted by Bullington [11] smooth earth (12 dB) or that measured by Okumura [10].

In order to understand better the Longley-Reasoner empirical equation, it is important to look at the data for the 84 paths they used as the basis for their empirical equation. As shown in Table 7 of their report [5], the minimum effective receive height of these 84 line-of-sight paths ranged from 1.5 to 286 m. Of particular relevance to land mobile communications, however, is the fact that only 10 percent of these 84 line-of-sight paths had effective receive heights less than 5.8 m, whereas 50 percent had effective receive heights greater than 38 m. Moreover, these paths covered a frequency range from 40 MHz to nearly 10 000 MHz. From the cumulative distribution information presented only 20 percent of the paths were in the frequency range from 516 to 1310 MHz. As readily apparent from this information, the primary focus of their work was for point-to-point communication paths. In such cases, the effective transmitter height may at times be less than the effective receive height; thus Longley-Reasoner utilized the minimum of h_1 or h_2. For point-to-point communication paths utilizing transmit and receive antennas wherein either may be closer to the ground, the effective elevation of the antenna closest to the ground can reasonably be expected to be a more important parameter than the effective elevation of the taller antenna. Thus the Longley–Reasoner empirical equation did *not* utilize any statistically significant amount of data relevant to 900 MHz land mobile communication systems. Accordingly, use of this Longley–Reasoner equation for 900 MHz land mobile communication systems is a use far beyond the context for which the equation was developed. One consequence of this out-of-context use is the lack of dependence on the transmitter height noted above.

The fact that the median transmission loss relative to the free-space transmission loss is dependent primarily on the effective receiver height for all cases where $h/r \leqslant 0.5$ is most significant. This means that a significant part of the entire impact of the path geometry is embodied in the determination of the numerical value of the effective height of the receiver. The balance of the impact is through the terrain roughness parameter. Thus, any change in the method of determining the effective height of the receiver can significantly change the results. For example, the change in computation methods from those specified in the original Longley–Reasoner report to those actually used in the TIREM11 program does change the actual results. The impact of these changes on correlations between predictions and actual experimental data is unknown. Also, any modification that assumes a minimum effective receive height other than the actual structural height can have a significant impact on the results, as can the use of approximate conversion factors to convert the receiver height from feet to meters. Furthermore, depending upon the accuracy of the digital terrain information utilized, sizeable variations in the effective height of the receiver can often be due to digital data base factors, such as whether the elevation information is rounded to the nearest 10 ft or the nearest 10 m, rather than due to actual physical situations.

4) Illustrative TIREM11 Results: In order to understand better how the TIREM11 model functions in predicting median transmission loss for 900 MHz mobile communication systems, a series of smooth earth test cases have been run using program code made available by NTIA during the fall of 1986. The program code was adapted so that all test cases predict the median transmission loss relative to the free-space transmission loss. No corrections have been applied for signal variabilities; the prediction of signal variabilities will be discussed separately.

The first set of test cases we will consider are predictions of median transmission loss relative to the free-space transmission loss over smooth earth as a function of path length. Fig. 18 shows results obtained for a transmitting elevation of 500 ft (152 m) and a receiver elevation of 6 ft (1.8 m). Results obtained for a transmitting elevation of 200 ft (61 m) and a receiver elevation of 6 ft (1.8 m) are shown in Fig. 19.

Examination of these two curves shows that each has two sizeable discontinuities. As path length increases, the first discontinuity occurs when the TIREM11 program switches from the mode 7 computation using the Reasoner rough-earth equation to the mode 10 computation of diffraction-scatter. The next discontinuity occurs within the mode 10 diffraction-scatter computations.

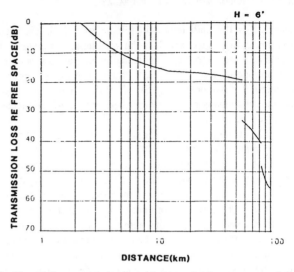

Fig. 18.   Median transmission loss relative to the free-space transmission loss for TIREM over smooth earth at 900 MHz; H_T = 500 ft (152 m); H_R = 6 ft (1.8 m).

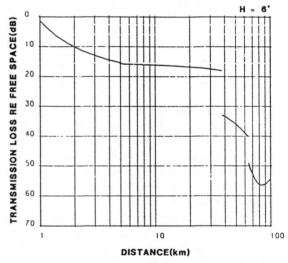

Fig. 19.   Median transmission loss relative to the free-space transmission loss for TIREM over smooth earth at 900 MHz; H_T = 200 ft (61 m); H_R = 6 ft (1.8 m).

The next set of test cases we will consider are predictions of median transmission loss relative to the free-space transmission loss over smooth earth as a function of path length for a series of receive heights. Fig. 20 shows results obtained for a transmitting elevation of 500 ft (152 m) and receiver elevations of 4, 6, 10, and 30 ft. Similar results obtained for a transmitting elevation of 200 ft (61 m) are shown in Fig. 21. The curves in these two figures also contain the two sizeable discontinuities found in the previous two figures.

Examination of the computer program shows that there is no attempt to "blend" the line-of-sight results into the beyond line-of-sight results. Accordingly, a discontinuity can be expected when the program switches from mode 7 to mode 10. In order to rule out program problems, the same smooth earth test cases were run by the NTIA program authors and the same two discontinuities were found. Test runs made by the NTIA

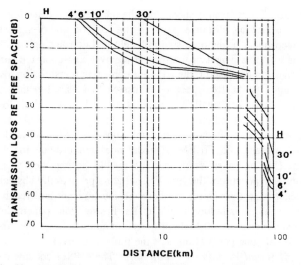

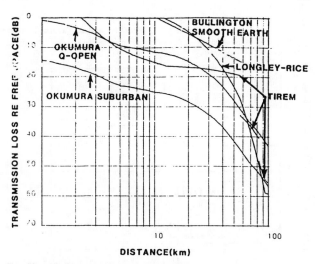

Fig. 20. Median transmission loss relative to the free-space transmission loss for TIREM over smooth earth.at 900 MHz; H_T = 500 ft (152 m); H_R = 4, 6, 10, and 30 ft.

Fig. 22. Median transmission loss relative to the free-space transmission loss for TIREM, Longley-Rice, Bullington smooth earth, Okumura suburban, and Okumura quasi-open at 900 MHz; H_T = 500 ft (152 m), H_R = 6 ft (1.8 m).

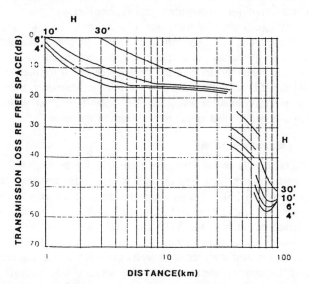

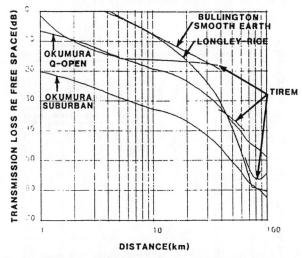

Fig. 21. Median transmission loss relative to the free-space transmission loss for TIREM over smooth earth at 900 MHz; H_T = 200 ft (61 m); H_R = 4, 6, 10, and 30 ft.

Fig. 23. Median transmission loss relative to the free-space transmission loss for TIREM, Longley-Rice, Bullington smooth earth, Okumura suburban, and Okumura quasi-open at 900 MHz; H_T = 200 ft (61 m), H_R = 6 ft (1.8 m).

program authors on TIREM11 version number 1.3 showed the same discontinuity at the point of transition to the beyond line-of-sight region. Results in the beyond line-of-sight region, however, were somewhat different. Rather than the second discontinuity that existed previously, a series of oscillation-like discontinuities were found that started at the point of transition to the beyond line-of-sight region.

The results of Figs. 18 and 19 can be directly compared with the same test results obtained for Okumura [10], Longley-Rice, and Bullington [11]. For easy comparison, the results for a 6 ft mobile receive antenna are shown in Figs. 22 and 23. Examination of these figures shows that TIREM11 predicts a sizeable region wherein the median transmission loss relative to the free-space transmission loss is nearly constant, followed by the two discontinuities discussed above.

The second set of test cases we will consider are TIREM11

predictions of median transmission loss relative to the free-space transmission loss over smooth earth as single knife-edge obstacles are introduced into specific paths. The results give insight into how TIREM11 handles the introduction of single knife-edge obstacles into the Fresnel zone. Fig. 24 shows results obtained for a transmitting elevation of 500 ft (152 m), a receiver elevation of 6 ft (1.8 m), and a path distance of 12 km. A knife-edge obstacle is introduced at a distance of 6 km from the receiver (the midpoint of the path) and the height of the knife-edge obstacle is increased from 0 to 80 m. This figure shows that there is a 6.2 dB discontinuity in the median transmission loss at the precise point where the TIREM11 model changes from the line-to-sight mode to the beyond line-of-sight diffraction/scatter mode. This occurs at the point where the knife-edge obstacle intersects the central ray connecting the transmit and receive antennas.

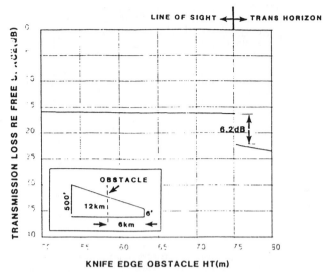

Fig. 24.   Median transmission loss relative to the free-space transmission loss for TIREM over a 12 km smooth earth path as a function of the height of a knife-edge obstacle located at 6 km (the midpoint of the path) from a 6 ft (1.8 m) receiver at 900 MHz; H_T = 500 ft (152 m).

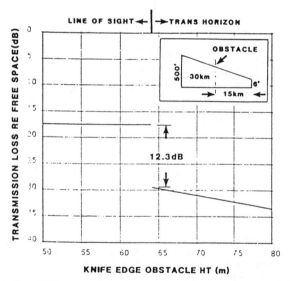

Fig. 25.   Median transmission loss relative to the free-space transmission loss for TIREM over a 30 km smooth earth path as a function of the height of a knife-edge obstacle located at 15 km (the midpoint of the path) from a 6 ft (1.8 m) receiver at 900 MHz; H_T = 500 ft (152 m).

Fig. 25 shows results obtained for the same antenna heights, but for a path distance of 30 km. A knife-edge obstacle is introduced at a distance of 15 km from the receiver (the midpoint of the path) and the height of the knife-edge obstacle is increased from 0 to 80 m. Now there is a 12.3 dB discontinuity at the point where the knife-edge obstacle inters^ ts the central ray connecting the transmit and receive antennas.

As is readily apparent from the preceding results, the TIREM11 model provides a discontinuous transition from the line-of-sight region to the beyond line-of-sight region *both* when the transition is caused by the curvature of the earth and when the transition is caused by single knife-edge obstacles.

Study of these results shows that the median transmission loss for a sizeable distance prior to the loss of line-of-sight is independent of the base station height and depends only upon the effective height of the mobile receiver and upon the path length. In this region, the increase in median transmission loss relative to the free-space transmission loss as a function of distance is very small. At the point where line-of-sight is lost due to the curvature of the earth, there is a sizeable and discontinuous increase in transmission loss. When obstacles are present, the TIREM11 model does not provide for the increase in transmission loss that would be expected as an obstacle penetrates the first Fresnel zone prior to the obstacle penetrating the central ray between the transmit and receive antennas. This failure to utilize first Fresnel zone penetration is significant as 900 MHz propagation almost always has first Fresnel zone penetration near the mobile receiver.

5) Signal Variability Predictions: The MPS11 documentation does not discuss the computation of signal variabilities. Also, the computer code of the MPS11 Master Propagation System obtained from NTIA in the fall of 1986 does not contain routines to compute signal variabilities.

6) Significant Factors not Considered: The MPS11 documentation and associated computer programs do not contain any provisions for determining corrections due to local environmental factors near the mobile receiver or for predicting the effects of either buildings or foliage. Moreover, there are no provisions for predicting the "short-term" variability (loss deviation) that is due to multipath propagation.

E. Okumura's Model

Full details of the Okumura technique for predicting the median transmission loss of land mobile radio systems are presented in [10]. Okumura's paper presents the results of extensive propagation tests for land mobile radio service carried out in both the VHF and UHF bands over various situations of irregular terrain and environmental clutter. The results are then analyzed statistically to determine the distance and frequency dependences of the median field strength, location variabilities, and antenna height gain factors for the base and vehicular station in urban, suburban, quasi-open, and open areas over quasi-smooth terrain. Path-specific correction factors are also developed to consider irregular terrain, such as rolling hills, isolated mountains, general sloped terrain, and mixed land-sea paths. Okumura techniques are applicable for distances of 1 to 100 km and for base station effective antenna heights of 30 to 1000 m. Because of the way the Okumura propagation results were derived and analyzed, a number of significant points must be remembered when utilizing the Okumura procedures to predict median transmission losses.

First, it is important to recognize that Okumura techniques are based on detailed propagation measurements for land-mobile radio systems; they do *not* utilize theoretical models and/or theoretical predictions concerning how radio waves should behave.

Second, Okumura techniques provide the basis for making not only area-to-area predictions, but also path-specific predictions. Many applications of Okumura overlook the path-specific corrections. In order for comparisons of Okumura with other point-to-point models to be valid, all applicable path-specific corrections must be applied before the comparisons are made. This is particularly true for those who use Hata's empirical formulas [12]. Hata's empirical formulas were developed from Okumura's in order to make the use of Okumura procedures simpler (it is much easier to compute median transmission losses from equations rather than by reading curves). With computation simplicity, however, also come some restrictions. Most notable, Hata's equations are severely limited in the range of applicability—they consider only the frequency range 100–1500 MHz, distance 1–20 km, base station antenna height 30–200 m, and vehicular antenna height 1–10 m. In addition, Hata's empirical formulas do *not* consider any of the path-specific corrections.

Third, Okumura techniques adopt as their "standard" propagation curves the propagation curves for the urban area environment. This was done *not* because the most typical propagation situation is urban, but because of computation considerations and the fact that the greatest degree of accuracy could be obtained if all other results were based on the use of the urban propagation curves as the "standard." The most typical situation in the United States is far from the urban situation.

Fourth, caution must be used when applying the Okumura definitions of environmental types to United States situations. Okumura's measurements are valid only for the building types found in Tokyo. Experience with comparable measurements in the United States has shown that the "typical" United States suburban situation is often somewhere between Okumura's suburban and open areas. Okumura's suburban definition is more representative of residential metropolitan areas with large groups of "row" houses.

The CCIR Report 567-3 (1986) has adopted the Okumura urban curve as its basic curve for land mobile 900 MHz propagation. Accordingly, all of the above cautions apply when using these curves to predict median transmission loss.

1) Path-Specific Corrections: Okumura's path-specific corrections are not easily implemented for use on digital computers. A number of assumptions need to be made. One such implementation has been done by Allsebrook and Parsons [13]. They considered and introduced a set of definitions applicable to computer codes for the following quantities:

- effective base station antenna height;
- rolling hilly terrain correction;
- vehicular ground slope correction.

It should be noted that the assumptions and definitions in their computer implementation are not necessarily unique; other definitions could be used and the predictions would differ accordingly.

Okumura also includes path-specific corrections for the following additional effects:

- isolated mountain ridge (shadow loss);
- mixed land-sea paths;
- orientation of urban streets relative to transmitter location;
- effects of vehicular antenna height gain.

In order to use these path-specific corrections on digital computers, a number of assumptions must necessarily be made and the resultant predictions would differ accordingly.

The European Broadcasting Union has developed a very useful empirical way of applying path-specific corrections to median field strength prediction. This technique is described in the CCIR Study Group Recommendation 370-3, dated February 20, 1980. This method was adopted by CCIR and appears in the 1982 version of CCIR Report 239-5 (sec. 4.2, pp. 237–238). It defines and uses a "terrain clearance angle" together with an empirically developed correction factor to "correct" the median field strength prediction for the specific path under consideration. The terrain clearance angle is defined as the angle measured between the horizontal at the receiving antenna and the line which clears all obstacles within 16 km in the direction of the transmitter. This technique considers both negative angles (when the line to the obstacles is above the horizontal) and positive angles (when the line to the obstacles is below the horizontal). The empirical correction curve utilized by this method was developed from measurements of over 200 paths in the Federal Republic of Germany, Finland, France, and the United Kingdom.

2) Okumura Results re Theoretical Models: Walfisch and Bertoni [14] and Ikegami *et al.* [16] have developed models for predicting how buildings influence the signal at vehicular receivers. They both view 900 MHz land mobile propagation as taking place over the tops of the buildings, with diffraction at the rooftops being the mechanism by which the fields reach street level. The diffraction loss for such a process depends primarily on building height and the separation between the rows of buildings and secondarily on the shape and construction materials of the roof and upper building walls.

The model of Walfisch and Bertoni [14] is discussed in detail in Appendix V. It considers the path loss to consist of three parts:

- the path loss between antennas in free space;
- reduction of the fields above the roof tops due to forward diffraction past many rows of buildings;
- diffraction of the roof top fields down to ground level.

This theoretical model proposes a wave propagation mechanism to explain how the presence of buildings influences the distance dependence of the median transmission loss. The model applies only within the radio horizon of an elevated base station antenna. Within the radio horizon, the measurements of Okumura [10] show a distance dependence that varies with base station antenna height within the range of about 31 to 35 dB per decade [10], [12]. The theoretical model also exhibits a variation of range dependence with antenna height, but the variation over typical ranges will be smaller and centered about 38 dB per decade. Measurements made in U.S. and English cities on flat terrain [13], [15] more typically suggest range dependence between 35 and 40 dB per decade.

In the Okumura model, the buildings in the vicinity of the mobile are accounted for via a correction factor that corrects the basic urban model to suburban, quasi-open, and open environments. This factor can be over 20 dB, and there is currently no objective way of selecting its value based on the measurable properties of the buildings near the mobile. Moreover, it is not clear how the correction factors obtained from measurements in Tokyo might apply in other cities with different building styles and construction materials. Okumura's classification of environmental areas is discrete. Moreover, it applies only to the location of the mobile unit. The path-loss slope is independent of the type of environment, except in those few close-in situations wherein the environmental correction would result in a path-loss less than the free-space loss. Several authors [17], [18] have expanded this concept to include a continuum of values through the introduction of a ground occupancy density parameter.

Because the theoretical model of Walfisch and Bertoni and the work of Ikagami et al. [16] give a mechanism by which the buildings influence the signal at the mobile, it may be possible to use them to predict the environmental correction for the Okumura model from the physical properties of the buildings. Their mechanism makes use of building height and the separation between rows of buildings as the key parameters for determining the environmental correction factor. These parameters are significantly different from those used for a ground occupancy density factor [17], [18].

3) Comparison of Path-Specific Prediction Methods: Several recent articles have been published that compare land mobile path-specific prediction techniques. Aurand and Post [19] published a study that gives a practical comparison of 12 prediction techniques with an emphasis on three essential criteria to consider in the selection of a model: 1) type of terrain or geography covered by each method; 2) the form of prediction provided (transmission loss estimates, field strength contour maps, etc.); and 3) the implementation difficulty and degree of sophistication. This study compares the features of Okumura techniques to the 11 other models. They conclude that the typical accuracy that can be expected from various methods is determined primarily by the degree to which a method accounts for terrain characteristics. This study, however, does not compare predictions with measured data on specific paths.

Delisle et al. [20] published a study that compared Okumura/Hata predictions with those of four other models. In addition, they made comparisons with experimental data they took in Ottawa at 910 MHz from a base transmitting antenna at 33.5 m AGL and a mobile antenna at 3.8 m AGL. They found general agreement between the various predictions provided that they were compared "in the conditions where they are applicable and provided they are corrected for specific conditions—such as losses over hilly terrain or due to buildings or vegetation."

Paunovic et al. [21] measured actual field strengths over a series of well defined path profiles over irregular terrain and compared them with the predictions of various models for VHF frequencies. This study also considered the "clearance angle" method developed by the European Broadcasting Union. Based on a comparison of predicted and experimentally determined field strength levels along different path profiles over irregular terrain, they found that the clearance angle method yielded the most reliable prediction values. They also concluded that the clearance angle method was very

operative, not complex, deprived of subjective factors, and thus was exceptionally well suited to computer applications. Accordingly, they chose the clearance angle method as the most suitable method for engineering land mobile communications systems.

4) Applicability: Okumura methods are uniquely different than most other propagation prediction techniques for land mobile communication systems. They are based on measurement data obtained from actual land mobile communication systems rather than on abstract theoretical models. Their use, however, requires that considerable engineering judgement be used, particularly in the selection of the appropriate environmental factor. More research is needed in order to be able to predict the environmental factor from the physical properties of the buildings surrounding a mobile receiver. In addition to the appropriate environmental factor, path-specific corrections must also be applied to convert Okumura median transmission loss predictions to predictions that apply to the specific path under study. Okumura's techniques for correcting for irregular terrain and other path-specific features require engineering interpretations and are thus not readily adaptable for computer use. In contrast, the European Broadcasting Union's Clearance Angle approach is a relatively simple approach for correcting for irregular terrain features; it is easily adaptable for computer use. Accordingly, clearance angle techniques provide a most useful way of determining the path-specific corrections required to make path-specific Okumura median transmission loss predictions.

F. Lee's Model

Lee's model is described in detail in [22, ch. 2], and is also summarized in Appendix VI of this report. Review of these documents shows that this model is applicable only for mobile receivers when they are located within the radio horizon of the base station antenna. This model does *not* consider the additional losses that are encountered as the mobile unit approaches the radio horizon of the base station antenna. Accordingly, the various curves and examples presented in Lee's book consider only distances out to 10 mi.

Lee's model has two components. The first component is an area-to-area path-loss prediction model. The second component uses the area-to-area path-loss prediction as a base and develops a point-to-point prediction therefrom. Three parameters are needed to make an area-to-area path-loss prediction:

- the power at the 1-mi point;
- the path-loss slope;
- an adjustment factor.

The power at the 1-mi point is the way Lee's model takes into account the environment of the transmitting antenna. Lee's model presumes that this parameter is determined empirically and no procedures are given for calculating it. He does give some representative values for different cities.

The path-loss slope is the way Lee's model takes into account the characteristics of the environment over which the signal is propagating. Typical ranges are from 30 dB/dec to 40 dB/dec, and he gives representative values to aid in the selection of this value. The path-loss slope is usually presumed constant over the entire propagation path, but procedures are given for treating changes in the path characteristics through changes in the path-loss slope at specific points along the path. Lee states that Okumura found 30 dB/dec. Close study of Okumura's paper [10], however, shows that this is an approximation. Okumura found 30 dB/dec up to distances of approximately 15 km; at greater distances, this value increased significantly. Okumura (fig. 12 of Okumura's paper) also found a dependence upon the effective height of the base station. For a 30 m antenna height he found approximately 34 dB/dec, while for a 250 m antenna height he found approximately 30 dB/dec, and for a 700 m antenna height he found approximately 26 dB/dec. It is important to note that the Okumura model has the same path-loss slope for all environments, urban, suburban, etc., provided the path-loss exceeds free-space, whereas Lee's model takes this parameter to be a characteristic of the environment.

The adjustment factor is the way Lee's model accounts for a number of different factors—specifically:

- base-station antenna height (standard given was 30 m or 100 ft);
- mobile-unit antenna height (standard given was 3 m or 10 ft);
- transmitter power (standard given was 10 W);
- base station antenna gain (standard given was 6 dB above dipole gain);
- mobile-unit antenna gain (standard given was 0 dB above dipole gain).

The area-to-area prediction gives a median path loss, or signal level, over flat terrain in general. In order to apply this to specific paths, corrections for the nature of the actual path must be made. Lee's model considers two specific types of situations: 1) nonobstructed paths; and 2) obstructed paths. Nonobstructed paths are defined to be those paths within the radio horizon that have no obstacles between the base station and the mobile unit. Lee's model presents a technique for correcting the area-to-area prediction by an effective base station antenna height that is determined by using the slope of the terrain in the vicinity of the mobile unit. This technique is relatively easy to apply to specific paths when they are being analyzed by engineers. However, it is not readily amenable for computer analysis. Any realistic implementation of this concept as a general computer algorithm requires a number of assumptions to be made so that the resultant computer algorithm can realistically analyze all possible types of terrain situations that can be encountered.

Obstructed paths are defined to be those paths that have a single terrain feature that obstructs the line-of-sight from the base station antenna to the mobile unit. Lee's model utilizes knife-edge diffraction to compute the shadow loss from the obstacle, which is computed as if the basic propagation were free space. No corrections are made to take into account the reflections that occur in the foreground of both the base antenna and the mobile unit. Bullington has presented techniques for computing shadow losses that consider the reflections that occur when the antennas are relatively close to the ground; Lee's model does not utilize these techniques.

Comparison with Okumura: Both Lee's model and Okumura use empirical data to predict signal strengths. Lee's model is restricted to distances well within the radio horizon of the base station antenna whereas Okumura applies out to distances of 100 km.

Lee's model explicitly considers the effects of the slope of the terrain in the vicinity of the mobile unit and also the effects of terrain obstacles. Okumura also has provisions for the treatment of a series of path-specific corrections, e.g., slope of the terrain in the vicinity of the mobile, mixed land-sea paths, isolated terrain obstacles. In order for a proper comparison to be made, it is necessary for all appropriate path-specific corrections to be made to both techniques. Okumura without path-specific corrections must never be compared to Lee's model with its path-specific corrections.

Both Lee's model and Okumura consider the effects of the environment, but there is a very important difference in the way they do so. Lee's model considers the effect of the environment surrounding the base station antenna in the determination of the power at the 1-mi point. The environment along the receive path is treated by the path-loss slope. Changes in the environment along the receive path are treated by introducing changes in the path-loss slope. It must be recognized, however, that in Lee's model the overall impact of environmental changes along the path are relatively small. The model applies only over distances up to approximately 10 mi— 1 decade from the 1-mi point. The changes in path-loss slope as a function of environment are relatively small (usually several decibels). Okumura's results, on the other hand, show no change in path-loss slope as a function of environment. Lee's model does not consider the effect of the environment in which the mobile unit is actually located except through changes in the path-loss slope. Thus, Lee's model does not predict significant changes in signal strength as the environment surrounding the mobile unit changes; the environment of the base station is the primary controlling factor.

In contrast, Okumura utilizes the environment of the mobile unit to predict signal levels. Okumura has no provision for treating the environment surrounding the base station antenna. Also, there is no change in the path-loss slope as a function of environment. All environmental effects are determined solely by the environment surrounding the mobile unit. Thus, Okumura predicts significant changes in signal strength as the environment surrounding the mobile unit changes.

This difference in treatment of the environment between Lee's model and Okumura is significant only when the mobile unit is in an environment significantly different than the base station antenna. When the base station antenna and the mobile unit are in the same type of environment, this difference essentially disappears.

REFERENCES

[1] P. L. Rice, A. G. Longley, K. A. Norton, and A. P. Barsis, "Transmission loss predictions for tropospheric communication circuits," NBS Tech. Note 101; two volumes; issued May 7, 1965; revised May 1, 1966; revised Jan, 1967.
[2] A. G. Longley and P. L. Rice, "Prediction of tropospheric radio transmission loss over irregular terrain; A computer method—1968," ESSA Tech. Rep. ERL 79-ITS 67.
[3] G. A. Hufford, A. G. Longley, and W. A. Kissick, "A guide to the use

of the ITS irregular terrain model in the area prediction mode," NTIA Rep. 82-100, Apr. 1982.

[4] G. A. Hufford, "Memorandum to users of the ITS irregular terrain model," Jan. 30, 1985.

[5] A. G. Longley and P. K. Reasoner, "Comparison of propagation measurements with predicted values in the 20 to 10,000 MHz range," ESSA Tech. Rep. ERL 148-ITS 97, Jan. 1970.

[6] A. G. Longley, "Radio propagation in urban areas," OT Rep. 78-144, Apr. 1978.

[7] ——, "Local variability of transmission loss—land mobile and broadcast systems," OT Rep., May 1976.

[8] *Telecommunications Analysis Services; Reference Guide*, NTIA.

[9] *Master Propagation System (MPS11); User's Manual.* (User's Manual available from NTIS under No. NTIS-PB83-178624. Computer program tape available under No. NTIS-PB83-173971.)

[10] Y. Okumura, E. Ohmori, T. Kawano, and K. Fukuda, "Field strength and its variability in VHF and UHF land-mobile radio service," *Rev. Elec. Commun. Lab.*, vol. 16, pp. 825–873, 1968.

[11] K. Bullington, "Radio propagation variations at VHF and UHF," *Proc. IRE*, vol. 38, pp. 27–32, 1950.

[12] M. Hata, "Empirical formula for propagation loss in mobile radio services," *IEEE Trans. Veh. Technol.*, vol. VT-29, pp. 317–325, 1980.

[13] K. Allsebrook and J. D. Parsons, "Mobile radio propagation in British cities at frequencies in the VHF and UHF bands," *IEEE Trans. Veh. Technol.*, vol. VT-26, pp. 313–322, 1977.

[14] J. Walfisch and H. L. Bertoni, "A theoretical model of UHF propagation in urban environments," *IEEE Trans. Antennas Propagat.*, to be published.

[15] G. D. Ott and A. Plitkins, "Urban path-loss characteristics at 820 MHz," *IEEE Trans. Veh. Technol.*, vol. VT-27, pp. 189–197, 1978.

[16] F. Ikegami, S. Yoshida, T. Takeuchi, and M. Umehira, "Propagation factors controlling mean field strength on urban streets," *IEEE Trans. Antennas Propagat.*, vol. AP-32, pp. 822–829, 1984.

[17] A. Akeyama, T. Nagatsu, Y. Ebine, "Mobile radio propagation characteristics and radio zone design method in local cities," *Rev. Elec. Commun. Lab.*, vol. 30, pp. 308–317, 1982.

[18] S. Kozono and K. Watanabe, "Influence of environmental buildings on UHF land mobile radio propagation," *IEEE Trans. Commun.*, vol. COM-25, pp. 1133–1143, 1977. (Correction: IEEE Trans. Commun., vol. com-26, pp. 199–200, 1978.)

[19] J. F. Aurand and R. E. Post, "A comparison of prediction methods for 800 MHz mobile radio propagation," *IEEE Trans. Veh. Technol.*, vol. VT-34, pp. 149–153, 1985.

[20] G. Delisle, J. P. Lefevre, M. Lecours, J. Y. Chouinard, "Propagation loss prediction: A comparative study with application to the mobile radio channel," *IEEE Trans. Veh. Technol.*, vol. VT-34, pp. 86–96, 1985.

[21] D. S. Paunovic, Z. D. Stojanovic, and I. S. Stojanovic, "Choice of a suitable method for the prediction of the field strength in planning land mobile radio systems," *IEEE Trans. Veh. Technol.*, vol. VT-33, pp. 259–265, 1984.

[22] W. C. Y. Lee, *Mobile Communications Design Fundamentals.* Indianapolis, IN: Howard W. Sams and Co., 1986.

Some of these references are reprinted in *Land-Mobile Communications Engineering*, D. Bodson, G. F. McClure, and S. R. McConoughey, Eds. New York: IEEE Press, 1984. The table of contents of this book is listed in Appendix I.

Indoor Radio Communications for Factories of the Future

Theodore S. Rappaport

Reprinted from *IEEE Communications Magazine*, pp. 15-24, May, 1989.

THE BOOM IN FACTORY AUTOMATION HAS created a need for reliable real-time communications. In 1984, the Manufacturing Automation Protocol (MAP) networking standard was established by industrial leaders to encourage commercialization and standardization of high data rate communications hardware for use in computer-controlled manufacturing [1]. In late 1985, the Technical and Office Protocol (TOP) was developed for computer communications in office buildings [2]. Both MAP and TOP are capable of supporting 10 Mb/s data rates for short periods of time and rely on coaxial cable or fiber optic cable to interconnect users. Twisted-pair interconnection of computer terminals is also commonly used as a communications channel in modern factories.

Methods for transporting parts-in-process in a futuristic, but realistic, JIT manufacturing environment is one of the thrust research areas at the ERC for Intelligent Manufacturing Systems.

To address concerns about the international competitiveness of the United States, the National Science Foundation (NSF) established large interdisciplinary Engineering Research Centers (ERCs) at ten universities in 1985. The charter of these ERCs is to conduct pioneering research aimed at vastly improving some of the manufacturing technologies and methodologies currently used in U.S. industry [3] [4]. Work at the ERC for Intelligent Manufacturing Systems at Purdue University revealed that Just-In-Time (JIT) manufacturing techniques can result in cost-savings of several orders of magnitude when applied to small and medium batch manufacturing processes. Industries that produce small-quantities of light-weight (< 25 kg) metal parts, medium-sized (< 1,000 units) production runs of electronic parts, and durable goods are particularly well-suited to JIT techniques [5].

Methods for transporting parts-in-process in a futuristic, but realistic, JIT manufacturing environment is one of the thrust research areas at the ERC for Intelligent Manufacturing Systems. Analysis has shown that an inexpensive, agile, mobile robot fleet, capable of navigating without any type of track, could easily accommodate a majority of the material flow required for a JIT manufacturing system. Navigational techniques and corresponding navigational error analyses have been conducted for a variety of autonomous guided, mobile robot systems [6–8].

A truly Autonomous Guided Vehicle (AGV) that does not use a tether requires a radio system for control. Optical systems are viable, but become inoperable when obstructed. Furthermore, radio systems are useful for providing fast and inexpensive connections for often-moved manufacturing equipment and computer terminals. Radio will also accommodate reconfigurable voice/data communications for other facets of factory operation. For example, RF tags, which can be easily attached to parts-in-process, provide a paperless data storage medium and are becoming an increasingly popular replacement for paper work order cards [9]. Eventually, radio communications may be used in homes and offices to provide universal, digital, portable communications [10]. Leading telecommunications firms, such as Bell Communications Research, AT&T Bell Laboratories, and Motorola are now exploring the viability of indoor radio communication systems for homes and offices [10–14].

Narrow-band (RF bandwidths < 25 kHz) VHF digital radio systems are presently marketed by a number of companies and are beginning to find use in data communications for inventory management (bar code readers, shelving dispatch) and dedicated control (for overhead cranes, wireguided vehicles, etc.). Most of these systems employ binary Frequency Shift Keying (FSK), a form of FM modulation. By retrofitting commercially available FM transceivers, many factory radio system vendors provide digital radios by using Audio Frequency Shift Keying (AFSK). In AFSK transmission, a data bit is mapped into a particular baseband tone having a duration equal to the reciprocal of the bit rate. There are two tones for each of the two possible bit values. The resulting baseband analog audio signal is then applied to the input of a standard FM transmitter. At the receiver, bandpass filters or tone decoders, which follow the FM demodulator, are sampled to determine which audio frequency (i.e., bit) was sent. Although AFSK is not as efficient as other forms of FSK, it is very simple to install on readily available analog FM transceivers since only audio circuitry must be added. Some factory radio system vendors use true FSK or Minimum Shift Keying (MSK), which is the most spectrally efficient form of binary FSK [15, p. 326].

Voice radio systems, which are presently used for paging and personnel dispatching inside buildings, employ FM at VHF and UHF frequencies. In many applications, analog FM transceivers are used in factory environments that contain both electromagnetic and acoustic noise. In the metal-working and textile industries particularly, operators in noisy environments use headphones that contain a microphone element. The ear-mounted microphone detects aural vibrations that pass through the eardrum and jawbone, while background noise is shielded from the operator and the microphone by the headset. At the receiving base station, signal processing is performed on the received speech to filter baseband background noise and to undo distortion induced by the throat/ear channel.

With modern speech coding algorithms, it is possible to greatly enhance voice communications in noisy, multipath radio channels. Several vendors have recently demonstrated powerful speech coding algorithms, which code speech at rates of between 6.5 kb/s and 13 kb/s, for next-generation U.S. digital cellular telephones. The Jet Propulsion Laboratory, in conjunction with the University of California and the Georgia Institute of Technology, have shown the viability of 4.8 kb/s speech codecs for satellite voice channels. Advancements in speech coding are discussed in the February 1988 issue of the *IEEE Journal on Selected Areas in Communications*. Information on research in mobile satellite systems is available in the *MSAT-X Quarterly,* which is published by the Jet Propulsion Laboratory, Pasadena, California.

It is important to note that voice systems can tolerate much greater bit error rates than data systems. Where an uncoded data transmission system may require bit error rates on the order of 10^{-5}, speech communications can tolerate bit error rates on the order of 10^{-2}. This is due to inherent redundancies in human speech and the brain's ability to interpret language.

While existing indoor factory radio systems are well suited for human operation and simple digital communication tasks, it is anticipated that for a moderately sized AGV fleet (> 20 vehicles) employing multiple-access digital radio communications within a Computer Integrated Manufacturing (CIM) environment, data rates of several hundred kilobits per second will be needed to accommodate real-time computer control and navigation of the vehicles.

In the U.S., the Federal Communications Commission (FCC) has allocated spectrum for narrow-band industrial radio communications in the VHF (450 MHz) and UHF (900 MHz) bands. More recently, the FCC authorized the use of suitably designed spread-spectrum systems for 900 MHz, 2,400 MHz, and 5,725 MHz [16]. Provided that transmitters meet FCC approval, unlicensed 1 W transmitter power levels may be used over bandwidths greater than 25 MHz. In Japan, spectrum has been set aside for 300 mW, 4,800 b/s indoor radio systems operating in the 400 MHz and 2,450 MHz bands [17]. Federal regulatory agencies have recently recognized the potential of wideband radio Local Area Networks (LANs), and radio propagation inside factories and office buildings is becoming increasingly understood. However, the current method for installing factory radio links requires lengthy surveys and often involves redundant equipment. Robust signalling techniques and system designs for wideband indoor radio systems do not currently exist, although this is an active area of research [i.e., 18–20].

Accurate characterization of the operating channels is a mandatory prerequisite for the development of reliable indoor radio systems. Radio channel propagation data from factory buildings have recently been made available through research at the ERC for Intelligent Manufacturing Systems, and a statistical impulse response channel model, based upon the details being developed at Virginia Tech. As is shown in this article, it is not environmental noise, but rather multipath propagation that limits the capacity of radio links operating above 1 GHz. The severity of multipath largely depends on factory inventory, building structure, and surrounding topography.

Factory Noise

Although much of the radio noise encountered in factories arises from weakly emitting sources, measurements have revealed that some types of industrial equipment produce harmonic RF energy and can radiate substantial noise (up to several hundred megahertz) [21]. Equipment such as RF-stabilized arc welders, induction heaters, and plastic bonders are acute sources of noise. Although interference is significant at HF and VHF, noise signatures of such equipment fall off rapidly above 1 GHz [21]. This trend has also been found in urban mobile-radio channels [22, p. 297]. Recent measurements have confirmed that typical machine-generated noise levels in operational factories are much less severe at higher frequencies [23]. Figure 1 shows results of peak noise power spectrum measurements (measured 4 m from engine cylinder machining line [23]) made along an engine manufacturing transfer line in full operation. When compared with the VHF spectrum, worst-case noise levels are 40 dB lower at UHF/microwave frequencies and are only a few dB above the thermal noise floor of the spectrum analyzer receiver. Although a careful study of impulsive noise has not been conducted, these results are encouraging and indicate machine-generated noise will not severely hamper most factory radio systems operating at UHF and above.

Multipath Propagation

Due to the large metal content of a factory, multipath interference is created by multiple reflections of the transmitted signal from the building structure and surrounding inventory. The resultant received waveform is a sum of time and frequency shifted versions of the original transmission, and, depending on parameters of the signal and the channel, the received signal may be greatly distorted.

Multipath has historically been identified as the most important factor limiting mobile and portable radio communication systems [22] [24]. For narrow-band systems (where the baseband digital symbol duration is several times greater than the extent of the multipath-induced propagation delays, i.e., flat-fading conditions), multipath causes large fluctuations (fading) in the received signal voltage due to the changing phasor sum of signal components arriving at the receiver antenna via different paths. Temporal variations of the channel as well as changing multipath geometry seen by a mobile user are the mechanisms for the fading. Additional signal loss will occur when an AGV or portable user is shadowed by inventory and equipment. In wideband systems, the scatterers create intersymbol interference and cause the channel to be frequency selective. Consequently, the maximum data rate supported by a multipath channel is limited. Typical flat-fading channels require 30 dB more transmitter power to achieve low bit-error rates (10^{-4}) than do systems operating over ideal channels [25].

In order to determine radio propagation characteristics inside factory buildings, radio wave propagation experiments at 1,300 MHz were conducted by the author in five operational factories in the spring and summer of 1987 [26]. Over 30,000 narrow-band fading measurements and 950 wideband impulse response measurements were made in a diverse collection of

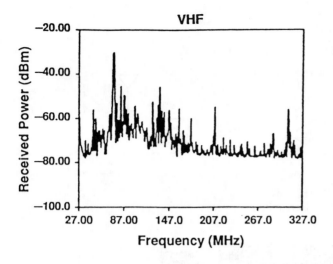

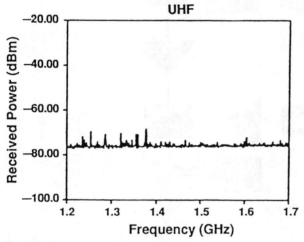

Fig. 1. Peak noise spectrum measurements.

industries and building structures. Factories that participated in the research included engine and automobile parts manufacturers and dry-foods producers. Detailed descriptions of the experiments and the measurement equipment are given in [26] [32] [40]. Briefly, four distinct topographical scenarios were identified in each of the five factories. These ranged from Line-Of-Sight (LOS) transmission paths along lightly cluttered aisles to heavily cluttered obstructed paths between adjacent aisles. In each topography, three measurement locations were selected having graduated transmitter-receiver separations between 10 and 80 m. The transmitter was positioned (in the clear) in the center of the particular topography while the receiver was moved along a 1 m track at each measurement location. Wideband channel impulse responses were measured in the time domain by repetitively transmitting a 10 ns pulse (7.8 ns rms duration) and receiving on a digital storage oscilloscope the attenuated, distorted, and delayed versions of the pulse.

The factory channel measurement apparatus consisted of a periodic pulse-modulated transmitter with a peak output power of 1 W. The receiver consisted of a low-noise amplifier followed by a square law envelope detector and a 350 MHz digital storage oscilloscope. A directional coupler allowed CW envelope measurements to be made simultaneously by a modified communications receiver. The measurement system block diagram is shown in Figure 2. Wideband discone antennas [28]

were used at both transmitter and receiver and were mounted at heights of 2 m above the floor. Channel measurement results obtained using this apparatus are given subsequently.

The term "path loss" is used to describe the relative power attenuation (in dB) seen by a receiver with respect to a convenient (close) reference distance between transmitter and receiver. In HF communication systems, a reference distance of 1 km is often used [29, ch. 2]. In recent wideband propagation measurements inside office buildings, a reference distance of 1 m was used [30]. With a reliable path loss model, one merely needs to know the transmitter power to accurately estimate the received power level for a particular distance from the transmitter, provided that the transmitting antenna used in the field is comparable to the type used to arrive at the path loss model.

CW measurements inside factory buildings revealed that, on the average, path loss increases according to a power law (in other words, received power falls off as an inverse power of distance) and is statistically described by a log-normal distribution about a mean value given by:

$$Path\ loss\ (d)\ \alpha\ d^n \qquad (1)$$

where d is the distance between transmitter and receiver in meters and n is the mean path loss exponent. For free space, $n = 2$. This model, which is sometimes called a large scale attenuation model, has been found to describe UHF channels inside and around houses and office buildings [10] [11] [31]. The term large-scale indicates that the model holds for a large range of distances between transmitter and receiver and accounts for the gross attenuation characteristics of the channel with distance. Large-scale statistics are derived by measuring average or median received signal levels at many local areas throughout a desired coverage area. This is different from small-scale fading models, which statistically describe the signal level seen by a mobile receiver while it is moved about a small area (typically on the order of a few square meters) at a typical operating velocity [32] [33]. For small-scale fading models, the large-scale effects are ignored since the distance between transmitter and receiver is virtually constant over the local area, and the model describes fading due to temporal variation of the receiver position. This phenomenon is often referred to as fast-fading, as the signal statistics over relatively short time intervals are characterized.

It is interesting to note that inside buildings, where surrounding objects in the channel move very slowly, fading is primarily due to motion of the receiver. Historically, ionospheric and tropospheric communications systems were used (and are still in use) between fixed stations, and it was the temporal variation of the propagating medium (i.e., the ionosphere) that induced signal fading. Consequently, fading statistics based on long-term and short-term temporal variations of the channels are used to determine link performance of many systems [29, ch. 6,7]. A definite distinction exists between models that describe signal strength as a function of distance as opposed to a function of time. The former is used in determining coverage areas and co-channel interference levels whereas the latter is valuable in determining bit-error rates and outage probabilities. By assuming that path loss (a distance-related phenomenon) and fast fading (a time-related phenomenon) are independent, predictions of instantaneously received signal levels for particular receiver locations and times can be computed through use of a probability density function that is the product of the two individual densities [31]. Preliminary work by the author indicates that such a modeling approach, when conditioned on the topography surrounding the receiver, accurately describes fading conditions inside factories.

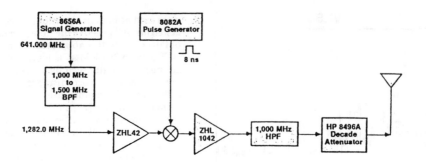

Transmitter

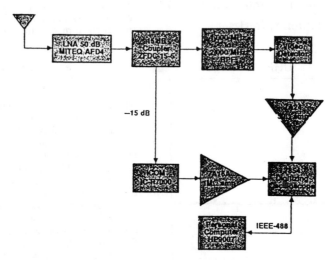

Receiver

Fig. 2. Factory multipath measurement apparatus.

To date, only two works have dealt with the temporal fading statistics of indoor radio channels [32] [34]. Using stationary transmitters and receivers, we found that ambient motion of people and inventory in factory buildings causes Rician fading for narrow-band communication systems. The Rician distribution has a probability density function given by:

$$f(x) = \frac{x}{\sigma_n^2} e^{-(x^2 + A^2)/2\sigma_n^2} I_o\left\{\frac{Ax}{\sigma_n^2}\right\} \quad 0 \leq x < \infty, A \geq 0$$

$$K = \frac{A^2}{2\sigma_n^2} \quad (2)$$

In (2), I_o (·) is the modified bessel function of the first kind and zero order. The parameter A represents the peak-to-zero value of the specular radio signal (comprised of the superposition of the dominant LOS signal and the time-invariant scattered signals reflected from walls, ceiling, and stationary inventory). σ_n^2 denotes the average power that is received over paths that vary with time due to objects moving inside the building [32]. The parameter K completely specifies

the Rician distribution. As K approaches negative infinity, the Rician distribution becomes the Rayleigh distribution given by:

$$f(x) = \frac{x}{\sigma_n^2} e^{-x^2/2\sigma_n^2} \quad 0 \leq x \leq \infty \quad (3)$$

The Rayleigh distribution is a classical one for communication and propagation studies, as it statistically describes the envelope of a Gaussian channel. The Gaussian assumption has been shown to accurately describe many physical time-varying channels, including the ionosphere [24], the troposphere [29], and urban mobile radio channels [22].

From data in [25] and [32], it appears that for open-plan style buildings with few internal partitions, temporal fading for fixed communication systems is Rician having a value of K equal to 10 dB. Bultitude has conducted measurements which indicate that in older office buildings that have many partitions, the fading is still Rician but with a value of K on the order of 5 dB [25]. This is not nearly as severe as Rayleigh-fading ionospheric or tropospheric channels.

Figure 3 shows the received signal strengths relative to a reference measurement made at 10 wavelength distance (2.3 m) in five factories. The figure is useful for computing large scale attenuation models for different factory topographies. Tables I and II indicate that in all cases, received power and Transmitter-Receiver (T-R) separation are highly correlated. In the tables, the mean path loss exponent $\bar{n}$ has been estimated from the slope of a linear least square error fit to the empirical data, where both distance and path loss are expressed logarithmically [31] [35]. In the tables, σ denotes standard deviation about the mean power law. Although path loss increases with distance more rapidly in obstructed topographies, large-scale attenuation in factory buildings does not appear to be as severe as in partitioned homes and office buildings, where path loss exponents typically range from 4 to 6 [13] [30] [31]. This is most likely due to the large ceiling expanses, wide aisles, and metal ceiling truss work and inventory that support, rather than impede, radio wave propagation.

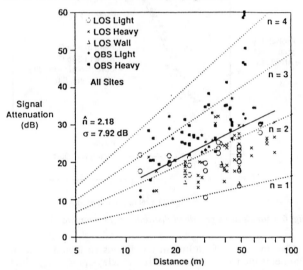

Fig. 3. Large-scale attenuation.

Because accurate descriptions of path obstacles were kept during the measurements, it is possible to extract from the data the RF signal loss caused by typical factory surroundings. By comparing the received signal levels for shadowed locations with the ensemble average of the factory measurements in a particular topography, average shadowing losses have been computed. Table III indicates typical shadowing losses that occur when a receiver is placed directly behind an obstruction (deep shadowing). Diffraction theory is used to predict the amount of RF energy that is received by an object that does not have a direct LOS path to the source. In many instances, an obstruction can be modeled as a knife-edge having a very thin width and a height equal to the physical height of the obstruction [27] [29]. Narrow-band measurements in factories reveal that path loss predicted by knife-edge diffraction theory is pessimistic; deeply shadowed locations experience received signal levels consistently 5 to 20 dB larger than predicted by diffraction theory. For geometries where obstructions are located towards the middle of the direct path, knife-edge diffraction is in closer agreement with the empirical data. This indicates that, just as in urban radio channels, multipath from surrounding structures can illuminate receivers when a direct LOS does not exist.

Measurements made with a moving CW receiver over many small areas (1 m long tracks in the middle and sides of the aisles) reveal that fading is usually Rayleigh in heavily clut-

TABLE I. Path Loss Exponent as a Function of Factory Building (1,300 MHz)

Factory Site	$\hat{n}$	σ (dB)	No. of Points	Corr. Coef.
Site B	2.39	10.20	33	.94
Site C	1.89	5.55	41	.98
Site D	2.43	7.94	34	.96
Site E	2.12	8.03	18	.96
Site F	1.92	4.79	17	.98

TABLE II. Path Loss Exponent as a Function of Factory Topography (1,300 MHz)

Path Loss Exponents in Different Factory Geographies

Factory Geography	$\hat{n}$	σ (dB)	No. of Points	Corr. Coef.
LOS light clutter	1.79	4.55	26	.98
LOS heavy clutter	1.79	4.42	43	.98
LOS along wall	1.49	3.9	8	.98
OBS light clutter	2.38	4.67	23	.99
OBS heavy clutter	2.81	8.09	43	.97

TABLE III. Shadowing Effects of Common Factory Equipment (1,300 MHz)

Shadowing Effects of Common Factory Equipment

Obstacle Description	Attenuation (dB)
2.5 m storage rack with small metal parts (loosely packed)	4-6
4 m metal box storage	10-12
5 m storage rack with paper products (loosely packed)	2-4
5 m storage rack with paper products (tightly packed)	6
5 m storage rack with large metal parts (tightly packed)	20
Typical N/C machine	8-10
Semi-automated Assembly Line	5-7
0.6 m square reinforced concrete pillar	12-14
Stainless Steel Piping for Cook-Cool Process	15
Concrete wall	8-15
Concrete floor	10

tered LOS and lightly cluttered obstructed topographies; Rician for paths along perimeter walls and on lightly cluttered LOS paths; and log-normal for paths that traverse heavily cluttered obstructed topographies. Figure 4 illustrates some of these typical fading distributions and their fit to some of the observed fading data. Rician distributed channels are highly desirable as they require a much smaller fading margin in system design. There is also evidence that suggests Rician fading channels support significantly larger bandwidths than do Rayleigh fading channels [34]. LOS paths along factory aisleways were consistently found to be Rician ($K = 6$ dB), thus suggesting that for radio communications wideband radio LAN terminals should be mounted along and at the ends of aisles.

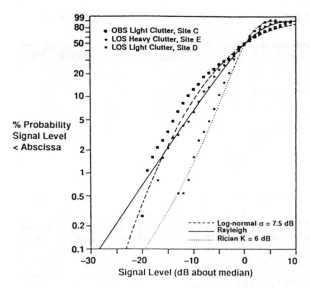

Fig. 4. Cumulative distributions of typical measurements and their fit to various distributions.

Antenna diversity is a well-known technique for combating fading on narrow-band channels [36–38]. The basic principal is to use several antennas separated in distance so that when one antenna receives a deeply faded signal, the other antenna(s) receive only a slightly faded version of the signal. The receiver then requires some level sensing circuitry that selects the best signal or various combining techniques to be used [38]. Diversity reception can provide several dB of gain against multipath when the antennas are positioned to achieve highly uncorrelated signal strengths. In manufacturing environments, measurements made over identical paths with different receiver antenna heights show that received signal strengths are often highly correlated (not independent) for vertical separations of two wavelengths (0.5 m). As seen in Figure 7, however, close-spaced antenna diversity may be useful when antennas are located with at least a quarter-wavelength separation along horizontal planes parallel to the ground. This can be seen by observing that the received signal levels on a particular antenna, at displacement differentials of approximately 0.06 m, consistently differ by more than 10 dB. Energy density antennas that couple both electric and magnetic fields are also useful in combating multipath fading in flat channels [39].

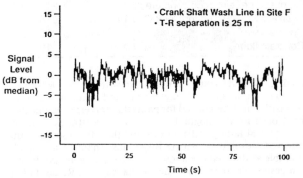

Fig. 5. Temporal fading measurement in engine assembly.

Factory channel impulse response measurements (also called power delay profiles) reveal that for LOS paths there typically exist only a few specular multipath components, with the direct signal having a significantly larger signal level than the latter components. Over obstructed paths, however, when either the transmitter or receiver is shadowed by large equipment or by stacks of inventory, the predominant energy arrives 50 to 150 ns after the first observable signal [40]. To determine how individual signal components change with receiver motion, 19 equally-spaced power impulse response measurements were made along 1 m tracks throughout various topographies in five factories. Figure 8 illustrates how specular reflections from perimeter walls, etc. are easily distinguishable. In Figure 9, typical spatially-averaged multipath power delay profiles from various factories and topographies are shown.

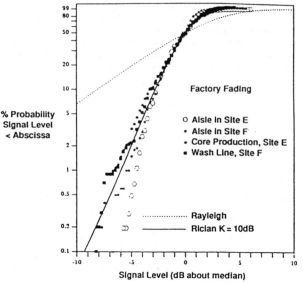

Fig. 6. Cumulative signal level distributions for temporal fading.

One measure of multipath conditions in a mobile radio channel is the root mean square (rms) delay-spread (σ_τ) which is inversely proportional to the maximum usable data rate of a channel [22] [41]. In [41], Bello characterized frequency selective channels in terms of a Taylor-series expansion of the average multipath power spectrum about the carrier frequency. He showed that for Wide Sense Stationary Uncorrelated Scattering (WSSUS) channels, the rms delay-spread describes the ratio of power in the first (linear) frequency term of the series expansion to the power in the (flat fading) constant term. Underlying assumptions are that the frequency selectivity (the slope of the channel transfer function with frequency) changes slowly over the operating bandwidth.

Root-mean-square delay-spread is computed as the square root of the second central moment of an averaged power delay profile. This average is usually computed over time [41], but for a slowly varying channel such as is found inside factories, this average may be computed over space by using many profile measurements in a local area [30] [40]. Analysis of the propagation database from inside factories has revealed that multipath characteristics can be correlated with building structure and type of inventory. Measurements in a food processing factory that manufactures dry-goods and has considerably less metal inventory than most factories had σ_τ values consistently half of those observed in factories that produce metal products. Newer factories which incorporate steel beams and steel-reinforced concrete in the building structure have larger delay spreads than older factories which use wood and brick for perimeter walls. Summarizing the results of the wideband measurements, the worst-case σ_τ value was 300 ns in a modern engine plant. Typical values ranged between 100 ns and 200 ns.

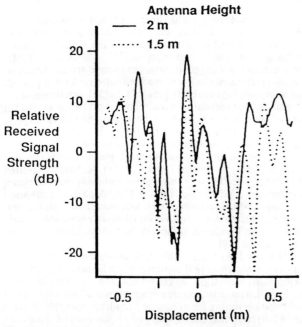

Antenna Height
— 2 m
······ 1.5 m

Relative
Received
Signal
Strength
(dB)

Fig. 7. *Received signal levels using antenna height diversity.*

Unlike in office buildings [30], delay-spread values in factories do not appear to depend on whether or not there exists a LOS path [40]. This is due to the open expanses and vast amount of reflecting material that readily supports multipath propagation.

In [22], DPSK bit-error-rate analyses for mobile radio channels undergoing both Rayleigh envelope (flat) fading and frequency selective fading indicate that as signaling bandwidth increases, multipath-induced intersymbol interference becomes the main cause of performance degradation. Our measurements show a worst-case rms delay-spread of 300 ns inside modern factory buildings. From Figure 4.2.9 in [22], it is easy to show that in order to achieve an irreducible bit error rate of better than 10^{-3}, baseband data rates must be limited to below

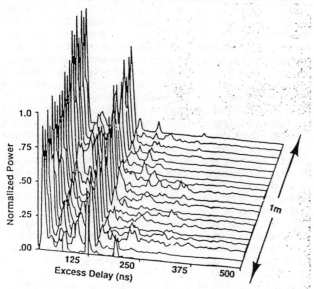

Fig. 8. *Closely-spaced power impulse response measurements.*

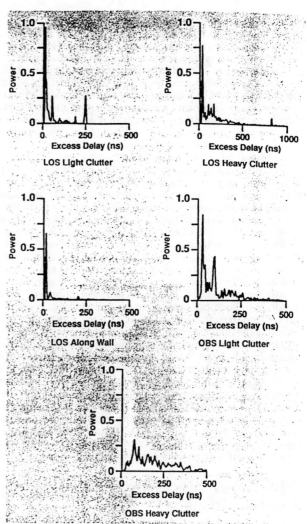

Fig. 9. *Power impulse responses.*

150 kb/s for rectangular pulse shapes (this is found from equation 4.2.71 in [22]). Raised-cosine pulse shapes can increase the worst-case throughput to 250 kb/s for the same channel. It must be stressed that measurements in [40] are the only known wideband measurements from within factories, and it may be that some buildings impose more severe frequency-selective fading conditions. In [30], Devasirvatham found an instance where rms delay-spread outside of a building was approximately 600 ns. It will be possible to improve the capacity of indoor digital radio communication systems through robust modulation techniques [42] [43], distributed antennas [44], and antenna diversity techniques [36] [37].

Important issues in the area of wideband indoor radio propagation measurements include measurement dynamic range and the effects of operating frequency on multipath. To date only three wideband indoor radio propagation works have been reported [11] [26] [30]. In [11], an apparatus that uses both a frequency swept technique and a pulsed transmitter was used to measure average multipath power delay profiles. In [26] and [40], a similar but simpler apparatus was used, and the average power profile was computed from a spatial average of individual profile measurements over several wavelengths. In [30], a spread-spectrum system was used to achieve greater dynamic range at the expense of temporal resolution. σ_τ values reported in [11] suggest that indoor radio channels could support bandwidths several times as large as those indicated by measurements in [26] and [30]. Consequently, Bit Error Rate (BER) calculations based on measurements in [11] are probably optimistic. It is interesting to note that measurements in [11] were made at 1.5 GHz, measurements in [26] were made at 1.3 GHz, and measurements in [30] were made at 850 MHz. Because of the limited amount of wideband indoor propagation data, it is unknown at this point what effect carrier frequency has on indoor radio propagation. One would think that in office buildings the delay spread would decrease with frequency due to increased attenuation by the structure. In factory buildings, however, this may not be the case.

Recently, rapid changes in channel group delay have been found to cause burst error in digital communication systems due to shifts in eye pattern timing [14] [45] [62]. As described in [14], it is necessary to use a phase-lock loop to accurately track the mean delay of the channel in order to detect digital symbols at the point where the signal-to-noise ratio is largest. In high-data-rate indoor radio systems, where channels vary slowly, the phase-lock loop will track the instantaneous centroid (first moment) of the impulse response. A phenomenon known as jitter occurs when the centroid of the impulse response changes rapidly, causing loss of bit synchronization and resulting in burst errors as the receiver attempts to reacquire bit synchronization. Differential-delay jitter, which is the rate at which the impulse-response centroid changes over time (or distance for a mobile moving at a constant velocity), is a useful design parameter since it indicates the required slew rate for a bit synchronization circuit. When the duration of a data bit is much larger (at least a factor of four or five greater) than the temporal extent of the indoor channel impulse response, it has been shown in [45] [62] that burst errors are caused primarily by envelope fades and not by timing jitter. In [18], empirical measurements indicate that inside factories, the impulse response centroid can change by as much as 180 ns with just a few centimeters of movement at the portable.

Multiple-Access Networking Considerations

Future factory radio communication systems will rely on multiple-access techniques to accommodate many fixed and mobile terminals. Multiple-access techniques such as Frequency Division Multiple-Access (FDMA), Time-Division Multiple-Access (TDMA), Code Division Multiple-Access (CDMA, also called spread-spectrum), and Carrier-Sense Multiple-Access (CSMA, also called packet radio) partition channel users into non-overlapping signal spaces [46–51]. Because of non-idealities, however, there is inevitable overlapping of signals and the resulting co-channel interference can appear as lengthy message delays or as degradation of the desired received signal [47–51].

Random access (CSMA, CDMA) radio LANs are attractive because they have relatively few synchronization requirements and support a distributed architecture. Unlike fixed assignment networks (FDMA, TDMA), which assume all users require their own channel, random access techniques rely on "bursty users" and assume that the likelihood of many users using the network at one time is small. For AGVs using reliable dead-reckoning systems, infrequent position updates suggest the use of random access networking for AGV control. In fact, for mobile robot systems which navigate by buried wire or paint strip, CSMA is a popular choice because communications traffic is generally very light. On the other hand, direct control of a fleet of vehicles by a central dispatching station (such as would be warranted for cleaning a contaminated nuclear reactor) might require a fixed assignment scheme. There also exist many types of demand-assignment access techniques that appear particularly suited to indoor radio networking [49] [53].

Considerable progress has been made in determining realistic limitations on the delay characteristics of packet networks [54], and a powerful product-form solution model for packet radio systems has been developed [55]. CSMA and CDMA strategies can be merged to enhance multiple-access communication performance in a multipath environment while providing some ranging capability [56] [57]. In [56], fundamental expressions that permit the calculation of BERs in packet spread-spectrum systems have been provided; these expressions permit system designers to analyze throughput and delay as a function of number of simultaneous users.

Work on indoor radio communication systems has revealed that by using a slotted packet radio technique (slotted ALOHA), overall system throughput and delay characteristics can be improved through capture [58]. A packet reservation technique, which accommodates both voice and data, is an example of a slotted, packetized multiple-access technique that exploits the burstiness of speech and gives priority to data transmission [59].

The July 1987 issue of the *IEEE Journal on Selected Areas in Communications* [49] contains several works on recent developments in multiple-access networks and their performances. In [60], a modeling approach has been used to include exact propagation timing for events on broadcast channels. Results from [60] show that the topology of a linear (bus-like) network should be considered when analyzing the performance of packet radio protocols, since previous protocol models underestimate performance. This indicates that for futuristic wideband indoor packet radio systems, the locations of repeaters may be an important design consideration not only from the standpoint of link budget and minimization of multipath, but also from the standpoint of channel capacity due to the inherent physical structure of the network. In [61], a CSMA protocol that uses a tone-modulated preamble has been developed to resolve contentions on packet radio channels.

There has been much interest in the use of TDMA for mobile and portable radio communications. In Europe, a Pan-European TDMA standard has been agreed upon for cellular radiotelephone, and several vendors are proposing TDMA for U.S. digital cellular systems. TDMA is also a viable technique for multiple-user indoor radio communications, and TDMA communication system prototypes are being developed at major research laboratories [12] [13] [19] [62]. TDMA offers many advantages, including communications flexibility and cheaper hardware. Some presently available factory radio systems employ low data-rate TDMA and use some form of priority scheduling to reallocate time slots that would normally go unused. A subtle but potentially important advantage of TDMA is that a mobile user can listen to the base as it transmits to other users within the frame. During this listening period, the mobile can use small-scale antenna diversity to choose the antenna for best reception (and transmission) for its upcoming time slot. A further advantage is that TDMA implies that only one user transmits at a time. Thus, class-C amplifiers, which are more efficient but have rich harmonic and intermodulation components, can be used. In FDMA, class-C amplifiers can create adjacent channel interference if users are in close proximity to one another (although feedback techniques exist which can reduce intermodulation emissions of class-C amplifiers).

FDMA is advantageous because the multiple-access system can be built up around existing and field-proven narrow-band technology. Although FDMA requires more hardware for a given number of users (because of a greater number of communication channels), the data rates in each of these channels is small. Thus, users in an FDMA system are not subjected to multipath-induced intersymbol interference, which can affect high bit-rate TDMA channels. Equalization, which becomes an issue for TDMA systems, is usually not needed for FDMA because each channel undergoes flat fading.

As the number of users increases, the real-time communications capability of random access techniques diminish, and a fixed assignment approach is required. Furthermore, if central computers using parallel processing architectures are required to simultaneously communicate, navigate, and control many simultaneous users on a virtually continuous basis, TDMA or FDMA approaches may be desired. Portable/mobile users transmitting large blocks of data (i.e., MIS, video transmissions, high resolution graphics, maps, etc.) are accommodated best by a fixed assignment network. Selection of networking strategies for radio links inside buildings will depend heavily upon the number of users, the duration of transmissions, the limit of sophistication at each terminal, the importance of real-time control, and whether or not it is desired to use radio for AGV navigation.

Conclusion

As part of the research mission of the NSF Engineering Research Center for Intelligent Manufacturing Systems, measurement, characterization, and modeling of indoor factory radio channels have been carried out. The work reveals that man-made noise is not a serious problem for indoor factory radio systems at frequencies greater than 1 GHz, and that fading characteristics are highly dependent upon local topography in the workplace. Shadowing data and large scale path loss models have been developed and form the basis for designing reliable narrow-band indoor radio LANs for portable communications and AGV control. Wideband measurements reveal that commercially available technology currently limits data rates to on the order of 150 kb/s in typical factory channels. While this accommodates current needs, it is anticipated that greater capacity will be required for the highly automated and flexible factories of the future. Ongoing work at Virginia Tech is aimed at developing robust wideband multi-access communication system designs and signaling techniques for indoor radio communications.

Acknowledgments

The author wishes to thank Clare McGillem and Heidi Peterson of Purdue University, and Charles Bostian, Tim Pratt, and Warren Stutzman of the Virginia Tech Satellite Communications Group for their helpful suggestions during the preparation of this article. The thoughtful comments of the reviewers are appreciated.

References

[1] M. Kaminski, Jr., "Protocols for Communicating in the Factory," *IEEE Spectrum*, vol. 23, no. 4, pp. 56–62, Apr. 1986.

[2] S. Farowich, "Communicating in the Technical Office," *IEEE Spectrum*, vol. 23, no. 4, pp. 63–67, Apr. 1986.

[3] National Academy of Engineers, *Engineering Research Centers: Leaders in Change*, Washington, D.C.: National Academy Press, 1987.

[4] L. Mayfield, "NSF's ERC Program—How it Developed," *Engineering Education*, pp. 130–132, Nov. 1987.

[5] J. Solberg and C. McGillem, "Automated Guided Vehicle Systems," *International Trends in Manufacturing Technology Series*, R. Holler, Edi., United Kingdom: IFS Publications Ltd., 1986.

[6] T. Rappaport, "The Design and Development of a Mobile Robot Location System," Masters Thesis in Electrical Engineering, Purdue University, West Lafayette, IN, Dec. 1984.

[7] C. McGillem and T. Rappaport, "An Infra-red Navigation Technique for Mobile Robots," *Proc. 1988 IEEE Intnl. Conf. on Robotics and Automation*, Philadelphia, PA, Apr. 26, 1988.

[8] C. McGillem, J. Callison, and T. Rappaport, "Autonomous Roving Vehicle Navigation, Control, and Communication," *9th Intnl. Conf. on Automation in Warehousing/6th Intnl. Conf. on Automated Guided Vehicle Systems*, Brussels, Belgium, Oct. 25, 1988.

[9] G. Schettler, "Developments in Radio Frequency Transponders," *1988 Materials Handling FOCUS Conference*, Materials Handling Research Center, Georgia Institute of Technology, Atlanta, GA, Sept. 15, 1988.

[10] D. Cox, H. Arnold, and P. Porter, "Universal Digital Portable Communications—A System Perspective," *IEEE Trans. J. Sel. Areas Comm.*, vol. SAC-5, no. 5, pp. 764–773, June 1987.

[11] A. A. M. Saleh and R. Valenzuela, "A Statistical Model for Indoor Multipath Propagation," *IEEE Trans. J. Sel. Areas Comm.*, vol. SAC-5, no. 2, pp. 138–146, Feb. 1987.

[12] A. Saleh, A. Rustako, Jr., and L. Cimini, Jr., "A TDMA Indoor Radio Communications System using Cyclical Slow Frequency Hopping and Coding—Experimental Results and Implementation Issues, *IEEE Globecom '88 Proc.*, pp. 41.3.1–6.

[13] R. Bernhardt, "Two-Way Transmission Quality in Portable Radio Systems," *IEEE Globecom '88 Proc.*, pp. 41.2.1–6.

[14] J. C-I Chuang, "The Effects of Multipath Delay Spread on Timing Recovery," *IEEE Trans. Veh. Tech.*, vol. VT-36, no. 3, pp. 135–140, Aug. 1987.

[15] L. Couch II, *Digital and Analog Communication Systems*, MacMillan, Second Ed., 1987.

[16] M. Marcus, "Regulatory Policy Considerations for Radio Local Area Networks," *IEEE Communications Magazine*, vol. 25, no. 7, pp. 95–99, July 1987.

[17] H. Nogami and M. Wakao, "Premises Radio Systems in Japan," *IEEE Communications Magazine*, vol. 25, no. 10, pp. 49–55, Oct. 1987.

[18] T. Rappaport, "Delay Spread and Time Delay Jitter in Manufacturing Environments," *1988 IEEE Veh. Tech. Conf. Record*, Philadelphia, PA, pp. 186–189, June 15, 1988.

[19] J. C-I Chuang and N. Sollenberger, "Burst Coherent Detection with Robust Frequency and Timing Estimation for Portable Radio Communications," *IEEE Globecom '88 Proc.* pp. 26.1.1–6.

[20] T. Takeuchi, F. Ikegami, S. Yoshida, and N. Kikuma, "Comparison of Multipath Delay Characteristics with BER Performance of High Speed Digital Mobile Transmission," *1988 IEEE Veh. Tech. Conf. Proc.*, pp. 119–126, June 15, 1988.

[21] E. Skomal, *Man-Made Radio Noise*, NY: Van Nostrand Reinhold Co., Ch. 3 and 4, 1978.

[22] W. Jakes, *Microwave Mobile Communications*, Wiley, 1974.

[23] L. Duran, "Characteristics of Electromagnetic Interference in Industrial Environments," Masters Thesis in Electrical Engineering, Purdue University, Dec. 1987.

[24] R. S. Kennedy, *Fading Dispersive Communication Channels*, Wiley, Ch. 1–3, 1969.

[25] R. Bultitude, "Measurement, Characterization and Modelling of Indoor 800/900 MHz Radio Channels for Digital Communications," *IEEE Communications Magazine*, vol. 25, no. 6, pp. 5–12, June 1987.

[26] T. Rappaport, "Characterizing the UHF Factory Multipath Channel," Ph.D. Dissertation, Purdue University, Dec. 1987. Also available as Tech. Report TR-ERC88-12, NSF Engineering Research Center for Intelligent Manufacturing Systems, Purdue University, June 1988.

[27] R. Colin, *Antennas and Radiowave Propagation*, McGraw-Hill, pp. 369–374, 1985.

[28] T. Rappaport, "Simple-to-Build Wide-Band Antennas," *RF Design*, pp. 35–41, Apr. 1988.

[29] J. Griffiths, *Radio Wave Propagation and Antennas*, Englewood Cliffs, NJ: Prentice-Hall International, 1987.

[30] D. Devasirvatham, "Time Delay Spread and Signal Level Measurements of 850 MHz Radio Waves in Building Environments, *IEEE Trans. Ant. Prop.*, vol. AP-34, no. 11, pp. 1,300–1,308, Nov. 1986.

[31] D. Cox, R. Murray, and A. Norris, "800-MHz Attenuation Measured In and Around Suburban Houses," *AT&T Bell Laboratories Technical Journal*, vol. 63, no. 6, pp. 921–954, July–Aug. 1984.

[32] T. Rappaport and C. McGillem, "UHF Fading in Factories," *IEEE J. Sel. Areas Comm.*, Jan. 1989.

[33] D. Cox, R. Murray, and A. Norris, "Measurements of 800-MHz Radio Transmission into Buildings with Metallic Walls," *Bell System Tech. J.*, vol. 62, no. 9, pp. 2,695, 2,716, Nov. 1983.

[34] R. Bultitude, "Measurement, Characterization and Modelling of 800/900 MHz Mobile Radio Channels for Digital Communications, Ph.D. Thesis, Carleton University, Ottawa, Feb. 1987.

[35] W. Dixon and F. Massey, Jr., *Introduction to Statistical Analysis*, McGraw-Hill, Fourth Edition, Ch. 11, 1983.

[36] R. G. Vaughan and J. B. Andersen, "Antenna Diversity in Mobile Communications," *IEEE Trans. Veh. Tech.*, vol. VT-36, no. 4, pp. 149–172, Nov. 1987.

[37] J. Winters, "Optimum Combining for Indoor Radio Systems with Multiple Users," *IEEE Trans. Comm.*, vol. COM-35, no. 11, pp. 1,222–1,230, Nov. 1987.

[38] S. Stein, "Fading Channel Issues in System Engineering (Invited Tutorial)," *IEEE J. Sel. Areas Comm.*, vol. SAC-5, no. 2, pp. 68–89, Feb. 1987.

[39] K. Ito and S. Sasaki, "A Small Printed Antenna Composed of Slot and Wide Strip for Indoor Communication Systems," *1988 IEEE Antennas and Propagation Symposium*, pp. 716–719, June 12, 1988.

[40] T. Rappaport, "Characterization of UHF Multipath Radio Channels in Manufacturing Environments," *IEEE Trans. Ant. Prop.*, Aug. 1989. Presented in part at *IEEE GLOBECOM '88*, Hollywood, FL, pp. 26–5.1, 26–5.7, Nov. 29, 1988.

[41] P. Bello, "Characterization of Randomly Time-variant Linear Channels," *IRE Trans. Comm. Sys*, vol. CS-11, pp. 361–393, Dec. 1963.

[42] R. Valenzuela, "Performance of Quadrature Amplitude Modulation for Indoor Radio Communications," *IEEE Trans. Comm.*, vol. COM-35, no. 11, Dec. 1987.

[43] M. Kavehrad and G. Bodeep, "Design and Experimental Results for Direct-Sequence Spread-Spectrum Radio using Differential Phase-Shift Keying Modulation for indoor Wireless Communications," *IEEE J. Sel. Areas Comm.*, vol. SAC-5, no. 5, pp. 815–823, June 1987.

[44] A. A. M. Saleh, A. J. Rustako, Jr., and R. Roman, "Distributed Antennas for Indoor Radio Communications," *IEEE Trans. Comm.*, vol. COM-35, no. 12, pp. 1,245–1,251, Dec. 1987.

[45] S. Ariyavisitakul et al., "Fractional-bit Differential Detection of MSK: A Scheme to Avoid Outages Due to FRequency-selective Fading," IEEE Trans. Veh. Tech., vol. VT-36, no. 1, pp. 36–42, Feb. 1987.

[46] W. Stallings, *Data and Computer Communications,* MacMillan, Second Ed., 1985.

[47] I. Rubin, "Message Delays in FDMA and TDMA Communication Channels," *IEEE Trans. Comm.*, vol. COM-27, no. 5, pp. 769–777, May 1979.

[48] L. Kleinrock and F. Tobagi, "Packet Switching in Radio Channels: Part I: Carrier Sense Multiple Access Modes and Their Throughput-delay Characteristics," *IEEE Trans. Comm.*, vol. COM-23, no. 12, pp. 1,400–1,416, Dec. 1975.

[49] Special Issue on Performance Evaluation of Multiple-Access Networks, *IEEE J. Sel. Areas Comm.*, V. O. K. Li, Edi., vol. SAC-5, no. 6, July 1987.

[50] M. Pursley, "Performance Evaluation for Phase-coded Spreead-spectrum Multiple-access Communication-Part I: System Analysis," *IEEE Trans. Comm.*, vol. COM-25, no. 8, pp. 795–799, Aug. 1977.

[51] L. Merakos, G. Exley, and C. Bisdikian, "Interconnection of CSMA Local Area Networks: The Frequency Division Approach," *IEEE Trans. Comm.*, vol. COM-35, pp. 730–738, July 1987.

[52] L. Lee and C. Un, "A Code-division Multiple-access Local Area Network," *IEEE Trans. Comm.*, vol. COM-35, no. 6, pp. 667–671, June 1987.

[53] S. Sachs, "Alternative Local Area Network Access Protocols," IEEE Communications Magazine, vol. 26, no. 3, pp. 25–45, Mar. 1988.

[54] S. Beuerman and E. Coyle, "The Delay Characteristics of CSMA/CD Networks," IEEE Trans. Comm., vol. 36, no. 5, pp. 553–563, May 1988.

[55] R. L. Hamilton, Jr. and E. Coyle, "A Two-Hop Packet Radio Network with Product Form Solution," Proc. 20th Annual Conf. on Info. Sciences and Systems, Princeton Univ., pp. 871–876, Mar. 1986.

[56] R. Morrow, Jr. and J. Lehnert, "Bit-to-Bit Error Dependence in Slotted DS/SSMA Packet Systems with Random Signature Sequences," accepted to *IEEE Trans. Comm.*, to appear early 1989.

[57] J. Fischer, J. Cafarella, C. Bouman, G. Flynn, V. Dolat, and R. Boisvert, "Wide-Band Packet Radio for Multipath Environments," IEEE Trans. Comm., vol. 36, no. 5, pp. 564–573, May 1988.

[58] D. Goodman and A. Saleh, "The Near/Far Effect in Local ALOHA Radio Communications," *IEEE Trans. Veh. Tech.*, vol. VT-36, no. 1, pp. 19–27, Feb. 1987.

[59] D. Goodman, R. Valenzuela, K. Gayliard, and B. Ramamurthi, "Packet Reservation Multiple Access for Local Wireless Communications," *1988 IEEE Veh. Tech. Conf. Record*, Philadelphia, PA, pp. 701–707, June 15, 1988.

[60] M. Molle, K. Sohraby, and A. Venetsanopoulos, "Space-Time Models of Asynchronous CSMA Protocols for Local Area Networks," *IEEE J. Sel. Areas Comm.*, vol. SAC-5, no. 6, pp. 956–968, July 1987.

[61] W. Lo and H. Mouftah, "Collision Detection and Multitone Tree Search for Multiple-Access Protocols on Radio Channels," *IEEE J. Sel. Areas Comm.*, vol. SAC-5, no. 6, pp. 1,035–1,040, July 1987.

[62] J. C-I Chuang, "The Effects of Time Delay Spread on Portable Radio Communications Channels with Digital Modulation," *IEEE J. Sel. Areas Comm.*, vol. SAC-5, no. 5, pp. 879–889, June 1987.

Biography

Theodore S. Rappaport was born on November 26, 1960. He received his B.S.E.E., M.S.E.E., and Ph.D. degrees from Purdue University, West Lafayette, IN, in 1982, 1984, and 1987, respectively. During his undergraduate career, he received three national scholarships from the Foundation for Amateur Radio, including the Radio Club of America Scholarship during his senior year. As a graduate student, he received the GTE Graduate Fellowship in Electrical Engineering and the EXXON Electrical Engineering Fellowship. He was with the Government Aerospace Systems Division of Harris Corporation, Melbourne, FL. during the summers of 1983 and 1986. From 1987 to 1988 he was manager of the Autonomous Guided Vehicle Project at the NSF Engineering Research Center for Intelligent Manufacturing Systems, Purdue University, West Lafayette, IN. There he managed the development of a prototype AGV fleet and conducted research in indoor radio communications for manufacturing environments. He is presently an Assistant Professor with the Bradley Department of Electrical Engineering, Virginia Polytechnic Institute and State University, Blacksburg, VA where he is teaching and conducting research in the areas of digital and satellite communications, radio wave propagation, and antennas. Dr. Rappaport is a consultant to the AGV and cellular radiotelephone industries, and is a reviewer for the IEEE.

914 MHz Path Loss Prediction Models for Indoor Wireless Communications in Multifloored Buildings

Scott Y. Seidel, *Student Member, IEEE,* and Theodore S. Rappaport, *Senior Member, IEEE*

Abstract—Quantitative models are presented that predict the effects of walls, office partitions, floors, and building layout on path loss at 914 MHz. The site-specific models have been developed based on the number of floors, partitions and concrete walls between the transmitter and receiver, and provide simple prediction rules which relate signal strength to the log of distance. The standard deviation between measured and predicted path loss is 5.8 dB for the entire data set, and can be as small as 4 dB for specific areas within a building. Average floor attenuation factors (FAF), which describe the additional path loss (in decibels) caused by floors between transmitter and receiver are found for as many as four floors in a typical office building. Average floor attenuation factors are found to be 12.9 and 16.2 dB for one floor between the transmitter and receiver in two different office buildings. For same-floor measurements, attenuation factors (AF) are found to be 1.4 dB for each cloth-covered office partition and 2.4 dB for each concrete wall between transmitter and receiver. Path loss contour plots for measured data are presented. In addition, contour plots for the path loss prediction error indicate that the prediction models presented in this paper are accurate to within 6 dB for a majority of locations in a building.

I. INTRODUCTION

OBJECTS that surround transmitters and receivers severely affect the propagation characteristics of any radio channel. The performance of in-building high capacity wireless communications is limited by the propagation characteristics. Thus, it is important to understand how the physical surroundings impact the propagation environment. Several researchers have measured radio waves in buildings and statistically modeled their results [1]–[9]. A summary of path loss, narrow-band fading statistics, and root mean square (rms) delay spread in many different building types is given in [10].

In this paper, we first present statistical analyses of 914 MHz narrow-band path loss measurements inside four buildings, and then classify the measurements based on the physical surroundings. The measured buildings include a grocery store, a retail department store, and two multistory office buildings. A statistical model of the simple form d^n (see (2)) is used to relate average path loss to the log of distance where

d is the distance between the transmitter and receiver measured in three dimensions, and n is the mean path loss exponent, which indicates how fast path loss increases with distance ($n = 2$ for free space) [1]–[6], [8], [9], [12], [13]. Values of the mean path loss exponent n are found for each building and all four buildings combined. In multifloored buildings, we find that more accurate prediction is possible when the parameter n is viewed as a function of the number of floors between transmitter and receiver. We also develop an alternative path loss model ((5)) for multifloor buildings to quantify the additional path loss caused by multiple floors between transmitter and receiver.

For measurements in the office buildings when the transmitter and receiver are located on the same floor, we quantify average path loss caused by cloth-covered plastic office partitions and concrete walls between the transmitter and receiver. We assume free space propagation with distance and consider additional path loss to be caused by the physical obstructions that lie directly between the transmitter and receiver ((8)). This provides a tractable physical propagation model that is shown to be more accurate than a generic statistical model ((2)) that considers only the distance between the transmitter and receiver terminals, and not site-specific information. Contour plots of locations of equal path loss for a fixed base transmitter are presented for the two office buildings. The models in this paper are used to predict the path loss at the measurement locations. The differences between the measured and predicted path loss are used to develop path loss prediction error contour regions for the office buildings.

Section II describes the measurement procedure and the four measured buildings used to produce the models in this paper. Section III describes the important site-specific building parameters used to develop the models. Section IV presents the measured data and shows the best fit for three different in-building path loss models. Section V shows path loss contour for measured data and error contour regions for the difference between the measured path loss and the path loss predicted from site-specific models given in Section IV. Sections VI and VII conclude with a summary of results and future work.

II. MEASUREMENT PROCEDURE AND LOCATIONS

A. Measurement Procedure

Narrow-band (CW) signal strength measurements were made at 914 MHz with a system nearly identical to the one

Manuscript received May 23, 1991; revised October 21, 1991. This work was supported by an NSF Graduate Fellowship, the MPRG Industrial Affiliates Program, and a grant from NCR Corporation.

The authors are with the Mobile and Portable Radio Research Group, Bradley Department of Electrical Engineering, Virginia Polytechnic Institute and State University, Blacksburg, VA 24061.

IEEE Log Number 9105955.

used in [3]. A 1 W CW signal was transmitted by an omnidirectional quarter-wave monopole antenna at a height of either 1.0 or 1.5 m above the floor [11]. The mobile receiver omnidirectional discone antenna was either 1.0 or 1.8 m above the floor. The receiver can instantaneously measure signals between 0 and -91 dBm over a 15 kHz bandwidth. With $+29$ dBm transmitter power, our maximum system path loss is 120 dB. This is on the order of the maximum dynamic range expected for emerging personal communications networks (PCN) which will be deployed within buildings during the next several years.

The stationary transmitter was placed at several locations within each of the four buildings. Locations which are potential candidates for future microcellular base stations, such as centrally located areas and perimeter areas within a wing of a building were chosen for most transmitter sites. At some other locations, the base antenna was placed within a partitioned office cubicle to determine the effects of office partitions between the transmitter and receiver. For each transmitter location, the mobile receiver thoroughly canvassed the building at transmitter–receiver (T–R) separations that ranged between 1.5 and 90 m. During each measurement, the mobile receiver moved at constant walking velocity along a straight path which varied in length between 2.4 and 60 m, depending on surroundings. The mobile's position was continuously recorded so that site-specific propagation models can be developed from the data.

B. Measurement Locations

1) Grocery Store: The open-plan shopping area of the grocery store has dimensions of 46 m × 67 m, and a ceiling height of 5 m. This area consists of metal shelves that are 22 m long and 2.5 m high. The aisles next to the shelves are each 2.3 m wide. For most measurements, the transmitter was located at the far side of the store while the receiver canvassed the entire shopping area. For several other measurements, the base transmitter was located between two checkout lanes in the front of the store.

2) Retail Store: The single story open-plan retail store has inside dimensions of 61 m × 52 m, and a ceiling height of 4.5 m. The store is divided into several departments. Movable 1.8 m high cloth-covered plastic partitions partially separate some of the departments. We call these soft partitions since they can be relocated easily, and are not a fixed part of the building structure. Other departments are divided by metal shelves ranging from 1.5 to 2 m high. Main aisles are 2 m wide and secondary aisles are 1.2 m wide. The three transmitter locations were near the checkout lanes, in the lawn furniture department, and in the center of the store near small appliances. The receiver moved throughout the shopping area.

3) Office Building 1: Office building 1 has five floors and two wings where office areas are boxed in with soft partitions. These partitions are cloth-covered plastic dividers which divide each large open-plan office area into several smaller individual cubicles. Typical cubicles cover 2.4 m × 2.4 m of floor space and are surrounded by 1.5 m tall soft partitions. Aisles between the cubicles are about 1.2 m wide.

In the center of the West wing of the building are conference rooms with concrete block walls which span from the floor to the ceiling but are typically 5–10 m in width (they do not block off the entire building). On the fifth floor of office building 1, the transmitter was located on the perimeter of the West wing. The direct path between the transmitter and receiver passed outside the building when the receiver was in the Central wing. The schematic drawing for the West wing of the fifth floor can be seen in the contour plots in Figs. 7 and 11. Fig. 1 is a photograph of a typical soft partitioned area in the West wing on the fifth floor.

The fourth floor of the West wing of office building 1 is similar in layout to the fifth floor. The transmitter location on the fourth floor of office building 1 was near the center of a partitioned office cubicle. The transmitter was at least 0.5 m from each of the surrounding partitions. A schematic drawing for the fourth floor in the West wing can be seen in the contour plots in Figs. 8 and 12. For the transmitter locations on the fourth and fifth floors, the mobile receiver moved throughout the West wing on the fourth and fifth floors, and also in the Central wing on the fifth floor. Additional multifloor measurements were made with the transmitter on the first floor near the edge of the building while the mobile receiver traversed a straight path along the edge of the building on the second through fifth floors. Both the length and width are greater than the height of office building 1.

4) Office Building 2: Office building 2 has four stories and office area layouts that are similar to those in office building 1. At one location, the transmitter was placed directly behind a soft partition in an office cubicle on the second floor of office building 2. Several multifloor measurements were made with the transmitter next to an elevator in the basement. The mobile receiver traversed aisles on the second and third floors. A schematic drawing of the second floor of office building 2 can be seen in the contour plots in Figs. 9, 10, and 13. Office building 2 is longer and wider than it is high.

III. Data Processing

Fig. 2 shows a typical measurement run where the receiver was moved along a 12 m track in an aisle of the retail store. The abscissa represents the T–R separation, which only changes by 10 m along the 12 m measurement run since the receiver was not moved radially away from the transmitter. The median signal strength over a distance of 20 λ (6.56 m) was computed at 20 λ intervals for each measurement run and is considered a discrete measurement location for the development of path loss models and contour plots. A 20 λ distance was chosen so that the fast-fading of the envelope caused by multipath would not influence the large-scale path loss for a given measurement track [13]. Preliminary work at MPRG shows that in buildings, large-scale path loss is uncorrelated at 20 λ spacings. Thus, each path loss measurement represents the median path loss measured over a 6.56 m section of a straight path. Median signal strengths were converted to absolute path loss by subtracting the median received signal strengths from the $+29.0$ dBm transmitter power. Building blueprints and path loss measurements were imported to a computer-aided design program (AutoCAD).

Fig. 1.   Picture of a soft partitioned environment in office building 1. Notice the 1.5 m high cloth covered plastic dividers (soft partitions) that separate office cubicles.

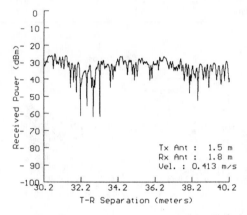

Fig. 2.   An example of typical 914 MHz CW signal fading as the receiver is moved in a retail department store. Fade depths which dip 30 dB below the median can occur.

For each path loss measurement, we have the following information:

median path loss (over 6.56 m track segment)
blueprints for the measured building
exact transmitter location
exact receiver location
transmitter–receiver separation (measured as a straight line in three dimensions)
number of floors between transmitter and receiver
number of soft partitions between transmitter and receiver
number of concrete walls between transmitter and receiver.

With the above information, we have developed models for path loss as a function of distance, number of floors, soft partitions, and concrete walls between the transmitter and receiver, and wing of the building that the terminals are in. In addition, we have produced contour plots of locations with equal path loss in buildings.

IV. PATH LOSS PREDICTION MODELS

Although the path loss models presented here are for CW measurements, [2] showed that when individual multipath

component amplitudes are uncorrelated, or phases of individual multipath components are independent and identically distributed over $[0, 2\pi)$, CW and wide-band (250 MHz RF bandwidth) path loss measurements are equivalent when averaged over distances of a few wavelengths. Thus, since our modeling uses local spatial path loss values averaged over 20 λ our models may be used to describe average wide band path loss for these environments, as well. Such models are appropriate for emerging systems that use wide bandwidth (i.e., spread spectrum) to mitigate small-scale fades. Furthermore, work in [5], [8], [9] showed virtually no statistical difference in path loss from 900 MHz to 4.0 GHz for same floor measurements in several buildings. From [2], [5], [8], [9], one could logically infer that the same floor path loss models described in this paper could be applied throughout the low microwave bands (1–5 GHz). In [14], it is shown that the attenuation caused per floor was 6 dB higher at 1700 MHz than at 900 MHz. As a first approximation, this result could be used to scale floor loss for different frequencies. However, more work is required to explicitly determine the effect of multiple floors on path loss for different frequencies.

A. Distance-Dependent Path Loss Model

A model used in [1]–[6], [8], [9], [12] indicates that mean path loss increases exponentially with distance. That is, the mean path loss is a function of distance to the n power

$$\overline{PL}(d) \propto \left(\frac{d}{d_0} \right)^n \qquad (1)$$

where $\overline{PL}$ is mean path loss, n is the mean path loss exponent which indicates how fast path loss increases with distance, d_0 is a reference distance, and d is the transmitter-receiver separation distance. When plotted on a log–log scale, this power-distance relationship is a straight line. Absolute mean path loss, in decibels, is defined as the path loss from the transmitter to the reference distance d_0, plus the additional path loss described by (1) in decibels

$$\overline{PL}(d)[\text{dB}] = PL(d_0)[\text{dB}] + 10 \times n \times \log_{10}\left(\frac{d}{d_0} \right). \qquad (2)$$

For these data, a 1 m reference distance was chosen and we assume $\text{PL}(d_0)$ is due to free space propagation from the transmitter to a 1 m reference distance. Assuming antenna gains equal system cable losses, which is valid for our system, this leads to 31.7 dB path loss at 914 MHz over a 1 m free space path. Measurements show this is accurate to within a decibel nominally [3].

In [1], [2] path loss was shown to be log-normally distributed about (2). Assuming the distribution of large-scale path loss about (2) is log-normal for our data, we determine the mean path loss exponent n and standard deviation σ (in decibels), which are viewed as parameters that are a function of building type, building wing, and number of floors between transmitter and receiver. Even though our measured data show the distribution is not always strictly log-normal, the standard deviation provides a quantitative measure of the accuracy of the model used to predict the path loss for a given environment. Further, when a measurement database is large, the distribution of path loss values over a wide range of distances tends to a log-normal distribution. The path loss at a T–R separation of d meters is then given by

$$PL(d)[\text{dB}] = \overline{PL}(d)[\text{dB}] + X_\sigma[\text{dB}] \qquad (3)$$

where X_σ is a zero mean log-normally distributed random variable with standard deviation σ in decibels. Linear regression was used to compute values of the parameters n and σ in a minimum mean square error (MMSE) sense for the measured data. The data have been grouped by building type, building wing, and the number of floors between the transmitter and receiver to provide smaller standard deviations. As is shown subsequently, this model more accurately predicts path loss as a function of distance when the model parameters n and σ are determined as a function of the general surroundings.

Table I summarizes the mean path loss exponents, standard deviations about the mean for different environments, and the number of measurement locations (20 λ track segments) used to compute the statistics for each category. From Table I, it can be seen that the parameters for path loss prediction for the entire data set are $n = 3.14$ and a large standard deviation of 16.3 dB. This large value of σ is typical for data collected from different building types, and indicates that only 68% of actual measurements will be within ±16.3 dB of the predicted mean path loss. These parameters may be used in the model for a first-order prediction of mean signal strength when only T–R separation but no specific building information is known, but is clearly unsatisfactory for site layout or capacity prediction. For measurements where the transmitter and receiver are on the same floor, the path loss is less severe, and the standard deviation is reduced only slightly. We found $n = 2.76$ and $\sigma = 12.9$ dB using data from four buildings.

The best fit exponent value for all measurement runs in the grocery store is less than two ($n = 1.81$) with a standard deviation of 5.2 dB. In the retail department store, mean path loss increases with distance slightly greater than free space ($n = 2.18$) and there is a spread of 8.7 dB about the mean value. The path loss results for the grocery and retail stores closely agree with those found in open-plan factory buildings [2], [3], [5].

Scatter plots of path loss versus T–R separation for office building measurements are given in Figs. 3 and 4. The dotted lines indicate the distance-dependent mean path loss model ((2)) for $n = 1$ through $n = 6$ and a 1 m reference distance. The dashed line indicates the best mean path loss model in a MMSE sense for the data presented in the scatter plot. Different symbols are used to indicate data from different environments, and overall n and σ are given on the left side of each graph. Multifloor measurements were possible in the two office buildings, and nearly all measurements had multiple obstructions such as concrete walls, windows, and soft partitions between the transmitter and receiver. From Fig. 3,

TABLE I
THE PARAMETERS MEAN PATH LOSS EXPONENT n AND STANDARD DEVIATION σ FOR USE IN THE DISTANCE-DEPENDENT PATH LOSS MODEL IN (2) BASED ON MEASUREMENTS AT A CARRIER FREQUENCY OF 914 MHz

	n	σ(dB)	Number of Locations
All Buildings:			
All Locations	3.14	16.3	634
Same Floor	2.76	12.9	501
Through 1 Floor	4.19	5.1	73
Through 2 Floors	5.04	6.5	30
Through 3 Floors	5.22	6.7	30
Grocery Store	1.81	5.2	89
Retail Store	2.18	8.7	137
Office Building 1:			
Entire Building	3.54	12.8	320
Same Floor	3.27	11.2	238
West Wing 5th Floor	2.68	8.1	104
Central Wing 5th	4.01	4.3	118
West Wing 4th Floor	3.18	4.4	120
Office Building 2:			
Entire Building	4.33	13.3	100
Same Floor	3.25	5.2	37

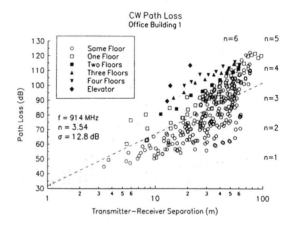

Fig. 3. Scatter plot of CW path loss as a function of distance in office building 1. The symbols represent the number of floors between the transmitter and receiver. Notice the large spread of data about the mean path loss predicted by the simple distance-dependent statistical model with $n = 3.54$.

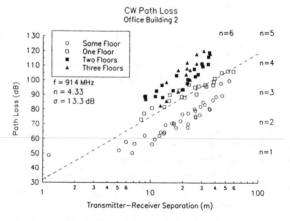

Fig. 4. Scatter plot of CW path loss as a function of distance in office building 2. The symbols represent the number of floors between the transmitter and receiver. Notice the clustering of data as a function of the number of floors between transmitter and receiver.

mean path loss increases with distance to the 3.54 power with a large standard deviation of 12.8 dB in office building 1. The simple d^n path loss model in Fig. 3 does not use knowledge of office partitions or the number of floors between the transmitter and receiver. Transmission between more obstructions leads to higher path loss.

In office building 2, mean path loss increases with distance to the 4.33 power as shown in Fig. 4. The number of floors between the transmitter and receiver can be seen to severely influence the path loss for a given T–R separation. Thus, the number of floors has an impact on the parameter n in the path loss model, and should be quantified for accurate path loss prediction.

The standard deviations about the mean path loss model for individual buildings and for site-specific classifications of measurements in Table I are smaller than those for data from all four buildings. For example, the standard deviation for data from all four buildings is 16.3 dB. If we consider only office building 1, σ drops to 12.8 dB. For same floor measurements only, $\sigma = 11.2$ dB. Further classification of same floor transmitter and receiver measurements into West wing fifth floor, Central wing fifth floor, and West wing fourth floor reduces the standard deviations to 8.1, 4.3, and 4.4 dB for each of the areas, respectively. Thus, more accurate signal strength prediction must be based on building information.

In multifloor environments, (4) is used to describe the mean path loss as a function of distance. Equation (4) is identical to (2) and emphasizes that the mean path loss exponent is a function of the number of floors between transmitter and receiver. The values of n(multifloor) are given in Table I for use in (4).

$$\overline{PL}(d)[\text{dB}] = PL(d_0)[\text{dB}]$$
$$+ 10.0 \times n(\text{multifloor}) \times \log_{10}\left(\frac{d}{d_0}\right). \quad (4)$$

B. Floor Attenuation Factor (FAF) Path Loss Model

In Section IV-A, the path loss in multifloored environments was predicted by a mean path loss exponent that was a function of the number of floors between transmitter and receiver. Alternatively, a constant floor attenuation factor (in decibels), which is a function of the number of floors and building type, may be added to the mean path loss predicted by a path loss model which uses the *same floor* path loss exponent for the particular building type ((5)).

$$\overline{PL}(d)[\text{dB}] = PL(d_0)[\text{dB}] + 10.0 \times n(\text{same floor})$$
$$\times \log_{10}\left(\frac{d}{d_0}\right) + \text{FAF}[\text{dB}] \quad (5)$$

where d is in meters and $PL(d_0)[\text{dB}] = 31.7$ dB at 914 MHz.

Table II gives the floor attenuation factors, the standard deviations (in decibels) of the difference between the mea-

TABLE II
AVERAGE FLOOR ATTENUATION FACTOR IN DECIBELS FOR ONE, TWO, THREE, AND FOUR FLOORS BETWEEN THE TRANSMITTER AND RECEIVER IN THE TWO OFFICE BUILDINGS; ALSO PRESENTED ARE THE STANDARD DEVIATION IN DECIBELS AND THE NUMBER OF LOCATIONS USED TO COMPUTE THE STATISTICS

	FAF (dB)	σ(dB)	Number of Locations
Office Building 1:			
Through 1 floor	12.9	7.0	52
Through 2 floors	18.7	2.8	9
Through 3 floors	24.4	1.7	9
Through 4 floors	27.0	1.5	9
Office Building 2:			
Through 1 floor	16.2	2.9	21
Through 2 floors	27.5	5.4	21
Through 3 floors	31.6	7.2	21

sured and predicted path loss, and the number of discrete measurement locations used to compute the statistics. Values for the floor attenuation factor in Table II are an average (in decibels) of the difference between the path loss observed at multifloor locations and the mean path loss predicted by the simple d^n model ((2)) where n is the *same floor* exponent given in Table I for the particular building structure and d is the shortest distance measured in three dimensions, between the transmitter and receiver. This is similar to the procedure used in [9], [14] to determine the attenuation caused by floors between transmitter and receiver.

The average floor attenuation factors for an identical number of floors between the transmitter and receiver for the two buildings differ by 3–8 dB. All floors in the two office buildings were made of reinforced concrete. Office building 1 was built within the past ten years, and office building 2 is 20 to 30 years old. Both buildings are longer and wider than they are high. Presently, it is unclear what causes the difference between the two buildings. It is interesting to note that in these buildings the average FAF is not a linear function of the number of floors between the transmitter and receiver as was found in [9], [14]. It is possible different floors cause different amounts of path loss, and there may be other factors such as multipath reflections from surrounding buildings that affect the path loss. More measurements are underway in many multifloored buildings to quantify the floor attenuation factors as a function of frequency and building materials, with the goal that eventually, the loss between different floors in buildings may be predicted without measurements.

As an example of how to use the two different models to predict the mean path loss through three floors of office building 1, assume the mean path loss exponent for *same floor* measurements in a building is $n = 3.27$, the mean path loss exponent for three-floor measurements is $n = 5.22$, and the average floor attenuation factor is FAF = 24.4 dB for three floors between transmitter and receiver. These parameters are found from Tables I and II for office building 1. Then, at a T–R separation of $d = 30.0$ m, the predicted mean path loss using (4) is

$$\overline{PL}(30 \text{ m})[\text{dB}] = PL(1 \text{ m})[\text{dB}] + 10.0$$
$$\times 5.22 \times \log_{10}(30) = 108.8 \text{ dB} \quad (6)$$

or, using (5),

$$\overline{PL}(30\ m)[dB] = PL(1\ m)[dB] + 10.0$$
$$\times\ 3.27 \times \log_{10}(30) + 24.4 = 104.4\ dB. \quad (7)$$

C. Soft Partition and Concrete Wall Attenuation Factor Model

The previous models include the effects of T–R separation, building type, and the number of floors between the transmitter and receiver, and are a first step for including site information to improve propagation prediction. Although Table I shows standard deviations can be reduced by accounting for the number of floors between the transmitter and receiver, the standard deviations given in Table I for all *same floor* measurements in the two office buildings are still as large as 11.2 and 13.3 dB. This indicates that building features around transmitter and receiver locations must be considered for a more accurate propagation model. This was done in [7] for hallways and rooms adjacent to a main corridor and in [9] for floors and plasterboard partitioned walls.

There are often obstructions between the transmitter and receiver even when the terminals are on the same floor. We consider the path loss effects of soft partitions and concrete walls between the transmitter and receiver for same floor measurements on the fourth and fifth floor of the West wing of office building 1 and the second floor of office building 2. These areas are classified as soft partitioned environments and are all the measured locations where soft partitions and concrete walls were the only obstructions between the transmitter and receiver. For a physical model that will apply in general, we assume path loss increases with distance as in free space ($n = 2$), so long as there are no obstructions between the transmitter and receiver. Then, we include attenuation factors for each soft partition and concrete wall that lie directly between the transmitter and receiver. For simplicity, any kind of concrete support column that wholly or partially blocks the direct path between the transmitter and receiver is labeled a concrete wall. Let p be the number of soft partitions and q be the number of concrete walls between the transmitter and receiver. The mean path loss predicted by the attenuation factor path loss model is then given by

$$\overline{PL}(d)[dB] = 20.0 \times \log_{10}\left(\frac{4\pi d}{\lambda}\right)$$
$$+ p \times AF(\text{soft partition})[dB]$$
$$+ q \times AF(\text{concrete wall})[dB]. \quad (8)$$

Notice that no reference distance is used since free space propagation is assumed for all distances. This model is similar in form to one proposed in [9] for floors and walls.

For each of the discrete path loss measurements in the soft partitioned environments, we computed the difference between the measured path loss and the path loss that would occur due to free space propagation for a transmitter and receiver at the same separation distance. We also recorded the number of soft partitions and the number of concrete walls (or concrete building support columns) between the transmitter and receiver. Linear regression was used to find the best fit, in a minimum mean square error sense, for the attenuation factors of each soft partition and each concrete wall between the transmitter and receiver, where we assume all soft partitions induce identical path loss, and all concrete walls induce identical path loss. We also found the standard deviation in decibels of the difference between the measured path loss and the path loss predicted by (8).

When all path loss measurements in soft partitioned environments are considered, the AF is 1.39 dB for each soft partition and 2.38 dB for each concrete wall between the transmitter and receiver. When each soft partitioned environment is considered separately, the attenuation factors range from 0.92 to 1.57 dB for each soft partition and from 1.99 to 2.45 dB for each concrete wall. Fig. 5 shows a scatter plot of the actual measured path loss versus the path loss predicted by (8) with AF (soft partition) = 1.39 dB and AF (concrete wall) = 2.38 dB for the three soft partitioned environments. The diagonal straight line in Fig. 5 shows where measured and predicted path loss are identical. The standard deviation of the difference between measured and predicted path loss is 4.1 dB. Compare the 4.1 dB standard deviation of the attenuation factor model for the data from three Soft Partitioned environments to the standard deviations for the distance-dependent path loss model in (2) for each area considered separately. In office building 1, the standard deviations are 4.4 and 8.1 dB for same floor measurements in the West wing on the fourth and fifth floor, respectively, and 5.2 dB for the second floor of office building 2. The soft partition and concrete wall attenuation factor model in (8) explains the deviation of the mean path loss exponent from free space ($n = 2$) based on a physical model that assumes free space propagation with distance and attributes additional path loss to identifiable physical obstructions between the transmitter and receiver. The first such attenuation factors were presented in [2] for open-plan factory buildings.

Although the 4.1 dB standard deviation of the attenuation factor model is greater than the 2.94 dB for in-building path loss models reported in [7], the building areas modeled here are much different propagation environments than those in [7]. For our data, multiple obstructing objects are present between the transmitter and receiver whereas in [7], measurements were made in open hallways and rooms adjacent to hallways. In [7], the effects of obstructions between the transmitter and receiver used a received power model of the form $a - b \log(\text{distance})$ where the a and b are found by a minimum mean square error fit similar to the method used here. The parameter a has no physical significance and is essentially an offset used to fit the data. It would be possible to model our data with an offset to reduce the standard deviation. However, an offset which varies from building to building that cannot be associated with the physical surroundings is not useful for a propagation prediction tool that could be used to predict path loss contours in a particular building *a priori*. In [9], standard deviations were 3–4 dB after correcting for floors and walls. However, no values for the attenuation these obstructions caused are given.

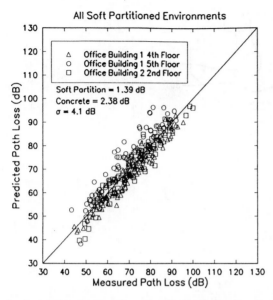

Fig. 5. Scatter plot of measured versus predicted path loss for soft partitioned environments. The standard deviation of the prediction error is 4.1 dB.

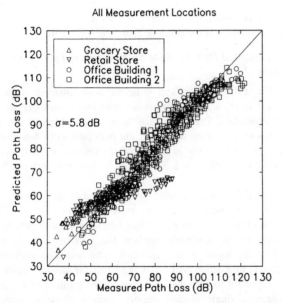

Fig. 6. Scatter plot of measured versus predicted path loss for all measurement locations. The prediction error is log-normally distributed with a standard deviation of 5.8 dB.

Fig. 6 shows a scatter plot of measured and predicted path loss for all measured locations in the four buildings. The predicted values were found using the distance-dependent path loss model ((2)) for all locations in the grocery and retail stores, and all co-floor transmitter and receiver combinations in the office building areas that cannot be classified as soft partitioned environments. The floor attenuation factor model ((5)) was used for multifloor locations in the two office buildings. In the soft partitioned environments in the office buildings, the soft partition and concrete wall attenuation

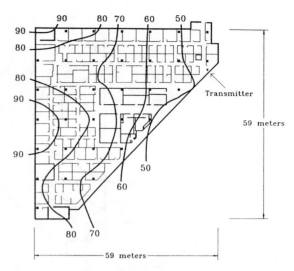

Fig. 7. Contour plot of locations with equal path loss in 10 dB steps for the fifth floor of the West wing of office building 1. The transmitter was located in the upper right corner of the figure.

factor model ((8)) was used to predict path loss. We have found that the distribution of errors between measured and predicted path loss is log-normal with a standard deviation of 5.8 dB. Path loss was also shown to be log-normally distributed after the attenuation caused by floors and walls was corrected for in [9]. By including the effects of floors, soft partitions, and concrete walls between the transmitter and receiver, we have reduced the standard deviation of our path loss prediction error from 16.3 to 5.8 dB.

V. PATH LOSS CONTOUR PLOTS

Building blueprints and both measured and predicted path loss data were imported to a computer-aided design program (AutoCAD). The data have been used to form contour plots of locations of equal path loss for a given transmitter location. In each figure, the transmitter location is indicated by an arrow pointing to an "X" at the transmitter location. Curved solid lines indicate locations of equal path loss from the transmitter in 10 dB steps. The amount of path loss is indicated at the end of the lines on the perimeter of the blueprints.

A. Measured Path Loss Contours

The contour plot of locations with equal measured path loss for the West wing on the fifth floor of office building 1 is given in Fig. 7. The transmitter was located in the upper right-hand corner of the figure as indicated. Curved solid lines represent contours of equal path loss from the transmitter in 10 dB steps. The thin lines on the drawing indicate 1.5 m high soft partition cubicle dividers. In the center of the wing are conference rooms with concrete block walls which span from the floor to the ceiling. These walls are indicated by thick lines on the drawing. In Fig. 7, notice that when the thick conference room walls are between the receiver and the transmitter, the signal is attenuated much more rapidly than at other locations. Along the diagonal hallway along the edge

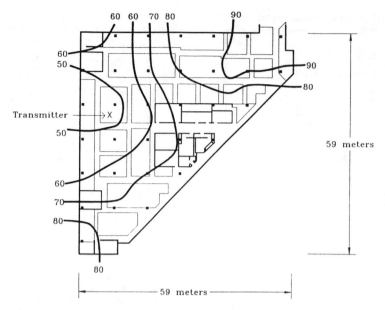

Fig. 8.   Contour plot of locations with equal path loss in 10 dB steps for the fourth floor of the West wing of office building 1. The transmitter was located on the left side of the figure.

of the building, the radio coverage is quite good, and obeys better than free space propagation. Propagation better than in free space was also observed in [2], [3], [7] for open hallways that can guide signal energy.

The contour plot for the fourth floor West wing of office building 1 is shown in Fig. 8. The transmitter was located inside an office cubicle on the left side of the figure. Thick lines indicate concrete block walls that span from floor to ceiling. The thin lines which surround rectangular areas represent the perimeters of soft-partitioned office cubicles. Both the transmitter and the receiver antenna were 1.0 m above the floor. The path loss seems to be consistent with T–R separation and is not greatly affected by the direction of propagation.

Fig. 9 shows the contour plot for the second floor of office building 2. Notice that the contour lines for equal path loss extend farther from the transmitter toward the top of the figure. This is because the geometry in the favorable direction is similar to a line-of-sight aisle where path loss generally falls off slower than free space due to guiding of signal energy. The 80–100 dB path loss lines near the top of the figure indicate that path loss increases much more rapidly through the concrete walls than through the soft partitions.

Fig. 10 gives the contour plot in 5 dB steps for identical receiver locations on the second floor of office building 2 with the transmitter located in the basement of the building directly below the location indicated by an "X" in the figure. The path loss is primarily due to the concrete floors between the transmitter and receiver. These floors cause the mean path loss exponent to be 5.31 for measurements on the second floor with the transmitter in the basement when (4) is used to predict path loss. The nearly circular contours indicate that a simple d^n model can be used to accurately predict site-specific path loss in multifloored environments. A d^n

path loss model predicts circular contours since the path loss depends only on the distance from the transmitter.

B. Predicted Path Loss for Soft Partitioned Environments Using (8)

The soft partition and concrete wall attenuation factor model in (8) was used to predict path loss in soft partitioned environments. The attenuation factors for soft partitions and concrete walls used to predict the path loss were determined from measurements in three soft partitioned environments. AF (soft partition) = 1.39 dB and AF(concrete wall) = 2.38 dB.

The differences between measured and predicted path loss have been used to generate error contours. The absolute value of the error has been plotted to show regions where the model in (8) accurately predicts the path loss and regions where the model is less accurate. The path loss prediction error is proportional to the darkness of the shaded region. It is important to note that the error can be positive or negative within a single shaded region. This is done to indicate regions where the models have difficulty predicting the path loss. Thus, the use of a correction factor for the shaded regions is not appropriate.

The error contours for the West wing of the fifth floor of office building 1 and the contours of measured path loss for the same transmitter location and building wing are given in Fig. 11. The prediction error is less than 3 dB for about half the area in this location. There are several places where the prediction error is greater than 9 dB. This may be partly explained in that the Attenuation Factors for the fifth floor in the West wing of office building 1 were the lowest measured and differ by about 0.4 dB from the attenuation factors used to predict path loss. Thus, on average, we predict 0.4 dB more path loss than was actually observed for each partition

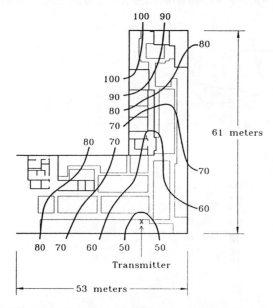

Fig. 9. Contour plot of locations with equal path loss in 10 dB steps for the second floor of office building 2 with the transmitter on the same floor. The transmitter was in the lower right corner of the figure.

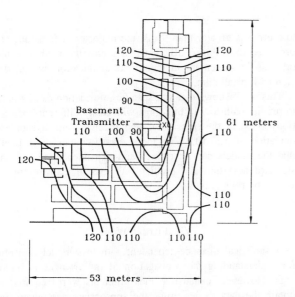

Fig. 10. Contour plot of locations with equal path loss in 5 dB steps for the second floor of office building 2 with the transmitter in the basement.

and concrete wall in this wing. When many partitions and concrete walls are between the transmitter and receiver, the error can be 4–5 dB.

Fig. 12 shows the measured and error contours for the fourth floor West wing of office building 1. The prediction error is less than 3 dB for over half the locations and less than 6 dB for nearly all of this wing. The error contours for this portion of the building show that (8) can be used to accurately predict path loss based on site-specific information.

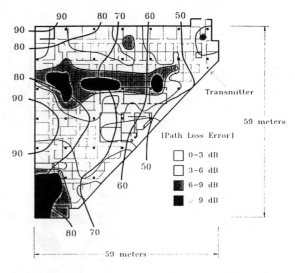

Fig. 11. Error contour plot for the fifth floor of office building 1. Path Loss error is proportional to the darkness of the shaded region. The path loss was predicted using (8) with AF(soft partition) = 1.39 dB and AF(concrete wall) = 2.38 dB.

Fig. 13 shows the error contour plot for the second floor of office building 2. The contours of measured path loss are given in Fig. 13. As in the case of the fourth floor West wing of office building 1, the path loss error is less than 3 dB in over half of the area and less than 6 dB for nearly all of the floor. Once again, error contours indicate that (8) can be used to accurately predict path loss in soft partitioned environments.

VI. CONCLUSION

Path loss models based on measured data at 914 MHz have been presented for four different buildings. The models are based on a simple d^n exponential path loss vs. distance relationship. In open-plan buildings such as the grocery store and the retail store, the path loss exponent n is close to 2. For environments with many more obstructions between the transmitter and receiver, the path loss exponent can be much higher. The models have been shown to be more accurate when different buildings and dissimilar areas within the same building are considered separately.

A floor-dependent path loss exponent (Table I) may be used to model the effects of the number of floors between the transmitter and receiver in conjunction with (4). Alternatively, the path loss exponent for co-floor propagation, along with a floor attenuation factor (Table II) to account for the additional path loss due to floors, may be used to predict path loss in conjunction with (5).

Attenuation factors for plastic covered office partitions and concrete walls that are located between the transmitter and receiver have been found. These factors are useful since they allow us to predict path loss in terms of free space path loss and physically identifiable objects that are part of the propagation environment. With the model in (8), it is possible to predict path loss contours for a transmitter and receiver located in the same wing of an office building where individ-

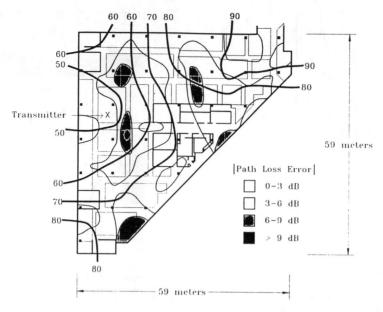

Fig. 12. Error contour plot for the fourth floor of office building 1. The large unshaded region indicates where the path loss prediction error is less than 3 dB.

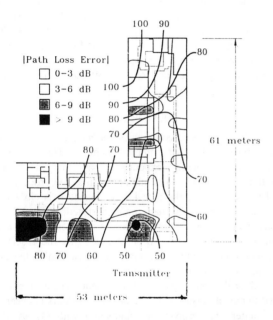

Fig. 13. Error contour plot for the second floor of office building 2 with the transmitter on the same floor. The prediction error is less than 6 dB for nearly the entire floor.

ual office cubicles are separated by cloth-covered plastic dividers. For all of our measured data, AF(soft partitions) = 1.39 dB/partition and AF (concrete walls) = 2.38 dB/wall.

Examination of error contours which show the difference between measured and predicted path loss for the buildings given here show how simple path loss models that use site-specific information, such as the locations of soft partitions and concrete walls, can be used to predict path loss in buildings. Although there are some locations where the path

loss error is greater than 9 dB, the majority of locations are predicted to within 6 dB. Thus, it seems the method could offer an improvement in indoor communication system design and installation.

This work shows that site-specific information can be used to predict path loss in buildings with many different obstructions between the transmitter and receiver with a standard deviation of 5.8 dB. With more measurements, it is likely that the models can be refined and expanded to reduce the standard deviation and predict path loss in many more diverse environments.

VII. FUTURE WORK

A statistical distance-dependent path loss model is useful for understanding the propagation of radio waves in buildings. However, exhaustive measurements were required to obtain the data to determine the appropriate parameters for the models for these particular buildings. Models that allow a system designer to predict path loss contours for all types of buildings without measurements would be extremely cost-effective and time-efficient. In order to develop these models, attenuation factors for many different types of common objects in buildings must be determined for different frequency ranges. More measurements in similar and different building types at different frequencies are required to determine the effects of multipath propagation must be considered for extremely accurate propagation prediction.

ACKNOWLEDGMENT

The authors would like to thank Mike Keitz and Ken Blackard for their help in the data collection. The thoughtful comments of the reviewers that helped improve the quality of this paper are gratefully appreciated.

REFERENCES

[1] D. C. Cox, R. R. Murray, and A. W. Norris, "800 MHz attenuation measured in and around suburban houses," *AT&T Bell Lab. Tech. J.*, vol. 63, pp. 921–954, July/Aug. 1984.

[2] T. S. Rappaport, "Characterization of UHF multipath radio channels in factory buildings," *IEEE Trans. Antennas Propagat.*, vol. 37, pp. 1058–1069, Aug. 1989.

[3] T. S. Rappaport and C. D. McGillem, "UHF fading in factories," *IEEE J. Select. Areas Commun.*, vol. 7, pp. 40–48, Jan. 1989.

[4] F. C. Owen and C. D. Pudney, "Radio propagation for digital cordless telephone at 1700 MHz and 900 MHz," *Electron. Lett.*, vol. 25, no. 1, pp. 52–53, 1989.

[5] D. A. Hawbaker and T. S. Rappaport, "Indoor wideband radiowave propagation measurements at 1.3 GHz and 4.0 GHz," *Electron. Lett.*, vol. 26, no. 21, pp. 1800–1802, 1990.

[6] A. A. M. Saleh and R. A. Valenzuela, "A statistical model for indoor multipath propagation," *IEEE J. Select. Areas Commun.*, vol. SAC-5, pp. 128–137, Feb. 1987.

[7] J-F. Lafortune and M. Lecours, "Measurement and modeling of propagation losses in a building at 900 MHz," *IEEE Trans. Veh. Technol.*, vol. 39, pp. 101–108, May 1990.

[8] D. M. J. Devasirvatham, M. J. Krain, and D. A. Rappaport, "Radio propagation measurements at 850 MHz, 1.7 GHz, and 4.0 GHz inside two dissimilar office buildings," *Electron. Lett.*, vol. 26, no. 7, pp. 445–447, 1990.

[9] A. J. Motley and J. M. Keenan, "Radio coverage in buildings," *British Telecom Tech. J.*, Special Issue on Mobile Communications, vol. 8, no. 1, pp. 19–24, Jan. 1990.

[10] D. Molkdar, "Review on radio propagation into and within buildings," *Inst. Elec. Eng. Proc.*, pt. H, vol. 138, no. 1, pp. 61–73, Feb. 1991.

[11] S. Y. Seidel and T. S. Rappaport, "900 MHz path loss measurements and prediction techniques for in-building communication system design," presented at *41st IEEE Veh. Technol. Conf.*, St. Louis, MO, May 21, 1991, pp. 613–618.

[12] T. S. Rappaport, S. Y. Seidel, and K. Takamizawa, "Statistical channel impulse response models for factory and open plan building radio communication system design," *IEEE Trans. Commun.*, vol. 39, pp. 794–807, May 1991.

[13] W. C. Y. Lee, *Mobile Communications Engineering.* New York: McGraw-Hill, 1982.

[14] A. J. Motley and J. M. P. Keenan, "Personal communication radio coverage in buildings at 900 MHz and 1700 MHz," *Inst. Elec. Eng. Electron. Lett.*, vol. 24, no. 12, pp. 763–764, 1989.

Scott Y. Seidel (S'89) was born in Falls Church, VA, in 1966. He received the B.S. and the M.S. degrees in electrical engineering from Virginia Polytechnic Institute and State University, Blacksburg, in 1988 and 1989, respectively. He is pursuing the Ph.D. degree in electrical engineering at Virginia Polytechnic Institute and State University under the support of an NSF Graduate Research Fellowship.

Since 1989, he has been a member of the Mobile and Portable Radio Research Group at Virginia Tech. He has been involved in the development of both urban microcellular and indoor radio channel models and is interested in site-specific propagation prediction and system design. He is co-inventor of SIRCIM, an indoor radio channel simulator that has been adopted by over 40 companies and universities.

Mr. Seidel is a member of Tau Beta Pi, Eta Kappa Nu, and Phi Kappa Phi.

Theodore S. Rappaport (S'83–M'84–S'85–M'87–SM'90) was born in Brooklyn, NY, on November 26, 1960. He received the B.S.E.E, M.S.E.E, and Ph.D. degrees from Purdue University, West Lafayette, IN, in 1982, 1984, and 1987, respectively.

From 1984 to 1988 he was with the NSF Engineering Research Center for Intelligent Manufacturing Systems at Purdue University. In 1988, he joined the Electrical Engineering faculty of Virginia Tech where he is an Assistant Professor and Director of the Mobile and Portable Radio Research Group. He conducts research in mobile radio communication system design and RF propagation prediction through measurements and modeling, and consults often in these areas. He guides a number of graduate and undergraduate students in mobile radio communications, and has authored or coauthored more than 50 technical papers in the areas of mobile radio communications and propagation, vehicular navigation, ionospheric propagation, and wide-band communications. He holds a U.S. patent for a wide-band antenna, and is co-inventor of SIRCIM, an indoor radio channel simulator that has been adopted by over 40 companies and universities.

Dr. Rappaport received the Marconi Young Scientist Award in 1990 for his contributions in indoor radio communications. He serves as senior editor of the IEEE JOURNAL ON SELECTED AREAS IN COMMUNICATIONS. He is a member of IEEE COMSOC Radio Committee, and is a Fellow of the Radio Club of America. He is a member of ASEE, Eta Kappa Nu, Tau Beta Pi, Sigma Xi, and is a life member of the ARRL.

Reprinted from *IEEE Transactions on Vehicular Technology* , Vol. 2, VT-23, pp. 1-20, Nov. 1974.

Properties of Mobile Radio Propagation Above 400 MHz

DOUGLAS O. REUDINK, MEMBER, IEEE

Abstract—This paper begins with a discussion of multipath interference. The spatial description of the field impinging upon a mobile radio antenna is derived and the power spectrum and other properties of the signal envelope are considered. Next, large scale variations of the average signal are discussed. Measurements of observed attenuation on mobile paths over both smooth and irregular terrain are summarized. The paper concludes with a discussion of methods of predicting the area of coverage from a base station.

I. INTRODUCTION

THIS PAPER reviews mobile radio transmission in the frequency ranges above 400 MHz. The emphasis is entirely on CW propagation and the results should thus be applied only to relatively narrow-band systems (< 100-kHz channels, for example). The rapid and extreme amplitude fluctuations of the mobile signal are discussed in Section II. The next section first considers factors that affect transmission such as diffraction, rain, and the atmosphere. Then observed attenuation on mobile radio paths are summarized. In Section IV methods are given for predicting field strength.

Manuscript received December 6, 1973; revised July 5, 1974. This paper was presented at the 24th IEEE Vehicular Technology Conference, Cleveland, Ohio, December 4–5, 1973.
The author is with Bell Telephone Laboratories, Inc., Crawford Hill Laboratory, Holmdel, N.J. 07733.

II. SHORT TERM FADING

One readily accessible property of the signal transmitted over a mobile radio propagation path is the amplitude variation of its envelope as the position of the mobile terminal is moved. This information is generally presented in the form of time recordings of the signal level; with uniform vehicle motion there is, of course, a 1:1 correspondence between distance measured on the recording and distance traveled in the street. A typical recording [1] is shown in Fig. 1 for a run made at 836 MHz in a suburban environment. The occasional deep fades and quasi-periodic occurrence of minima are clearly evident in the expanded section of the record.

Recordings such as these made by many workers in the field over the frequency range from 50 MHz to 11 200 MHz have shown that the envelope of the mobile radio signal is Rayleigh distributed [2]-[5] when measured over distances of a few tens of wavelengths where the mean signal is sensibly constant. This suggests the assumption [6], reasonable on physical grounds, that at any point the received field is made up of a number of horizontally traveling plane waves with random amplitudes and angles of arrival for different locations. The phases of the waves are uniformly distributed from zero to 2π. The amplitudes and phases are assumed to be statistically independent.

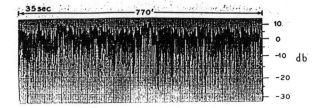
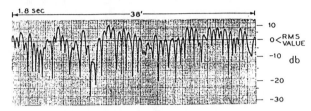

Fig. 1. Typical received signal variations at 836 MHz measured at mobile speed of 15 mi/h records taken on same street with different recording speeds.

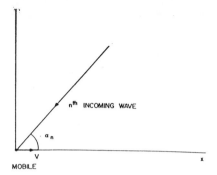

Fig. 2. Typical component wave incident on mobile receiver.

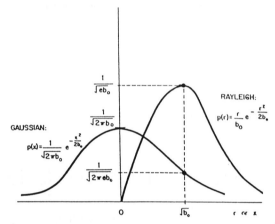

Fig. 3. Gaussian and Rayleigh probability density functions.

A diagram of this simple model is shown in Fig. 2, with plane waves from stationary scatterers incident on a mobile traveling in the x direction with velocity v. The xy plane is assumed to be horizontal. The vehicle motion introduces a Doppler shift in every wave

$$\omega_n = \beta v \cos \alpha_n \qquad (2.1)$$

where $\beta = 2\pi/\lambda$, λ equals wavelength of the transmitted carrier frequency. If the transmitted signal is vertically polarized the E-field seen at the mobile can thus be written

$$E_z = E_0 \sum_{n=1}^{N} C_n \cos (\omega_c t + \theta_n) \qquad (2.2)$$

where

$$\theta_n = \omega_n t + \phi_n \qquad (2.3)$$

and ω_c is the carrier frequency of the transmitted signal, $E_0 C_n$ is the (real) amplitude of the nth wave in the E_z field. The ϕ_n are random phase angles uniformly distributed from 0 to 2π. Furthermore, the C_n are normalized so that the ensemble average

$$\sum_{n=1}^{N} C_n^2 = 1.$$

We note from (2.1) that the Doppler shift is bounded by the values $\pm \beta v$, which, in general, will be very much less than the carrier frequency. For example, for $f_c = \omega_c/2\pi = 1000$ MHz, $v = 96$ km/h (60 mi/h)

$$\frac{1}{2\pi} \beta v = \frac{v}{\lambda} = 90 \text{ Hz}. \qquad (2.4)$$

The field component may thus be described as narrowband random processes. Furthermore, as a consequence of the central limit theorem, for large values of N they are approximately Gaussian random processes, and the considerable body of literature devoted to such processes may be utilized. It must be kept in mind that this is still an approximation; for example, (2.2) implies that the mean

signal power is constant with time, whereas it actually undergoes slow variations as the mobile moves distances of hundreds of feet. Nevertheless, the Gaussian model is successful in predicting the measured statistics of the signal to good accuracy in most cases over the ranges of interest for the variables involved, thus its use is justified. In Section III we shall consider in detail the large-scale variations of this average.

Following Rice [7] we can express E_z as

$$E_z = T_c(t) \cos \omega_c t - T_s(t) \sin \omega_c t \qquad (2.5)$$

where

$$T_c(t) = E_0 \sum_{n=1}^{N} C_n \cos (\omega_n t + \phi_n) \qquad (2.6)$$

$$T_s(t) = E_0 \sum_{n=1}^{N} C_n \sin (\omega_n t + \phi_n) \qquad (2.7)$$

are Gaussian random processes, corresponding to the in-phase and quadrature components of E_z, respectively. We denote by T_c and T_s the random variables corresponding to $T_c(t)$ and $T_s(t)$ for fixed t. They have zero mean and equal variance

$$\langle T_c^2 \rangle = \langle T_s^2 \rangle = \frac{E_0^2}{2} = \langle | E_z |^2 \rangle. \qquad (2.8)$$

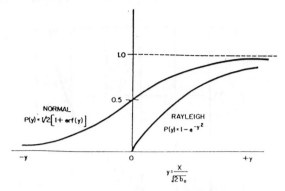

Fig. 4. Normal and Rayleigh cumulative distributions.

The brackets indicate an ensemble average over the ϕ_n and C_n. T_c and T_s are also uncorrelated (and therefore independent)

$$\langle T_c T_s \rangle = 0. \qquad (2.9)$$

Since T_c and T_s are Gaussian, they have probability densities of the form

$$p(x) = \frac{1}{(2\pi b_0)^{1/2}} \exp\left(-\frac{x^2}{2b_0}\right) \qquad (2.10)$$

where $b_0 = E_0^2/2 =$ mean power, and $x = T_c$ or T_s.

The envelope of E_z is given by

$$r = (T_c^2 + T_s^2)^{1/2} \qquad (2.11)$$

and Rice [7] has shown that the probability density of r is

$$p(r) = \begin{cases} \dfrac{r}{b_0} \exp\left(-\dfrac{r^2}{2b_0}\right), & r \geq 0 \\[2ex] 0, & r < 0 \end{cases} \qquad (2.12)$$

which is the Rayleigh density formula. The Gaussian and Rayleigh densities are shown in Fig. 3 for illustration.

The cumulative distribution functions of T_c (or T_s) are also of interest

$$P[x \leq X] = \int_{-\infty}^{X} p(x)\, dx = \frac{1}{2}\left[1 + \operatorname{erf}\left(\frac{X}{(2b_0)^{1/2}}\right)\right] \qquad (2.13)$$

where the error function is defined by

$$\operatorname{erf}(y) = \frac{2}{\pi^{1/2}} \int_0^y \exp(-t^2)\, dt. \qquad (2.14)$$

Similarly for the envelope

$$P[r \leq R] = \int_{-\infty}^{R} p(r)\, dr = 1 - \exp\left(-\frac{R}{2b_0}\right). \qquad (2.15)$$

These distribution functions are illustrated in Fig. 4.

RF Power Spectrum

From the viewpoint of an observer on the mobile unit, the signal received from a CW transmission as the mobile moves with constant velocity may be represented as a carrier whose phase and amplitude are randomly varying, with an effective bandwidth corresponding to twice the maximum Doppler shift of βV. Many of the statistical properties of this random process can be determined from its moments, which, in turn are most easily obtained from the power spectrum [6].

We assume that the field may be represented by the sum of N waves, as in (2.2). As $N \rightarrow \infty$ we would expect to find that the incident power included in an angle between α and $\alpha + d\alpha$ would approach a continuous, instead of discrete, distribution. Let us denote by $p(\alpha)\, d\alpha$ the fraction of the total incoming power within $d\alpha$ of the angle α, and also assume that the receiving antenna is directive in the horizontal plane with power gain pattern $G(\alpha)$. The differential variation of received power with angle is then $b_0 G(\alpha) p(\alpha)\, d\alpha$; we equate this to the differential variation of received power with frequency by noting the relationship between frequency and angle of (2.2)

$$f(\alpha) = f_m \cos \alpha + f_c \qquad (2.16)$$

where $f_m = \beta v/2\pi = v/\lambda$, the maximum Doppler shift. Since $f(\alpha) = f(-\alpha)$, the differential variation of power with frequency may be expressed

$$S(f)\,|df| = b_0[p(\alpha)G(\alpha) + p(-\alpha)G(-\alpha)]\,|d\alpha|. \qquad (2.17)$$

However,

$$|df| = f_m\,|-\sin\alpha\, d\alpha| = [f_m^2 - (f - f_c)^2]^{1/2}\,|d\alpha|$$

thus

$$S(f) = \frac{b_0}{[f_m^2 - (f - f_c)^2]^{1/2}}\big[p(\alpha)G(\alpha) + p(-\alpha)G(-\alpha)\big] \qquad (2.18)$$

where

$$\alpha = \cos^{-1}\left(\frac{f - f_c}{f_m}\right) \quad \text{and} \quad S(f) = 0, \quad \text{if } |f - f_c| > f_m.$$

We assume the transmitted signal is vertically polarized. The electric field will then be in the z direction and may be sensed by a vertical whip antenna on the mobile, with $G(\alpha) = 1$. Substituting in (2.18), the power spectrum of the electric field is

$$S_{E_z}(f) = \frac{b_0}{[f_m^2 - (f - f_c)^2]^{1/2}}[p(\alpha) + p(-\alpha)]. \qquad (2.19)$$

The simplest assumption for the distribution of power with arrival angle α is a uniform distribution

$$p(\alpha) = \frac{1}{2\pi}, \qquad -\pi \leq \alpha \leq \pi. \qquad (2.20)$$

Assuming no additional antenna directivity, the power spectrum is

$$S_{E_z}(f) = \frac{2b_0}{\omega_m}\left[1 - \left(\frac{f - f_c}{f_m}\right)^2\right]^{-1/2}. \qquad (2.21)$$

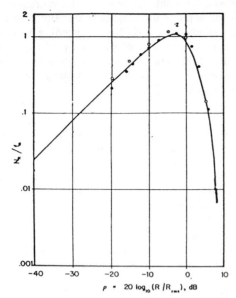

Fig. 5. Normalized level crossing rates of envelope. Measured values: ○—11 215 MHz, ●—836 MHz (E_z).

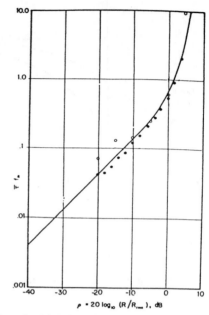

Fig. 6. Normalized durations of fade of envelope. Measured values: ○—11 215 MHz, ●—836 MHz (E_z).

From the relations between correlations and power spectra the moments can also be obtained [6]

$$b_n = (2\pi)^n \int_{f_c-f_m}^{f_c+f_m} S(f)(f-f_c)^n \, df. \qquad (2.22)$$

Level Crossing Rates

As illustrated in Fig. 1 the signal envelope experiences very deep fades only occasionally; the shallower the fade the more frequently it is likely to occur. A quantitative expression of this property is the level crossing rate N_R, which is defined as the expected rate at which the envelope crosses a specified signal level R in the positive direction. In general, it is given by [7]

$$N_R = \int_0^\infty \dot{r} p(R, \dot{r}) \, d\dot{r} \qquad (2.23)$$

where the dot indicates the time derivative and $p(R, \dot{r})$ is the joint density function of r and $\dot{r}$ at $r = R$. Rice [7] gives the joint density function in the four random variables $r, \dot{r}, \theta, \dot{\theta}$ of a Gaussian process

$$p(r, \dot{r}, \theta, \dot{\theta}) = \frac{r^2}{4\pi^2 b_0 b_2} \exp\left[-\frac{1}{2}\left(\frac{r^2}{b_0} + \frac{\dot{r}^2}{b_2} + \frac{r^2\dot{\theta}^2}{b_2}\right)\right] \quad (2.24)$$

where $\tan\theta = -T_s/T_c$. Integrating this expression over θ from 0 to 2π and $\dot{\theta}$ from $-\infty$ to $+\infty$ we get

$$p(r, \dot{r}) = \underbrace{\frac{r}{b_0} \exp\left(-\frac{r^2}{2b_0}\right)}_{p(r)} \underbrace{\frac{1}{(2\pi b_2)^{1/2}} \exp\left(-\frac{\dot{r}^2}{2b_2}\right)}_{p(\dot{r})} \quad (2.25)$$

since $p(r, \dot{r}) = p(r)p(\dot{r})$, r and $\dot{r}$ are independent and uncorrelated. Substituting (2.25) into (2.23) we get the

level crossing rate

$$N_R = \frac{p(R)}{(2\pi b_2)^{1/2}} \int_0^\infty \dot{r} \exp\left(-\frac{\dot{r}^2}{2b_2}\right) dr = \left(\frac{b_2}{b_0}\right)^{1/2} \rho \exp(-\rho^2)$$

$$(2.26)$$

where

$$\rho = R/(\langle r^2 \rangle)^{1/2} = R/(2b_0)^{1/2} = R/R_{rms}. \quad (2.27)$$

Substituting the appropriate values of the moments b_0 and b_2 we get the expressions for the level crossing rate

$$E_z : N_R = (2\pi)^{1/2} f_m \rho \exp(-\rho^2). \qquad (2.28)$$

This expression is plotted in Fig. 5 along with some measured values [1]. The rms level of E_z is crossed at a rate of 0.915 f_m; for example: at $f = 1000$ MHz and $V = 96$ km/h, $f_m = 90$ Hz, thus $N_R = 82/s$ at $\rho = 0$ dB. Lower signal levels are crossed less frequently, as shown by the curves. The maximum level crossing rate occurs at $\rho = -3$ dB.

Duration of Fades

The average duration of fades below $r = R$ is also of interest. Let τ_i be the duration of the ith fade. Then the probability that $r \leq R$ for a total time interval of length T is

$$P[r \leq R] = \frac{1}{T}\sum \tau_i. \qquad (2.29)$$

The average fade duration is

$$\bar{\tau} = \frac{1}{T N_R} \sum \tau_i = \frac{1}{N_R} P[r \leq R]. \qquad (2.30)$$

Since

$$P[r \leq R] = \int_0^R p(r)\, dr = 1 - \exp(-\rho^2). \quad (2.31)$$

Then

$$\bar{\tau} = \left(\frac{\pi b_0}{b_2}\right)^{1/2} \frac{1}{\rho} [\exp(\rho^2) - 1]. \quad (2.32)$$

Substituting the appropriate values of the moments

$$E_z: \bar{\tau} = \frac{\exp(\rho^2) - 1}{\rho f_m (2\pi)^{1/2}}. \quad (2.33)$$

This expression is plotted in Fig. 6 along with some measured values [1].

III. LARGE SCALE VARIATIONS OF THE AVERAGE SIGNAL

The principal methods by which energy is transmitted to a mobile, namely, reflection and diffraction are often indistinguishable, thus it is convenient to lump the losses together and call them scatter (or shadow) losses. In Section II it was shown that this scattering gives rise to fields whose amplitudes are Rayleigh distributed in space and the Rayleigh model led to very powerful results. In this section it is shown that while the "local statistics" may be Rayleigh, the "local mean" varies because of the terrain and the effects of other obstacles. We shall begin by reviewing briefly the more deterministic factors that influence propagation and conclude with a look at the statistical nature of signal strength.

Free Space Transmission Formula

The power received by an antenna separated from a radiating antenna is given by a simple formula, provided there are no objects in the region which absorb or reflect energy. This free space transmission formula [8] depends upon the inverse square of the antenna separation distance d and is given by

$$P_0 = P_t \left(\frac{\lambda}{4\pi d}\right)^2 g_b g_m \quad (3.1)$$

where

P_0 received power,
P_t transmitted power,
λ wavelength,
g_b power gain of the base station antenna,
g_m power gain of the mobile station antenna.

Thus the received radiated power decreases 6 dB for each doubling of the distance. On first inspection, one might conclude that higher frequencies might be unsuitable for mobile communications because the transmission loss increases with the square of the frequency. However, this usually can be compensated for by increased antenna gain. In mobile communications, it is often desirable to have antennas whose patterns are omnidirectional in the azimuthal plane, thus the increase in gain is required in

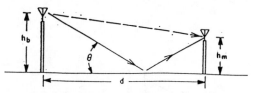

Fig. 7. Propagation paths over plane earth.

the elevation plane. In the limit of the higher microwave frequencies, this additional gain may become impractical to realize for effective communications between an elevated base station and a mobile.

Knowing the propagation characteristics over a smooth conducting flat earth provides a starting point for estimating the effects of propagation over actual paths. Analytical results for propagation over a plane earth have been derived by Norton [9]-[11] and simplified by Bullington [12], [13]. For base station and mobile antennas elevated heights h_b and h_m, respectively, above ground as shown in Fig. 7 the power received is

$$P_r = 4P_0 \sin^2 \frac{2\pi h_b h_m}{\lambda d} \quad (3.2)$$

where P_0 is the expected power over a free space path. In most mobile radio applications, except very near the base station antenna, $\sin(\Delta/2) \approx \Delta/2$, thus the transmission over a plane earth is given by the approximation

$$P_r = P_t g_b g_m \left(\frac{h_b h_m}{d^2}\right)^2 \quad (3.3)$$

yielding an inverse fourth power relationship of received power with distance from the base station antenna. However, at the higher microwave frequencies the assumption of a plane earth may no longer be valid due to surface irregularities. A measure of the surface "roughness," which provides an indication of the range of validity of (3.2), is given by the Rayleigh criterion, which is

$$C = \frac{4\pi\sigma\theta}{\lambda} \quad (3.4)$$

where σ is the standard deviation of the surface irregularities relative to the mean height of the surface, λ is the wavelength, and θ is the angle of incidence measured in radians from the horizontal. Experimental evidence shows that for $C < 0.1$, specular reflection results and the surface may be considered smooth. Surfaces are considered "rough" for values of C exceeding 10 and under these conditions the reflected wave is very small in amplitude. Bullington [12] has found experimentally that most practical paths at microwave frequencies are relatively "rough" with reflection coefficients in the range of 0.2–0.4.

Knife Edge Diffraction

Very often in the mobile ratio environment a line-of-sight path to the base station is obscured by obstructions

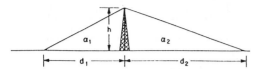

Fig. 8.   Geometry for propagation over knife edge.

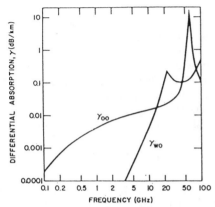

Fig. 9.   Signal attenuation from oxygen and water vapor in atmosphere.

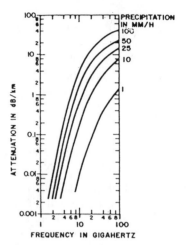

Fig. 10.   Signal attenuation for various rainfall rates.

such as hills, trees, and buildings. When the shadowing is caused by a single object such as a hill, it is instructive to treat the object as a diffracting knife edge to estimate the amount of signal attenuation. The exact solution to the problem of diffraction over a knife edge is well known and is discussed in many textbooks ([14], for example).

For most microwave mobile radio applications several assumptions can be made to simplify the calculations. Consider an infinite completely absorbing (rough) half-plane which divides space into two parts as in Fig. 8. When the distances d_1 and d_2 from the half-plane to the transmitting antenna and the receiving antenna are large compared to the height h, and h itself is large compared with the wavelength λ, i.e.,

$$d_1, d_2 \gg h \gg \lambda \qquad (3.5)$$

then the diffracted power can be given by the expression

$$P = P_0 \frac{1}{2\pi^2 h^2}. \qquad (3.6)$$

This result can be considered independent of polarization as long as the conditions of (3.5) are met. In cases where the earth's curvature has an effect there can be up to four paths. A simplified method of computing knife edge diffraction for such cases is treated by Anderson and Trolese [15]. Closer agreement with data over measured paths has been obtained by calculations, which better describe the geometry of the diffracting obstacle [16]–[18].

Effects of Rain and the Atmosphere

Microwave mobile radio signals are attenuated by the presence of rain, snow, and fog. Losses depend upon the

frequency and upon the amounts of moisture in the path. At the higher microwave frequencies, frequency selective absorption results because of the presence of oxygen and water vapor in the atmosphere [19]. The first peak in the absorption due to water vapor occurs around 24 GHz, while for oxygen the first peak occurs at about 60 GHz as shown in Fig. 9.

The attenuation due to rain has been studied experimentally [20] by a number of workers. Fig. 10 provides an estimate of the attenuating effect of rainfall as a function of frequency for several rainfall rates. It should be noted that very heavy rain showers are usually isolated and not large in extent. Nevertheless, at frequencies above 10 GHz the effects of rain cannot be neglected.

Signal attenuation over actual mobile radio paths result from a complicated dependence upon the environment, and the formulas derived here apply only in the special cases where the propagation paths can be clearly described. In the remainder of this section we will concentrate our attention on propagation effects over irregular terrain.

Propagation Over Irregular Terrain

Some classification of the environment is also necessary since signal attenuation varies depending upon the type of objects that obstruct the path. Rather than attempt to precisely define many types of environments and then describe propagation characteristics for each we will restrict our definitions of environment to three types:

open areas—areas where there are very few obstacles such as tall trees or buildings in the path, for instance, farm land or open fields;
suburban areas—areas with houses, small buildings and trees, often near the mobile unit;
urban areas—areas which are heavily built up with tall buildings and multistory residences.

Extensive measurements of radio transmission loss over various terrains in unpopulated areas have been made in the frequency ranges from 20 to 10 000 MHz by workers

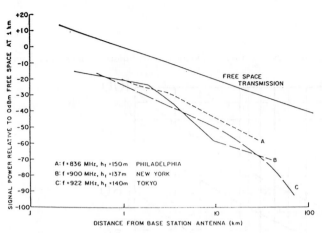

Fig. 11. Examples of transmission loss with distance.

at the Institute for Telecommunications Sciences, and this data has been tabulated in several ESSA reports [21]–[23]. Mobile radio measurements in urban and suburban areas in the microwave region have been made by a number of workers [1]–[5], [24]–[28]. To date, the most extensive work in the field has been reported by Okumura *et al.* [5]. We will rely heavily upon their results to generate prediction curves that are provided in the latter portion of this paper.

Field Strength Variation in Urban Areas

Within the urban environment the received field strength is found to vary with the base station and mobile antenna heights, transmitting frequency, the distance from the transmitter, and the width and orientation of the streets. The median field strengths in a quasi-smooth urban area show a relatively continuous change with frequency, antenna height, and distance, while other effects appear less simply related.

Distance and Frequency Dependence

One of the fundamental problems in the study of radio propagation is to describe the manner in which the signal strength attenuates as the receiving unit moves away from a transmitting base station. Obviously the signal level will fluctuate markedly (even when the Rayleigh fading is averaged out) since building heights, street widths, and terrain features are not constant. However, if we consider for a moment the behavior of the median values of the received signals, we find that there is a general trend for the signal levels to decrease more rapidly the further the mobile is separated from the base station. Fig. 11 is a plot of the received signal power versus distance as measured by independent workers in three different cities—New York [2], Philadelphia [24], and Tokyo [5]. All the measurements were made at approximately 900 MHz from relatively high base station locations. This remarkably consistent trend of both the falloff of the median signal value with distance and the excess

attenuation relative to free space leads one to hope that signal strength parameters in different cities will likewise exhibit consistent traits.

For antenna separation distances between 1 km and 15 km the attenuation of median signal power with distance changes from nearly an inverse fourth power decrease for very low base station antenna heights to a rate only slightly faster than the free space decrease for extremely high base station antennas. For antenna separation distances greater than 40 km, the signal attenuation is very rapid. Fig. 12 shows predictions derived by Okumura [5] for the basic median signal attenuation relative to free space in a quasi-smooth urban areas as it varies with both distance and frequency. These curves provide the starting point for predicting signal attenuation, which will be discussed in Section IV. The curves assume a base station antenna height of 200 m and a mobile antenna height of 3 m. Adjustments to these basic curves for different base station antenna heights and mobile antenna heights are considered in the paragraphs that follow next.

Effect of Base Station Antenna Height

Okumura [5] has found that the variation of received field strength with distance and antenna height remains essentially the same for all frequencies in the range from 200 to 2000 MHz. For antenna separation distances less than 10 km the received power varies very nearly proportional to the square of the base station antenna height (6 dB per octave). For very high base station antennas and for large separation distances (greater than 30 km) the received power tends to be proportional to the cube of the height of the base station antenna (9 dB per octave). Fig. 13 is a set of prediction curves that give the change in received power (often called the height-gain factor) realized by varying the base station antenna height. The curves are plotted for various antenna separation distances and predict the median received power relative to a 200-m base station antenna and a 3-m mobile antenna. They may be used for frequencies in the range 200 to 2000 MHz.

Effect of Mobile Antenna Height

For obvious reasons mobile antenna heights are generally limited to no more than 4 m. For a large range of frequencies and for several base station antenna heights Okumura observed a height-gain advantage of 3 dB for a 3-m high mobile antenna compared to a 1.5-m high mobile antenna. For special cases where antenna height can be above 5 m the height-gain factor depends upon the frequency and the environment. In a medium sized city where the transmitting frequency is 2000 MHz the height-gain factor may be as much as 14 dB per octave, while for a very large city and a transmitting frequency under 1000 MHz the height-gain factor may be as little as 4 dB per octave for antennas above 5 m. Prediction curves for the vehicle height-gain factor in urban areas are given in Fig. 14.

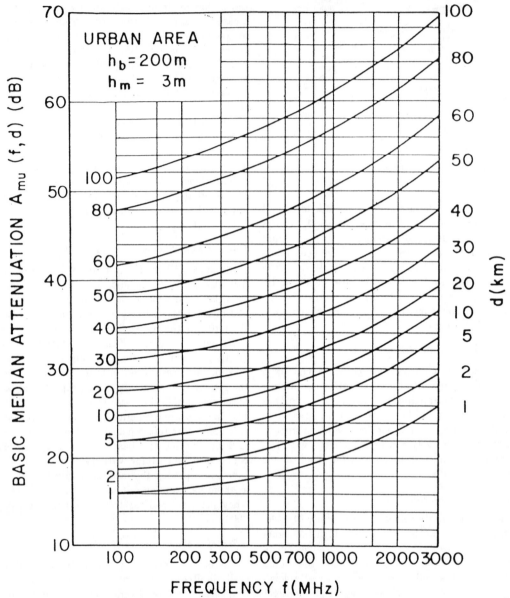

Fig. 12.   Prediction curve for basic median attenuation relative to free space in urban area over quasi-smooth terrain referred to $h_b = 200$ m, $h_m = 3$ m.

Correction Factor for Suburban and Open Areas

Suburban areas are generally characterized by lower buildings and generally less congestion of obstacles than in cities. Consequently, one should expect that radio signals would propagate better in such environments. Okumura has found that there is practically no change in the difference between urban and suburban median attenuation (suburban correction factor) with changes in base station antenna height or with separation distances between the base and mobile antennas. The suburban propagation does depend somewhat on the frequency and increases to some extent at the higher frequencies. A plot of the suburban correction factor is shown by the solid curve in Fig. 15 for frequencies in the range of 100 to 3000 MHz. Recent data shows a 10-dB difference between urban and suburban values of the median received signal strength at a frequency of 11 200 MHz, slightly less than that extrapolated in Fig. 15. Open areas, which occur rather infrequently, tend to have significantly better propagation paths than urban and suburban areas and typical received signal strengths run nearly 20 dB greater for the same antenna height and separation distance. The upper curve shown in Fig. 15 provides a correction in dB that may be added directly to the prediction value for the urban case. Rural areas or areas with only slightly built up sections have median signal strength somewhere between the two curves.

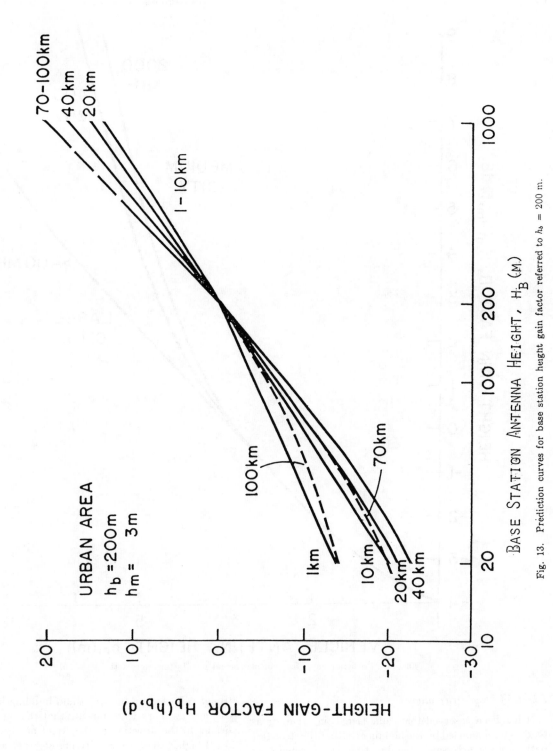

Fig. 13. Prediction curves for base station height gain factor referred to $h_b = 200$ m.

147

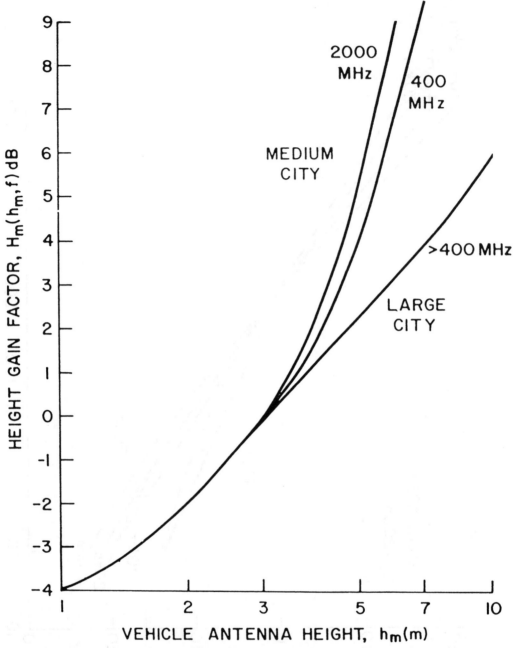

Fig. 14. Prediction curve for vehicular height gain factor referred to $h_m = 3$ m.

Effects of Street Orientation

It has been observed that radio signals in urban areas tend to be channeled by the buildings so that the strongest paths are not necessarily the direct paths diffracted over the edge of nearby obstructing buildings, but are found to be from directions parallel to the streets. Streets that run radially or approximately radially from the base station are most strongly affected by this channeling phenomena. This causes the median received signal strength to differ by as much as 20 dB at locations nearby the transmitter [24].

The distribution of the signal paths in the horizontal plane as seen from the mobile vehicle in an urban area is strongly affected by the street and building layout. Tests [27] in New York City indicate that the signals arriving parallel to the direction of the street are typically 10 to 20 dB higher than waves arriving at other angles. These tests were carried out at 11.2 GHz by scanning with a highly directive antenna (beamwidth of 5°) at various locations in the city.

Effects of Foliage

There are a great many factors that affect propagation behind obstacles such as a grove of trees. Precise estimates of attenuation are difficult because tree heights are not uniform; also, the type, shape, density, and distribution

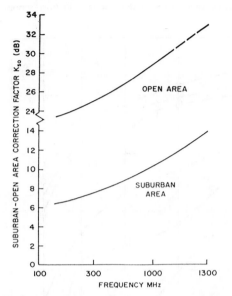

Fig. 15. Prediction curves for suburban and open area correction factor K_{so}.

of the trees influence the propagation. In addition, the density of the foliage depends upon the season of the year. However, some success has been obtained by treating trees as diffracting obstacles with an average effective height.

An experimental study of propagation behind a grove of live oak and hackberry trees in Texas for several frequencies has been reported by Lagrone et al. [29]. At frequencies from about 0.5 to 3 GHz and for distances greater than about five times the tree height, the measurements were in good agreement with the theoretical predictions of diffraction over an ideal knife edge, assuming distances and heights the same as those in the measurements. At shorter distances some propagation takes place through the trees, which acts to reduce the effective height of the diffracting edge and at the same time increases the apparent distance from the diffracting edge to the antenna.

Measurements at 836 MHz and 11.2 GHz were made, which compared the signal strength measured on the same streets in summer and in winter in suburban Holmdel, N. J. [30]. At the UHF frequency the average received signal strength in the summer when the trees were in full leaf was roughly 10 dB lower than for the corresponding locations in later winter. At X band frequencies the losses during the summer appear to be greater in the areas where the signal levels were previously low.

Signal Attenuation in Tunnels

It is well known that frequencies in the VHF region commonly used for mobile communications are severely attenuated in tunnel structures [31]. Only by using special antennas are these frequencies usable in long (over 300-m) tunnels. However, at microwave frequencies tunnels are effective guiding or channeling mechanisms and can offer significant improvement over VHF for communications.

A test [32] was performed in the center tube of the

Lincoln Tunnel (3000 m long), which connects midtown Manhattan to New Jersey under the Hudson River. The inside of the tunnel is roughly rectangular in cross section with a height of 4 m and a width of 8 m. Seven test frequencies roughly an octave apart were used to make signal attenuation measurements.

The average loss of signal strength in dB against the antenna separation for the seven frequencies is plotted in Fig. 16. For convenience in plotting the data an arbitrary reference level of 0 dB at 300-m antenna separation was chosen. It is worth noting that the 153- and 300-MHz attenuation rates are nearly straight lines implying that the signal attenuation has an exponential relationship to the separation. At 153 MHz the loss is extremely high (in excess of 40 dB in 300 m, where at 300 MHz the rate of attenuation is of the order of 20 dB in 300 m. At higher frequencies a simple exponential attenuation rate is not evident. For the major portion of the length of the tunnel the received signal level at 900 MHz has an inverse fourth power dependence upon the antenna separation, while at 2400 MHz the loss has an inverse or square dependence. At frequencies above 2400 MHz, dependence of the signal strength with antenna separation is less than the free space path loss (throughout most of the length of the tunnel). Roughly, the attenuation rates appear to be only 2–4 dB in 300 m for frequencies in the 2400 to 11 000 MHz range.

Effects of Irregular Terrain

The traditional approach of predicting attenuation from propagation over irregular terrain has been to approximate the problem to one that is solvable in closed form. This is usually done by solving problems dealing with smooth regular boundaries such as planes or cylinders [33]–[42]. To some extent propagation over more complicated obstacles can be determined by constructing models and performing laboratory experiments at optical frequencies or frequencies in the millimeter wavelength region [43]–[45].

A computer program has been published by Longley and Rice [46] that predicts the long term median radio transmission loss over irregular terrain. The method predicts median values of attenuation relative to the transmission loss in free space and requires the following: the transmission frequency, the antenna separation, the height of the transmitting and receiving antennas, the mean surface refractivity, the conductivity and dielectric constant of the earth, polarization, and a description of the terrain. This program was based upon thousands of measurements and compares well with measured data [47].

The computer method of Longley and Rice provides both point-to-point and area predictions which agree well with the experiment, but a description of the calculation of the many parameters used in their method would be rather lengthy. The reader is referred to their work for precise calculations. We shall adopt a somewhat less accurate method in which we obtain correction factors

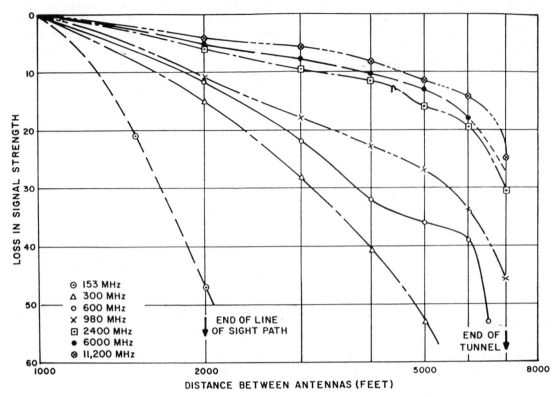

Fig. 16. Signal loss versus log of antenna separation for seven frequencies.

to our basic median curves (Figs. 12–14) for various terrain effects.

Correction Factors for Undulating, Sloping, and Land-Sea Terrain

An approximate prediction curve for undulating terrain is given in Fig. 17, which is based upon work reported by the CCIR [19] and Okumura [5]. This estimates the correction factor to the basic median attenuation curves obtained previously for quasi-smooth urban terrain. More exact predictions would probably have some dependence upon frequency and on antenna separation distance. If the location of the vehicle is known to be near the top of the undulation the correction factor in Fig. 17 can be ignored. On the other hand, if the location is near the bottom of the undulation, the attenuation is higher and is indicated on the lower curve in Fig. 17.

In cases where the median height of the ground is gently sloping for distances of the order of 5 km, a correction factor may be applied. Let us define the average slope θ_m measured in milliradians as illustrated in Fig. 18. Depending upon the antenna separation distance d, the terrain slope correction factor K_{sp} is given in Fig. 19 in terms of the average slope θ_m. (It should be noted that these curves are based upon rather scant amounts of data and should be considered as estimates applying in the frequency range of 450 to 900 MHz.)

Usually on propagation paths where there is an expanse of water between the transmitting and receiving stations the received signal strength tends to be higher than for cases where the path is only over land. The change in signal strength depends upon the antenna separation distance and whether the water lies closer to the mobile receiver or the base station transmitter, or somewhere in between. Let us define a ratio β, which represents the fraction of the path that consists of propagation over water. Two path geometries and the definition of β are illustrated in Fig. 20. It has been observed that when the latter portion of the path from the base station antenna to the mobile antenna is over water, the signal strength is typically 3 dB higher than for cases where the latter portion of the base mobile path is over land. Prediction curves for mixed land-sea paths have been obtained experimentally by Okumura [5] as shown in Fig. 20, which provides correction factors in terms of the percentage of the path over water.

Statistical Distribution of the Local-Mean Signal

Thus far we have obtained results based primarily upon experimental evidence, which has provided us with the behavior of median signal levels obtained by averaging received signals over a distance of 10 to 20 m. Smooth curves were obtained relating the variation of the median received signal with distance, base station antenna height, and frequency in urban, suburban, and rural areas. Consistent but less accurate predictions were found for dependence on street orientation, isolated ridges, rolling hills, and land-sea paths. Considerably less data is avail-

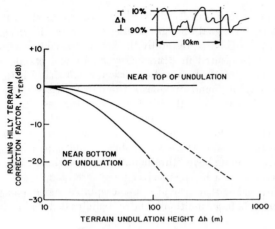

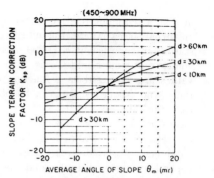

Fig. 17. Rolling hilly terrain correction factor K_{ter}.

Fig. 19. Measured value and prediction curves for "slope terrain correction factor" K_{sp}.

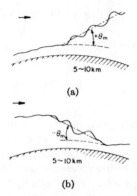

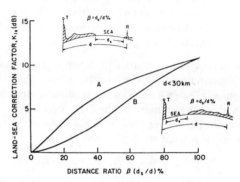

Fig. 18. Definition of average angle of general terrain slope. (a) Positive slope $(+\theta_m)$. (b) Negative slope $(-\theta_m)$.

Fig. 20. Prediction curves for land-sea correction factor K_{ls}.

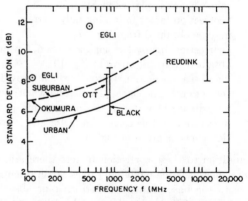

Fig. 21. Prediction curves for standard deviation of median field strength variation in urban, suburban, and rolling hilly terrain.

able that describes the fluctuations of the received signal about the median value. The dependence of the signal distribution upon the parameters mentioned requires a good deal more investigation before definitive results are available.

One consistent result that has been observed is that the distribution of the received signals at fixed base and mobile antenna heights, frequency, and separation distance from the base station within the same environment class (urban, for example) have very nearly a normal distribution when the distribution is plotted for the received signal measured in dB. Such a probability distribution is often referred to as log-normal [48]. Also the excess path loss, that is, the difference (in dB) between the computed value of the received signal strength in free space (3.1) and the actual measured value of the local mean received signal has been observed to be log-normally distributed.

Fig. 21 is a prediction curve for the standard deviation σ of the log-normal distribution, which describes the variation of the median signal strength values in suburban areas as given by Okumura [5]. The data spreads at 850 MHz are from Black [24] and Ott [26]; at 11.2 GHz the data is from Reudink [25]. The data points at 127 and 510 MHz are from the work of Egli [49].

IV. PREDICTION OF FIELD STRENGTH

In order to predict the median power received by a mobile unit from a base station antenna in a basic urban environment we may use the following equation (all quantities in dB):

$$P_p = P_0 - A_m(f,d) + H_b(h_b,d) + H_m(h_m,f) \quad (4.1)$$

where

P_p value of the predicted received power;
P_0 power received for free space transmission (3.1);

$A_m(f,d)$ median attenuation relative to free space in an urban area where the effective base station antenna height h_b is 200 m and the vehicle antenna height h_m is 3 m; these values are expressed as a function of distance and frequency and can be obtained from the curves of Fig. 12;

$H_b(h_b,d)$ base station height-gain factor expressed in dB relative to a 200-m high base station antenna in an urban area; this function is dependent upon distance and has been plotted in Fig. 13;

$H_m(h_m,f)$ is the vehicle station height-gain factor expressed in dB relative to a 3-m high vehicle station antenna in an urban area; this factor is dependent upon frequency and has been plotted in Fig. 14.

If the particular propagation path happens to be over a different environment type or involve terrain that is not "quasi-smooth" we may amend our prediction formula for P_p by adding one or more of the correction factors that were described in earlier portions of the chapter. Thus the "corrected" predicted received power P_c is

$$P_c = P_p + K_{so} + K_{ter} + K_{sp} + K_{ls} \qquad (4.2)$$

where

K_{so} "correction factor for suburban and open terrain," which is plotted in Fig. 15;

K_{ter} "correction factor for rolling hilly terrain," which may be obtained from Fig. 17;

K_{sp} "correction factor for sloping terrain," which is obtained from the curve in Fig. 19;

K_{ls} "land-sea correction factor," which provides a correction to the signal attenuation when there is an expanse of water in the propagation path; a description of this correction factor and a prediction curve is given in Fig. 20.

In addition to these correction factors there are other factors such as isolated mountain ridges, street orientation relative to the base station, the presence or absence of foliage, effects of the atmosphere, and in the case of undulating terrain the position of the vehicle relative to the median height. These additional effects together with the fact that the correction factors and indeed the basic transmission factors A_m, H_b, and H_m are average values based upon empirical data should indicate that discrepancies between measured and predicted values are still possible. It is reassuring to point out, however, that these prediction curves, which are essentially the same as those of Okumura, have been compared to measured data with a great deal of success [5]. This was done over a variety of environments, antenna heights and separation, and for a number of frequencies.

Determination of Signal Coverage in a Small Area

Let us assume that the local mean signal strength in an area at a fixed radius from a particular base station antenna is log-normally distributed. Let the local mean (that is, the signal strength averaged over the Rayleigh fading) in dB be expressed by the normal variable x with mean $\bar{x}$ (measured in dBm for example) and standard deviation σ (dB). To avoid confusion, recall that $\bar{x}$ is the median value found previously from (4.1). As we have seen, $\bar{x}$ depends upon the distance r from the base station as well as several other parameters. Let x_0 be the receiver "threshold." We shall determine the fraction of the locations (at $r = R$) wherein a mobile would experience a received signal above "threshold." The "threshold" value chosen need not be the receiver noise threshold, but may be any value that provides an acceptable signal under Rayleigh fading conditions. The probability density of x is

$$p(x) = \frac{1}{\sigma(2\pi)^{1/2}} \exp\left[\frac{-(x-\bar{x})^2}{2\sigma^2}\right]. \qquad (4.3)$$

The probability that x exceeds the threshold x_0 is

$$P_{x_0}(R) = P(x \geq x_0) = \int_{x_0}^{\infty} p(x)\, dx$$

$$= \tfrac{1}{2} + \tfrac{1}{2} \operatorname{erf}\left(\frac{x_0 - \bar{x}}{\sigma\sqrt{2}}\right). \qquad (4.4)$$

If we have measured or theoretical values for $\bar{x}$ and for σ in the area of interest we can determine the percent of the area for which the average signal strength exceeds x_0. For example, at a radius where the median and hence the mean of the log-normal signal strength is -100 dBm ($\bar{x} = -100$ dBm at some particular separation distance R and for some radiated power) assume our system threshold happens to be -110 dBm, then if we assume $\sigma = 10$ dB we have

$$P_{x_0}(R) = \tfrac{1}{2} + \tfrac{1}{2}\operatorname{erf}\frac{1}{\sqrt{2}} = 0.84.$$

Determination of the Coverage Area from a Base Station

It is also of interest to determine the percent of locations *within* a circle of radius R in which the received signal strength from a radiating base station antenna exceeds a particular threshold value. Let us define the fraction of useful service area F_u as that area, within a circle of radius R, for which the signal strength received by a mobile antenna exceeds a given threshold x_0. If P_{x_0} is the probability that the received signal x exceeds x_0 in an incremental area dA, then

$$F_u = \int P_{x_0}\, dA.$$

In a real-life situation one would probably be required to break the integration into small areas in which P_{x_0} can be estimated and then sum over all such areas. For purposes of illustration let us assume that the behavior of the mean value of the signal strength $\bar{x}$ follows an r^{-n} law. Thus

$$\bar{x} = \alpha - 10n \log_{10}\frac{r}{R} \qquad (4.6)$$

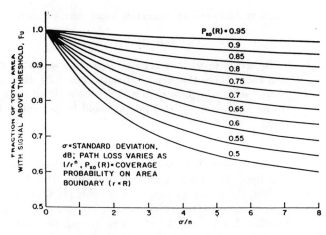

Fig. 22.   Fraction of total area with signal above threshold F_u.

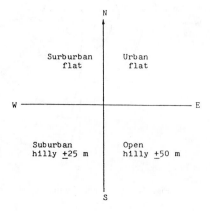

Fig. 23.   Environment surrounding base station.

TABLE I

Distance (km)	P_0 (dBm)	A_m (dB)	H_b (dB)	H_m (dB)	P_p (dBm)
2	−47	−23	−4	−2	−76
5	−55	−27	−5	−2	−89
10	−61	−30	−5	−2	−98
15	−65	−31	−6	−2	−104
20	−67	−33	−6	−2	−108
25	−69	−35	−6	−2	−112
30	−71	−37	−6	−2	116
40	−73	−41	−7	−2	123
50	−75	−50	−7	−2	134

where α, expressed in dB, is a constant determined from the transmitter power, antenna heights and gains, etc. Then

$$P_{x_0} = \tfrac{1}{2} - \tfrac{1}{2}\, \mathrm{erf}\, \frac{x_0 - \alpha + 10n \log r/R}{\sigma\sqrt{2}}. \qquad (4.7)$$

Then letting $a = (x_0 - \alpha)/\sigma\sqrt{2}$ and $b = 10n \log_{10} e/\sigma\sqrt{2}$ we get

$$F_u = \frac{\pi R^2}{2} + \int_0^R e\, \mathrm{erf}\left(a - b \ln \frac{r}{R}\right) dr. \qquad (4.8)$$

The integral can be evaluated by substituting $t = a - b \ln (r/R)$, so that

$$F_u = \frac{\pi R^2}{2} + \frac{\pi R^2 \exp (2a/b)}{b} \int_a^\infty \exp\left(-\frac{2t}{b}\right) \mathrm{erf}\,(t)\, dt.$$

$$(4.9)$$

From [50, p. 6, #1]

$$F_u = \frac{R^2}{2}\left\{ 1 + \mathrm{erf}\,(a) + \exp\left(\frac{2ab+1}{b^2}\right) \right.$$
$$\left. \cdot \left[1 - \mathrm{erf}\, \frac{ab+1}{b}\right]\right\}. \qquad (4.10)$$

For example let us choose α such that $\bar{x} = x_0$ at $r = R$, then $a = 0$ and

$$F_u = R^2 \left\{ \tfrac{1}{2} + \exp\left(\frac{1}{b^2}\right)\left[1 - \mathrm{erf}\left(\frac{1}{b}\right)\right]\right\}. \qquad (4.11)$$

Let us further assume that $n = 3$ and $\sigma = 9$, then $F_u = 0.71$, or about 71 percent of the area within a circle of radius R has signal above threshold when 1/2 the locations on the circumference have a signal above threshold.

For the case where the propagation follows a power law, the important parameter is σ/n. Fig. 22 is a plot of the fraction of the area within a circle of radius R, which has a received signal above a threshold for various fractions of coverage on the circle.

V. AN EXAMPLE

Suppose that as a young entrepreneur you have decided to go into the radio controlled pizza delivery business. You received a license to operate at a frequency of 1.0 GHz transmitting 10 W into a 10-dB gain antenna from the top of your home office, the Leaning Tower of Pizza, located in suburban LaPimento. Since it is essential that your product be delivered promptly and orders received correctly, you want to estimate the regions where there is reliable radio coverage.

Looking at a terrain map you determine that your effective base station height is 100 m above the surrounding terrain. The pizza trucks have 2-dB gain dipole antennas 2 m above the ground. With this information you go to (4.1) and calculate basic predictions of received signal power. Table I gives some representative values of P_p for various distances.

Suppose that the environment around the base station can be described as shown in Fig. 23.

The values calculated in Table I apply directly to the NE section but for the remaining areas you need to apply the correction factors of (4.2). For the other areas you add correction factors to P_p to obtain

$$P_c: \quad \mathrm{NW} = P_p + K_{so} = P_p + 10$$

$$P_c: \quad \mathrm{SW} = P_p + K_{so} + K_{ter} = P_p + 10 - 4.$$

$$P_c: \quad \mathrm{SE} = P_p + K_{so} + K_{ter} = P_p + 29 - 7.$$

With Table I and the corrected values for P_p we can plot contours of equal signal strength as shown in Fig. 24.

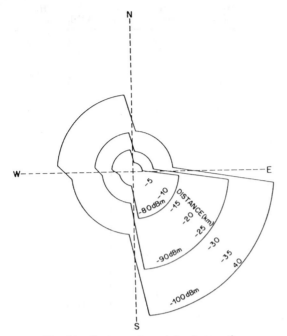

Fig. 24.   Contours of equal signal strength.

Now you look at Fig. 1 and determine subjectively that you want your minimum signal for reliable communications to be 10 dB above the rms of the Rayleigh fluctuations. If the radio receivers have 7-dB noise figures and transmit FM in a 50-kHz bandwidth, then the minimum signal desired is

kTB	−127 dBm.
noise figure	+7 dB
FM threshold	10 dB
subjective threshold	10 dB
minimum signal	−100 dBm.

Thus you obtain the radius (different in each quadrant) at which half the locations can be expected to have signal strengths above −100 dBm and half below this value.

We can carry this example another step to determine the radii, say where 90 percent of the locations have signals above −100 dBm. To do this we go to Fig. 19 to find σ and then invert (4.4) to find the extra margin in signal strength required to get 90 percent coverage instead of 50 percent. Choosing 6.5 and 8.5 dB for σ for the urban and suburban areas, respectively, we then have from (4.4)

urban:

$$0.9 = 0.5 + 0.5 \text{ erf}\left[\frac{100 - x}{6.5\sqrt{2}}\right]$$

$$x = -91.5 \text{ dBm}$$

suburban:

$$0.9 = 0.5 - 0.5 \text{ erf}\left[\frac{100 - x}{8.5\sqrt{2}}\right]$$

$$x = -89 \text{ dBm}.$$

Referring to the coverage contours of Fig. 23 we find the radii where the signal strengths are −89 dBm and −91.5 dBm. On these radii we expect the signal strength to be above our −100 dBm threshold 90 percent of the time.

VI. CONCLUSIONS

Experiments have verified that the Rayleigh model is excellent for predicting short-term fading. Based on this model, RF power spectra, level crossing rates, duration of fades, and other statistical parameters can be accurately predicted. Likewise, the effects on propagation of certain geometric obstacles, rain, and the atmosphere are well known. However, attempts to determine propagation over terrain that is cluttered with trees, buildings, hills and valleys, and a variety of other obstructions are necessarily less successful. This paper has shown that the local mean value of received signal strength is a relatively continuous function of frequency, antenna heights, and distance. Further, it was shown that correction factors can account for building density, rolling hilly terrain, sloping terrain, and land-sea paths. Finally, it was shown that variations in the local-mean signal strength have a lognormal distribution. Using this as a model, we can predict the expected coverage from a base station for a large range of antenna heights and frequencies and for many types of terrain.

ACKNOWLEDGMENT

I would like to thank W. C. Jakes, Jr. for allowing me to use material that will be used in his forthcoming book, *Microwave Mobile Communications*.

REFERENCES

[1] W. C. Jakes, Jr. and D. O. Reudink, "Comparison of mobile radio transmission at UHF and X band," *IEEE Trans. Veh. Technol.*, VT-16, pp. 10–14, Oct. 1967.
[2] W. R. Young, Jr., "Comparison of mobile radio transmission at 150, 450, 900, and 3700 MHz," Bell Syst. Tech. J., vol. 31, pp. 1068–1085, Nov. 1952.
[3] P. M. Trifonov, V. N. Budko, and V. S. Zotov, "Structure of USW field strength spatial fluctuations in a city," *Trans. Telecomm. Radio Eng.*, vol. 9, pp. 26–30, Feb. 1964.
[4] H. W. Nyland, "Characteristics of small-area signal fading on mobile circuits in the 150 MHz band," *IEEE Trans. Veh. Technol.*, vol. VT-17, pp. 24–30, Oct. 1968.
[5] Y. Okumura et. al., "Field strength and its variability in VHF and UHF land-mobile radio service," *Rev. Elec. Commun. Lab.*, 16, pp. 825–873, Sept. 1968.
[6] M. J. Gans, "A power-spectral theory of propagation in the mobile-radio environment," *IEEE Trans. Veh. Technol.*, vol. VT-21, pp. 27–38, Feb. 1972.
[7] S. O. Rice, "Mathematical analysis of random noise," *Bell Syst. Tech. J.*, vol. 23, pp. 282–332, July 1944; vol. 24, pp. 46–156, Jan. 1945, and S. O. Rice, "Statistical properties of a sine wave plus random noise," *Bell Syst. Tech. J.*, vol. 27, pp. 109–157, Jan. 1948.
[8] H. T. Friis, "A note on a simple transmission formula," *Proc. IRE*, vol. 34, pp. 254–256, May, 1946.
[9] K. A. Norton, "The propagation of radio waves over the surface of the earth in the upper atmosphere, Part I," *Proc. IRE*, vol. 24, pp. 1367–1387, Oct. 1936.
[10] ———, "The propagation of radio waves over the surface of the earth in the upper atmosphere, Part II," *Proc. IRE*, vol. 25, pp. 1203–1236, Sept. 1937.
[11] ———, "The calculation of ground wave field intensity over a finitely conducting spherical earth," *Proc. IRE*, vol. 29, pp. 623–639, Dec. 1941.

[12] K. Bullington, "Radio propagation at frequencies above 30 megacycles," *Proc. IRE*, vol. 35, pp. 1122–1136, Oct. 1947.

[13] ——, "Radio propagation fundamentals," *Bell Syst. Tech. J.* vol. 36, p. 593, May 1957.

[14] D. S. Jones, *The Theory of Electro-magnetism*, New York: McMillian 1964, ch. 9.

[15] L. J. Anderson and L. G. Trolese, "Simplified method for computing knife edge defraction in the shadow region," *IRE Trans. Antennas Propagat.*, vol. AP–6, pp. 281–286, July 1958.

[16] N. P. Bachynski and M. G. Kingsmill, "Effect of obstacle profile on knife-edge diffraction," *IRE Trans. Antennas Propagat.*, vol. 10, pp. 201–205, Mar. 1962.

[17] G. Millington, "A Note on diffraction round a sphere or cylinder," *Marconi Rev.*, vol. 23, p. 170, 1960.

[18] H. T. Dougherty and L. J. Maloney, "Application of diffraction by convex surfaces to irregular terrain situations," *Radio Phone*, vol. 68B, no. 2, Feb. 1964.

[19] CCIR Rep. 370–1, Oslo, 1966, II.

[20] D. C. Hogg, "Statistics on attenuation of microwaves by intense rain," *Bell Syst. Tech. J.*, vol. 48, no. 9, Nov. 1969.

[21] P. L. McQuate, J. M. Harman, and A. P. Barsis, "Tabulations of propagation data over irregular terrain in the 230 to 9200 MHz frequency range, part 1: Gun Barrel Hull receiver site," ESSA Tech. Rep. ERL65–ITS52, Mar. 1968.

[22] P. L. McQuate, J. M. Harman, M. E. Johnson, and A. P. Barsis, "Tabulations of the propagation data over irregular terrain in the 230 to 9200 MHz frequency range, part 2: Fritz Peal receiver site," ESSA Tech. Rep. ERL65–ITS58–2, Dec. 1968.

[23] P. L. McQuate, J. M. Harman, M. E. McClamaham, and A. P. Barsis, "Tabulations of propagation data over irregular terrain in the 230 to 9200 MHz frequency range, part 3: North Table Mountain-Golden," ESSA Technical Report ERL65–ITS58–3, July, 1970.

[24] D. M. Black and D. O. Reudink, "Some characteristics of mobile radio propagation at 836 MHz in the Philadelphia area," *IEEE Trans. Veh. Technol.*, vol. VT–21, pp. 45–51, May 1972.

[25] D. O. Reudink, "Comparison of radio transmission at X-band frequencies in suburban and urban areas," *IEEE Trans. Antennas Propagat.*, vol. AP–20, pp. 470–473, July 1972.

[26] G. D. Ott, "Data processing summary and path loss statistics for Philadelphia HCMTS measurements program," unpublished.

[27] D. O. Reudink, "Preliminary investigation of mobile radio transmission at X-band in an urban area," presented at the 1967 Fall URSI Meeting at Ann Arbor, Mich.

[28] W. C. Y. Lee and R. H. Brandt, "The elevation angle of mobile radio signal arrival," *IEEE Trans. Veh. Technol.*, vol. VT–22, pp. 110–113, Nov. 1973.

[29] A. H. LaGrone and C. W. Chapman, "Some propagation characteristics of high UHF signals in the immediate vicinity of trees," *IRE Trans. Antennas Propagat.*, vol. 9, pp. 487–491, Sept. 1961.

[30] D. O. Reudink and M. F. Wazowicz, "Some propagation experiments relating foliage loss and defraction loss at X-band at UNF frequency," *IEEE Trans. Veh. Technol.*, vol. VT–22, pp. 114–122, Nov. 1973.

[31] R. A. Farmer and N. H. Shepnera, "Guided radiation.... the key to tunnel talking," *IEEE Trans. Veh. Commun.*, vol. VC–14, pp. 93–102, Mar 1965.

[32] D. O. Reudink, "Mobile radio propagation in tunnels," *IEEE Vehicular Technology Group Conf.*, San Francisco, Calif., Dec. 2–4, 1968.

[33] S. O. Rice, "Diffraction of plane radio waves by prarbolic cylinder," *Bell Syst. Tech. J.*, 33(2), 417–504, (1954).

[34] J. R. Wait and A. M. Conda, "Diffraction of Electromagnetic Waves by Smooth Obstacles for Grazing Angles," *J. Res. Nat. Bur. Stand.*, vol. 63D, pp. 181–197, 1959.

[35] J. C. Schelleng, C. R. Burrows, and E. B. Ferrell, "Ultra shortwave propagation," *Proc. IRE*, vol. 21, pp. 427–463, 1933.

[36] H. Selvidge, "Diffraction Measurements at Ultra-High frequencies," *Proc. IRE*, vol. 29, pp. 10–16, Jan. 1941.

[37] T. B. A. Senior, "The diffraction of a dipole field by a perfectly conduction half-plane," *Quart. J. Mech. Appl. Math.*, vol. 6, pp. 101–114, 1953.

[38] D. E. Kerr, *Propagation of Short Radio Waves.* New York: McGraw-Hill, p. 728, 1951.

[39] H. T. Dougherty and L. J. Maloney, Application of diffraction by convex surfaces to irregular terrain situations." *Radio Sci.*, vol. 68D 284–305, 1964.

[40] G. Millington, "A note on diffraction round a sphere or cylinder," *Marconi Rev.*, vol. 23, pp. 170–182, 1960.

[41] M. H. L. Pryce, "The diffraction of radio waves by the curvature of the earth," *Advances Phys.*, vol. 2, pp. 67–95, 1953.

[42] J. R. Wait and A. M. Conda, "Pattern of an antenna on a curved lossy surface," *IRE Trans. Antennas Propagat.*, vol. AP–6, pp. 187–188, Oct. 1958; and vol. AP–8, p. 628, Nov. 1960.

[43] K. Hacking, "UHF propagation over rounded hills," *Proc. Inst. Elec. Eng.*, vol. 117, Mar. 1970.

[44] M. P. Bachynski, "Scale model investigations of electromagnetic wave propagation over natural obstacles," *RCA Rev.*, vol. 25, pp. 105–44, 1963.

[45] K. Hacking, "Optical diffraction experiments simulating propagation over hills at UHF," in *Inst. Elect. Eng.*, London, *Conf. Pub.* #48, 1968.

[46] A. G. Longley and P. L. Rice, "Prediction of tropospheric radio transmission loss over irregular terrain, a computer method–1968," ESSA Research Laboratories ERL79–ITS67. (U.S. Government Printing Office, Washington, D.C.), 1968.

[47] A. G. Longley and R. K. Reasoner, "Comparison of propagation measurements with predicted values in the 20 to 10,000 MHz range," ESSA Tech. Rep. ERL148–ITS97, Jan. 1970.

[48] J. Aitchison and J. A. C. Brown, *Lognormal Distribution With Special Reference to Its Uses in Economics.* Cambridge, England: Cambridge University Press. 1957.

[49] J. Egli, "Radio propagation above 40 MC over irregular terrain," *Proc. IRE*, vol. 45, pp. 1383–1391, Oct. 1957.

[50] Ng, W. Edward and Murray Geller, "A Table of Integrals of the Error Functions," *J. Res. Nat. Bur. Stand.*, vol. 73B, Jan.–Mar. 1969.

Contributors

David S. Eddy (M'62) was born in Kalamazoo, Mich., on February 10, 1936. He received the B.S.E.E. and M.S.E.E. degrees in electrical engineering from the University of Michigan, Ann Arbor, in 1959 and 1960, respectively.

He joined the staff of the General Motors Research Laboratories, Warren, Mich., in 1960. His major fields of research activities are the development of electronic ceramic materials and processes, and the development of solid-state transducer devices and applications.

Mr. Eddy is a member of Eta Kappa Nu, Sigma Xi, the American Ceramic Society, and the American Association for the Advancement of Science.

John H. McMahon was born in 1915 in Minneapolis, Minn. He received the B.S. degree in electrical engineering from the University of Minnesota, Minneapolis, in 1938.

He has been employed since in AM, FM, and TV broadcasting and consulting and in antenna and transmitter design for use in the United States and overseas for Page Communications Engineers and Sanders Associates, as also (on a domestic basis only) for Airborne Instruments Laboratory, Station KSTP in St. Paul, Weldon and Carr, and DECO Incorporated. In 1970, he entered government service with the Federal Communications Commission and is currently Chief, Systems Engineering Group, Spectrum Management Task Force, Office of the Chief Engineer, Washington, D. C.

Louis L. Nagy (S'60–M'65) was born in Detroit, Mich., on January 15, 1942. He received the B.S. degree in electrical engineering from the University of Michigan, Dearborn, in 1965, and the M.S. and Ph.D. degrees in electrical engineering from the University of Michigan, Ann Arbor, in 1969 and 1973, respectively.

From 1965 to 1969, he was a faculty member of the University of Michigan's Willow Run Research Laboratories where he specialized in radar and electronic counter-measures systems. In 1969 he joined the staff of the General Motors Proving Grounds as a Project

Engineer where he investigated and applied radar and electro technology to vehicular performance evaluation systems. In 19 he was transferred to the newly formed Microwave Group of Electronics Department at the General Motors Research Laboratories. He is currently an Associated Senior Research Engineer charge of the microwave engineering section of the Microwave Grou His current research is directed at automotive radar systems, a the automotive environment as it is related to these systems.

Dr. Nagy is a member of Tau Beta Pi, Eta Kappa Nu, ar Sigma Xi.

Edward A. Neham (M'66) was born on Ma 24, 1943, in New York City, N. Y. He re ceived the B.S.E.E. degree from Drexel Uni versity, Philadelphia, Pa., in 1965.

From 1965 to 1970, while working at IIT Research Institute, Annapolis, Md., he wa involved in the analysis of various communications and radar systems with regard to electromagnetic compatibility and spectrum engineering. During the period from 1970 to 1972 he was employed by National Scientific Laboratories as a Senior Engineer, working primarily on the application of computer technology to the solution of various satellite communications problems. Upon joining the technical staff of Kelly Scientific Corporation, Washington, D. C., in 1972, he concentrated his attention on land mobile radio systems with particular regard to spectrum engineering, and was integrally involved with the FCC's Spectrum Management Task Force. He is now with PWC/PMS, McLean, Va.

Douglas O. Reudink (M'66) was born in West Point, Nebr., on May 6, 1939. He received the B.A. degree from Linfield College, McMinnville, Oreg., in 1961, and the Ph.D. degree in mathematics from Oregon State University, Corvallis, in 1965.

Since joining Bell Telephone Laboratories, Holmdel, N.J., in 1964, he has been engaged in electronic systems research with particular emphasis in the field of mobile communications. His recent work has been concerned with fundamentals of mobile radio propagation, diversity techniques, and the configuration and control of mobile systems. He is currently head of the Satellite Systems Research Department.

Chapter 4

MODULATION TECHNIQUES

THE CHOICE OF A DIGITAL MODULATION SCHEME

IN A MOBILE RADIO SYSTEM

Pritpal Singh Mundra *, T. L. Singal * and Rakesh Kapur **

*   Punjab Wireless Systems Limited, SAS Nagar, INDIA.
**   Electronics Research & Development Centre, SAS Nagar, INDIA.

ABSTRACT - The modulation scheme that should be adopted in a mobile radio communication system must primarily satisfy the objective of achieving high spectral efficiency and narrow power spectrum. The paper describes the emerging modulation technologies in the family of linear modulation and constant envelope techniques. The multilevel modulation, spectral efficiency, power spectrum and inter symbol interference are discussed. The effect of coding on transmission rate and improvement in SNR requirement is considered.

I. INTRODUCTION

A modulation scheme used for mobile environment should utilise the transmitted power and RF channel bandwidth as efficiently as possible. This is because the mobile radio channel is both power and band limited. To conserve power, efficient source encoding schemes are generally used but this is at the cost of bandwidth. Whereas to save the spectrum in band limited systems, spectrally efficient modulation techniques are used. The objective of spectrally efficient modulation is to maximize the bandwidth efficiency. It is also desirable to achieve the bandwidth efficiency at a prescribed BER with minimum transmitted power.

Coming to the cost aspect - To satisfy the requirement of power economy, class C amplifiers are preferred. Hence, a trade off between power spectrum efficiency and spectral efficiency has to be made keeping in view the system complexity. Care has to be taken to ensure that the hardware complexity and the cost are properly weighed.

II. DIGITAL MODULATION TECHNIQUES

Among the many digital modulation techniques in use, some are responsive primarily to the goal of spectral efficiency, while others focus on the objective of achieving a narrow power spectrum. The large variation in signal amplitude due to Rayleigh fading encountered, render the digital amplitude modulation schemes almost inoperative. There are basically two broad modulation strategies emerging which emphasize the use of PSK or MSK derived modulation schemes.

A. Linear Modulation Techniques

The family of linear -modulation techniques require a high degree of linearity in modulating the carrier by baseband signal and RF power amplification before transmission. Linear modulation is more spectral efficient, but requires a linear power amplifier so as to avoid the signal amplitude variations which may result in intermodulation products. The most important linear modulation methods are QPSK, OK-QPSK, Π/4 shift QPSK and higher -level PSK.

0-7803-1266-x/93/53.00©1993 IEEE

Quaternary Phase Shift Keying (QPSK) : A generalised model of QPSK modulator is shown in Figure 1(a). QPSK is characterised by two parts of the baseband data signal the inphase signal $I(t)$ and the quadrature signal $Q(t)$. For this reason the data rate of individual signals, $I(t)$ and $Q(t)$, is half that of the base band signal. This also cuts the bandwidth requirement of QPSK to 1/2 as compared to BPSK. The modulated signal has 180 and ±90 degree phase shifts. The signal transitions are abrupt and unequal and this causes large spectrum dispersion. QPSK also needs more power and higher CNR (approx. 3 dB) than BPSK, to obtain the same performance for same probability of error.

Offset Keyed QPSK (OK-QPSK) : The difference between QPSK and OK-QPSK is in the alignment of the in phase signal $I(t)$ and the quadrature signal $Q(t)$. OK-QPSK is obtained by introducing a shift or offset equal to one bit delay (Tb) in the quadrature signal $Q(t)$. See Figure 1(b). This ensures that the $I(t)$ and $Q(t)$ signals have signal transitions at the time instants separated by Ts/2. The modulated signal transitions are ±90 degree maximum. It has no 180 degree phase shift and this results in reduction of out of band radiations. However, the abrupt phase transitions still remain.

Π/4 Phase Shift QPSK : Π/4 -QPSK is a compromise between QPSK and OK-QPSK. It can be regarded as a modification to QPSK and has carrier phase transitions which are restricted to ±45 degree and ±135 degree. A block diagram of Π/4 shift QPSK modulator is shown in figure 1(c). Like OK-QPSK the carrier phase does not undergo instantaneous 180 degree phase transition. Thus, the main advantage is that of the reduced envelope fluctuation as compared to that of QPSK. The BER performance of the Π/4 shift QPSK is controlled by CCI rather than fading. The fact which is very important for cellular mobile radio systems which work on frequency reuse concept and where CCI is the major source of interference. [5].

B. Continous Phase Modulation Techniques

Continuous-phase modulation schemes avoids the linearity requirements which reduces the cost of amplification. Modulation techniques derived from the CPM family have quite narrow power spectra. On the other hand, the spectral efficiency is somewhat lower. Among the important constant envelope or continuous phase modulation schemes being explored currently are MSK and GMSK.

Minimum Shift Keying (MSK) : MSK can be regarded either as a special case of OK-QPSK or as a form of FSK modulation. The baseband signal is filtered sinusoidaly to produce a graceful transition from one binary state to another. MSK is a binary modulation with symbol interval Tb and frequency deviation = ±1/4Tb. And there is phase

1

Reprinted from *IEEE Vehicular Technology Conference*, pp. 1-4, 1993.

159

continuity of the modulated RF carrier at the bit transitions. RF phase varies linearly exactly ±90 degree with respect to the carrier over one bit period Tb.

Gaussian Minimum Shift Keying (GMSK): The use of a pre-modulation low-pass filter with a Gaussian characteristics with the MSK approach achieves the requirement of uniform envelope, in addition to spectral containment. This modulation scheme is known as GMSK (Gaussian MSK). The Gaussian filter is used to suppress out of band noise and adjacent channel interference. GMSK provides high spectrum efficiency and a constant amplitude signal that allows class C power amplifiers to be used minimising power consumption, weight and cost. [2].

Tamed Frequency Modulation (TFM) : In MSK the phase continuity is achieved, but the derivative of the phase is still discontinuous. If the phase change is made still smoother, a much narrower spectrum can be achieved. A scheme involving pre-filtering combined with an algorithm for selecting the carrier phase shift, according to the original data values, has been developed. This scheme, known as Tamed Frequency Modulation (TFM), has spectral containment characteristics similar to GMSK. [1].

III. SPECTRAL EFFICIENCY

Spectral efficiency is defined as the number of bits per second per hertz (b/s/Hz). Spectral efficiency influences the spectrum occupancy in a mobile radio system. Theoretically, an increase in the number of modulation levels results into higher spectral efficiency. But the precision required at the demodulator to detect the phase and frequency changes also increases exponentially. Which results into higher S/N requirement to achieve same BER performance. Table I shows the comparison of spectral efficiency and the required S/N (for BER of 1 in 10^6) for PSK and MSK modulation systems.

The QPSK-type linear modulation schemes have recently drawn more attention, since they offer higher spectral efficiency and are considerably simpler to be implemented. The north American (ADC) and Japanese (JDC) digital cellular systems currently under development employ $\Pi/4$ phase shift QPSK.

TABLE I - COMPARISON OF MODULATION SYSTEMS

MODULATION	SPECTRAL EFFICIENCY	REQUIRED S/N
BPSK	1 b/s/Hz	11.1 dB
QPSK	2 b/s/Hz	14.0 dB
PSK (16 level)	4 b/s/Hz	26.0 dB
MSK (2 level)	1 b/s/Hz	10.6 dB
MSK (4 level)	2 b/s/Hz	13.8 dB

Table II summarises various modulation schemes adopted in second generation cellular and cordless telephone systems. It is observed that though linear modulation schemes offer better spectral efficiency, GMSK is also considered as a promising modulation scheme. The European GSM (Group Speciale Mobile) system is based on GMSK with a bit rate of 271 kbps and Bb.T = 0.3. The DECT (Digital European Cordless Telephone) system uses GMSK with Bb.T = 0.5 at a data rate of 1.152 Mb/sec..

The choice between linear modulation and constant

envelope modulation technique is influenced by the fact that how much maximum spectral efficiency must be stressed in comparison with cost and size, and also the requirement of adjacent channel selectivity.

TABLE II - SPECTRAL EFFICIENCY OF DIGITAL CELLULAR and CORDLESS TELEPHONE SYSTEMS

SYSTEM	MODULATION TECHNIQUE	CHANNEL B.W. KHz	DATA RATE Kbps	SPECTRAL EFFICIENCY b/s/Hz
JDC	Π/4 QPSK	25	42.0	1.68
ADC	Π/4 QPSK	30	48.6	1.62
GSM	GMSK (BbT = 0.3)	200	270.8	1.35
CT-2	GMSK	100	72.0	0.72
DECT	GMSK(BbT = 0.5)	1728	1572.0	0.67

Bb is the pre-modulation filter bandwidth.

IV. INTERSYMBOL INTERFERENCE

One of the effects produced by mobile environment is the distortion of the signal due to delay spread, which results in intersymbol interference. It is known that if the effects of delay spread are ignored or equalized, the spectral efficiency improves for higher level modulation. And this is realised at the cost of high SNR. Theoretically, the impact of delay spread can be reduced by adopting higher level modulation techniques. For example, a 4-level modulation signal transmitted at the data rate of 200 kbps would have symbol duration of about 10us as against 20 us in 16-level modulation. However, under the influence of delay spread, the simulations indicate that no significant performance improvement can be achieved as the level of modulation exceeds 4, even if SNR approaches infinity. It is indicated that BER performance depends strongly on the rms value of delay spread. Thus a 4 level modulation scheme seems to be reasonable when delay spread is significant. [4].

V. POWER SPECTRUM EFFICIENCY

The energy of a modulated carrier is distributed in the main lobe and side lobes. The power spectrum is a determinent of adjacent channel interference.

A power efficient digital communication system will have to be non linearly amplified for cost effective utilization of power. When a band limited linearly modulated carrier undergoes non linear amplification, the filtered side lobes re-appear. This causes severe adjacent channel and co-channel interference. The BER performance is also degraded. Hence, such systems may not efficiently utilize the available frequency spectrum. Power efficiency gained is lost as a result of non linear amplification.

The power spectrum density for QPSK is given by eq. (1)

$$W(f-fc) = \frac{A}{2} \cdot Ts \cdot \left[\frac{\sin\{\pi \, Ts \, (f-fc)\}}{\{\pi \, Ts \, (f-fc)\}} \right]^2 \qquad (1)$$

Where fc is the unmodulated carrier frequency, A is a constant proportional to the total transmitted power and Ts is the symbol period. The equation (1) is valid for BPSK, QPSK and multilevel PSK as long as the symbols are mutually independent.

Constant phase modulation techniques can be non linearly amplified without significant spectral regeneration. The power spectrum density for MSK is given by equation (2).

2

160

$$W(f - fc) = \frac{A^2 \cdot Ts}{4\pi^2} \left[\frac{\cos(\pi f Ts)}{\{(f Ts - fc Ts)^2 - 1/4\}} \right]^2 \qquad (2)$$

Figure 2 shows the power spectrum distribution of QPSK and MSK. The difference in time alignment of the bit streams for QPSK and OK-QPSK does not change the power spectral density distribution. The difference in power efficiency using QPSK or Π/4 shift QPSK is very small. We see that QPSK has a side lobe level about 12 dB and MSK has about 24dB down from the carrier. The power spectrum of MSK signal concentrates more around the main lobe and decreases more quickly in the other side lobe areas i.e. the power spectrum of MSK signal decreases with f(-4) while that of QPSK decreases with f(-2). MSK is more spectrum efficient than QPSK but it has a wider main lobe.

Figure 3 shows the computed results of power spectra of GMSK for various values of BbT. [6]. It is observed that the spectrum occupancy can be controlled by the normalised 3dB BbT of the Gaussian premodulation filter.

If BbT is infinite this gives a phase change of 90 degree in each interval, which is MSK modulation. The power spectral density of GMSK with BbT = 0.2 is nearly equal to that of TFM. The value of BbT can be selected considering overall spectrum efficiency requirements. The spectral spread is nearly the same till 1/T but beyond that both QPSK and GMSK differ significantly. Among all these schemes the complexity of QPSK is considered to be high.

VI. OUT OF BAND POWER

In a channel without any bandwidth restriction, the QPSK modulated carrier has a uniform envelope but a wide spread of sideband energy. To contain this, post filtering (bandwidth equal to 1.1RL) can be adopted. The filtering is acceptable to receivers, but the effect of filtering has been to introduce carrier amplitude variations. If the signal is transmitted through a nonlinear power amplifier in the transmitter, then the amplitude variations result in the regrowth of the side lobes, which effectively expand the bandwidth again.

Figure 4 shows the computed results of fractional power outside the desired normalised channel bandwidth for various values of BbT. [6]. It is seen that for a non filterd MSK signal 99% of the power is contained within a channel of bandwidth equal to 1.2RL. Where as for GMSK (BbT = 0.5) it is contained within a channel of bandwidth equal to 1.04RL and for GMSK (with BbT = 0.2) in a channel of bandwidth equal to 0.79RL.

The out-of-band emission performance of QPSK and Π/4 shift QPSK is almost the same when a reasonably linear amplifier is used. [8]. For QPSK 99% of the power is contained in a channel bandwidth equal to 8RL as compared to 1.2RL for non filterd MSK signal. Thus less post modulation filtering will be required for MSK than for QPSK to reduce the band power to a given value. The first null of the MSK signal occurs at 0.75, while that of QPSK occurs at 0.5 of the bit rate. A narrow post modulation filter removes more energy from the first lobe of MSK than that of PSK and introduces more intersymbol interference. The literature shows different values for an optimum channel bandwidth from the implementation point. Ishizuka and Hirade [7] claim an optimum for Bc = 0.59RL. In a realistic radio channel using class C power amplifier a Bc = 0.75RL is a realistic optimum assumption.

Minimum spectral dispersion and a uniform envelope for the modulated carrier are, therefore, important targets to aim for. QPSK appears to be non ideal on both counts.

VII. EFFECT OF CODING

The carrier modulation and coding though are logically independent process but are strongly interrelated. The improvement in either of them is towards achievement of a common goal of higher spectral efficiency. Higher level modulation offers higher spectral efficiency but higher SNR is required to achieve a given BER objective which is difficult to achieve in a mobile environment. The required S/N can be reduced by using low bit rate voice coding and efficient error correction techniques prior to modulation.

The selection of a coding technique is mainly driven by achieving the desired voice quality by adding minimum overhead while protecting the information bits to fight the channel noise. This strategy is assisted by treating those bits within a message which decide the speech quality, separately from those which are less critical.

LPC-RPE speech coder used in GSM cellular system operates at 13 kbps data rate. The most sensitive bits are protected by a CRC code with rate = 1/2 convolutional code with constraint length 5. The overall coded bit rate per speech signal is 22.8 kbps. To combat the channel noise, 30% overhead (10.1 kbps) as guard time, synchronization, etc and 0.95 kbps as SACCH is added before transmission. Thus the gross bit rate is 33.85 kbps per channel and 270.8 kbps for 8 channels. The spectral efficiency of GSM system employing GMSK modulation (BT = 0.3) operating with a 200 KHz channel spacing is 1.35 b/s/Hz.

The speech coding technique adopted for ADC system is VSELP operating at 7.95 kbps. The channel coding used is a convolutional code with constraint length 6. The overall coded bit rate is 13 kbps only. With 16% overhead and 0.6 kbps SACCH, the overall bit rate is 16.2 kbps per channel and 48.6 kbps for 3 channels. The spectral efficiency (1.6 b/s/Hz) is better than that of GSM system. The ADC system uses Π/4-QPSK modulation with 30 KHz channel spacing.

Traditional LPC-type coders are capable of handling very low bit rates (1.2 - 4.8 kbps), but tend to be extremely vulnerable to errors. RELP coder operating at 8 - 16 kbps and Subband coders operating at 16 kbps have also proved to be fairly robust.

VIII. CONCLUSION

Both presented digital modulation schemes have been used extensively in mobile communication systems. The selection of one over the other depends on the priorities set in the system requirements. If most efficient bandwidth utilisation and moderate hardware complexity is the key note - QPSK (Π/4 QPSK) will be a better choice. Whereas continuous phase modulation schemes offer constant envelope, narrow power spectra, good error rate performance, etc. Therefore when out of band signal power, tolerance against filter parameter and non-linear power amplifiers are important features, compromise in channel separation is permissible and higher circuit complexity is of less consideration- GMSK is the solution. Spectral efficiency can be further improved by using suitable coding technique. Digital modulation scheme combined with robust coder which can inherently withstand higher bit error rates needing less channel coding will be the ultimate choice for a spectrum efficient mobile system.

3

161

IX. REFERENCES

[1] F. de Jager and C. B. Dekker, "Tamed Frequency Modulation, A Novel Method to Achieve Spectrum Economy in Digital Transmission," IEEE Trans. on Communications, pp 534-542, May, 1978.

[2] G..D. Aria, F.Muratore and V. Palestini," Performance of GMSK and Comparisons with the Modulation Methods of 12PM3, " Proceedings 37th IEEE Vehicular Technology Conference VTC-87, Tampa, Florida, USA, pp 246-252, June 1-3, 1987.

[3] H. Robert Mathwich, Joseph F. Balcewicz and Martin Hecht, "The Effect of Tandem Band and Amplitude Limiting on Eb/No Performance of Minimum (frequency) Shift Keying (MSK)," IEEE Trans. on Communications, Vol. COM-22, No. 10, pp 1525 - 1540, October 1974

[4] Justin C.I. Chuang, "The Effects of Delay Spread on 2-PSK, 4-PSK, 8-PSK and 16-QAM in a Portable Radio Environment," IEEE Transactions on Vehicular Technology, Vol. 38, No. 2, pp 43-45, May 1989.

[5] Kamilo Feher, "Modems for Emerging Digital Cellular Mobile Radio Systems, " IEEE Transactions on Vehicular Technology, Vol. 40, No. 2, pp 355 - 365, May 1991.

[6] K. Murota and K. Hirade, "GMSK Modulation for Digital Mobile Radio Telephony, " IEEE Trans. on Communications, Vol COM-29, No. 7, pp 1044 - 1050, July 1981.

[7] M. Ishizuka and K. Hirade, "Optimum Gaussian Filter and Deviated Frequency Locking Scheme for Coherent Detection of MSK," IEEE Trans. on Communications, Vol. COM-28, No. 6, pp 850 - 857, June 1980.

[8] S. Ariyavisitakul and Ting Ping Liu, "Characterizing the Effects of Nonlinear Amplifiers on Linear Modulation for Digital Portable Radio Communications," IEEE Transactions on Vehicular Technology, Vol. 39, No. 4, pp 383-389, November 1990.

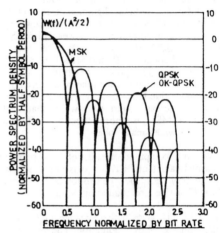

Fig.2: Power spectra of QPSK and MSK

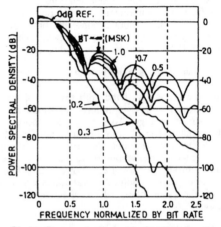

Fig. 3: Power spectra of GMSK signals

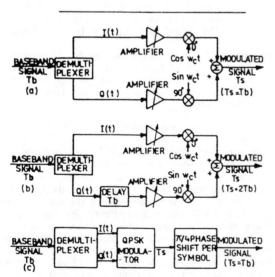

Fig. 1: Generalised Model of (a) QPSK;
(b) OK-QPSK; (c) π/4-QPSK
Tb: Baseband data Rate, Ts: Symbol Rate

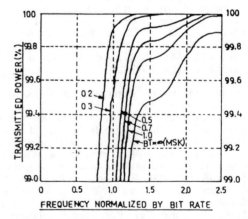

Fig.4: Fractional Power Ratio –GMSK signals

162

Modulation Methods for PCNs

For optimal efficiency, macrocellular and microcellular personal communications systems may require different modulation schemes.

William T. Webb

Reprinted from *IEEE Communications Magazine*, pp. 90-95, Dec. 1992.

ithin future personal communications networks there is likely to be a shortage of bandwidth to cope with the large user base and high-bit-rate services that will be required. Many procedures have been adopted to improve spectral efficiency, including low-bit-rate speech coders, a variety of multiple access techniques, the deployment of microcells, and an increase in the network intelligence. The spectral occupancy required by the modulation scheme is an important quality parameter. Binary and quatenary modulation schemes are simple to implement and can tolerate low signal-to-interference ratios (SIRs) but only encode one or two bits per symbol. Higher-level modulation schemes such as quadrature amplitude modulation (QAM) encode more bits per symbol. However, they require a higher SIR, and in non-code-division multiple access (CDMA) systems this often necessitates an increase in cluster size and a concomitant reduction in the number of available channels. We are interested here in whether the gain in bits per symbol offsets the reduction in number of channels.

We may anticipate that the spectral congestion will cause operators to deploy microcells in urban areas. While conventional (macrocellular) coverage areas can be modelled as hexagons, this representation is invalid for microcells. Here we consider a range of modulation schemes for both macro- and microcellular communications systems to determine efficient modulation schemes for use in future personal communication systems.

For cellular communications there are two main measures of efficiency, channel efficiency and spectrum efficiency, where channel efficiency is the number of channels that can be provided within a given bandwidth, and spectrum efficiency is the number of channels that can be provided with frequency reuse. For cellular systems, system capacity is directly related to spectrum efficiency. In order to derive equations detailing spectrum efficiency, Lee considered an idealized hexagonal cell structure that has six significant interferers, regardless of the cluster size [1]. The SIR is the received signal power from the wanted base station (BS) divided by the summation of the interference powers from the cochannel BSs.

Based on a logarithmic path loss model, neglecting noise, and assuming that the signal level from all the interferers is approximately equal, the SIR becomes $R^{-\gamma}/(6D^{-\gamma})$ where R is the radius of a cell, D the distance between interfering cells, and γ the path loss exponent. As

$$D/R = \sqrt{3K},$$

where K is the number of cells in a frequency reuse pattern, and the number of radio channels available in each cell, m, is $B_t/(B_c.K)$ where B_t is the total bandwidth available and B_c the bandwidth required per channel, then for $\gamma = 4$ we may represent m in terms of SIR, viz:

$$m = \frac{B_t}{B_c\sqrt{\frac{2}{3}SIR}}. \qquad (1)$$

Now higher spectrum efficiency may be achievable by increasing the number of bits per symbol. This has the effect of reducing B_c but also requires a higher SIR. Gejji tried to formulate an equation that would make this trade-off explicit [2]. To do this he assumed that the transmission was limited by the SIR, and replaced signal-to-noise ratio (SNR) in Shannon's capacity equation with SIR, having assumed the interference was Gaussian-like. A parameter α was introduced to represent the

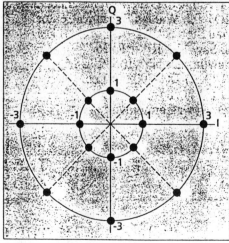
■ **Figure 1.** *16-level star QAM constellation.*

WILLIAM T. WEBB is technical director at Multiple Access Communications Ltd., Southampton, UK, working in the field of modulation techniques, computer simulation, and propagation modeling.

 0163-6804/92/$03.00 1992© IEEE

closeness of fit to the Shannon limit. This parameter will be dependent on the modulation, bit rate, technology used, etc. No attempt has been made to estimate α for the various schemes; it is left within the equation and assumed to be a constant. Replacing B_c in equation 1 by the bandwidth in Shannon's equation gives the number of channels as

$$m = 1.224a \frac{B_t \log_2(1+SIR)}{R_b \sqrt{SIR}} = (1.224a \frac{B_t}{R_b})M \quad (2)$$

where R_b is the bit rate and M is a system parameter that links the number of channels with the SIR. Gejji went on to consider mathematical models of practical modulation schemes, and came to the conclusion that four-level modulation was most appropriate for cellular communications.

Spectrum Efficiency in Conventional Cells

We have reevaluated the results obtained by Gejji using a comprehensive simulation of frequency-shift keying (FSK), phase-shift keying (PSK), and fixed-level and variable-level QAM, as we judged these types of modulation to be suitable for personal communication services (PCS) systems. The QAM simulations are for differential star QAM [3], and the variable-level QAM schemes are for variable-rate star QAM [4, 5].

The star QAM scheme was introduced when work showed the square constellation to be unsuitable for Rayleigh fading channels due to carrier recovery and automatic gain control (AGC) problems [3]. A 16-level constellation is shown in Fig. 1. It has eight possible lock positions, which are resolved by differential coding. Of the four bits in each symbol, the first is differentially encoded onto the QAM phasor amplitude so that a "1" causes a change to the amplitude ring that was not used in the previous symbol, and a "0" causes the current symbol to be transmitted at the same amplitude as the previous symbol. The remaining three bits are differentially Gray encoded onto the phase so that, for example, "000" would cause the current symbol to be transmitted with the same phase as the previous one, "001" would cause a 45-degree phase shift relative to the previous symbol, "011" a 90-degree shift relative to the previous symbol, and so on. Decoding data is reduced to a comparison test between the previous and current received symbols. This system considerably improves the bit error rates (BERs) compared to the square constellation because it eliminates the long error bursts that occurred when a false lock was made. Of considerable importance is that with differential amplitude encoding we may dispense with AGC, which thereby removes errors caused by an inability of the AGC to follow the fading envelope.

The variable-rate QAM varies the number of modulation levels according to the integrity of the channel, so that when the receiver is not in a fade we increase the number of constellation points, and as the receiver enters a fade we decrease them down to a value that provides an acceptable BER. Successful variable-rate transmission requires that the fast fading channel changes slowly compared to the symbol rate. For time-division multiple access

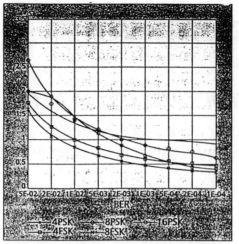

■ **Figure 2.** *Spectrum efficiency for PSK/FSK.*

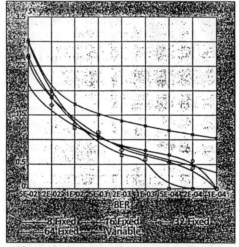

■ **Figure 3.** *Spectrum efficiency for QAM schemes.*

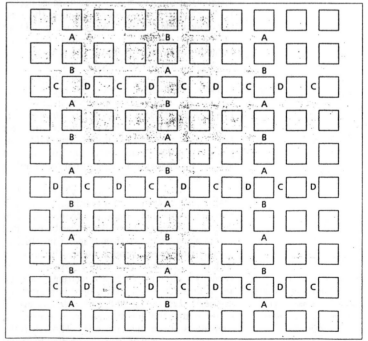

■ **Figure 4.** *Microcellular BS pattern at 1.8 GHz for a four-cell cluster.*

(TDMA) systems the transmission received by the MS is used to estimate the channel integrity, which then dictates the number of QAM levels to be used by the MS transmitter. Similarly, the BS uses the signal it receives to determine the number of QAM levels to be used in the subsequent BS transmissions. The QAM constellation changes as the number of levels is varied maintaining the principles of star QAM and alternately doubling amplitude levels and phase points as more bits per symbol are required, up to a maximum of 6 b/symbol. For the fixed-rate star QAM schemes the 8- and 16-level star QAM constellations have two amplitude levels, whereas the 32-and 64-level constellations have 4 levels, and for all these fixed schemes the data rate is constant.

The simulation program used includes clock and carrier recovery, cochannel and adjacent-channel interference, Rayleigh fading, and the use of convolutional and block coding. Near-optimal filtering at transmitters, receivers, and cochannel transmitters was used. A carrier frequency of 2 GHz, mobile speed of 15 m/s, and data rate of 25 kBaud was assumed.

Graphs of the BER against SIR performance for a range of modulation schemes, each using a range of channel coding techniques, was established. This led to over 60 performance curves. Using these curves, spectrum efficiency for a variety of BERs was established. To do this, 9 BER values spanning the range 5×10^{-2} to 1×10^{-4} were selected. For each modulation scheme and coding scheme, the SIR at which these error rates were achieved was read off our performance curves. The SIR was then used to calculate M in equation 2, which was multiplied by the throughput (this takes the redundancy required for the error coding into account) to give a figure of merit, s, by which the various systems can be judged. This number is proportional to the number of channels available in each cell with the particular modulation schemes under consideration.

The graphs of s versus BER for hexagonal cells are given in Figs. 2 and 3. In these figures, "nPSK," where n is an integer, refers to n-level PSK. A similar notation is used for FSK. For QAM, "n fixed" refers to n-level star QAM where the number of levels are fixed over the duration of the transmission, whereas the variable-level scheme employed an average of 4 b/symbol. From these graphs we can see that the performance of all the schemes is similar. The variable-rate scheme is superior for all values of BER. If the variable level scheme is not available, then for BERs above 1×10^{-2} the fixed star QAM schemes with 64 and 32 levels are preferred. For BERs below this, then, as Gejji postulated, four-level PSK is most appropriate, followed by eight-level PSK. At a BER of 1×10^{-3}, 16-level star QAM can only provide 60 percent of the channels that 4-level PSK can and is more complex to implement.

Spectrum Efficiency in Microcells

While the above equations hold for idealized hexagonal cellular systems, they are not applicable to microcellular systems where the microcellular boundaries are controlled by the buildings and roads in the vicinity of the BS. There is an enormous variety of street patterns, but for simplicity we considered microcellular clusters based on an idealized grid pattern. The SIR resulting from these clusters was established and the figure of merit, s, computed. Conclusions were then drawn as to the most appropriate modulation schemes for microcellular systems.

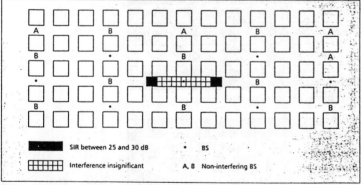

■ Figure 5. *Microcellular BS pattern at 900 MHz.*

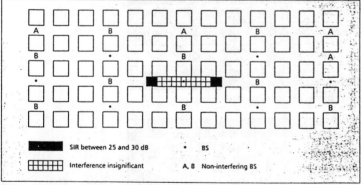

■ Figure 6. *Interference with K=4 at 1.8 GHz.*

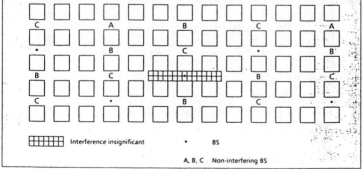

■ Figure 7. *Six-cell interference at 1.8 GHz.*

Microcellular Clusters

In a microcell, the interference from any given cochannel interferer varies significantly throughout the microcell, due to the rapid decrease in signal level as the mobile moves away from its BS. To simplify our determination of SIR values we resorted to a microcellular prediction tool developed at Multiple Access Communications (MAC), known as the Microcellular Design System (MIDAS). This tool can predict both signal and interference levels for any microcellular cluster in both practical and idealized situations. In real streets it must be supplied with a digital map. Here we simply fed into the package the rectilinear grid pattern of roads having 100 m blocks and 20-m-wide streets. A range of predictions were performed at both 900 MHz and 1.8 GHz. In order to establish the microcell size, the microcell boundary was based approximately on the minimum received power of a CT-2 type system. With a 10-mW transmitter and 0-dBi gain antennas, the effective radiated power was 10 dBm. Given a minimum signal level of –95 dBm at which data can be recovered and a fading margin of 20 dB, the minimum received signal level was –75 dBm, which relates to a path loss of 85 dB. This minimum signal level will be modulation-dependent; here we assume that the transmit power is adjusted to maintain a constant microcell size regardless of modulation type. If all microcells in the cluster use the same transmitter power, the SIRs to be calculated are independent of this power. With microcell BSs placed midway along the blocks, which was the best way to produce a cluster with a regular frequency assignment pattern, this path loss gave a range of 180 m at 1.8 GHz, giving a microcell length of 360 m. This required a microcell every third block, as shown in Fig. 4, which also illustrates the BS pattern for a four-cell cluster. In this figure, which is not drawn to scale, the square areas represent city blocks and the letters represent a BS location and its specific frequency assignment.

At 900 MHz, the same path loss led to a range of 300 m, giving a microcell length of 600 m and a BS every 5th block, as shown in Fig. 5, which also shows the frequency assignment for a 6-cell cluster, where the same representation as in Fig. 4 applies. Using these BS patterns we considered the SIRs. In Figs. 6 through 9, the small square blocks represent BSs using the same frequencies that are significant interferers, the letters represent insignificant interferers and their frequency assignments, while the shaded area represents a microcell. This shaded area does not represent the standalone coverage area; instead, it indicates the area of a microcell within a cluster. The interference has been assumed to be insignificant when the SIR was greater than 30 dB. A minimum of four frequencies are required to avoid SIRs of 0 dB or less, so we do not consider clusters with fewer than four frequencies.

At 1.8 GHz the use of four frequencies leads to the interference contour shown in Fig. 6. Here two frequencies have been assigned to the vertical streets and two to the horizontal, as shown in Fig. 4. The staggered pattern of the interfering BSs on neighboring streets leads to the minimum interference levels for four-cell clusters for this particular street pattern. The interfering BSs are both on the same street as the main BS and on parallel streets. At 1.8 GHz, only the BSs on the same

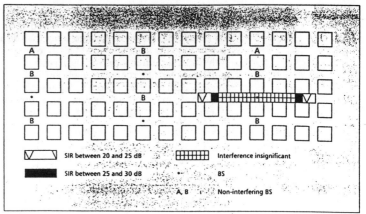

Figure 8. *Interference with K=4 at 900 MHz.*

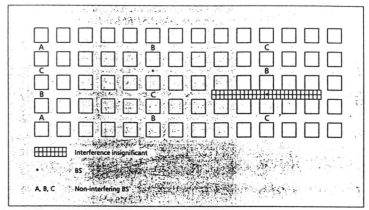

Figure 9. *Interference with K=6 at 900 MHz.*

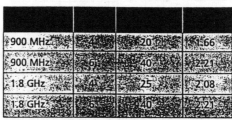

900 MHz	4	20	1.66
900 MHz	6	40	2.21
1.8 GHz	4	25	2.08
1.8 GHz	6	40	2.21

Table 1. *Microcell capacity results.*

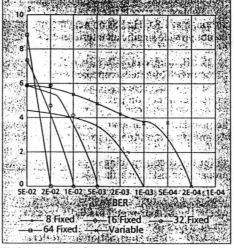

Figure 10. *Performance of QAM schemes in microcells with 20-dB SNR, K=4.*

166

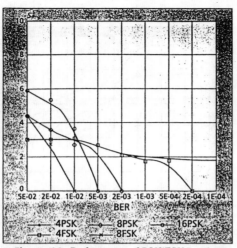

■ **Figure 11.** *Performance of PSK/FSK in microcells with 20-dB SNR, K=4.*

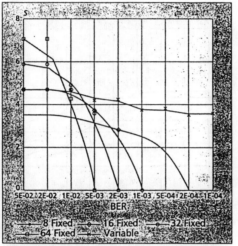

■ **Figure 12.** *Performance of QAM schemes in microcells with 30-dB SNR, K=4.*

Although microcells only experience significant interference in certain places, a communications system should be designed to cope with this level of interference.

street cause significant interference. The use of six-cell clusters leads to the interference plot shown in Fig. 7. With three frequency sets available for the horizontal streets, BSs can be spaced further apart on the same street and also staggered on neighboring streets to cause less interference, although at 1.8 GHz BSs on neighboring streets do not provide significant interference. As can be seen, with $K = 4$, there are areas where the SIR drops to 25 dB, whereas with $K = 6$ the microcell becomes SNR-limited as opposed to SIR-limited.

At 900 MHz, the interference from BSs using the same frequency on parallel adjacent streets becomes more significant due to the increase in reflection and diffraction around corners at this frequency. A similar method to that used to construct the cluster at 1.8 GHz was used at 900 MHz, but with the wider microcell spacing caused by increased propagation distances at the lower frequency. Interference plots for $K = 4$ and $K = 6$ are shown in Figs. 8 and 9, respectively. With $K = 4$, areas where the SIR dropped to 20 dB were experienced. This SIR was lower than that for 1.8 GHz, due to the higher diffraction and reflection of energy around corners. With $K = 6$, the system again became SNR-limited.

System Design for Microcells

Communications systems are generally designed for worst-case propagation. So, despite the fact that microcells only experience significant interference in certain places, a communications system should nevertheless be designed to cope with this level of interference. It is possible that future designers will find ways of overcoming or reducing the interference, even in four-cell clusters, by a combination of dynamic channel allocation and switching the frequencies of mobiles experiencing significant interference to different frequencies within the same set. However, here we only consider a simple system with fixed allocation and investigate its results. More intelligent systems will tend to reduce the cluster size.

Microcellular Spectrum Capacity

In the hexagonal cellular structures a simple relationship exists between K and the SIR; hence, we were able to eliminate the cluster size, K, in forming our figure of merit. No such simple relationship for K exists in microcells, but manipulation of the previous equations allows us to substitute our values of SIR for the various cluster sizes and frequencies considered, leading to values for M' reported in Table 1. We have assumed that where the interference was insignificant the SNR was used instead of the SIR, where the channel was noise-limited to 40-dB SNR. Our figure of merit, s, is M' multiplied by the throughput.

As we wish M' to be high, a cluster size of six is preferred for this particular SNR. The points at which four- and six-cell clusters have identical ratings is when the assumed SNR is 30 dB for the 900-MHz case and 37.5 dB for the 1.8-GHz case. This means that if the channel SNR within the microcell falls below 37.5 dB, four-cell clusters become more efficient than six-cell clusters for 1.8 GHz, and if the SNR falls below 30 dB, four-cell clusters become more efficient for 900 MHz. The microcellular SNR is of great importance in deciding upon the modulation scheme and depends on transmitter power and receiver sensitivity for any given system. Here we give results for a range of SNRs, always using the most efficient cluster size for each SNR value to derive the number of available channels.

Modulation Schemes for Microcells

Given SNRs of 20, 30, and 40 dB and assuming an SIR of 20 dB for the four-cell cluster and an infinite SIR for the six-cell cluster, we consider the most efficient modulation schemes to use within the microcellular environment. Graphs showing the BER against number of available channels for a variety of modulation schemes are given in Figs. 10 and 11 for the 20-dB case, in Figs. 12 and 13 for the 30-dB case, and Figs. 14 and 15 for the 40-dB case. These graphs have been derived in the same way as the graphs for the conventional cells, but with the appropriate SIRs assumed. The QAM schemes were star QAM, using two concentric circles with four and eight amplitude points per ring for the eight- and sixteen-level schemes, respectively. In all cases a Rayleigh fading channel was assumed with transmission at 1.8 GHz, a mobile speed of 15 m/s, and a data rate of 25 kbaud. The results at 900 MHz are similar because the same value of M' applies to both frequencies with

94

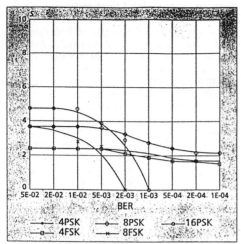

■ **Figure 13.** *Performance of PSK/FSK in micro-cells with 30-dB SNR, K=4.*

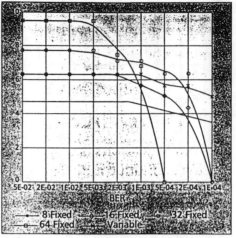

■ **Figure 14.** *Performance of QAM schemes in microcells with 40-dB SNR, K=6.*

the cluster size $K = 6$. For a cluster size of four, the value of s in the results reported for 1.8 GHz should be scaled by the ratio of M' at 900 MHz to M' at 1.8 GHz. Differential transmission and reception have been used along with near-optimal filtering and modified early-late clock recovery [5], and with the interferer using the same modulation scheme as the transmitter. The nomenclature used in Figs. 2 and 3 applies to these figures.

For the 20-dB SNR case, variable QAM and the higher-level schemes are most effective at high BERs. For BERs below 5×10^{-4} only 4PSK is suitable. When the SNR is 30 dB, the variable-level QAM is superior for all but the highest BERs. If only fixed-level schemes are considered, then 8PSK proves slightly better than 4PSK and 4FSK for low BERs. For the higher BERs, 16PSK and 16-level star QAM provide similar performances, while for the very highest BERs the higher-level modulation schemes prove best. Finally, for an SNR of 40 dB it can be seen that for all but the lowest BERs, the 64- and 32-level star QAM schemes are the most successful. At BERs below 2×10^{-4}, the variable-rate QAM schemes are superior to the other schemes simulated, with the four- and eight-level PSK and FSK schemes giving substantially worse performance for the range of SNR and SIR values examined.

Conclusions

We have considered the interference levels that might be expected for a range of conventional cellular and microcellular clusters that can be used in PCN/PCS. Simulations suggested that for conventional size cells, 32- and 64-level QAM schemes are preferred for BERs above 1×10^{-2}, while 4PSK or variable-rate QAM schemes are better for lower BERs. For microcells with communications at both 900 MHz and 1.8 GHz, four- or six-cell clusters are advocated, depending on the SNR expected. Based on the expected SNRs and interference levels, we surmise that variable-rate QAM schemes are superior to the other modulation schemes considered here. For low BERs, 4PSK may often provide the best performance, whereas for high BERs, particularly when the SNR is high, 32- and 64-level star QAM are the most suitable.

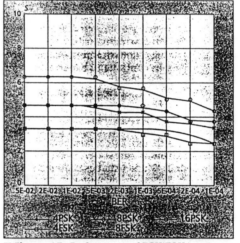

■ **Figure 15.** *Performance of PSK/FSK in micro-cells with 40-dB SNR, K=6.*

Acknowledgment

Some of this work was performed on behalf of the Radiocommunications Agency as part of a project researching multi-level modulation for the mobile radio environment.

References

[1] W. C. Y. Lee, "Spectrum Efficiency in Cellular," *IEEE Trans. on Vehicular Tech.*, Vol. 38, No. 2, pp. 69-75, May 1989.
[2] R. R. Gejji, "Channel Efficiency in Digital Cellular Communication Systems," *Proc. 42nd IEEE Vehicular Tech. Conf.*, Denver, CO, 1992.
[3] W. T. Webb, L. Hanzo, and R. Steele, "Bandwidth-Efficient QAM Schemes for Rayleigh Fading Channels," *IEE Proc.*, Pt. I, Vol. 138, No. 3, pp. 169-175, June 1991.
[4] R. Steele and W. T. Webb, "Variable Rate QAM for Data Transmissions over Mobile Radio Channels," keynote paper, Wireless '91, Calgary, Alberta, Canada, July 1991.
[5] W. T. Webb, "QAM—The Modulation Scheme for Future Mobile Radio Communications?," *IEE Elect. and Commun. J.*, Vol. 4, No. 4, Aug. 1992, pp. 167-176.

Biography

WILLIAM T. WEBB spent three years at Southampton University, sponsored by Ferranti Computer Systems Ltd. While at Southampton he was awarded the top year prize for each of his three years of study and graduated in 1989 with a first-class honors degree. He gained his Ph.D., with a thesis entitled "QAM Techniques for Digital Mobile Radio," in 1992, also from Southampton University. Since 1989 he has been working for Multiple Access Communications Ltd. in the field of modulation techniques, computer simulation, and propagation modelling, and was promoted to the position of technical director in March 1992.

Based on the expected SNRs and interference levels, we surmise that variable-rate QAM schemes are superior to the other modulation schemes considered here.

Defining, Designing, and Evaluating Digital Communication Systems

A tutorial that emphasizes the subtle but straightforward relationships we encounter when transforming from data-bits to channel-bits to symbols to chips.

Bernard Sklar

Reprinted from *IEEE Communications Magazine*, pp. 92-101, Nov. 1993.

BERNARD SKLAR is the head of advanced systems at Communications Engineering Services, an adjunct professor at the University of Southern California, and a visiting professor at the University of California at Los Angeles.

*T*he design of any digital communication system begins with a description of the channel (received power, available bandwidth, noise statistics, and other impairments such as fading), and a definition of the system requirements (data rate and error performance). Given the channel description, we need to determine design choices that best match the channel and meet the performance requirements. An orderly set of transformations and computations has evolved to aid in characterizing a system's performance. Once this approach is understood, it can serve as the format for evaluating most communication systems.

In subsequent sections, we shall examine the following four system examples, chosen to provide a representative assortment: a bandwidth-limited uncoded system, a power-limited uncoded system, a bandwidth-limited and power-limited coded system, and a direct-sequence spread-spectrum coded system. The term coded (or uncoded) refers to the presence (or absence) of error-correction coding schemes involving the use of redundant bits.

Two primary communications resources are the received power and the available transmission bandwidth. In many communication systems, one of these resources may be more precious than the other, and hence most systems can be classified as either bandwidth limited or power limited. In bandwidth-limited systems, spectrally-efficient modulation techniques can be used to save bandwidth at the expense of power; in power-limited systems, power-efficient modulation techniques can be used to save power at the expense of bandwidth. In both bandwidth- and power-limited systems, error-correction coding (often called channel coding) can be used to save power or to improve error performance at the expense of bandwidth. Recently, trellis-coded modulation (TCM) schemes have been used to improve the error performance of bandwidth-limited channels without *any* increase in bandwidth [1], but these methods are beyond the scope of this tutorial.

The Bandwidth Efficiency Plane

*F*igure 1 shows the abscissa as the ratio of bit-energy to noise-power spectral density, E_b/N_0 (in decibels), and the ordinate as the ratio of throughput, R (in bits per second), that can be transmitted per hertz in a given bandwidth, W. The ratio R/W is called bandwidth efficiency, since it reflects how efficiently the bandwidth resource is utilized. The plot stems from the Shannon-Hartley capacity theorem [2-4], which can be stated as

$$C = W \log_2 \left(1 + \frac{S}{N} \right) \qquad (1)$$

where S/N is the ratio of received average signal power to noise power. When the logarithm is taken to the base 2, the capacity, C, is given in b/s. The capacity of a channel defines the maximum number of bits that can be reliably sent per second over the channel. For the case where the data (information) rate, R, is equal to C, the curve separates a region of practical communication systems from a region where such communication systems cannot operate reliably [3,4].

M-ary Signaling

Each symbol in an M-ary alphabet is related to a unique sequence of m bits, expressed as

$$M = 2^m \quad \text{or} \quad m = \log_2 M \qquad (2)$$

where M is the size of the alphabet. In the case

0163-6804/93/$03.00 © 1993 IEEE

of digital transmission, the term "symbol" refers to the member of the M-ary alphabet that is transmitted during each symbol duration, T_s. In order to transmit the symbol, it must be mapped onto an electrical voltage or current waveform. Because the waveform represents the symbol, the terms "symbol" and "waveform" are sometimes used interchangeably. Since one of M symbols or waveforms is transmitted during each symbol duration, T_s, the data rate, R in b/s, can be expressed as

$$R = \frac{m}{T_s} = \frac{\log_2 M}{T_s} \quad \text{bit/s} \quad (3)$$

Data-bit-time duration is the reciprocal of data rate. Similarly, symbol-time duration is the reciprocal of symbol rate. Therefore, from Equation (3), we write that the effective time duration, T_b, of each bit in terms of the symbol duration, T_s, or the symbol rate, R_s, is

$$T_b = \frac{1}{R} = \frac{T_s}{m} = \frac{1}{m R_s} \quad (4)$$

Then, using Equations (2) and (4) we can express the symbol rate, R_s, in terms of the bit rate, R, as follows.

$$R_s = \frac{R}{\log_2 M} \quad (5)$$

From Equations (3) and (4), any digital scheme that transmits $m = \log_2 M$ bits in T_s seconds using a bandwidth of W Hz operates at a bandwidth efficiency of

$$\frac{R}{W} = \frac{\log_2 M}{W T_s} = \frac{1}{W T_b} \quad \text{(bit/s)/Hz} \quad (6)$$

where T_b is the effective time duration of each data bit.

Bandwidth-Limited Systems

From Equation (6), the smaller the $W T_b$ product, the more bandwidth efficient will be any digital communication system. Thus, signals with small $W T_b$ products are often used with bandwidth-limited systems. For example, the new European digital mobile telephone system known as groupe special mobile (GSM) uses Gaussian minimum-shift keying (GMSK) modulation having a $W T_b$ product equal to 0.3 Hz/(b/s), where W is the bandwidth of a Gaussian filter [5].

For uncoded bandwidth-limited systems, the objective is to maximize the transmitted information rate within the allowable bandwidth, at the expense of E_b/N_0 (while maintaining a specified value of bit-error probability, P_B). The operating points for coherent M-ary phase-shift keying (MPSK) at $P_B = 10^{-5}$ are plotted on the bandwidth-efficiency plane (Fig. 1). We assume Nyquist (ideal rectangular) filtering at baseband [6]. Thus, for MPSK, the required double-sideband (DSB) bandwidth at an intermediate frequency (IF) is related to the symbol rate as follows.

$$W = \frac{1}{T_s} = R_s \quad (7)$$

where T_s is the symbol duration, and R_s is the symbol rate. The use of Nyquist filtering results in the minimum required transmission bandwidth that yields zero intersymbol interference; such ideal filtering gives rise to the name Nyquist minimum bandwidth.

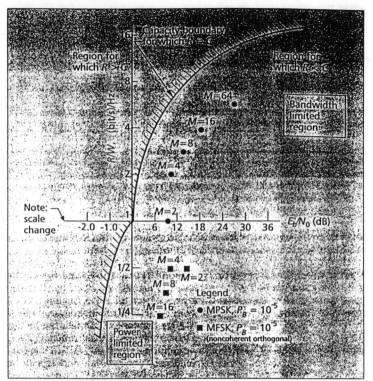

■ **Figure 1.** *Bandwidth-efficiency plane.*

From Equations (6) and (7), the bandwidth efficiency of MPSK modulated signals using Nyquist filtering can be expressed as

$$\frac{R}{W} = \log_2 M \quad \text{(bit/s)/Hz} \quad (8)$$

The MPSK points in Fig. 1 confirm the relationship shown in Equation (8). Note that MPSK modulation is a bandwidth-efficient scheme. As M increases in value, R/W also increases. MPSK modulation can be used for realizing an improvement in bandwidth efficiency at the cost of increased E_b/N_0. Although beyond the scope of this article, many highly bandwidth-efficient modulation schemes are under investigation [7].

Power-Limited Systems

Operating points for noncoherent orthogonal M-ary frequency-shift keying (MFSK) modulation at $P_B = 10^{-5}$ are also plotted (Fig. 1). For MFSK, the IF Nyquist minimum bandwidth is as follows [4]:

$$W = \frac{M}{T_s} = M R_s \quad (9)$$

where T_s is the symbol duration, and R_s is the symbol rate. With MFSK, the required transmission bandwidth is expanded M-fold over binary FSK since there are M different orthogonal waveforms, each requiring a bandwidth of $1/T_s$. Thus, from Equations (6) and (9), the bandwidth efficiency of noncoherent orthogonal MFSK signals using Nyquist filtering can be expressed as

$$\frac{R}{W} = \frac{\log_2 M}{M} \quad \text{(bit/s)/Hz} \quad (10)$$

The MFSK points in Fig. 1 confirm the relationship shown in Equation (10). Note that MFSK modulation is a bandwidth-expansive scheme. As M increases, R/W decreases. MFSK modulation can be

M	m	R (b/s)	R_s (symb/s)	MPSK minimum bandwidth (Hz)	MPSK R/W	MPSK E_b/N_0 (dB) $P_B = 10^{-5}$	Noncoherent orthog MFSK min. bandwidth (Hz)	MFSK R/W	MFSK E_b/N_0 (dB) $P_B = 10^{-5}$
2	1	9600	9600	9600	1	9.6	19,200	1/2	13.4
4	2	9600	4800	4800	2	9.6	19,200	1/2	10.6
8	3	9600	3200	3200	3	13.0	25,600	1/3	9.1
16	4	9600	2400	2400	4	17.5	38,400	1/4	8.1
32	5	9600	1920	1920	5	22.4	61,440	1/8	7.4

■ Table 1. *Symbol rate, Nyquist minimum bandwidth, bandwidth efficiency, and required E_b/N_0 for MPSK and noncoherent orthogonal MFSK signaling at 9600 b/s.*

used for realizing a reduction in required E_b/N_0 at the cost of increased bandwidth.

In Equations (7) and (8) for MPSK, and Equations (9) and (10) for MFSK, and for all the points plotted in Fig. 1, Nyquist (ideal rectangular) filtering has been assumed. Such filters are not realizable! For realistic channels and waveforms, the required transmission bandwidth must be increased to account for realizable filters.

In the examples that follow, we will consider radio channels that are disturbed only by additive white Gaussian noise (AWGN) and have no other impairments and, for simplicity, we will limit the modulation choice to constant-envelope types, i.e., either MPSK or noncoherent orthogonal MFSK. For an uncoded system, MPSK is selected if the channel is bandwidth limited, and MFSK is selected if the channel is power limited. When error-correction coding is considered, modulation selection is not so simple, because coding techniques can provide power-bandwidth tradeoffs more effectively than would be possible through the use of any M-ary modulation scheme considered in this article [8].

In the most general sense, M-ary signaling can be regarded as a waveform-coding procedure, i.e., when we select an M-ary modulation technique instead of a binary one, we in effect have replaced the binary waveforms with better waveforms — either better for bandwidth performance (MPSK), or better for power performance (MFSK). Even though orthogonal MFSK signaling can be considered a coded system, i.e., a first-order Reed-Muller code [9], we restrict our use of the term "coded system" to those traditional error-correction codes using redundancies, e.g., block codes and convolutional codes.

Nyquist Minimum Bandwidth Requirements for MPSK and MFSK Signaling

The basic relationship between the symbol (or waveform) transmission rate, R_s, and the data rate, R, was shown in Equation (5) to be

$$R_s = \frac{R}{\log_2 M}$$

Using this relationship together with Equations (7-10) and R = 9600 b/s, a summary of symbol rate, Nyquist minimum bandwidth, and bandwidth efficiency for MPSK and noncoherent orthogonal MFSK was compiled for M = 2, 4, 8, 16, and 32 (Table 1). Values of E_b/N_0 required to achieve a bit-error probability of 10^{-5} for MPSK and

MFSK are also given for each value of M. These entries (which were computed using relationships that are presented later in this paper) corroborate the trade-offs shown in Fig. 1. As M increases, MPSK signaling provides more bandwidth efficiency at the cost of increased E_b/N_0, while MFSK signaling allows a reduction in E_b/N_0 at the cost of increased bandwidth.

Example 1: Bandwidth-limited Uncoded System

Suppose we are given a bandwidth-limited AWGN radio channel with an available bandwidth of W = 4000 Hz. Also, suppose that the link constraints (transmitter power, antenna gains, path loss, etc.) result in the received average signal-power to noise-power spectral density, S/N_0 being equal to 53 dB-Hz. Let the required data rate, R, be equal to 9600 b/s, and let the required bit-error performance, P_B, be at most 10^{-5}. The goal is to choose a modulation scheme that meets the required performance. In general, an error-correction coding scheme may be needed if none of the allowable modulation schemes can meet the requirements. However, in this example, we shall find that the use of error-correction coding is not necessary.

Solution to Example 1

For any digital communication system, the relationship between received S/N_0 and received bit-energy to noise-power spectral density, E_b/N_0, is as follows [4].

$$\frac{S}{N_0} = \frac{E_b}{N_0} R \tag{11}$$

Solving for E_b/N_0 in decibels, we obtain

$$\frac{E_b}{N_0}(\text{dB}) = \frac{S}{N_0}(\text{dB - Hz}) - R \, (\text{dB - bit / s})$$

$$= 53 \, \text{dB - Hz} - (10 \times \log_{10} 9600) \, \text{dB - bit / s} \tag{12}$$

$$= 13.2 \, \text{dB (or 20.89)}$$

Since the required data rate of 9600 b/s is much larger than the available bandwidth of 4000 Hz, the channel is bandwidth limited. We therefore select MPSK as our modulation scheme. We have confined the possible modulation choices to be constant-envelope types; without such a restriction, we would be able to select a modulation type with greater bandwidth-efficiency. To conserve power, we compute the *smallest possible* value of M such that the MPSK minimum bandwidth does not exceed the available bandwidth of 4000 Hz.

171

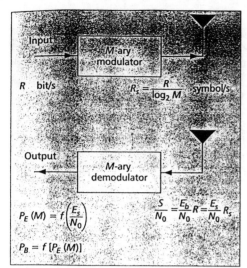

■ Figure 2. *Basic modulator/demodulator (MODEM) without channel coding.*

Table 1 shows that the smallest value of M meeting this requirement is $M = 8$. Next, we determine whether the required bit-error performance of $P_B \leq 10^{-5}$ can be met by using 8-PSK modulation alone, or whether it is necessary to use an error-correction coding scheme. Table 1 shows that 8-PSK alone will meet the requirements, since the required E_b/N_0 listed for 8-PSK is less than the received E_b/N_0 derived in Equation (12). Let us imagine that we do not have Table 1, however, and evaluate whether or not error-correction coding is necessary.

Figure 2 shows a basic modulator/demodulator (MODEM) block diagram that summarizes the functional details of this design. At the modulator, the transformation from data bits to symbols yields an output symbol rate, R_s, that is a factor $\log_2 M$ smaller than the input data-bit rate, R, as is seen in Equation (5). Similarly, at the input to the demodulator, the symbol-energy to noise-power spectral density E_s/N_0 is a factor $\log_2 M$ larger than E_b/N_0, since each symbol is made up of $\log_2 M$ bits. Because E_s/N_0 is larger than E_b/N_0 by the same factor that R_s is smaller than R, we can expand Equation (11), as follows.

$$\frac{S}{N_0} = \frac{E_b}{N_0} R = \frac{E_s}{N_0} R_s \qquad (13)$$

The demodulator receives a waveform (in this example, one of $M = 8$ possible phase shifts) during each time interval T_s. The probability that the demodulator makes a symbol error, $P_E(M)$, is well approximated by the following equation [10].

$$P_E(M) \cong 2Q\left[\sqrt{\frac{2E_s}{N_0}} \sin\left(\frac{\pi}{M}\right) \right] \text{ for } M > 2 \quad (14)$$

where $Q(x)$, sometimes called the complementary error function, represents the probability under the tail of a zero-mean unit-variance Gaussian density function. It is defined as follows [11].

$$Q(x) = \frac{1}{\sqrt{2\pi}} \int_x^\infty \exp\left(-\frac{u^2}{2}\right) du \qquad (15)$$

A good approximation for $Q(x)$, valid for $x > 3$, is given by the following equation [12].

$$Q(x) \cong \frac{1}{x\sqrt{2\pi}} \exp\left(-\frac{x^2}{2}\right) \qquad (16)$$

In Fig. 2 and all the figures that follow, rather than show explicit probability relationships, the generalized notation f(x) has been used to indicate some functional dependence on x.

A traditional way of characterizing communication efficiency in digital systems is in terms of the received E_b/N_0 in decibels. This E_b/N_0 description has become standard practice, but recall that there are no bits at the input to the demodulator; there are only waveforms that have been assigned bit meanings. The received E_b/N_0 represents a bit-apportionment of the arriving waveform energy.

To solve for $P_E(M)$ in Equation (14), we need to compute the ratio of received symbol-energy to noise-power spectral density, E_s/N_0. Since from Equation (12)

$$\frac{E_b}{N_0} = 13.2 \text{ dB} \quad \text{(or } 20.89)$$

and because each symbol is made up of $\log_2 M$ bits, we compute the following using $M = 8$.

$$\frac{E_s}{N_0} = (\log_2 M) \frac{E_b}{N_0} \qquad (17)$$
$$= 3 \times 20.89 = 62.67$$

Using the results of Equation (17) in Equation (14), yields the symbol-error probability, $P_E = 2.2 \times 10^{-5}$. To transform this to bit-error probability, we use the relationship between bit-error probability P_B, and symbol-error probability P_E, for multiple-phase signaling [9], as follows:

$$P_B \cong \frac{P_E}{\log_2 M} = \frac{P_E}{m} \quad \text{(for } P_E \ll 1) \qquad (18)$$

which is a good approximation when Gray coding is used for the bit-to-symbol assignment [10]. This last computation yields $P_B = 7.3 \times 10^{-6}$, which meets the required bit-error performance. No error-correction coding is necessary and 8-PSK modulation represents the design choice to meet the requirements of the bandwidth-limited channel (as we had predicted by examining the required E_b/N_0 values in Table 1).

Example 2: Power-limited Uncoded System
Now, suppose that we have exactly the same data rate and bit-error probability requirements as in Example 1, but let the available bandwidth, W, be equal to 45 kHz, and the available S/N_0 be equal to 48 dB-Hz. The goal is to choose a modulation or modulation/coding scheme that yields the required performance. We shall again find that error-correction coding is not required.

Solution to Example 2
The channel is clearly not bandwidth limited since the available bandwidth of 45 kHz is more than adequate for supporting the required data rate of 9600 b/s. We find the received E_b/N_0 from Equation (12) as follows.

$$\frac{E_b}{N_0} \text{ (dB)} = 48 \text{ dB-Hz}$$
$$- (10 \times \log_{10} 9600) \text{ dB-bit /s} \qquad (19)$$
$$= 8.2 \text{ dB} \quad \text{(or } 6.61)$$

For an uncoded system, we select MPSK if the channel is bandwidth limited, and we select MFSK if the channel is power limited.

The code
should be
as simple
as possible.
Generally,
the shorter
the code, the
simpler will
be its imple-
mentation.

Since there is abundant bandwidth but a relatively small E_b/N_0 for the required bit-error probability, we consider that this channel is power limited and choose MFSK as the modulation scheme. To conserve power, we search for the *largest possible M* such that the MFSK minimum bandwidth is not expanded beyond our available bandwidth of 45 kHz. A search results in the choice of $M = 16$ (Table 1). Next, we determine whether the required error performance of $P_B \leq 10^{-5}$ can be met using 16-FSK alone, i.e., without error-correction coding. Table 1 shows that 16-FSK alone meets the requirements, since the required E_b/N_0 for 16-FSK is less than the received E_b/N_0 derived in Equation (19). Let us imagine again that we do not have Table 1 and evaluate whether or not error-correction coding is necessary.

The block diagram in Fig. 2 summarizes the relationships between symbol rate R_s and bit rate R, and between E_s/N_0 and E_b/N_0, which is identical to each of the respective relationships in Example 1. The 16-FSK demodulator receives a waveform (one of 16 possible frequencies) during each symbol time interval T_s. For noncoherent orthogonal MFSK, the probability that the demodulator makes a symbol error, $P_E(M)$, is approximated by the following upper bound [13].

$$P_E(M) \leq \frac{M-1}{2} \exp\left(-\frac{E_s}{2N_0}\right) \qquad (20)$$

To solve for $P_E(M)$ in Equation (20), we compute E_s/N_0, as in Example 1. Using the results of Equation (19) in Equation (17), with $M = 16$, we get

$$\frac{E_s}{N_0} = (\log_2 M)\frac{E_b}{N_0} \qquad (21)$$
$$= 4 \times 6.61 = 26.44$$

Next, using the results of Equation (21) in Equation (20) yields the symbol-error probability, $P_E = 1.4 \times 10^{-5}$. To transform this to bit-error probability, P_B, we use the relationship between P_B and P_E for orthogonal signaling [13], given by

$$P_B = \frac{2^{m-1}}{2^m - 1} P_E \qquad (22)$$

This last computation yields $P_B = 7.3 \times 10^{-6}$, which meets the required bit-error performance. We can meet the given specifications for this power-limited channel by using 16-FSK modulation, without any need for error-correction coding (as we had predicted by examining the required E_b/N_0 values in Table 1).

Example 3: Bandwidth-limited and Power-limited Coded System

We start with the same channel parameters as in Example 1 ($W = 4000$ Hz, $S/N_0 = 53$ dB-Hz, and $R = 9600$ b/s), with one exception. In this example, we specify that P_B must be at most 10^{-9}. Table 1 shows that the system is both bandwidth limited and power limited, based on the available bandwidth of 4000 Hz and the available E_b/N_0 of 13.2 dB, from Equation (12). (8-PSK is the only possible choice to meet the bandwidth constraint; however, the available E_b/N_0 of 13.2 dB is certainly insufficient to meet the required P_B of 10^{-9}). For this small value of P_B, we need to consider the performance improvement that error-

n	k	t
7	4	1
15	11	1
	7	2
	5	3
31	26	1
	21	2
	16	3
	11	5
63	57	1
	51	2
	45	3
	39	4
	36	5
	30	6
127	120	1
	113	2
	106	3
	99	4
	92	5
	85	6
	78	7
	71	9
	64	10

■ Table 2. *BCH codes (partial catalog).*

correction coding can provide within the available bandwidth. In general, one can use convolutional codes or block codes.

The Bose, Chaudhuri, and Hocquenghem (BCH) codes form a large class of powerful error-correcting cyclic (block) codes [14]. To simplify the explanation, we shall choose a block code from the BCH family. Table 2 presents a partial catalog of the available BCH codes in terms of n, k, and t, where k represents the number of information (or data) bits that the code transforms into a longer block of n coded bits (or channel bits), and t represents the largest number of incorrect channel bits that the code can correct within each n-sized block. The rate of a code is defined as the ratio k/n; its inverse represents a measure of the code's redundancy [14].

Solution to Example 3

Since this example has the same bandwidth-limited parameters given in Example 1, we start with the same 8-PSK modulation used to meet the stated bandwidth constraint. However, we now employ error-correction coding so that the bit-error probability can be lowered to $P_B \leq 10^{-9}$.

To make the optimum code selection from Table 2, we are guided by the following goals:
- The output bit-error probability of the combined modulation/coding system must meet the system error requirement.
- The rate of the code must not expand the required transmission bandwidth beyond the available channel bandwidth.
- The code should be as simple as possible. Generally, the shorter the code, the simpler will be its implementation.

The uncoded 8-PSK minimum bandwidth requirement is 3200 Hz (Table 1) and the allowable channel bandwidth is 4000 Hz, so the uncoded signal bandwidth can be increased by no more

'96

than a factor of 1.25 (i.e., an expansion of 25 percent). The very first step in this (simplified) code selection example is to eliminate the candidates in Table 2 that would expand the bandwidth by more than 25 percent. The remaining entries form a much reduced set of "bandwidth-compatible" codes (Table 3).

A column designated "Coding Gain, G" has been added for MPSK at $P_B = 10^{-9}$ (Table 3). Coding gain in decibels is defined as follows.

$$G \text{ (dB)} = \left(\frac{E_b}{N_0}\right)_{uncoded} \text{(dB)} - \left(\frac{E_b}{N_0}\right)_{coded} \text{(dB)} \quad (23)$$

G can be described as the reduction in the required E_b/N_0 (in decibels) that is needed due to the error-performance properties of the channel coding. G is a function of the modulation type and bit-error probability, and it has been computed for MPSK at $P_B = 10^{-9}$ (Table 3). For MPSK modulation, G is relatively independent of the value of M. Thus, for a particular bit-error probability, a given code provides about the same coding gain when used with any of the MPSK modulation schemes. Coding gains were calculated using a procedure outlined in the "Calculating Coding Gain" section below.

A block diagram summarizes this system which contains both modulation and coding (Fig. 3). The introduction of encoder/decoder blocks brings about additional transformations. The relationships that exist when transforming from R b/s to R_c channel-b/s to R_s symbol/s are shown at the encoder/modulator. Regarding the channel-bit rate, R_c, some authors prefer the units of channel-symbol/s (or code-symbol/s). The benefit is that error-correction coding is often described more efficiently with nonbinary digits. We reserve the term "symbol" for that group of bits mapped onto an electrical waveform for transmission, and we designate the units of R_c to be channel-b/s (or coded-b/s).

We assume that our communication system cannot tolerate any message delay, so the channel-bit rate, R_c, must exceed the data-bit rate, R, by the factor n/k. Further, each symbol is made up of $\log_2 M$ channel bits, so the symbol rate, R_s, is less than R_c by the factor $\log_2 M$. For a system containing both modulation and coding, we summarize the rate transformations as follows.

$$R_c = \left(\frac{n}{k}\right)R \quad (24)$$

$$R_s = \frac{R_c}{\log_2 M} \quad (25)$$

At the demodulator/decoder in Fig. 3, the transformations among data-bit energy, channel-bit energy, and symbol energy are related (in a reciprocal fashion) by the same factors as shown among the rate transformations in Equations (24) and (25). Since the encoding transformation has replaced k data bits with n channel bits, then the ratio of channel-bit energy to noise-power spectral density, E_c/N_0, is computed by decrementing the value of E_b/N_0 by the factor k/n. Also, since each transmission symbol is made up of $\log_2 M$ channel bits, then E_s/N_0, which is needed in Equation (14) to solve for P_E, is computed by incrementing E_c/N_0 by the factor $\log_2 M$. For a system containing both modulation and coding, we summarize

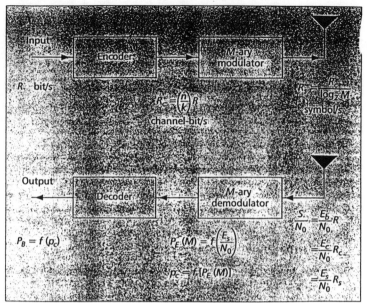

■ **Figure 3.** *MODEM with channel coding.*

n	k	t	Coding Gain, G (dB) MPSK, $P_B = 10^{-9}$
31	26	1	2.0
63	57	1	2.2
	51	2	3.1
127	120	1	2.2
	113	2	3.3
	106	3	3.9

■ **Table 3.** *Bandwidth-compatible BCH codes.*

the energy to noise-power spectral density transformations, as follows.

$$\frac{E_c}{N_0} = \left(\frac{k}{n}\right)\frac{E_b}{N_0} \quad (26)$$

$$\frac{E_s}{N_0} = (\log_2 M)\frac{E_c}{N_0} \quad (27)$$

Using Equations (24) through (27), we can now expand the expression for S/N_0 in Equation (13), as follows (Appendix A).

$$\frac{S}{N_0} = \frac{E_b}{N_0}R = \frac{E_c}{N_0}R_c = \frac{E_s}{N_0}R_s \quad (28)$$

As before, a standard way of describing the link is in terms of the received E_b/N_0 in decibels. However, there are no data bits at the input to the demodulator, and there are no channel bits; there are only waveforms that have bit meanings, and thus the waveforms can be described in terms of bit-energy apportionments.

Since S/N_0 and R were given as 53 dB-Hz and 9600 b/s, respectively, we find as before, from Equation (12), that the received $E_b/N_0 = 13.2$ dB. The received E_b/N_0 is fixed and independent of n, k, and t (Appendix A). As we search Table 3 for the ideal code to meet the specifications, we can iteratively repeat the computations suggested in Fig. 3. It might be useful to program on a PC (or calculator) the following four steps as a function of n, k, and t. Step 1 starts by combining Equations (26) and (27).

Step 1:

$$\frac{E_s}{N_0} = (\log_2 M)\frac{E_c}{N_0} = (\log_2 M)\left(\frac{k}{n}\right)\frac{E_b}{N_0} \quad (29)$$

Step 2:

$$P_E(M) \cong 2Q\left[\sqrt{\frac{2E_s}{N_0}}\sin\left(\frac{\pi}{M}\right)\right] \quad (30)$$

which is the approximation for symbol-error probability, P_E, rewritten from Equation (14). At each symbol-time interval, the demodulator makes a symbol decision, but it delivers a channel-bit sequence representing that symbol to the decoder. When the channel-bit output of the demodulator is quantized to two levels, 1 and 0, the demodulator is said to make hard decisions. When the output is quantized to more than two levels, the demodulator is said to make soft decisions [4]. Throughout this paper, we assume hard-decision demodulation.

Now that we have a decoder block in the system, we designate the channel-bit-error probability out of the demodulator and into the decoder as p_c, and we reserve the notation P_B for the bit-error probability out of the decoder. We rewrite Equation (18) in terms of p_c as follows.

Step 3:

$$p_c \cong \frac{P_E}{\log_2 M} = \frac{P_E}{m} \quad (31)$$

relating the channel-bit-error probability to the symbol-error probability out of the demodulator, assuming Gray coding, as referenced in Equation (18).

For traditional channel-coding schemes and a given value of received S/N_0, the value of E_s/N_0 with coding will always be less than the value of E_s/N_0 without coding. Since the demodulator with coding receives less E_s/N_0, it makes more errors! When coding is used, however, the system error-performance doesn't only depend on the performance of the demodulator, it also depends on the performance of the decoder. For error-performance improvement due to coding, the decoder must provide enough error correction to more than compensate for the poor performance of the demodulator.

The final output decoded bit-error probability, P_B, depends on the particular code, the decoder, and the channel-bit-error probability, p_c. It can be expressed by the following approximation [15].

Step 4:

$$P_B \cong \frac{1}{n}\sum_{j=t+1}^{n} j\binom{n}{j}p_c^j(1-p_c)^{n-j} \quad (32)$$

where t is the largest number of channel bits that the code can correct within each block of n bits. Using Equations (29) through (32) in the above four steps, we can compute the decoded bit-error probability, P_B, as a function of n, k, and t for each of the codes listed in Table 3. The entry that meets the stated error requirement with the largest possible code rate and the smallest value

of n is the double-error correcting (63,51) code. The computations are

Step 1:

$$\frac{E_s}{N_0} = 3\left(\frac{51}{63}\right)20.89 = 50.73$$

where $M = 8$, and the received $E_b/N_0 = 13.2$ dB (or 20.89).

Step 2:

$$P_E \cong 2Q\left[\sqrt{101.5}\times\sin\left(\frac{\pi}{8}\right)\right]$$

$$= 2Q(3.86) = 1.2\times10^{-4}$$

Step 3:

$$p_c \cong \frac{1.2\times10^{-4}}{3} = 4\times10^{-5}$$

Step 4:

$$P_B \cong \frac{3}{63}\binom{63}{3}(4\times10^{-5})^3(1-4\times10^{-5})^{60}$$

$$+\frac{4}{63}\binom{63}{4}(4\times10^{-5})^4(1-4\times10^{-5})^{59}+\ldots$$

$$= 1.2\times10^{-10}$$

where the bit-error-correcting capability of the code is $t = 2$. For the computation of P_B in Step 4, we need only consider the first two terms in the summation of Equation (32) since the other terms have a vanishingly small effect on the result. Now that we have selected the (63, 51) code, we can compute the values of channel-bit rate, R_c, and symbol rate, R_s, using Equations (24) and (25), with $M = 8$.

$$R_c = \left(\frac{n}{k}\right)R = \left(\frac{63}{51}\right)9600 \cong 11{,}859 \text{ channel-bit /s}$$

$$R_s = \frac{R_c}{\log_2 M} = \frac{11859}{3} = 3953 \text{ symbol /}$$

Calculating Coding Gain

Perhaps a more direct way of finding the simplest code that meets the specified error performance is to first compute how much coding gain, G, is required in order to yield $P_B = 10^{-9}$ when using 8-PSK modulation alone; then we can simply choose the code that provides this performance improvement (Table 3). First, we find the uncoded E_s/N_0 that yields an error probability of $P_B = 10^{-9}$ by writing from Equations (18) and (31) the following.

$$P_B \cong \frac{P_E}{\log_2 M} \cong \frac{2Q\left[\sqrt{\frac{2E_s}{N_0}}\sin\left(\frac{\pi}{M}\right)\right]}{\log_2 M} = 10^{-9} \quad (33)$$

At this low value of bit-error probability, it is valid to use Equation (16) to approximate $Q(x)$ in Equation (33). By trial-and-error (on a programmable calculator), we find that the uncoded $E_s/N_0 = 120.67 = 20.8$ dB, and since each symbol is made up of $\log_2 8 = 3$ bits, the required $(E_b/N_0)_{uncoded} = 120.67/3 = 40.22 = 16$ dB. From the given parameters and Equation (12), we know that the received $(E_b/N_0)_{coded} = 13.2$ dB. Using Equation (23), the

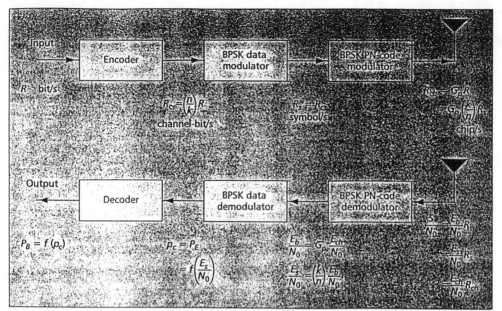

■ **Figure 4.** *Direct-sequence spread-spectrum MODEM with channel coding.*

required coding gain to meet the bit-error performance of $P_B = 10^{-9}$ is

$$G\ (dB) = \left(\frac{E_b}{N_0}\right)_{uncoded} (dB) - \left(\frac{E_b}{N_0}\right)_{coded} (dB)$$

$$= 16\ dB - 13.2\ dB = 2.8\ dB$$

To be precise, each of the E_b/N_0 values in the above computation must correspond to exactly the same value of bit-error probability (which they do not). They correspond to $P_B = 10^{-9}$ and $P_B = 1.2 \times 10^{-10}$, respectively. However, at these low probability values, even with such a discrepancy, this computation still provides a good approximation of the required coding gain. In searching Table 3 for the simplest code that will yield a coding gain of at least 2.8 dB, we see that the choice is the (63, 51) code, which corresponds to the same code choice that we made earlier.

Example 4: Direct Sequence (DS) Spread Spectrum Coded System

Spread-spectrum systems are not usually classified as being bandwidth- or power-limited. However, they are generally perceived to be power-limited systems because the bandwidth occupancy of the information is much larger than the bandwidth that is intrinsically needed for the information transmission. In a direct-sequence spread-spectrum (DS/SS) system, spreading the signal bandwidth by some factor permits lowering the signal-power spectral density by the same factor (the total average signal power is the same as before spreading). The bandwidth spreading is typically accomplished by multiplying a relatively narrowband data signal by a wideband spreading signal. The spreading signal or spreading code is often referred to as a pseudorandom code, or PN code.

Processing Gain — A typical DS/SS radio system is often described as a two-step BPSK modulation process. In the first step, the carrier wave is modulated by a bipolar data waveform having a value +1 or -1 during each data-bit duration; in the

second step, the output of the first step is multiplied (modulated) by a bipolar PN-code waveform having a value +1 or -1 during each PN-code-bit duration. In reality, DS/SS systems are usually implemented by first multiplying the data waveform by the PN-code waveform and then making a single pass through a BPSK modulator. For this example, however, it is useful to characterize the modulation process in two separate steps — the outer modulator/demodulator for the data, and the inner modulator/demodulator for the PN code (Fig. 4).

A spread-spectrum system is characterized by a processing gain, G_p, that is defined in terms of the spread-spectrum bandwidth, W_{ss}, and the data rate, R, as follows [16].

$$G_p = \frac{W_{ss}}{R} \tag{34}$$

For a DS/SS system, the PN-code bit has been given the name "chip," and the spread-spectrum signal bandwidth can be shown to be about equal to the chip rate. Thus, for a DS/SS system, the processing gain in Equation (34) is generally expressed in terms of the chip rate, R_{ch}, as follows.

$$G_p = \frac{R_{ch}}{R} \tag{35}$$

Some authors define processing gain to be the ratio of the spread-spectrum bandwidth to the symbol rate. This definition separates the system performance due to bandwidth spreading from the performance due to error-correction coding. Since we ultimately want to relate all of the coding mechanisms relative to the information source, we shall conform to the most usually accepted definition for processing gain, as expressed in Equations (34) and (35).

A spread-spectrum system can be used for interference rejection and multiple access (allowing multiple users to access a communications resource simultaneously). The benefits of DS/SS signals are best achieved when the processing gain is very large; in other words, the chip rate of the spreading (or PN) code is much larger than the data

176

rate. In such systems, the large value of G_p allows the signaling chips to be transmitted at a power level well below that of the thermal noise. We will use a value of $G_p = 1000$. At the receiver, the despreading operation correlates the incoming signal with a synchronized copy of the PN code, and thus accumulates the energy from multiple (G_p) chips to yield the energy per data bit. The value of G_p has a major influence on the performance of the spread-spectrum system application. However, the value of G_p has no effect on the received E_b/N_0. In other words, spread spectrum techniques offer no error-performance advantage over thermal noise. For DS/SS systems, there is no disadvantage either! Sometimes such spread-spectrum radio systems are employed only to enable the transmission of very small power-spectral densities, and thus avoid the need for FCC licensing [17].

Channel Parameters for Example 4 — Consider a DS/SS radio system that uses the same (63, 51) code as in the previous example. Instead of using MPSK for the data modulation, we shall use BPSK. Also, we shall use BPSK for modulating the PN-code chips. Let the received $S/N_0 = 48$ dB-Hz, the data rate $R = 9600$ b/s, and the required $P_B \leq 10^{-6}$. For simplicity, assume that there are no bandwidth constraints. Our task is simply to determine whether or not the required error performance can be achieved using the given system architecture and design parameters. In evaluating the system, we will use the same type of transformations used in previous examples.

Solution to Example 4

A typical DS/SS system can be implemented more simply than the one shown in Fig. 4. The data and the PN code would be combined at baseband, followed by a single pass through a BPSK modulator. We assume the existence of the individual blocks in Fig. 4, however, because they enhance our understanding of the transformation process. The relationships in transforming from data bits, to channel bits, to symbols, and to chips (Fig. 4) have the same pattern of subtle but straightforward transformations in rates and energies as previous relationships (Figs. 2-3). The values of R_c, R_s, and R_{ch} can now be calculated immediately since the (63,51) BCH code has already been selected. From Equation (24)

$$R_c = \left(\frac{n}{k}\right)R = \left(\frac{63}{51}\right)9600 \cong 11{,}859 \text{ channel-bit/s}$$

Since the data modulation considered here is BPSK,

$$R_s = R_c \cong 11{,}859 \text{ symbol/s}$$

and from Equation (35), with an assumed value of $G_p = 1000$,

$$R_{ch} = G_p R = 1000 \times 9600 = 9.6 \times 10^6 \text{ chip/s}$$

Since we have been given the same S/N_0 and the same data rate as in Example 2, we find the value of received E_b/N_0 from Equation (19) to be 8.2 dB (or 6.61). At the demodulator, we can now expand the expression for S/N_0 in Equation (28) and Appendix A, as follows.

$$\frac{S}{N_0} = \frac{E_b}{N_0}R = \frac{E_c}{N_0}R_c = \frac{E_s}{N_0}R_s = \frac{E_{ch}}{N_0}R_{ch} \quad (36)$$

Corresponding to each transformed entity (data bit, channel bit, symbol, or chip) there is a change in rate, and similarly a reciprocal change in energy-to-noise spectral density for that received entity. Equation (36) is valid for any such transformation when the rate and energy are modified in a reciprocal way. There is a kind of *conservation of power (or energy)* phenomenon in the transformations. The total received average power (or total received energy per symbol duration) is fixed regardless of how it is computed — on the basis of data-bits, channel-bits, symbols, or chips.

The ratio E_{ch}/N_0 is much less in value than E_b/N_0. This can seen from Equations (36) and (35), as follows.

$$\frac{E_{ch}}{N_0} = \frac{S}{N_0}\left(\frac{1}{R_{ch}}\right) = \frac{S}{N_0}\left(\frac{1}{G_p R}\right) = \left(\frac{1}{G_p}\right)\frac{E_b}{N_0} \quad (37)$$

But, even so, the despreading function (when properly synchronized) accumulates the energy contained in a quantity G_p of the chips, yielding the same value, $E_b/N_0 = 8.2$ dB, as was computed earlier from Equation (19). Thus, the DS spreading transformation has no effect on the error performance of an AWGN channel [4], and the value of G_p has no bearing on the value of P_B in this example. From Equation (37), we can compute

$$\begin{aligned}\frac{E_{ch}}{N_0} \text{ (dB)} &= \frac{E_b}{N_0} \text{ (dB)} - G_p \text{ (dB)}\\ &= 8.2 \text{ dB} - (10 \times \log_{10} 1000) \text{ dB}\\ &= -21.8 \text{ dB}\end{aligned} \quad (38)$$

The chosen value of processing gain ($G_p = 1000$) enables the DS/SS system to operate at a value of chip energy *well below the thermal noise*, with the same error performance as without spreading.

Since BPSK is the data modulation selected in this example, each message symbol therefore corresponds to a single channel bit, and we can write

$$\frac{E_s}{N_0} = \frac{E_c}{N_0} = \left(\frac{k}{n}\right)\frac{E_b}{N_0} = \left(\frac{51}{63}\right) \times 6.61 = 5.35 \quad (39)$$

where the received $E_b/N_0 = 8.2$ dB (or 6.61). Out of the BPSK data demodulator, the symbol-error probability, P_E, (and the channel-bit error probability, p_c) is computed as follows [4].

$$p_c = P_E = Q\left(\sqrt{\frac{2E_c}{N_0}}\right) \quad (40)$$

Using the results of Equation (39) in Equation (40) yields

$$p_c = Q(3.27) = 5.8 \times 10^{-4}$$

Finally, using this value of p_c in Equation (32) for the (63, 51) double-error correcting code yields the output bit-error probability of $P_B = 3.6 \times 10^{-7}$. We can therefore verify that, for the given architecture and design parameters of this example, the system does in fact achieve the required error performance.

Conclusion

The goal of this tutorial has been to review fundamental relationships in defining, designing, and evaluating digital communication system performance. First, we examined the concept of bandwidth-limited and power-limited systems and how such conditions influence the design when the choices are confined to MPSK and MFSK modulation. Most important, we focused on the definitions and computations involved in transforming from data bits to channel bits to symbols to chips. In general, most digital communication systems share these concepts; thus, understanding them should enable one to evaluate other such systems in a similar way.

References

[1] G. Ungerboeck, "Trellis-Coded Modulation with Redundant Signal Sets," Part I and Part II, *IEEE Commun. Mag.*, vol. 25, pp. 5-21, Feb. 1987.
[2] C. E. Shannon, "A Mathematical Theory of Communication," *BSTJ*, vol. 27, pp. 379-423, 623-657, 1948.
[3] C. E. Shannon, "Communication in the presence of Noise," *Proc. IRE*, vol. 37, no. 1, pp. 10-21, Jan. 1949.
[4] B. Sklar, "*Digital Communications: Fundamentals and Applications*," Prentice-Hall Inc., Englewood Cliffs, N.J., 1988.
[5] M. R. L. Hodges, "The GSM Radio Interface," *British Telecom Technol. J.*, vol. 8, no. 1, pp. 31-43, Jan. 1990.
[6] H. Nyquist, "Certain Topics on Telegraph Transmission Theory," *Trans. AIEE*, vol. 47, pp. 617-644, April 1928.
[7] J. B. Anderson and C-E. W. Sundberg, "Advances in Constant Envelope Coded Modulation," *IEEE Commun., Mag.*, vol. 29, no. 12, pp. 36-45, Dec. 1991.
[8] G. C. Clark, Jr. and J. B. Cain, "*Error-Correction Coding for Digital Communications*," (Plenum Press, New York, 1981).
[9] W. C. Lindsey, and M. K. Simon, "*Telecommunication Systems Engineering*," (Prentice-Hall, Englewood Cliffs, NJ, 1973).
[10] I. Korn, "*Digital Communications*," (Van Nostrand Reinhold Co, New York, 1985).
[11] H. L. Van Trees, "*Detection, Estimation, and Modulation Theory*," Part I, (John Wiley and Sons, Inc., New York, 1968).
[12] P. O. Borjesson and C.E. Sundberg, "Simple Approximations of the Error Function Q(x) for Communications Applications," *IEEE Trans. Comm.*, vol. COM-27, pp. 639-642, March 1979.
[13] A.J. Viterbi, "*Principles of Coherent Communication*," McGraw-Hill Book Co., New York, 1966.
[14] S. Lin and D. J. Costello, Jr., "*Error Control Coding: Fundamentals and Applications*," (Prentice-Hall Inc., Englewood Cliffs, N.J., 1983).
[15] J. P. Odenwalder, "*Error Control Coding Handbook, Linkabit Corporation*," San Diego, CA, July 15, 1976.
[16] A. J. Viterbi, "Spread Spectrum Communications — Myths and Realities," *IEEE Commun. Mag.*, pp. 11-18, May 1979.
[17] Title 47, Code of Federal Regulations, Part 15 Radio Frequency Devices.

Appendix A
Received E_b/N_0 Is Independent of the Code Parameters

Starting with the basic concept that the received average signal power, S, is equal to the received symbol or waveform energy, E_s, divided by the symbol-time duration T (or multiplied by the symbol rate, R_s), we write

$$\frac{S}{N_0} = \frac{E_s/T}{N_0} = \frac{E_s}{N_0} R_s \qquad (A1)$$

where N_0 is noise-power spectral density.

Using Equations (27) and (25), rewritten below,

$$\frac{E_s}{N_0} = (\log_2 M) \frac{E_c}{N_0} \quad \text{and,} \quad R_s = \frac{R_c}{\log_2 M}$$

let us make substitutions into Equation (A1), which yields

$$\frac{S}{N_0} = \frac{E_c}{N_0} R_c \qquad (A2)$$

Next, using Equations (26) and (24), rewritten below,

$$\frac{E_c}{N_0} = \left(\frac{k}{n}\right)\frac{E_b}{N_0} \quad \text{and} \quad R_c = \left(\frac{n}{k}\right) R$$

let us now make substitutions into Equation (A2), which yields the relationship expressed in Equation (11).

$$\frac{S}{N_0} = \frac{E_b}{N_0} R \qquad (A3)$$

Hence the received E_b/N_0 is only a function of the received S/N_0 and the data rate, R. It is independent of the code parameters, n, k, and t. These results are summarized in Fig. 3.

Biography

BERNARD SKLAR received a B.S. in math and science from the University of Michigan, an M.S. in electrical engineering from the Polytechnic Institute of Brooklyn, and a Ph.D. in engineering from the University of California, Los Angeles. He has more than 35 years experience in a wide variety of technical development positions at Republic Aviation Corp., Hughes Aircraft Co., Litton Industries, Inc., and The Aerospace Corporation. Currently, he is the head of advanced systems at Communications Engineering Services, a consulting company that he founded in 1984; an adjunct professor at the University of Southern California; and a visiting professor at the University of California at Los Angeles, where he teaches communications. He is the author of the book, *Digital Communications*. He is a Fellow of the Institute for the Advancement of Engineering, and a past Chairman of the Los Angeles Council IEEE Education Committee.

178

mission is currently being studied [1]–[7]. While digital transmission can surely bring many advantages, some technical problems must be solved. This paper is concerned with a digital modulation for future mobile radio communications.

From the viewpoint of mobile radio use, the out-of-band radiation power in the adjacent channel should be generally suppressed 60–80 dB below that in the desired channel. So as to satisfy this severe requirement, it is necessary to manipulate the RF output signal spectrum. Such a spectrum manipulation cannot usually be performed at the final RF stage in the multichannel SCPC transceivers because the transmitted RF frequency is variable. Therefore, intermediate-frequency (IF) or baseband filtering with frequency up conversion is mostly used. However, when such a spectrum-manipulated signal is translated up and passed through a nonlinear class-C power amplifier, the required spectrum manipulation should not be violated by the nonlinearities. In order to mitigate the impairments, some narrow-band digital modulation schemes with constant or less fluctuated envelope property have been researched [8]–[10].

In this paper, premodulation Gaussian filtered minimum shift keying (GMSK) with coherent detection is proposed as an effective digital modulation for the present purpose, and its fundamental properties are analyzed with the aid of machine computation. The relationship between out-of-band radiation suppression and bit-error-rate (BER) performance is made clear. Constitution of the modulator and demodulator is then discussed. The superiority of this modulation is supported by some experimental test results.

GMSK Modulation for Digital Mobile Radio Telephony

KAZUAKI MUROTA, MEMBER, IEEE, AND KENKICHI HIRADE, MEMBER, IEEE

Abstract—This paper is concerned with digital modulation for future mobile radio telephone services. First, the specific requirements on the digital modulation for mobile radio use are described. Then, premodulation Gaussian filtered minimum shift keying (GMSK) with coherent detection is proposed as an effective digital modulation for the present purpose, and its fundamental properties are clarified with the aid of machine computation. The constitution of modulator and demodulator is then discussed from the viewpoints of mobile radio applications. The superiority of this modulation is supported by some experimental test results.

I. INTRODUCTION

It is well known that voice transmission in many VHF and UHF mobile radio telephone systems has usually been made by using a single-channel-per-carrier (SCPC) analog FM transmission technique. However, in order to provide highly secure voice and/or high-speed data transmission by the use of large-scale integrated (LSI) transceivers, digital mobile radio trans-

Paper approved by the Editor for Communication Theory of the IEEE Communications Society for publication after presentation at 29th IEEE Vehicular Technology Conference, Chicago, IL, March 1979. Manuscript received May 28, 1980; revised January 5, 1981.

The authors are with the Yokosuka Electrical Communication Laboratory, Nippon Telegraph and Telephone Public Corporation, Kanagawa-Ken, Japan.

II. GMSK MODULATION

A. Spectrum Manipulation of MSK

Minimum shift keying (MSK), which is binary digital FM with a modulation index of 0.5, has the following good properties: constant envelope, relatively narrow bandwidth, and coherent detection capability [11]–[13]. However, it does not satisfy the severe requirements with respect to out-of-band radiation for SCPC mobile radio. MSK can be generated by direct FM modulation. As is easily found, the output power spectrum of MSK can be manipulated by using a premodulation low-pass filter (LPF), keeping the constant envelope property, as shown in Fig. 1. To make the output power spectrum compact, the premodulation LPF should have the following properties:

1) narrow bandwidth and sharp cutoff
2) lower overshoot impulse response
3) preservation of the filter output pulse area which corresponds to a phase shift $\pi/2$.

Condition 1) is needed to suppress the high-frequency components, 2) is to protect against excessive instantaneous frequency deviation, and 3) is for coherent detection to be applicable as simple MSK.

Generally, the introduction of the premodulation LPF violates the minimum frequency spacing constraint and the fixed-phase constraint of MSK. However, the above two constraints are not intrinsic requirements for effective coherent binary FM with modulation index 0.5. Such a premodulation-filtered MSK signal can be detected coherently because its

Reprinted from *IEEE Transactions on Communications*, Vol. COM-29, No. 7, July, 1981.

179

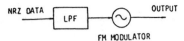

Fig. 1.   Premodulation baseband-filtered MSK.

pattern-averaged phase-transition trajectory does not deviate from that of simple MSK.

B. Fundamental Properties of GMSK

A Gaussian LPF satisfies all the above-described characteristics. Consequently, the modified MSK modulation using a premodulation Gaussian LPF can be expected to be an excellent digital modulation technique for the present purpose. Such a modified MSK is named Gaussian MSK or GMSK in connection with Gaussian low-pass filtering. Let us now investigate the GMSK modulation from various aspects.

Output Power Spectrum: Fig. 2 shows the machine-computed results of the output power spectrum of the GMSK signal versus the normalized frequency difference from the carrier center frequency $(f - f_c)T$ where the normalized 3 dB-down bandwidth of the premodulation Gaussian LPF $B_b T$ is a parameter. The spectrum for GMSK with $B_b T = 0.2$ is nearly equal to that of TFM.

The effective variable parameter $B_b T$ can be selected by the system designer considering overall spectrum efficiency of the cellular zone structure.

Fig. 3 shows the machine-computed results of the fractional power in the desired channel versus the normalized bandwidth of the predetection rectangular bandpass filter (BPF) $B_i T$. Table I shows the occupied bandwidth for the prescribed percentage of power where $B_b T$ is also a variable parameter. For comparison, the occupied bandwidth of TFM is also shown in Table I.

Fig. 4 shows the machine-computed results of the ratio of the out-of-band radiation power in the adjacent channel to the total power in the desired channel where the normalized channel spacing $f_s T$ is taken as the abscissa and both channels are assumed to have the ideal rectangular bandpass characteristics with $B_i T = 1$. The situation of $f_s T = 1.5$ and $B_i T = 1$ corresponds to the case of $f_s \cong 25$ kHz and $B_i = 16$ kHz when $f_b = 1/T = 16$ kbits/s. From Fig. 4, it is found that the GMSK with $B_b T = 0.28$ can be adopted as the digital modulation for conventional VHF and UHF SCPC mobile radio communications without carrier frequency drift where the ratio of out-of-band radiation power in the adjacent channel to the total power in the desired channel must be lower than −60 dB. When a certain amount of carrier frequency drift (for example $\Delta f = \pm 1.5$ kHz) exists, $B_b T = 0.2$ is needed.

BER Performance: Let us now consider the theoretical BER performance of GMSK modulation using coherent detection in the presence of additive white Gaussian noise.

Since the GMSK modulation of interest is a certain kind of binary digital modulation, its BER performance bound in the high SNR condition is approximately represented as

$$P_e = \frac{1}{2} \operatorname{erfc}\left(\frac{d_{min}}{2\sqrt{N_0}}\right) \tag{1}$$

where N_0 is the power spectrum density of the additive white Gaussian noise and erfc( ) is the complementary error func-

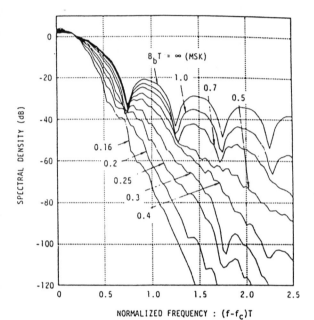

Fig. 2.   Power spectra of GMSK.

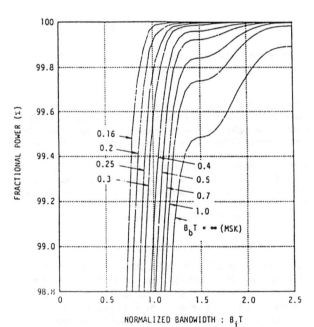

Fig. 3.   Fractional power ratio of GMSK.

TABLE I
OCCUPIED BANDWIDTH CONTAINING A GIVEN PERCENTAGE
POWER

$B_b T$ \ %	90	99	99.9	99.99
0.2	0.52	0.79	0.99	1.22
0.25	0.57	0.86	1.09	1.37
0.5	0.69	1.04	1.33	2.08
MSK	0.78	1.20	2.76	6.00
TFM	0.52	0.79	1.02	1.37

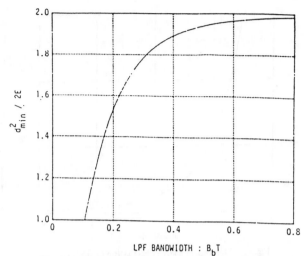

Fig. 5. Normalized minimum signal distance of GMSK.

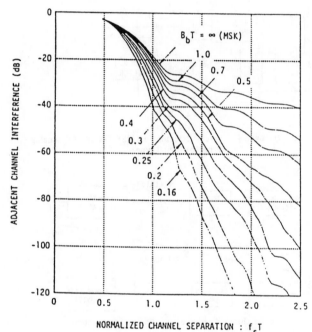

Fig. 4. Adjacent channel interference of GMSK.

tion given by

$$\text{erfc}(x) = \frac{2}{\sqrt{\pi}} \int_x^\infty \exp(-u^2) \, du. \qquad (2)$$

Furthermore, d_{min} is the minimum value of the signal distance d between mark and space in Hilbert space observed during the time interval from t_1 to t_2 and d is defined by

$$d^2 = \frac{1}{2} \int_{t_1}^{t_2} |u_m(t) - u_s(t)|^2 \, dt \qquad (3)$$

where $u_m(t)$ and $u_s(t)$ are the complex signal waveforms corresponding to the mark and the space transmissions, respectively.

While the BER performance bound given by (1) is attained only when the ideal maximum likelihood detection is adopted, it gives an approximate solution for the ideal BER performance of GMSK modulation with coherent detection.

Fig. 5 shows the machine-computed results for d_{min} of the GMSK signal versus $B_b T$ where E_b denotes the signal energy per bit defined by

$$E_b = \frac{1}{2} \int_0^T |u_m(t)|^2 \, dt = \frac{1}{2} \int_0^T |u_s(t)|^2 \, dt. \qquad (4)$$

In the case $B_b T \to \infty$, which corresponds to the simple MSK signal, Fig. 5 yields $d_{min} = 2\sqrt{E_b}$, which is that of antipodal transmission. It is noticed that the meaningful observation time interval for the GMSK signal $t_2 - t_1$ may be made longer than $2T$, which corresponds to that for the simple MSK signal, due to the intersymbol interference (ISI) effect on the phase transitions.

Substituting the machine-computed results of d_{min} into (1), the BER performance of the GMSK modulation with coherent detection is obtained. Fig. 6 shows the performance degradation of GMSK from antipodal transmission due to the ISI effect of the premodulation LPF. This figure shows that the performance degradation is small and that the required E_b/N_0 of GMSK with $B_b T = 0.25$ does not exceed more than 0.7 dB compared to that of antipodal transmission.

III. IMPLEMENTATION

A. Modulator

The simple and easy method is to modulate the frequency of VCO directly by the use of baseband Gaussian pulse stream, as shown in Fig. 1. However, this modulator has the weak point that it is difficult to keep the center frequency within the allowable value under the restriction of maintaining the linearity and the sensitivity for the required FM modulation. Such a weak point can be removed by the use of an elaborate PLL modulator with a precisely designed transfer characteristics or an orthogonal modulator with digital waveform generators [14]. Instead of such a modulator, a $\pi/2$-shift binary PSK (BPSK) modulator followed by a suitable PLL phase smoother, as shown in Fig. 7, is considered to be a prominent alternative where the transfer characteristics of this PLL are also designed for the output power spectrum to satisfy the required condition.

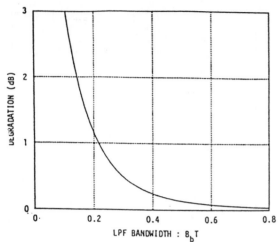

Fig. 6.   Theoretical E_b/N_0 degradation of GMSK.

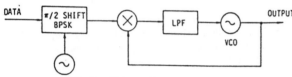

Fig. 7.   PLL-type GMSK modulator.

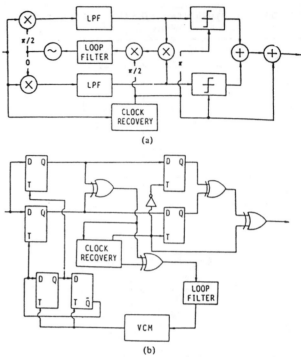

Fig. 8.   Orthogonal coherent detector for MSK/GMSK. (a) Analog type. (b) Digital type.

B. Demodulator

Similar to the simple MSK or TFM system, the orthogonal coherent detector is also applicable for the GMSK system. When realizing such an orthogonal coherent detector, one of the most important and difficult problems is how to recover the reference carrier and the timing clock. The most typical method is de Buda's one [12]. In his method, the reference carrier is recovered by dividing by four the sum of the two discrete frequencies contained in the frequency doubler output and the timing clock is directly recovered by their difference. Remembering that the action of the well-known Costas loop as a carrier recovery circuit for BPSK systems is equivalent to that of a PLL with a frequency doubler [15], de Buda's method is realized by the equivalent one shown in Fig. 8(a). This modified method can easily be implemented by conventional digital logic circuits and its configuration is also shown in Fig. 8(b). In this configuration, two D flip-flops act as the quadrature product demodulators and both of the Exclusive-Or logic circuits are used for the baseband multipliers. Furthermore, the mutually orthogonal reference carriers are generated by the use of two D flip-flops, and the VCO center frequency is then set equal to the four times carrier center frequency. This configuration is considered to be especially suitable for the mobile radio unit which must be simplified, miniaturized, and economized.

IV. EXPERIMENTS

A. Test System

Fig. 9 shows the block diagram of the experimental test system where the carrier frequency and the bit rate are $f_c = 70$ MHz and $f_b = 16$ kbits/s, respectively. A pseudonoise (PN) pulse sequence with a repetition period of $N = (2^{15} - 1)$ bits is generated by the 15-stage feedback shift register (FSR) and is used as a test pattern signal. After passing through a pre-

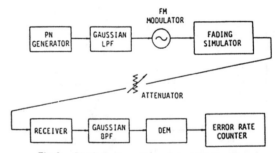

Fig. 9.   Block diagram of experimental test system.

modulation Gaussian LPF having a variable bandwidth B_b, the PN sequence is put into the synthesized RF signal generator having an external FM modulation capability. The frequency deviation of the RF signal generator is set equal to $\Delta f_d = \pm 4$ kHz, which corresponds to the MSK condition for the 16 kbits/s transmission. Then the GMSK signal of our choice is obtained as the RF signal generator output, and is transmitted into the receiver via the Rayleigh fading simulator [16]. Predetection bandpass filtering in the receiver is performed by the precisely designed Gaussian bandpass crystal filter. The bandpass-filtered output is demodulated by the digital orthogonal coherent detector shown in Fig. 8. The regenerated output is fed into the error-rate counter for the BER measurement.

B. Power Spectrum and Eye Pattern

Fig. 10 shows the measured power spectra of the RF signal generator output when $B_b T$ is a variable parameter. It is clearly seen that the measured results agree well with the machine-computed ones shown in Fig. 2. Moreover, GMSK with $B_b T = 0.25$ is shown to satisfy the severe requirements

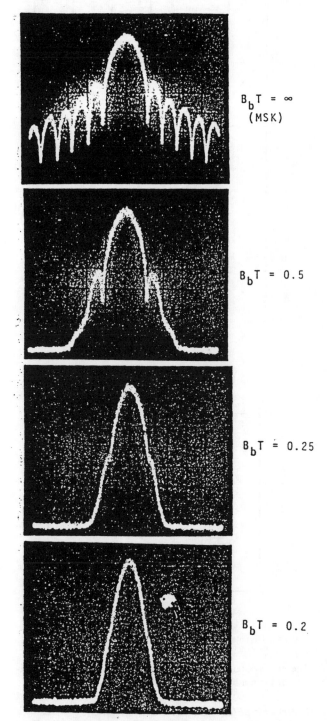

Fig. 10.   Measured power spectra of GMSK (V: 10 dB/div., H: 10 kHz/div.).

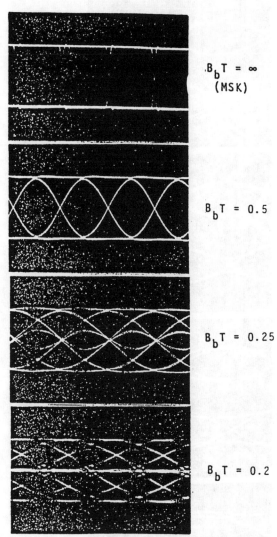

Fig. 11.   Instantaneous frequency variation of GMSK.

of the out-of-band radiation of SCPC mobile radio communications. The corresponding eye pattern measured at the premodulation Gaussian LPF output is shown in Fig. 11. This figure shows that the above satisfactory performance of the out-of-band radiation can only be attained by the sacrifice of introducing severe ISI effects into the baseband waveform of the FM modulator input. It might be feared that such a severe

ISI effect causes inferior transmission performance. However, this misgiving is happily unwarranted because the demodulator output of GMSK with $B_b T = 0.25$ degrades only slightly from that of simple MSK. It is easily found from Fig. 12 which shows the respective eye patterns measured by the analog-type orthogonal coherent detector shown in Fig. 8(a). It is also certified from the BER performance test results described later.

C. Static BER Performance

Fig. 13 shows experimental test results for static BER performance in the nonfading environment where the normalized 3 dB-down bandwidth of the premodulation Gaussian LPF, $B_b T$, is a variable parameter and the normalized 3 dB-down bandwidth of the predetection Gaussian BPF is $B_i T \cong 0.63$, i.e., $B_i = 10$ kHz for $f_b = 1/T = 16$ kbits/s. The condition $B_i T \cong 0.63$ is nearly optimum, as shown in Fig. 14. From Fig. 13, performance degradation of GMSK with $B_b T = 0.25$ relative to simple MSK is found to be only 1.0 dB. Moreover, the measured static BER performance of simple MSK degrades by 0.7 dB from the theoretical one of ideal antipodal binary

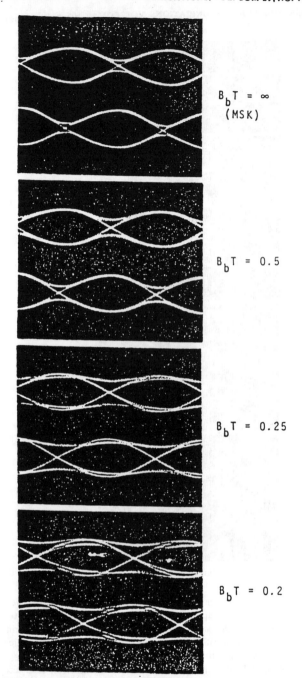

Fig. 12.   GMSK eye patterns demodulated by orthogonal coherent detector.

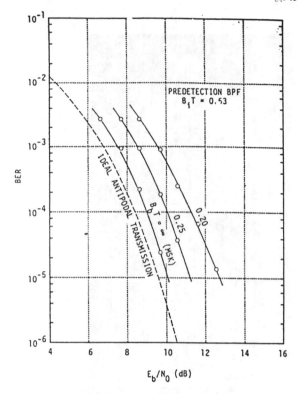

Fig. 13.   Static BER performance.

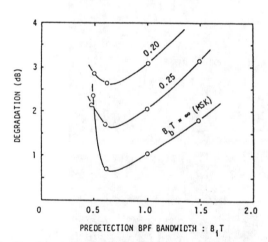

Fig. 14.   Degradation of required E_b/N_0 for obtaining BER of 10^{-3}.

transmission system. If γ denotes the received signal energy-to-noise density ratio, i.e., E_b/N_0, the measured static BER performance in the nonfading environment can be approximated as

$$P_e(\gamma) \cong \tfrac{1}{2} \operatorname{erfc}(\sqrt{\alpha\gamma}) \tag{5}$$

where erfc( ) is the complementary error function given by (2) and α is a constant parameter determined as

$$\alpha \cong \begin{cases} 0.68 & \text{for GMSK with } B_b T = 0.25 \\ 0.85 & \text{for simple MSK } (B_b T \to \infty). \end{cases} \tag{6}$$

The above-obtained results can be estimated by the degradation of the minimum signal distance shown in Figs. 5 and 6.

D. Dynamic BER Performance

In the practical V/UHF land mobile radio environment, signal transmission between a fixed base station and a moving vehicle is usually performed via random multiple propagation routes. Consequently, fast and deep multipath fading, which can generally be treated by the well-known Rayleigh fading model, appears on the received signals of both stations and degrades the signal transmission performance severely.

In particular, when a quasi-stationary slow Rayleigh fading

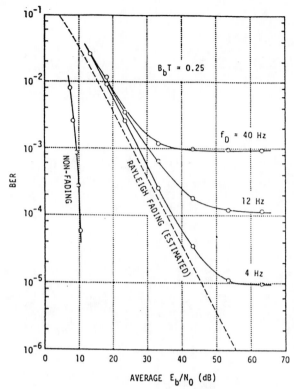

Fig. 15. Dynamic BER performance.

model is assumed, dynamic BER performance is given by

$$P_e(\Gamma) = \int_0^\infty P_e(\gamma)p(\gamma)\,d\gamma \tag{7}$$

where Γ is the average E_b/N_0 and $p(\gamma)$ is the probability density function (pdf) of γ given by

$$p(\gamma) = \frac{1}{\Gamma} \exp\left(-\frac{\gamma}{\Gamma}\right). \tag{8}$$

Substituting (5) and (8) into (7) yields

$$P_e(\Gamma) \cong \frac{1}{2}\left(1 - \sqrt{\frac{\alpha\Gamma}{\alpha\Gamma + 1}}\right) \cong \frac{1}{4\alpha\Gamma} \tag{9}$$

where α is the constant parameter given by (6).

However, the dynamic BER performance in the fast Rayleigh fading environment, where the temporal variation effect of the fading cannot be neglected, has not yet been theoretically estimated because the tracking performance of the carrier recovery circuit in such environment cannot be analyzed. Fig. 15 shows the experimental test results of dynamic BER performance of the GMSK with $B_bT = 0.25$ in the simulated fast Rayleigh fading environment where the maximum Doppler frequency, i.e., the fading rate f_D, is a variable parameter. For comparison, theoretically estimated dynamic BER performance in the quasi-stationary slow Rayleigh fading environ-

ment, i.e., $f_DT \to 0$, is also shown by the dashed line in the same figure.

V. CONCLUSION

As an effective digital modulation for mobile radio use, premodulation Gaussian-filtered minimum shift keying (GMSK) modulation with coherent detection has been proposed. The fundamental properties have been analyzed with the aid of machine computation. The constitution of modulator and demodulator has also been discussed. The superiority of this modulation has been supported by experimental results.

ACKNOWLEDGMENT

The authors wish to thank Dr. K. Miyauchi, S. Ito, K. Izumi, and Dr. S. Seki for their helpful guidance. They also are grateful to Dr. M. Ishizuka and H. Suzuki for their fruitful discussions.

REFERENCES

[1] O. Bettinger, "Digital speech transmission for mobile radio service," *Elec. Commun.*, vol. 47, pp. 224–230, 1972.
[2] J. S. Bitler and C. O. Stevens, "A UHF mobile telephone system using digital modulation: Preliminary study," *IEEE Trans. Vehic. Technol.*, vol. VT-22, pp. 78–81, Aug. 1973.
[3] N. S. Jayant, R. W. Schafer, and M. R. Karim, "Step-size-transmitting differential coders for mobile telephony," in *Proc. IEEE Int. Conf. Commun.*, June 1975, pp. 30/6–30/10.
[4] D. L. Duttweiler and D. G. Messerschmitt, "Nearly instantaneous companding and time diversity as applied to mobile radio transmission," in *Proc. IEEE Int. Conf. Commun.*, June 1975, pp. 40/12–40/15.
[5] J. C. Feggeler, "A study of digitized speech in mobile telephony," presented at the Symp. on Microwave Mobile Commun., session V-3, Boulder, CO, Sept.–Oct. 1976.
[6] H. M. Sachs, "Digital voice considerations for the land mobile radio services," in *Proc. IEEE 27th Vehic. Technol. Conf.*, Mar. 1977, pp. 207–219.
[7] K. Hirade and M. Ishizuka, "Feasibility of digital voice transmission in mobile radio communications," *Paper Tech. Group. IECE Japan*, vol. CS78-2, Apr. 1978.
[8] F. G. Jenks, P. D. Morgan, and C. S. Warren, "Use of four-level phase modulation for digital mobile radio," *IEEE Trans. Electromagn. Compat.*, vol. EMC-14, pp. 113–128, Nov. 1972.
[9] P. K. Kwan, "The effects of filtering and limiting a double-binary PSK signal," *IEEE Trans. Aerosp. Electron. Syst.*, vol. AES-5, pp. 589–594, July 1969.
[10] S. A. Rhodes, "Effects of hardlimiting on bandlimited transmission with conventional and offset QPSK modulation," in *Proc. IEEE Nat. Telecommun. Conf.*, 1972, pp. 20F/1–20F/7.
[11] H. C. van den Elzen and P. van der Wurf, "A simple method of calculating the characteristics of FSK signals with modulation index 0.5," *IEEE Trans. Commun.*, vol. COM-20, pp. 139–147, Apr. 1972.
[12] R. de Buda, "Coherent demodulation of frequency shift keying with low deviation ratio," *IEEE Trans. Commun.*, vol. COM-20, pp. 466–470, June 1972.
[13] H. Miyakawa *et al.*, "Digital phase modulation scheme using continuous-phase waveform," *Trans. IECE Japan*, vol. 58-A, pp. 767–774, Dec. 1975.
[14] F. de Jager and C. B. Dekker, "Tamed frequency modulation, a novel method to achieve spectrum economy in digital transmission," *IEEE Trans. Commun.*, vol. COM-20, pp. 534–542, May 1978.
[15] R. L. Didday and W. C. Lindsey, "Subcarrier tracking methods and communication system design," *IEEE Trans. Commun. Technol.*, vol. COM-16, pp. 541–550, Aug. 1968.
[16] K. Hirade *et al.* "Fading simulator for land mobile radio communications," *Trans. IECE Japan*, vol. 58-B, pp. 449–459, Sept. 1975.

MODEMS for Emerging Digital Cellular-Mobile Radio System

Kamilo Feher, *Fellow, IEEE*

Abstract—Digital MODEM (modulation–demodulation) techniques for emerging digital cellular telecommunications–mobile radio system applications are described and analyzed in this paper. In particular, theoretical performance, experimental results, principles of operation and various architectures of "$\pi/4$-QPSK" ($\pi/4$-shifted coherent or differential QPSK) modems for second generation U.S. digital cellular radio system applications are presented. The spectral/power efficiency and performance of the $\pi/4$-QPSK modems (American and Japanese digital cellular emerging standards) is studied and briefly compared to GMSK modems (proposed for European DECT and GSM cellular standards). Improved filtering strategies and digital pilot aided (digital channel sounding) techniques are also considered for $\pi/4$-QPSK and for other digital modems. These techniques could significantly improve the performance of digital cellular and of other digital land mobile and satellite mobile radio systems. Research results and publications of members of the Digital Communications Research Laboratory (DCRL) cellular/mobile radio group, University of California, Davis have been used in most of the performance prediction and review (tutorial) sections of this paper. More spectrally efficient modem trends for future cellular/mobile (land mobile) and satellite communication systems applications are also highlighted.

I. INTRODUCTION

A. Business and Capacity Challenge/Standards

DIGITAL cellular and digital mobile radio communication systems will be introduced and become operational during the 1990's in the U.S., Canada, Europe, Japan, and many other countries around the world. These "second generation" national and international land mobile and satellite mobile systems will have to have an increased capacity. The available frequency spectrum for analog and/or for digital mobile radio systems is basically the same. Systems are being developed which will have

- increased capacity—10 times;
- reduced cost;
- improved performance;
- new innovative services (in addition to telephone);
- the same or similar basestation (cell site) configurations as currently operational analog systems (as real estate is very expensive);
- dual mode operation–smooth transition from analog to digital [28], [29].

Manuscript received September 1, 1990; revised November 2, 1990. This work was supported by MICRO, NASA, and Ericsson.

The author is with the Department of Electrical Engineering and Computer Sciences, University of California, Davis, CA 95616.

IEEE Log Number 9144474.

TABLE I
SPECTRAL EFFICIENCY OF THE CURRENTLY USED AMPS STANDARD, OF THE AMERICAN DIGITAL CELLULAR (ADC) SECOND GENERATION (TIA 45.3) STANDARD AND OF POSSIBLE FUTURE SYSTEMS. THE SIGNALING RATE OF THE AMPS SYSTEM IS AT 10 KB/S

	Late 1980's and Early 1990's AMPS	Mid-1990's ADC-TIA 45.3	For late 1990's Research Objectives
RF bandwidth	Analog 30 kHz	Digital 30 kHz	
Modulation	FM-FDMA	$\pi/4$-QPSK-TDMA	
Voice capacity	1 channel	3–6 channels	
Efficiency	0.33 b/s/Hz	1.62 b/s/Hz	3–b/s/Hz
Bit rate	10 kb/s (signaling)	48.6 kb/s	
Subscribers (1990)	3.5 million	15–20 million	

Obviously many other "system criteria" could be listed above. Numerous references, including books by Lee [1], [3] describe new and emerging capacity and system service requirements.

Design Challenge for the 1990's and Beyond: The MODEM (modulation–demodulation) and, in general, the mobile radio "physical layer" design challenge for the 1990's and beyond is illustrated in Table I.

The currently operational "Advanced Mobile Phone System" AMPS [26] serves approximately *3.5 million customers* in the U.S., toward the end of the 1980's (December 1989). For cellular telephone it uses analog FM modulation in a 30-kHz radio frequency (RF) bandwidth, i.e., 30-kHz per voice channel. For inter- and intrachannel signaling binary FM, i.e., FSK signal at a rate of 10 kb/s, is transmitted in the same 30-kHz RF band. The 30-kHz individual channel spacing, channel no. 1 is at 825.03 MHz for the mobile units and at 870.03 MHz land base station. The 20-MHz FCC-authorized band is used for 666 channels [15], [28], [29]. In order to accommodate the *15–20 million users* toward the end of this century (many market forecasts are even more optimistic, i.e., suggest an even higher demand), in the same frequency spectrum a *fivefold increase* in spectral efficiency as compared to the currently operational analog systems will be required. Additional capacity could be achieved by bit rate reduced FEC-coded voice (e.g., good voice quality below 10 kb/s) and by introduction of more microcells. In this paper we focus on modem/radio techniques and the capacity increases attained by new generations of modems.

Radio propagation caused *delay spread* may require adaptive modem/channel equalization. This challenging modem/

Reprinted from *IEEE Transactions on Vehicular Technology*, Vol. 40, No. 2, pp. 355-365, May, 1991.

radio R&D task is described by Proakis [35] and in numerous recent references.

The spectral efficiency of the AMPS interchannel signaling (10 kb/s signal) and of comparable bit rate digitized voice signals will have to increase from the present 0.33–1.6 b/s/Hz, thus a five-times (500%) increase in the spectral efficiency requirement. To maintain the overall system capacity advantage and power efficiency, the robustness or immunity to noise and interference (CCL: cochannel interference; ACI: adjacent channel interference) will have to be comparable and/or better than that of the AMPS systems.

For systems requiring even higher capacity, further increase in the spectral efficiency from 2 to 5 b/s/Hz may be required. Research trends will be briefly reviewed in Section VI.

B. Impact of Mobile-Cellular Standards (U.S. and International) on Modem Selection/Design

Several standards are being developed and/or finalized for various mobile and cellular communications applications. Some of the most important world-wide standard systems and committees are shown below.

Abbreviation	Standard for Application in Country and Committee Name
ADC	American Digital Cellular—second generation of US and Canadian cellular radio standards developed and specifiedZ by the Telecommunications Industry Association (US–Canada) TIA 45.3 committee [15], [27], [28],
GSM	Pan-European (Group Speciale Mobile) System specified by the European Telecommunications Standards Institute (ETSI) [30],
DECT	Digital European Cordless Telephone System specified by the DECT committee for applications throughout Europe [31],
CT-2	Cordless Telephone—Second Generation Digital, developed by the British Post Office and committee of CT-2 for applications in Britain and in Europe,
JDC	Japanese Digital Cellular currently being developed by NTT Japan in cooperation with the Ministry of Post, Telegraph and Telecommunications (MPT) of Japan of applications in Japan [27],
APCO-25	Standardization of the APCO committee, Project 25, for new emerging digital U.S./Canadian services serving APCO and NASTD markets.

All of these standards could have worldwide impact on mobile radio systems applications. Various market forecasts indicate a tremendous growth rate for all of these systems. The number of users (served by just one standard) could be *50–70 million* by 1995. Table II illustrates some of the systems parameters, modulation formats, and the various market forecasts. (Evidently we are not in a position to guarantee the accuracy of these market forecasts!)

The MODEM techniques recommended and/or already adopted for these standards are $\pi/4$-QPSK, GMSK (or GFSK), and 4 PAM-FM, see Table II.

The American Digital Cellular standard (ADC) and the Japanese Digital Cellular (JDC) systems will use $\pi/4$-QPSK modems. In this paper we describe and study the performance of the $\pi/4$-QPSK systems which will serve these major U.S./Canadian and Japanese systems.

For a detailed description of the GMSK and related PAM-FM systems, see [1]–[4], numerous IEEE TRANSACTIONS and conference papers, and documents prepared by the DECT and GSM committees, including [30] and [31].

C. Technical Modem Background

Power and bandwidth efficient digital communications systems traditionally use coherent detection [2], [4], [21]. Although coherent systems perform well in stationary additive white Gaussian noise (AWGN) environments, and also have a theoretical power efficiency advantage in Rayleigh and Rician faded mobile systems, their performance may degrade significantly when disturbances such as multipath fading, Doppler shifts and other excessive forms of phase noise are present. These effects have been taken more and more into consideration in the design of new relatively narrow-band digital cellular and digital mobile radio communication systems, i.e., systems which operate in the 1.2 to 50-kb/s bit rate range and have a Doppler shift in the range of 15–100 Hz. Differential detection avoids the need for carrier recovery, and therefore achieves fast synchronization. Therefore, differentially detected systems may be more suitable, not only for narrow-band fading mobile time division multiplex access channels, but also for burst operated (TDMA) systems for burst operated (TDMA) systems requiring fast synchronization and resynchronization [4].

Coherent systems, i.e., coherent demodulators, have a 1–3-dB power efficiency advantage compared to noncoherent demodulators and are more suitable for adaptive channel equalization and fade-countermeasure designs, assuming the bit rate is several orders of magnitude larger than the Doppler shift. A power efficient digital communications system may have to be nonlinearly amplified [12], [16] for cost effective utilization of the available power and may require a nonlinear AGC (automatic gain control circuit). However, when a bandlimited linearly modulated carrier with nonconstant envelope undergoes nonlinear amplification, the filtered sidelobes are restored and in-phase to quadrature in-band crosstalk is generated. This causes severe adjacent channel and in-band cochannel interference. Hence, such systems may not efficiently utilize the available frequency spectrum. The BER performance is also degraded as a result of nonlinear amplification. Therefore, any spectral efficiency gained by using a linear modulation scheme may be lost after nonlinear amplification [18], [20], [21].

A well-known method of reducing envelope fluctuations, and hence spectral regeneration in linear modulation schemes is offset (staggered) QPSK and QAM [2], [14]. However, this type of modulated signal may require coherent demodulators and, due to large Doppler shifts of the mobile channel, may not be suitable for relatively low bit rate system applications, e.g., systems having a bit rate of less than 10 kb/s.

Constant envelope modulation techniques such as MSK, GMSK, TFM, and digital multilevel PAM-FM and/or PRS-FM can be nonlinearly amplified without significant spectral regeneration, and can be differentially or discriminator detected [1]–[4]. The disadvantage of these modulation techniques is that they may have a low spectral efficiency which does not always meet the requirements of emerging standards for digital mobile systems. Recently, therefore, attention has

TABLE II
RADIO/MODEM ILLUSTRATIVE DATA FOR SELECTED EUROPEAN/AMERICAN SYSTEMS AND JAPAN. MOST OF THE SYSTEM PARAMETERS ARE PRELIMINARY. "ILLUSTRATIVE DATA" CHANGES AND MODIFICATIONS ARE ANTICIPATED FOR MOST OF THE DATA. SEVERAL ENTRIES HAVE NOT BEEN FINALIZED

		Europe DECT	Europe Cellular Mobile GSM	England CT-2	ADC-U.S. Dig. Cellular and Canada TIA 45.3	APCO/ NASTD U.S./Canada Project 25	JDC Japan
1.	RF frequency band (MHZ)	1880–1900	890–915 935–960	864.1–868.1 40 channels	880–900? verify exact number	several RF 150–800 MHz FCC projection	
2.	Channel bandwidth	1728 kHz	200 kHz (interleaving?)	100 kHz (?)	30 kHz (interleaving)	12.5 kHz (?)	25 kHZ (interleaving)
3.	Bit rate	1152 kb/s	270.833 kb/s	72 kb/s	48.6 kb/s	9.6 kb/s Motorola propr	42 kb/s
4.	Spectral efficiency	0.67 b/s/Hz	1.35 b/s/Hz	0.72 b/s/Hz	1.6 b/s/Hz	0.77 b/s/Hz	1.6 b/s/Hz
5.	Access Method	TDMA	TDMA	TDMA (?)	TDMA	FDMA	TDMA/ 3-slots
6.	Mobile RF Power/Out-of-Band	250 mW / − 35 dBR	20 W (43 dBm) / − 33 dBr	/ − _ dBr (?)	0.6–4 W − 26–45 dBr	/ − 65 dBr (?)	− 30 dBr (?)
7.	Modulation	GFSK BT = 0.5	GMSK BT = 0.3	GMSK BT =	π/4-QPSK	4-FM (4-PAM-FM)	π/4-QPSK α = 0.5
8.	Demodulation	Coherent or Noncoher. (?)	Coherent (most prob.)		Different or Coherent (?)	FM Noncoher. (?)	
9.	Adaptive Equalization	No (?)	Yes	(?)	Yes	no (?)	yes (?)
10.	Performance	not too stringent	stringent	(?)	stringent	stringent (?)	stringent (?)
11.	Market size (quantity by year)	50–70 million 1995 ??	15 million (?) 1995	??	10–15 million 1997 (?)	1–3 million 1995 ??	

been focused again on the use of linear modulation techniques for nonlinearly amplified systems to meet the simultaneous power and spectral efficiency requirements.

Linear modulation techniques which are suitable for fast fading mobile channels having a significant delay spread and may have to be used as burst operated TDMA systems should satisfy the following requirements.

1) A power efficient amplifier may be used without introducing significant spectral regeneration (spectral restoration, regrowth or splatter) of the transmitted carrier.

2) Extremely low out-of-band radiation can be achieved after nonlinear power efficient amplification without the need for postamplification filtering. (Such a filter may not be practical in most frequency agile TDMA applications and in narrow-band mobile systems.) The out-of-band integrated power has to be in the 5–35-dB region for interleaved cellular radio applications and in the 55–80-dB region for mobile (adjacent RF channel) system applications.

3) Synchronization may be achieved fast and immunity against fast fading may be provided. These may be achieved by noncoherent (differential or discriminator) detection.

4) Adaptive equalization or other fading countermeasure techniques to mitigate the effect of excessive radio propagation caused delay spread and of Doppler shift can be implemented.

In the search for alternative unstaggered linear modulation systems which have low envelope fluctuation, Akaiwa and Nagata [12] and many others, including [13], [16], [18] have studied π/4-QPSK, a technique first introduced by Baker in 1962 [11]. This modification to QPSK has carrier phase transitions which are restricted to $\pm \pi/4$ and $\pm 3\pi/4$. As the phase does not undergo instantaneous $\pm \pi$ transitions as in

QPSK, the envelope fluctuation is significantly reduced. Also, coherent as well as noncoherent detection can be applied to π/4-QPSK. It has been shown in [2] that the spectral efficiency obtained by QPSK is double that obtained by comparable two-level digital FM, GMSK, or TFM, constant envelope modulation techniques.

Due to these attractive features of π/4-QPSK, it has been chosen as the standard modulation technique for the emerging American Digital Cellular (ADC) second generation standard radio system and is also considered for the second generation Japanese standard [27]. The π/4-QPSK modulator, in the case of the second generation American standard, is differentially encoded. The demodulation architecture was not defined by the standardization committee. Manufacturing companies may use coherent, differential or discriminator techniques for signal demodulation [15], [28], [29]. Due to the same attractive properties that lead to its selection for land mobile radio systems, we observe that π/4-QPSK is also an attractive modulation scheme for use in burst operated TDMA satellite systems [22].

A somewhat more detailed literature (however noncomprehensive) search of π/4-QPSK systems indicates the significant amount of research and development related to these systems. Differentially encoded π/4-QPSK was first proposed for data transmission via telephone lines by Baker of Bell Telephone Laboratories [11]. In continued search for alternative nonstaggered linear modulation systems, which have reduced envelope fluctuations and are suitable for coherent as well as differential detection, the application of π/4-QPSK for land mobile channels was proposed in [32] and 15 years later in [12]. Discriminator detection of π/4-QPSK in land mobile channels is studied in [13]. Dif-

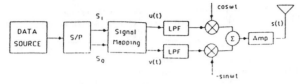

Fig. 1. Block diagram of the transmitter of the $\pi/4$-QPSK modem. The binary data are first converted to two parallel data streams. The parallel data are differentially encoded to a symbol denoted by the carrier phase. The carrier phase can also be characterized by the complex envelope of the carrier. In baseband, the complex envelope is represented by in-phase and quadrature components. The baseband signals are filtered by square-root raised-cosine filters with $x/\sin x$ aperture equalizers. The bandlimited signals are used to quadrature modulate a carrier [22].

ferentially detected $\pi/4$-QPSK ($\pi/4$-DQPSK) in an AWGN-CCI channel is studied in [16]. Comprehensive performance evaluations of $\pi/4$-DQPSK in Rayleigh channels are given in [16]–[24]. Fade-compensated coherent $\pi/4$-QPSK in Rayleigh channels is studied in [18]. Improved baseband pulse shaping techniques which reduce the envelope fluctuation of $\pi/4$-QPSK have been studied in [19], [21]. Power amplifier linearization techniques suitable for $\pi/4$-QPSK and other linear modulation schemes have also been extensively described in recent IEEE Vehicular Technology Conference Proceedings. In this paper, both differentially and coherently detected $\pi/4$-QPSK modems operated in a stationary as well as in mobile radio Rayleigh channels, i.e., channels with AWGN 5 are described. For higher bit rate digital communication systems coherent systems with carrier recovery and/or low-redundancy fade compensation [18]–[23] are suitable. Coherent demodulation is desirable since it improves the power efficiency by 3 dB. A novel coherent demodulator of $\pi/4$-QPSK that uses exclusively two-level decision is described in [22] and briefly reviewed in this paper.

II. $\pi/4$-QPSK-MODULATOR TRANSMITTER MODEL

Although not essential, $\pi/4$-QPSK signals are frequently differentially encoded. This is a desirable property for differential detection and coherent demodulation with phase ambiguity in the recovered carrier. It is also desirable if both coherent and differential receivers are allowed. The transmitter model of the $\pi/4$-QPSK system is shown in Fig. 1. Let $u(t)$ and $v(t)$ denote the unfiltered baseband non-return-to-zero (NRZ) pulses in the in-phase (I) and quadrature (Q) channel, respectively [22]. The signal levels of u_k and v_k which are the pulse amplitudes of the I- and Q-channel for $kT \leq t < (k+1)T$ are determined by the signal levels of the previous pulses and the current information symbol denoted by θ_k as follows:

$$u_k = u_{k-1} \cos \theta_k - v_{k-1} \sin \theta_k \quad (1a)$$

$$v_k = u_{k-1} \sin \theta_k + v_{k-1} \cos \theta_k \quad (1b)$$

In (1a) and (1b), θ_k is in turn determined by the current symbol denoted by (s_I, s_Q) of the information source. The relationship between θ_k and the input symbol is given in Table III. Note that $u(t)$ and $v(t)$ can take the amplitudes of ± 1, 0, and $\pm 1/\sqrt{2}$. The coherently demodulated signals

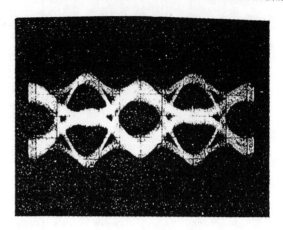

1 μs/div

Fig. 2. The "five-level" eye-diagram of the coherent demodulated $\pi/4$-QPSK signals. At every other sampling instant the signals are two-level. In between, the signals are three-level. Only the in-phase channel is shown. The bit rate used in this experimental setup is 800 kb/s. The receiver low-pass filter is a Butterworth filter with cutoff frequency 256 kHz [23].

TABLE III
PHASE SHIFT AS A FUNCTION OF INFORMATION SYMBOL $\pi/4$-DQPSK

Information	θ	$\cos \theta$	$\sin \theta$
11	$\pi/4$	+	+
01	$3\pi/4$	−	+
00	$-3\pi/4$	−	−
10	$-\pi/4$	+	−

(eye diagrams) are either two-level or three-level at the odd and even sampling instants. A "five-level" eye diagram of the $\pi/4$-QPSK signal is shown in Fig 2. The block diagram shown in Fig. 1 is only a conceptual one, not necessarily a hardware implementation. For hardware design alternatives of $\pi/4$-QPSK modems, see [16], [17], [24].

Let us assume that the low-pass filter is absent first, i.e., an infinite-bandwidth channel is assumed. Assume that the phase of the carrier is 0 during $0 \leq t < T$, i.e., $u_0 = 1$, $v_0 = 0$. At $t = T$, symbol (1, 1) is sent from the information source, then θ_1 is $\pi/4$. From (1a) and (1b) we have $u_1 = 1/\sqrt{2}$ and $v_1 = 1/\sqrt{2}$, the phase of the carrier "jumps" to $\pi/4$. Equations (1a) and (1b) is actually a linear transformation with output a rotation of input in the u-v (complex envelope) plane. The θ_k is the angle rotated and the angle between the complex envelope and the u-axis is the phase of the carrier. From Table III and (1a) and (1b) it follows that if the carrier is at one of the four states denoted by "*" in Fig. 3(a) during the present symbol duration, it shifts (or rotates) to one of the four states denoted by "0" in Fig. 3(a) during next symbol duration and vice versa. Hence the carrier always shifts phase between two symbols, but the phase shift can only be $k\pi/4$ where k is ± 1 or ± 3. If the pulses are bandlimited, the phase transition becomes smoother. However, if ISI-free filters are used, the phase of the carrier is preserved at the sampling instant. A "constellation" of a

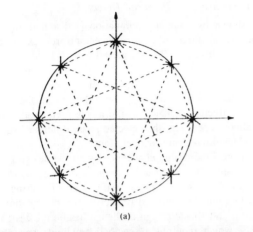

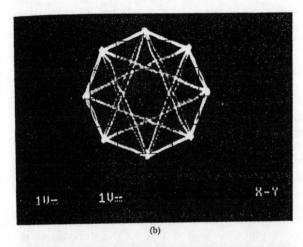

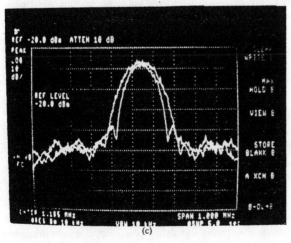

(a)

(b)

(c)

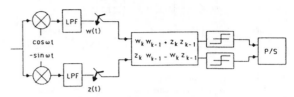

Fig. 4. Block diagram of the baseband differential detector. The low-pass filters are assumed to be square-root raised-cosine.

Fig. 3. Possible phase states of the $\pi/4$-QPSK modulated carrier at sampling instants. The links between two states indicate the allowed phase transition. (b) The constellation of the $\pi/4$-QPSK signals. In this hardware experiment, sinewave shaping $\pi/4$-QPSK (SP-QPSK) is used [24]. (c) A new generation of improved performance $\pi/4$-QPSK modems developed at C. Davis has a reduced out-of-band spectrum in nonlinearly amplified power efficient radio system applications. In the illustrated spectral measurement we use an $f_b = 250$-kb/s rate. These "$\pi/4$-QPSK" (upper trace) MQAM modems (lower trace) are described in [2], [14], [19], [21]. Horizontal scale: 100 kHz/div; vertical: 10 dB/div.

new sinewave shaped $\pi/4$-QPSK (SP-QPSK) [24] system is shown in Fig. 3(b). Measured spectra are illustrated in Fig. 3(c).

From the block diagram and the description, it follows that the information is completely contained in the phase difference "θ_k" of the carrier between two sampling instants. In the receiver only the phase difference between two sampling instants is needed for detection of information. Thus differential detection can be employed. Differential detection has the advantage of hardware simplicity and in relatively low bit rate fast Rayleigh faded channels the error floor, caused by Doppler shift, may be lower than in coherent systems. However, coherent detection is desirable when higher power efficiency is required.

In the following sections we describe four detection schemes for $\pi/4$-QPSK. They are:

1) baseband differential detection;
2) conventional IF band differential detection;
3) FM-discriminator detection;
4) a novel coherent demodulator structure based on [22]. The noncoherent demodulators have equivalent bit-error rate (BER) performance [16], [20].

III. DQPSK Differential Detection Techniques

A. Baseband Differential Detection

The block diagram of a baseband differential detector is shown in Fig. 4. The local oscillator (LO) is assumed to have the same frequency as the unmodulated carrier. It is not necessarily phase coherent [13]. However, the phase error is cancelled by differential detection. Since the assumed "ideal" channel is a Nyquist channel, $w_k = u_k$, $z_k = v_k$ in the sampling instants. In the interference- and noise-free environment we have $\theta = 0$. In general, $\theta \neq 0$ and we have

$$w_k = \cos(\phi_k - \theta) \qquad (2a)$$

$$z_k = \sin(\phi_k - \theta) \qquad (2b)$$

where $\phi_k = \tan^{-1} v_k / u_k$ is the phase of the carrier at $t = kT$ which is preserved through the Nyquist channel [13], [22].

After the detection operation we have

$$x_k = w_k w_{k-1} + z_k z_{k-1} = \cos(\phi_k - \theta_{k-1}) \qquad (3a)$$

$$y_k = z_k w_{k-1} - w_k z_{k-1} = \sin(\phi_k - \theta_{k-1}). \qquad (3b)$$

From Table III the detector decides

$$S_I = 1, \quad \text{if } x_k > 0; S_I = 0, \quad \text{if } x_k < 0 \qquad (4a)$$

$$S_Z = 1, \quad \text{if } y_k > 0; S_Q = 0, \quad \text{if } y_k < 0. \qquad (4b)$$

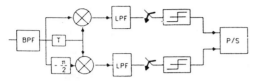

Fig. 5.   Block diagram of the IF-band differential detector employing delay line and mixers. The BPF is assumed to have square-root raised-cosine rolloff. The LPF is assumed to be an ideally brick-wall filter with $2(1 + \alpha)$ times Nyquist bandwidth.

Fig. 6.   Block diagram of the FM-discriminator detector. The modulo 2π operation is used in the decision. The BPF is also assumed to have square-root raised-cosine rolloff.

In general, for M-ary DPSK, x_k and y_k can be used to jointly decide the transmitted symbol. This differential scheme is often called ''optimal'' differential detector since matched Nyquist filters are used. The BER, $P(e)$ versus bit-energy to noise–power–density ratio E_b/N_0 is given as

$$P_e(E_b/N_0) = e^{-2E_b/N_0} \sum_{\infty} 0 (\sqrt{2} - 1)^k l_k(\sqrt{2} \, E_b/N_0)$$

$$- \frac{1}{2} l_0(\sqrt{2} \, E_b/N_0) e^{-2E_b/N_0} \qquad (5)$$

where l_k is the kth-order modified Bessel function of first kind [20].

B. IF Band Differential Detection

In Fig. 5 we illustrate a conventional differential detector employing delay line and phase detector (mixer) components. The advantage of this detection scheme is that no local oscillator is needed. In conventional differential detection, Butterworth BPF and LPF are assumed. The bandwidth of the filters are chosen as $0.57/T$ to minimize the effect of ISI and noise [13]. However if a square-root raised-cosine rolloff BPF is used to match the transmitted signals, the carrier phase is preserved and the noise power is minimized under ISI-free conditions. After differential detection the signal bandwidth becomes twice that of the baseband in the transmitter end. An ideal brickwall LPF with bandwidth $(1 + \alpha)f_s$ is used to filter out the bandpass signal around $2f_c$ without introducing ISI.

C. FM-Discriminator

The block diagram of the FM-discriminator detector is shown in Fig. 6. We also propose the use of the square-root raised-cosine rolloff BPF to preserve the carrier phase at $t = nT$. The hard-limiter does not change the phase of the carrier. After FM-discriminator and integrate-sample-and-dump (ISD), the phase difference between two sampling instants are detected by the modulo 2π phase detector. The modulo 2π decision of the phase difference is employed to improve the BER performance. With the modulo 2π decision, the effect of click noise is removed since click noise causes 2π phase shift.

D. Equivalence of Differential Detectors

It is shown by computer simulation [13] and theoretical analysis that all three differential detectors proposed in this section are equivalent. Both baseband and IF-band differential detectors detect the cosine and sine functions of the phase difference first and decide the phase difference accordingly. The FM-discriminator followed by ISD detects the phase difference directly. If matched filtering is performed before any nonlinear operation, the detected phase difference should not be affected by nonlinear operations.

In Scheme I the design challenge lies on the design of the local oscillator. If the LO has a frequency difference of Δf relative to the CR, the phase drifts $2\pi\Delta fT$ during one symbol duration. This phase drift causes BER degradation. In Schemes II and III the design challenge lies on the design of the BPF, which required a specified amplitude and phase responses. In general the BPF causes ISI and has a wider noise bandwidth than the Nyquist bandwidth.

IV.   COHERENT DETECTION OF $\pi/4$-QPSK SYSTEMS

Differential or noncoherent detection has advantages of hardware simplicity and robustness against random FM in fading channels. However, differential detection has 2–3-dB degradation in Gaussian and in slowly fading channels compared with coherent detection. In a Rician-fading channel with strong line-of-sight (LOS) signal or Rayleigh-fading channels with fade compensation, coherent demodulation is desirable for improved power efficiency. This is particularly useful for higher bit rate digital mobile radio applications, e.g., bit rates higher than the 40-kb/s range. In [22] we describe a novel coherent $\pi/4$-QPSK demodulator having a block diagram illustrated in Fig. 7. This demodulator requires only two-level detection and achieves improved performance.

V.   PERFORMANCE OF $\pi/4$-DQPSK SYSTEMS IN FREQUENCY-SELECTIVE FAST RAYLEIGH-FADING CHANNELS

A. System Model

The general model of a digital cellular mobile communication system considered in this paper is depicted in Fig. 8 [18]. The block TX1 is the transmitter of the desired signal that transmits the signal

$$s_T(t) = A(t)\cos \omega t - B(t)\sin \omega t \qquad (6)$$

where $A(t)$ and $B(t)$ are the bandlimited in-phase and quadrature baseband signals, respectively, and ω is the angular carrier frequency. The signal $s_T(t)$ is randomly modulated by a Rayleigh envelope $R_1(t)$ and uniform phase $\phi_1(t)$. A delayed signal $s_T(t - \tau)$ is randomly modulated by an independent second Rayleigh envelope $R_2(t)$ and uniform phase $\phi_2(t)$. The combination of $s_T(t)$ and $s_T(t - \tau)$ accounts for the frequency-selective fading. The two-ray model is recommended by the Telecommunications Industry Association (TIA) standard committee [15] to evaluate the tolerance of delay spread in digital cellular systems. Another transmit-

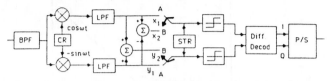

Fig. 7. Block diagram of a $\pi/4$-QPSK coherent demodulator. Analog summers are used to rotate the complex envelope by $\pi/4$. By properly clocking the switch, the input to the threshold detectors are always two-level. For a detailed description of this improved performance demodulator, see [22].

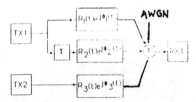

Fig. 8. Block diagram of the system model of a frequency-selective fast Rayleigh-fading channel corrupted by CCI and AWGN. (1) TX1-RX1: the desired channel (2) TX2: the transmitter of the cochannel (3) $R_i(t)$: the Rayleigh envelope (4) $\phi i(t)$: the uniform phase.

ter TX2 transmits the cochannel "signal"

$$s_c(t) = C(t)\cos \omega t - D(t)\sin \omega t. \qquad (7)$$

The cochannel signal is also randomly modulated by a Rayleigh envelope $R_3(t)$ and uniform phase $\phi_3(t)$. We assume that $R_i(t)$, $i = 1, 2, 3$ and $\phi_i(t)$, $i = 1, 2, 3$ are statistically mutually independent. The white Gaussian noise

$$n(t) = n_c(t)\cos \omega t - n_s(t)\sin \omega t \qquad (8)$$

is added to the three received faded signals in the front end of the receiver.

In linear static Gaussian channels, matched ISI-free filters are used for optimum BER performance [4]. In matched ISI-free systems, ISI is introduced by the transmitter filter; the receiver filter removes the ISI so that the cascade system is ISI-free. In a Rayleigh-fading channel the receiver filter does not remove the ISI completely because of the time-varying nature of the channel. To avoid ISI we assume that a raised-cosine filter is used in the transmitter and a brickwall filter with bandwidth equal to $(1 + \alpha)F_N + f_{D,\max}$ is used in the receiver. Here f_N is the Nyquist frequency and $f_{D,\max}$ is the maximum Doppler frequency and α is the filter "rolloff" parameter [2]. With this filtering strategy the output of the transmitter filter is ISI-free. The channel is modeled as a time-varying but memoryless system, hence no ISI is introduced. The receiver filter passes the received signal without distortion. Therefore the cascaded system is free of ISI. This causes 0.57 dB $(10\log[(1 + \alpha)(1 - 0.25\alpha)])$ E_b/N_0 degradation for $\alpha = 0.2$ compared to matched filtering with ISI neglected.

If CCI is the dominant interference as in cellular systems, the performance is the same as in the case of matched filtering systems. Full raised-cosine transmitter filters with $\sin x$ aperture equalizers have other advantages. For IF-band differential detector and FM-discriminator detector to be ISI-free raised-cosine filtering must be performed before any nonlinear operation [13]. If the raised-cosine filtering is

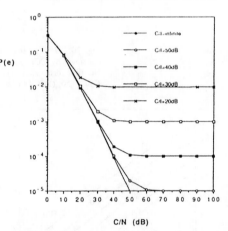

Fig. 9. The BER performance of $\pi/4$-DQPSK in a flat slow fading channel corrupted by AWGN and CCI. $f_c = 850$ MHz, $f_s = 24$ kBd, $\alpha = 0.2$. No Doppler spread and time dispersion.

equally separated between transmitter and receiver, the ideal receiver BPF may be required to have a square-root raised-cosine amplitude and linear phase response. It is impractical to design and implement a narrowband BPF with such a stringent specification. If the raised-cosine filtering is performed in the transmitter, a maximum-flat narrow-band filter can be used as the receiver BPF. Although the noise bandwidth is wider than the minimum Nyquist bandwidth, the degradation is negligible in CCI controlled systems [19]–[22].

B. $\pi/4$-DQPSK Performance Results

Detailed theoretical analysis, computer simulation studies and experimental results of our model have been derived and presented in [13], [16], [20]. In the following section we will highlight results from these references. *Note:* Additional practical degradations and deviations from the theoretical model are not reported in this section.

The BER of *uncoded and of unequalized* (i.e., no adaptive channel equalizer) $\pi/4$-DQPSK systems in various Rayleigh-faded mobile environments is highlighted in the following figures.

In Fig. 9 the theoretical BER $= f(C/N)$, in a Rayleigh-faded system model is presented. Note that for the Rayleigh-faded environment within 1-dB accuracy we found [20] that

$$\text{BER} = f(C/N) = f(C/I).$$

That is, the in-band interference, or cochannel interference,

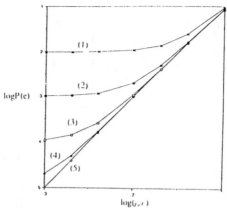

Fig. 10. $P(e)$ versus $f_D T$ of $\pi/4$-DQPSK in a flat fast fading channel corrupted by CCI. $f_c = 850$ MHz, $f_s = 24$ kB, $\alpha = 0.2$, $C/N = \infty$ dB. No time dispersion. (1) $C/I = 20$; (2) $C/I = 30$; (3) $C/I = 40$; (4) $C/I = 50$; (5) $C/I = \infty$ dB.

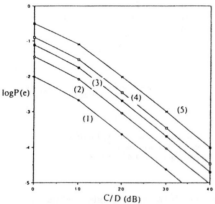

Fig. 11. $P(e)$ versus C/D of $\pi/4$-DQPSK in a frequency-selective slow fading channel. $C/N = \infty$ dB. No Doppler spread. $f_c = 850$ MHz, $f_s = 24$ kB, $\alpha = 0.2$, $C/I = \infty$ dB, (1) $\tau = 0.1$; (2) $\tau = 0.2$; (3) $\tau = 0.3$; (4) $\tau = 0.4$; (5) $\tau = T$.

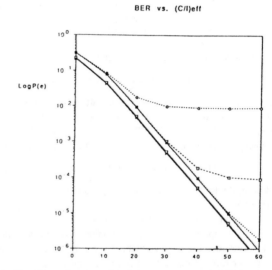

Fig. 12.   The performance of the fade compensated $\pi/4$-QPSK as functions of $(C/I)_{\text{eff}}$, the effective carrier-to-interference power ratio. The performances of "ideal" coherent $\pi/4$-QPSK and "ideal" $\pi/4$-DQPSK are also depicted for [18]. (1) Dashed line without symbol: ideal coherent demodulation (2) Dashed–dotted line: ideal differential detection (3) Real lines: fade compensated coherent demodulation (a) white circle: $f_D T = 3 \times 10^{-2}$ (b) white square: $f_D T = 3 \times 10^{-3}$ (c) cross: $f_D T = 3 \times 10^{-4}$ (4) Dashed lines: differential detection (a) white circle: $f_D T = 3 \times 10^{-2}$ (b) white square: $f_D T = 3 \times 10^{-3}$ (c) cross: $f_D T = 3 \times 10^{-4}$.

assuming it is independently Rayleigh faded, gives us practically the same results as Gaussian noise.

The Doppler shift may cause severe error floors, i.e., residual BER. From Fig. 10 we note as an example that

$$\log f_D T = -2, \ i.e., \ f_d T = 0.01$$

is the cause of a BER floor of BER $= 10^{-3}$ if there is no cochannel interference and a BER $= 10^{-2}$ with CCI = 20 dB. At an increased Doppler, e.g., $f_D T = 0.1$ the residual BER floor is as high as 10^{-1}, i.e., BER $= 10^{-1}$. This error floor is independent of the available C/I.

The impact of delay spread caused degradations in a frequency selective fading environment is shown in Fig. 11. The C/D or C/I parameter in this figure represents the carrier power (C) to delayed carrier power (D) ratio used in the two-ray Rayleigh-faded system model. Note that for a reasonable worst case example $C/D = 0$ dB and that the delayed Rayleigh-faded signal "D" is delayed by 0.3 times the symbol duration, i.e., $\tau = 0.3T$. This could cause a residual BER floor of BER $= 10^{-1}$. Adaptive equalization

would be required for this demodulator, operated in such a severe propagation-caused delay spread environment.

The performance results, illustrated in Figs. 9–11 indicate that for digital mobile systems, operating below 10 kb/s, the Doppler-caused BER floor may be of major concern while for systems operating in the 30–50 kb/s range and at higher rates, e.g., 300 kb/s, the radio propagation-caused random delay spread may be of major concern. Differentially demodulated $\pi/4$-QPSK outperforms coherent $\pi/4$-QPSK in high Doppler shift environment. On the other hand it is more complex to adaptively equalize noncoherent systems. Adaptive equalization may be essential for cellular/mobile cover-

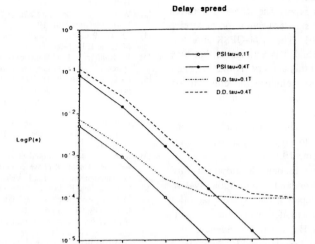

Fig. 13. Error-floors of the fade compensated $\pi/4$-QPSK and $\pi/4$-DQPSK in a frequency-selective fading (delay spread) channel as functons of C/D for $\tau = 0.1\ T$ and $\tau = 0.4\ T$. The fading rate is assumed to be $f_D T = 3 \times 10^{-3}$. Pilot symbol insert (PSI) and conventional differentially detected (DD) $\pi/4$-DQPSK results are illustrated.

TABLE IV
COMPARISON OF BER PERFORMANCE OF SEVERAL SPECTRALLY EFFICIENT MODULATION SCHEMES IN A MOBILE CHANNEL

		$\pi/4$-DQPSK	Coded 16-QAM with Channel State Derived Decoding [25]	Pilot Symbol Aided 16-QAM [33]
Spectral efficiency (based on source data rate)		1.6 b/s/Hz (typically)	2.2–3.2 b/s/Hz	2.9 b/s/Hz
E_b/N_0 required for $P_e = 10^{-3}$ $4 \cdot 10^{-3} < f_D T_S < 1 \cdot 10^{-2}$		27 dB	19 dB	30 dB
Error Floor (no CCI)	$f_D T_S = 4 \cdot 10^{-3}$	$1 \cdot 10^{-4}$	$< 10^{-4}$	
	$f_D T_S = 1 \cdot 10^{-2}$	$6 \cdot 10^{-4}$	$< 10^{-4}$	
Error Floor with CCI	$C/I = 20$ dB	$1 \cdot 10^{-2}$	$1 \times 10^{-4}\ 2 \cdot 10^{-2}$	
	$C/I = 30$ dB	$1 \cdot 10^{-3}$	$< 10^{-4}$	$2 \cdot 10^{-2}$
$f_D T_S = 4 \cdot 10^{-3}$	$C/I = 40$ dB	$2 \cdot 10^{-4}$	$< 10^{-4}$	$2 \cdot 10^{-3}$
Error Floor with CCI	$C/I = 20$ dB	$1 \cdot 10^{-2}$	5×10^{-4}	
	$C/I = 30$ dB	$1 \cdot 10^{-3}$	$< 10^{-4}$	
$f_D T_S = 1 \cdot 10^{-2}$	$C/I = 40$ dB	$6 \cdot 10^{-4}$	$< 10^{-4}$	

age areas where the delay spread is in excess of 0.2 of the symbol duration, i.e., 0.2 T.

C. Digital Pilot Symbol Aided Coherent Modems Reduce the Residual BER Floors

In [18] and [23] it has been demonstrated that digital pilot symbol aided coherent PSK systems, including $\pi/4$-QPSK systems, have a significantly reduced error floor in large Doppler spread-fast Rayleigh- and Rician-faded environments. Illustrative results are given in Figs. 12 and 13. Both coherent $\pi/4$-QPSK and differentially detected (DD) $\pi/4$-QPSK residual BER floors are highlighted. In Fig. 12 the improvements due to digital pilot sounding are very significant [18]. For example, for $f_D T = 3.10^{-2}$ the BER floor of a coherent system is 10^{-2}. This floor is reduced by more than three orders of magnitude to 10^{-6} with Doppler phase

compensated digital channel (pilot symbol) sounding techniques. The performance improvement, in frequency-selective fading environments attained by digital channel sounding, is approximately 3 dB, as illustrated in Fig. 13.

VI. SPECTRALLY EFFICIENT (MORE THAN 2 b/s/Hz) MODEM RESEARCH FOR DIGITAL MOBILE RADIO SYSTEMS

Several organizations throughout the world are active in "spectrally efficient" digital mobile radio system research. We define spectrally efficient systems which meet the following criteria:

Spectral efficiency	2-b/s/Hz minimum
Out-of-band integrated power	−60 dBr
Power efficiency	comparable to $\pi/4$-DQPSK

Future generations of modems are expected to be even more power- and spectral-efficient than the $\pi/4$-QPSK mobile radio modems described in this paper. The second generation U.S. digital cellular network standardization procedure is nearing completion during 1991. The $\pi/4$-QPSK modems will be on the market by 1992 and it is anticipated that they will carry traffic until the end of this century. However, even before the end of the 1990's more efficient modems will be required together with several spectrally efficient digital mobile systems with an efficiency of 2–5 b/s/Hz. For this reason digital modems having performance, such as highlighted in Table IV, will have to be developed and implemented during the later part of the 1990's.

Some of the spectral efficient modem developments are highlighted in this paragraph. For land mobile radio system applications Sampei [33] developed a 64-kb/s digital pilot aided "channel sounding coded" modem with fade compensation which achieves 2.9-b/s/Hz spectral efficiency. Researchers at the Jet Propulsion Laboratory [34] have been very active in trellis coded modem research. At the University of California, Davis, we have developed a 16-QAM modem which has state-derived decoding algorithms and has a very promising performance in the 2.2–3-b/s/Hz range. Illustrative results are shown in Table IV [25]. For mobile satellite communications systems digital pilot aided compensation is used in trellis-coded QAM and in M-ary PSK systems [23].

This is only a brief partial list of spectrally efficient modem R&D achievement. Conference proceedings such as the IEEE-VTC, ICC, GLOBECOM, and the IEEE TRANSACTIONS ON VEHICULAR TECHNOLOGY as well as the IEEE TRANSACTIONS ON COMMUNICATIONS and the IEEE TRANSACTIONS ON INFORMATION THEORY contain many papers which describe new developments. I am confident that, before the year 2000, we will see practical, operational digital mobile cellular systems which have spectral efficiencies in the 3–5 b/s/Hz range.

ACKNOWLEDGMENT

The author wishes to express his gratitude to members of the research staff of our Digital Communications Research Laboratory (DCRL) Mobile–Cellular Radio Group, University of California, Davis for their suggested improvements of this manuscript and their permission to use jointly published research results in review sections of this paper. In particular the assistance of C. L. Liu, D. Subasinghe-Dias, and S. Gurunathan is acknowledged. The support and cooperation of MICRO, NASA, of our industrial sponsors, and in particular, of Ericsson (K. Raith, Dr. J. Uddenfeldt) enabled us the completion of numerous computer simulation hardware and system/experimental projects highlighted in this paper.

REFERENCES

[1] W. C. Y. Lee, *Mobile Communications Design Fundamentals*. Indianapolis, IN: Howard W. Sams, 1986.

[2] K. Feher, Ed., *Advanced Digital Communications: Systems and Signal Processing Techniques*. Englewood Cliffs, NJ: Prentice-Hall, 1987.

[3] W. C. Y. Lee, *Mobile Cellular Telecommunication Systems*. New York: McGraw-Hill, 1989.

[4] J. G. Proakis, *Digital Communications*, 2nd ed. New York: McGraw-Hill, 1989.

[5] K. Raith and J. Uddenfeldt, "Capacity of digital cellular TDMA systems," pp. 323–332, this issue.

[6] D. E. Borth and P. D. Rasky, "An experimental RF link system to permit evaluation of the GSM air interface standard," in *Proc. Third Nordic Seminar on Digital Land Mobile Radio Communication*, Copenhagen, Denmark, Sept. 13–15, 1988.

[7] J. P. Weck *et al.*, "Power delay profiles measured in mountainous terrain," presented at the 39th IEEE Veh. Technol. Conf., Philadelphia, PA.

[8] S. Ariyavisitakul, "Equalization of hard limited slowly fading multipath signal using a phase equalizer with a time reversal structure," presented at the 40th IEEE Veh. Technol. Conf., Orlando, FL.

[9] C. L. Liu and K. Feher, "Pilot symbol aided coherent M-ary PSK in frequency-selective fast Rayleigh fading channels," Univ. California, Davis, Digital Commun. Res. Lab. Rep. DCRL-S-68, Sept. 1990. (Also *IEEE Trans. Commun.*, submitted).

[10] C. Y. Weng and K. Feher, "Anti multipath design strategy for excessive delay spread fast Rayleigh faded digital FM (CPM) mobile radio systems," Univ. California, Davis, Digital Commun. Res. Lab. Rep. DCRL-SC-8, Oct. 12, 1990. (Also, *IEEE-ICC-91*, submitted).

[11] P. A. Baker, "Phase modulation data sets for serial transmission at 2000 and 2400 bits per second, Part 1," *AIEE Trans. Commun. Electron.*, July 1962.

[12] Y. Akaiwa and Y. Nagata, "Highly efficient digital mobile communications with a linear modulation method," *IEEE J. Select. Areas Commun.*, vol. SAC-5, June 1987.

[13] C. L. Liu and K. Feher, "Noncoherent detection of $\pi/4$-shift QPSK systems in a CCI-AWGN combined interference environment," in *Proc. IEEE 40th Veh. Technol. Conf.*, San Francisco, CA, May 1989.

[14] J. S. Seo and K. Feher, "SQUAM: a new superposed QAM modem technique," *IEEE Trans. Commun.*, vol. COM-33, Mar. 1985.

[15] Electronic Industries Association Specification IS-54, "Dual-mode subscriber equipment compatibility specification," EIA Project No. 2215, Dec. 1989.

[16] S. H. Goode, H. L. Kazecki, and D. W. Dennis, "A comparison of limiter-discriminator, delay and coherent detection for $\pi/4$-QPSK," in *Proc. 40th IEEE Veh. Technol. Conf.*, Orlando, FL, May 1990, pp. 687–694.

[17] Y. Guo and K. Feher, "Performance evaluation of differential $\pi/4$-shift QPSK systems in a Rayleigh fading/delay spread/CCI/AWGN environment," in *Proc. 40th IEEE Veh. Technol. Conf.*, Orlando, FL, May 1990, pp. 420–424.

[18] C. L. Liu and K. Feher, "A new generation of Rayleigh fade compensated $\pi/4$-QPSK coherent modems," in *Proc. 40th IEEE Veh. Technol. Conf.*, Orlando, FL, May 1990, pp. 482–486.

[19] K. Feher and C. L. Liu, "A modified Nyquist filtering strategy for digital cellular mobile radio systems," Univ. California, Davis, Digital Communications Research Lab. Rep. S-12, Aug. 27, 1990.

[20] C. L. Liu and K. Feher, "Bit-error-rate performance $\pi/4$-DQPSK in a frequency-selective fast Rayleigh fading channel," Univ. California, Davis, Digital Commun. Res. Lab. Rep. S-62, Sept. 1990.

[21] D. Subasinghe-Dias and K. Feher, "Baseband pulse shaping for $\pi/4$-DQPSK for nonlinearly amplified digital satellite and mobile channels," Univ. California, Davis, Digital Commun. Res. Lab. Rep. S-64, May 31, 1990.

[22] C. L. Liu and K. Feher, "$\pi/4$-QPSK modems for satellite sound/data broadcast systems," *IEEE Trans. Broadcasting*, Mar. 1991.

[23] M. L. Moher and J. H. Lodge, "TCMP—A modulation and coding strategy for Rician fading channels," *IEEE J. Select. Areas Commun.*, vol. 7, Dec. 1989.

[24] S. Katoh and K. Feher, "SP-QPSK: A new modulation technique for satellite and land-mobile digital broadcasting," *IEEE Trans. Broadcasting*, vol. 36, pp. 195–202, Sept. 1990.

[25] D. Subasinghe-Dias and K. Feher, "Coded 16 QAM with channel state-derived decoding for fast fading mobile channels," Univ. California, Davis, Digital Commun. Res. Lab. Rep. SC-7, Oct. 9, 1990.

[26] R. Arredondo, "AMPS—Advanced mobile phone system," *Bell Syst. Tech. J.*, Jan. 1979.

[27] N. Nakajima and K. Kinoshita, "A system design for TDMA mobile radios," in *Proc. IEEE Veh. Technol. Conf.*, Orlando, FL, May 6-9, 1990.

[28] EIA/TIA, "Cellular system: Recommended minimum performance standards for 800 MHz dual-mode mobile stations," PN2216, incorporating EAI/TIA IS-19B Draft; TR.45.3, Project 2216 Electronic Industries Association (EIA), Washington, DC, Oct. 1990.

[29] EIA/TIA, "Cellular system: Recommended minimum performance standards for 800 MHz base stations supporting dual-mode mobile stations," TR.45.3, Project Number 2217, incorporating EIA/TIA IS-20 Draft, Electronic Industries Association (EIA), Washington, DC, Oct. 1990.

[30] G. S. M., "Physical layer on the radio-path," Vol. "G"; *GSM 05.04*; Release GSM/PN; ETSI—European Telecommun. Standards Inst., Nice, France, July 1988.

[31] DECT, "Digital European cordless telecommunication system—common interface specifications," Code: RES-3(89), DECT, 1989.

[32] F. G. Jenks *et al.*, "Use of four level phase modulation for digital mobile radio," *IEEE Trans Electromagn. Compat.*, vol. EMC-14, Nov. 1972.

[33] S. Sampei and T. Sunaga, "Rayleigh fading compensation method for 16-QAM in digital land-mobile channels," IEEE Veh. Technol. Conf. 1989.

[34] D. Divsalar and M. K. Simon, "Multiple trellis coded modulation (MTCM)," *IEEE Trans. Commun.*, vol. 36, Apr. 1988.

[35] J. G. Proakis, "Adaptive equalization for TDMA digital mobile radio," *IEEE Trans. Veh. Technol.*, pp. 333-341, this issue.

Kamilo Feher (S'73–M'73–SM'77–F'89) is a professor of Electrical Engineering at the University of California, Davis, and Director, Consulting Group, DIGCOM, Inc. He has 25 years of industrial and academic R&D, teaching, management, and consulting experience. He has been a consultant to U.S., Canadian, and many overseas corporations and to governments and has presented numerous short courses. He holds six U.S. and international patents. He has coauthored over 230 original research papers and is the author of five books: *Advanced Digital Communications: Systems and Signal Processing Techniques* (Englewood Cliffs, NJ: Prentice-Hall, 1987); *Telecommunications Measurements, Analysis and Instrumentation* (Englewood Cliffs, NJ: Prentice-Hall, 1983); *Digital Demodulations Techniques in an Interference Environment*, (Gainesville, VA: Don White Consultants, 1977). He has supervised large digital communications (university and industry based) research teams. His inventions are used by internationally acclaimed major corporations throughout the world. His major discoveries of emerging digital cellular and mobile radio (modulation–demodulation) communications and digital satellite mobile/broadcasting systems are expected to have most significant contributions to the ultimate objective of communications—to enable anyone to communicate instantly with anyone else from anywhere.

For his contribution to Digital Communications R&D (as a professor at the University of Ottawa, Canada), Dr. Feher has been awarded the Steacie Memorial Fellowship, by the Natural Sciences and Engineering Research Council of Canada (NSERC), and has received the "Engineering Medal, Ontario, Canada."

Spread Spectrum for Mobile Communications

Raymond L. Pickholtz, *Fellow, IEEE*, Laurence B. Milstein, *Fellow, IEEE*, and
Donald L. Schilling, *Fellow, IEEE*

Abstract—The characteristics of spread spectrum that make it advantageous for mobile communications are described. The parameters that determine both the performance and the total capacity are introduced, and an analysis is presented which yields (approximately) the number of users that can simultaneously communicate while maintaining a specified level of performance. Spread-spectrum overlay, wherein a code division multiple access (CDMA) network shares a frequency band with narrow-band users, is analyzed, and it is seen that excision of the narrow-band signals from the CDMA receivers before despreading can improve both performance and capacity.

I. INTRODUCTION

THE use of spread spectrum techniques for military communications has become quite commonplace over the past decade, and it was the potential use of spread spectrum in hostile environments that motivated most of the basic research in the subject for the past 20 years or more. In recent years, however, there has been an even greater interest in understanding both the capabilities and the limitations of spread-spectrum techniques for commercial applications. Entirely new opportunities such as wireless LAN's, personal communication networks, and digital cellular radios have created the need for research on how spread spectrum systems can be "reoptimized" for most efficient use in these nonhostile environments.

Indeed, in this new decade we will experience a revolution in telecommunication, which might be even greater than the PC revolution of the past decade. With this increased need to communicate, and with a "fixed" available spectrum, it becomes necessary to use this spectrum more effectively. Spread spectrum is a technique for efficiently using spectrum by allowing additional (spread spectrum) users to use the same band as other, existing users.

The new applications of spread-spectrum communications have characteristics which are quite different from those of interest in the past. That is, up until fairly recently, most interest in spread spectrum was dominated by the classical military scenario of intentional, smart jamming [1]. However, once the threat of having to design a system which is capable of combatting an intelligent adversary is removed, one can consider ways of improving the receiver design to make the system more efficient and more practical for commercial applications. Indeed, the new design philosophy will emphasize spectral efficiency, cost, reliability and complexity reduction.

Of the many potential uses for spread-spectrum communications in civilian applications, code division multiple access (CDMA) appears to be the most popular. This is especially true in what is arguably the hottest topic in communications today, mobile communications. In such situations, multipath is often a fundamental limitation to system performance, and spread spectrum is a well-known technique to combat multipath.

In addition to multiple accessing capability and multipath rejection, one of the major opportunities that arise when using spread spectrum communications is the possibility of overlaying low level direct sequence (DS) waveforms on top of existing narrow-band users, and hence increasing the overall spectrum capacity even more so than just through the use of the CDMA network. However, such a procedure must be done very carefully so as to not cause intolerable interference for either the existing narrow-band users or the spread spectrum users. One means of helping to ensure this is to use signal processing techniques to suppress those narrow-band users that occupy the spread bandwidth; this will clearly lessen the interference level that the DS spread-spectrum signals have to live with. However, it will also help the narrow-band users because the power levels of the CDMA users can now be correspondingly decreased precisely because they do not have to compete with the preexisting signals.

Finally, since spread sprectrum is obtained by the use of noise-like signals, where each user has a uniquely addressable code, privacy is inherent.

II. REVIEW OF SPREAD SPECTRUM CHARACTERISTICS

The idea behind spread spectrum is to transform a signal with bandwidth B_s into a noise-like signal of much larger bandwidth B_{ss}. This is illustrated in Fig. 1, where the ordinate is power spectral density (watts/hertz) and the abscissa is the frequency axis (hertz). Assume the total power transmitted by the spread-spectrum signal is the same as that in the original signal. In this case, the power spectral density of the spread-spectrum signal is $P_s(B_s/B_{ss})$; the ratio B_{ss}/B_s, called the *processing gain*, is denoted by N, and is typically 10–30 dB. Hence the power of the radiated spread spectrum signal is spread over 10–1000 times the original bandwidth, while its power spectral density is correspondingly reduced by the same amount. It is this feature that gives

Manuscript received May 1, 1990; revised September 17, 1990. This paper was supported in part by the NSF I/UCRC for Ultra-High Speed Integrated Circuits and Systems, and by SCS Telecom, Inc.

R. L. Pickholtz is with the Department of Electrical Engineering and Computer Science, George Washington University, Washington, DC 20052.

L. B. Milstein is with the Department of Electrical and Computer Engineering, University of California at San Diego, La Jolla, CA 92093.

D. L. Schilling is with the Department of Electrical Engineering, City College of New York, New York, NY 10031.

IEEE Log Number 9144469.

Reprinted from *IEEE Transactions on Vehicular Technology*, Vol. 40, No. 2, pp. 313-321, May, 1991.

199

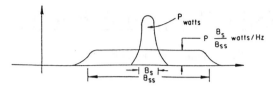

Fig. 1.   Spectra of signal before and after spreading.

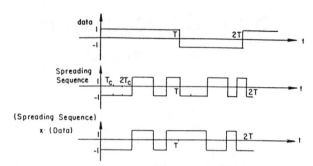

Fig. 2.   Time waveforms involved in generating a direct sequence signal.

the spread spectrum signal the characteristic of causing little interference to a narrow-band user and is the basis of proposed *overlay* systems that operate spread-spectrum concurrently with existing narrow-band systems. The other requirement of the spread-spectrum signal is that it be "noise-like." That is, each spread-spectrum signal should behave as if it were "uncorrelated" with every other spread signal using the same band. In practice, the correlation used need not be zero (i.e., the signals are not completely orthogonal), and thus there can be many more such signals with this property than if they were all required to be truly orthogonal. The consequence of this will be seen in the following.

III. Basic Principles and Features

There are several ways to implement a spread spectrum system. Each requires:

- signal spreading by means of a code;
- synchronization between pairs of users;
- care to insure that some of the signals do not overwhelm the others (near-far problem);
- source and channel coding to optimize performance and total throughput.

The two most popular signal spreading schemes are:

- direct sequence spreading;
- frequency hopping.

Spread spectrum is a "second modulation technique." For example, in direct sequence, one starts with a standard digital modulation, such as binary phase shift keying (BPSK) and then applies the spreading signal. The despreading at the front end of the receiver then delivers the BPSK signal to the standard processor for such signals. In frequency hopping, the modulated signal is first generated. Then the spread spectrum technique consists of changing the center frequency of the transmitted signal every T_H seconds, so that the *hop rate* is $f_H = (1/T_H)$ hops/s. This could be done *slowly* (one hop per many symbols) or *fast* (many hops per symbol). The total frequency spread B_{ss} is equal to the total number N of distinct frequencies used for hopping times the frequency bin occupied for each such frequency. For slow hoppers, this is $N/T = B_{ss}$, or $N = B_{ss}/B_s$, where $B_s \sim 1/T$ and T is the duration of a data symbol. We observe that N is the processing gain defined above.

To make the FH signal "noise-like," the frequency hopping pattern is driven by a "pseudo-random" number generator having the property of delivering a uniform distribution for each frequency that is "independent" on each hop. So long as the intended receiver knows this pseudo-random

sequence, and can arrange to synchronize with the transmitter, the frequency hopped (FH) pattern may be dehopped (or despread) to receive the original signal, which is then processed normally. FH systems are very popular in military as well as commercial communications, and they enjoy certain advantages over DS (such as not requiring a contiguous band).

Returning to direct sequence, we see from Fig. 2 that the spreading is accomplished by multiplying the modulated information-bearing signal by a (usually) binary $\{\pm 1\}$ baseband code sequence waveform, $PN_i(t)$. The code sequence waveform may be thought of as being (pseudo) randomly generated so that each binary chip can change (with probability = 1/2) every T_c s. Thus the signal for the ith transmitter is

$$S_i(t) = PN_i(t) \, A d(t)(\cos \omega_0 t + \phi), \qquad (1)$$

where $d(t)$ is the data modulation (assumed to be ± 1 for BPSK signaling), A is the amplitude of the BPSK waveform, and ϕ is a random phase.

Since T_c is less than T (and frequently *much* less than T), the ratio of the spread bandwidth, B_{ss}, to the unspread bandwidth, B_s, is given by $B_{ss}/B_s = T/T_c = N$, the processing gain. It is clear that a receiver with access to $PN_i(t)$, and synchronized to the spread spectrum transmitter, can receive the data signal, $d(t)$, by a simple correlation. That is, in the interval $[0, T]$, if the data symbol is d_1, which can take on values ± 1, then

$$\frac{2}{T} \int_0^T PN_i(t)(\cos \omega_0 t + \phi) S_i(t) \, dt = A d_1 = \pm A. \quad (2)$$

If the spreading sequence is properly designed, it will have many of the randomness properties of a fair coin toss experiment where "1" = heads and "−1" = tails. These properties include the following:

1) in a long sequence, about 1/2 the chips will be +1 and 1/2 will be −1;
2) a run of length r chips of the same sign will occur about $2^{-r}l$ times in a sequence of l chips;
3) the autocorrelation of the sequence $PN_i(t)$ and $PN_i(t + \tau)$ will be very small except in the vicinity of $\tau = 0$;
4) the cross correlation of any two sequences $PN_i(t)$ and $PN_j(t + \tau)$ will be small.

An important class of sequences called *maximal length linear feedback shift register sequences* are well known to exhibit properties 1), 2), and 3). In particular, the autocorrelation function

$$R_i(\tau) = \frac{1}{T_p} \int_0^{T_p} PN_i(t) PN_i(t + \tau) \, dt \qquad (3)$$

is given by

$$R_i(\tau) = \begin{cases} 1 - \dfrac{\tau}{T_p}\left(1 + \dfrac{T_c}{T_p}\right), & 0 \le \tau \le T_c \\[2mm] -\dfrac{T_c}{T_p}, & T_c \le \tau \le (N-1)T_c \\[2mm] \tau - \dfrac{T_p - T_c}{T_c}\left(1 + \dfrac{T_p}{T_c}\right) - \dfrac{T_p}{T_c}, & \\[2mm] & (N-1)T_c \le \tau \le NT_c, \end{cases} \qquad (4)$$

where T_p is the period of the sequence and $R_i(\tau)$ is also periodic with period T_p.

Each of these properties can be quite significant in a mobile communications system. For example, if $T = T_p$ and $N = T/T_c = 255$, then (1) tells us that a signal due to multipath, arriving τ s after the first signal, is attenuated by $R_i(\tau)$. In particular, if $T_c \le \tau \le T - T_c$, then the power of the multipath signal is reduced by $(T_c/T)^2 = (1/255)^2$ (or about 48 dB).

Note, however, that (3) refers to a full correlation (i.e., a correlation over the complete period of the spreading sequence). Since data are usually present on the signal, and since data transitions typically occur 50% of the time, (3) should really be replaced by

$$\pm \frac{1}{T_p} \int_0^{\tau} PN_i(t) PN_i(t + \tau) \, dt$$

$$\pm \frac{1}{T_p} \int_{\tau}^{T_p} PN_i(t) PN_i(t + \tau) \, dt, \qquad (5)$$

where the independent $\pm$ signs on the two terms of (5) correspond to the fact that they are due to different data symbols. When either both signs are plus or both signs are minus, (3) applies and, for the example presented above, the attenuation would indeed be 48 dB. However, when the two signs differ, (5) applies, indicating that we now have the sum of two partial correlations, rather than one total correlation. In particular, if $\tau = 63T_c$ (i.e., if, in our example, the multipath is delayed by about one quarter of the symbol duration), then, for one specific maximal length shift register sequence, the attenuation of the multipath can be shown to be reduced from 48 to 16 dB.

Finally, we focus on two issues which are crucial to deploying a successful spread spectrum system. The first is the requirement for acquisition of synchronization. The synchronization between code generators is essential either at initialization or if a dropout has occurred long enough for the clock to slip. No communication is possible until acquisition of synchronization has been achieved. This is usually done by means of a search for the chip epoch that results in a large

correlation. Furthermore, a loop must keep the code generators in synchronization after acquisition has taken place. This is usually done with a *delay locked loop*, which takes advantage of the correlation property 3) above.

The second issue is *power control*. This is required because a close-in undesired transmitter can swamp a remote, desired transmission. While the powers are additive, the close-in transmitter has a $(d_d/d_u)^r$ advantage in power, where d_u is the distance to the undesired transmitter, d_d is the distance to the desired transmitter, and r is the propagation exponent. For example, for free space, if $r = 2$ and if $d_d/d_u = 20$, then the undesired signal has an advantage of $(20)^2$, or 26 dB. This intolerable condition may be mitigated by using power control to reduce the transmission power of a close-in user. For example, in a mobile cellular environment, one would attempt to separately control the forward power from the cell to each mobile and the reverse power from each mobile to the cell site.

IV. CODE DIVISION MULTIPLE ACCESS

Certainly the fundamental reason for the interest in spread spectrum communications for cellular radio is the fact that CDMA can allow many users to access the channel simultaneously, just as TDMA and FDMA can. The distinction, however, between CDMA and either TDMA or FDMA is that CDMA provides, in addition to the basic multiple accessing capability, the other attributes described above (such as privacy, multipath tolerance, etc.). These latter attributes are either not available with the use of the narrow-band waveforms which are employed with TDMA or FDMA, or are much more difficult to achieve. For example, one can typically implement a narrowband digital communication link that is tolerant of multipath interference by including in the receiver an adaptive equalizer, but this increases the complexity of the receiver, and may affect the ability to perform a smooth handover. Indeed, since the equalizer must continually adapt to an ever changing channel, it is a high-risk component of a TDMA system. Furthermore, CDMA degrades gracefully. No more than N users can simultaneously access a TDMA or FDMA system. However, if more than N users simultaneously access a CDMA system, the noise level, and hence the error rate, increases in proportion to the percent overload.

There are, of course, a number of disadvantages associated with CDMA, the two most obvious of which are the problem of "self-jamming," and the related problem of the "near–far" effect. The self-jamming arises from the fact that in an asynchronous CDMA network, the spreading sequences of the different users are not orthogonal, and hence in the despreading of a given user's waveform, nonzero contributions to that user's test statistic arise from the transmissions of the other users in the network. This is as distinct from either TDMA or FDMA, wherein for reasonable time or frequency guardbands, respectively, orthogonality of the received signals can be (approximately) preserved.

Given that such orthogonality cannot be preserved in CDMA, the obvious point of interest is to determine how much degradation in system performance results. From a

IEEE TRANSACTIONS ON VEHICULAR TECHNOLOGY, VOL. 40, NO. 2, MAY 1991

quantitative point of view, there are many analyses and bounding techniques available to answer this question [2]–[6]. Qualitatively, we can easily see two major areas of concern for the specific application of digital cellular radio. The first is the propagation law; because these channels typically have multiple reflections associated with them, the net propagation law is not that which would be observed over a free-space channel. Measurements have indicated that the received power falls off roughly as the inverse of the distance between the transmitter and the receiver raised to a power that is somewhere between two and four, and because of the potentially large number of users causing the multiple access interference, there can be a noticeable difference in performance depending on which power law is used for performance computations.

Associated with this is the near-far problem, that is, signals closer to the receiver of interest are received with smaller attenuation than are signals located further away. This means that power control techniques must be used in the cell of interest. However, this still does not guarantee that interference from neighboring cells might not arrive with power levels higher than can be tolerated, especially if the waveforms in different cells are undergoing independent fading.

One final concern is the following: in CDMA, a smooth handover from one cell to the next requires acquisition by the mobile of the new cell before it relinquishes the old cell.

It is seen then that while the use of spread spectrum techniques offers some unique opportunities to system designers of digital cellular radios, there are issues to be concerned about as well, and some of these issues are addressed below. In particular, in what follows, we try to discuss some aspects of spread spectrum which are at least somewhat particular to portable communication networks.

A. Voice Activity Effects in CDMA and Capacity Estimates

CDMA systems tend to be self-interference limited. That is, in attempting to have many users communicate simultaneously, the mutual interference sets a limit on the number of simultaneously active users. To the extent that not every user in the network is always transmitting, the capacity of the system is increased. Note that this argument can also be applied to FDMA and TDMA systems, but there is a fundamental difference. In these latter systems, some type of central control is needed to implement a demand assignment protocol; in CDMA, no such control is necessary. Alternately, if such control is indeed used, one still achieves a performance advantage in CDMA because of this effect that is not achieved with either FDMA or TDMA. This is because with CDMA, if a given user stops talking, and no new user wants to access the channel at that particular instant of time, then all the remaining users on the channel experience less interference.

In light of the above, it is of interest to determine the effect of voice activity on the performance of a CDMA network ([7]). Toward that end, consider starting with the following approximate expression for the bit error rate (BER) of a CDMA system, which was originally derived in [8]:

$$P_e = \phi\left(-\left[\frac{\eta_0}{2E} + \frac{m}{3N}\right]^{-1/2}\right), \quad (6)$$

where E/η_0 is the ratio of energy-per-bit-to-noise power spectral density, $\phi(\cdot)$ is the cumulative normal function, N is the number of chips-per-bit, and m is the number of additional active users (i.e., the total number of active users is $m + 1$). Equation (6) is based upon the so-called "Gaussian approximation," i.e., the despread multiple access interference is approximated as a Gaussian random variable. This approximation has been studied in a variety of references (see, e.g., [9]–[11]), and has been found to hold reasonably well when long spreading sequences are used (i.e., when the period of the spreading sequence spans many data symbols) and when $N \gg 1$, $m \gg 1$, and P_e is not too small. This latter assumption is reasonable when forward error correction is used, since (6) corresponds to the channel BER, not the decoded BER. In particular, the Gaussian approximation is shown in [10] to hold, for fixed N, as K becomes large when thermal noise can be neglected, and in what follows, we will indeed be interested in the situation where the performance is limited by the multiple access interference, not by the thermal noise.

Note that m is a random variable, since it depends upon how many users are "on," and how many of the "on" users are talking. Therefore, (6) is more accurately designated as $P(e \mid m)$. To find the unconditional BER, which we will denote by $\bar{P}_e$, we observe the following: If $L + 1$ is the total number of users, if p is the probability that one of them is "on," and if α is the probability that one of the "on" users is talking, then

$$\bar{P}_e = \sum_{k=0}^{L} \sum_{m=0}^{k} \phi\left(-\left[\frac{\eta_0}{2E} + \frac{m}{3N}\right]^{-1/2}\right)\binom{L}{k}p^k(1-p)^{L-k}$$
$$\cdot \begin{bmatrix} k \\ m \end{bmatrix}\alpha^m(1-\alpha)^{k-m}. \quad (7)$$

In order to avoid having to evaluate (7) (since L might be a large number), we can consider the following approximation: Define

$$M \triangleq \alpha L p, \quad (8)$$

i.e., M is the average number of "on" users who are talking. Then we approximate P_e by $\tilde{P}_e$, where

$$\tilde{P}_e \triangleq \phi\left(-\left[\frac{\eta_0}{2E} + \frac{M}{3N}\right]^{-1/2}\right). \quad (9)$$

Note that while we have arrived at (9) as an *approximation* to (7), in reality, (9) is, at times, a lower bound to (4). To see this, note that the second derivative of the $\phi(\cdot)$ function of (7) with respect to m is given by

$$\frac{d^2\phi\left(\left[\frac{\eta_0}{2E} + \frac{m}{3N}\right]^{-1/2}\right)}{dm^2} = \frac{\exp\left[-\frac{1}{2}\left[\frac{\eta_0}{2E} + \frac{m}{3N}\right]^{-1}\right]}{\sqrt{2}\,\pi(6N)^2\left[\frac{\eta_0}{2E} + \frac{m}{3N}\right]^{5/2}}$$

$$\cdot \left[\left[\frac{\eta_0}{2E} + \frac{m}{3N} \right]^{-1} - 3 \right]$$

$$\triangleq \beta \left[\left[\frac{\eta_0}{2E} + \frac{m}{3N} \right]^{-1} - 3 \right] \quad (10)$$

where β is a function of many parameters, but is always positive. Therefore, (10) will be greater than zero for all m if

$$\frac{L}{N} < 1 - \frac{3}{2} \left(\frac{\eta_0}{E} \right). \quad (11)$$

In turn, this implies that

$$\phi \left(- \left[\frac{\eta_0}{2E} + \frac{m}{3N} \right]^{-1/2} \right) \quad (12)$$

is convex U in m, and hence that (9) lower bounds (7) [12].

Another way to look at this problem is from an operational point of view. A user wanting to access a cell in order to place a call must first ask permission to do so. If the cell has a receiver available to accept the call, the mobile user and receiver, employing the same code, communicate. If no receiver is "free," a busy signal is returned to the user, who must keep trying until a receiver becomes available.

To obtain an initial (although certainly approximate) estimate of the number of receivers, say $\hat{m}$, allotted to a cell, it is necessary to make some modifications to (6), since (6) applies to an AWGN channel, whereas mobile channels experience fading. To account for the fading, we model the channel as a two-ray multipath channel with each ray undergoing independent Rayleigh fading and having equal average signal-to-noise ratio. We further assume that these two paths can be resolved and coherently summed to yield twofold maximal-ratio combining diversity. Then, assuming for simplicity a scenario where the interference does not fade, we can use standard techniques to show that the average probability of error is given by (see [13]) $P_e \approx 3/(16\Gamma^2)$, where Γ is the average signal-to-noise ratio on each ray and is assumed to be $\gg 1$. Note that this corresponds to somewhat of a worst-case situation, since we are assuming that the desired signal experiences fading but that the multiple access interference does not. Letting $\Gamma = 1/(\eta_0/E + 2\hat{m}/3N)$, for a given P_e, we can solve for $\hat{m}$. For example, suppose $P_e = 10^{-2}$. (Note that this is the channel error rate; through the use of forward error correction coding, the error rate can then be reduced to a satisfactory level.) Then Γ should be about 6 dB, and, assuming $E/\eta_0 \gg 1$, the number of users, $\hat{m}$, is approximately $\hat{m} \simeq 0.38N$, or about one-third of the processing gain.

To refine this estimate, as well as account for other sources of degradation, such as interference from adjacent cells, consider the following: In a digital communication system operating in Gaussian noise, the usual performance measure is E/η_0, the ratio of the energy/bit to the noise power spectral density. We may calculate this number by examining the output of a matched filter receiver (matched to one of the PN coded waveforms) whose input consists of the desired signal, thermal white noise with one side power density η_0

W/Hz, and interference consisting of m *other* PN coded waveforms with an activity factor $\alpha < 1$ and possibly other interference, such as users in adjacent cells, that increases the measured interference by $(1 + K)$. Applying the matched filter integration over one symbol time, $T = NT_c = E_s$ for a ± 1 signal, we get the *effective* energy/symbol to noise spectral density

$$\left(\frac{E_s}{\eta_0} \right)_{\text{eff}} = \frac{\frac{1}{2}(NT_c)^2}{NT_c \frac{\eta_0}{2} + \frac{NT_c^2}{3} \hat{m}(1 + K)\alpha}$$

$$= \frac{1}{\frac{\eta_0}{E_s} + \frac{2}{3N} \hat{m}(1 + K)\alpha}, \quad (13)$$

where $\hat{m} \triangleq pL$.

Let $rE_b = E_s$ (we now use E_b as the energy per bit to distinguish it from E_s), where $r < 1$ is the dimensionless code-rate of an error correcting code (if used). Dividing both sides by r gives

$$\left(\frac{E_b}{\eta_0} \right)_{\text{eff}} = \frac{1}{\frac{\eta_0}{E_b} + \frac{2r}{3N} \hat{m}(1 + K)\alpha}. \quad (14)$$

In order to guarantee a given performance, this quantity must be at least equal to the *required* value $(E_b/\eta_0)_{\text{req}}$. If we define the required $(E_b/\eta_0)_{\text{req}} \triangleq \lambda_{\text{req}}$, and $E_b/\eta_0 \triangleq \lambda_0$ (this is $(E_b/\eta_0)_{\text{eff}}$ without interference), and solve for $\hat{m}$, we obtain

$$\hat{m} \leq \left\lfloor \frac{3}{2\alpha(1 + K)} \frac{N}{r\lambda_{\text{req}}} \left(1 - \frac{\lambda_{\text{req}}}{\lambda_0} \right) \right\rfloor. \quad (15)$$

A typical calculation might be based on the following parameters:

Signal-to-thermal noise ratio	$\lambda_0 = 30$ dB
Signal-to-noise ratio required before decoding	$\lambda_{\text{req}} = 6$ dB
FEC code rate	$r = \frac{1}{2}$
Uncoded bit rate	$R_b = 4800$ b/ps
Transmitted bit rate after encoding	$R_s = 9600$ b/ps
Spread bandwidth	$B_{ss} = 12.5$ MHz[1]
Adjacent channel spill over factor	$K = 1/2$
Speaker activity factor	$\alpha = 1/2$
Processing gain	$N = \dfrac{B_{ss}}{2R_s} = 640 \; (\simeq 28$ dB$)$

For these numbers, we get from (15) $\hat{m} \leq 640$ active users

[1] 12.5 MHz is the bandwidth available for cellular transmission and for reception.

(with no frequency reuse). It should be pointed out, however, that this number may be off by a significant factor in any application, since the result is very sensitive both to the assumptions used in the modeling and to the values used for the illustrative parameters. Further, notice that although the number of *active* users is 640, there may be many more potential users who will (statistically) share the channel without the need to reassign codes or spectral space. For example, if the probability of a mobile being turned on is $p = 0.01$, there may be as many as 64 000 such potential users at any single cell.

V. OVERLAY CONSIDERATIONS AND NARROW-BAND INTERFERENCE REJECTION

As indicated above, one of the most attractive features of using spread spectrum in a mobile scenario is the ability to overlay the spread spectrum waveforms on top of existing narrow-band users. In a sense, this allows a different type of frequency reuse, or frequency sharing. Frequency sharing is a consequence of the fact that a DS waveform, spread over a sufficiently wide bandwidth, has a very low power spectral density, and hence the additional degradation its presence causes to a narrow-band user located somewhere in its bandwidth, over and above that caused by the thermal noise, can be made quite small.

This assumes, of course, that the power of the DS waveform is not too large. Since, in a "shared" environment, each DS user must be able to withstand the interference caused by the presence of the narrow-band users, it is of interest to make that interference as small as possible. That is, the larger the level of the narrow-band-to-spread spectrum interference, the larger must be the transmitted power of the spread spectrum signal, and hence the larger will be the spread spectrum-to-narrow-band interference.

As one means of potentially alleviating this seemingly vicious cycle, one can consider employing any one of a number of narrow-band interference suppression techniques [14]. These are techniques that are based upon signal processing schemes, and which attempt to make use of the key difference between a DS waveform and a narrow-band interfering signal, namely the great disparity in bandwidth. In the remainder of this section, we briefly describe two narrow-band interference rejection schemes and show some typical performance results for both the narrow-band-to-spread spectrum interference and the spread spectrum-to-narrow-band interference.

Let us assume that we are employing frequency sharing. Then each spread spectrum receiver receives wide-band spread spectrum waveforms in the presence of wide-band thermal noise and narrow-band interference (e.g., narrow-band microwave signals). Assume that, prior to despreading, the receiver attempts to accumulate, say, μ samples of the received waveform and, on the basis of those μ samples, predict the value of the $(\mu + 1)$st sample. The only component of the received signal that can be accurately predicted is the interference component. This is because both the DS signal and the thermal noise are wide-band processes, and hence their past histories are not of much use in predicting

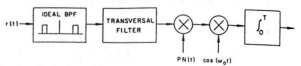

Fig. 3.   Block diagram of direct sequence receiver which employs an interference suppression filter.

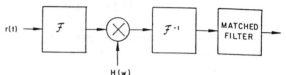

Fig. 4.   Block diagram of transform domain processing receiver.

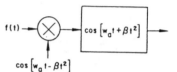

Fig. 5.   Implementation used to generate a real-time Fourier transform.

future values. On the other hand, the interference, being a narrow-band process, *can* have its future values predicted from past values.

Therefore, if a prediction of the $(\mu + 1)$st value of the received waveform is made, and if this estimate is then subtracted from the actual received value of the $(\mu + 1)$st sample, the interference component will be significantly attenuated, whereas the DS signal and the thermal noise will have their values only slightly altered. A block diagram of a receiver employing this type of suppression filter is shown in Fig. 3 [15]. Similarly, if a two-sided transversal filter is used instead of a single-sided filter (sometimes referred to as a prediction-error filter), both past and future values of the received waveform can be used to predict the current value of the process. In reality, since the interfering narrow-band waveforms will not be at fixed locations (in frequency) as the mobile moves, some means of implementing an adaptive interference rejection filter is needed. Such a filter can be designed using any one of a number of adaptive algorithms, and analyses of system performance when the well-known LMS algorithm is used are presented in [16] and [17].

An alternate approach to narrow-band interference rejection is to use the technique of transform domain processing. This latter scheme is illustrated in Fig. 4. It consists of a Fourier transformer, a multiplier, an inverse Fourier transformer, and a matched filter. In essence, the filtering by the transfer function $H(\omega)$ is performed by multiplication followed by an inverse Fourier transformation, rather than by convolution. This multiplication, while ostensibly being performed in the "frequency domain," is, of course, accomplished by the Fourier transform device in real-time.

Clearly, the essence of the scheme is the ability to perform a real-time Fourier transform, and this operation, for example, can be accomplished using the block diagram of Fig. 5. The signal to be transformed, say $f(t)$, is multiplied by the waveform $\cos(\omega_a t - \beta t^2)P_T(t)$ (i.e., it is multiplied by a linear FM or chirp waveform), and the product is used as the input to a time-invariant linear filter whose impulse response

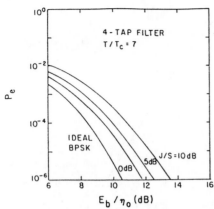

Fig. 6. Bit error rate results of direct sequence receiver when narrow-band interference occupying 10% of spread bandwidth is present.

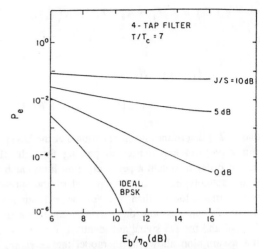

Fig. 7. Bit error rate results of direct sequence receiver when narrow-band interference occupying 50% of spread bandwidth is present.

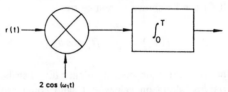

Fig. 8. Block diagram of BPSK receiver.

is $\cos(\omega_a t + \beta t^2) P_{T_1}(t)$, where $P_a(x)$ is 1 for $0 \le x \le a$ and zero elsewhere (see, e.g., [18], [19] for details on this type of signal processing). That this type of processing does, indeed, result in the Fourier transform of a time-limited input signal $f(t)$ can be seen as follows. Denoting the output of the system shown in Fig. 5 by $f_0(t)$, we have, for $t \in [T, T_1]$,

$$f_0(t) = \int_0^T f(\tau) \cos\left(\omega_a \tau - \beta \tau^2\right)$$
$$\cdot \cos\left[w_a(t-\tau) + \beta(t-\tau)^2\right] d\tau$$
$$= \frac{1}{2} \cos\left(w_a t + \beta t^2\right) \int_0^T f(\tau) \cos 2\beta t\tau \, d\tau$$
$$+ \frac{1}{2} \sin\left(w_a t + \beta t^2\right) \int_0^T f(\tau) \sin 2\beta t\tau \, d\tau$$
$$+ \frac{1}{2} \int_0^T f(\tau) \cos\left[2 w_a \tau - 2\beta \tau^2\right]$$
$$+ 2\beta t\tau - w_a t - \beta t^2\right] d\tau. \qquad (16)$$

Ignoring the third term of (16) (it is a double frequency term), we have

$$f_0(t) = \frac{1}{2} F_R(2\beta t) \cos\left(w_a t + \beta t^2\right)$$
$$+ \frac{1}{2} F_I(2\beta t) \sin\left(w_a t + \beta t^2\right), \qquad (17)$$

where $F_R(w)$ and $F_I(w)$ are the real and imaginary part of the Fourier transform of $f(t)$, respectively.

To obtain a perspective on how these systems perform, consider Figs. 6 and 7, both taken from [20], and corresponding to the receiver of Fig. 3. These figures show curves of average probability of error versus the ratio of energy-per-bit-to-noise spectral density. The curves are parameterized by the ratio of interference power-to-signal power (denoted J/S), and the results shown in Fig. 6 correspond to the interference (modeled as a narrow-band Gaussian noise process) occupying 10% of the spread spectrum bandwidth, while those of Fig. 7 correspond to an interference bandwidth equal to 50% of the spread bandwidth. From these figures, it

is evident that the rejection scheme works well if the narrow-band user does not occupy more than 10% of the bandwidth of the CDMA user, although the performance of the technique degrades very rapidly as the interference bandwidth increases beyond about 10%. However, we note that the results shown are uncoded error rates. Using forward error correction, the error rate can be significantly reduced.

Up to this point, we have concentrated on the interference that the narrow-band users induce onto the spread spectrum users. It is at least as important to quantify the opposite effect, the interference on the narrow-band waveforms caused by the spread spectrum overlay. Toward this end, consider the following simplified system shown in Fig. 8, which is used to model a BPSK receiver. The received waveform $r(t)$ is given by the sum of a conventional BPSK signal, a DS waveform, and AWGN. That is,

$$r(t) = A_1 d_1(t) \cos \omega_1 t + A_2 d_2(t-\tau) PN(t-\tau)$$
$$\cdot \cos\left(\omega_2 t + \theta\right) + n_w(t), \qquad (18)$$

where $d_1(t)$ and $d_2(t)$ are independent random binary sequences. The bit rate of $d_1(t)$ is $1/T$, and that of $d_2(t)$ is $1/T_2$. Each of the bit streams is assumed to be bipolar, taking values ± 1, $PN(t)$ is the spreading sequence for the DS waveform and consists of rectangular pulses of duration T_c s, A_1 and A_2 are constant amplitudes, τ is a random time delay, θ is a random phase, and $n_w(t)$ is AWGN of two-sided spectral density $\eta_0/2$.

If we assume the receiver of Fig. 8 is perfectly synchronized to the BPSK waveform, then the test statistic at the output can be shown to be given by

$$g_1(T_1) = A_1 T_1 + A_2 I(T_1) + N(T_1), \qquad (19)$$

where

$$I(T_1) \triangleq \int_0^{T_1} d_2(t - \tau) PN(t - \tau)$$
$$\cdot \cos\left[(\omega_2 - \omega_1)t + \theta\right] dt. \qquad (20)$$

Note that $I(T_1)$ determines the attenuation that the DS spread spectrum waveform experiences in passing through BPSK receiver. In order to obtain a perspective on how much that attenuation actually is, assume the period of the spreading sequence is much longer than T_1, assume τ is an integer multiple of T_c, and assume the carrier frequencies of the BPSK signal and the DS signal are identical (i.e., $\omega_1 = \omega_2$). The first assumption allows us to model the spreading sequence as a random binary sequence, and the second assumption in conjunction with the first allows us to ignore $d_2(t)$ (i.e., we can combine the effects of $d_2(t)$ and $PN(t)$). Hence $I(T)$ can be shown to reduce to

$$I(T_1) = \sum_{i=1}^{N_1} c_i T_c \cos\theta, \qquad (21)$$

where the $\{c_i\}$ are independent, identically distributed random variables taking on values ± 1 with equal probability, and where N_1 is given by

$$N_1 \triangleq \frac{T_1}{T_c}. \qquad (22)$$

Note that N_1 corresponds to the ratio of the duration of a data symbol of the narrow-band user to the duration of a chip of the spread spectrum user. The phase θ is a random variable uniformly distributed in $[-\pi, \pi]$.

If we assume that $N_1 \gg 1$, then $\sum_{i=1}^{N_1} c_i$ can be approximated as a zero-mean Gaussian random variable with variance equal to N_1. In turn, $I(T_1)$, conditioned upon θ, can be approximated as a conditional Gaussian random variable, having zero-mean and variance $N_1 T_c^2 \cos^2\theta$. Therefore, the conditional probability of error of the system can be approximated by

$$P(e\,|\,\theta) \simeq \phi\left(\frac{-A_1 T_1}{\left[\eta_0 T_1 + T_c^2 A_2^2 N_1 \cos^2\theta\right]^{1/2}}\right)$$

$$= \phi\left[-\left[\frac{\eta_0}{2E_1} + \left(\frac{A_2}{A_1}\right)^2 N_1 \left(\frac{T_c}{T_1}\right)^2 \cos^2\theta\right]^{-1/2}\right]$$

$$= \phi\left(-\left[\frac{\eta_0}{2E_1} + \frac{S_2}{N_1 S_1}\cos^2\theta\right]^{-1/2}\right), \qquad (23)$$

and the average probability of error can be approximated by

$$P_e = \frac{1}{2\pi}\int_0^{2\pi} P(e\,|\,\theta)\, d\theta. \qquad (24)$$

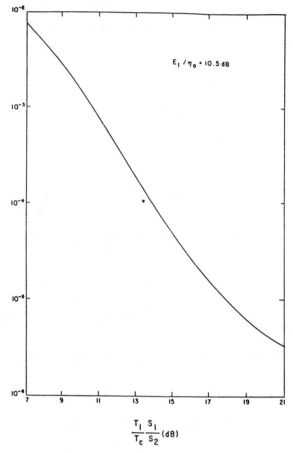

Fig. 9. Bit error rate performance of BPSK receiver when operating in presence of direct sequence waveform.

In (23), $E_1 \triangleq A_1^2 T_1 / 2$ is the energy-per-bit of the BPSK signal, $S_1 \triangleq A_1^2 / 2$ is the average power of the BPSK signal, and $S_2 \triangleq A_2^2 / 2$ is the average power of the PN signal. Interestingly, if these results are compared to those of [21], it will be seen that this problem is the same as that of detecting a DS waveform in the presence of sinusoidal interference, except in our problem it is the DS signal that represents the interference. However, the key point is that the mathematical expressions for average probability of error are the same for both problems.

To obtain some perspective as to how much interference the BPSK signal can withstand, consider Fig. 9, which shows the results of evaluating (24) for $E_1 / \eta_0 = 10.5$ dB (i.e., for an average probability of error of the BPSK receiver 10^{-6} in the absence of interference). It can be seen that, if an error rate of 10^{-5} is desired in the presence of the spread spectrum overlay, then the value of $N_1 S_1 / S_2$ must be about 18 dB. For example, assume the narrow-band user occupies 1% of the spread bandwidth (i.e., assume $N_1 = 100$). Then the ratio of power in the BPSK waveform to power in the DS waveform can be as small as -2 dB, (i.e., the level of the BPSK signal can actually be less than that of the DS waveform and the performance of the BPSK receiver will still not degrade to a BER greater than 10^{-5}).

VI. CONCLUSION

As more people demand access to communication media, system planners will be required to inaugurate more efficient techniques of allocating spectrum than are currently used today. One such technique is spread spectrum. The natural attributes of spread spectrum waveform design, such as inherent security, multipath rejection, and multiple accessing capability, make spread spectrum very competitive for a variety of commercial applications, in particular, those involving mobile units.

In this paper, we have tried to emphasize that, in addition to the above properties, spread spectrum CDMA can also be used to share the spectrum with existing narrow-band users. Thus, rather than reallocating spectrum and disrupting existing users, spread spectrum "layers" can be allocated. As pointed out in the paper, there is a limit to how many layers can be allocated and that this limit can be extended by using interference rejection techniques. (Remember that these interference rejection techniques reduce the interference seen by each of the "layered" users, thereby allowing them to decrease their powers. This, in turn, further reduces their interference on the existing narrow-band users.)

Over the past five years, the use of spread spectrum has increased significantly, and it appears that this trend will increase even more dramatically during the coming decade.

REFERENCES

[1] R. L. Pickholtz, D. L. Schilling, and L. B. Milstein, "Theory of spread-spectrum communications—A tutorial," *IEEE Trans. Commun.*, vol. COM-30, pp. 855–884, May 1982.

[2] D. E. Borth and M. B. Pursley, "Analysis of direct-sequence spread spectrum multiple-access communications over Rician fading channels," *IEEE Trans. Commun.*, vol. COM-27, pp. 1566–1577, Oct. 1979.

[3] M. B. Pursley, D. V. Sarwate, and W. Stark, "Error probability for direct sequence spread spectrum multiple-access communications—Part I: Upper and lower bounds," *IEEE Trans. Commun.*, vol. COM-30, pp. 975–984, May 1982.

[4] K. J. Wu and D. L. Neuhoff, "Average error probability for direct sequence spread spectrum multiple access communication systems," in *Proc. 18th Annu. Allerton Conf. Commun., Cont., and Comput.*, Oct. 1980, pp. 359–380.

[5] K. Yao, "Error probability of asynchronous spread spectrum multiple access communication systems," *IEEE Trans. Commun.*, vol. COM-25, pp. 803–809, Aug. 1977.

[6] R.-H. Dou and L. B. Milstein, "Error probability bounds and approximations for DS spread spectrum communication systems with multiple tone or multiple access interference," *IEEE Trans. Commun.*, vol. COM-32, pp. 493–502, May 1984.

[7] K. S. Gilhousen, I. M. Jacobs, R. Padovani, and L. A. Weaver, Jr., "Increased capacity using CDMA for mobile satellite communication," *IEEE J. Select. Areas Commun.*, vol. 8, pp. 503–514, May 1990.

[8] M. B. Pursley, "Performance evaluation for phase-coded spread spectrum multiple-access communication, Part I, System analysis," *IEEE Trans. Commun.*, vol. COM-25, pp. 795–799, Aug. 1977.

[9] J. S. Lehnert and M. B. Pursley, "Error probabilities for binary direct-sequence spread-spectrum communications with random signature sequences," *IEEE Trans. Commun.* vol. COM-35, pp. 87–98, Jan. 1987.

[10] R. K. Morrow, Jr. and J. S. Lehnert, "Bit-to-bit dependence in slotted DS/SSMA packet systems with random signature sequences," *IEEE Trans. Commun.*, vol. 37, pp. 1052–1061, Oct. 1989.

[11] M. B. Pursley, "The role of spread spectrum in packet radio networks," *Proc. IEEE*, vol. 75, pp. 115–13, Jan. 1987.

[12] R. G. Gallager, *Information Theory and Reliable Communication*. New York: Wiley, 1968, ch. 4.

[13] M. Schwartz, W. R. Bennett, and S. Stein, *Communication Systems & Techniques*. New York: McGraw-Hill, 1966, ch. 10.

[14] L. B. Milstein, "Interference rejection techniques in spread spectrum communications," *Proc. IEEE* vol. 76, pp. 657–671, June 1988.

[15] R. A. Iltis and L. B. Milstein, "Performance analysis of narrowband interference rejection techniques in DS spread spectrum systems," *IEEE Trans. Commun.*, vol. COM-32, pp. 1169–1177, Nov. 1984.

[16] ——, "An approximate statistical analysis of the Widrow LMS algorithm with application to narrowband interference rejection," *IEEE Trans. Commun.*, vol. COM-33, pp. 121–130, Feb. 1985.

[17] N. J. Bershad, "Error probabilities for DS spread-spectrum systems using an ALE for narrowband interference rejection," *IEEE Trans. Commun.*, vol. 36, pp. 588–595, May 1988.

[18] L. B. Milstein and P. K. Das, "An analysis of a real-time transform domain filtering digital communication system, Part I: Narrowband interference rejection," *IEEE Trans. Commun.*, vol. COM-28, pp. 816–824, June 1980.

[19] ——, "An analysis of a real-time transform domain filtering digital communication system—Part II: Wideband interference rejection," *IEEE Trans. Commun.*, vol. COM-31, pp. 21–27, Jan. 1983.

[20] E. Masry and L. B. Milstein, "Performance of DS spread-spectrum receivers employing interference-suppression filter under a worst-case jamming condition," *IEEE Trans. Commun.*, vol. 34, pp. 13–21, Jan. 1988.

[21] L. B. Milstein, S. Davidovici, and D. L. Schilling, "The effect of tone interfering signals on a direct sequence spread spectrum communication system," *IEEE Trans. Commun.*, vol. COM-30, pp. 436–336, Mar. 1982.

Raymond L. Pickholtz (S'54–A'55–M'60–SM'77–F'82) received the Ph.D. degree in electrical engineering from the Polytechnic Institute of Brooklyn (now New York) in 1966.

He is a Professor and former Chairman of the Department of Electrical Engineering and Computer Science at the George Washington University, Washington, DC. He is also President of Telecommunications Associates, a research and consulting firm specializing in communication system disciplines. He was a researcher at RCA Laboratories and at ITT Laboratories. He has been on the faculty of the Polytechnic Institute of Brooklyn and of Brooklyn College, and has been a Visiting Professor at the Université du Quebec and the University of California. He is the Editor of the Telecommunication Series for Computer Science Press. He was an Editor of the IEEE the TRANSACTIONS ON COMMUNICATIONS and Guest Editor for special issues on computer communication military communications, and spread spectrum. He has published scores of papers and holds six U.S. patents.

Dr. Pickholtz is a fellow of the American Association for Advancement of Science (AAAS). In 1990, he was elected President of the IEEE Communications Society.

Laurence B. Milstein (S'66–M'68–SM'77–F'85) received the B.E.E. degree from the City College of New York, New York, NY, in 1964, and the M.S. and Ph.D. degrees in electrical engineering from the Polytechnic Institute of Brooklyn, Brooklyn, NY, in 1966 and 1968, respectively.

From 1968 to 1974 he was employed by the Space and Communications Group of Hughes Aircraft Company, and from 1974 to 1976 he was a member of the Department of Electrical and Systems Engineering, Rensselaer Polytechnic Institute, Troy, NY. Since 1976 he has been with the Department of Electrical and Computer Engineering, University of California at San Diego, La Jolla, where he is a Professor and former Department Chairman, working in the area of digital communication theory with special emphasis on spread-spectrum communication systems. He has also been a consultant to both government and industry in the areas of radar and communications.

Dr. Milstein was an Associate Editor for Communication Theory for the IEEE TRANSACTIONS ON COMMUNICATIONS, and an Associate Technical Editor for the *IEEE Communications Magazine*. He is the Vice President for Technical Affairs of the IEEE Communications Society, is a member of the Board of Governors of the IEEE Information Theory Society, and is a member of Eta Kappa Nu and Tau Beta Pi.

Donald L. Schilling (S'56–M'58–SM'69–F'75) is the Herbert G. Keyser Distinguished Professor of Electrical Engineering at the City College of the City University of New York, New York, NY, where he has been a Professor since 1969. Prior to that, he was a Professor at the Polytechnic Institute of New York, Brooklyn, NY. He is also President of SCS Telecom, Inc. In this capacity, he directs programs dealing with research and development, and training, in the military and commercial aspects of telecommunications. He co-authored eight international bestselling texts, *Electronic Circuits: Discrete and Integrated* (1969); *Introduction to Systems, Circuits and Devices* (1973); *Principles of Communications Systems* (1971); *Digital Integrated Electronics* (1977); *Electronic Circuits: Discrete and Integrated* (2nd edition—1979); *Principles of Communications Systems* (2nd edition 1986); *Electronic Circuits: Discrete and Integrated*, (3rd edition); and *Dynamic Project Management: A Practical Guide for Managers and Engineers* (1989). He has published more than 140 papers in the telecommunications field. He is a well-known expert in the field of military communications systems and has made many notable contributions in spread spectrum communications systems, FM and phase locked systems and has directed research efforts in the performance of HF and Meteor Burst Communications. His algorithm for an adaptive DM is used on the space shuttle.

Dr. Schilling was President of the IEEE Communications Society from 1979–1981 and a member of the Board of Directors of the IEEE from 1981–1983. He was Editor of the IEEE TRANSACTIONS ON COMMUNICATIONS from 1968–1978. He was nominated and accepted as a member of the U.S. Army Science Board in November 1987. He is a member of Sigma Xi and has been an international representative for the IEEE in the Soviet Union where he was part of a Popov Society exchange, and in China where he led an IEEE delegation.

Bit Error Simulation for $\pi/4$ DQPSK Mobile Radio Communications using Two-Ray and Measurement-Based Impulse Response Models

Victor Fung, Theodore S. Rappaport, *Senior Member, IEEE,* and Berthold Thoma

Abstract— A combination hardware and software simulation technique that allows real-time bit-by-bit error simulation for mobile radio systems is described in this paper. The technique simulates mobile radio communication links and generates average bit error rate (BER) and bit-by-bit error patterns. The hardware simulates bit errors between a data source and sink in real time using the error patterns. Various communication system parameters (e.g., modulation scheme, data rate, signal-to-noise ratio, and receiver speed) and different channel environments (i.e., outdoor and indoor multipath fading channels) may be specified and permit performance comparison. Additive white Gaussian noise and cochannel interference effects are also simulated by the software. Using the simulation tool, we studied average BER results for $\pi/4$ DQPSK with Nyquist pulse shaping in indoor and outdoor, flat, and frequency-selective fading channels. BER results for high data rate (> 450 kb/s) transmission in channels generated by a measurement-based indoor channel model, SIRCIM [1], are compared with results in channels generated by the classic two-ray Rayleigh fading model. Simulation results show that when the ratio of rms delay spread to symbol duration is greater than about 0.04, the irreducible BER is not only a function of rms delay spread but is also a function of the temporal and spatial distribution of multipath components. In addition, an example of bit-by-bit error simulation of the transmission of a video image in a mobile radio fading channel is shown. This simulation methodology, which has been implemented in a program called BERSIM, allows subjective evaluation of link quality between a source and sink in laboratory in real time without requiring any radio frequency hardware.

I. INTRODUCTION

THE accurate prediction of average and instantaneous BER in multipath channels will become increasingly important in system design as demand grows for digital wireless communication systems. BER predictions allow designers to determine acceptable modulation methods, coding techniques, and receiver implementations in the operating environments. However, since there are numerous system and channel parameters (e.g., signal-to-noise ratio, data rate, modulation type, cochannel interference, impulsive noise, mobile speed) that can affect BER in mobile communications, it is extremely difficult to evaluate the performance of mobile communication

systems using analytical techniques alone. Also, due to the complexity and time-varying nature of mobile radio channels, it is often too complicated to design and optimize the parameters by analysis. Thus, computer simulation becomes a viable tool in performance evaluation and tradeoff analysis in the design of mobile communication systems. This paper describes a simulator methodology that can predict average BER and generate bit-by-bit error patterns for real-time bit error simulation. The bit error rate simulator (BERSIM) can simulate various popular modulation schemes (i.e., BPSK, FSK, and $\pi/4$ DQPSK), can accommodate data rates up to 15 Mb/s (for $\pi/4$ DQPSK), E_b/N_0 from 0 to 100 dB, receiver speeds up to 150 km/h, square root raised cosine pulse shaping, and cochannel interference in indoor and outdoor, flat, and frequency-selective fading channels. A two-ray Rayleigh fading channel model described by Clarke [2] and implemented by Smith [3] in software is used as one of the indoor and outdoor channel models in the simulator, where the delay between the two rays can be specified. Also, a measurement-based statistical indoor channel model, SIRCIM (Simulation of Indoor Radio Channel Impulse Response Models), developed by Rappaport and Seidel [1] is used to provide channel impulse responses for indoor channels. This paper presents BER results for both indoor and outdoor channels and illustrates how the type of channel model can dramatically impact BER. Different methods of timing recovery in dispersive fading channels impact the BER of a transmission are not considered in this study, but are implemented in BERSIM and studied in [20]. Simulations in this paper use the centroid of the spatially averaged power delay profile to determine symbol timing.

Flat and frequency-selective fading [4] are two distortion phenomena due to multipath that can cause bit errors. This paper focuses on frequency-selective fading for indoor and outdoor channels. BER results for $\pi/4$ DQPSK, BPSK, and FSK in indoor flat fading channels can be found in [5]. For high data rate indoor systems (e.g., where the ratio of rms delay spread to symbol period is greater than 10^{-2} [16]), frequency-selective fading will introduce intersymbol interference (ISI) in the data stream and cause bit errors. This research evaluates the performance of $\pi/4$ DQPSK in indoor channels at data rates of 450 kbs, the data rate proposed by Bellcore [6], 1.152 Mb/s, which is the data rate proposed for DECT (Digital European Cordless Telecommunications) [7], and a higher data rate of 2 Mb/s for comparison. For outdoor channels, the data rate of

Manuscript received February 12, 1992; revised August 1, 1992.
V. Fung is with Bell Northern Research, Richardson, TX 75081.
T. S. Rappaport and B. Thoma are with the Mobile and Portable Radio Research Group, Bradley Department of Electrical Engineering, Virginia Polytechnic Institute and State University, Blacksburg, VA 24061.
IEEE Log Number 9207827.

Reprinted from *IEEE Journal on Selected Areas in Communications*, Vol. 11, No. 3, Apr., 1993.

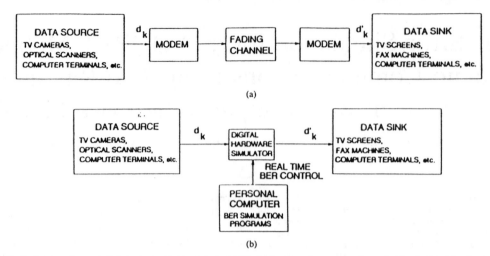

(a)

(b)

Fig. 1. (a) Block diagram of actual digital communication system. (b) Block diagram of a system using a baseband digital hardware BER simulator with software simulation as a driver for real-time BER control.

24,300 symbols/second, specified in the U.S. Digital Cellular Standard IS-54, is used [15]. $\pi/4$ DQPSK has been chosen as the modulation scheme for U.S. Digital Cellular, and it has been shown to perform well in indoor flat fading channels [5]. It is used exclusively in simulations presented in this paper. However, the simulation methods described can be applied to other modulation schemes as well.

For data transmission systems, error bursts due to signal nulls or intersymbol interference are a primary concern, and understanding the temporal distribution of errors is necessary to implement successful antenna diversity or coding techniques. With accurate BER computer simulations in fading environments, it becomes possible to test digital radio communication systems by using a simple and inexpensive baseband digital hardware BER simulator between the data source and sink as shown in Fig. 1. Digital hardware BER simulators, which operate on an applied digital data stream, can be programmed in real time to insert bit errors based on the bit-by-bit error patterns generated by the software simulator. In that case, an entire communication system operating in mobile fading environments with adjustable parameters can be simulated in baseband using the software and hardware simulator. Subjective evaluating between data source and data sink can then be performed conveniently in a laboratory.

In this paper, the communication systems used for the simulation, the two-ray channel model, the SIRCIM channel model, and the methods used to predict average BER and to generate bit-by-bit error patterns for $\pi/4$ DQPSK are described. Earlier work on these topics have already appeared in [11], [1], [5]. BER results for various system parameters in outdoor channels using a two-ray channel model, and BER results for indoor channels using SIRCIM with three different data rates are presented. The indoor channel (SIRCIM) results are then compared with the results of the classic two-ray channel model, with the same rms delay spread, to determine whether irreducible BER is dependent only on rms delay spread or if

the actual distribution of multipaths also impact BER. Finally, an example of real-time digital video transmission in mobile radio channels is used to demonstrate the utility of real-time bit-by-bit error simulation and the burstiness of errors in wireless data communications.

It should be noted that the two-ray Rayleigh fading channel model has been used extensively in the literature [1], [16]–[19]. With the recent advent of channel sounders that can resolve multipath components to resolutions of a few ns, however, it has become clear that individual multipath components do not fade as a Rayleigh distribution but rather log-normal or Ricean [1]. The motivation for the present paper is to explore how different channel models and data rates affect the bit errors in wireless systems.

II. COMMUNICATION SYSTEM MODEL

A block diagram of the $\pi/4$ DQPSK system used in the simulation is shown in Fig. 2. All bandpass signals and channels described in this paper are represented and simulated using the corresponding baseband in-phase and quadrature forms [8], [5]. Referring to Fig. 2, a pseudorandom binary bit stream $d(t)$ is sent through the communication system for BER simulation. The binary bit stream $d(t)$ is stored in computer memory and is eventually compared with the received bits $d'(t)$ at the output of the simulated receiver for determination of bit errors. The number of bits sent can be determined by the bounded binomial sampling method [9], or a brute force method can be used where a large number of bits (typically several millions) are generated and the resulting errors counted.

By placing a square root raised cosine filter in both the transmitter and the receiver, both Nyquist pulse shaping and matched filtering (in flat fading channels) are achieved. Specifically, [5, eq. (15)–(17)] were used to simulate the $\pi/4$ DQPSK modem at baseband. The ideal transfer function of the square

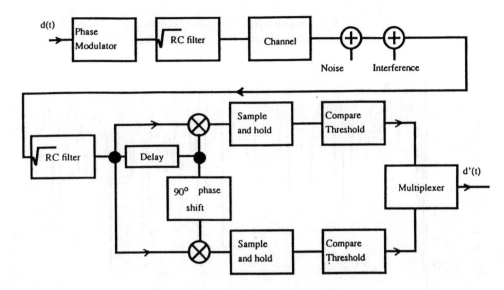

Fig. 2.   Block diagram of the $\pi/4$ DQPSK system.

root raised cosine filter is defined as [10]

$$H_{\sqrt{RC}} = \begin{cases} 1 & 0 \le |f| \le \dfrac{r_b}{2}(1-\alpha) \\ \cos\left(\dfrac{\pi}{2\alpha r_b}\left[|f| - \dfrac{r_b}{2}(1-\alpha)\right]\right) \\ & \dfrac{r_b}{2}(1-\alpha) < |f| < \dfrac{r_b}{2}(1+\alpha) \\ 0 & \dfrac{r_b}{2}(1+\alpha) < |f| \end{cases} \quad (1)$$

where α is the rolloff factor of the filter.

The impulse response of the filter is used for convolution in the time domain, and is obtained by performing an inverse Fourier transform [8]. In BERSIM, the impulse response of the square root raised cosine filter is truncated to 12 symbols in duration and is stored as a collection of amplitudes as a function of time.

The in-phase $I(t)$ and quadrature $Q(t)$ parts of the modulated signal are convolved with the impulse response of the square root raised cosine filter in the simulation [5]. The signals at the output of the filter are convolved with the time-varying complex impulse response of the simulated channel. However, the channel is considered to be stationary for an integer number of symbol durations. The generation of impulse responses for indoor and outdoor channels is described in Section III. As shown in Fig. 2, white Gaussian noise is added to the signal at the output of the channel. In the simulation, the mean signal level is held constant while the added noise power is changed for BER analysis as a function of signal-to-noise ratio. Cochannel interference is simulated by passing a different pseudorandom bit stream through an independent simulated channel and adding it to the desired signal at the input of the receiver. The cochannel interference level is set by the ratio of the desired average signal power (C) to the average interference signal power (I), which is denoted as C/I. The matched filter in the receiver is a square root raised cosine filter identical to that of the transmitter.

Different types of detectors for $\pi/4$ DQPSK modulation are known and have been shown to offer similar performance and various design advantages [11], [18], [19], [21]. In this work, IF differential detection is assumed [5], [11].

After detection, $d'(t)$ is written to computer memory in sequential order and is compared with the original bit stream $d(t)$ to determine bit errors. This technique enables one to determine the exact number and time correlation of bit errors. The errors can be recorded in their exact order of occurrence for later replay with the hardware portion of the simulator.

III. CHANNEL DESCRIPTION

Channel models which have accurate first- and second-order fading statistics (i.e., level crossing rate, fading distributions of multipath components, and instantaneous delay spread) are vital for the accurate prediction of burst errors in real channels and are needed for accurate bit-by-bit error simulation. A two-ray Rayleigh fading model is one of the channel models used by BERSIM to generate both outdoor and indoor channels. Time correlation of the flat fading behavior in the two-ray model is preserved by the Doppler fading spectrum used to generate the two-ray impulse responses [8]. Also, a more accurate measurement-based statistical channel simulation model, SIRCIM [1], is used to provide the channel impulse responses for indoor channels.

A. Two-Ray Channel Model

The discrete impulse response of the flat and frequency-selective fading channels is given by the two-ray model

$$h(kT_s) = \alpha_1 e^{j\theta_1(kT_s)}\delta(kT_s) \\ + \alpha_2 e^{j\theta_2(kT_s)}\delta(kT_s - \tau) \quad (2)$$

where T_s is the fractional symbol sampling period described in [5], α_1 and α_2 are independent and Rayleigh distributed, θ_1

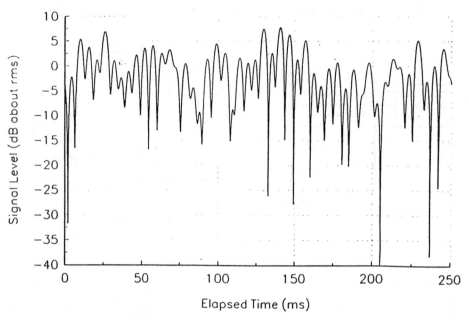

Fig. 3. A typical simulated Rayleigh fading envelope.

and θ_2 are independent and uniformly distributed over $(0, 2\pi]$, and τ is the time delay between the two rays. The power sum of $E\{\alpha_1^2\}$ and $E\{\alpha_2^2\}$ is set to unity in the simulation, so the channel has unity average gain over a simulation run. The ratio of $E\{\alpha_1^2\}$ to $E\{\alpha_2^2\}$ is the power ratio of the main ray (C) to the delayed ray (D) and is denoted as C/D. A flat fading channel is formed by setting α_2 to zero. The rate of change of phases θ_1 and θ_2 over time (or equivalently over the travel distance of the mobile) is dictated by the Doppler fading spectrum.

A software fading simulator similar to [3] has been used to generate the amplitude and phase of each ray in the two-ray model. The power spectrum of the fading envelope is given in [2]

$$s(f) = \begin{cases} \dfrac{E_0^2}{2\pi f_m \sqrt{1 - \left(\dfrac{f}{f_m}\right)^2}} & |f| < f_m \\ 0 & \text{elsewhere} \end{cases} \quad (3)$$

where E_0^2 is related to the received signal power, and $f_m = v/\lambda$ is the Doppler frequency in Hertz corresponding to a receiver speed v (m/s) and a carrier wavelength λ (in meters). The fading spectrum is used to simulate the Rayleigh fading found in mobile radio [2], [4]. Specifically, the spectra of two independent complex Gaussian processes are shaped to the fading spectrum in (3), and then an IFFT is used to provide a complex fading envelope for a mobile travel distance of 24 wavelengths. In a manner similar to [3], 8192 samples spaced by the appropriate frequency sampling interval are used to represent each fading spectrum in the frequency domain.

The value of the frequency sampling interval depends on the doppler shift, and is always chosen to provide sufficient resolution to resolve even short deep fades in the time domain. Since the fading envelope resulting from the IFFT spans a fixed travel distance, smaller sampling intervals are used for slower receiver speeds. In order to provide a real waveform from the inverse Fourier transform of each of the two complex Gaussian spectra, the real part of each complex Gaussian spectrum is made an even function by mirroring values about the carrier frequency. Similarly, the imaginary part is made an odd function about the carrier before performing an inverse Fourier transform. The real outputs of those two IFFT's are used as in-phase and quadrature baseband fading signals in the time domain [8]. The distribution of the envelope of the resulting signals in the time domain is Rayleigh, and the phase is uniformly distributed from 0 to 2π over several hundred samples. The envelope and phase generated are used as a time series to determine the amplitude and phase of the rays in (2), at a particular sample time. Since 8192 samples of interval $1/T$ in the frequency domain correspond to 8192 samples equally spaced over the duration T in the time domain, linear interpolation is used to determine the amplitude and phase of each sample point within each bit sent through the channel. The frequency-selective fading is simulated by adding a delay τ between the two rays, and is implemented in software. A typical fading envelope generated by the simulator with a Doppler frequency of 91 Hz (receiver speed = 120 km/h, λ = 0.35 m) is shown in Fig. 3. The two-ray model simulator has been thoroughly tested [8], [20], and the cumulative distribution and normalized level crossing rate agree with the theoretical results given in [1], [4], [12].

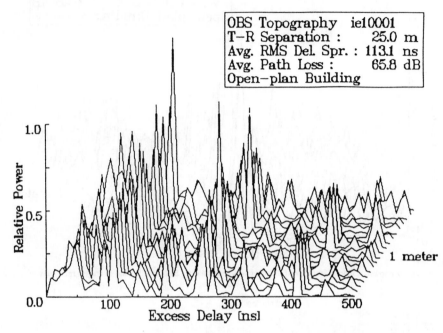

OBS Topography	ie10001
T—R Separation :	25.0 m
Avg. RMS Del. Spr. :	113.1 ns
Avg. Path Loss :	65.8 dB
Open—plan Building	

Fig. 4. Typical wideband impulse response profile for OBS channels generated by SIRCIM. Spatial averged rms delay spread is 113.1 ns.

B. SIRCIM Channel Model

SIRCIM is an indoor UHF multipath radio channel software simulator designed from over 50,000 wideband and narrowband measurements from many different buildings. SIRCIM recreates the statistics of measured wideband impulse responses in both LOS (Line of Sight) and OBS (Obstructed Sight) topographies, and generates 19 baseband complex impulse responses for indoor mobile radio channels when traveling over a small area (a 1 m path). Statistical models for the number of multipath components, the variation of the multipath components, the amplitudes, phases, and fading statistics of individual multipath components, and the excess delays are all based on measurements reported in [1], [13], and more recent data. A detailed description of the channel simulator SIRCIM can be found in [1], [13], [5]. Extensive tests have been made to verify SIRCIM before using it for the BER analysis presented here [1], [8], [20].

Examples of multipath profiles generated by SIRCIM, which were used for simulations in this paper, are shown in Figs. 4 and 5. These are typical spatially-varying OBS channel multipath delay profiles (squared magnitude impulse responses–phases are not shown for brevity) and are realistic representations of field measured multipath profiles reported in [1].

The average BER results presented in this paper are based on the channel characteristics of simulated measurement locations seen by a mobile at each location. At each location, the mobile is assumed to move along a short track at a specified velocity. The angle between the line of motion of the receiver and the line drawn between transmitter and receiver is randomly distributed from 0° to 360° for each simulated measurement track. The transmitter/receiver (T/R) separation

is assumed to be 25 m, and the speed of the moving receiver is assumed to be 1 m/s for all indoor simulations, which roughly corresponds to the walking speed of a portable radio user. SIRCIM assumes that the scatterers are aligned along aisles located to each side of the transmitter, and the width between these aisles in our analysis is arbitrarily chosen to be 7 m for each simulation, although this can be varied easily [13].

SIRCIM produces 19 spatially-varying complex impulse responses spaced by quarter-wavelength increments for any carrier frequency from 900 MHz to 60 GHz. A single multipath component is assumed within a multipath delay bin of 7.8 ns. In this paper, we have used a carrier frequency of 1300 MHz for all simulations, which implies a 1 m measurement track. The simulator BERSIM can read SIRCIM channel files, and also allows the user to load in other complex channel files (either measured or computer generated) using a similar, but more general, data format as described later. Fig. 6 illustrates how time-varying impulse responses are represented as snapshots which vary over the travel distance of the mobile. BERSIM accepts complex impulse response profiles consisting of D equal-width multipath delay bins and I equally spaced profiles of separation Δs_{ch}. SIRCIM files use $D = 64$, a sampling interval $\Delta \tau_{ch}$ of 7.8125 ns (corresponding to 500 ns of excess delay for each complex impulse response snapshot), and Δs_{ch} of a quarter wavelength. However, BERSIM allows any value of $\Delta \tau_{ch}$, any Δs_{ch} less than a half wavelength of the carrier frequency, and allows D to be as large as 128.

Since mobile radio measurements always involve spatial sampling of the complex impulse response, BERSIM uses cubic spline interpolations on the amplitudes and phases of each multipath delay bin over space to generate hundreds of interpolated impulse responses between the primary profiles

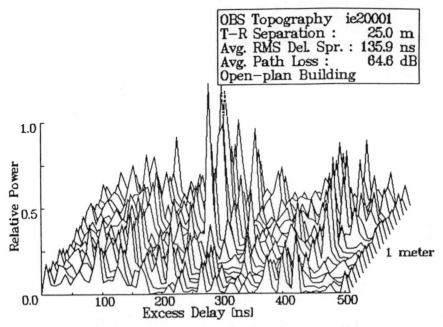

Fig. 5.   Typical wideband impulse response profile for OBS channels generated by SIRCIM. Spatial averaged rms delay spread is 135.9 ns.

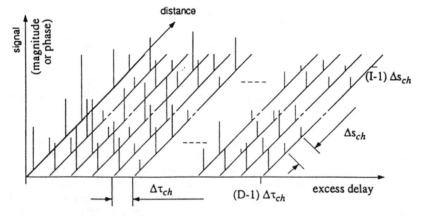

Fig. 6.   Format of a BERSIM channel impulse response file. For SIRCIM files, $\Delta\tau_{ch}$ is 7.8 ns, $D = 64$, and Δs_{ch} is the spatial sampling of a quarter wavelength of the carrier frequency. For simulations in this paper, $I = 19$ for a 1 m travel distance and carrier frequency of 1300 MHz.

as shown in Fig. 7. For SIRCIM channels, this interpolation procedure can produce thousands of impulse responses from the 19 original profiles. The spatial separation Δs_{symb} of each new complex impulse response defines discrete spatial locations at which data are convolved with a static complex impulse response. The value of Δs_{symb} is determined by the velocity of the mobile, and M, the number of impulse responses after interpolation over space, is a function of the number of symbols sent during simulation and the mobile velocity [20]. Note that BERSIM interpolates each original multipath component before convolving the channel with the digital signal. The user must be sure that the spatial samples and time delay resolution of the channel data is adequate to ensure there is no aliasing over space. The high temporal resolution and quarter-wavelength spatial samples of SIRCIM

ensures that each multipath component can be accurately regenerated over space.

In order to use each impulse response component for BER simulations, the time delay bins of each impulse response profile must be decimated into larger time delay bins of time period T_s, where T_s is the sampling period of a symbol. As described in [5], BERSIM uses 13 samples to represent each transmitted data symbol. Assuming that the channel is static over T_s, all multipath components within T_s are vectorially summed to form a new multipath component as shown in Fig. 8. To reduce computation time, no splining is performed over time delay at discrete spatial locations.

In our indoor simulations, the data rate is high (i.e., > 450 kb/s) and the receiver is moving slowly (i.e., 1 m/s) so that we may assume the channel remains unchanged during the

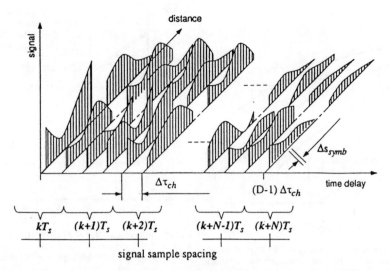

Fig. 7. Format of a BERSIM impulse response file after interpolating the amplitudes and phases of individual multipath components over space using a cubic spline technique. Note the fading behavior of individual multipath components is preserved. As shown in Fig. 8, the sample period of an applied symbol encompasses several consecutive time delay bins. The spatial separation Δs_{symb} is a function of mobile velocity.

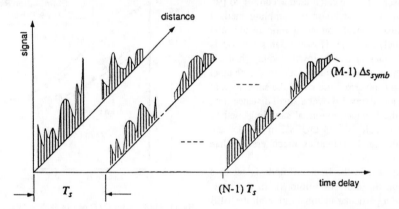

Fig. 8. Format of a BERSIM impulse response file after matching the impulse response time delay samples to the symbol samples. As described in [5], BERSIM convolves an integer number of symbols with each complex impulse response at discrete distances. M denotes the number of specific spatial locations used during a simulation run, and is a function of the velocity and the amount of data transmitted during simulation.

transmission of hundreds of symbols. For a data rate of 450 kb/s, the channel is assumed to be stationary for 38 symbols. For a data rate of 1.152 Mb/s and receiver speed of 1 m/s, the channel is assumed to be stationary during the transmission of 96 symbols. For a data rate of 2 Mb/s, the channel is assumed to be stationary for 200 symbols. These assumptions are reasonable, since the impulse response variation between two successive impulse responses over space is very small for the considered speeds.

IV. BER CALCULATION

Two different methods are used to determine the number of bits sent in a BER simulation and to calculate the average BER. One method is the "bounded binomial sampling" method described in [9], and the other is a brute-force "free run" method. Bounded binomial sampling tells how many bits must be sent in a BER simulation from the parameters provided. It

also provides double boundaries in simulated BER results so that the three BER lies both within a certain relative precision (e.g., 50%) and a certain absolute precision (e.g., 10^{-2}). The relative precision is more appropriate to low BER, while the absolute precision is more appropriate to high BER [8]. In our simulation, the confidence level was 99%, relative precision was 25%, and absolution precision was 10^{-2}. This means that there is 99% confidence that the true bit error rate is both within 25% and 10^{-2} of the BER found by our simulation. To use the bounded binomial sampling method, the maximum number of bits sent in a BER simulation depends on the BER estimated before the actual run. We assumed the BER would be on the same order of magnitude of analytical results given in [14], or would give pessimistic estimates when analytical results were not available. We set the maximum number of bits to be approximately 10 times the number of bits required to produce one bit error using the bounded binomial sampling technique. Our BER simulation for a particular run terminates

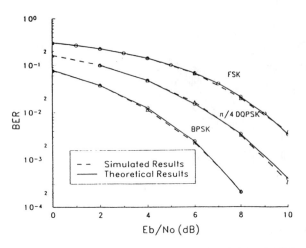

Fig. 9. BER versus E_b/N_o for an AWGN channel for $\pi/4$ DQPSK, BPSK, and FSK compared with theoretical results. The symbols are shaped using a raised cosine filter with rolloff factor $\alpha = 1$.

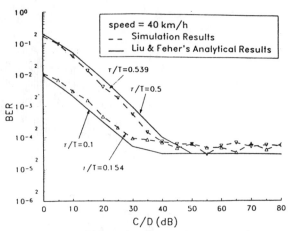

Fig. 10. BER versus C/D of $\pi/4$ DQPSK in a frequency-selective two-ray Rayleigh fading channel for two different signal delays. $E_b/N_o = 100$ dB, $C/I = 100$ dB, $fc = 850$ MHz, $fs = 24.300$ kBd, $\alpha = 0.2$, $v = 40$ km/h.

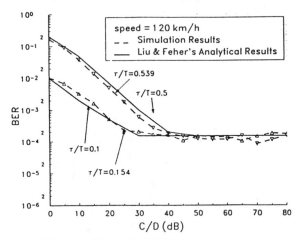

Fig. 11. BER versus C/D of $\pi/4$ DQPSK in a frequency-selective two-ray Rayleigh fading channel for two different signal delays. $E_b/N_o = 100$ dB, $C/I = 100$ dB, $fc = 850$ MHz, $fs = 24.300$ kBd, $\alpha = 0.2$, $v = 120$ km/h.

due to one of the three possible outcomes described in [9], and the median unbiased BER are calculated according to [9]. The advantages of the bounded binomial sampling method are that it sets confidence levels on simulation results and also requires less simulation time. However, this method is not desirable in high data rate and slowly fading channels (indoor channels), where a sudden deep fade in the channel might generate numerous bit errors and cause the simulation to terminate before the mobile traveled over a small distance (i.e., a 1 m track). Thus, the bounded binomial sampling method was used in finding the BER for $\pi/4$ DQPSK in AWGN and two-ray channels with mobile velocities much greater than walking speeds.

The number of bits sent in the free-run mode is set to a very large number (on the order of millions). In this way, the BER is typical for a particular channel although the BER has no confidence bounds. The free-run method requires a longer run time; however, this ensures that the simulation will cover the channels intended and the simulation will not terminate prematurely. The free-run method was used in the BER simulations in SIRCIM channels.

V. SIMULATION RESULTS

A. Simulation Results in AWGN Channels

Simulation results for various modulation techniques, including $\pi/4$ DQPSK in AWGN channels, are compared with theoretical results in Fig. 9 to test the accuracy of the simulator. It can be seen that the simulation results match well with the theoretical results presented in the literature [5], [8], [9], [18], [19], [21]. This indicates that the implementations of different modulation and detection techniques within the simulator are correct.

B. Simulation Results Using Two-Ray Channel Model

Figs. 10 and 11 show BER versus C/D (average power ratio of line-of-sight ray to delayed ray) in two-ray Rayleigh

fading channels for two different receiver speeds and τ/T ratios, where τ denotes the time delay between the two rays and T is the symbol duration. As C/D increases, the channels become less frequency-selective and the random FM (fast fading) places a lower limit on the BER. By comparing the bit error floors in Figs. 10 and 11, the lower BER limit is determined by the receiver speed. The BER limits are about $2 * 10^{-5}$ and $2 * 10^{-4}$ for receiver speeds of 40 and 120 km/h, respectively. For C/D below 40 dB, the BER is dominated by frequency-selective fading (ISI). By comparing the results for C/D below 25 dB in Figs. 10 and 11, the variation of the receiver speed appears to have little effect on BER for the same C/D ratio. We have compared our simulation results with Liu and Feher's [14] analytical results in the figures for the same receiver speeds and slightly different τ/T. As shown in Figs. 10 and 11, our simulation results agree closely with results given in [14], where we have used only Monte-Carlo simulation without any analytical techniques.

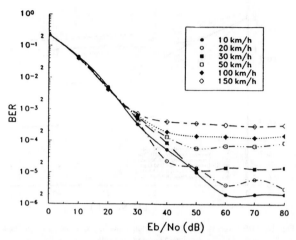

Fig. 12. BER versus E_b/N_o in a flat Rayleigh fading channel for various mobile speeds. $fc = 850$ MHz, $fs = 24.300$ kBd, $\alpha = 0.2$, $C/I = 100$ dB.

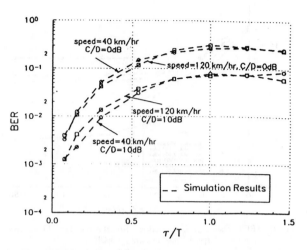

Fig. 13. BER versus t/T in frequency-selective fading channels for various receiver speeds and C/D ratios. $E_b/N_o = 100$ dB.

Fig. 12 shows the BER simulation results for $\pi/4$ DQPSK in flat fading channels. Results are found for various receiver speeds of 10 to 150 km/h. An irreducible bit error floor exists at E_b/N_0 above 60 dB.. The irreducible bit error floor is caused by random FM and is a function of the receiver speed. For low $E_b/N_o (E_b/N_o < 30\,\text{dB})$, bit errors are caused by the additive white Gaussian noise and changing the receiver speed has minimal effect on the BER.

BER performance as a function of τ/T in frequency-selective fading channels is shown in Fig. 13. The influence of receiver speed and C/D are also shown in the figure. The results show that BER increases with the increase of τ/T from 0.077 to 0.77 for receiver speeds of 40 and 120 km/h and both C/D ratios. In our simulation, τ/T ranged from 0.077 to 1.46, and Fig. 13 shows that BER is not a strong function of the receiver speed for τ/T in that range. There is a BER ceiling of about 0.25 for τ/T greater than 1, which is determined by the C/D ratio. The BER ceiling is due to saturation of the bit errors caused by frequency-selective fading.

The effect of cochannel interference in flat fading channels is shown in Fig. 14. The BER decreases as C/I increases from 10 to 40 dB. Three different receiver speeds (40, 70 and 120 km/h) were used in our C/I simulations. Simulation results show that the variation of receiver speed makes no difference on the BER performance for a given C/I ratio. This agrees with the observation by Malupin and McNair [22]. Experimental results found in [22] from actual field measurements and analytical results for flat fading channels [14] are also shown in Fig. 14 for comparison. Results in [22] probably have a higher irreducible bit error at large C/I due to the frequency-selective fading nature of real-world channels, although it is unclear why [14] portrays better (i.e., lower BER) at E_b/N_o below 30 dB.

C. Simulation Results Using SIRCIM for Indoor Channels

Simulation results for indoor channels are based on the two channel impulse responses shown in Figs. 4 and 5. These two impulse responses are typical impulse responses for OBS

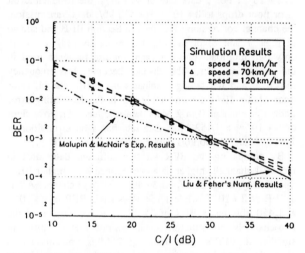

Fig. 14. BER versus C/I in frequency-selective fading channels for various τ/T. Receiver speed = 40 and 120 km/h. $E_b/N_o = 100$ dB, $C/D = 0$ dB.

channels for open-plan buildings. While we realize these two channels may not be the worst case in a particular environment, one of the goals of this work was to develop a tool that can use measurement-based models and field measurements to determine BER within a local area. We call this small-scale BER behavior, due to the small travel distances experienced by the receiver in simulation. BER results using channels in Figs. 4 and 5 should be taken as typical and are used to illustrate the difference in BER for different channels. Indeed, BERSIM may be used to produce real-time error patterns or BER results for any channel that is applied in the manner described in Section III. Since the receiver speed was chosen to be 1 m/s in simulations, one second of the time-varying channel impulse response is shown for two different channels in Figs. 4 and 5. The rms delay spreads for channels A and B are 113.1 and 135.9 ns, respectively. BER performance for $\pi/4$ DQPSK in these two channels is shown in Fig. 15. For a data rate

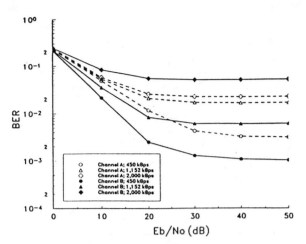

Fig. 15. BER versus E_b/N_o in simulated channel A and channel B, receiver speed 1 m/s, frequency 1.3 GHz.

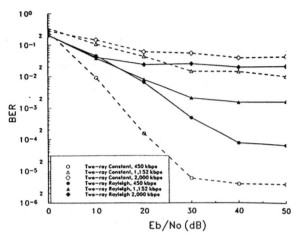

Fig. 16. BER versus E_b/N_o in Rayleigh two-ray channels and constant amplitude two-ray channels.

of 1.152 Mb/s in channel A, the irreducible bit error floor is $2 * 10^{-2}$. For a data rate of 450 kb/s, the irreducible bit error floor drops to $3 * 10^{-3}$. For 450 kb/s data transmission, the channel conversion (described in Section III-B and shown in Figs. 7 and 8) of the impulse responses [20] results in only two multipath bins for each impulse response used in the simulation. Thus, the channel becomes less frequency selective, and flat fading dominates the BER. The bit error floor of $3 * 10^{-2}$ shown in Fig. 15 is the flat fading error floor. For channel B, a 2 Mb/s data rate has a bit error floor of $6 * 10^{-2}$; a 1.152 Mb/s data rate has a bit error floor of $7 * 10^{-3}$; and a 450 kb/s data rate has a floor of 10^{-3}.

By comparing the BER for two different data rates in identical channels, higher data rates indeed have higher BER. Specifically, at an E_b/N_o of 30 dB, 2 Mb/s in channel A has a BER of $3 * 10^{-2}$, while 1.152 Mb/s has a BER of $2 * 10^{-2}$ and 450 kb/s has a BER of $5 * 10^{-3}$. A comparison of the two channels shows that the same data rate produces different small-scale BER's. The simulated 2 Mb/s transmission in channel A has a lower bit error floor than the same simulation for channel B. This is due to the fact that channel A has a lower τ/T than channel B. However, for the lower data rates and flat fading conditions ($\tau/T < 0.08$), deeper fades cause channel A to offer a higher BER floor than channel B.

Fig. 16 shows the BER performance for data rates of 450 kb/s, 1.152 Mb/s, and 2 Mb/s in an independent Rayleigh two-ray fading channel model where each ray is assumed to fade over space for a moving receiver. The purpose of this simulation was to compare the BER performance between a two-ray channel and a SIRCIM channel (channel B), where both channels have virtually identical rms delay spreads. Due to the discretization of time delays in computer simulation and the fact that BERSIM uses 13 samples per symbol and convolves an integer number of symbols with each impulse response, the rms delay spread of the two-ray channel can only be set to 170.9 ns (i.e., $1/(2 * 13(450k/2))$) for 450 kb/s, 133.5 ns (i.e., $2/(2 * 13(1.152M/2))$) for 1.152 Mb/s, and 153.8 ns (i.e., $4/(2 * 13(2M/2))$) for 2 Mb/s. The factor of two results from the fact that for a two-ray model with

equal average power in each ray, the rms delay spread σ_{rms}, as computed in [1], is half of the temporal separation between the two rays. The data rate and a fixed number of samples per symbol determine the discrete values of rms delay spread that can be simulated in a two-ray model. The 170.9, 133.5, and 153.8 ns values of rms delay spread for the three two-ray channels are comparable to the 135.9 ns rms delay spread of channel B.

We also simulated two-ray multipath channels with constant (nonfading) ray amplitudes at a mobile receiver in order to compare fading channels with stationary channels. Thus, the amplitude of each ray was constant but the phase was changing based on the Doppler fading mechanism and free-space propagation. This experiment was conducted in order to determine how important the multipath amplitudes are in causing bit errors in $\pi/4$ DQPSK. The BER for 450 kb/s, 1.152 Mb/s, and 2 Mb/s in a constant amplitude two-ray model are also shown in Fig. 16 for comparison. The phase of each ray in the constant amplitude two-ray model was generated exactly the same way as for the Rayleigh two-ray channel using the method described in Section III.

By comparing Fig. 15 with Fig. 16, it can be seen that the irreducible BER in a Rayleigh two-ray fading channel is very close to the irreducible BER in channel B for Mb/s. At a data rate of 1.152 Mb/s, the Rayleigh two-ray channel has a *lower* BER than the simulated channels. The difference between errors in the Rayleigh two-ray channel and channels A and B is even more pronounced for a data rate of 450 kb/s. Since almost flat fading conditions exist for 450 kb/s, the difference in the BER performance is that the resulting narrowband fading envelope in a real-world indoor channel is not necessarily Rayleigh distributed within a local area. It can also be seen that the BER is sensitive to the shape of the profile if one considers almost equal rms delay spreads. As a result, the Rayleigh two-ray channel should *not* be used to model indoor fading channels, but rather field measurements or models based on such data should be used for modem and system design [1].

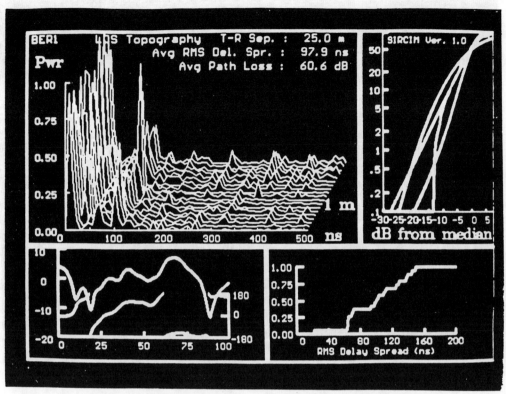

Fig. 17. An example of video image transmitted through the hardware simulator.

In summary, the irreducible BER for $\pi/4$ DQPSK is not simply a function of rms delay spread for $0.2 > \sigma_{rms}/T > 0.02$ but is also a function of the temporal variations of the multipath components and their phases. Extensive work that has studied the impact of timing recovery, jitter, and multipath phase shifts on BER is available in [20].

By examining the results in the Rayleigh fading and constant amplitude two-ray models, it can be seen that the constant amplitude two-ray model has a higher bit error floor than the Rayleigh two-ray model at the data rates of 1.152 and 2 Mb/s, but has a lower BER at the data rate of 450 kb/s. This shows that the constant amplitude channel model yields worse BER results when compared with Rayleigh two-ray models for $\sigma_{rms}/T > 0.08$ (a frequency-selective fading channel) because the instantaneous rms delay spread is always at its maximum (delay between the two rays/2). It also shows that the bit errors in the Rayleigh two-ray model are dominated by the delay spread, and not Rayleigh fading of multipath signals at $\sigma_{rms}/T > 0.08$. Fig. 16 shows that, for σ_{rms}/T less than 0.04, the channel is less frequency selective. Thus, the Rayleigh fading in the Rayleigh two-ray channel generates more bit errors than the constant amplitude two-ray channel. As mentioned in [16], envelope fading is the most important mechanism for small delay spreads. It is interesting to note from Figs. 15 and 16 that both Rayleigh and constant envelope channels give much more optimistic BER's than the real-world-based channel models (channels A and B) produced by SIRCIM. From these results, we conclude that future computer simulations and system design should use channel models based on the real-world behavior of amplitude and phase, such as provided by SIRCIM.

VI. REAL-TIME BIT ERROR SIMULATION

BERSIM provides real-time BER simulation, and can play back the corrupted data at a specified rate in real time through hardware on a personal computer. Similar hardware for simulation on UNIX-based workstations is under development. The user may provide baseband data from actual hardware data sources or may read in stored data files such as video or voice data. The bit-by-bit error patterns used in the real-time simulation are generated using the method described in [5]. Specifically, a binary one is written to an external file if a bit error is encountered, and a binary zero is written if no error occurs. Once the off-line simulation is done, real-time simulation can be performed. Data streams from the data source (either externally provided or read from the computer disk) are sent to the data sink through the hardware simulator. The hardware simulator has a fast DMA controller on board that reads the bit-by-bit error pattern in real time from the source and inserts errors into the data stream according to the error pattern. The BERSIM hardware card clocks the corrupted data stream using TTL voltage levels at the specified data rate of the simulation up to 15 Mb/s. In this manner, digital communications in mobile radio channels are simulated at the baseband and a subjective evaluation of data links can be

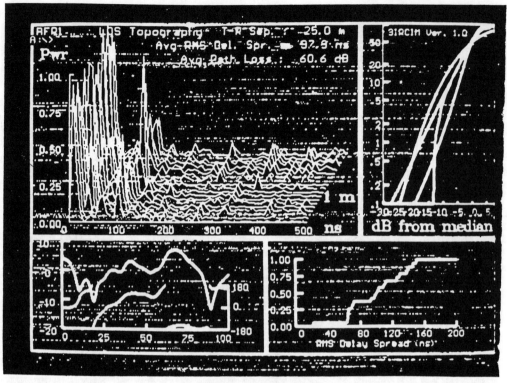

Fig. 18. Corrupted video image received at data sink.

performed easily in the laboratory. As an example, a test video image shown in Fig. 17 was transmitted through the hardware simulator. The simulation was for $\pi/4$ DQPSK operating at 48.6 Kb/s, and the video image was represented by binary pixels. The simulated digital cellular radio channel was a frequency-selective fading channel of 6 μs rms delay spread, E_b/N_0 of 20 dB, and C/I of 30 dB. The corrupted image received at the data sink is shown in Fig. 18. The bursty nature of bit errors in mobile fading channels can be seen clearly in the image. Subjective quality evaluation of wireless data links using speech and image coding in a wide range of channels, modulations, or systems can be performed easily with BERSIM.

VII. Summary

An accurate software/hardware bit-by-bit error simulator for mobile radio communications has been described. Simulation results in indoor and outdoor channels are shown and compared with theoretical results. BER results in simulated frequency-selective fading channels generated by several channel models such as two-ray, constant amplitude, and simulated indoor radio channel impulse models (SIRCIM) are presented and compared. We found that BER is not only dependent on the rms delay spread, but also on the distribution of temporal and spatial multipath components in local areas. An important result is that a two-ray Rayleigh fading model is a poor fit for indoor wireless channels and, if used, can underestimate the BER by orders of magnitude. A real-time bit error simulation

of video transmission using the bit-by-bit error simulator has also been shown. The simulator, called BERSIM, has been demonstrated to be a useful tool for evaluating emerging data transmission products for digital mobile communications.

Acknowledgment

The authors wish to thank B. Tisdale of the Contel Technology Center for his comments concerning this work, and M. Keitz of Virginia Tech for designing and building the hardware simulator. BERSIM is patent-pending and is available from Virginia Tech.

References

[1] T. S. Rappaport, S. Y. Seidel, and K. Takamizawa, "Statistical channel impulse response models for factory and open plan building radio communication system design," *IEEE Trans. Commun.*, vol. 39, no. 5, pp. 794–807, May 1991.
[2] R. H. Clarke, "A statistical theory of mobile-radio reception," *Bell Syst. Tech. J.*, pp. 957–1000, July/Aug. 1968.
[3] J. I. Smith, "A computer generated multipath fading simulation for mobile radio," *IEEE Trans. Vehic. Technol.*, vol. VT-24., no. 3, pp. 39–40, Aug. 1975.
[4] W. C. Jakes, *Microwave Mobile Communications*. New York: Wiley, 1974.
[5] T. S. Rappaport and V. Fung, "Simulation of bit error performance of FSK. BPSK, and $\pi/4$ DQPSK in flat fading indoor radio channels using a measurement-based channel model," *IEEE Trans. Vechic. Technol.*, vol. VT-40, no. 4, pp. 731–741, Nov. 1991.
[6] D. C., Cox, A radio system proposal for widespread low-power tetherless communications," *IEEE Trans. Commun.*, vol. 39, no. 2, pp. 324–335, Feb. 1991.
[7] D. J. Goodman, "Trends in cellular and cordless communications," *IEEE Commun. Mag.*, vol. 29, no. 6, June 1991.

[8] V. Fung, "Simulation of BER performance of FSK, BPSK, $\pi/4$ DQPSK in flat and frequency-selective fading channels," Masters Thesis, Virginia Polytech. Instit. and State Univ., Aug. 1991.

[9] E. L. Crow and M. J. Miles, "A minimum cost, accurate statistical method to measure bit error rates," *Int. Conf. Comput. Commun. Rec.*, pp. 631–635, 1976.

[10] K. Feher, and Engineers of Hewlett-Packard, *Telecommunications Measurements, Analysis, and Instrumentation.* Englewood Cliffs, NJ: Prentice-Hall, 1987.

[11] C. L. Liu, and K. Feher, "Noncoherent detection of $\pi/4$-DQPSK systems in a CCI–AWGN combined interference environment," in *Proc. IEEE Vehic. Technol. Conf.*, May 1989, pp. 83–94.

[12] J. D. Parsons, and J. G. Gardiner *Mobile Communication Systems.* Blackie, 1989.

[13] S. Y. Siedel, "UHF indoor radio channel models for manufacturing environments," Masters Thesis, Virginia Polytech. Instit. and State Univ., Aug. 1989.

[14] C-L. Liu, and K. Feher, "Performance of non-coherent $\pi/4$-QPSK in a frequency-selective fast Rayleigh fading channel," in *Proc. IEEE Int. Conf. Commun.*, Atlanta, GA, Apr. 1990, pp. 335.7.1–335.7.5.

[15] "Cellular system dual-mode mobile station-base station compatibility specification," Electron. Industry Assoc./Telecommun. Industry Assoc. Interim Stand. IS-54, May 1990.

[16] J. C.-I. Chuang, "The effects of time delay spread on portable radio communications channels with digital modulation," *IEEE J. Select. Areas Commun.*, vol. SAC-5, pp. 879–889, June 1987.

[17] C-L. Liu, and K. Feher, "Bit error rate performance of $\pi/4$-QPSK in a frequency-selective fast Rayleigh fading channel," *IEEE Trans. Vehic. Technol.*, vol. VT-40, no. 3, pp. 558–568, Aug. 1991.

[18] S. H. Goode, H. L. Kazecki, and Y. Shimazaki, "A comparison of limiter-discriminator, delay and coherent detection for $\pi/4$ DQPSK," in *Proc. IEEE Vehic. Technol. Conf.*, Orlando, FL, May 1990, pp. 687–694.

[19] S. Chennakeshu and G. J. Saulnier, "Differential detection of $\pi/4$-shift-DQPSK for digital cellular radio," in *Proc. IEEE Vehic. Technol. Conf.*, St. Louis, May 1991, pp. 186–191.

[20] B. Thoma, "Bit error rate simulation enhancement and outage prediction in mobile communication systems," Masters Thesis, Virginia Polytech. Instit. and State Univ., July 1992.

[21] Y. Akaiwa and Y. Nagata, "Highly efficiency digital mobile communications with a linear modulation method," *IEEE J. Select. Areas Commun.*, vol. SAC-5, pp. 890–895, June 1987.

[22] R. P. Malupin and I. M. McNair, "Bit error rate characteristics in a suburban fading environment," presented at *IEEE Vehic. Technol. Conf.*, Orlando, FL, May 1990.

Theodore S. Rappaport (S'83-M'84-S'85-M'87-SM'91) was born in Brooklyn, NY, on November 26, 1960. He received the B.S.E.E., M.S.E.E., and Ph.D. degrees from Purdue University in 1982, 1984, and 1987, respectively.

In 1988, he joined the Electrical Engineering faculty of Virginia Tech, Blacksburg, where he is an Associate Professor and Director of the Mobile and Portable Radio Research Group. He conducts research in mobile radio communication system design and RF propagation prediction through measurements and modeling. He guides a number of graduate and undergraduate students in mobile radio communications, and has authored or coauthored more than 70 technical papers in the areas of mobile radio communications and propagation, vehicular navigation, ionospheric propagation, and wideband communications. He holds several U.S. patents and is coinventor of SIRCIM, an indoor radio channel simulator that has been adopted by over 75 companies and universities. In 1990, he received the Marconi Young Scientist Award for his contributions in indoor radio communications, and was named a National Science Foundation Presidential Faculty Fellow in 1992. He serves as Senior Editor of the IEEE JOURNAL ON SELECTED AREAS IN COMMUNICATIONS and coedited *Wireless Personal Communications* (Kluwer Academic). He is a Registered Professional Engineer in the State of Virginia and is a Fellow of the Radio Club of America. He is also President of TSR Technologies, a cellular radio and paging test equipment manufacturer.

Berthold Thoma was born in Wenkheim, Germany, on June 29, 1963. He received the Diplom Ingenieur (FH) degree from the Fachhochschule Heilbronn, Germany in 1990.

Before entering the Fachhochschule Heilbronn in 1985, he was on a three and a half year apprenticeship with the A.W.d.H. Neckarzimmern, Germany to qualify as a trained radio engineer. During his studies in Germany, he received a Carl Duisberg scholarship to work for six months as a Research Assistant at Middlesex Polytechnic, London in the Robotics Laboratory. For his thesis at the Fachhochschule Heilbronn, he joined S-TEAM Elektronik Untereiseisheim where he developed a direct-sequence spread spectrum-based LAN transceiver. In 1990, he entered Virginia Tech as a Fulbright Scholar and later joined their Mobile and Portable Radio Research Group. His graduate research focused on simulations of mobile radio communication systems to analyze bit error rate mechanisms in those systems.

Victor Fung (S'91) was born in Hong Kong in 1961. He received the B.S. degree in electrical engineering from Michigan State University, East Lansing, in 1984 and the M.S. degree in electrical engineering from Virginia Polytechnic Institute and State University, Blacksburg, in 1991.

From 1984 to 1986, he worked as a Project Electronics Engineer at Lutron Electronics Limited. In 1986, he joined Commodore in Hong Kong, where he worked on the testing of Amiga personal computers. Between 1987 to 1989, he was a Senior Engineer at MiniScribe (HK) Limited, where he worked on high-capacity hard disk testing. As a master's student, he worked with Dr. Rappaport in the Mobile and Portable Radio Research Group. He is now with BNR, Richardson, TX. His research interests are communication system simulation in mobile radio environments. He is currently developing software tools for simulating bit error rate performance for various modulation techniques in mobile fading channels.

Chapter 5

CHANNEL CODING, EQUALIZATION, AND DIVERSITY

The Application of Error Control to Communications

Elwyn R. Berlekamp
Robert E. Peile
Stephen P. Pope

In any system that handles large amounts of data, uncorrected and undetected errors can degrade performance, response time, and possibly increase the need for intervention by human operators

Reprinted from *IEEE Communications Magazine*, Vol. 25, No. 4, Apr., 1987.

Error Control is an area of increasing importance in communications. This is partly because the issue of data integrity is becoming increasingly important. There is downward pressure on the allowable error rates for communications and mass storage systems as bandwidths and volumes of data increase. Certain data cannot be wrong; for example, no one can be complacent about the effect of an undetected data error on a weapons control system. More generally, in any system which handles large amounts of data, uncorrected and undetected errors can degrade performance, response time, and possibly increase the need for intervention by human operators.

Just as important as the data integrity issue is the increasing realization that error control is a system design technique that can fundamentally change the trade-offs in a communications system design. To take some examples:

1) In satellite communications, high-integrity, low redundancy coding can reduce the required transmit power, reduce the hardware costs of earth stations, and allow closer orbital spacing of geosynchronous satellites.
2) Relative to uncoded modulation, Trellis Coded Modulation allows more data to be transmitted over a limited bandwidth.
3) In situations where data is transmitted on an auxiliary carrier over a preexisting channel, such as data-over-voice modems and subcarrier data transmission within existing broadcast FM and TV signals, coding results in increased channel utilization without altering the existing mode in which the channel is used.

One aspect of coding is unchanged: coding can change data quality from problematic to acceptable. However, the above examples illustrate a new aspect; if a communications system has no problem with data quality, the designers ought to review the design, and consider discarding or downgrading the most expensive and troublesome elements while using use error control techniques to overcome the resulting loss in performance. (A similar strategy also applies to mass memory applications. In these applications, the pay-off is increased storage density which provides both increased capacity and increased throughput.)

From this position, sophisticated coding is increasingly able to offer a unique competitive edge to many diverse areas.

The opening statement that "error control is an area of increasing importance to communications" is also a statement on the history of error control. In fact, the history of error control is very rich and colorful [29]. Coding theory effectively started in 1948 with the appearance of Shannon's classic paper [1]. Early activity was intense and rapidly split between Information Theory, that is what was theoretically possible, and Coding Theory, that is how coding gains could be achieved.

In essence, Shannon's paper proved that a stationary channel could be made arbitrarily reliable given that a fixed fraction of the channel was used for redundancy. Conversely, he showed that, if the fixed fraction was not

0163-6804/87/0004-0044$01.00 © 1987 IEEE

used, reliable performance was not possible. This raised several immediate questions:

1) How could this theoretical performance be translated into practical benefits?
2) What expectations did Shannon's somewhat subtle result arouse?

These questions are discussed below.

Realizing Theoretical Performance

In relation to the first point, it soon became apparent that the practical problem of achieving anything like the performance promised by Shannon was extremely difficult. Moreover, determining the performance limit hinged upon a complete knowledge of the noise statistics on the channel. The philosophy of the early Coding Theorists was roughly as follows: Given that we are informed of, or are allowed to study and classify, the exact noise statistics of your channel, we can prove that your present methods are hopeless and superb performance is possible. However, we do not know how to obtain this performance but we are working on techniques that obtain some small fraction of this performance.

It is not hard to see that the above philosophy is difficult to pursue in a nonresearch environment. In fact, there has been a history of skepticism toward coding from the very start [2]. In terms of the practical application of coding, the Shannon philosophy has fallen out of favor. The real problem is that the noise distribution on a channel is very rarely known and might be impossible to define. For example, the authors recently looked at the noise afflicting data transmitted on a sub-carrier within the bandwidth of an FM station. The noise was primarily related to program content; Pat Benatar was 4.2 times noisier than Count Basie, drum solos excepted. Clearly any absolute definition of noise would involve predictions of trends in popular music. Fortunately, this is not necessary.

The alternative approach (which we credit to Jacobs and Viterbi) was to take a particular code, analyze it under one or more simple types of noise, and present the results in comparison to those obtained with uncoded data communications. Further, in this approach, other codes may be analyzed, shown to be inferior and presented as a "strawmen." The real impact of this approach is that, by offering a tangible product, it throws questions about the appropriateness of a code back to the communications system designer and his knowledge of the channel.

The success of this approach led to the first large scale application of Forward Error Correction to communication, the use of the (2,1) K=7 convolutional code on satellite communications. This application is discussed in the section titled "half-rate coding against predominantly Gaussian noise." One result of this approach is a growing consensus as to which codes are of practical importance, and a growing awareness of the pros and cons of the various classes of codes. To expand on this last point, there are literally hundreds of different codes and decoding algorithms. It is probably possible to take any one of these codes and devise noise conditions and

constraints that will make this code optimal. However, these conditions and constraints will be extremely far-fetched in the majority of cases. There is a much smaller class of codes that are applicable to a large number of practical problems and a still smaller number of codes that can meet other practical constraints, such as high-speed and/or low complexity implementation.

Expectations of Coding

In relation to the second question, the early expectations of coding theoreticians were high. It was widely held that, if certain problems could be solved, a major revolution in communications would ensue. The problems were seen as threefold:

1) To find good codes.
2) To find decoding algorithms for the codes.
3) To find ways of implementing the decoding algorithms.

In fact, these problems were solved but the revolution did not happen. This was largely for practical reasons. Communications has undergone several revolutions since 1948, mostly of a much less mathematical and more direct nature. This had led coding to be declared to be of no practical importance on innumerable occasions. (Incredibly, entire workshops have been held to promulgate this negative conclusion [3]!). In spite of such declarations, practical acceptance of coding is probably higher now than at any previous time. What has changed?

As was pointed out in the introduction, error control can offer significant gains in systems where other components would be expensive to upgrade. If you accept that a new technology offers a period of high gains followed by diminishing returns, many elements of a communications chain have improved enormously in the last twenty years and are now in the diminishing return phase. Coding is now at the stage of being able to offer an attractive alternative to improving performance.

Perhaps this transformation is best illustrated by the evolution of Reed-Solomon (RS) codes. The existence of the codes was published in 1960 [4]. Their status for the next decade was almost totally academic. In 1966 G.D. Forney [5] established that RS codes could be concatenated to supply extremely good performance but, at the time, the existence and implementation of decoding algorithms were daunting problems. Discovery of good decoding algorithms was a gradual task, solved by the efforts of several people in the mid to late 60s, a breakthrough occurring with the publication of Berlekamp's algorithm [6]. In fact, the search for more efficient decoding algorithms has continued ever since; the definition of "efficient" depending upon the technology being used for implementation.

Reed-Solomon codes were still regarded as of academic interest at the start of the 70s. (The late 60s and early 70s mark the lowest point in the history of practical forward error correction; expectations were low. Conversely, the theoretical side of coding was in good shape). By 1982, this was not so. RS codes were incorporated in a Deep-Space standard [7], used in the JTIDS system and commercially available in popular Compact Disc systems. What were the causes of this turnabout?

Partly, it was the explosion in digital circuitry. RS codes are highly suited to digital implementation and, as the capabilities of digital electronics grew, their implementation became possible. It was also due to their performance. In many ways RS codes have complementary properties to convolutional codes. If a system designer has doubts that his channel will exhibit noise conditions suited to convolutional codes, there is strong likelihood that an RS code will function under these conditions. (See the section titled "Half-rate Coding for Predominantly Gawsian Noise.")

Note how this fits in with the changing strategy described in "Realizing Theoretical Performance." The communications designer is being offered a limited number of coding options and, depending upon the channel noise conditions and other constraints, the "best" option will change. It is the purpose of this article to review the properties of various error control schemes in order to clarify the properties of the available techniques.

Error Control Techniques

There are several divisions between types of error control. The first major division is between Automatic Request for Retransmission (ARQ) and Forward Error Correction (FEC). In the communications context, both techniques add redundancy to data prior to transmission in order to reduce the effect of errors that occur during transmission. However, the philosophy is very different.

Forward Error Correction utilizes redundancy so that a decoder can correct the errors at the receiver. There does not need to be a return path. ARQ utilizes redundancy to detect errors and, upon detection, to request a repeat transmission. A return path is necessary.

Adaptive techniques such as fault tolerance could be regarded as another form of error control. For example, in a network it is possible to avoid errors by routing traffic around damaged parts of the network (adaptive routing).

The remainder of this section introduces the fundamental properties of ARQ and FEC techniques. The section titled "Comparison of Techniques" compares techniques and their properties in greater depth. (Reflecting the author's bias, the emphasis is on Forward Error Correction.)

It should be apparent that hybrid strategies involving combinations of the above techniques are possible. These are discussed in the section "ARQ and Hybrid ARQ/FEC Strategies." The section titled "Applications" gives some practical applications of FEC techniques.

Forward Error Correction Techniques

The strategy of Forward Error Correction is to get it right the first time.

The underlying tenet of FEC (in a communications system) is to take user data, add some redundancy, and transmit both the user data and the redundant data. At the receiver, the possibly corrupted signal is processed by the decoder which utilizes the redundancy to extract correct data. An analogy may be made to a person speaking slowly and repetitively over a noisy phone line, adding

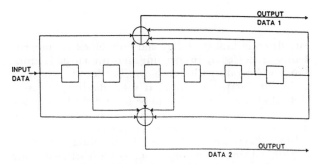

Fig. 1.   A (2,1) K=7 Odenwalter Encoder.

more redundancy for the listener to process into the correct message.

Note also that FEC does not need a return path; in our analogy, the speaker could be dictating into an answering machine.

Forward error correction codes come in many diverse forms with many diverse properties. One immediate division is into *convolutional codes* and *block codes*. These are described in outline in the next two sections.

convolutional codes; the (2,1) K=7 Odenwalter code

Figure 1 shows a flow chart for an encoder for the (2,1) $K=7$ convolutional code generally used in commercial satellite applications. The encoder processes a continuous stream of data. One bit of user data enters the register. By adding certain stages of the register, two bits of data are obtained and transmitted. Note that the value of the output bits depends on seven user data bits; this accounts for the $K=7$ notation. K is referred to as the constraint length. Note also that a particular bit of user data affects the output for seven clock cycles; the encoder is adding time diversity.

Decoding FEC codes is normally more complex than encoding. For this code, decoding is most often done by the Viterbi Algorithm [8], although other alternatives exist [9]. Whichever method is used, it should be clear that the decoder has to know the history of the decoded stream, that is the values held in the shift register, before being able to decode a particular bit. Furthermore, it has to look at the subsequent history of the stream to examine the total influence of that bit on the transmitted stream.

This suggests that with convolutional codes:

1) If the decoder loses or makes a mistake in the history of the stream, errors will propagate. Such events can be caused by a fading or burst interference channel.
2) The decoder has to perform a large number of operations per decoded bit.

These questions are examined further in "Code Comparison for Gaussian Noise Environments."

block codes; Reed-Solomon codes

Block codes do not process data in the same continuous fashion as convolutional codes. The user data is split into discrete blocks of data and each block is independently processed by an encoding algorithm to add redundancy and produce a longer block. The decoder works on a similar basis; each block is individually processed. Note that the decoder has to be informed of the block boundaries; synchronization is normally more of a problem for block codes than convolutional.

Block codes are very diverse. The encoding and decoding algorithms often employ many sophisticated finite algebraic concepts. Although this makes block coding into a specialist area, the use of finite algebra is very suited to implementation in digital electronics, allowing for high-speed operation. (Of course, this does not imply every digital implementation is well-suited; the authors have observed different implementations of the same code that differ by a factor of 8-10 in IC count!)

This article intends to emphasize practical uses of Reed-Solomon codes. There are many reasons for the practical importance of RS codes, some of which are described in "Code Comparison for Gaussian Noise Environments." This section confines itself to describing the parameters and correction power of the code.

RS codes operate on multi-bit symbols rather than individual bits. Consider a RS(64,40) code on 6-bit symbols. The encoder groups the user data into blocks of 240 bits. These blocks are treated as 40 symbols where each symbol has 6 bits.

An encoding algorithm (see [6] for details) expands these 40 symbols to 64 symbols. In a systematic RS encoder this is achieved by appending 24 redundant symbols (non-systematic RS encoders exist which scramble and expand the data; there are many incompatible RS codes with the same parameters).

In general, RS codes exist on b bit symbols for every value of b. RS(n,k) codes on b bit symbols exist for all n and k for which:

$$0<k<n<2^b+2$$

Clearly, there is a lot of choice.
(A popular value for b is 8; in this case, the symbols are bytes. 8 bit RS codes are extremely powerful.)

It remains to consider the correction power of RS codes. The general formula is that, if there are $r=n-k$ redundant symbols, t symbol errors can be corrected in a codeword provided that $2t$ does not exceed r. In the RS(64,40) example, every pattern of up to 12 symbol errors is correctable.

A mnemonic is often used for the above formula. The decoder has r redundant symbols "to spend." The decoder has to "spend" one redundant symbol to locate an error, and another redundant symbol to find the correct value of the symbol in that location. Given that the total expense can not exceed r, the formula results.

This mnemonic is also useful in introducing erasure decoding. Suppose some external agent were to inform the decoder where some of the errors were located. In this case, the decoder would not have to "spend" symbols locating these errors. More technically, if the decoder is instructed that s symbols are unreliable (the symbols are referred to as "erased") and, in addition, there are t errors that the decoder is not informed about, the correct data can be extracted provided that $2t+s\leq r$.

Even on a channel with random errors, it is not imme-

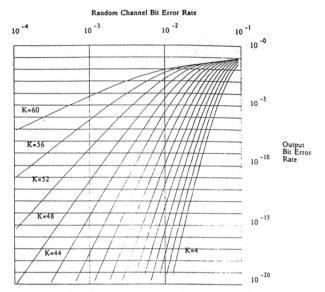

Fig. 2. RS(64,k) Random Digital Error Performance.

diately obvious how these formulas translate into performance. To give some appreciation, Fig. 2 shows the performance of RS(64,k) codes against random digital errors for all values of k divizable by 4. The performance is for decoding of errors only (no erasures).

Automatic Request for retransmission

There are many different types of ARQ. ARQ techniques include, amongst several others, "Selective-Repeat" and "Go-back N" systems. In addition, there is a logical split between positive and negative acknowledgment systems. The differences are conceptually not difficult.

positive and negative acknowledgment

In a negative acknowledgment system, the data source re-transmits data only upon request; no news is good news. In a positive acknowledgement system, the data source expects confirmation from the data sink on the correct receipt of every block. In a positive acknowledgement system, the throughput can be low, particularly if the round-trip delay between the data source and the data sink is long, that is, over a satellite link. The advantage of positive acknowledgement is, of course, superior data security. In practice, the majority of commercial ARQ systems default to negative acknowledgement with positive acknowledgement as an option.

"go back N" and selective-repeat protocols

In a "Go back N" system, the ARQ is normally negative and the data sink requests that all of the last N blocks are repeated whenever a block is received in error (popular values of N are 4 or 6). A "Go back 4" system can be found in the EUROCOM D/1 tactical area system trunk-group protocols.

The advantage of a "Go back N" system is that the blocks do not have to be individually labeled and that the algorithms are accordingly simpler than a comparable selective repeat scheme. In fact, the lack of numbering leads to less overhead and shorter blocks which, in turn, means that the performance in random noise can be slightly superior. However, there are disadvantages with "Go back N" systems. In any "Go back N" system, in order to make sure the erroneous block is repeated, the maximum round-trip delay on the link and, hence, the maximum range, is limited. The delay can not exceed the time it takes to transmit N blocks. For satellite links, this restriction is a major problem. Another criticism of "Go Back N" systems is that re-transmissions send more data than necessary.

In a selective repeat system, the ARQ is normally negative and the data-sink asks for specific blocks to be repeated. The advantage of Selective-Repeat Protocols is that only blocks with errors are repeated. In many noise conditions, these ARQ protocols are best in achieving good throughput. In a Selective-Repeat protocol, blocks have to be individually labeled. This indicates that the protocol overhead is either higher than a comparable "Go Back N" system or that the blocks are longer. While long blocks solve many problems, they are more likely to contain errors than a shorter block length. This presents a trade-off between minimizing the loss of throughput due to protocol overhead (long blocks) and minimizing the loss of throughput due to repeats (short blocks). Selective-Repeat protocols are more complex than "Go Back N" protocols but they have much less restriction as to range. However, some limitations still exist. The blocks have to be numbered individually and the length of the counter upper bounds the maximum range. In practice, the range is more likely to be determined by the amount of memory that the data source has available to store transmitted blocks. Note that as the speed of satellite links increases, the time delay of a satellite link is translated into an increased number of blocks for the data source to store.

Comparison of techniques

In this section we examine and cross-compare the various techniques that have been introduced in the previous section. In accordance with the major divisions, the comparison falls into two parts.

1) A comparison of Convolutional codes to block codes. This is discussed in terms of a predominantly Gaussian Noise channel.
2) A discussion of the relative merits of FEC, ARQ, and hybrid schemes.

half-rate coding for predominantly Gaussian noise

The use of half-rate codes to combat Gaussian noise has a long history in satellite technology. The impact of the $(2,1)$ $K=7$ convolutional code described earlier has been profound. In pacticular, the success of this code has been an important element in changing industry's perspective of coding from being a mathematical curiosity to being an important and practical element in many communication systems.

The nature of satellite communications appears to be changing, both in the amount of traffic (increasing) and type of traffic (from digitized voice to intermachine data).

Both of these changes present more challenging requirements on the technology. For example, whole new families of modulation schemes have appeared to make better use of the available bandwidth and transponder power [10],[11]. On the coding level the change has been away from conditions for which the Odenwalter code is suited and towards conditions for which other coding strategies are more apt. The reasons for this change in code suitability are outlined in this section.

Convolutional Codes

This section confines itself to discussing the $(2,1)$ $K=7$ code and decoders using the Viterbi Algorithm. Gaussian noise performance curves for this code are shown in Fig. 3.

The code is extremely good when all of the following conditions exist:

1) The noise is white and Gaussian.
2) The demodulator provides reliable *soft decision* information (probablistic information on the likelihood of a received symbol being a 0 or 1).
3) The output data (user data) only needs a low level of integrity such as an output BER of 10^{-3} to 10^{-7}.
4) The transmission speed is low enough to allow the decoder to perform a relatively large number of operations per bit.

These conditions, and what happens when they are contravened, are now discussed in more detail.

Gaussian noise—The Viterbi Algorithm is very good at processing soft decision data into an estimate of the transmitted data. However, the complexity of the Viterbi Algorithm confines its application to codes with a small constraint length. $K=7$ is appropriate at conventional speeds; $K=11$ is about as large as feasible even on very slow channels, and even $K=5$ may be difficult on chan-

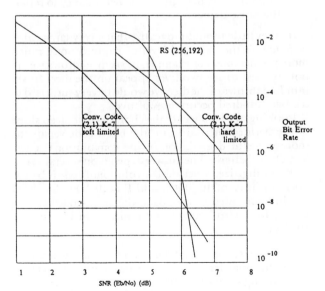

Fig. 3. *RS Code versus Convolutional Code.*

nels whose throughput is hundreds of millions of bits per second.

This means that the decoder processes the data within a fairly small window before forming a decision as to the transmitted bit. If the data is badly corrupted or the soft decision information is unreliable within that window, the decoder is likely to go wrong and output a burst of errors. Such conditions can exist both for interference channels and for fading channels e.g. non-white Gaussian noise. It is worth noting that the world is getting less white and less Gaussian. Adjacent channel interference, accidental or deliberate interference, antenna misalignment all give conditions not intrinsically suited to convolutional codes.

Reliable soft decision information—For white Gaussian noise, a decoder can be supplied with reliable soft decision data relatively easily. The maximum likelihood metric simplifies to the inner product of the received data and the hypothesised transmitted data [12]. For fading channels and/or high-power interference, this simplification is not valid. For fading channels, the received SNR must appear in the metric. This at the very least complicates the decoder, and also raises questions as to whether the demodulator/AGC can supply the received SNR with adequate precision.

Low output data-integrity—The performance of a convolutional code depends upon the constraint length. Because this length is normally $K=7$, the performance is limited. It is possible (if not too helpful) to speak of the constraint length of block codes. Typically, this length is very much larger (for example, $K=2000$). How does this difference appear? The key to appreciating the different performance characteristics of block and convolutional codes lies in splitting the effect of the code into two parts:

1) The code adds diversity which averages the effect of the noise over a number of bits that depend on K.
2) Provided the averaged received noise does not deviate too badly from the expected level of average noise the decoder can use the redundancy to repair the noise damage.

It takes little thought to realize that a very large value of K makes the averaged level of received noise much more predictable. For example, in six dice throws we are not shocked to see two fives appear on the dice. In six hundred throws of the dice, we would be very surprised to see two hundred occurrences of five.

Looking at Fig. 3 we see the characteristic steep slope of the performance curve for the block code versus the more graceful degradation of the convolutional code. Which is "better"? The answer depends on several variables. For digitized voice the convolutional code is better. Such traffic is intelligible in a BER of 10^{-3}. For blocks of machine-oriented control data, the block-coded data is better; BER requirements are typically in the 10^{-10} to 10^{-14} range.

Transmission speed and decoder complexity—Convolutional decoders have a high complexity in terms of decoding operations per output bit. Furthermore, a majority of these operations involve addition and com-

parison of real numbers (or, at least, digitized approximations to real numbers).

By contrast, block codes have a much lower number of decoding operations per output bit. Moreover, Reed-Solomon decoders normally operate directly on bits and real number arithmetic is avoided.

Equally important is the observation that the convolutional decoder complexity increases as the redundancy decreases. For block codes, the complexity decreases as the redundancy decreases. For high code-rates (such as, 14/15 rate) this greatly favors block codes.

In practical terms, the above considerations limit the speed of convolutional decoders. The authors are aware of paper studies into single un-multiplexed convolutional decoders that can work at 120 Mb/s but are unaware of any existing machines that work beyond 20-30 Mb/s.

Conversely, powerful RS block decoders have been built and delivered that operate at rates above 120 Mb/s. RS decoders that operate at channel rates in excess of 2.0 Gb/s are currently being developed [28].

The speed issue is becoming increasingly important as the switch to multiplexed traffic continues. It is becoming less efficient to protect individual slow traffic channels and more economic to protect the grouped channels at the aggregate rate.

Block Codes

There are many different families of block codes. Of major importance for practical applications is the family of Reed-Solomon codes. The reason for their preeminence is that they can combat combinations of both random and burst errors. They also can have a long block-length, assuring a sharp performance curve. The major drawback with RS codes (for satellite use) is that the present generation of decoders do not make full use of bit-based soft decision information. They can handle erasure-information, an indication of when a multi-bit character is unreliable, but not intermediary levels of confidence.

In fact decoding algorithms have been implemented which make use full-use of soft decision information on the byte level [20], but most presently available decoders do not have this advantage.

Interleaving

The section "Code Comparison for Gaussian Noise Environments" indicates that convolutional codes are weak when it comes to burst noise and that Reed-Solomon codes are superior [15]. A common counterargument is that the codes can be interleaved so that the un-interleaved received data is more or less random and, therefore, more or less optimal for the convolutional decoder. After all, Reed-Solomon codes are often interleaved.

This argument is appealing but fallacious. In a very precise information-theoretic sense, the very worst type of noise is random noise. If the noise has structure, the structure can be exploited. In less high-brow terms, the effect of short interference bursts on a telephone line disturbs reception less than a constant level of white noise. The reason for this is not profound. The receiver can locate when a burst occurs and make good use of the location information. In an interleaved code, the de-interleaver carefully tries to convert a less damaging type of noise (bursts) into a more damaging type of noise (random). Given that interleaving is wrong on a philosophical level but practically necessary in many applications, a useful design methodology is to start with a code that has considerable burst-error ability and interleave to a much lesser extent.

In more detail, system delay constraints typically preclude interleaving to a depth so great as to make the channel appear memoryless; there is often a nonnegligible probability that a channel noise burst will have length which is a significant fraction of the system delay constraint. In such cases, the proper comparison is not between idealized coder-interleaver systems with infinite interleaving depth, but rather between specific schemes that meet the delay constraint.

When this delay constraint is tens of thousands or hundreds of thousands of bits, and bursts occasionally approach such lengths, then long RS codes with well-designed interleavers enjoy an enormous performance advantage over interleaved convolutional codes. However, when burst length and delay constraints are both only tens or a few hundreds of bits, then long RS codes are precluded. In this case, interleaved convolutional codes with Viterbi decoding may be the only viable solution.

Concatenated Codes

In many applications concatenated codes offer a way of obtaining the best of two worlds. The idea is to use two codes in series, as shown in Fig. 4. Why, it can be asked, use two codes? If one can not work, why use two? An answer can be made on many levels, but for the satellite case, the following summarizes why this approach works well.

The inner decoder processes the incoming data. This decoder uses all available soft decision data to obtain the best performance in Gaussian conditions. However, the inner decoder is normally limited in performance and is vulnerable to interference and fading. The effect of the inner decoder is to clean-up the majority of Gaussian noise and to indicate stretches of data with which it could not cope. The output of the inner decoder is predominantly either correct data or stretches of burst-errors that are indicated as being unreliable. This output then becomes the input to the RS decoder. From the "Comparison of Techniques" section it should be clear that the RS decoder is now being presented with its ideal input

that is, mostly short burst errors accompanied by helpful erasure information.

The effect of using the two decoders, each combating its preferred type of noise, results in strong performance. The exponential nature of code performance comes into play. The inner decoder reduces poor quality data to medium quality data and the outer decoder reduces medium quality data to very good quality data. The performance of the sum is much greater than the sum of the performance either part could achieve alone.

However, the above description is deliberately vague about the inner code selection. Should this be block or convolutional? The choice depends on the application, the data rates and the channel conditions.

For very high data rates, the choice is for block codes. An inner convolutional decoder would fail to keep up. For predominantly Gaussian slower channels, the inner convolutional decoder is very attractive. Under these conditions, the convolutional inner code tends to outperform most short block codes of interest (ignoring some types of very low rate spread-spectrum systems). The section "The CCSDS Standard" Concatenated Block and Convolutional Code" discusses the performance obtained by concatenating a low redundancy (8 percent) RS code with the (2,1) $K=7$ Odenwalter code. If the noise is predominantly not Gaussian, the choice is for an inner block code. Each inner codeword can protect one character of a Reed-Solomon codeword; the effect of a burst can not propagate through the inner decoder and corrupt more characters than necessary. The section "A Concatenated Reed-Solomon Binary Block Code System" discusses the performance obtained by concatenating a short binary block code with an RS code.

Summary of Comparison for Half-Rate Coding

The continuing growth of data communications over satellite presents new challenges and reappraisals to the communications engineer. In particular, the characteristics of convolutional codes are much less suited for the demands of high-speed data needing high integrity than for their traditional application of low-speed digitized voice traffic. The required performance tends to favor Reed-Solomon codes. RS codes exhibit a very sharp improvement of block error-rate with an improvement of channel quality making their use ideal for data. However, RS codes used in isolation fail to make good use of the soft decision data which is both available and reliable in commercial satellite applications. In order to exploit all the available information and to obtain the RS performance at even worse signal-to-noise ratios, concatenated codes can be used to great effect in this application. If either extremely fast data-rates or very little redundancy is available for coding, concatenated codes are less suitable; RS codes alone may be preferable.

comparison of high-rate codes under Gaussian noise conditions

"Half-rate Coding for Predominantly Gaussian Noise" discussed the role that half-rate coding has played and is likely to continue to play in Gaussian noise environments, such as satellite communications.

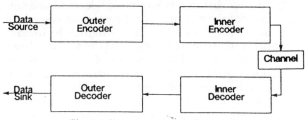

Fig. 4. Concatenated Code Schematic.

However, the pressure for bandwidth and the increased performance of modern codecs raises questions about the need for so much redundancy. This section addresses what can be done with a minimum of redundancy, for example, 5-25 percent.

The question falls into two overlapping portions: modulation and coding. The introduction of trellis-coding into commercially-available modems has finally shown that combined modulation and coding provides gains which far outweigh the traditional operational and maintenance advantages of modularity. Trellis coding uses convolutional coding but "buries" the redundancy into a more complex signal constellation. This constellation presents worse performance in isolation, but this is more than compensated for by the coding gain. The net result is that the gain can be used to send more bits over the same bandwidth, seemingly without redundancy. TCM is discussed in [13]. Research is now extending coding further into the communications chain, for example, the decoding/demodulation process is being extended to interact with or include the channel equalization process.

"Half-rate Coding for Predominantly Gaussian Noise" discussed the drawbacks in using convolutional codes alone. Unfortunately, the same criticisms can be applied to trellis coding used alone. The coding is vulnerable to fades and interference bursts. In very-high speed applications, above 20 Mb/s, it will be difficult to implement the decoders/demodulators. In such applications, high-performance, low-redundancy block codes present a cost-effective option used either with simpler modulations or concatenated with trellis codes.

"Half-rate Coding for Predominantly Gaussian Noise" presented the advantages in concatenating RS codes with conventional convolutional codes. The same reasons apply to concatenating with trellis codes. However, just as trellis codes can be regarded as an unconventional form of convolutional codes, the most efficient form of block code to concatenate is probably not a conventional block code. (In fact, the authors have developed a class of block codes that are very closely matched to the noise statistics produced by a trellis decoder.) The remainder of this section discusses what can be achieved with simple modulation schemes and codes that use only a small percentage of the available bandwidth.

If the redundancy is to combat the noise, the noise has to be a correspondingly small percentage of the codeword. In order to ensure that this occurs most of the time, the received noise has to be averaged over a long period of time, reducing the effect of a freak streak of bad luck; the code must have a large block length or constraint length. This immediately rules out convolutional codes. The Viterbi Algorithm is only economic for small constraint lengths. Sequential decoding is often promoted as a way of increasing the constraint length, but sequential algorithms typically have the undesirable property that the variance in the number of decoding operations per bit is infinite [14]. Thus, in order to achieve any reasonable rate of continuous throughput, a sequential decoder may require an average operating speed much faster than the channel, as well as a very large buffer and a very large decoding delay.

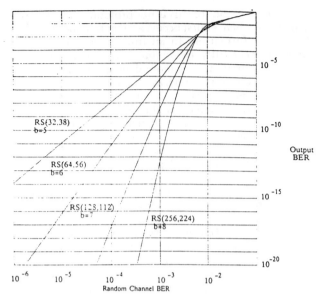

Fig. 5.   RS Codes of 7/8-th Rate

In order to meet the long block length requirement, one attractive alternative is to use RS codes. The effect of block length is illustrated by Fig. 5. Figure 5 shows RS codes working at rate 7/8. The first code is a 5-bit code that is block length 31*5=155 bits. The other codes are 6,7, and 8 bit codes of the same rate. Notice that the 8-bit code's (block length 255*8=2040 bits) performance is much more "knife-edged". For data communication this is to be preferred; BER performance in the 10^{-20} region is obtained at much lower SNRs.

There are other reasons why RS codes are suited to this role:

1) RS codes are symbol or character based codes rather than binary codes. For example, the RS(255,223) code takes 223 bytes of data and encodes them into 255 bytes of data. The decoder and encoder logic works with byte-based arithmetic. The decoder processes 255 items of data, not 255*8 items of binary data. This reduces the complexity of the logic as compared to a "true" binary code of the same length. In addition, decoding algorithms exist for RS codes that are extremely efficient at reducing the number of operations per output bit. This allows the increase in block length to be implementable. In fact, the number of operations per decoded bit grows with the number of redundant symbols; low redundancy RS codes are less complex than high redundancy RS codes.

2) RS codes have a very low misdecode rate. If the noise exceeds the amount that the code can correct, the decoder will almost always recognize the impossibility of decoding the data and can flag the data as being corrupt. The error-rate of data that is both undecodable and NOT recognized as such by the decoder is typically 5 orders of magnitude below the overall BER. This property is very valuable for inter-machine data. Furthermore, this bad data signaling can be used in other parts of the communi-

cations chain, for example, for ARQ purposes, for Automatic Power Control purposes and for channel assessment purposes.

3) RS codes work on a block basis. One of the consequences of working with data is that the data often has a natural frame or packet size. For such blocked data, the statistic of how many bits per block are wrong, the BER, is much less important than the probability of getting a block right. This latter quantity can be computed accurately if the block code is designed to synchronize with the frame or packet boundaries. Exactly the same comments apply to the misdecode rate.

4) The symbol structure of RS codes tends to absorb short bursts. The point is that, as far as the decoder is concerned, having several bits in the symbol wrong is no worse than having one bit wrong. If the noise tends to "bunch," as, for example, in multi-bit baud modulation, the performance can be better than for purely random noise.

5) Finally, RS codes can often be easily accommodate within a multiplexed traffic stream. The codes can be systematic, with the data untouched and check characters appended. This allows for a multiplexed frame where the majority of the channels are dedicated to user data, frame-sync or control channels, and a minority of the channels are dedicated to code check characters. Note that the code delay is divided equally amongst the individual channels. A practical example is discussed in the section titled "Block Codes on a TDMA System."

In summary, high-rate coding against Gaussian noise tends to indicate:

1) Convolutional coding for low-speed, low-integrity applications.
2) Concatenated convolutional and block coding for low-speed, high-integrity applications.
3) Block codes alone for high-speed and high-integrity applications.

Of these three areas, the last two seem to be of increasing interest to satellite communications.

ARQ and Hybrid ARQ/FEC Strategies

ARQ at the link level

Consider first the effect of using ARQ on a link. The throughput of ARQ alone (often called Type 0 ARQ) is very dependent on the channel conditions. Figure 6 shows the throughput of 1000-bit blocks of ARQ-ed data when faced with Gaussian noise. The throughput collapses when the expected number of errors per block becomes sizable.

Periodic interference can hurt ARQ techniques for the same reason. In one example, an airport radar was found to be corrupting data communications on each sweep. This translated to a burst error on a once-per-block basis, giving virtually zero throughput.

One adaptive technique is to shorten the ARQ block length as the noise worsens. This does help the throughput, but there are drawbacks. Firstly, the amount

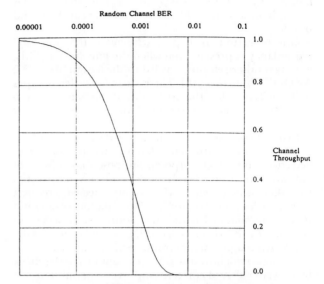

Fig. 6. ARQ Throughput for 1000-Bit Blocks.

of protocol overhead per block is normally fixed. Hence, as the blocks are shortened, less data is transferred. Secondly, the technique only helps within fairly limited bounds against random errors. Even with optimally sized blocks, the low throughput of ARQ under non-benign conditions remains a major disadvantage.

In summary, Type 0 ARQ is probably most applicable to channels which are mostly error-free but subject to infrequent bursts of interference of limited duration.

The most important advantage of ARQ is that the delivered data from ARQ has predictable quality; the greatest disadvantage is that throughput is dependent on channel conditions. FEC behaves in a complimentary fashion: throughput is constant, with data quality depending on the channel. This immediately suggests combinations of FEC and ARQ.

One common hybrid FEC/ARQ scheme is called Type 1 ARQ. In Type 1 schemes, the initial aim is to correct data using FEC. If this fails, the error-correcting decoder can be designed to have a high degree of residual error-detection, and this detection can be used to trigger an ARQ mechanism.

Although this sounds logical, the effectiveness of Type 1 depends, as ever, on the type of channel noise. If the link is uniformly noisy, there are drawbacks with Type 1. Figure 5 shows the performance of several RS codes. Notice how steep the curves are. The addition of ARQ under these random error conditions is of mainly psychological benefit. If the code is working, the ARQ is redundant. If the code is not working, the ARQ will not help as the possibility of getting a correct repeat is remote. The combined FEC/ARQ is only helpful in the narrow descent region. For levels of noise that slowly vary in intensity between reasonable bounds, adaptive FEC techniques can be more apt than Type 1 FEC/ARQ, that is the protocols adapt the FEC code, redundancy and interleaving to the prevailing noise conditions. Conversely, if the link is mostly uniformly noisy but prone to infrequent onsets of severe bursts of interference or fading, Type 1 FEC/ARQ can be very effective. An example of a Type 1 FEC/ARQ scheme can be found in [15].

One objection to Type 1 schemes is that the FEC redundancy is present even when the noise is not. This, amongst other reasons, has led to the design of Type 2 ARQ/FEC schemes [16]. There are many different types of Type 2 ARQ but the predominant characteristic is to use ARQ first and FEC second. This is best explained by example.

In one Type 2 system [16] data is sent out without FEC protection but with redundancy attached for error-detection and ARQ purposes. Suppose that n bytes of data are sent in a packet. If the data is received correctly, then there is no problem. If errors are detected, a request is sent back to the data source. The data source does not repeat the data. Instead the source uses a $(2n,n)$ systematic code to obtain n check bytes and transmits the check bytes, again with error-detection attached.

The data sink now has a choice of options. If the check bytes were received correctly, an inverse permutation is applied to extract the correct data. If the receiver detects errors in the check bytes, the receiver has an erroneous block of message bytes and an erroneous block of check bytes. The two blocks are then passed to a $(2n,n)$ error-correcting decoder which attempts to extract the correct data. If this succeeds, the process is over. If not, another request is sent to the data source and the process is repeated, that is, the data bytes or a new set of n check bytes may be repeated.

network level

The application of ARQ to networks (packet-switched networks in particular) is extensive.

The premise of packet-switched networks is that there is much more bandwidth in a network than in any single link. Data can be sent quickly if it is packetized and the packets are transmitted over several routes in parallel.

This approach fits in well with selective repeat ARQ; if one or more of the packets hit a bad or noisy link, then this packet can be repeated and, with high probability, the repeat will be sent on a good route.

Currently there is considerable work and interest in adding adaptive routing techniques to packet-switched nets, i.e. the network isolates faulty links and does not route data over these links. Note that the data communications equipment at each end of a faulty link can continue to send test data over the faulty link, monitoring for any improvement.

The motive for adaptive routing is straightforward; if a link is not working why bother to use it? The only point of debate is about the definition of when a link is "working" or not. Suppose no link is working? Can we make more links work with FEC and ARQ? How quickly can we route around new centers of damage and bring improved links back?

These points are not academic; the success of packet-switched technology in local area networks has led to their use over military radio networks where there is a threat of Electronic Counter Measures (ECM).

The implications of ECM are now discussed in greater detail.

Obviously, adaptive routing strategies fail if every single link is not "working". Furthermore, Fig. 6 indicates that Gaussian noise can render Type 0 ARQ strategies ineffective on the link level. Furthermore, the levels of Gaussian noise required to do so would be considered mild in the ECM radio environment: an obvious jamming strategy is to corrupt every link with relatively low-levels of noise. This can be protected against by adding FEC, as well as or in place of ARQ, to each link, for example, Type 1 FEC/ARQ.

Assuming these measures have been taken, this presents the jammer with three choices. He can:

1) Give up (the ideal solution from the communicator's position).
2) Up the stakes and jam all the channels at much higher power levels.
3) Concentrate his resources on a few of the links.

In fact the latter might be forced on him by geographical considerations.

In either of the last two cases, the most likely result is that some of the links will be operative and some will not (the dichotomy being strengthened by the knife-edged performance of good codes, see Fig. 5). In these circumstances, an adaptive routing strategy can be effective.

However, adaptive routing does not solve every problem. When an ECM attack starts, the re-routing must take some finite time to notice, diagnose and re-route traffic around areas of damage. This time can not be made too short; it is not desired to close links at the slightest defect. Unfortunately, in the time period immediately after an attack the following is likely to be true of the network:

1) It is most necessary to transmit command and control data quickly and efficiently.
2) It is in this time-period that the delay caused by the attack is most severe.

To quantify these points, Curve 1 of Fig. 7 shows the normalized time delay that results from an ECM attack in

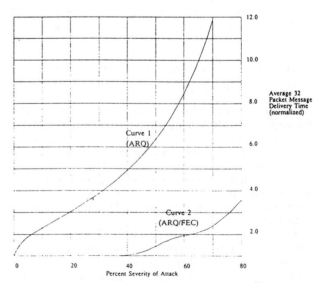

Fig. 7. Average Time-Delay in an Attacked Network.

the absence of re-routing. The x-axis shows the severity of the attack in terms of the percentage of non-working routes from A to B. The y-axis shows the average delay in sending 32 packages of data. In the absence of damage the time is normalized to 1 unit. Even with adaptive rerouting, in the period immediately after the attack, Fig. 7 could still be valid due to the time taken to re-route.

Another concern with adaptive re-routing is spoofing. The ECM can jam some percentage of the channels until the re-routing closes them down. The jammer can then jam some other set of channels, safe in the knowledge that there will be a time lag before the closed channels can be re-opened.

The above discussion indicates that for a packet-switched network to be survivable against sustained ECM attacks, error-control must be present on several levels. FEC/ARQ is necessary at the link level to make it as difficult as possible for a link to be rendered inoperative. In addition, FEC/ARQ techniques on an end-to-end basis are necessary to protect against the detrimental effect of inoperative links.

Suitable end-to-end ARQ techniques are already supplied by packet-switching and adaptive routing techniques, but suitable end-to-end forward error correction network techniques have received relatively little attention. The authors have developed one example of packet-coding; another example is the "code-combining" method of D. Chase [17].

Curve 2 of Fig. 7 shows the performance of a packet-coded ARQ scheme in transferring 32 packets of data. Note that it is uniformly superior to the uncoded case and that negligible delay is caused if less than 30 percent of the channels are jammed. When this is added to the gains due to coding on the link level (that is, when it is difficult to jam one link, let alone 30 percent), the task of the jammer is seen to be much harder.

The above discussion and examples are designed to show how adaptation offers an effective solution to a wide range of communication problems. The key to adaptive systems is a thorough knowledge of the parameters that should be adapted and the properties of the different techniques that can be deployed. From this point of view, FEC offers a set of properties that are complimentary to the techniques that have been most often used to date. As the demands on communication systems become more severe, the co-design of systems using FEC as an ingredient is expected to bring about a rapprochement of modulation, ARQ, multiplexing, packet-switched and protocol design.

Applications

This section looks at some particular examples where error control has been applied. The examples have been chosen to illustrate some of the points discussed in the previous section.

Block Codes on a TDMA System

A high rate (16/17) Reed-Solomon codec has recently been integrated into a satellite TDMA system [18]. The advantage of using such a high-rate code in this application was not only that the coding overhead is 6 percent, but also that the existing satellite transponder equip-

ment could be readjusted to accommodate the slightly higher bandwidth without modification. This allowed integration of the RS code into the TDMA environment in a very transparent fashion.

The Reed-Solomon code used is a RS(204,192) code on 8 bit symbols. For Gaussian noise conditions, this code will improve a raw bit-error rate of 10^{-1} to a corrected bit-error rate of 10^{-11}. The channel data rate is 51 megabits/second, with the user data rate being 48 megabits/second.

The development model of this system is constructed of standard ECL components. Encoding is performed in real-time using straightforward techniques. Decoding is performed by a dedicated, microcoded Galois field processor [24].

At these data rates, a decoder that operates in real-time on worst case data is not always the most cost-effective implementation. Instead, an approach known as "Best Case" or "Average-Case" Optimization can be used [19]. In this approach, the decoding engine can operate in real time under the expected channel conditions, but will not keep up with the incoming data if the channel becomes exceptionally noisy. A data-buffer (154 codewords in this case) is used to absorb the data-dependency of the decoder throughput, thus simulating real-time operation for the specified channel conditions.

The buffered decoder approach takes advantage of the fact that the complexity of decoding a received RS codeword is a function of the number of errors—the more errors, the more processing time is required. The performance of a buffered decoder is identical to that of a true real-time decoder except in those worst-case scenarios where buffer overflow occurs. In this event, some amount of uncorrected channel data will be passed to the user. The incidence of buffer overflow for a given system can be analyzed using queueing theory. As a practical illustration, the results obtained from the TDMA system codec are given in Fig. 8. The lower curve on this graph gives the theoretical relationship between raw BER and corrected BER for the RS code. The circles are data points measured during field testing of the codec unit. The difference between the theoretical and measured curves in the region between 10^{-2} and 10^{-3} raw BER is due to the occurrence of buffer overflow events. For raw BER less than about 7×10^{-1} the measured data agrees very closely with the theoretical performance.

The CCSDS Standard: Concatenated Block and Convolutional Code

This section reports on the Consultative Committee for Space Data Systems (CCSDS) Blue Book standard for Telemetry Channel Coding [7]. A schematic for the system is shown in Fig. 9. A space platform supplies a data source for which errors would prove damaging, for example, compressed video data. With such data, if any errors are not corrected, the de-compression algorithms will propagate the effect normally to the detriment of several successive frames. The data source is first encoded by a RS(255,223) encoder on 8 bit symbols. The output of the RS encoder is then interleaved on a symbol basis and input into the (2,1) $K=7$ convolutional encoder shown in Fig. 1.

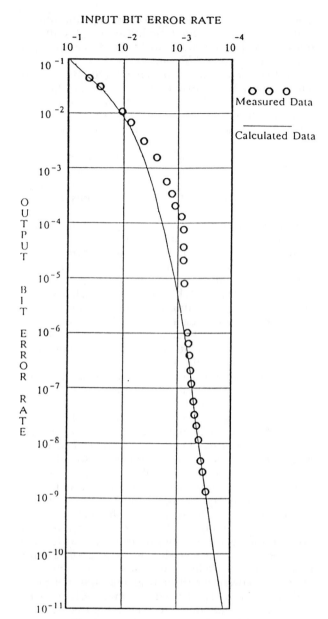

Fig. 8. *Performance of the TDMA Codec.*

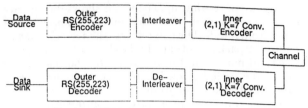

Fig. 9. *CCSDS Concatenated Code Schematic.*

code is offset from the theoretical by about 0.25 to 0.5 dB. This is suspected to be due to non-Gaussian effects on the channel.

Fortunately, the output from the convolutional decoder is well-suited to interleaved RS codes. The interleaving slices the bursts into shorter lengths that a codeword can handle with high probability. The symbol structure of the RS code effectively compresses the error bursts. For example, an error burst of 8 bits can affect 2 symbols at most. Finally, the RS decoder attempts to correct the errors.

Figures 10 and 11 show the theoretical and practical effect of the RS code. In the practical case, the curve is near vertical in the 3.25 to 3.4 dB region. This type of curve presents the system designer with a power budget, that is, if the designer provides a signal above a pre-set power, the data quality can be regarded as perfect.

In terms of complexity, several innovative techniques were used (and standardized) to reduce the RS encoder complexity [22,23]. The encoder, being on the space platform, is the critical component in terms of reducing hardware. It is interesting to note that these encoders require less hardware and less power than encoders for much "simpler" binary codes of the same redundancy but of lesser performance. The symbol nature of the RS codes and the algebraic nature of the encoding and decoding algorithms is well-matched to digital implementation.

The data rate is not defined in the standard. The RS code has been implemented to work at T1 (1.544 Mb/s) speeds and as high as 2 Mb/s.

The decoding reverses the above chain and is more complex. Figure 10 shows the theoretical performance of the convolutional code alone as opposed to uncoded. In fact, practical experiments have been performed for the scheme [21]. Figure 11, reproduced by kind permission of NASA JPL, shows some of the measured results.

The convolutional decoder uses the Viterbi Algorithm to process all the available soft decision data. This results in improved performance but, because of the small value of constraint length, the rate of descent is not that steep in either the practical or theoretical measurements. It is noticeable that in the measured case, the convolutional

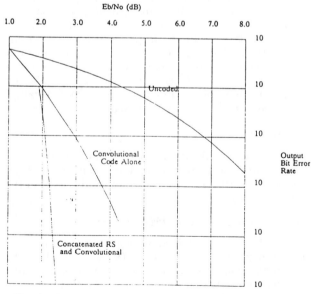

Fig. 10. *Performance of RS(255,223) and (2,1), K=7 Conv. Code*

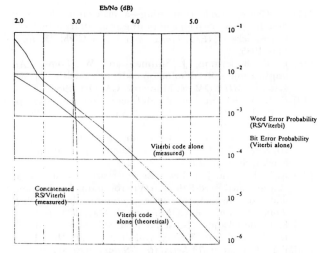

Fig. 11. *Performance of the RS/Viterbi Concatenated Coding System. (By permission of the Jet Propulsion Laboratory [22])*

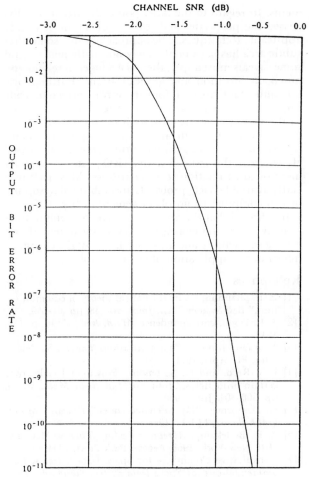

Fig. 12. *Performance of the Model 7 Codec.*

The RS decoder consists of a sync acquisition subsystem and a Galois-field processor [24]. It is also interesting to note that CCSDS is now considering a Green Book recommendation for coding Space Station communications [25]. In this application, the links are a lot faster (300 Mb/s). At these speeds, convolutional decoding becomes difficult. The Green Book recommendation is to retain the RS code and drop the convolutional.

A Concatenated Reed-Solomon/ Binary Block Code System

In tactical military communications systems, a variety of error-control techniques are often combined into a single system.

One such system which has been recently developed [26] combines the use of a Reed-Solomon outer code, pseudo-random interleaving, and a binary inner code which is designed to impart certain spectral characteristics to the modulated signal.

The design of the binary inner code for this system is fairly involved. The system specifications impose a bias requirement on this code: the frequency of one's in the encoded data should be approximately one third of the total number of bits. Bias requirements such as this can arise when it is necessary for the receiver to recover the signal under very severe noise and interference. This restriction was satisfied by using the (24,12) extended Golay code, shortened to length (22,10), and then expurgated so as to include only 128 codewords, each of which has seven one's. It follows that the inner code has rate 7/22 and satisfies the bias requirement.

The overall inner coding technique can be categorized as a spread-spectrum method. Maximum likelihood estimation is used in the soft-decision inner code decoding algorithm, and an erasure indication is supplied to the outer Reed-Solomon decoder. The outer code is over GF(128) with 27 check symbols. The overall code rate of the concatenated codec is 0.25.

Pseudorandom interleaving is used to improve the

performance in the presence of interference. This form of interleaving is less suspectable to counter-measures than the more typical block or helical interleaving methods, for which effective periodic jamming signals can be designed.

The performance of this codec is shown in Fig. 12, which relates decoded BER to channel signal-to-noise ratio. The decoder provides usable output when the channel SNR is as bad as -1 dB. This sort of performance could also have been obtained using a low-rate, soft-decision convolutional code as the inner code; however, such a code would not satisfy the bias requirement. The use of the modified Golay code to meet this requirement represents part of the growing trend in which channel coding, modulation and synchronization are co-designed, rather than treated as separate system functions.

Conclusions

This article has introduced the changing strategy and tactics of error control. Certain classes of forward error correction codes and re-transmit protocols can fundamentally change system performance provided they are matched to the channel conditions and user requirements. In regard to the latter point, the article has discussed the properties and features of the most widely applicable techniques. The approach to this somewhat subtle area has been deliberately nonmathematical and some caveats must apply; there is a danger in relying too much on intuition and generalizations. Furthermore, the important area of forward error correction code implementation and algorithm design is only discussed in the widest of terms (although references are given). Some other important areas (for example, redundant signal sets) are only mentioned briefly; they are discussed elsewhere in this issue. Finally, ARQ protocols are only mentioned in relation to error control. Most commercially available ARQ protocols are embedded in sophisticated computer protocols of general utility.

It is hoped that the article will give the communications system designer a clearer view of the potential that coding can achieve and an appreciation of the changing trends in the application of error control.

References

[1] C. E. Shannon, "A mathematical theory of communication," *Bell System Tech. Jour.*, vol. 38, pp. 61–656.

[2] J. L. Doob, correspondence, *Math. Rev.*, V. 10, p. 133, 1949.

[3] *New Directions in Communications Theory*, St. Petersburg, FL, April 1971.

[4] I. S. Reed, and G. Solomon, "Polynomial codes over certain finite fields," *Jour. Soc. Ind. Appl. Math.*, vol. 8, pp. 300–304, June 1960.

[5] G. D. Forney, "Concatenated codes," *Research Monograph no. 37*, M.I.T., 1966.

[6] E. R. Berlekamp, *Algebraic Coding Theory*, McGraw-Hill, New York, 1968; Aegean Park Press, 1984.

[7] Consultative Committee for Space Data Systems, *Recommendations for Space Data System Standards: Telemetry Channel Coding "Blue Book,"* May 1984.

[8] A. J. Viterbi, "Error bounds for convolutional codes and an asymptotically optimum decoding algorithm," *IEEE Trans. Info. Theory*, vol. IT-13, pp. 260–269, Apr. 1967.

[9] G. D. Forney, "Convolutional codes III: sequential decoding," *Tech. Report 7004-1*, Information Systems Laboratory, Stanford University.

[10] T. Le-Ngoc, K. Feher, and H. P. Van, "New modulation techniques for low-cost power and bandwidth efficient satellite earth stations," *IEEE Trans. Comm.*, vol. COM-30, no. 1, Jan. 1982.

[11] G. D. Forney, R. G. Gallager, G. R. Lang, F. M. Longstaff, and S. U. Qureshi, "Efficient modulation for band-limited channels," *IEEE Jou. of Selected Areas in Comm.* vol. SAC-2, no. 5, Sept. 1984.

[12] A. J. Viterbi, and J. Omura, *Principles of Digital Communications and Coding*, McGraw-Hill, 1979.

[13] G. Ungerboeck, "Trellis-coded modulation with redundant signal sets—an overview," *IEEE Communications Magazine*, vol. 25, no. 2, Feb. 1987.

[14] I. M. Jacobs and E. R. Berlekamp, "A lower bound to the distribution of computation for sequential decoding," *IEEE Trans Info. Theory*, vol. IT-13, pp. 167–174, Apr. 1967.

[15] T. Kasami, T. Fujiwara and S. Lin, "A concatenated coding scheme for error control," *IEEE Trans. Comm.*, vol. COM-34, no. 5, May 1986.

[16] S. Lin and D. J. Costello, "A survey of various ARQ and hybrid ARQ schemes, and error detection using linear block codes," *IEEE Communications Magazine*, vol. 22, no. 12, Dec. 1984.

[17] D. Chase, "Code combining—a maximum-likelihood decoding approach for combining an arbitrary number of noisy packets," *IEEE Trans. Comm.*, vol. COM-33, no. 5, May 1985.

[18] E. R. Berlekamp, J. Shifman and W. Toms, "An application of Reed-Solomon codes to a satellite TDMA system," *MILCOM 86*, Monterey, CA, Oct. 1986.

[19] E. R. Berlekamp and R. McEliece, "Average-case optimized buffered decoders," NATO Advanced Study Institute, Paris, July 11–22, 1983.

[20] L. R. Welch and E. R. Berlekamp, *Error-Correction for Algebraic Block Codes*, U.S. Patent Application No. 536951, Sept. 28, 1983.

[21] K. Y. Liu and J. J. Lee, "Recent results on the use of concatenated Reed-Solomon/Viterbi channel coding and data compression for space communications," *IEEE Trans. Comm.*, vol. COM-32, no. 5, May 1984.

[22] E. R. Berlekamp, "Bit-serial Reed-Solomon encoders," *IEEE Trans. Info. Theory*, vol. IT-28, no. 6, Nov. 1982.

[23] E. R. Berlekamp, *Bit-Serial Reed-Solomon Encoders*, U.S. Patent No. 4,410,989, Oct. 18, 1983.

[24] E. R. Berlekamp, *Galois Field Computer*, U.S. Patent No. 4,162,480, July 24, 1979.

[25] Consultative Committee for Space Data Systems, *Space Station: Application of CCSDS Recommendations for Space Data System Standards to the Space Station Information System (SSIS) Architecture "Green Book,"* Oct. 1985.

[26] E. R. Berlekamp, P. Tong, R. McEliece, R. J. Currie, and C. K. Rushforth, "An error-control code with an imbalance of ones and zeroes to provide a residual carrier component," *MILCOM 86*, Monterey, CA, Oct. 1986.

[27] CCITT Recommendation X.25, "Interface between data terminal equipment (DTE) and data circuit termination equipment (DCE) for terminals operating in the packet mode on public data networks."

[28] E. R. Berlekamp, "Hypersystolic computers," JASON Workshop on Advanced Computer Architectures, La Jolla, CA, July 1986.

[29] E. R. Berlekamp, (ed), *Key Papers in the Development of Coding Theory*, IEEE Press, 1973.

Elwyn R. Berlekamp received B.S., M.S., and Ph.D. degrees from the Massachusetts Institute of Technology in 1962, 1962, and 1964, respectively.

In 1973, Dr. Berlekamp founded Cyclotomics, Inc., an organization dedicated to the research, study, design, and implementation of high-performance encoding and decoding systems. Since 1982, he has been devoting most of his time to this enterprise.

Robert Peile received a B.Sc. degree in Mathematics with First Class Honors from the University of London 1976. In 1977 he received his MSc in Mathematics from Oxford University and in 1979 his D. Phil. in Mathematics from Oxford University.

At Cyclotomics, Dr. Peile is a member of the Senior Scientific Staff. He is involved in the development and configuration of specialized systems that utilize error correction and detection.

Stephen P. Pope received the B.S. degree with majors in Mathematics and Engineering from the California Institute of Technology, Pasadena, California, in 1978, and the Ph.D. degree in Electrical Engineering and Computer Science from the University of California, Berkeley, in February 1985.

After completing his studies at the University of California, Dr. Pope joined the staff of Cyclotomics, where he is the manager of VLSI design activities.

■

Adaptive Equalization

SHAHID U. H. QURESHI, SENIOR MEMBER, IEEE

Invited Paper

Bandwidth-efficient data transmission over telephone and radio channels is made possible by the use of adaptive equalization to compensate for the time dispersion introduced by the channel. Spurred by practical applications, a steady research effort over the last two decades has produced a rich body of literature in adaptive equalization and the related more general fields of reception of digital signals, adaptive filtering, and system identification. This tutorial paper gives an overview of the current state of the art in adaptive equalization. In the first part of the paper, the problem of intersymbol interference (ISI) and the basic concept of transversal equalizers are introduced followed by a simplified description of some practical adaptive equalizer structures and their properties. Related applications of adaptive filters and implementation approaches are discussed. Linear and nonlinear receiver structures, their steady-state performance and sensitivity to timing phase are presented in some depth in the next part. It is shown that a fractionally spaced equalizer can serve as the optimum receive filter for any receiver. Decision-feedback equalization, decision-aided ISI cancellation, and adaptive filtering for maximum-likelihood sequence estimation are presented in a common framework. The next two parts of the paper are devoted to a discussion of the convergence and steady-state properties of least mean-square (LMS) adaptation algorithms, including digital precision considerations, and three classes of rapidly converging adaptive equalization algorithms; namely, orthogonalized LMS, periodic or cyclic, and recursive least squares algorithms. An attempt is made throughout the paper to describe important principles and results in a heuristic manner, without formal proofs, using simple mathematical notation where possible.

I INTRODUCTION

The rapidly increasing need for computer communications has been met primarily by higher speed data transmission over the widespread network of voice-bandwidth channels developed for voice communications. A modulator-demodulator (modem) is required to carry digital signals over these analog passband (nominally 300- to 3000-Hz) channels by translating binary data to voice-frequency signals and back (Fig. 1). The thrust toward common carrier digital transmission facilities has also resulted in application of modem technology to line-of-sight terrestrial radio and satellite transmission, and recently to subscriber loops.

Analog channels deliver corrupted and transformed versions of their input waveforms. Corruption of the waveform

Manuscript received July 16, 1984; revised March 6, 1985.
The author is with Transmission Products, Codex Corporation, Mansfield, MA 02048, USA.

—usually statistical—may be additive and/or multiplicative, because of possible background thermal noise, impulse noise, and fades. Transformations performed by the channel are frequency translation, nonlinear or harmonic distortion, and time dispersion.

In telephone lines, time dispersion results when the channel frequency response deviates from the ideal of constant amplitude and linear phase (constant delay). Equalization, which dates back to the use of loading coils to improve the characteristics of twisted-pair telephone cables for voice transmission, compensates for these nonideal characteristics by filtering.

A synchronous modem transmitter collects an integral number of bits of data at a time and encodes them into symbols for transmission at the signaling rate. In pulse amplitude modulation (PAM), each signal is a pulse whose amplitude level is determined by the symbol, e.g., amplitudes of -3, -1, 1, and 3 for quaternary transmission. In bandwidth-efficient digital communication systems, the effect of each symbol transmitted over a time-dispersive channel extends beyond the time interval used to represent that symbol. The distortion caused by the resulting overlap of received symbols is called intersymbol interference (ISI) [54]. This distortion is one of the major obstacles to reliable high-speed data transmission over low-background-noise channels of limited bandwidth. In its broad sense, the term "equalizer" applies to any signal processing device designed to deal with ISI.

It was recognized early in the quest for high-speed (4800-bit/s and higher rate) data transmission over telephone channels that rather precise compensation, or equalization, is required to reduce the intersymbol interference introduced by the channel. In addition, in most practical situations the channel characteristics are not known beforehand. For medium-speed (up to 2400-bit/s) modems, which effectively transmit 1 bit/Hz, it is usually adequate to design and use a compromise (or statistical) equalizer which compensates for the average of the range of expected channel amplitude and delay characteristics. However, the variation in the characteristics within a class of channels as in the lines found in the switched telephone network, is large enough so that automatic adaptive equalization is used nearly universally for speeds higher

Reprinted from *Proceedings of the IEEE*, Vol. 73, No. 9, pp. 1340-1387, Sept., 1985

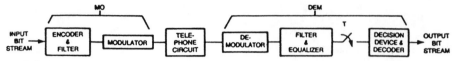

Fig. 1. Data transmission system.

than 2400 bits/s. Even 2400-bit/s modems now often incorporate this feature.

Voice-band telephone modems may be classified into one of three categories based on intended application: namely, for two-wire public switched telephone network (PSTN), four-wire point-to-point leased lines, and four-wire multipoint leased lines. PSTN modems can achieve 2400-bit/s two-wire full-duplex transmission by sending 4 bits/symbol and using frequency division to separate the signals in the two directions of transmission. Two-wire full-duplex modems using adaptive echo cancellation are now available for 2400- and 4800-bit/s transmission. Adaptive echo cancellation in conjunction with coded modulation will pave the way to 9600-bit/s full-duplex operation over two-wire PSTN circuits in the near future. At this time, commercially available leased-line modems operate at rates up to 16.8 kbits/s over conditioned point-to-point circuits, and up to 9.6 kbits/s over unconditioned multipoint circuits. An adaptive equalizer is an essential component of all these modems. (See [27] for a historical note on voice-band modem development.)

In radio and undersea channels, ISI is due to multipath propagation [87], [98], which may be viewed as transmission through a group of channels with differing relative amplitudes and delays. Adaptive equalizers are capable of correcting for ISI due to multipath in the same way as ISI from linear distortion in telephone channels. In radio-channel applications, an array of adaptive equalizers can also be used to perform diversity combining and cancel interference or jamming sources [6], [72]. One special requirement of radio-channel equalizers is that they be able to track the time-varying fading characteristics typically encountered. The convergence rate of the adaptation algorithm employed then becomes important during normal data transmission [87]. This is particularly true for 3-kHz-wide ionospheric high-frequency (HF), 3- to 30-MHz, radio channels which suffer from severe time dispersion and relatively rapid time variation and fading. Adaptive equalization has also been applied to slowly fading tropospheric scatter microwave digital radios, in the 4- to 11-GHz bands, at rates up to 200 Mbits/s [82].

In the last decade there has been considerable interest in techniques for full-duplex data transmission at rates up to 144 kbits/s over two-wire (nonloaded twisted-copper pair) subscriber loops [2], [22], [69], [106], [112]. Two competing schemes for achieving full-duplex transmission are time-compression multiplex or burst mode and adaptive echo cancellation. Some form of adaptive equalization is desirable, if not indispensible, for these baseband modems due to a number of factors: high transmission rates specially for the burst-mode scheme, attenuation distortion based on the desired range of subscriber loop lengths and gauges, and the presence of bridged taps, which cause additional time dispersion.

The first part of this paper, intended primarily for those

not familiar with the field, is a simplified introduction to intersymbol interference and transversal equalizers, and an overview of some practical adaptive equalizer structures. In the concluding sections of the first part, we briefly mention other related applications of adaptive filters (such as echo cancellation, noise cancellation, and prediction) and discuss past and present implementation approaches.

Before presenting the introductory material, however, it seems appropriate to summarize the major areas of work in adaptive equalization, with reference to key papers and to sections of this article where these topics are discussed. (The interested reader should refer to Lucky [59] and Price [83] for a comprehensive survey of the literature and extensive bibliographies of work up to the early 1970s.) Unfortunately, use of some as yet undefined technical terms in the following paragraphs is unavoidable at this stage.

Nyquist's telegraph transmission theory [115] in 1928 laid the foundation for pulse transmission over band-limited analog channels. In 1960, Widrow and Hoff [109] presented a least mean-square (LMS) error adaptive filtering scheme which has been the workhorse adaptive equalization algorithm for the last decade and a half. However, research on adaptive equalization of PAM systems in the early 1960s centered on the basic theory and structure of zero-forcing transversal or tapped-delay-line equalizers with symbol interval tap spacing [55], [56]. In parallel, the theory and structure of linear receive and transmit filters [116], [117] were developed which minimize mean-square error for time-dispersive additive Gaussian noise channels [31]. By the late 1960s, LMS adaptive equalizers had been described and understood [33], [54], [84]. It was recognized that over highly dispersive channels even the best linear receiver falls considerably short of the matched filter performance bound, obtained by considering the reception of an isolated transmitted pulse [54]. Considerable research followed on the theory of optimum nonlinear receiver structures under various optimality criteria related to error probability [1], [59], [87]. This culminated in the development of the maximum-likelihood sequence estimator [24] using the Viterbi algorithm [25] and adaptive versions of such a receiver [17], [51], [61], [62], [88], [89], [103]. Another branch of research concentrated on a particularly simple suboptimum receiver structure known as the decision-feedback equalizer [3], [4], [12], [32], [71], [83], [93]. Linear feedback, or infinite impulse response (IIR), adaptive filters [47] have not been applied as adaptive equalizers due to lack of guaranteed stability, lack of a quadratic performance surface, and a minor performance gain over transversal equalizers [87]. As the advantages of double-sideband suppressed-carrier quadrature amplitude modulation (QAM) over single-sideband (SSB) and vestigial-sideband (VSB) modulation were recognized, previously known PAM equalizers were extended to complex-valued structures suitable for joint equalization of the in-phase and quadrature signals in a QAM receiver [15], [16], [49], [84], [119]. Transversal and decision-feedback equalizers with forward-filter tap spacing that is less than the symbol interval were suggested in the late 1960s and early 1970s [6], [58], [71]. These fractionally spaced equalizers were first used in commercial telephone line modems [26], [53] and military tropospheric scatter radio systems [114] in the mid 1970s. Their theory and many performance advantages over conventional "symbol-spaced" equalizers have been the subject of several articles [39], [45], [91], [104]. The timing phase sensitivity of the mean-square error of symbol-spaced [64], fractionally spaced [45], [91], [104], and decision-feedback [94] equalizers has also been a research topic in the 1970s. Recently, interest in a nonlinear decision-aided receiver structure [85] now known as an ISI canceller has been revived by using a fractionally spaced equalizer as a matched filter [34], [80].

In the second part of the paper we develop the various receiver structures mentioned above and present their important steady-state properties. The first two sections are devoted to the definition of the baseband equivalent channel model and the development of an optimum receive filter which must precede further linear or nonlinear processing at the symbol rate. The next section on linear receivers shows that while the conventional, matched-filter plus symbol-spaced equalizer, and fractionally spaced forms of a linear receiver are equivalent when each is unrestricted (infinite in length), a finite-length fractionally spaced equalizer has significant advantages compared with a practical version of the conventional linear receiver. Nonlinear receivers are presented in the fourth section with a discussion of decision-feedback equalizers, decision-aided ISI cancellation, and adaptive versions of the maximum-likelihood sequence estimator. The final section of this part of the paper addresses timing phase sensitivity. A few important topics which have been excluded due to space limitations are: adaptive equalization of nonlinearities [5], [?], diversity-combining adaptive equalizer arrays to combat selective fades and interference in radio channels [6], [?], [114], and a particular passband equalizer structure [11], [?].

Until the early 1970s most of the equalization literature was devoted to equalizer structures and steady-state analysis [59], partly due to the difficulty of analyzing the transient performance of practical adaptive equalization algorithms. Since then some key papers [38], [65], [102], [111] have contributed to the understanding of the convergence of the LMS stochastic update algorithm for transversal equalizers, including the effect of channel characteristics on the rate of convergence. The third part of this paper is devoted to this subject and a discussion of digital precision considerations [?], [13], [36], [38]. The important topic of decision-directed convergence [60], [66] and self-recovering adaptive equalization algorithms [42], [95] has been omitted.

The demand for polled data communication systems using multipoint modems [26] which require fast setup at the central site receiver has led to the study of fast converging equalizers using a short preamble or training sequence. The fourth part of this paper summarizes three classes of fast-converging equalization algorithms. Some of the early work on this topic was directed toward orthogonalized LMS algorithms for partial response systems [8], [75], [88]. Periodic or cyclic sequences for equalizer training and methods for fast startup based on such sequences have been widely used in practice [43], [70], [76], [90], [91]. The third class of fast converging algorithms are self-orthogonalizing [37]. In 1973, Godard [41] described how the Kalman filtering algorithm can be used to estimate the LMS equalizer coefficient vector at each symbol interval. This was later recognized [20] to be a form of recursive least squares (RLS) estimation problem. Development of computationally efficient RLS algorithms has recently been a subject of intense research activity [46], [73], [78], [79], [87] leading to transversal [10], [11], [20] and lattice [52], [63], [74], [96], [97], [127] forms of the algorithm. Some of these algorithms have been applied to adaptive equalizers for HF radio modems [87], [126] which need to track a relatively rapidly time-varying channel. However, the extra complexity of these algorithms has so far prevented application to the startup problem of telephone line modems where periodic equalization [43] and other cost-effective techniques, e.g., [26], [125], are applicable.

Block least squares methods are widely used in speech coding [46], [122] to derive new adaptive filter parameters for each frame of the nonstationary speech waveform. Block implementations of adaptive filters have been suggested [130, and references therein] where the filter coefficients are updated once per block, and the output samples are computed a block at a time using transform-domain "high-speed convolution." Such implementations generally reduce the number of arithmetic operations at the expense of a more complex control structure, additional memory requirements, and a greater processing delay.

A brief view of the general direction of future work in adaptive equalization is given in the final part of the paper.

A. Intersymbol Interference

Intersymbol interference arises in all pulse-modulation systems, including frequency-shift keying (FSK), phase-shift keying (PSK), and quadrature amplitude modulation (QAM) [54]. However, its effect can be most easily described for a baseband pulse-amplitude modulation (PAM) system. A model of such a PAM communication system is shown in Fig. 2. A generalized baseband equivalent model such as this can be derived for any linear modulation scheme. In this model, the "channel" includes the effects of the transmitter filter, the modulator, the transmission medium, and the demodulator.

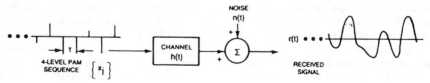

Fig. 2. Baseband PAM system model.

A symbol x_m, one of L discrete amplitude levels, is transmitted at instant mT through the channel, where T seconds is the signaling interval. The channel impulse response $h(t)$ is shown in Fig. 3. The received signal $r(t)$ is

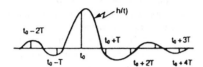

Fig. 3.   Channel impulse response.

the superposition of the impulse responses of the channel to each transmitted symbol and additive white Gaussian noise $n(t)$

$$r(t) = \sum_j x_j h(t - jT) + n(t).$$

If we sample the received signal at instant $kT + t_0$, where t_0 accounts for the channel delay and sampler phase, we obtain

$$r(t_0 + kT) = x_k h(t_0)$$
$$+ \sum_{j \neq k} x_j h(t_0 + kT - jT) + n(t_0 + kT).$$

The first term on the right is the desired signal since it can be used to identify the transmitted amplitude level. The last term is the additive noise, while the middle sum is the interference from neighboring symbols. Each interference term is proportional to a sample of the channel impulse response $h(t_0 + iT)$ spaced a multiple iT of symbol intervals T away from t_0 as shown in Fig. 3. The ISI is zero if and only if $h(t_0 + iT) = 0$, $i \neq 0$; that is, if the channel impulse response has zero crossings at T-spaced intervals.

When the impulse response has such uniformly spaced zero crossings, it is said to satisfy Nyquist's first criterion. In frequency-domain terms, this condition is equivalent to

$$H'(f) = \sum_n H(f - n/T) = \text{constant for } |f| \leq 1/2T.$$

$H(f)$ is the channel frequency response and $H'(f)$ is the "folded" (aliased or overlapped) channel spectral response after symbol-rate sampling. The band $|f| \leq 1/2T$ is commonly referred to as the Nyquist or minimum bandwidth. When $H(f) = 0$ for $|f| > 1/T$ (the channel has no response beyond twice the Nyquist bandwidth), the folded response $H'(f)$ has the simple form

$$H'(f) = H(f) + H(f - 1/T), \quad 0 \leq f \leq 1/T.$$

Fig. 4(a) and (d) shows the amplitude response of two linear-phase low-pass filters: one an ideal filter with Nyquist bandwidth and the other with odd (or vestigial) symmetry around $1/2T$ hertz. As illustrated in Fig. 4 (b) and (e), the folded frequency response of ech filter satisfies Nyquist's first criterion. One class of linear-phase filters, which is commonly referred to in the literature [54], [87], [113] and is widely used in practice [23], [118], is the raised-cosine family with cosine rolloff around $1/2T$ hertz.

In practice, the effect of ISI can be seen from a trace of the received signal on an oscilloscope with its time base

synchronized to the symbol rate. Fig. 5 shows the outline of a trace (eye pattern) for a two-level or binary PAM system. If the channel satisfies the zero ISI condition, there are only two distinct levels at the sampling time t_0. The eye is then fully open and the peak distortion is zero. Peak distortion (Fig. 5) is the ISI that occurs when the data pattern is such that all intersymbol interference terms add to produce the maximum deviation from the desired signal at the sampling time.

The purpose of an equalizer, placed in the path of the received signal, is to reduce the ISI as much as possible to maximize the probability of correct decisions.

B. Linear Transversal Equalizers

Among the many structures used for equalization the simplest is the transversal (tapped-delay-line or nonrecursive) equalizer shown in Fig. 6. In such an equalizer, the current and past values $r(t - nT)$ of the received signal are linearly weighted by equalizer coefficients (tap gains) c_n and summed to produce the output. If the delays and tap-gain multipliers are analog, the continuous output of the equalizer $z(t)$ is sampled at the symbol rate and the samples go to the decision device. In the commonly used digital implementation, samples of the received signal at the symbol rate are stored in a digital shift register (or memory), and the equalizer output samples (sums of products) $z(t_0 + kT)$ or z_k are computed digitally, once per symbol according to

$$z_k = \sum_{n=0}^{N-1} c_n r(t_0 + kT - nt)$$

where N is the number of equalizer coefficients, and t_0 denotes sample timing.

The equalizer coefficients c_n, $n = 0, 1, \cdots, N - 1$ may be chosen to force the samples of the combined channel and equalizer impulse response to zero at all but one of the N T-spaced instants in the span of the equalizer. This is shown graphically in Fig. 7. Such an equalizer is called a zero-forcing (ZF) equalizer [55].

If we let the number of coefficients of a ZF equalizer increase without bound, we would obtain an infinite-length equalizer with zero ISI at its output. The frequency response $C(f)$ of such an equalizer is periodic, with a period equal to the symbol rate $1/T$ because of the T-second tap spacing. After sampling, the effect of the channel on the received signal is determined by the folded frequency response $H'(f)$. The combined response of the channel, in tandem with the equalizer, must satisfy the zero ISI condition or Nyquist's first criterion

$$C(f)H'(f) = 1, \quad |f| \leq 1/2T.$$

From the above expression we see that an infinite-length zero-ISI equalizer is simply an inverse filter, which inverts the folded-frequency response of the channel. A finite-length ZF equalizer approximates this inverse. Such an inverse filter may excessively enhance noise at frequencies where the folded channel spectrum has high attenuation. This is undesirable, particularly for unconditioned telephone connections, which may have considerable attenuation distortion, and for radio channels, which may be subject to frequency-selective fades.

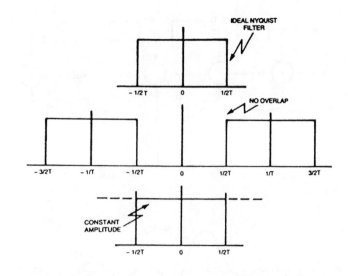

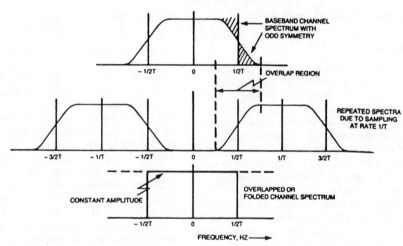

fig. 4. Linear phase filters which satisfy Nyquist's first criterion.

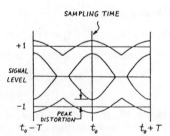

fig. 5. Outline of a binary eye pattern.

Clearly, the ZF criterion neglects the effect of noise altogether. Also, a finite-length ZF equalizer is guaranteed to minimize the peak distortion or worst case ISI only if the peak distortion before equalization is less than 100 percent [55], i.e., if a binary eye is initially open. However, at high speeds on bad channels this condition is often not met.

The least mean-square (LMS) equalizer [54] is more robust. Here the equalizer coefficients are chosen to minimize the mean-square error—the sum of squares of all the ISI terms plus the noise power at the output of the equalizer. Therefore, the LMS equalizer maximizes the signal-to-distortion ratio at its output within the constraints of the equalizer time span and the delay through the equalizer.

The delay introduced by the equalizer depends on the position of the main or reference tap of the equalizer. Typically, the tap gain corresponding to the main tap has the largest magnitude.

If the values of the channel impulse response at the sampling instants are known, the N coefficients of the ZF and the LMS equalizers can be obtained by solving a set of N linear simultaneous equations for each case.

Most current high-speed voice-band telephone-line modems use LMS equalizers because they are more robust in the presence of noise and large amounts of ISI, and superior to the ZF equalizers in their convergence properties. The same is generally true of radio-channel modems [72], [82], [114] except in one case [23] where the ZF equalizer was selected due to its implementation simplicity.

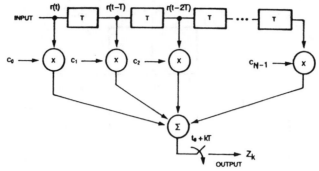

Fig. 6. Linear transversal equalizer.

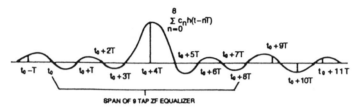

$$\sum_{n=0}^{8} c_n h(t - nT)$$

SPAN OF 9 TAP ZF EQUALIZER

Fig. 7. Combined impulse response of a channel and zero-forcing equalizer in tandem.

C. Automatic Synthesis

Before regular data transmission begins, automatic synthesis of the ZF or LMS equalizers for unknown channels, which involves the iterative solution of one of the above-mentioned sets of simultaneous equations, may be carried out during a training period. (In certain applications, such as microwave digital radio systems, and remote site receivers in a multipoint telephone modem network [43], the adaptive equalizers are required to bootstrap in a decision-directed mode (see Section I-E) without the help of a training sequence from the transmitter.)

During the training period, a known signal is transmitted and a synchronized version of this signal is generated in the receiver to acquire information about the channel characteristics. The training signal may consist of periodic isolated pulses or a continuous sequence with a broad, uniform spectrum such as the widely used maximum-length shift-register or pseudo-noise (PN) sequence [9], [54], [76], [118], [119]. The latter has the advantage of much greater average power, and hence a larger received signal-to-noise ratio (SNR) for the same peak transmitted power. The training sequence must be at least as long as the length of the equalizer so that the transmitted signal spectrum is adequately dense in the channel bandwidth to be equalized.

Given a synchronized version of the known training signal, a sequence of error signals $e_k = z_k - x_k$ can be computed at the equalizer output (Fig. 8), and used to adjust the equalizer coefficients to reduce the sum of the squared errors. The most popular equalizer adjustment method involves updates to each tap gain during each symbol interval. Iterative solution of the coefficients of the equalizer is possible because the mean-square error (MSE) is a quadratic function of the coefficients. The MSE may be envisioned as an N-dimensional paraboloid (punch bowl) with a bottom or minimum. The adjustment to each tap gain is in a direction opposite to an estimate of the gradient of the MSE with respect to that tap gain. The idea is to move the set of equalizer coefficients closer to the unique optimum set corresponding to the minimum MSE. This symbol-by-symbol procedure developed by Widrow and Hoff [109] is commonly referred to as the continual or stochastic update method because, instead of the true gradient of the mean-square error,

$$\partial E\left[e_k^2\right]/\partial c_n(k)$$

a noisy but unbiased estimate

$$\partial e_k^2/\partial c_n(k) = 2e_k r(t_0 + kT - nT)$$

is used. Thus the tap gains are updated according to

$$c_n(k + 1) = c_n(k) - \Delta e_k r(t_0 + kT - nT),$$
$$n = 0, 1, \cdots, N - 1;$$

where $c_n(k)$ is the nth tap gain at time k, e_k is the error signal, and Δ is a positive adaptation constant or step size.

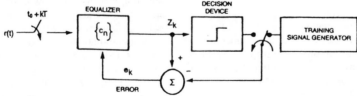

Fig. 8. Automatic adaptive equalizer.

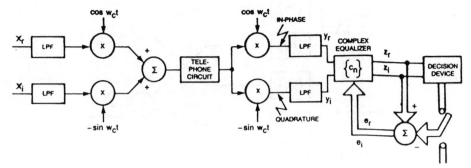

fig. 9. QAM system with baseband complex adaptive equalizer.

D. Equalizer Convergence

The exact convergence behavior of the stochastic update method is hard to analyze (see Section III-B). However, for a small step size and a large number of iterations, the behavior is similar to the steepest descent algorithm, which uses the actual gradient rather than a noisy estimate.

Here we list some general convergence properties: a) fastest convergence (or shortest settling time) is obtained when the (folded)-power spectrum of the symbol-rate sampled equalizer input is flat, and when the step size Δ is chosen to be the inverse of the product of the received signal power and the number of equalizer coefficients; b) the larger the variation in the above-mentioned folded-power spectrum, the smaller the step size must be, and therefore the slower the rate of convergence; c) for systems where sampling causes aliasing (channel foldover or spectral overlap), the convergence rate is affected by the channel-delay characteristics and the sampler phase, because they affect the aliasing. This will be explained more fully later.

E. Adaptive Equalization

After the initial training period (if there is one), the coefficients of an adaptive equalizer may be continually adjusted in a decision-directed manner. In this mode, the error signal $e_k = z_k - \hat{x}_k$ is derived from the final (not necessarily correct) receiver estimate $\{\hat{x}_k\}$ of the transmitted sequence $\{x_k\}$. In normal operation, the receiver decisions are correct with high probability, so that the error estimates are correct often enough to allow the adaptive equalizer to maintain precise equalization. Moreover, a decision-directed adaptive equalizer can track slow variations in the channel characteristics or linear perturbations in the receiver front end, such as slow jitter in the sampler phase.

The larger the step size, the faster the equalizer tracking capability. However, a compromise must be made between fast tracking and the excess mean-square error of the equalizer. The excess MSE is that part of the error power in excess of the minimum attainable MSE (with tap gains frozen at their optimum settings). This excess MSE, caused by tap gains wandering around the optimum settings, is directly proportional to the number of equalizer coefficients, the step size, and the channel noise power. The step size that provides the fastest convergence results in an MSE which is, on the average, 3 dB worse than the minimum

achievable MSE. In practice, the value of the step size is selected for fast convergence during the training period and then reduced for fine tuning during the steady-state operation (or data mode).

F. Equalizers for QAM Systems

So far we have only discussed equalizers for a baseband PAM system. Modern high-speed voice-band modems almost universally use phase-shift keying (PSK) for lower speeds, e.g., 2400 to 4800 bits/s, and combined phase and amplitude modulation or, equivalently, quadrature amplitude modulation (QAM) [54], for higher speeds, e.g., 4800 to 9600 or even 16 800 bits/s. At the high rates, where noise and other channel distortions become significant, modems using coded forms of QAM such as trellis-coded modulation [27], [105], [107] are being introduced to obtain improved performance. QAM is as efficient in bits per second per hertz as vestigial or single-sideband amplitude modulation, yet enables a coherent carrier to be derived and phase jitter to be tracked using easily implemented decision-directed carrier recovery techniques [49]. A timing waveform with negligible timing jitter can also be easily recovered from QAM signals. This property is not shared by vestigial sideband amplitude-modulation (AM) systems [128].

Fig. 9 shows a generic QAM system, which may also be used to implement PSK or combined amplitude and phase modulation. Two double-sideband suppressed-carrier AM signals are superimposed on each other at the transmitter and separated at the receiver, using quadrature or orthogonal carriers for modulation and demodulation. It is convenient to represent the in-phase and quadrature channel low-pass filter output signals in Fig. 9 by $y_r(t)$ and $y_i(t)$, as the real and imaginary parts of a complex valued signal $y(t)$. (Note that the signals are real, but it will be convenient to use complex notation.)

The baseband equalizer [84], with complex coefficients c_n, operates on samples of this complex signal $y(t)$ and produces complex equalized samples $z(k) = z_r(k) + jz_i(k)$, as shown in Fig. 10. This figure illustrates more concretely the concept of a complex equalizer as a set of four real transversal filters (with cross-coupling) for two inputs and two outputs. While the real coefficients $c_{rn}, n = 0, \cdots, N - 1$, help to combat the intersymbol interference in the in-phase and quadrature channels, the imaginary coefficients $c_{in}, n = 0, \cdots, N - 1$, counteract the cross-interference between the two channels. The latter may be caused by

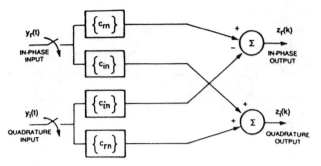

Fig. 10. Complex transversal equalizer for QAM modems.

asymmetry in the channel characteristics around the carrier frequency.

The coefficients are adjusted to minimize the mean of the squared magnitude of the complex error signal, $e(k) = e_r(k) + je_i(k)$, where e_r and e_i are the differences between z_r and z_i, and their desired values. The update method is similar to the one used for the PAM equalizer except that all variables are complex-valued

$$c_n(k+1) = c_n(k) - \Delta e_k y^*(t_0 + kT - nT),$$
$$n = 0, 1, \cdots, N-1$$

where y^* is the complex conjugate of y. Again, the use of complex notation allows the writing of this single concise equation, rather than two separate equations involving four real multiplications, which is what really has to be implemented.

The complex equalizer can also be used at passband [6], [16] to equalize the received signal before demodulation, as shown in Fig. 11. Here the received signal is split into its in-phase and quadrature components by a pair of so-called phase-splitting filters, with identical amplitude responses and phase responses that differ by 90°. The complex passband signal at the output of these filters is sampled at the symbol rate and applied to the equalizer delay line in the same way as at baseband. The complex output of the equalizer is demodulated, via multiplication by a complex exponential as shown in Fig. 11, before decisions are made and the complex error computed. Further, the error signal is remodulated before it is used in the equalizer adjustment algorithm. The main advantage of implementing the equalizer in the passband is that the delay between the

demodulator and the phase error computation circuit is reduced to the delay through the decision device. Fast phase jitter can be tracked more effectively because the delay through the equalizer is eliminated from the phase correction loop. The same advantage can be attained with a baseband equalizer by putting a jitter-tracking loop after the equalizer.

G. Decision-Feedback Equalizers

We have discussed placements and adjustment methods for the equalizer, but the basic equalizer structure has remained a linear and nonrecursive filter. A simple nonlinear equalizer [3], [4], [32], [71], [93], which is particularly useful for channels with severe amplitude distortion, uses decision feedback to cancel the interference from symbols which have already been detected. Fig. 12 shows such a decision-feedback equalizer (DFE). The equalized signal is the sum of the outputs of the forward and feedback parts of the equalizer. The forward part is like the linear transversal equalizer discussed earlier. Decisions made on the equalized signal are fed back via a second transversal filter. The basic idea is that if the value of the symbols already detected are known (past decisions are assumed to be correct), then the ISI contributed by these symbols can be canceled exactly, by subtracting past symbol values with appropriate weighting from the equalizer output. The weights are samples of the tail of the system impulse response including the channel and the forward part of the equalizer.

The forward and feedback coefficients may be adjusted simultaneously to minimize the MSE. The update equation

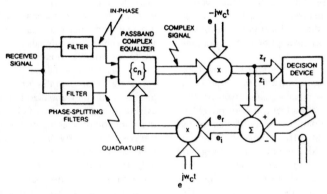

Fig. 11. Passband complex adaptive equalizer for QAM system.

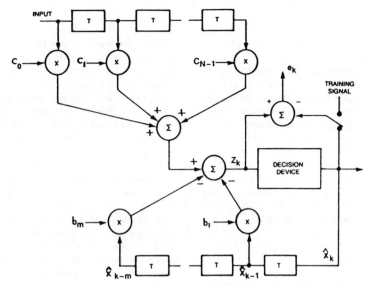

Fig. 12. Decision-feedback equalizer.

for the forward coefficients is the same as for the linear equalizer. The feedback coefficients are adjusted according to

$$b_m(k+1) = b_m(k) + \Delta e_k \hat{x}_{k-m}, \qquad m = 1, \cdots, M$$

where $\hat{x}_k$ is the kth symbol decision, $b_m(k)$ is the mth feedback coefficient at time k, and there are M feedback coefficients in all. The optimum LMS settings of $b_m, m = 1, \cdots, M$, are those that reduce the ISI to zero, within the span of the feedback part, in a manner similar to a ZF equalizer. Note that since the output of the feedback section of the DFE is a weighted sum of noise-free past decisions, the feedback coefficients play no part in determining the noise power at the equalizer output.

Given the same number of overall coefficients, does a DFE achieve less MSE than a linear equalizer? There is no definite answer to this question. The performance of each type of equalizer is influenced by the particular channel characteristics and sampler phase, as well as the actual number of coefficients and the position of the reference or main tap of the equalizer. However, the DFE can compensate for amplitude distortion without as much noise enhancement as a linear equalizer. The DFE performance is also less sensitive to the sampler phase [94].

An intuitive explanation for these advantages is as fol-

lows: The coefficients of a linear transversal equalizer are selected to force the combined channel and equalizer impulse response to approximate a unit pulse. In a DFE, the ability of the feedback section to cancel the ISI, because of a number of past symbols, allows more freedom in the choice of the coefficients of the forward section. The combined impulse response of the channel and the forward section may have nonzero samples following the main pulse. That is, the forward section of a DFE need not approximate the inverse of the channel characteristics, and so avoids excessive noise enhancement and sensitivity to sampler phase.

When a particular incorrect decision is fed back, the DFE output reflects this error during the next few symbols as the incorrect decision traverses the feedback delay line. Thus there is a greater likelihood of more incorrect decisions following the first one, i.e., error propagation. Fortunately, the error propagation in a DFE is not catastrophic. On typical channels, errors occur in short bursts that degrade performance only slightly.

H. Fractionally Spaced Equalizers

A fractionally spaced transversal equalizer [6], [39], [45], [58], [71], [91], [104] is shown in Fig. 13. The delay-line taps

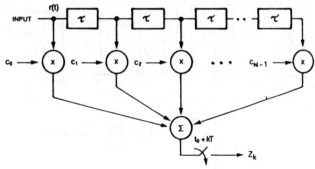

Fig. 13. Fractionally spaced equalizer.

of such an equalizer are spaced at an interval τ which is less than, or a fraction of, the symbol interval T. The tap spacing τ is typically selected such that the bandwidth occupied by the signal at the equalizer input is $|f| < 1/2\tau$, i.e., τ-spaced sampling satisfies the sampling theorem. In an analog implementation, there is no other restriction on τ, and the output of the equalizer can be sampled at the symbol rate. In a digital implementation τ must be KT/M, where K and M are integers and $M > K$. (In practice, it is convenient to choose $\tau = T/M$, where M is a small integer, e.g., 2.) The received signal is sampled and shifted into the equalizer delay line at a rate M/T and one output is produced each symbol interval (for every M input samples). In general, the equalizer output is given by

$$z_k = \sum_{n=0}^{N-1} c_n r(t_0 + kT - nKT/M).$$

The coefficients of a KT/M equalizer may be updated once per symbol based on the error computed for that symbol, according to

$$c_n(k + 1) = c_n(k) - \Delta e_k r(t_0 + kT - nKT/M),$$
$$n = 0, 1, \cdots, N - 1.$$

It is well known (see Section II) that the optimum receive filter in a linear modulation system is the cascade of a filter matched to the actual channel, with a transversal T-spaced equalizer [14], [24], [31]. The fractionally spaced equalizer, by virtue of its sampling rate, can synthesize the best combination of the characteristics of an adaptive matched filter and a T-spaced equalizer, within the constraints of its length and delay. A T-spaced equalizer, with symbol-rate sampling at its input, cannot perform matched filtering. An FSE can effectively compensate for more severe delay distortion and deal with amplitude distortion with less noise enhancement than a T equalizer.

Consider a channel whose amplitude and envelope delay

characteristics around one band edge $f_c - 1/2T$ hertz differ markedly from the characteristics around the other band edge $f_c + 1/2T$ hertz in a QAM system with a carrier frequency of f_c hertz. Then the symbol-rate sampled or folded channel-frequency response is likely to have a rapid transition in the area of spectral overlap. It is difficult for a typical T equalizer, with its limited degrees of freedom (number of taps), to manipulate such a folded channel into one with a flat frequency response. An FSE, on the other hand, can independently adjust the signal spectrum (in amplitude and phase) at the two band-edge regions before symbol-rate sampling (and spectral overlap) at the equalizer output, resulting in significantly improved performance.

A related property of an FSE is the insensitivity of its performance to the choice of sampler phase. This distinction between the conventional T-spaced and fractionally spaced equalizers can be heuristically explained as follows. First, symbol-rate sampling at the input to a T equalizer causes spectral overlap or aliasing, as explained in connection with Fig. 4. When the phases of the overlapping components match they add constructively, and when the phases are 180° apart they add destructively, which results in the cancellation or reduction of amplitude as shown in Fig. 14. Variation in the sampler phase or timing instant corresponds to a variable delay in the signal path; a linear phase component with variable slope is added to the signal spectrum. Thus changes in the sampler phase strongly influence the effects of aliasing; i.e., they influence the amplitude and delay characteristics in the spectral overlap region of the sampled equalizer input. The minimum MSE achieved by the T equalizer is, therefore, a function of the sampler phase. In particular, when the sample phase causes cancellation of the band-edge ($|f| = 1/2T$ hertz) components, the equalizer cannot manipulate the null into a flat spectrum at all, or at least without significant noise enhancement (if the null is a depression rather than a total null).

In contrast, there is no spectral overlap at the input to an

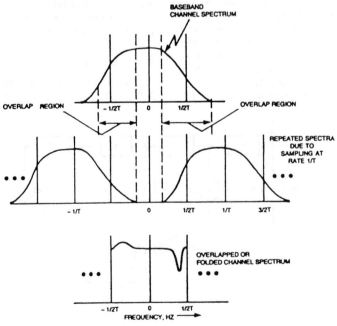

Fig. 14. Spectral overlap at the input to a T equalizer.

FSE. Thus the sensitivity of the minimum MSE, achieved with an FSE with respect to the sampler phase, is typically far smaller than with a T equalizer.

Comparison of numerical performance results of T and $T/2$ equalizers for QAM systems operating over representative voice-grade telephone circuits [91] has shown the following properties: a) a $T/2$ equalizer with the same number of coefficients (half the time span) performs almost as well or better than a T equalizer; b) a pre-equalizer receive shaping filter is not required with a $T/2$ equalizer; c) for channels with severe band-edge delay distortion, the T equalizer performs noticeably worse than a $T/2$ equalizer regardless of the choice of sampler phase.

I Other Applications

While the primary emphasis of this paper is on adaptive equalization for data transmission, a number of the topics covered are relevant to other applications of adaptive filters. In this section, generic forms of adaptive filtering applications are introduced to help in establishing the connection between the material presented in later sections and the application of interest. But first let us briefly mention applications of automatic or adaptive equalization in areas other than data transmission over radio or telephone channels.

One such application is generalized automatic channel equalization [57], where the entire bandwidth of the channel is to be equalized without regard to the modulation scheme or transmission rate to be used on the channel. The tap spacing and input sample rate are selected to satisfy the sampling theorem, and the equalizer output is produced at the same rate. During the training mode, a known signal is transmitted, which covers the bandwidth to be equalized. The difference between the equalizer output and a synchronized reference training signal is the error signal. The tap gains are adjusted to minimize the MSE in a manner similar to that used for an automatic equalizer for synchronous data transmission.

Experimental use of fixed transversal equalizers has also been made in digital magnetic recording systems [99]. The recording method employed in such a case must be linearized by using an ac bias. Having linearized the "channel," equalization can be employed to combat intersymbol interference at increased recording densities, using a higher symbol rate or multilevel coding.

Fig. 15 shows a general form of an adaptive filter with input signals x and y, output z, and error e. The parameters of an LMS adaptive filter are updated to minimize the mean-square value of the error e. In the following paragraphs, we point out how the adaptive filter is used in different applications by listing how x, y, z, and e are interpreted for each case.

Equalization:

y Received signal (filtered version of transmitted data signal) plus noise uncorrelated with the data signal.
x Detected data signal.
z Equalized signal used to detect received data.
e Residual intersymbol interference plus noise.

Echo Cancellation: Echo cancellation is a form of a general system identification problem, where the system to be identified is the echo path linear system. The coeffi-

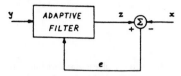

Fig. 15. General form of an adaptive filter.

cients of a transversal echo canceler converge in the mean to the echo path impulse response samples.

i) Voice:

y Far-end voice signal plus uncorrelated noise.
x Echo of far-end voice plus near-end voice plus noise.
z Estimated echo of far-end voice.
e Near-end voice plus residual echo plus noise.

Adaptation is typically carried out in the absence of the near-end voice signal. When double talk is detected (both near- and far-end signals present), update of the echo canceler coefficients is inhibited [69].

ii) Data:

y Transmitted data signal.
x Echo of transmitted data signal plus received signal plus noise.
z Estimated echo of transmitted data signal.
e Received signal plus residual echo plus noise.

Filter adaptation is typically required to be continued in the presence of a large interfering received signal which is uncorrelated with the transmitted data [69], [108]. A method proposed in [22] involves locally generating a delayed replica of the received signal and subtracting it from e before using the residual for echo canceler update.

Noise Cancellation:

y Noise source correlated with noise in x.
x Desired signal plus noise.
z Estimate of noise in x.
e Desired signal plus residual noise.

One example is that of canceling noise from the pilot's speech signal in the cockpit of an aircraft [110]. In this case, y may be the pickup from a microphone in the pilot's helmet, and x is the ambient noise picked up by another microphone placed in the cockpit. See [110] for a number of other interesting applications of noise and periodic interference cancellation, e.g., to electrocardiography.

Prediction:

y Delayed version of original signal.
x Original signal.
z Predicted signal.
e Prediction error or residual.

A well-known example is linear predictive coding (LPC) of speech where the end result is the set of estimated LPC coefficients [46]. Due to the nonstationary nature of the speech signal, LPC coefficients are typically obtained separately for each new frame (10 to 25 ms) of the speech signal.

In adaptive differential pulse code modulation (ADPCM) of speech, the purpose of adaptive prediction is to generate a residual signal with less variance so that it can be quantized and represented by fewer bits for transmission [46]. In this case:

- y Reconstructed speech signal = quantized residual plus past prediction.
- x Original speech signal.
- z Prediction
- e Residual to be quantized for transmission.

Note that the reconstructed speech signal is used for y instead of a delayed version of the original speech signal, and the predictor coefficients are updated using the quantized residual instead of e. Both the reconstructed speech signal and the quantized residual signal are also available at the ADPCM decoder so that the predictor coefficients at the decoder can be adapted in a manner identical to that used at the ADPCM encoder.

Adaptive Arrays: A further generalization of the adaptive filter of Fig. 15 is shown in Fig. 16 where a number of input signals are processed through an array of adaptive

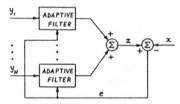

Fig. 16. Adaptive filter array.

filters whose outputs are summed together. Such adaptive arrays are useful in diversity combining [6], [114] and in dealing with jamming or spatially distributed interference [72], [110].

J. Implementation Approaches

One may divide the methods of implementing adaptive equalizers into the following general categories: analog, hardwired digital, and programmable digital.

Analog adaptive equalizers, with inductor–capacitor (LC) tapped-delay lines and switched ladder attenuators as tap gains, were among the first implementations for voice-band modems. The switched attenuators later gave way to field-effect transistors (FETs) as the variable gain elements. Analog equalizers were soon replaced by digitally implemented equalizers for reduced size and increased accuracy. Recently, however, there is renewed interest in large-scale integrated (LSI) analog implementations of adaptive filters based on switched capacitor technology [2], [112]. Here the equalizer input is sampled but not quantized. The sampled analog values are stored and transferred as charge packets. In one implementation [2], the adaptation circuitry is digital. The variable tap gains are typically stored in digital memory locations and the multiplications between the analog sample values and the digital tap gains take place in analog fashion via multiplying digital-to-analog converters. In another case [112], a five-tap adaptive transversal filter has been fabricated on a single integrated circuit (IC) using an all-analog implementation approach combining switched capacitor and charge-coupled device (CCD) technologies. The IC consists of a five-tap CCD delay line, a convolver, five correlators (one for each tap gain), an offset error canceler, and an error signal generator. Integrators and four-quadrant analog multipliers are implemented in switched capacitor technology. The IC can be configured for use as an echo canceler, a linear equalizer, or a decision-feedback equalizer at sampling rates up to 250 kHz. Analog or mixed analog–digital implementations have significant potential in applications, such as digital radio and digital subscriber loop transmission, where symbol rates are high enough to make purely digital implementations difficult.

The most widespread technology of the last decade for voice-band modem adaptive equalizer implementation may be classified as hardwired digital technology. In such implementations, the equalizer input is made available in sampled and quantized form suitable for storage in digital shift registers. The variable tap gains are also stored in shift registers and the formation and accumulation of products takes place in logic circuits connected to perform digital arithmetic. This class of implementations is characterized by the fact that the circuitry is hardwired for the sole purpose of performing the adaptive equalization function with a predetermined structure. Examples include the early units based on metal–oxide–semiconductor (MOS) shift registers and transistor–transistor logic (TTL) circuits. Later implementations [53], [100] were based on MOS LSI circuits with dramatic savings in space, power dissipation, and cost.

A hardwired digital adaptive filtering approach described in [106] for a 144-kbit/s digital subscriber loop modem is based on a random-access memory (RAM) table lookup structure [124]. Both an echo canceler (EC) and a decision-feedback equalizer (DFE), where inputs are the transmit and receive binary data sequences, respectively, can be implemented in this way. An output signal value is maintained and updated in the RAM for each of the 2^N possible states of an N-tap transversal filter with a binary input sequence. Such a structure is not restricted to be linear and, therefore, can adapt to compensate for nonlinearities [123]. A two-chip realization of a digital subscriber loop modem based on a joint EC-DFE RAM structure is described in [121].

The most recent trend in implementing voice-band modem adaptive equalizers is toward programmable digital signal processors [43], [81], [101], [120]. Here, the equalization function is performed in a series of steps or instructions in a microprocessor or a digital computation structure specially configured to efficiently perform the type of digital arithmetic (e.g., multiply and accumulate) required in digital signal processing. The same hardware can then be time-shared to perform functions such as filtering, modulation, and demodulation in a modem. Perhaps the greatest advantage of programmable digital technology is its flexibility, which permits sophisticated equalizer structures and training procedures to be implemented with ease.

For microwave digital radio systems, adaptive equalizers have been implemented both in the passband at the intermediate frequency (IF) stage and at baseband [98]. Passband equalizers are analog by necessity, e.g., an amplitude slope equalizer [23] at 70-MHz IF, and a dynamic resonance equalizer using p-i-n and varactor diodes [82] at 140-MHz IF. Three-tap $T/2$ transversal equalizers have been implemented in the passband (at 70-MHz IF) using quartz surface acoustic wave filters, analog correlators, and tap multipliers for a 4-PSK 12.6-Mbit/s digital radio [114]. Baseband transversal equalizers have been implemented using a combination of analog and digital or all-digital circuitry. A five-tap zero-forcing equalizer using lumped delay elements, hybrid integrated circuits for variable-gain and buffer amplifiers, and emitter-coupled logic for tap control is described in [13] for a 16-QAM 90-Mbit/s digital radio. A five-tap LMS transversal equalizer [82], and all digital DFEs have also been reported [98], [114].

II. RECEIVER STRUCTURES AND THEIR STEADY-STATE PROPERTIES

A. Baseband Equivalent Model

To set a common framework for discussing various receiver configurations, we develop a baseband equivalent model of a passband data transmission system. We start from a generic QAM system (Fig. 9) since it can be used to implement any linear modulation scheme.

The passband transmitted signal can be compactly written as

$$s(t) = \text{Re}\left[\sum_n x_n a(t - nT) \exp(j2\Pi f_c t)\right]$$

where $\{x_n\}$ is the complex sequence of data symbols with in-phase (real) and quadrature (imaginary) components, x_r and x_i, respectively, such that $x_n = x_{rn} + jx_{in}$, $a(t)$ is the transmit pulse shape, and f_c is the carrier frequency. We shall assume that the baseband transmit spectrum $A(f)$ is band-limited to $|f| \leq (1 + \alpha)/2T$ hertz where the rolloff factor α, between 0 and 1, determines the excess bandwidth over the minimum $|f| \leq 1/2T$. (Note that greater than 100-percent excess bandwidth is sometimes used in radio and baseband subscriber loop transmission systems.)

The received signal is

$$r(t) = s(t) * h_p(t) + n_p(t)$$

where $h_p(t)$ is the passband channel impulse response, $n_p(t)$ is "passband" Gaussian noise, and the operator $*$ represents convolution. The in-phase and quadrature outputs of the receive low-pass filters, $y_r(t)$ and $y_i(t)$, may be represented in complex notation as $y(t) = y_r(t) + jy_i(t)$

$$y(t) = g(t) * [r(t) \exp(-j2\Pi f_c t)]$$

where $g(t)$ is the impulse response of the receive filter. Assume that the receive filter completely rejects the double-frequency signal components produced by demodulation and centered around $2f_c$. Then the baseband received signal may be written as

$$y(t) = \sum_n x_n h(t - nT) + n(t) \qquad (1)$$

where

$$h(t) = g(t) * h_b(t) * a(t) \qquad (2)$$

is the complex-valued impulse response of the baseband equivalent model (Fig. 17). The real-valued passband channel impulse response $h_p(t)$ and the complex-valued baseband channel impulse response, $h_b(t)$, are related according to

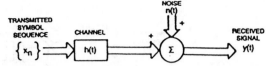

Fig. 17. General complex-valued baseband equivalent channel model.

$$h_p(t) = \text{Re}\left[h_b(t) \exp(j2\Pi f_c t)\right].$$

The corresponding frequency-domain relationship is

$$H_p(f) = H_b\left[(f + f_c) + H_b^*(f - f_c)\right]/2.$$

Thus $h_b(t)$ is the impulse response of the positive frequency part of the channel spectral response translated to baseband.

The noise waveform $n(t)$ in (1) is also complex-valued, i.e.,

$$n(t) = g(t) * \left[n_p(t) \exp(-j2\Pi f_c t)\right].$$

The receive low-pass filters (in Fig. 9) typically perform two functions: rejection of the "double-frequency" signal components, and noise suppression. The latter function is accomplished by further shaping the baseband signal spectrum. In the baseband equivalent model, only the first of these functions of the receive filters is performed by $g(t)$ and absorbed in $h(t)$ given in (2). For simplicity, the noise $n(t)$ in the baseband equivalent model is assumed to be white with jointly Gaussian real and imaginary components.

B. Optimum Receive Filter

Given the received signal $y(t)$, what is the best receive filter? This question has been posed and answered in different ways by numerous authors. Here, we follow Forney's development [24]. He showed that the sequence of T-spaced samples, obtained at the correct timing phase, at the output of a matched filter is a set of sufficient statistics for estimation of the transmitted sequence $\{x_n\}$. Thus such a receive filter is sufficient regardless of the (linear or nonlinear) signal processing which follows the symbol-rate sampler.

For the baseband equivalent model derived in the previous section, the receive filter must have an impulse response $h^*(-t)$, where the superscript $*$ denotes complex conjugate. The frequency response of this matched filter is $H^*(f)$, where $H(f)$ is the frequency response of the channel model $h(t)$.

If the data sequence $\{x_n\}$ is uncorrelated with unit power, i.e.,

$$E\left[x_n x_m^*\right] = \delta_{nm}$$

then the signal spectrum at the matched filter output is $|H(f)|^2$, and the noise power spectrum is $|H(f)|^2$. After T-spaced sampling, the aliased or "folded" signal spectrum is

$$S_{hh}(f) = \sum_n |H(f - n/T)|^2, \qquad 0 \leq f \leq 1/T$$

and the noise power spectrum is $N_0 S_{hh}(f)$.

If the transmission medium is ideal, then the baseband signal spectrum $H(f)$ at the matched filter input is determined solely by the transmit signal-shaping filters. From the discussion in Section I-A it is clear that if ISI is to be avoid-

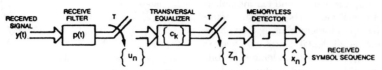

Fig. 18. Conventional linear receiver.

ed, the composite of the transmit filter and receive matched filter response must satisfy the Nyquist criterion, i.e.,

$$S_{hh}(f) = R_0, \qquad 0 \leqslant f \leqslant 1/T$$

where, in general,

$$R_0 = T\int_0^{1/T} S_{hh}(f)\, df.$$

Therefore, the overall Nyquist amplitude response $|H(f)|^2$ must be equally divided between the transmit and receive filters. For instance, each filter may have an amplitude response which is a square root of a raised-cosine characteristic [54], [113]. For such an ideal additive white Gaussian noise (AWGN) channel, the matched filter, symbol-rate sampler, and a memoryless detector comprise the optimum receiver.

Let us now consider the more interesting case of an AWGN channel with linear distortion. For such a channel, the simple linear receiver described above is no longer adequate. The symbol-rate sampled sequence, though still providing a set of sufficient statistics, now contains intersymbol interference in addition to noise. The current received symbol is distorted by a linear combination of past and future transmitted symbols. Therefore, a memoryless symbol-by-symbol detector is not optimum for estimating the transmitted sequence. Nonlinear receivers which attempt to minimize some measure of error probability are the subject of Section II-D, where the emphasis is on techniques which combine nonlinear processing with adaptive filters.

It is instructive to first study linear receivers, which attempt to maximize signal-to-noise ratio (SNR) (minimize MSE) prior to memoryless detection.

C. Linear Receivers

We begin by reviewing the conventional linear receiver comprising a matched filter, a symbol-rate sampler, an infinite-length symbol-spaced equalizer, and a memoryless detector. In the following section we show that the matched-filter, sampler, symbol-spaced equalizer combination is a special case of a more general infinite-length fractionally spaced transversal filter/equalizer. In fact, this general filter may be used as the receive filter for any receiver structure without loss of optimality. Sections II-C3 and II-C4 present a contrast between practical forms of the conventional and fractionally spaced receiver structures.

1) Matched Filter and Infinite-Length Symbol-Spaced Equalizer: If further processing of the symbol-rate sampled sequence at the output of a matched receive filter is restricted to be linear, this linear processor takes the general form of a T-spaced infinite-length transversal or nonrecursive equalizer followed by a memoryless detector [31], [54]. Let the periodic frequency response of the transversal equalizer be $C(f)$. The equalized signal spectrum is

$S_{hh}(f)C(f), 0 \leqslant f \leqslant 1/T$. The optimum $C(f)$ is one which minimizes the MSE at its output. The MSE is given by

$$\epsilon = T\int_0^{1/T} |1 - S_{hh}(f)C(f)|^2 + N_0 S_{hh}(f)|C(f)|^2\, df \quad (3)$$

where the first term is the residual ISI power, and the second term is the output-noise power. Differentiating the integrand with respect to $C(f)$ and equating the result to zero, one obtains the minimum MSE (MMSE) equalizer frequency response

$$C(f) = 1/[N_0 + S_{hh}(f)]. \quad (4)$$

Thus the best (MMSE) matched-filtered equalized signal spectrum is given by

$$S_{hh}(f)/[N_0 + S_{hh}(f)]. \quad (5)$$

Substituting (4) in (3), one obtains the following expression for the minimum MSE achievable by a linear receiver:

$$\epsilon_{min}\,(\text{linear}) = T\int_0^{1/T} N_0/[N_0 + S_{hh}(f)]\, df. \quad (6)$$

The receiver structure (Fig. 18) comprising a matched filter, a symbol-rate sampler, an infinite-length T-spaced equalizer, and a memoryless detector is referred to as the conventional linear receiver.

If the equalizer response $C(f)$ is designed to satisfy the zero-ISI constraint then $C(f) = 1/S_{hh}(f)$.

The overall frequency response of the optimum zero-forcing (ZF) linear receiver is given by $H^*(f)/S_{hh}(f)$, which forces the ISI at the receiver output to zero. The MSE achieved by such a receiver is

$$\epsilon_{ZF}\,(\text{linear}) = T\int_0^{1/T} N_0/S_{hh}(f)\, df. \quad (7)$$

$\epsilon_{ZF}\,(\text{linear})$ is always greater than or equal to $\epsilon_{min}\,(\text{linear})$ because no consideration is given to the output-noise power in designing the ZF equalizer. At high signal-to-noise ratios the two equalizers are nearly equivalent.

2) Infinite-Length Fractionally Spaced Transversal Filter: In this section, we derive an alternative form of an optimum linear receiver.

Let us start with the conventional receiver structure comprising a matched filter, symbol-rate sampler, T-spaced transversal equalizer, and a memoryless detector. As a first step, linearity permits us to interchange the order of the T-spaced transversal equalizer and the symbol-rate sampler. Next, the composite response $H^*(f)C(f)$ of the matched filter in cascade with a T-spaced transversal equalizer may be realized by a single continuous-time filter. Let us assume that the received signal spectrum $H(f) = 0$ outside the range $|f| \leqslant 1/2\tau, \tau \leqslant T$, and the flat noise spectrum is also limited to the same band by an ideal antialiasing filter with cutoff frequency $1/2\tau$. The composite matched-filter equalizer can then be realized by an infinite-length continuous-time transversal filter with taps spaced at τ-second intervals and frequency response $H^*(f)C(f), |f| \leqslant 1/2\tau$, which is periodic with period $1/\tau$ hertz. The continuous output of this composite transversal filter may be sampled at the symbol rate without any further restriction on the tap-spacing τ. If the frequency response $C(f)$ is selected according to (4), the symbol-rate output of the composite filter is identical to the corresponding output (5) in the conventional MMSE linear receiver.

To implement this composite matched-filter equalizer as a fractionally spaced digital nonrecursive filter, the effective tap-spacing must be restricted to KT/M, where K and M are relatively prime integers, $K < M$, and the fraction $KT/M \leqslant \tau$. The desired frequency response $H^*(f)C(f)$ of this fractionally spaced digital transversal filter is periodic with period M/KT hertz, and limited to the band $|f| \leqslant 1/2\tau < M/2KT$ since $H(f)$ is limited to the same bandwidth. An ideal antialiasing filter with a cutoff frequency $M/2T$ hertz is assumed before the rate M/T sampler.

The operation of the digital filter may be visualized as follows. Each symbol interval, M input samples are shifted into digital shift register memory, every Kth sample in the shift register is multiplied by a successive filter coefficient, and the products summed to produce the single output required per symbol interval.

Note that the M/T rate input of the fractionally spaced digital filter has the signal spectrum

$$
\begin{aligned}
&H(f), && |f| \leqslant M/2KT \\
&0, && M/2KT < |f| \leqslant M/2T.
\end{aligned}
$$

The frequency response of the digital filter is $H^*(f)C(f), |f| \leqslant M/2KT$. Thus if the output of this filter was produced at the rate M/T, the output signal spectrum would be

$$
\begin{aligned}
&H(f)H^*(f)C(f), && |f| \leqslant 1/2KT \\
&0, && M/2KT < |f| \leqslant M/2T.
\end{aligned}
$$

When the filter output is produced at the symbol rate, it has the desired aliased signal spectrum

$$
\sum_n |H(f - n/T)|^2 C(f - n/T), \qquad 0 \leqslant f \leqslant 1/T.
$$

Noting that $C(f)$ is periodic with period $1/T$, this output-signal spectrum is recognized as $S_{hh}(f)C(f)$, which is the same as for the conventional MMSE linear receiver (5), provided $C(f)$ is selected according to (4).

This simple development proves the important point that an infinite-length fractionally spaced digital transversal filter is at once capable of performing the functions of the matched filter and the T-spaced transversal equalizer of the conventional linear receiver.

Let us further show that the symbol-rate sampled outputs of a fractionally spaced digital filter form a set of sufficient statistics for estimation of the transmitted sequence under the following conditions. The digital filter with tap spacing KT/M has the frequency response $H^*(f)C(f), |f| \leqslant M/2KT$, where the received signal spectrum $H(f)$ is zero outside the band $|f| \leqslant M/2KT$, $C(f)$ is periodic with period $1/T$, and $C(f)$ is information lossless. A sufficient condition for $C(f)$ to be information lossless is that $C(f)$ is invertible, i.e., $C(f) \neq 0$, $0 \leqslant f \leqslant 1/T$. However, it is only necessary that $C(f)/S_{hh}(f)$ is invertible, i.e., $C(f)/S_{hh}(f) \neq 0, 0 \leqslant f \leqslant 1/T$. In words, this condition implies that $C(f)$ may not

introduce any nulls or transmission zeros in the Nyquist band, except at a frequency where the signal (and the noise power spectrum at the matched-filter output) may already have a null.

The above result shows that with an appropriately designed $C(f)$, the symbol-rate outputs of a fractionally spaced filter, with frequency response $H^*(f)C(f)$, may be used without loss of optimality, for any linear or nonlinear receiver, regardless of the criterion of optimality.

In a linear receiver, where a memoryless detector operates on the symbol-rate outputs of the fractionally spaced filter, the function $C(f)$ may be designed to minimize the MSE at the detector input. The optimum $C(f)$ is then obtained using the same procedure as outlined in the previous section for the conventional linear receiver. Thus the MMSE KT/M-spaced filter frequency response is

$$
H^*(f)/[N_0 + S_{hh}(f)], \qquad |f| \leqslant M/2KT. \tag{8}
$$

The MMSE achieved by this filter is, of course, the same as ϵ_{min} (linear) derived earlier (6) for the conventional receiver structure.

If the criterion of optimality used in designing $C(f)$ is to force the intersymbol interference at the detector input to zero, one obtains the overall filter response

$$
H^*(f)/S_{hh}(f), \qquad |f| \leqslant M/2KT. \tag{9}
$$

The MSE achieved in this case is the same as ϵ_{ZF} (linear) derived earlier (7).

3) Fixed-Filter and Finite-Length Symbol-Spaced Equalizer: The conventional MMSE linear receiver is impractical for two reasons. First, constraints of finite length and computational complexity must be imposed on the matched filter as well as the T-spaced transversal equalizer. Secondly, in most applications, it is impractical to design, beforehand, a filter which is reasonably matched to the variety of received signal spectra resulting from transmissions over different channels or a time-varying channel. Thus the most commonly used receiver structure comprises a fixed filter, symbol rate sampler, and finite-length T-spaced adaptive equalizer (Fig. 18). The fixed-filter response is either matched to the transmitted signal shape or is designed as a compromise equalizer which attempts to equalize the average of the class of line characteristics expected for the application. For the present discussion, let us assume that the fixed filter has an impulse response $p(t)$ and a frequency response $P(f)$. Then, the T-spaced sampled output of this filter may be written as

$$
u_k = \sum_n x_n q(kT - nT) + v_k \tag{10}
$$

where

$$
q(t) = p(t) * h(t)
$$

and

$$
v_k = \int n(t)p(kT - t)\, dt.
$$

Denoting the N equalizer coefficients at time kT by the column vector c_k, and the samples stored in the equalizer delay line by the vector u_k, the equalizer output is given by

$$
z_k = c_k^T u_k
$$

where the superscript T denotes transpose. Minimizing the MSE $E[|z_k - x_k|^2]$ leads to the set of optimum equalizer coefficients

$$c_{opt} = A^{-1}\alpha \qquad (11)$$

where A is an $N \times N$ Hermitian covariance matrix $E[u_k^* u_k^T]$, and α is an N-element cross-correlation vector $E[u_k^* x_k]$.

Using the assumption that the data sequence $\{x_k\}$ is uncorrelated with unit power, it can be shown that the elements of the matrix A and vector α are given by

$$a_{i,j} = \sum_k q^*(kT) q(kT + iT - jT)$$

$$+ N_0 \int p^*(t) p(t + iT - jT)\, dt \qquad (12)$$

and

$$\alpha_i = q^*(-iT). \qquad (13)$$

The MMSE achieved by this conventional suboptimum linear receiver is given by

$$\epsilon_{min}(con) = 1 - \alpha^{*T}A^{-1}\alpha. \qquad (14)$$

Alternatively, the N equalizer coefficients may be chosen to force the samples of the combined channel and equalizer impulse response to zero at all but one of the N T-spaced instants in the span of the equalizer. The zero-forcing equalizer coefficient vector is given by

$$c_{ZF} = Q^{-1}\delta \qquad (15)$$

where Q is an $N \times N$ matrix with elements

$$q_{i,j} = q(iT - jT) \qquad (16)$$

and δ is a vector with only one nonzero element, that element being unity. The MSE achieved by the zero-forcing suboptimal receiver is given by

$$\epsilon_{ZF}(con) = \epsilon_{min}(con) + (c_{ZF} - c_{opt})^{*T} A (c_{ZF} - c_{opt})$$

where the quadratic form is the excess MSE over the LMS solution.

It is instructive to derive expressions for the equalizer and its performance as the number of coefficients is allowed to grow without bound. Since the $N \times N$ matrices A and Q are Toeplitz, their eigenvalues can be obtained by the discrete Fourier transform (DFT) of any row or column as $N \to \infty$ [44]. Thus by taking the DFT of (11) for c_{opt} we obtain the frequency spectrum of the infinite-length T-spaced LMS equalizer

$$C_{opt}(f) = \frac{Q_{eq}^*(f)}{|Q_{eq}(f)|^2 + N_0 S_{pp}(f)} \qquad (17)$$

where

$$Q_{eq}(f) = \sum_n Q(f - n/T) \qquad (18)$$

is the aliased spectrum of $q(t)$, and

$$S_{pp}(f) = \sum_n |P(f - n/T)|^2 \qquad (19)$$

is the aliased power spectrum of $p(t)$. The minimum achievable MSE is given by

$$\epsilon_{min}(con) = T \int_0^{1/T} 1 - \frac{|Q_{eq}(f)|^2}{N_0 S_{pp}(f) + |Q_{eq}(f)|^2}\, df$$

$$= T \int_0^{1/T} \frac{N_0}{N_0 + |Q_{eq}(f)|^2 / S_{pp}(f)}\, df. \qquad (20)$$

The corresponding expressions for the zero-forcing equalizer are

$$C_{ZF}(f) = 1/Q_{eq}(f) \qquad (21)$$

and

$$\epsilon_{ZF}(con) = T \int_0^{1/T} N_0 S_{pp}(f)/|Q_{eq}(f)|^2\, df. \qquad (22)$$

When $P(f)$ is a matched filter, i.e., $P(f) = H^*(f)$, the above expressions reduce to those given in Section II-B1 because $S_{pp}(f) = Q_{eq}(f) = S_{hh}(f)$.

The smallest possible MSE (zero-ISI matched filter bound) is achieved when in (20) we have

$$|Q_{eq}(f)|^2/S_{pp}(f) = S_{hh}(f) = R_0, \qquad 0 \leqslant f \leqslant 1/T.$$

This occurs when the channel amplitude characteristic is ideal and perfect equalization is achieved by the matched filter. The greater the deviation of $|Q_{eq}(f)|^2/S_{pp}(f)$ from its average R_0, the greater $\epsilon_{min}(con)$. The aliased power spectrum $S_{pp}(f)$, as defined in (19), is independent of the phase characteristics of $P(f)$ or the sampler phase. The value of the squared absolute value $|Q_{eq}(f)|^2$ of the aliased spectrum $Q_{eq}(f)$, on the other hand, is critically dependent on the sampler phase in the rolloff region due to aliasing. Thus the minimum MSE achieved by the conventional receiver is dependent on the sampler phase even when the number of T-spaced equalizer coefficients is unlimited ($N \to \infty$).

When N is finite, the value of $\epsilon_{min}(con)$ depends on the channel impulse response $h(t)$, the noise power spectral density N_0, the choice of the fixed receive filter $p(t)$, and the number N of T-spaced equalizer coefficients. Therefore, it is difficult to say much about the performance or degree of suboptimality of this receiver structure without resorting to numerical computations for particular examples. As noted above, one general characteristic of this receiver structure is the sensitivity of its performance to the choice of sampler phase. This point is discussed further in Section II-E.

4) Finite-Length Fractionally Spaced Transversal Equalizer: This suboptimum linear receiver structure is simply a practical form of the infinite-length structure discussed in Section II-C2. We shall restrict our attention to the digitally implemented fractionally spaced equalizer (FSE) with tap spacing KT/M (see Fig. 19). The input to the FSE is the received signal sampled at rate M/T

$$y(kT/M) = \sum_n x_n h(kT/M - nT) + n(kT/M). \qquad (23)$$

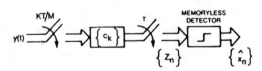

Fig. 19. Linear receiver based on a fractionally spaced transversal equalizer.

Each symbol interval, the FSE produces an output according to

$$z(kT) = \sum_{n=0}^{N-1} c_n y(kT - nKT/M). \quad (24)$$

Denoting the N equalizer coefficients at time kT by the vector c_k and the N most recently received samples (spaced kT/M seconds apart) by the vector y_k, the equalizer output may be written as

$$z_k = c_k^T y_k.$$

Minimizing the MSE $E[|z_k - x_k|^2]$ leads to the set of optimum equalizer coefficients

$$c_{opt} = A^{-1}\alpha \quad (25)$$

where A is an $N \times N$ covariance matrix $E[y_k^* y_k^T]$, and α is an N-element cross-correlation vector $E[y_k^* x_k]$.

Using the assumption that the data sequence $\{x_k\}$ is uncorrelated with unit power, it can be shown that the elements of the matrix A and vector α are given by

$$a_{i,j} = \sum_k h^*(kT - iKT/M) h(kT - jKT/M) + N_0 \delta_{ij} \quad (26)$$

$$\alpha_i = h^*(-iKT/M). \quad (27)$$

The MMSE achieved by the FSE is given by

$$\epsilon_{min}(FSE) = 1 - \alpha^{*T} A^{-1} \alpha. \quad (28)$$

On the surface, the FSE development is quite similar to that of the conventional T-spaced LMS equalizer given in the previous section. There are, however, significant differences. First, unlike the T equalizer, the FSE does not require a fixed receive shaping filter $p(t)$. Secondly, note that while the FSE input covariance matrix A is Hermitian, it is not Toeplitz. In fact, each diagonal periodically takes one of M different values. Due to the non-Toeplitz cyclostationary nature of the A matrix, it is no longer possible to obtain the eigenvalues of A by simply taking the DFT of one of its rows even as $N \to \infty$. However, it is possible to decompose the set of infinite equations

$$Ac = \alpha$$

into M subsets each with M Toeplitz submatrices. Using this procedure, it can be shown [39] that as $N \to \infty$, a fraction $(M - K)/M$ of the eigenvalue are equal to N_0, and the remaining eigenvalues are of the form $(M/K) S_{hh}(f) + N_0$, where

$$S_{hh}(iM/NKT) = \sum_n |H(iM/NKT - n/T)|^2,$$
$$i = 0,1,2,\cdots,(NK/M) - 1. \quad (29)$$

The frequency response of the optimum FSE approaches (8) as $N \to \infty$, and its MSE approaches ϵ_{min} (linear) given in (6).

As the noise becomes vanishingly small, an infinitely long FSE has a set of zero eigenvalues. This implies that there are an infinite number of solutions which produce the same minimum MSE. The nonunique nature of the infinite FSE is evident from the fact that when both signal and noise vanish in the frequency range $1/2T < |f| \leqslant M/2T$, the in-finite FSE spectrum $C(f)$ can take any value in this frequency range without affecting the output signal or MSE.

Gitlin and Weinstein [39] show that for transmission systems with less than 100-percent excess bandwidth, the matrix A is nonsingular for a finite-length FSE even as the noise becomes vanishingly small. Therefore, there exists a unique set of optimum equalizer coefficients c_{opt} given by (25).

Deviation of the coefficient vector c_k from the optimum results in the following excess MSE over $\epsilon_{min}(FSE)$ given in (28):

$$(c_k - c_{opt})^{*T} A(c_k - c_{opt}).$$

This quadratic form may be diagonalized to obtain $d_k^{*T} \Lambda d_k$, where the diagonal matrix Λ has the eigenvalues of A along its main diagonal, and d_k is the transformed coefficient deviation vector according to

$$d_k = V(c_k - c_{opt}) \quad (30)$$

where the columns of the diagonalizing matrix V are the eigenvectors of A. From the analysis of the infinite FSE, one would expect that when the number of FSE coefficients is "large," a significant fraction $(M - K)/M$ of the eigenvalues of A are relatively small. If the ith eigenvalue is very small, the ith element of the deviation vector d_k will not contribute significantly to the excess MSE. It is, therefore, possible for coefficient deviations to exist along the eigenvectors corresponding to the small eigenvalues of A without significant impact on the MSE. Thus many coefficient vectors may produce essentially the same MSE.

The most significant difference in the behavior of the conventional suboptimum receiver and the FSE is a direct consequence of the higher sampling rate at the input to the FSE. Since no aliasing takes place at the FSE input, it can independently manipulate the spectrum in the two rolloff regions to minimize the output MSE after symbol-rate sampling. Thus unlike the T-spaced equalizer, it is possible for the FSE to compensate for timing phase as well as asymmetry in the channel amplitude or delay characteristics without noise enhancement. This is discussed further in Section II-E.

D. Nonlinear Receivers

The MMSE linear receiver is optimum with respect to the ultimate criterion of minimum probability of symbol error only when the channel does not introduce any amplitude distortion, i.e., $S_{hh}(f) = R_0, 0 \leqslant f \leqslant 1/T$. The linear receive filter then achieves the matched filter (mf) bound for MSE

$$\epsilon_{min}(mf) = T \int_0^{1/T} N_0 / [N_0 + R_0] \, df$$

and a memoryless threshold detector is sufficient to minimize the probability of error. When amplitude distortion is present in the channel, a linear receive filter, e.g., FSE, can reduce ISI and output MSE by providing the output signal spectrum

$$S_{hh}(f) / [N_0 + S_{hh}(f)].$$

The corresponding output error power spectrum is $N_0 / [N_0 + S_{hh}(f)]$. Thus noise power is enhanced at those frequencies where $S_{hh}(f) < R_0$. A memoryless detector operating on the output of this received filter no longer minimizes symbol error probability.

Recognizing this fact, several authors have investigated optimum or approximately optimum nonlinear receiver structures subject to a variety of criteria [59]. Most of these receivers use one form or another of the maximum *a posteriori* probability rule to maximize either the probability of detecting each symbol correctly [1] or of detecting the entire transmitted sequence correctly. The classical maximum-likelihood receiver [54] consists of m^k matched filters, where k is the length of the transmitted sequence whose symbols are drawn from a discrete alphabet of size m.

The complexity of the classical receiver, which grows exponentially with the message length, can be avoided by using the Viterbi algorithm. This recursive algorithm which was originally invented to decode convolutional codes was recognized to be a maximum-likelihood sequence estimator (MLSE) of the state sequence of a finite-state Markov process observed in memoryless noise [25]. Forney [24] showed that if the receive filter is a whitened matched filter, its symbol rate outputs at the correct sampling times form a set of sufficient statistics for estimation of the information sequence. Thus the transmission system between the data source and the Viterbi algorithm (VA) can be considered as a discrete channel, as shown in Fig. 20. The state and hence the input sequence of the discrete channel can be estimated by the VA which observes the channel output corrupted by additive white Gaussian noise [24]. The computational complexity of the MLSE is proportional to m^{L-1}, the number of discrete channel states, where L is the number of terms in the discrete channel pulse response.

The MLSE maximizes the mean time between error (MTBE) events, a reasonable criterion for practical automatic repeat request (ARQ) data communications systems where efficiency is measured by throughput (i.e., the number of blocks of data correctly received versus the total number of blocks transmitted).

The symbol error probability of the MLSE [24] is estimated by an expression of the form

$$\Pr(e) \simeq KQ[d_{min}/2\sigma].$$

The minimum Euclidean distance between any two valid neighboring sequences is d_{min}^2, σ^2 is the mean square white Gaussian noise power at the input to the VA, and d_{min}^2/σ^2 is the effective SNR.

$$Q(x) = (1/2\pi) \int_x^\infty \exp(-y^2/2)\, dy$$

is the Gaussian probability of error function, and the error coefficient K may be interpreted as the average number of ways in which minimum distance symbol errors can occur.

The lower bound on the probability of error for binary transmission over an ideal AWGN channel is given by

$$Q\left[\left(2\epsilon_{min}(\mathrm{mf})\right)^{-1/2}\right].$$

The MLSE approaches this lower bound at high SNR for all channels except those with extremely severe ISI. For instance, for a class IV partial response system with a discrete channel model of the form $1 - D^2$, the symbol error rate achieved by the MLSE is about 4 times the lower bound given above. In decibels this difference is small (about 0.5 dB for a symbol error rate of 10^{-4}) and goes to zero as the effective SNR goes to infinity.

For unknown and/or slowly time-varying channels, the MLSE can be made adaptive by ensuring that both the whitened matched filter and the channel model used by the VA adapt to the channel response. Magee and Proakis [62] proposed an adaptive version of the VA which uses an adaptive identification algorithm to provide an estimate of the discrete channel pulse response. Structures with adaptive whitened matched filters (WMF) were proposed in [61] and [88].

Forney [24] derived the WMF as the cascade connection of a matched filter $H^*(f)$, a symbol-rate sampler, and a T-spaced transversal whitening filter whose pulse response is the anticausal factor of the inverse filter $1/S_{hh}(f)$. The noise at the output of the WMF is white and the ISI is causal. Price [83] showed that the WMF is also the optimum forward filter in a zero-forcing decision-feedback equalizer, where the feedback transversal filter can exactly cancel the causal ISI provided all past decisions are correct. (Earlier, Monsen [71] had derived the optimum forward filter for an LMS DFE.)

At this point it is helpful to discuss simpler forms of nonlinear receivers, i.e., decision feedback equalization and

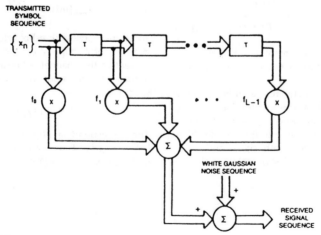

Fig. 20. Discrete channel model.

general decision-aided ISI cancellation, before returning to the topic of adaptive receiver filtering for maximum-likelihood sequence estimation.

1) Decision Feedback Equalizers: The MLSE unravels ISI by deferring decisions and weighing as many preliminary decision sequences as the number of states in the discrete channel model. Thus in most cases, the MLSE makes use of all the energy in the discrete channel impulse response to maximize the effective SNR. By contrast, a DFE makes memoryless decisions and cancels all trailing ISI terms. Even when the WMF is used as the receive filter for both the MLSE and the DFE, the latter suffers from a reduced effective SNR, and error propagation, due to its inability to defer decisions.

a) Infinite length: A decision-feedback equalizer (DFE) takes advantage of the symbols which have already been detected (correctly with high probability) to cancel the intersymbol interference due to these symbols without noise enhancement. An infinite-length DFE receiver takes the general form (Fig. 21) of a forward linear receive filter, symbol-rate sampler, canceler, and memoryless detector. The symbol-rate output of the detector is then used by the feedback filter to generate future outputs for cancellation.

As pointed out by Belfiore and Park [4], an equivalent structure to the DFE receiver of Fig. 21 is the structure shown in Fig. 22. The latter may be motivated from the point of view that given the MMSE forward filter, e.g., an infinite-length FSE, we know that the sequence of symbol-rate samples at the output of this filter form a set of sufficient statistics for estimating the transmitted sequence. Then what simple form of nonlinear processing could further reduce the MSE? To this end, let us examine the power spectra of the two components of distortion: noise and ISI. Noise at the output of the MMSE forward filter (see Section II-C1) has the power spectrum

$$N_0 S_{hh}(f)/[N_0 + S_{hh}(f)]^2, \qquad 0 \leqslant f \leqslant 1/T$$

and the residual ISI has the power spectrum

$$|1 - S_{hh}(f)/[N_0 + S_{hh}(f)]|^2 = N_0^2/[N_0 + S_{hh}(f)]^2,$$
$$0 \leqslant f \leqslant 1/T.$$

Since noise and ISI are independent, the power spectrum of the total distortion or error sequence is given by the sum of the noise and ISI power spectra, i.e.,

$$|\hat{E}(f)|^2 = N_0/[N_0 + S_{hh}(f)], \qquad 0 \leqslant f \leqslant 1/T. \quad (31)$$

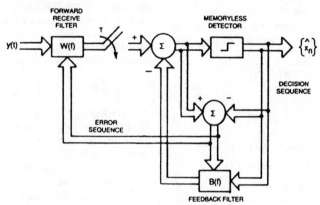

Fig. 21. Conventional decision-feedback receiver.

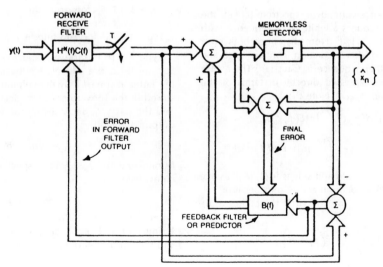

Fig. 22. Predictor form of decision-feedback receiver.

The error sequence is white if and only if $S_{hh}(f)$ is a constant; e.g., $S_{hh}(f) = R_0$, $|f| \leqslant 1/2T$, when the channel has no amplitude distortion. In this case, further reduction in MSE is not possible. However, for channels with amplitude distortion, the power of the error sequence at the output of the forward filter can be reduced further by linear prediction [122], provided past samples of the error sequence are available. An estimate of these past error samples can be obtained by decision feedback via a memoryless detector, as shown in Fig. 22. To complete the picture, it remains to derive the optimum predictor spectrum as the number of predictor coefficients grows without bound.

The error sequence at the predictor output has the spectrum

$$E(f) = \hat{E}(f) + \hat{E}(f)B(f)$$

where $B(f)$ is the desired spectrum of the infinitely long predictor, i.e.,

$$B(f) = \sum_{n=1}^{\infty} b_n e^{-j2\pi fnT}$$

The optimum $B(f)$ is one which minimizes the final MSE

$$\epsilon(\text{DFE}) = T \int_0^{1/T} |1 + B(f)|^2 |\hat{E}(f)|^2 \, df.$$

The solution is available from the theory of one-step predictors and Toeplitz quadratic forms [44]. There exists a factorization of the inverse power spectrum of the error sequence, such that

$$\Gamma(f)\Gamma^*(f) = 1/|\hat{E}(f)|^2$$

where

$$\Gamma(f) = \sum_{n=0}^{\infty} \gamma_n \exp(-j2\pi fnT)$$

i.e., $\{\gamma_n\}$ is causal. Then the optimum $B(f)$ is given by

$$B(f) = \Gamma(f)/\gamma_0 - 1, \qquad 0 \leqslant f \leqslant 1/T$$

where the normalizing factor γ_0 is the average value of $\Gamma(f)$ in the range $|f| \leqslant 1/2T$. The minimum achievable MSE is

$$\epsilon_{\min}(\text{DFE}) = 1/|\gamma_0|^2. \tag{32}$$

Given the error sequence with power spectrum $|\hat{E}(f)|^2$, the optimum predictor produces a white error sequence with power spectral density $1/|\gamma_0|^2$. Note that since both error sequences, before and after prediction, contain noise and ISI components, neither sequence is Gaussian. The MMSE given in (32) can be expressed directly in terms of the folded power spectrum $S_{hh}(f)$. Note that

$$T \int_0^{1/T} \ln |\hat{E}(f)|^2 \, df = T \int_0^{1/T} \ln |E(f)|^2 \, df$$
$$- T \int_0^{1/T} \ln |1 + B(f)|^2 \, df.$$

The first integral on the right-hand side is $\ln(1/|\lambda_0|^2)$ since $|E(f)|^2 = 1/|\lambda_0|^2$. The second integral is zero because $1 + B(f)$ has all its zeros inside the unit circle [122]. Thus using (32), we have

$$\epsilon_{\min}(\text{DFE}) = \exp \left[T \int_0^{1/T} \ln |\hat{E}(f)|^2 \, df \right]$$
$$= \exp \left[-T \int_0^{1/T} \ln (1 + S_{hh}(f)/N_0) \, df \right]. \tag{33}$$

This expression is identical to the MMSE for an infinite length conventionl DFE [93], which proves the equivalence of the predictor and conventional DFE structures.

We can write the following equivalence relationship by comparing the two DFE structures shown in Figs. 21 and 22:

$$W(f) = H^*(f)C(f)[1 + B(f)]. \tag{34}$$

After some inspection, it becomes evident that the infinite-length forward filter $W(f)$ in the conventional DFE structure is the cascade of a matched filter and the anticausal factor of the optimum $C(f)$ given in (4) [71]. So long as the length of the forward filter in each of the two DFE structures is unconstrained, the two structures remain equivalent even when the feedback (or prediction) filter is reduced to a finite length. Note, however, that while the forward filter $W(f)$ in the conventional structure depends on the number of feedback coefficients, the forward filter in the predictor structure is independent of the predictor coefficients.

An alternative to the optimum mean-square DFE receiver is the zero-forcing formulation [4], [83]. The forward filter is again of the form (34). However, $C(f)$ is designed to satisfy the zero-ISI constraint, i.e., $C(f) = 1/S_{hh}(f)$. Thus the forward filter $W(f)$ is the cascade of a matched filter and the anticausal factor of the inverse filter $1/S_{hh}(f)$. The noise at the output of $W(f)$ is white and the ISI is causal (as in the case of Forney's whitened matched filter). The causal ISI is completely canceled by an infinite-length decision-feedback filter. The final error sequence consists solely of noise which is white and Gaussian with power

$$\epsilon_{\text{ZF}}(\text{DFE}) = \exp \left[-T \int_0^{1/T} \ln (S_{hh}(f)/N_0) \, df \right]. \tag{35}$$

As expected, $\epsilon_{\min}(\text{mf}) \leqslant \epsilon_{\min}(\text{DFE}) \leqslant \epsilon_{\text{ZF}}(\text{DFE})$.

b) Finite length: Neglecting zero-forcing equalizers, there are four possible structures for finite-length DFE receivers, based on the conventional or FSE forward filters, and conventional or predictor forms of feedback filters. Let us denote these as

Type 1: Conventional forward filter + conventional feedback filter.

Type 2: Conventional forward filter + predictor feedback filter.

Type 3: FSE forward filter + conventional feedback filter.

Type 4: FSE forward filter + predictor feedback filter.

Type 1: To the forward equalizer structure described in Section II-C3, we add N_b feedback coefficients. Denoting the latter at time kT by the column vector $\boldsymbol{b}_k$, the samples stored in the forward equalizer delay line by the vector $\boldsymbol{u}_k$, and the past N_b decisions by the vector $\boldsymbol{x}_k$, the equalizer output is given by

$$z_k = \boldsymbol{c}_k^T \boldsymbol{u}_k - \boldsymbol{b}_k^T \boldsymbol{x}_k.$$

Minimizing the MSE with respect to the feedback coefficients leads to

$$\boldsymbol{b}_k = Q\boldsymbol{c}_k \tag{36}$$

where Q is an $N_b \times N$ matrix with elements given by (16). Using (36) and proceeding as in Section II-C1, it can be shown that the set of optimum forward coefficients is given by

$$\boldsymbol{c}_{\text{opt}} = \hat{A}^{-1}\boldsymbol{\alpha}. \tag{37}$$

The $N \times N$ matrix $\hat{A}$ has elements $\hat{a}_{i,j}$ similar to $a_{i,j}$ given

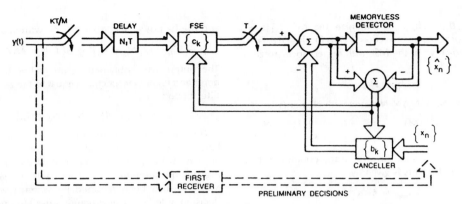

fig. 23. Decision-aided ISI cancellation.

by (12), except that the summation over k now excludes the set $1 \leqslant k \leqslant N_b$ (which is in the span of the feedback filter). Finally

$$b_{\text{opt}} = Qc_{\text{opt}}.$$

Type 2: In this case, the forward equalizer coefficients can be obtained from (11) independently of the predictor coefficients. Let $\hat{e}_k$ be an N_b-element vector at time kT consisting of the N_b most recent error signals before prediction at instants $k, k-1, \cdots, k - N_b - 1$, with the forward equalizer coefficients at their optimal values. Then, the set of optimum predictor coefficients b_k is the solution of the normal equations

$$E\left[\hat{e}_{k-1}^* \hat{e}_{k-1}^T \right] b_k = E\left[\hat{e}_k^* \hat{e}_k \right]. \tag{38}$$

Type 3: This case is similar to Type 1, except that the forward filter structure described in Section II-C4 is used. The set of optimum forward coefficients can be obtained using

$$c_{\text{opt}} = \hat{A}^{-1}\alpha$$

where the matrix $\hat{A}$ has elements $a_{i,j}$ similar to $\hat{a}_{i,j}$ given by (26), except that the summation over k now excludes the set $1 \leqslant K \leqslant N_b$. The optimum feedback coefficients are given by

$$b_{\text{opt}} = Hc_{\text{opt}}$$

where H is an $N_b \times N$ matrix with elements $h_{i,j} = h(iT - jKT/M)$.

Type 4: This type of DFE is similar to Type 2. The optimum forward coefficients given by (25) still apply and the predictor coefficients can be obtained by solving a set of normal equations similar to that given in (38).

For a direct comparison of the performance of the four types of DFE structures, resort must be made to numerical solution for particular channel characteristics, number of equalizer coefficients, etc. One general comment that can be made is that for an equal number of forward and feedback coefficients, the DFE structures of Type 1 and 3 will always achieve an output MSE at least as low or lower than the MSE achieved by structures of Type 2 and 4, respectively. This is true because unlike Type 1 and 3 structures, independent solution of the forward equalizer and the feedback predictor coefficients in Type 2 and 4 DFE structures does not in general guarantee joint minimization of the final MSE.

One important factor in the practical performance of DFE structures of all types is the effect of error propagation on the final error probability. Assuming correct decisions, the improvement in output signal-to-MSE ratio provided by a DFE reduces the probability of occurrence of the first error. However, once an error occurs it tends to propagate due to incorrect decision feedback. Bounds on the error multiplication factor have been developed by Duttweiler *et al.* [12] and by Belfiore and Park [4].

2) Decision-Aided ISI Cancellation: The concept of decision feedback of past data symbols to cancel intersymbol interference can theoretically be extended to include future data symbols. If all past and future data symbols were assumed to be known at the receiver, then given a perfect model of the ISI process, all ISI could be canceled exactly without any noise enhancement. Such a hypothetical receiver could, therefore, achieve the zero-ISI matched-filter bound on performance. In practice, the concept of decision-aided ISI cancellation can be implemented by using tentative decisions and some finite delay at the receiver. In the absence of tentative decision errors, the ISI due to these finite number of future data symbols (as well as past data symbols) can be canceled exactly before final receiver decisions are made. Such a receiver structure with a two-step decision process was proposed by Proakis [85]. Recently, Gersho and Lim [34] observed that the MSE performance of this receiver structure could be improved considerably by an adaptive matched filter in the path of the received signal prior to ISI cancellation. In fact, the zero-ISI matched-filter bound on MSE could be achieved in the limit assuming correct tentative decisions. A general theoretical treatment is available in [80].

Consider the block diagram of Fig. 23 where correct data symbols are assumed to be known to the canceler. Let the forward filter have N coefficients fractionally spaced at KT/M-second intervals. Then the final output of the structure is given by

$$z_k = c_k^T y_k - b_k^T x_k$$

where c_k and y_k are the forward filter coefficient and input vectors, respectively, with N elements as defined earlier in Section II-C2, b_k is the vector of the canceler coefficients spaced T seconds apart, and x_k is the vector of data symbols. Each of these vectors is of length $N_1 + N_2$. The data vector x_k has elements $x_{k+N_1}, \cdots x_{k+1}, x_{k-1}, \cdots, x_{k-N_2}$. Note that the current data symbol x_k is omitted. The canceler has N_1 noncausal and N_2 causal coefficients

$b_{-N_1}, \cdots, b_{-1}, b_1, \cdots, b_{N_2}$. To minimize the MSE $E[|z_k - x_k|^2]$, we can proceed as in Section II-D1b for the DFE. First, setting the derivative of the MSE with respect to b_k to zero we obtain

$$b_k = Hc_k$$

where H is an $(N_1 + N_2) \times N$ matrix with elements

$$h_{i,j} = h(iT - jKT/M), \qquad i = -N_1, \cdots, 1, 1, \cdots, N_2$$
$$j = 0, 1, \cdots, N - 1.$$

Using this result, it can be shown that the optimum set of forward filter coefficients is given by

$$c_{\mathrm{opt}} = \hat{A}^{-1}\alpha$$

where the matrix $\hat{A}$ has elements $\hat{a}_{i,j}$ similar to $a_{i,j}$ given by (26), except that the summation over k now excludes the set $k = -N_1, \cdots, 1, 1, \cdots, N_2$ (which is in the span of the canceler). The vector α is the same as in (27). Finally, the optimum canceler coefficient vector $b_{\mathrm{opt}} = Hc_{\mathrm{opt}}$ consists of T-spaced samples of the impulse response of the channel and the forward filter in cascade, omitting the reference sample.

The role of the forward filter in this structure may be better understood from the following point of view. Given an optimum (LMS) desired impulse response (DIR) for the canceler (with the reference sample forced to unity), the forward filter equalizes the channel response to this desired impulse response with least MSE. In fact both the DIR, modeled by the canceler coefficients, and the forward filter coefficients are being jointly optimized to minimize the final MSE. The same basic idea was used by Falconer and Magee [17] to create a truncated DIR channel for further processing by the Viterbi algorithm. In [17], a unit energy constraint was imposed on the DIR while in the ISI canceler structure the reference sample of the DIR is constrained to be unity. Another difference between the two structures is the use of a forward filter with fractional tap spacing rather than a predetermined "matched" filter followed by a T-spaced equalizer. The use of a fractionally spaced forward filter for creating a truncated DIR was proposed in [88] noting its ability to perform combined adaptive matched filtering and equalization.

As we allow the lengths of the forward filter and the canceler to grow without bound, the forward filter evolves into a matched filter with frequency response [34], [80]

$$C(f) = \frac{H^*(f)}{N_0 + R_0}.$$

The canceler models the T-spaced impulse response of the channel and forward filter in cascade, all except the reference sample. The frequency response of the canceler may be written as

$$B(f) = \frac{S_{hh}(f)}{N_0 + R_0} - T\int_0^{1/T} \frac{S_{hh}(f)}{N_0 + R_0}\, df$$
$$= \frac{S_{hh}(f) - R_0}{N_0 + R_0}.$$

Note that since all the quantities in the above expression for the canceler frequency response are real, the canceler coefficients must be Hermitian symmetric.

The output noise power spectrum is given by

$$N_0|C(f)|^2 = N_0 S_{hh}(f)/[N_0 + R_0]^2$$

and the ISI power spectrum may be written as

$$|1 - S_{hh}(f)/[N_0 + R_0]|^2 + |B(f)|^2 = N_0^2/[N_0 + R_0]^2.$$

The output MSE, obtained by integrating the sum of the noise and ISI power spectra, is equal to the zero-ISI matched filter bound, i.e.,

$$\epsilon_{\min}(\mathrm{mf}) = N_0/[N_0 + R_0].$$

The critical question regarding decision aided ISI cancellation is the effect of tentative decision errors on the final error probability of the receiver. Published results answering this question are not yet available. However, reduced MSE has also been reported [5] when a version of the decision-aided receiver structure is used to cancel nonlinear ISI.

3) Adaptive Filters for MLSE: Adaptive receive filtering prior to Viterbi detection is of interest from two points of view. First, for unknown and/or slowly time-varying channels the receive filter must be adaptive in order to obtain the ultimate performance gain from maximum-likelihood sequence estimation. Secondly, the complexity of the MLSE becomes prohibitive for practical channels with a larger number of ISI terms. Therefore, in a practical receiver, an adaptive receive filter may be used to limit the time spread of the channel as well as to track slow time variation in the channel characteristics [17], [89].

By a development similar to that given in Section II-C2, it can be shown that a fractionally spaced transversal filter can model the characteristics of the WMF proposed by Forney [24]. However, the constraint on this filter to produce zero anticausal ISI makes it difficult to derive an algorithm for updating the filter coefficients in an adaptive receiver.

The general problem of adaptive receive filtering for MLSE may be approached as follows. We know from Section II-C2 that an LMS FSE produces the composite response of a matched filter and an LMS T-spaced equalizer, when the MSE is defined with respect to a unit pulse DIR. In general, an LMS FSE frequency response can always be viewed as the composite of a matched filter and an LMS T-spaced equalizer response

$$H^*(f)G(f)/[N_0 + S_{hh}(f)] \qquad (39)$$

where $G(f)$ is the DIR frequency response with respect to which the MSE is minimized. After symbol-rate sampling, the signal spectrum at the output of the FSE is given by $S_{hh}(f)G(f)/[N_0 + S_{hh}(f)]$, $|f| \leq 1/2T$, and the output noise power spectrum may be written as

$$N_0 S_{hh}(f)|G(f)|^2/[N_0 + S_{hh}(f)]^2, \qquad |f| \leq 1/2T.$$

The selection of the DIR is therefore the crux of the problem. Fredericsson [30] has shown that from an effective MSE point of view, best performance is obtained when the DIR is selected such that its power spectrum is

$$|G(f)|^2 = [N_0 + S_{hh}(f)]/(R_0 + N_0), \qquad |f| \leq 1/2T. \qquad (40)$$

With this optimum DIR and the receive filter selected according to (39), the residual ISI power spectrum at the input to the Viterbi algorithm is given by

$$|G(f) - S_{hh}(f)G(f)/[N_0 + S_{hh}(f)]|^2$$
$$= N_0/[N_0 + S_{hh}(f)](R_0 + N_0).$$

Summing the residual ISI and noise power spectra we

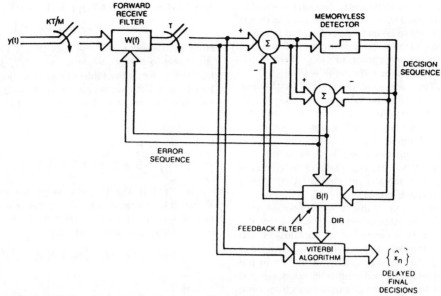

fig. 24. Adaptive MLSE receiver with causal desired impulse response (DIR) developed via decision-feedback equalization.

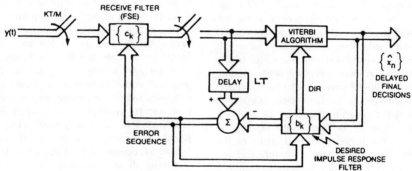

Fig. 25. General form of adaptive MLSE receiver with finite-length desired impulse response.

obtain the power spectrum of the combined error sequence at the input to the Viterbi algorithm. The error sequence is found to be white with mean-squared value $N_0/(R_0 + N_0)$, which is equal to the matched-filter bound. Note that the Gaussian noise component of the error sequence is approximately white at a moderately high SNR. However, the total error sequence while white is not Gaussian due to residual ISI.

Let us select the DIR such that $G(f)$ is the causal factor of the power spectrum given in (40). Then the resulting FSE (39) may be recognized as the optimum forward filter of an infinite-length conventional LMS DFE. At moderately high SNR this FSE approaches the WMF. An MLSE receiver using such a receive filter is shown in Fig. 24. This receiver can easily be made adaptive by updating both the FSE and the feedback filter coefficients to jointly minimize the mean-squared value of the error sequence. The feedback filter coefficients also provide all but the first DIR coefficient for use by the Viterbi algorithm. The first DIR coefficient is assumed to be unity.

Fredericsson [30] points out the difficulty of obtaining a general explicit solution for the optimum truncated DIR of a specified finite length. However, when the DIR is limited

to two or three terms [17], [30], it invariably takes the form of one or the other familiar class of partial response systems [48] with a null at one or both band edges. As mentioned in Section II-D, the Viterbi algorithm is able to recover most of the nearly 3-dB loss which otherwise results from use of bandwidth-efficient partial-response systems.

Several methods of jointly optimizing the fractionally spaced receive filter and the truncated DIR are available which minimize the MSE at the input to the Viterbi algorithm (VA). These methods differ in the form of constraint [17], [68] on the DIR which is necessary in this optimization process to exclude the selection of the null DIR corresponding to no transmission through the channel. The general form of such a receiver is shown in Fig. 25.

One such constraint is to restrict the DIR to be causal and to restrict the first coefficient of the DIR to be unity. In this case, the delay LT in Fig. 25 is equal to the delay through the VA and the first coefficient of $\{b_k\}$ is constrained to be unity.

If the causality constraint is removed (as in [34] for decision-aided ISI cancellation) but the reference (or center) coefficient of the DIR is constrained to unity, another form of the receiver shown in Fig. 25 is obtained [103]. In this case the delay LT is equal to the delay through the VA plus N_1T where the DIR has $2N_1 + 1$ coefficients with its center coefficient constrained to be unity. As before, this structure can easily be made adaptive by updating the FSE and DIR coefficients in a direction opposite to the gradient of the squared error with respect to each coefficient.

The least restrictive constraint on the DIR is the unit energy constraint proposed by Falconer and Magee [17]. This leads to yet another form of the receiver structure shown in Fig. 25. However, the adaptation algorithm for updating the desired impulse response coefficients $\{b_k\}$ is considerably more complicated (see [17]). Note that the fixed predetermined WMF and T-spaced prefilter combination of [17] has been replaced in Fig. 25 by a general fractionally spaced adaptive filter.

A common characteristic of the above mentioned truncated DIR suboptimum MLSE receiver structures is that the sample sequence at the input to the VA contains residual ISI (with respect to the DIR), and that the noise sequence is not white. Bounds on the performance of the VA in the presence of correlated noise and residual ISI are developed in [89]. In practice, so long as the DIR power spectrum more or less matches the nulls or high-attenuation regions of the channel folded power spectrum, a reasonable length FSE can manipulate the channel response to the truncated DIR without significant noise enhancement or residual ISI. However, the degree to which the noise sequence at the VA input is uncorrelated depends on the constraints imposed on the DIR. As mentioned earlier, as the causal DIR length is allowed to grow, the forward filter in the structure of Fig. 25 approaches the WMF at moderately high SNR, resulting in uncorrelated noise at the VA input. On the other hand, if the DIR is noncausal, the receive filter approaches a matched filter as the DIR length approaches the length of the original channel impulse response, resulting in the noise to be colored according to the folded power spectrum of the received signal. For such receiver structures, the modified VA [103], which takes the noise correlation into account, is more appropriate.

E. Timing Phase Sensitivity

As noted in earlier sections, conventional suboptimum receiver structures based on T-spaced equalizers suffer from extreme sensitivity to sampler timing phase. The inherent insensitivity of the performance of fractionally spaced equalizers to timing phase was heuristically explained in Section I-H. Here we present an overview of the influence of timing phase on the performance of various receiver structures.

Let us reconsider the conventional linear receiver of Section II-C1 in the presence of a timing phase offset t_0. Assume that $H(f) = 0$, $|f| > 1/T$, i.e., the channel has at most 100-percent excess bandwidth. The sampled noise sequence at the output of the matched filter is given by

$$n_k = n(kT + t_0) = \int n(t)h^*(t - kT - t_0)\, dt$$

where $h^*(-t)$ is the impulse response of the matched filter $H^*(f)$. The power spectrum, $N_0 S_{hh}(f)$, of the noise sequence $\{n_k\}$ is independent of the timing phase t_0.

The sampled signal spectrum, however, is a function of t_0 according to

$$S_{hh}(f, t_0) = \exp(-j2\pi f t_0)\big[\,|H(f)|^2 + |H(f - 1/T)|^2$$
$$\cdot \exp(j2\pi t_0/T)\big], \qquad 0 \leqslant f \leqslant 1/T$$

Note that when $t_0 = 0$, the alias $|H(f - 1/T)|^2$ adds constructively to $|H(f)|^2$, while for $t_0 = T/2$, destructive aliasing takes place in the foldover region around $1/2T$ hertz. In particular, if the channel power spectrum is the same at the two band edges, a null is created in the sampled signal spectrum at $f = 1/2T$ hertz when $t_0 = T/2$.

The MSE corresponding to (3) may be written as a function of timing phase, i.e.,

$$\epsilon(t_0) = T\int_0^{1/T} |1 - S_{hh}(f, t_0)C(f)|^2 + N_0 S_{hh}(f)|C(f)|^2\, df$$

where $C(f)$ is the periodic frequency response of the T-spaced equalizer which follows the sampler. Proceeding to optimize $C(f)$, as in Section II-B1, the minimum MSE may be derived as a function of t_0

$$\epsilon_{min}(t_0) = T\int_0^{1/T} N_0/\big[\,N_0 + S_{hh}^2(f, t_0)/S_{hh}(f)\big]\, df.$$

$$(41)$$

Note that if $S_{hh}^2(f, t_0)/S_{hh}(f)$ is small in the foldover region, due to poor choice of t_0, the intergrand becomes relatively large in that range of frequencies. This leads to a larger ϵ_{min}. Clearly, timing phase is unimportant when there is no excess bandwidth and therefore no aliasing.

The above development shows that for systems with excess bandwidth, the performance of the conventional MMSE linear receiver with a fixed matched filter is sensitive to choice of timing phase due to the inability of the T-spaced equalizer to invert a "null" in the sampled signal spectrum without excessive noise enhancement. This sensitivity can be avoided if the matched filter spectrum $H^*(f)$ is adjusted by a linear phase factor, $\exp(j2\pi f t_0)$, to explicitly compensate for the timing offset t_0. An infinite-length fractionally spaced equalizer obtains its insensitivity to timing phase in this way.

For instance, given a timing offset t_0, an infinite length $T/2$ equalizer synthesizes the frequency response

$$W(f) = H^*(f)\exp(j2\pi f t_0)/\big[N_0 + S_{hh}(f)\big], \qquad |f| \leqslant 1/T.$$

The signal spectrum at the equalizer input is the DFT of the $T/2$-sampled channel impulse response $\{h(kT/2 + t_0)\}$, i.e., $H(f, t_0) = H(f)\exp(-j2\pi f t_0), |f| \leqslant 1/T$. Aliasing, due to symbol-rate sampling, takes place at the equalizer output after $W(f)$ has explicitly compensated for the timing offset. This results in the equalized signal spectrum

$$H(f, t_0)W(f) + H(f - 1/T, t_0)W(f - 1/T)$$
$$= S_{hh}(f)/\big[N_0 + S_{hh}(f)\big], \qquad 0 \leqslant f \leqslant 1/T.$$

Thus the residual ISI, noise, and MSE at the $T/2$ equalizer output are all independent of the timing phase.

In a practical conventional receiver structure, the matched filter is typically replaced by a filter matched to the transmitted pulse or the received pulse over the average channel. Mazo [64] has shown that under some assumptions, the best timing phase for an infinite-length T-spaced equalizer is one which maximizes the energy in the band edge, $f = 1/2T$ hertz, component in the symbol-rate sampled signal at the equalizer input. However, on channels with severe delay distortion and moderately large excess bandwidth, e.g., 50-percent, even the timing phase which maximizes band-edge energy can lead to near nulls elsewhere in the folded spectrum. On such a channel, a T-spaced equalizer performs poorly regardless of the choice of timing phase. Conditions for amplitude depressions to occur in the folded spectrum are given in [91]. For instance, asymmetrical delay distortion in the two band-edge regions can cause the phase responses of the two aliasing components to jitter by π radians a few hundred hertz away from the band edge. Thus while the aliases add at the band edge, they subtract where they are opposite in phase. This cancellation may be accentuated by amplitude distortion in the channel causing the amplitudes of the canceling aliases to be nearly equal.

By contrast, a finite-length FSE maintains its ability to compensate for a timing offset in such a way as to equalize with the minimum of noise enhancement.

The results discussed above can be extended to decision-feedback receiver structures. Consider a conventional DFE consisting of a matched filter, a symbol-rate sampler, and infinite-length T-spaced forward and feedback filters. The MSE of such a receiver can be derived as a function of the timing phase (using the method given in [94])

$$\epsilon_{\min}(t_0) = \exp\left\{ -T\int_0^{1/T} \ln\left[1 + S_{hh}^2(f,t_0)/N_0 S_{hh}(f) \right] df \right\}.$$

(42)

As t_0 is varied, the greatest deviation in $\epsilon_{\min}(t_0)$ occurs for a channel such that

$$|H(f)|^2 = \begin{cases} 1, & |f| \leqslant (1-\alpha)/2T \\ 0.5, & (1-\alpha)/2T \leqslant |f| \leqslant (1+\alpha)/2T \end{cases}$$

where α is the rolloff factor. For this channel

$$S_{hh}(f,t_0)|_{t_0=0} = S_{hh}(f) = 1, \qquad 0 \leqslant f \leqslant 1/T$$

and

$$S_{hh}(f,t_0)|_{t_0=T/2} = \begin{cases} 1, & |f| \leqslant (1-\alpha)/2T \\ 0, & (1-\alpha)/2T \leqslant f \leqslant (1+\alpha)/2T. \end{cases}$$

Using this result in (42) we obtain for the DFE

$$\epsilon_{\min}(t_0)|_{t_0=0} = 1/(1 + 1/N_0)$$

and

$$\epsilon_{\min}(t_0)|_{t_0=T/2} = 1/(1 + 1/N_0)^{(1-\alpha)}.$$

For $N_0 = 0.01$ (20-dB SNR) and $\alpha = 0.1$, the MSE of the decision-feedback equalizer degrades by 2 dB when t_0 is varied from the best choice $t_0 = 0$ to the worse choice $t_0 = T/2$. The corresponding result for the conventional linear receiver is obtained by using (41)

$$\epsilon_{\min}(t_0)|_{t_0=0} = N_0/(1 + N_0)$$

and

$$\epsilon_{\min}(t_0)|_{t_0=T/2} = \alpha + (1-\alpha)N_0/(1 + N_0).$$

For $N_0 = 0.01$ and $\alpha = 0.1$, the MSE of the linear receiver degrades by 10.4 dB as t_0 is varied from 0 to $T/2$.

This example illustrates the ability of the conventional DFE to compensate for the spectral null created by poor choice of timing phase with much less noise enhancement

than a conventional linear receiver. However, this relative insensitivity may not translate into actual performance insensitivity from an error probability point of view due to error propagation in the DFE. In order to cancel the ISI due to a spectral null created by a bad timing phase, the feedback filter must develop a relatively long impulse response with large magnitude coefficients. Such a DFE is likely to suffer from severe error propagation.

The performance of a DFE with a fractionally spaced forward filter is, of course, insensitive to timing phase by virtue of explicit compensation of the timing offset by the forward filter before symbol-rate sampling. The same comment applies to a fractionally spaced receive filter used in conjunction with a maximum-likelihood sequence estimator.

III. Least Mean-Square Adaptation

In this section we expand upon the topics briefly introduced in Sections I-C through I-E, that is, LMS adaptation algorithms, their convergence properties, and excess MSE. The effect of finite precision in digital implementations is also discussed. The results of this section are applicable to other forms of adaptive filters, e.g., an echo canceler, with appropriate reinterpretation of terms.

The deterministic gradient algorithm, which is of little practical interest, is presented first to set the stage for a discussion of the LMS or stochastic gradient algorithm.

A. Deterministic Gradient Algorithm

When the equalizer input covariance matrix A and the cross-correlation vector α (see Sections II-C3 and II-C4) are known, one can write the MSE as a function of A, α, and the equalizer coefficient vector c_k according to

$$\epsilon_k = c_k^{*T} A c_k - 2\,\mathrm{Re}\left[c_k^{*T}\alpha \right] + 1.$$

Taking the gradient of the MSE with respect to c_k gives

$$\partial \epsilon_k / \partial c_k = 2(A c_k - \alpha).$$

(43)

Thus a deterministic (or exact) gradient algorithm for adjusting c_k to minimize ϵ_k can be written as

$$c_{k+1} = (I - \Delta A)c_k + \Delta\alpha$$

(44)

where Δ is the step-size parameter. (This update procedure is also known as the steepest descent algorithm.) Using the fact that $\alpha = A c_{\mathrm{opt}}$, from (11) or (25), and subtracting c_{opt} from both sides of (44), we obtain

$$c_{k+1} - c_{\mathrm{opt}} = (I - \Delta A)(c_k - c_{\mathrm{opt}}).$$

(45)

In order to analyze the stability and convergence of the deterministic gradient algorithm, we use coordinate transformation to diagonalize the set of equations (45) so that

$$d_{k+1} = (I - \Delta\Lambda)d_k$$

(46)

where the transformed coefficient deviation vector is defined in (30).

Since Λ is a diagonal matrix, it is clear from (46) that the ith element of d_k decays geometrically according to

$$d_{ik} = (1 - \Delta\lambda_i)^k d_{i0}, \qquad i = 0, 1, \cdots, N-1$$

(47)

where λ_i is the ith eigenvalue of A, and d_{i0} is the initial value of the ith transformed tap-gain deviation. Recall from

Section II-C2 that given d_k, the MSE at step k may be written as the sum of $\epsilon_{\min}$ and the excess MSE

$$\epsilon_k = \epsilon_{\min} + d_k^{*T} \Lambda d_k$$

$$= \epsilon_{\min} + \sum_{i=0}^{N-1} \lambda_i (1 - \Delta \lambda_i)^{2k} |d_{i0}|^2. \qquad (48)$$

Given finite initial deviations, the deterministic algorithm is stable and the MSE converges to $\epsilon_{\min}$ provided

$$0 < \Delta < 2/\lambda_{\max} \qquad (49)$$

where $\lambda_{\max}$ is the maximum eigenvalue of A. If all the eigenvalues of A are equal to λ and Δ is selected to be $1/\lambda$, the excess MSE will be reduced to zero in one adjustment step of the deterministic gradient algorithm (44). For a T-spaced equalizer, this condition corresponds to a flat folded power spectrum at the equalizer input.

When the eigenvalues of A have a large spread, i.e., the ratio $\rho = \lambda_{\max}/\lambda_{\min}$ is large, no single value of the step size Δ leads to fast convergence of all the tap-gain deviation components. When $\Delta = 2/(\lambda_{\max} + \lambda_{\min})$, the two extreme tap-gain deviation components converge at the same rate according to $[(\rho - 1)/(\rho + 1)]^k$ [33]. All other components converge at a faster rate with a time constant at most as large as $(\rho + 1)/2$ iterations. The impact of the large eigenvalue spread on the convergence of the excess MSE is somewhat less severe because tap-gain deviations corresponding to the small eigenvalues contribute less to the excess MSE (see (48)).

Two possibilities for speeding up the convergence of this deterministic algorithm can be devised. The first of these involves the use of a variable step size Δ_k in (44). This leads to the relationships

$$d_{ik} = d_{i0} \prod_{n=0}^{k-1} (1 - \Delta_n \lambda_i), \qquad i = 0, 1, \cdots, N-1$$

corresponding to (47). Observe that, if the eigenvalues are known beforehand, complete convergence can be obtained in N steps by using a variable step size provided the N values of the step size are selected such that $\Delta_n = 1/\lambda_n$, $n = 0, 1, \cdots, N-1$. Thus each step reduces one of the tap-gain deviation components to zero.

The second method of obtaining fast convergence is to replace the scalar Δ in (44) by the precomputed inverse matrix A^{-1}. This is tantamount to direct noniterative solution, since regardless of initial conditions or the eigenvalue spread of A, or the equalizer size N, optimum solution is obtained in one step.

B. LMS Gradient Algorithm

In practice, the channel characteristics are not known beforehand. Therefore, the gradient of the MSE cannot be determined exactly and must be estimated from the noisy received signal. The LMS gradient algorithm [109] is obtained from the deterministic gradient algorithm (44) by replacing the gradient

$$2(AC_k - \alpha) = 2E\left[y_k^* \left(y_k^T c_k - x_k \right) \right]$$

by its unbiased but noisy estimate $y_k^* (y_k^T c_k - x_k)$. The equalizer coefficients are adjusted once in every symbol interval according to

$$c_{k+1} = c_k - \Delta y_k^* \left(y_k^T c_k - x_k \right) = c_k - \Delta y_k^* e_k \qquad (50)$$

where y_k is the equalizer input vector, x_k is the received data symbol, and e_k is the error in the equalizer output.

Subtracting c_{opt} from both sides of (50) allows us to write

$$\left(c_{k+1} - c_{\text{opt}} \right) = \left(I - \Delta y_k^* y_k^T \right) \left(c_k - c_{\text{opt}} \right) - \Delta y_k^* e_{k \text{opt}}$$

where $e_{k \text{opt}}$ is the instantaneous error if the optimum coefficients were used. The transformed coefficient deviation vector d_k is now a random quantity. Neglecting the dependence of c_k on y_k, we see that the mean of d_k follows the recursive relationship (46), i.e.,

$$E[d_{k+1}] = (I - \Delta \Lambda) E[d_k]. \qquad (51)$$

The ensemble average of the MSE evolves according to

$$\epsilon_k = \epsilon_{\min} + E\left[d_k^{*T} \Lambda d_k \right]$$

where the second term on the right is the average excess MSE

$$\epsilon_{\Delta k} = \sum_{i=0}^{N-1} \lambda_i E\left[|d_{ik}|^2 \right].$$

This quantity is difficult to evaluate exactly in terms of the channel and equalizer parameters. Using the assumption that the equalizer input vectors y_k are statistically independent [38], [65], [84], [102], the following approximate recursive relationship can be derived:

$$E\left[|d_{k+1}|^2 \right] = M E\left[|d_k|^2 \right] + \Delta^2 \epsilon_{\min} \lambda \qquad (52)$$

where λ is the vector of eigenvalues of A and the $N \times N$ matrix M has elements

$$M_{ij} = (1 - 2\Delta \lambda_i) \delta_{ij} + \Delta^2 \lambda_i \lambda_j.$$

Let ρ be the ratio of the maximum to the effective average eigenvalue of A.

Three important results can be derived [38] using this line of analysis [102] and eigenvalue bounds [38]:

1) The LMS algorithm is stable if the step size Δ is in the range

$$0 < \Delta < 2/(N\rho\bar{\lambda}) \qquad (53)$$

where $\bar{\lambda}$ is the average eigenvalue of A and is equal to the average signal power at the equalizer input, defined according to $E[y_k^T y_k^*]/N$. When Δ satisfies (53), all eigenvalues of the matrix M in (52) are less than one in magnitude permitting mean-square coefficient deviations in (52) to converge.

2) The excess MSE follows the recursive relationship

$$\epsilon_{\Delta k+1} = \left[1 - 2\Delta\bar{\lambda} + \Delta^2 N\rho(\bar{\lambda})^2 \right] \epsilon_{\Delta k} + \Delta^2 \epsilon_{\min} N\rho(\bar{\lambda})^2.$$
$$(54)$$

If Δ is selected to minimize the excess MSE at each iteration, we would obtain

$$\Delta_k = \epsilon_{\Delta k}/(\epsilon_{\min} + \epsilon_{\Delta k})(N\rho\bar{\lambda}).$$

Initially $\epsilon_{\Delta k} \gg \epsilon_{\min}$, so that fastest convergence is obtained with an initial step size

$$\Delta_0 = 1/(N\rho\bar{\lambda}). \qquad (55)$$

Note that Δ_0 is half as large as the maximum permissible step size for stable operation.

3) The steady-state excess MSE can be determined from (54) to be

$$\epsilon_\Delta = \Delta \epsilon_{\min} N\rho\bar{\lambda}/(2 - \Delta N\rho\bar{\lambda}). \qquad (56)$$

A value of $\Delta = \Delta_0$ results in $\epsilon_\Delta = \epsilon_{\min}$, that is, the final MSE is 3 dB greater than the minimum achievable MSE. In order to reduce ϵ_Δ to $\gamma\epsilon_{min}$, where γ is a small fraction, Δ must be reduced to

$$\Delta = 2\gamma/[(1+\gamma)N\rho\bar\lambda] = 2\Delta_0\gamma/(1+\gamma). \quad (57)$$

For instance, a reduction of Δ to $0.1\Delta_0$ results in a steady-state MSE which is about 0.2 dB greater than the minimum achievable MSE.

Note that the impact of a distorted channel, with an eigenvalue ratio $\rho > 1$, on the excess MSE, its rate of convergence, and the choice of Δ is the same as if the number of equalizer coefficients N was increased to $N\rho$. For a T-spaced equalizer, some of the eigenvalues (and hence ρ) depend on the timing phase and the channel envelope delay characteristics in the band-edge regions.

The above results apply equally to T-spaced and fractionally spaced equalizers, except that the recursion for the excess MSE (54) for a KT/M-spaced equalizer is given by

$$\epsilon_{\Delta,k} = [1 - 2\Delta(M/K)\bar\lambda + \Delta^2 N(M/K)\rho(\bar\lambda)^2]$$
$$+ \Delta^2\epsilon_{min}N(M/K)\rho(\bar\lambda)^2.$$

As mentioned earlier, only about K/M of the eigenvalues of the correlation matrix A are significant for a KT/M FSE. For the same average signal power at the equalizer input, equal to the average eigenvalue $\bar\lambda$, the significant eigenvalues are generally M/K times larger for an N-coefficient KT/M FSE compared with the eigenvalues for an N-coefficient T-spaced equalizer. Thus the eigenvalue ratio ρ for an FSE should be computed only over the significant eigenvalues of A. Note that A and ρ for an FSE are independent of timing phase and channel envelope delay characteristics, and $\rho = 1$ when the unequalized amplitude shape is square-root of Nyquist.

For well-behaved channels (ρ approximately 1), $\epsilon_{\Delta k}$ for an N-coefficient KT/M FSE, using the best initial step size given by (55), initially converges faster as $(1 - M/KN)^k$ compared with $(1 - 1/N)^k$ for an N-coefficient T equalizer using the same best initial step size. Conversely, an MN/K-coefficient FSE, using a step size K/M times as large, generally exhibits the same behavior with respect to the convergence and steady-state value of excess MSE as an N-coefficient T-spaced equalizer.

As a rule of thumb, the symbol-by-symbol LMS gradient algorithm with the best initial Δ leads to a reduction of about 20 dB in MSE for well-behaved channels (ρ approximately 1) in about five times the time span T_{eq} of the equalizer. At this time it is desirable to reduce Δ by a factor of 2 for the next $5T_{eq}$ seconds to permit finer tuning of the equalizer coefficient. Further reduction in excess MSE can be obtained by reducing Δ to its steady-state value according to (57). (For distorted channels, an effective time span of ρT_{eq} should be substituted in the above discussion.)

C. Digital Precision Considerations

The above discussion may suggest that it is desirable to continue to reduce Δ in order to reduce the excess MSE to zero in the steady state. However, this is not advisable in a practical limited precision digital implementation of the adaptive equalizer. Observe [36] that as Δ is reduced, the coefficient correction terms in (50), on the average, become smaller than half the least significant bit of the coefficient; adaptation stalls and the MSE levels off. If Δ is reduced

further, the MSE increases if the channel characteristics change at all or if some adjustments made at peak errors are large enough to perturb the equalizer coefficients.

The MSE which can be attained by a digital equalizer of a certain precision can be approximated as follows [38]. Let the equalizer coefficients be represented by a uniformly quantized number of B bits (including sign) in the range $(-1, 1)$. Then the real and imaginary parts of the equalizer coefficients will continue to adapt, on the average, so long as

$$\Delta(\epsilon\bar\lambda/2)^{1/2} \geq 2^{-B}. \quad (58)$$

It is desirable to select a compromise value of Δ such that the total MSE $\epsilon = \epsilon_{min} + \epsilon_\Delta$, with ϵ_Δ predicted by infinite-precision (analog) analysis (56), is equal to the lower limit on ϵ determined by digital precision (58). The required precision can be estimated by substituting Δ from (57) into (58)

$$2^B \geq [N\rho(1+\gamma)/\sqrt2\,\gamma](\bar\lambda/\epsilon)^{1/2} \quad (59)$$

where $(\bar\lambda/\epsilon)$ can be recognized as the desired equalizer input-power-to-output-MSE ratio.

As an example, consider a 32-tap T-spaced equalizer ($N = 32$) for a well-behaved channel ($\rho = 1$). Select $\gamma = 0.25$ (corresponding to a 1-dB increase in output MSE over ϵ_{min}) and a desired equalizer output-signal-power-to-MSE ratio of 24 dB (adequate for 9.6-kbit/s transmission). Let the input-to-output power scale factor for this equalizer be 2, so that $(\bar\lambda/\epsilon)^{1/2} = 10^{27/20}$. Solving (59) for B, we find that 12-bit precision is required.

The required coefficient precision increases by 1 bit for each doubling of the number of coefficients, and for each 6-dB reduction in desired output MSE. However, for a given ϵ_{min}, each 6-dB reduction in excess MSE ϵ_Δ requires a 2-bit increase in the required coefficient precision. This becomes the limiting factor in some adaptive filters, e.g., an echo canceler, which must track slow variations in system parameters in the presence of a large uncorrelated interfering signal [108].

Note that the precision requirement imposed by the LMS gradient adaptation algorithm is significantly more stringent than a precision estimate based on quantization noise due to roundoff in computing the sum of products for the equalizer output. If each product is rounded individually to B bits and then summed, the variance of this roundoff noise is $N2^{-2B}/3$. Assuming an equalizer output signal power of $1/6$, the signal-to-output roundoff noise ratio is $2^{2B}/2N$, which is 54 dB for $B = 12$ and $N = 32$: 30 dB greater than the desired 24-dB signal-to-MSE ratio in our example.

Roundoff in the coefficient update process, which has been analyzed in [7], is another source of quantization noise in adaptive filters. This roundoff causes deviation of the coefficients from the values they take when infinite precision arithmetic is used. The MSE ϵ_r contributed by coefficient roundoff [7] is approximated by $\epsilon_r = N2^{-2B}/6\Delta$. Again using an output signal power value of $1/6$, the signal-to-coefficient roundoff noise ratio is $\Delta2^{2B}/N$. For our example, we obtain $\Delta = 0.0375$ from (59) using $\gamma = 0.25$, $N = 32$, $\rho = 1$, and $\bar\lambda = 1/3$. The signal-to-coefficient roundoff noise ratio is 43 dB for $B = 12$, which is 19 dB greater than the desired 24-dB signal-to-MSE ratio, suggesting that the effect of coefficient roundoff is insignificant.

265

Moreover, each time we reduce Δ by a factor of 2, the coefficient precision B must be increased by 1 bit in order to prevent adaptation from stalling according to (58). This reduction in Δ and the corresponding increase in B further reduces the MSE due to coefficient roundoff by 3 dB. Using the expression for ϵ_r given above, and Δ from (57), one can rewrite (58) as

$$\epsilon/\epsilon_r \geqslant 6\rho(1 + \gamma)/\gamma.$$

Since $\rho \geqslant 1$ and γ is typically a small fraction, the MSE due to coefficient roundoff, ϵ_r, is always small compared to ϵ provided coefficient precision is sufficient to allow adaptation to continue for small desired values of γ and Δ. Thus coefficient roundoff noise can be neglected in the process of estimating LMS adaptive filter precision requirements.

Simple update schemes using nonlinear multipliers to produce the coefficient correction terms have been devised to address the precision problem and reduce implementation complexity at some penalty in performance [13]. Two such schemes are:

1) Sign-bit multiplication update

$$c_{k+1} = c_k - \Delta \, \text{sgn}\left(y_k^*\right) \text{sgn}(e_k). \qquad (60)$$

2) Power-of-two multiplication update

$$c_{k+1} = c_k - \Delta f\left(y_k^*\right) f(e_k) \qquad (61)$$

where

$$f(x) = \text{sgn}(x) 2^{\lfloor \log_2 |x| \rfloor}$$

and $[\cdot]$ is used here to denote the greatest integer less than or equal to the argument.

In (60) and (61), the $\text{sgn}(\cdot)$ and $f(\cdot)$ functions are applied separately to each element of a vector and to the real and imaginary parts of a complex quantity. Analysis and simulation results show [13] that while sign-bit multiplication update (60) suffers a significant degradation in convergence time compared to the true LMS gradient update (50), the loss in performance is small when the power-of-two multiplication update (61) is used.

Whenever some of the eigenvalues of the input matrix A for an adaptive filter become vanishingly small, the output MSE is relatively insensitive to coefficient deviations corresponding to these eigenvalues. In some practical situations this may lead to numerical instability: when input signals are not adequately dense across the entire bandwidth of the adaptive filter (e.g., a signal consisting of a few tones at the input to a long adaptive equalizer or voice echo canceler), or an FSE which naturally has a set of relatively small eigenvalues, as mentioned in Section II-C4. In these cases, finite precision errors tend to accumulate along the eigenvectors (or frequency components) corresponding to small eigenvalues without significantly affecting the output MSE. If a natural converging force constraining these components (e.g., background noise) is weak, and the digital precision errors have a bias, these fluctuations may eventually take one or more filter coefficients outside the allowed numerical range, e.g., $(-1, 1)$. This overflow can be catastrophic unless saturation arithmetic is used in the coefficient update process to prevent it.

Another solution is to use a so-called tap-leakage algorithm [40]

$$c_{k+1} = (1 - \Delta\mu)c_K - \Delta y_k^* e_k$$

where a decay factor $(1 - \Delta\mu)$ has been introduced into the usual stochastic gradient algorithm (50). The tap-leakage algorithm seeks to minimize a modified mean-square cost function

$$\mu c_k^* c_k^T + E\left[|e_k|^2\right]$$

which is a sum of the MSE and an appropriate fraction μ of the squared length of the N-dimensional coefficient vector. The decay tends to force the coefficient magnitudes and, by Parseval's theorem, the equalizer power spectral response toward zero. For an FSE, a value of μ on the order of a small eigenvalue is suggested [40].

The tap-leakage algorithm is similar to an adaptive predictor update algorithm for an ADPCM system [129] where the predictor coefficients are forced to decay toward predetermined compromise values during silence intervals. This prevents the effect of channel errors on the decoder adaptive predictor coefficients from persisting.

IV. Fast Converging Equalizers

The design of update algorithms to speed up the convergence of adaptive filters has been a topic of intense study for more than a decade. Rapid convergence is important for adaptive equalizers designed for use with channels, such as troposcatter and HF radio, whose characteristics are subject to time variations [87]. In voice-band telephone applications, reduction of the initial setup time of the equalizer is important in polling multipoint networks [26] where the central site receiver must adapt to receive typically short bursts of data from a number of transmitters over different channels.

In this section we present an overview of three classes of techniques devised to speed up equalizer convergence.

A. Orthogonalized LMS Algorithms

Recall from Section III-A that for the deterministic gradient algorithm, no single value of the step size Δ leads to fast convergence of all the coefficient deviation components when the eigenvalues of the equalizer input covariance matrix A have a large spread. Using the independence assumption, the same is true regarding the convergence of the mean of the coefficient deviations for the LMS gradient algorithm (see (51) in Section III-B). The excess MSE is a sum of the mean-square value of each coefficient deviation weighted by the corresponding eigenvalues of A. Slow decay of some of these mean-square deviations, therefore, slows down the convergence of the excess MSE. Substituting the best initial Δ from (55) in (54), we obtain the recursion

$$\epsilon_{\Delta k+1} = (1 - 1/N\rho)\epsilon_{\Delta k} + \epsilon_{\min}/N\rho. \qquad (62)$$

Observe that the initial decay of $\epsilon_{\Delta k}$ is geometric with a time constant of approximately $N\rho$ symbol intervals. Thus for the same length equalizer, a severely distorted channel ($\rho = 2$) will cause the rate of convergence of the LMS gradient algorithm to be slower by a factor of 2 compared to that for a good channel. The inadequacy of the LMS gradient algorithm for fast start-up receivers becomes obvious if we consider a 9.6-kbit/s, 2400-Bd modem with an equalizer spanning 32 symbol intervals. For a severely distorted channel, more than 320 equalizer adjustments over a 133-ms interval would be required before data transmission could begin.

For partial-response systems [48], [50], where a controlled amount of intersymbol interference is introduced to obtain a desired spectral shape, the equalizer convergence problem is fundamental. It can be shown that $\rho = 2$ for an ideal cosine-shaped spectrum at the equalizer input for a Class IV SSB partial-response [8] or a Class I QAM partial-response [90] system. Noting this slow convergence, Chang [8] suggested the use of a prefixed weighting matrix to transform the input signals to the equalizer tap gains to be approximately orthonormal. All eigenvalues of the transformed equalizer input covariance matrix are then approximately equal resulting in faster equalizer convergence.

Another orthogonalized LMS update algorithm [75], [88] is obtained by observing that the decay of the mean of all the transformed coefficient deviation components could be speeded up by using a diagonal matrix $\text{diag}(\Delta_i)$ instead of the scaler Δ in (51), such that each element Δ_i of this matrix is the inverse of the corresponding element λ_i of Λ. Transforming $\text{diag}(\Delta_i)$ back to the original coordinate system, we obtain the orthogonalized LMS update algorithm

$$c_{k+1} = c_k - Py_k^* e_k. \tag{63}$$

As in Chang's scheme [8], the best value of the weighting matrix P is given by

$$P = V^{*T}\text{diag}(\Delta_i)V = V^{*T}\Lambda^{-1}V = A^{-1}$$

where the columns of the diagonalizing matrix V are eigenvectors of A (see (30)). In practice, A is not known beforehand, therefore, P can only approximate A^{-1}. For instance, in partial-response systems where the dominant spectral shape is known beforehand, we can use $P = S^{-1}$, where S is the covariance matrix of the partial-response shaping filter.

A practical advantage of this algorithm (63) over Chang's structure is that the weighting matrix is in the path of the tap-gain corrections rather than the received signal. The computation required, therefore, need not be carried out to as much accuracy.

As we shall see in Section IV-C, the fastest converging algorithms are obtained when P is continually adjusted to do the best job of orthogonalizing the tap-gain corrections.

8 Periodic or Cyclic Equalization

As mentioned in Section I-C, one of the most widely used [118], [119] methods of training adaptive equalizers in high-speed voice-band modems is based on PN training sequences with periods significantly greater than the time span of the equalizer. Here we discuss the techniques [70], [76], [90] which can be used to speed up equalizer convergence in the special case when the period of the training sequence is selected to be equal to the time span of the equalizer.

1) Periodic or Averaged Update: Consider a training sequence $\{x_k\}$ with period NT for a T-spaced equalizer with N coefficients. Let the equalizer coefficients be adjusted periodically, every N-symbol intervals, according to the following LMS algorithm with averaging:

$$c_n(k+1) = c_n(k) - \Delta \sum_{j=0}^{N-1} e_{kN+j} y^*(kNT + jT - nT),$$

$$n = 0, 1, \cdots, N-1.$$

Let us denote the error sequence during the kth period by

the vector e_k whose jth element is given by

$$e_{kN+j} = \sum_{n=0}^{N-1} c_n(k) y(kNT + jT - nT) - x_{kN+j},$$

$$j = 0, 1, \cdots, N-1.$$

The update algorithm can be written in matrix notation as

$$c_{k+1} = c_k - \Delta Y_k^* e_k \tag{64}$$

where the n,jth element of the $N \times N$ matrix Y_k is given by $y(kNT + jT - nT)$. Substituting $e_k = Y_k^T c_k - x_k$, we obtain

$$c_{k+1} = c_k - \Delta(Y_k^* Y_k c_k - Y_k^* x_k).$$

Neglecting noise and using the fact that the periodic training sequence can be designed to be white [53], i.e.,

$$\frac{1}{N} \sum_{n=0}^{N-1} x_n x_k^* = \delta_{nk}$$

we have

$$Y_k^* Y_k = NA_p \quad \text{and} \quad Y_k^* x_k = N\alpha_p.$$

The elements of the matrix A_p and vector α_p are given by

$$a_{i,j} = \sum_{n=0}^{N-1} h^*(nT - iT) h(nT - jT)$$

and

$$\alpha_i = h^*(-iT) \tag{65}$$

where $h(nT)$, $n = 0, 1, \cdots, N-1$, are the T-spaced samples of the periodic channel response to a sequence of periodic impulses spaced NT seconds apart. Thus the periodic update algorithm may be written as

$$c_{k+1} = (I - \Delta NA_p)c_k + \Delta N\alpha_p. \tag{66}$$

Note the similarity of (66) to the deterministic gradient algorithm (44). However, in this case the matrix A_p is not only Toeplitz but also circulant, i.e., each row of A_p is a circular shift of another row. The elements $a_{i,j}$ of the ith row are the coefficients R_{i-j} of the periodic autocorrelation function of the channel. The final solution to the difference equation (66) is the "optimum" set of periodic equalizer coefficients: $c_{p\text{opt}} = A_p^{-1}\alpha_p$. Substituting in (66), we obtain

$$c_{k+1} - c_{p\text{opt}} = (I - \Delta NA_p)(c_k - c_{p\text{opt}}). \tag{67}$$

For a noiseless ideal Nyquist channel $R_{i-j} = \delta_{ij}$, i.e., $A_p = I$. Therefore, a single averaged adjustment with $\Delta = 1/N$ results in perfect periodic equalization.

In general, the transformed coefficient deviation vector after the kth periodic update is given by

$$d_k = (I - \Delta N\Lambda_p)^k d_0$$

where Λ_p is a diagonal matrix with eigenvalues λ_{ip}, $i = 0, 1, \cdots, N-1$, equal to the coefficients of the discrete Fourier transform (DFT) of the channel periodic autocorrelation function R_n, $n = 0, 1, \cdots, N-1$. Thus

$$\lambda_{pi} = \sum_{n=0}^{N-1} R_n \exp(-j2\pi ni/N)$$

$$= \left| \sum_k H(i/NT - k/T) \right|^2, \quad i = 0, 1, \cdots, N-1.$$

$$\tag{68}$$

Using these results, it can be shown that in the absence of

noise the perfect periodic equalizer has a frequency response equal to the inverse of the folded channel spectrum at N uniformly spaced discrete frequencies

$$C(n/NT) = 1 \bigg/ \sum_k H(n/NT - k/T), \quad n = 0, 1, \cdots, N - 1.$$

The excess MSE after the kth update is given by

$$\epsilon_{p\Delta k} = d_k^{*T} \Lambda d_k$$

$$= \sum_{i=0}^{N-1} \lambda_{pi}(1 - \Delta N \lambda_{pi})^{2k} |d_{i0}|^2.$$

If the initial coefficients are zero, then $|d_{i0}|^2 = 1/\lambda_{pi}$, and

$$\epsilon_{p\Delta k} = \sum_{i=0}^{N-1} (1 - \Delta N \lambda_{pi})^{2k}. \tag{69}$$

Each component of this sum converges provided $0 < \Delta < 2/(N\lambda_{p\max})$. Fastest convergence is obtained when $\Delta = 2/[N(\lambda_{p\max} + \lambda_{p\min})]$.

2) Stochastic Update: So far we have examined the convergence properties of the periodic update or LMS steepest descent algorithm with averaging. It is more common, and as we shall see, more beneficial to use the continual or stochastic update method, where all coefficients are adjusted in each symbol interval according to (50). Proceeding as in Section III-B and noting that in the absence of noise $e_{k\,\text{opt}} = 0$ for a periodic input, we obtain

$$(c_{k+1} - c_{p\text{opt}}) = (I - \Delta y_k^* y_k^T)(c_k - c_{p\text{opt}}).$$

Let us define an $N \times N$ matrix

$$B_k \triangleq (I - \Delta y_k^* y_k^T).$$

Note that since y_k is periodic with period N, B_k is a circulant matrix and

$$B_{k+1} = U^{-1} B_k U$$

where the $N \times N$ cyclic shift matrix U is of the form

$$U = \begin{bmatrix} 0 & 1 & 0 & \cdot & \cdot & 0 \\ 0 & 0 & 1 & \cdot & \cdot & 0 \\ \cdot & \cdot & \cdot & \cdot & \cdot & \cdot \\ \cdot & \cdot & \cdot & \cdot & \cdot & \cdot \\ 0 & 0 & 0 & \cdot & \cdot & 1 \\ 1 & 0 & 0 & \cdot & \cdot & 0 \end{bmatrix}.$$

Note that $U^T U = U^N = I$. Consider the first N updates from time zero to $N - 1$. Then

$$(c_N - c_{p\text{opt}}) = B_{N-1} B_{N-2} \cdots B_0 (c_0 - c_{p\text{opt}})$$

$$= (U^{-N+1} B_0 U^{N-1})(U^{-N+2} B_0 U^{N-2}) \cdots B_0 (c_0 - c_{p\text{opt}})$$

$$= (U B_0)^N (c_0 - c_{p\text{opt}}).$$

In general

$$(c_{kN} - c_{p\text{opt}}) = (U B_0)^{kN} (c_0 - c_{p\text{opt}}). \tag{70}$$

The convergence of the coefficient deviations, therefore, depends on the eigenvalues of $U B_0$. Using the fact that $y_0^* y_0^T$ is singular with rank 1, after some manipulation, the characteristic equation $\det(\lambda I - U B_0) = 0$ can be reduced to the form

$$\lambda^N + \Delta N \sum_{n=0}^{N-1} \lambda^n R_n - 1 = 0. \tag{71}$$

Here

$$R_n = (y_i^T y_{k+n}^*)/N, \quad n = 0, 1, \cdots, N - 1$$

are the coefficients of the periodic autocorrelation function of the equalizer input (defined earlier in terms of the periodic impulse response of the channel). When $R_n = 0$ for $n \geq 1$, corresponding to an ideal Nyquist channel, all roots of (71) are equal to the Nth roots of $(1 - \Delta N R_0)$. Thus perfect equalization is obtained after N updates with $\Delta = 1/(N R_0)$.

Let $\lambda_{\max}$ be the maximum magnitude root of (71). Then the coefficient deviations after every N stochastic updates are reduced in magnitude according to (70) provided $|\lambda_{\max}^N| < 1$. Moreover, from the maximum modulus principle of holomorphic functions [92] we have the condition that

$$|\lambda_{\max}^N| \leq \max_{0 \leq i \leq N-1} \left| 1 - \Delta N \sum_{n=0}^{N-1} R_n \exp(-j2\pi ni/N) \right|$$

or

$$|\lambda_{\max}^N| \leq \max_{0 \leq i \leq N-1} |1 - \Delta N \lambda_{pi}| \tag{72}$$

where λ_{pi} are the eigenvalues of A_p in the periodic update method. Since (72) holds with equality only when $R_n = 0$ for $n \geq 1$, we reach the important conclusion for periodic training that for the ideal Nyquist channel the stochastic and averaged update algorithms converge equally fast, but for all other channels the stochastic update algorithm results in faster convergence. This behavior has also been observed for equalizer convergence in the presence of random data [71].

In the presence of noise, an exact expression for the excess MSE for the stochastic update algorithm is difficult to derive for periodic training sequences. However, assuming zero initial coefficients the following expression is a good approximation to results obtained in practice for moderately high SNR:

$$\epsilon_{\Delta kN} = \sum_{i=0}^{N-1} \left[(1 - \Delta N \lambda_i)^{2k} + \Delta^2 N \lambda_i^2 \epsilon_{p\min} \right] \tag{73}$$

where λ_i are the roots of (71) (the effect of noise can be included by defining $R_n = E[y_k^T y_{k+N}^*]/N$). For an ideal channel, $\lambda_i = R_0 = 1$ for all i, and the excess MSE for periodic training converges to $\epsilon_{p\min}$ after N adjustments with $\Delta = 1/N$, where $\epsilon_{p\min}$ is the minimum achievable MSE for periodic training.

The significant difference in the convergence behavior of the stochastic gradient algorithm for random data and periodic training is now apparent by comparing (54) and (73). The well-controlled correlation properties of periodic training sequences tend to reduce the average settling time of the equalizer by about a factor of two compared to the settling time in the presence of random data.

An important question regarding periodic equalization is that once the coefficients have been optimized for a periodic training sequence, how close to optimum is that set of coefficients for random data. The answer depends primarily on the selected period of the training sequence (and hence the equalizer span) relative to the length of the channel impulse response. When the period is long enough to contain a sufficiently large percentage (say 95 percent) of the energy of the channel impulse response, the edge effects in the periodic channel response and equalizer coefficients are small. In frequency-domain terms, the discrete tones of the periodic training sequence are adequately dense to obtain representative samples of the channel spectrum. Under these conditions, the excess MSE due to the periodicity of the training sequence is small compared to the excess MSE due to the large value of Δ which must be selected for fast initial convergence. After rapid initial convergence has been obtained in this manner, it may be desirable to make finer adjustments to the equalizer using a pseudo-random training sequence with a longer period, or begin decision-directed adaptation using randomized customer data.

3) Application to Fractionally Spaced Equalizers: The averaged and stochastic update methods of periodic training are also applicable to fractionally spaced equalizers [91]. The period of the training sequence is still equal to the time span of the equalizer. Thus a sequence with period NT can be used to train an equalizer with NM/K coefficients spaced KT/M seconds apart.

The equalizer coefficients may be adjusted periodically, every N symbol intervals, according to the averaged update algorithm

$$c_n(k+1) = c_n(k) - \Delta \sum_{j=0}^{N-1} e_{kN+j} y^*(kNT - jT - nKT/M),$$

$$n = 0, 1, \cdots, NM/K - 1.$$

In the absence of noise, the elements of the matrix A_p and vector α_p (given in (65) for a T-spaced equalizer) are given below for a KT/M-spaced equalizer

$$a_{i,j} = \sum_{n=0}^{N-1} h^*(nT - iKT/M) h(nT - jKT/M)$$

and

$$\alpha_i = h^*(-iKT/M).$$

The matrix A_p is no longer Toeplitz or circulant and, therefore, its eigenvalues cannot be obtained by DFT techniques (68). However, it can be shown that the N significant eigenvalues are samples of the channel folded power spectrum, that is,

$$\lambda_{pi} = (M/K) \sum_k |H(i/NT - k/T)|^2, \qquad i = 0, 1, \cdots, N - 1$$

$$(74)$$

and the remaining $N(M - K)/K$ eigenvalues are zero. The frequency response of the perfect periodic equalizer at N uniformly spaced discrete frequencies is given by

$$C(n/NT) = H^*(n/NT) \Big/ \sum_k |H(n/NT - k/T)|^2,$$

$$n = 0, 1, \cdots, NM/K - 1.$$

Starting with zero initial coefficients, the excess MSE after the kth averaged update is related to the eigenvalues (74) according to

$$\epsilon_{p\Delta k} = \sum_{i=0}^{N-1} (1 - \Delta N \lambda_{pi})^{2k}$$

If the channel folded power spectrum is flat, i.e., $\lambda_{pi} = M/K$, $i = 0, 1, \cdots, N - 1$, then a single averaged update with $\Delta = K/MN$ results in a matched filter which also

removes intersymbol interference. In contrast with T-spaced equalizers, the ideal amplitude shape of the unequalized system for fast convergence of a fractionally spaced equalizer is a square root of Nyquist rather than Nyquist. Moreover, the convergence of an FSE is not affected by sampler timing phase or channel delay distortion since the channel power spectrum and hence λ_{pi} are independent of phase-related parameters.

The stochastic update algorithm (50) for periodic training of fractionally spaced equalizers can be analyzed along the lines of Section IV-B2. For instance, for a $T/2$-spaced equalizer, neglecting noise, the coefficient deviation recursion (70) still applies. However, all matrices are now $2N \times 2N$ and the cyclic shift matrix U now produces a double shift. The eigenvalues of UB_0 are the roots of the characteristic equation $\det(\lambda I - UB_0) = 0$, or

$$(\lambda^N - 1)\left(\lambda^N + \Delta N \sum_{n=0}^{N-1} \lambda^n R_n - 1\right) = 0 \qquad (75)$$

where

$$R_n = (1/N) \sum_{k=0}^{2N-1} y(kT/2) y^*(kT/2 + nT),$$

$$n = 0, 1, \cdots, N - 1$$

are the T-spaced samples of the channel periodic autocorrelation. Note that half the roots of (75) are the Nth roots of unity. Coefficient deviation components corresponding to these roots do not converge. Appropriate selection of initial coefficients, e.g., $c_0 = 0$, ensures that these components are zero. The remaining roots of (75) are dependent on the properties of R_n, whose DFT may be recognized to be equal to the samples of the channel folded power spectrum (74). Therefore, like the averaged update algorithm, the convergence of the stochastic update algorithm is also independent of phase-related parameters. When the folded power spectrum is flat, $R_n = R_0 \delta_{n0}$, perfect equalization can be obtained in N adjustments with $\Delta = 1/NR_0$. When the channel power spectrum is not Nyquist ($R_n \neq 0, n \geq 1$), the maximum modulus principle (72) applies ensuring that the stochastic update algorithm will result in faster convergence than that obtained by periodic or averaged update. Comments with regard to the selection of an adequately long period of the training sequence given in Section IV-B2 still apply.

4) Accelerated Processing: One technique for reducing the effective settling time of an equalizer involves performing equalizer coefficient update iterations as often as permissible by the computational speed limitations of the implementation. For instance, for a periodic equalizer, one period of the received sequence of samples may be stored in the equalizer delay line and iterative updates using the averaged or stochastic update algorithm may be made at a rate faster than the usual [76]. This update rate may be selected to be independent of the modem symbol rate since the sequence of samples already stored in the equalizer delay line may be circularly shifted as often as required to produce new output samples. Based on each such output new coefficient correction terms can be computed using a circular shift of the locally stored periodic training sequence. Thus after the equalizer delay line has been filled with a set of received samples, the best periodic equalizer for that particular set of received samples can be determined almost instantly by accelerated processing given unlimited computational speed. If this set of received samples is representative of the channel response to the periodic training sequence then the equalizer obtained by such accelerated processing is a good approximation to the optimum periodic equalizer. However, all sources of aperiodicity, e.g., initial transients and noise, in these received samples degrade performance, since new received samples are not used in this method to reduce the effect of noise by averaging. Modified versions of the accelerated processing method are possible which reprocess some previously processed samples and then accept a new input sample as it becomes available, thus updating the equalizer coefficients several times per symbol interval.

5) Orthogonalized Periodic Equalizer: As discussed in Section IV-A, the convergence of the LMS gradient algorithm can be improved by inserting an orthogonalizing matrix P in the path of the coefficient corrections (63). The desired value of P is the inverse of the equalizer input correlation matrix A. For a periodic equalizer the average update algorithm (64) can be modified in a similar manner to

$$c_{k+1} = c_k - PY_k^* e_k. \qquad (76)$$

However, in this case the orthogonalizing matrix P can be replaced by a single inverse filter in the path of the periodic equalizer input sequence before it is used for coefficient adjustment [90]. This simplification is a consequence of the fact that, as discussed in Section IV-B1, the input correlation matrix A_p for a T-spaced periodic equalizer is Toeplitz and circulant. The inverse P of the circulant matrix NA_p is also circulant. For periodic input, the transformation performed by P is equivalent to a nonrecursive filtering operation (or periodic convolution) with coefficients equal to the elements of the first row of P.

The fast settling periodic equalizer structure with a single inverse filter is also applicable when the equalizer coefficients are updated symbol-by-symbol. The modified stochastic update algorithm is given by

$$c_n(k+1) = c_n(k) - \Delta e_k \sum_{j=0}^{N-1} p_j y^*(kT - jT - nT),$$

$$n = 0, 1, \cdots, N-1.$$

A block diagram of the equalizer structure and performance curves showing significant improvement in the settling time for partial response QAM systems are given in [90].

6) Discrete Fourier Transform Techniques: Throughout the discussion on periodic equalization, we have taken advantage of the circulant property of the equalizer input to use the discrete Fourier transform (DFT) for analysis. As pointed out in Section IV-B1, in the absence of noise the perfect periodic equalizer has a frequency response equal to the inverse of the folded channel spectrum at N uniformly spaced frequencies. Therefore, the equalizer coefficients can be directly obtained by transmitting a periodic training sequence and using the following steps at the receiver [43], [70], [90]:

1) Compute the DFT of one period of the equalizer input

$$Y_i = \sum_{k=0}^{N-1} y_k \exp(-j2\pi ik/N), \qquad i = 0, 1, \cdots, N-1.$$

2) Compute the desired equalizer spectrum according to

$$C_i = X_i Y_i^* / |Y_i|^2, \qquad i = 0, 1, \cdots, N-1$$

where X_i is the precomputed DFT of the training sequence.
3) Compute the inverse DFT of the equalizer spectrum to obtain the periodic equalizer coefficients

$$c_n = (1/N) \sum_{i=0}^{N-1} C_i \exp(j2\pi ni/N), \qquad n = 0, 1, \cdots, N-1.$$

A number of modifications may be made to improve performance of this direct computation method in the presence of noise and other distortions, such as frequency translation, which adversely affect periodicity. For instance, when the equalizer input is not strictly periodic with period NT due to channel-induced frequency translation, its DFT at any frequency suffers from interference from adjacent components. The effect of this interference can be minimized by windowing a longer sequence of input samples before taking the DFT. The window function should be selected such that its Fourier transform has reduced sidelobe energy while preserving the property of zero response at $1/NT$-hertz intervals. A $2NT$ second triangular window has both these properties [43]. A second minor modification can be made to step 2 by adding a constant estimate of the expected flat noise power spectral components to the denominator.

C. Recursive Least Squares (RLS) Algorithms

The orthogonalized LMS algorithms of Section IV-A can provide rapid convergence when the overall received signal spectral shape is known beforehand; for example, in partial-response systems. In certain voice-band modem applications special training sequences can be used to design fast equalizer startup algorithms, such as those in the last section. However, in general, a self-orthogonalizing method, such as one of the RLS algorithms described in this section, is required for rapidly tracking adaptive equalizers (or filters) when neither the reference signal nor the input (received) signal (or channel) characteristics can be controlled.

As discussed in earlier sections, the rate of convergence of the output MSE of an LMS gradient adaptive equalizer is adversely affected by the eigenvalue spread of the input covariance matrix. This slow convergence is due to the fundamental limitation of a single adjustable step size parameter Δ in the LMS gradient algorithm. If the input covariance matrix is known *a priori* then an orthogonalized LMS gradient algorithm can be derived, as in Section IV-A, where the scalar Δ is replaced by a matrix P. Most rapid convergence is obtained when P is the inverse of the equalizer input covariance matrix A, thus rendering the adjustments to the equalizer coefficients independent of one another.

In [41], Godard applied the Kalman filter algorithm to the estimation of the LMS equalizer coefficient vector under some assumptions on the equalizer output error and input statistics. The resulting algorithm has since been recognized to be the fastest known equalizer adaptation algorithm. It is an ideal self-orthogonalizing algorithm [37] in that the received equalizer input signals are used to build up the inverse of the input covariance matrix which is applied to the coefficient adjustment process. A disadvantage of the

Kalman algorithm is that it requires on the order of N^2 operations per iteration for an equalizer with N coefficients.

Falconer and Ljung [20] showed that the Kalman equalizer adaptation algorithm can be derived as a solution to the exact least squares problem without any statistical assumptions. An advantage of this approach is that the "shifting property" previously used for fast recursive least squares identification algorithms [73] can be applied to the equalizer adaptation algorithm. This resulted in the so-called fast Kalman algorithm [20] which requires on the order of N operations per iteration.

A third class of recursive least squares algorithms known as adaptive lattice algorithms [127] were first described for adaptive identification in [74] and for adaptive equalization in [63], [96], and [97]. Like the fast Kalman algorithm, adaptive lattice algorithms are recursive in time, requiring of the order of N operations per iteration. However, unlike the Kalman algorithms, adaptive lattice algorithms are order-recursive. That is, the number of equalizer coefficients (and the corresponding lattice filter sections) can be increased to $N + 1$ without affecting the already computed parameters of the Nth-order equalizer. Low sensitivity of the lattice coefficients to numerical perturbations is a further advantage.

In the remainder of this section, we shall briefly review the least square criterion and its variants, introduce the "shifting property" and the structure of the fast Kalman and adaptive lattice algorithms, and summarize some important results, complexity estimates, and stability considerations.

1) The Least Squares Criterion: The performance index for recursive least squares (RLS) algorithms is expressed in terms of a time average instead of a statistical or ensemble average as in LMS algorithms. The RLS equalizer adaptation algorithm is required to generate the N-coefficient vector c_n at time n which minimizes the sum of all squared errors as if c_n were used over all the past received signals, i.e., c_n minimizes

$$\sum_{k=0}^{n} |x_k - y_k^T c_n|^2. \tag{77}$$

This leads to the so-called prewindowed RLS algorithm, where the input samples y_k are assumed to be zero for $k < 0$.

In order to permit tracking of slow time variations, a decay factor w with a value slightly less than unity may be introduced. The resulting exponentially windowed RLS algorithm minimizes

$$\sum_{k=0}^{n} w^{n-k} |x_k - y_k^T c_n|^2. \tag{78}$$

The minimizing vector c_n is the solution of the discrete-time Wiener–Hopf equation obtained by setting the derivative of (78) with respect to c_n to zero. Thus we have the solution

$$c_n = A_n^{-1} \alpha_n$$

where the $N \times N$ estimated covariance matrix is given by

$$A_n = \sum_{k=0}^{n} w^{n-k} y_k^* y_k^T + \delta I = w A_{n-1} + y_n^* y_n^T \tag{79}$$

and the N-element estimated cross-correlation vector is given by

$$\alpha_n = \sum_{k=0}^{n} w^{n-k} y_k^* x_k = w \alpha_{n-1} + y_n^* x_k.$$

The parameter δ is selected as a small positive number to ensure that A_n is nonsingular. The matrix A_n and vector α_n are akin to the statistical autocorrelation matrix and cross-correlation vector encountered in the LMS analysis. However, in this case A_n is not a Toeplitz matrix even for T-spaced equalizers.

It can be shown [87] that given c_{n-1}, the coefficient vector for time n can be generated recursively according to

$$c_n = c_{n-1} + k_n e_n \tag{80}$$

where $e_n = x_n - y_n^T c_{n-1}$ is the equalizer output error and

$$k_n = A_n^{-1} y_n^* \tag{81}$$

is the Kalman gain vector. The presence of the inverse estimated covariance matrix in (81) explains the insensitivity of the rate of convergence of the RLS algorithms to the channel characteristics.

In the Kalman algorithm [41], the inverse matrix $P_n = A_n^{-1}$ and the Kalman gain vector are computed recursively according to

$$k_n = P_{n-1} y_n^* / [w + y_n^T P_{n-1} y_n^*]$$

and

$$P_n = [P_{n-1} - k_n y_n^T P_{n-1}] / w. \tag{82}$$

The order of N^2 complexity of this algorithm is due to the explicit recursive computation of P_n. This computation is also susceptible to roundoff noise.

Two other variations of the prewindowed and exponentially windowed least squares criteria are the "growing memory" covariance and "sliding window" covariance performance indices defined as

$$\sum_{k=N-1}^{n} w^{k-n} |x_k - y_k^T c_n|^2$$

and

$$\sum_{k=n-L+1}^{n} |x_k - y_k^T c_n|^2$$

respectively, where N is the number of equalizer coefficients and L is the fixed length of the sliding window. The sliding window method is equivalent to a block least squares approach, where the block is shifted by one sample and a new optimum coefficient vector is determined each sample time.

2) The Fast Kalman Algorithm: Consider the input vector y_{n-1} at time $n - 1$ for a T-spaced equalizer of length N. The vector y_n at time n is obtained by shifting the elements of y_{n-1} by one, discarding the oldest sample y_{n-N}, and adding a new sample y_n. This shifting property is exploited by using least square linear prediction. Thus an efficient recursive algorithm can be derived [20] for updating the Kalman gain vector k_n without explicit computation of the inverse matrix P_n. Here we shall briefly examine the role of forward and backward prediction in the fast Kalman algorithm. See [20] for a more detailed derivation generalized to fractionally spaced and decision-feedback equalizers.

Let F_{n-1} be a vector of N forward predictor coefficients which minimizes the weighted sum of squares of the forward prediction error f_n between the new input sample y_n and a prediction based on the vector y_{n-1}. That is, F_{n-1} minimizes

$$\sum_{k=0}^{n} w^{n-k} |f_k|^2$$

where

$$f_k = y_k - F_{n-1}^T y_{k-1}. \tag{83}$$

The least squares forward predictor coefficients can be updated recursively according to

$$F_n = F_{n-1} + k_{n-1} f_n. \tag{84}$$

Similarly, the vector B_{n-1} of N backward predictor coefficients permits prediction of the old discarded sample y_{n-N} given the vector y_n. Thus we have the backward error

$$b_n = y_{n-N} - B_{n-1}^T y_n \tag{85}$$

and the update equation

$$B_n = B_{n-1} + k_n b_n.$$

The updated Kalman gain vector k_n, which is not yet available, can be obtained as follows. Define

$$f_n' = y_n - F_n^T y_{n-1} \tag{86}$$

as the error between y_n and its prediction based on the updated forward predictor. Let

$$E_n = w E_{n-1} + f_n'^* f_n \tag{87}$$

be the estimated exponentially weighted squared prediction error. Then the augmented or extended Kalman gain vector with $N + 1$ elements is given by

$$\bar{k}_n = \left[\begin{array}{c} f_n'^* / E_n \\ \hline k_{n-1} - F_n f_n'^* / E_n \end{array} \right] = \left[\begin{array}{c} k_n' \\ \hline \mu_n \end{array} \right] \tag{88}$$

where the dashed lines indicate partitions of the vector k_n. Finally, the updated backward predictor vector and the Kalman gain vector are given by the recursive relationships

$$B_n = \left[B_{n-1} + k_n' b_n \right] / [1 - \mu_n b_n] \tag{89}$$

and

$$k_n = k_n' + B_n \mu_n \tag{90}$$

where k_n' and μ_n are the N-element and scalar partitions, respectively, of the augmented Kalman gain vector defined in (88).

The matrix computations (82) involved in the Kalman algorithm are replaced in the fast Kalman algorithm by the recursions (83) through (90) which use forward and backward predictors to update the Kalman gain vector as a new input sample y_n is received and the oldest sample y_{n-N} is discarded.

A fast exact initialization algorithm for the interval $0 \leqslant n \leqslant N$ given in [10] avoids the choice of a stabilizing δ in (79) and the resulting suboptimality of the solution at $n = N$.

3) Adaptive Lattice Algorithms: The Kalman and fast Kalman algorithms obtain their fast convergence by orthogonalizing the adjustments made to the coefficients of an ordinary linear transversal equalizer. Adaptive lattice (AL) algorithms, on the other hand, use a lattice filter structure to orthogonalize a set of received signal components [127].

The transformed received signal components are then linearly weighted by a set of equalizer coefficients and summed to produce the equalizer output. We shall briefly review the gradient [96] and least squares [87], [97] forms of adaptive lattice algorithms for linear T-spaced complex equalizers. See [78] and [52] for generalization of the least squares AL algorithm to fractionally spaced and decision-feedback equalizers.

The structure of an adaptive lattice gradient equalizer is shown in Fig. 26. An N-coefficient equalizer uses $N-1$

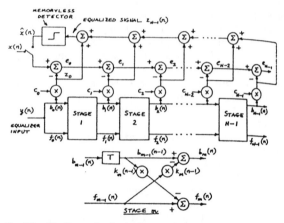

Fig. 26. Gradient adaptive lattice equalizer.

lattice filter stages. Each symbol interval a new received sample $y(n)$ enters stage 1. The mth stage produces two signals $f_m(n)$ and $b_m(n)$ which are used as inputs by stage $m + 1$. These signals correspond to the forward and backward prediction errors, respectively, of mth-order forward and backward linear LMS predictors. The two predictors have identical so-called reflection coefficients k_m for the mth stage. At time n, the prediction errors are updated according to

$$f_0(n) = b_0(n) = y(n) \tag{91}$$

and for $m = 1, \cdots, N - 1$

$$f_m(n) = f_{m-1}(n) - k_m(n-1) b_{m-1}(n-1) \tag{92}$$

$$b_m(n) = b_{m-1}(n-1) - k_m(n-1) f_{m-1}(n). \tag{93}$$

The reflection coefficients are updated to minimize the sum of the mean-square value of the forward and backward prediction errors. That is,

$$k_m(n) = k_m(n-1)$$
$$+ \left[f_{m-1}^*(n) b_m(n) + b_{m-1}^*(n-1) f_m(n) \right] / \nu_m \tag{94}$$

where

$$\nu_m = w \nu_m(n-1) + |f_{m-1}(n)|^2 + |b_{m-1}(n-1)|^2. \tag{95}$$

The equalizer output for each stage is computed according to

$$z_m(n) = z_{m-1}(n) + c_m(n-1) b_m(n),$$
$$m = 0, 1, \cdots, N - 1. \tag{96}$$

The final output $z_{N-1}(n)$ is used during data mode to compute the receiver decision $x(n)$. Initially, a reference signal $x(n)$ is substituted for the receiver decision in order to compute the error signals

$$e_m(n) = x(n) - z_m(n), \qquad m = 0, 1, \cdots, N - 1. \quad (97)$$

The last step in the algorithm is the equalizer coefficient update equation

$$c_m(n) = c_m(n-1) - 2e_m(n) b_m^*(n)/\nu_m,$$
$$m = 0, 1, \cdots, N - 1. \quad (98)$$

Equations (91)–(98) define the gradient AL equalizer algorithm.

One important property of the lattice structure is that as the reflection coefficients converge, the backward prediction errors $b_m(n)$, $m = 0, 1, \cdots, N - 1$, form a vector $\boldsymbol{b}(n)$ of orthogonal signal components, i.e.,

$$E\left[b_m(n) b_j(n) \right] = 0, \qquad \text{for } j \neq m.$$

The vector $\boldsymbol{b}(n)$ is a transformed version of the received vector $\boldsymbol{y}(n)$ with elements $y(n), y(n-1), \cdots, y(n-N+1)$. This transformation is performed by an $N \times N$ lower triangular matrix $\boldsymbol{L}$ according to $\boldsymbol{b}(n) = \boldsymbol{L}\boldsymbol{y}(n)$, where $\boldsymbol{L}$ is formed by the backward predictor coefficients of order m, $m = 1, \cdots, N - 1$. The prediction error $b_{m+1}(n)$ is given by

$$b_{m+1}(n) = y(n - m) - \sum_{j=1}^{m} B_m(j) y(n - m + j)$$

where the predictor coefficients B of order m and lower can all be derived from the first m reflection coefficients k_j, $l = 1, \cdots, m$.

The lower triangular form of the above transformation permits a simple way to increase the length or order of the lattice equalizer since the existing prediction errors and equalizer coefficients remain unchanged when another stage is added. The prediction errors at the mth stage are not functions of the reflection coefficients at succeeding stages as can be seen from Fig. 26, (92), and (93). The lattice equalizer is, therefore, order-recursive as well as time-recursive.

A computationally complex (requiring larger number of computations) but faster converging least squares form of the AL equalizer results when the performance index or cost function to be minimized is the exponentially windowed sum of squared errors given in (78) instead of the MSE. The structure of the least squares AL equalizer is shown in Fig. 27. Note that the lattice coefficients for forward and backward prediction for any of the lattice stages are no longer equal, each being independently up-

dated to minimize the weighted sum of squared forward and backward prediction errors, respectively. The least squares AL algorithm for a T-spaced complex equalizer is summarized below.

At time n, the inputs to the first lattice stage are set to the newly received sample, i.e.,

$$f_0(n) = b_0(n) = y(n) \quad (99)$$

and

$$E_0^f(n) = E_0^b(n) = wE_0^f(n-1) + y^*(n)y(n) \quad (100)$$

where $E_m^f(n)$ and $E_m^b(n)$ are the estimated sum of the squared forward and backward prediction errors, respectively, at stage m. Next, the order updates are performed for $m = 1, \cdots, N - 1$

$$K_m(n) = wK_m(n-1) + t_m(n-1) f_{m-1}(n) \quad (101)$$

the forward prediction errors

$$f_m(n) = f_{m-1}(n) - G_m(n-1) b_{m-1}(n-1) \quad (102)$$

the backward prediction error

$$b_m(n) = b_{m-1}(n-1) - H_m(n-1) f_{m-1}(n) \quad (103)$$

the lattice coefficient for forward prediction

$$G_m(n) = K_m(n)/E_{m-1}^b(n-1) \quad (104)$$

the lattice coefficient for backward prediction

$$H_m(n) = K_m^*(n)/E_{m-1}^f(n) \quad (105)$$

and

$$E_m^f(n) = E_{m-1}^f(n) - G_m(n) K_m^*(n) \quad (106)$$

$$E_m^b(n) = E_{m-1}^b(n) - H_m(n) K_m(n) \quad (107)$$

$$t_m(n) = \left[1 - \gamma_{m-1}(n) \right] b_{m-1}^*(n) \quad (108)$$

$$\gamma_m(n) = \gamma_{m-1}(n) + |t_m(n)|^2/E_{m-1}^b(n). \quad (109)$$

Now the equalizer output can be computed according to

$$z_m(n) = z_{m-1}(n) + \left[c_m(n-1)/E_{m-1}^b(n-1) \right] b_{m-1}(n),$$
$$m = 0, 1, \cdots, N - 1. \quad (110)$$

The final equalized signal is given by $z_{N-1}(n)$. The error signals are computed and the corresponding equalizer coefficient updates are performed for $m = 0, 1, \cdots, N - 1$ according to

$$e_m(n) = x(n) - z_m(n) \quad (111)$$

and

$$c_m(n) = wc_m(n-1) + t_m(n) e_{m-1}(n). \quad (112)$$

Equations (99) through (112) define the least squares AL algorithm.

4) Complexity and Numerical Stability: In the preceding sections, we have presented an overview of three basic forms of recursive least squares equalization algorithms. Fast RLS algorithms are still being actively studied to reduce computational complexity, specially for "multichannel" (fractionally spaced and decision-feedback) equalizers, and to improve stability when limited precision arithmetic is used. Some of the recent results are reported in [10], [11], and [52].

Accurate counts of the number of multiplications, additions/subtractions, and divisions are hard to summarize due to the large number of variations of the RLS algorithms

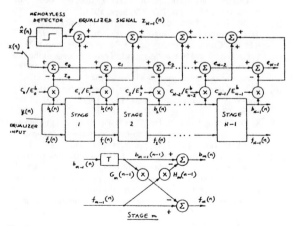

Fig. 27. Least squares adaptive lattice equalizer.

which have been reported in the literature. In the table below, the number of operations (multiplications and divisions) required per iteration is listed for T- and $T/2$-spaced transversal equalizers of span NT seconds for the LMS gradient, Kalman, fast Kalman/fast transversal, and RLS lattice algorithms. In each case, the smallest number of operations is given from the complexity estimates reported in [78], [10], and [52].

Algorithm	Number of Operations per Iteration	
	T Equalizer	$T/2$ Equalizer
LMS gradient	$2N$	$4N$
Kalman	$2N^2 + 5N$	$8N^2 + 10N$
Fast Kalman/fast transversal	$7N + 14$	$24N + 45$
RLS lattice	$15N - 11$	$46N$

The fast Kalman is the most efficient type of RLS algorithm. However, compared to the LMS gradient algorithm, the fast Kalman algorithm is still about four times as complex for T equalizers and six times for $T/2$ equalizers. The RLS lattice algorithm is still in contention due to its better numerical stability and order-recursive structure, despite a two-fold increase in computational complexity over the fast Kalman algorithm.

The discussion on RLS algorithms would not be complete without a comment on the numerical problems associated with these algorithms in steady-state operation. Simulation studies have reported the tendency of RLS algorithms implemented with finite precision to become unstable and the adaptive filter coefficients to diverge [10], [46], [52], [78], [126]. This is due to the long-term accumulation of finite precision errors. Among the different types of RLS algorithms, the fast Kalman or fast transversal type algorithms are the most prone to instability [46], [78]. In [78] instability was reported to occur when an exponential weighting factor $w < 1$ was used for the fast Kalman algorithm implemented with single-precision floating-point arithmetic. The Kalman and RLS lattice algorithms did not show this instability. A sequential processing dual-channel version of the RLS lattice algorithm for a DFE is reported to be stable even for fixed-point arithmetic with 10- to 12-bit accuracy [52]. However, in [10] an "unnormalized" RLS lattice algorithm is shown to become unstable. Normalized versions of fast transversal [10] and RLS lattice [127] algorithms are more stable but both require square roots, the lattice type having greater computational complexity.

The stability of RLS algorithms can be improved [10], [46], [126] by modification of the least squares criterion, and periodic reinitialization of the algorithm to avoid precision error buildup. The modified criterion takes into account the squared magnitude of the difference of the filter coefficients from their initial (or restart) values. The rationale for this so-called soft constraint [10] is the same as for the stochastic gradient algorithm with tap leakage discussed in Section III-C.

For a short transition period following each reinitialization, while the RLS algorithm is reconverging, an auxiliary LMS adaptive filter is used to compute outputs. Results of four variations of the periodic restart procedure for an adaptive decision feedback equalizer are given in [126].

V. CONCLUDING REMARKS

Adaptive equalization and the more general field of adaptive filtering have been areas of active research and development for more than two decades. It is, therefore,

tempting to state that no substantial further work remains to be done. However, this has not been the case in the last decade despite how mature the field appeared in 1973 [59]. In fact, a number of the topics covered in this paper, e.g., fractionally spaced equalizers, decision-aided ISI cancellation, and fast recursive least squares algorithms, were not yet fully understood or were yet to be discovered. Of course, tremendous strides have since been made in implementation technology which have spawned new applications, e.g., digital subscriber loops, and pushed existing applications toward their limits, e.g., 256-QAM digital radios and voice-band modems with rates approaching 19.2 kbits/s. Programmable digital signal processors now permit implementation of ever more sophisticated and computationally complex algorithms; and so the study and research must continue—in new directions. There is still more work to be done in adaptive equalization of nonlinearities with memory and in equalizer algorithms for coded modulation systems. However, the emphasis has already shifted from adaptive equalization theory toward the more general theory and applications of adaptive filters, and toward structures and implementation technologies which are uniquely suited to particular applications.

ACKNOWLEDGMENT

The author wishes to thank G. D. Forney, Jr., for his encouragement and guidance in preparing this paper. Thanks are also due to a number of friends and colleagues who have reviewed and commented on the paper.

REFERENCES AND BIBLIOGRAPHY

[1] K. Abend and B. D. Fritchman, "Statistical detection for communication channels with intersymbol interference," *Proc. IEEE*, vol. 58, pp. 779–785, May 1970.

[2] O. Agazzi, D. A. Hodges, and D. G. Messerechmitt, "Large scale integration of hybrid-method digital subscriber loops," *IEEE Trans. Commun.*, vol. COM-30, pp. 2095–2108, Sept. 1982.

[3] M. E. Austin, "Decision-feedback equalization for digital communication over dispersive channels," MIT Lincoln Lab., Lexington, MA, Tech. Rep. 437, Aug. 1967.

[4] C. A. Belfiore and J. H. Park, Jr., "Decision feedback equalization," *Proc. IEEE*, vol. 67, pp. 1143–1156, Aug. 1979.

[5] E. Biglieri, A. Gersho, R. D. Gitlin, and T. L. Lim, "Adaptive cancellation of nonlinear intersymbol interference for voiceband data transmission," *IEEE J. Selected Areas Commun.*, vol. SAC-2, pp. 765–777, Sept. 1984.

[6] D. M. Brady, "An adaptive coherent diversity receiver for data transmission through dispersive media," in *Proc. 1970 IEEE Int. Conf. Commun.*, pp. 21–35 to 21–39, June 1970.

[7] C. Caraiscos and B. Liu, "A roundoff error analysis of the LMS adaptive algorithm," *IEEE Trans. Acoust., Speech, Signal Process.*, vol. ASSP-32, pp. 34–41, Feb. 1984.

[8] R. W. Chang, "A new equalizer structure for fast start-up digital communication," *Bell Syst. Tech. J.*, vol. 50, pp. 1969–2014, July–Aug. 1971.

[9] R. W. Chang and E. Y. Ho, "On fast start-up data communication systems using pseudo-random training sequences," *Bell Syst. Tech. J.*, vol. 51, pp. 2013–2027, Nov. 1972.

[10] J. M. Cioffi and T. Kailath, "Fast, recursive-least-squares transversal filters for adaptive filtering," *IEEE Trans. Acoust. Speech, Signal Process.*, vol. ASSP-32, pp. 304–337, Apr. 1984.

[11] J. M. Cioffi and T. Kailath, "An efficient exact-least-squares fractionally spaced equalizer using intersymbol interpolation," *IEEE J. Selected Areas Commun.*, vol. SAC-2, pp. 743–756, Sept. 1984.

[12] D. L. Duttweiler, J. E. Mazo, and D. G. Messerschmitt, "An upper bound on the error probability in decision-feedback equalization," *IEEE Trans. Inform. Theory*, vol. IT-20, pp.

490–497, July 1974.

[13] D. L. Duttweiler, "Adaptive filter performance with nonlinearities in the correlation multiplier," *IEEE Trans. Acoust., Speech, Signal Process.*, vol. ASSP-30, pp. 578–586, Aug. 1982.

[14] T. Ericson, "Structure of optimum receiving filters in data transmission systems," *IEEE Trans. Inform. Theory* (Corresp.), vol. IT-17, pp. 352–353, May 1971.

[15] D. D. Falconer and G. J. Foschini, "Theory of minimum mean-square-error QAM system employing decision feedback equalization," *Bell Syst. Tech. J.*, vol. 53, pp. 1821–1849, Nov. 1973.

[16] D. D. Falconer, "Jointly adaptive equalization and carrier recovery in two-dimensional digital communication systems," *Bell Syst. Tech. J.*, vol. 55, pp. 317–334, Mar. 1976.

[17] D. D. Falconer and F. R. Magee, Jr., "Adaptive channel memory truncation for maximum likelihood sequence estimation," *Bell Syst. Tech. J.*, vol. 52, pp. 1541–1562, Nov. 1973.

[18] ____, "Evaluation of decision feedback equalization and Viterbi algorithm detection for voiceband data transmission–Parts I and II," *IEEE Trans. Commun.*, vol. COM-24, pp. 1130–1139, Oct. 1976, and pp. 1238–1245, Nov. 1976.

[19] D. D. Falconer and J. Salz, "Optimal reception of digital data over the Gaussian channel with unknown delay and phase jitter," *IEEE Trans. Inform. Theory*, vol. IT-23, pp. 117–126, Jan. 1977.

[20] D. D. Falconer and L. Ljung, "Application of fast Kalman estimation to adaptive equalization," *IEEE Trans. Commun.*, vol. COM-26, pp. 1439–1446, Oct. 1978.

[21] D. D. Falconer, "Adaptive equalization of channel nonlinearities in QAM data transmission systems," *Bell Syst. Tech. J.*, vol. 57, pp. 2589–2611, Sept. 1978.

[22] ____, "Adaptive reference echo cancellation," *IEEE Trans. Commun.*, vol. COM-30, pp. 2083–2094, Sept. 1982.

[23] G. L. Fenderson, J. W. Parker, P. D. Quigley, S. R. Shepard, and C. A. Siller, Jr., "Adaptive transversal equalization of multipath propagation for 16-QAM, 90-Mb/s digital radio," *AT&T Bell Lab. Tech. J.*, vol. 63, pp. 1447–1463, Oct. 1984.

[24] G. D. Forney, Jr., "Maximum-likelihood sequence estimation of digital sequences in the presence of intersymbol interference," *IEEE Trans. Inform. Theory*, vol. IT-18, pp. 363–378, May 1972.

[25] ____, "The Viterbi algorithm," *Proc. IEEE*, vol. 61, pp. 268–278, Mar. 1973.

[26] G. D. Forney, S. U. H. Qureshi, and C. K. Miller, "Multipoint networks: Advances in modem design and control," in *Nat. Telecom. Conf. Rec.*, pp. 50-1-1 to 50-1-4, Dec. 1976.

[27] G. D. Forney, Jr., R. G. Gallager, G. R. Lang, F. M. Longstaff, and S. U. Qureshi, "Efficient modulation for band-limited channels," *IEEE J. Selected Areas Commun.*, vol. SAC-2, pp. 632–647, Sept. 1984.

[28] G. J. Foschini and J. Salz, "Digital communications over fading radio channels," *Bell Syst. Tech. J.*, vol. 62, pp. 429–459, Feb. 1983.

[29] L. E. Franks, Ed., *Data Communication.* Stroudsburg, PA: Dowden, Hutchinson and Ross, 1974.

[30] S. A. Fredricsson, "Joint optimization of transmitter and receiver filters in digital PAM systems with a Viterbi detector," *IEEE Trans. Inform. Theory*, vol. IT-12, pp. 200–2210, Mar. 1976.

[31] D. A. George, "Matched filters for interfering signals," *IEEE Trans. Inform. Theory* (Corresp.), vol. IT-11, pp. 153–154, Jan. 1965.

[32] D. A. George, R. R. Bowen, and J. R. Storey, "An adaptive decision feedback equalizer," *IEEE Trans. Commun. Technol.*, vol. COM-19, pp. 281–293, June 1971.

[33] A. Gersho, "Adaptive equalization of highly dispersive channels," *Bell Syst. Tech. J.*, vol. 48, pp. 55–70, Jan. 1969.

[34] A. Gersho and T. L. Lim, "Adaptive cancellation of intersymbol interference for data transmission," *Bell Syst. Tech. J.*, vol. 60, pp. 1997–2021, Nov. 1981.

[35] R. D. Gitlin, E. Y. Ho, and J. E. Mazo, "Passband equalization of differentially phase-modulated data signals," *Bell Syst. Tech. J.*, vol. 52, pp. 219–238, February 1973.

[36] R. D. Gitlin, J. E. Mazo, and M. G. Taylor, "On the design of gradient algorithms for digitally implemented adaptive filters," *IEEE Trans. Circuit Theory*, vol. CT-20, pp. 125–136,

Mar. 1973.

[37] R. D. Gitlin and F. R. Magee, Jr., "Self-orthogonalizing algorithms for accelerated covergence of adaptive equalizers," *IEEE Trans. Commun.*, vol. COM-25, pp. 666–672, July 1977.

[38] R. D. Gitlin and S. B. Weinstein, "On the required tap-weight precision for digitally-implemented adaptive mean-squared equalizers," *Bell Syst. Tech. J.*, vol. 58, pp. 301–321, Feb. 1979.

[39] ____, "Fractionally-spaced equalization: An improved digital transversal equalizer," *Bell Syst. Tech. J.*, vol. 60, pp. 275–296, Feb. 1981.

[40] R. D. Gitlin, H. C. Meadors, Jr., and S. B. Weinstein, "The tap-leakage algorithm: An algorithm for the stable operation of a digitally implemented, fractionally-spaced adaptive equalizer," *Bell Syst. Tech. J.*, vol. 61, pp. 1817–1939, Oct. 1982.

[41] D. N. Godard, "Channel equalization using a Kalman filter for fast data transmission," *IBM J. Res. Develop.*, vol. 18, pp. 267–273, May 1974.

[42] ____, "Self-recovering equalization and carrier tracking in two-dimensional data communication systems," *IEEE Trans. Commun.*, vol. COM-28, pp. 1867–1875, Nov. 1980.

[43] ____, "A 9600 bit/s modem for multipoint communication systems," in *Nat. Telecomm. Conf. Rec.* (New Orleans, LA, Dec. 1981), pp. B3.3.1–B3.3.5.

[44] N. Grenander and G. Szego, *Toeplitz Forms and Their Application.* Berkeley, CA: Univ. Calif. Press, 1958.

[45] L. Guidoux, "Egaliseur autoadaptif a double echantillonnage," *L'Onde Electrique*, vol. 55, pp. 9–13, Jan. 1975.

[46] M. L. Honig and D. G. Messerschmitt, *Adaptive Filters; Structures, Algorithms, and Applications.* Boston, MA: Kluwer Academic Pub., 1984.

[47] C. R. Johnson, Jr., "Adaptive IIR filtering: Current results and open issues," *IEEE Trans. Inform. Theory*, vol. IT-30, pp. 237–250, Mar. 1984.

[48] P. Kabal and S. Pasupathy, "Partial-response signaling," *IEEE Trans. Commun.*, vol. COM-23, pp. 921–934, Sept. 1975.

[49] H. Kobayashi, "Simultaneous adaptive estimation and decision algorithm for carrier modulated data transmission systems," *IEEE Trans. Commun. Technol.*, vol. COM-19, pp. 268–280, June 1971.

[50] E. R. Kretzmer, "Binary data communication by partial response transmission," in *1965 ICC Conf. Rec.*, pp. 451–456; also, "Generalization of a technique for binary data communication," *IEEE Trans. Commun. Technol.*, vol. COM-14, pp. 67–68, Feb. 1966.

[51] W. U. Lee and F. S. Hill, "A maximum likelihood sequence estimator with decision feedback equalization," *IEEE Trans. Commun. Technol.*, vol. COM-25, pp. 971–979, Sept. 1977.

[52] F. Ling and J. G. Proakis, "A generalized multichannel least squares lattice algorithm based on sequential processing stages," *IEEE Trans. Acoust., Speech, Signal Process.*, vol. ASSP-32, pp. 381–389, Apr. 1984.

[53] H. L. Logan and G. D. Forney, Jr., "A MOS/LSI multiple configuration 9600 b/s data modem," in *Proc. IEEE Int. Conf. Commun.*, pp. 48-7 to 48-12, June 1976.

[54] R. W. Lucky, J. Salz, and E. J. Weldon, Jr., *Principles of Data Communication.* New York: McGraw-Hill, 1968.

[55] R. W. Lucky, "Automatic equalization for digital communication," *Bell Syst. Tech. J.*, vol. 44, pp. 547–588, Apr. 1965.

[56] ____, "Techniques for adaptive equalization of digital communication systems," *Bell Syst. Tech. J.*, vol. 45, pp. 255–286, Feb. 1966.

[57] R. W. Lucky and H. R. Rudin, "An automatic equalizer for general-purpose communication channels," *Bell Syst. Tech. J.*, vol. 46, pp. 2179–2208, Nov. 1967.

[58] R. W. Lucky, "Signal filtering with the transversal equalizer," in *Proc. 7th Ann. Allerton Conf. on Circuits and System Theory*, pp. 792–803, Oct. 1969.

[59] ____, "A survey of the communication theory literature: 1968–1973," *IEEE Trans. Inform. Theory*, vol. IT-19, pp. 725–739, Nov. 1973.

[60] O. Macchi and E. Eweda, "Convergence analysis of self-adaptive equalizers," *IEEE Trans. Inform. Theory*, vol. IT-30, pp. 161–176, Mar. 1984.

[61] L. R. MacKechnie, "Maximum likelihood receivers for channels having memory," Ph.D. dissertation, Dep. Elec. Eng., Univ. of Notre Dame, Notre Dame, Jan. 1973.

[62] F. R. Magee, Jr., and J. G. Proakis, "Adaptive maximum-likeli-

hood sequence estimation for digital signaling in the presence of intersymbol interference," *IEEE Trans. Inform. Theory* (Corresp.), vol. IT-19, pp. 120–124, Jan. 1973.

[63] J. Makhoul, "A class of all-zero lattice digital filters: properties and applications," *IEEE Trans. Acoust., Speech, Signal Process.*, vol. ASSP-26, pp. 304–314, Aug. 1978.

[64] J. E. Mazo, "Optimum timing phase for an infinite equalizer," *Bell Syst. Tech. J.*, vol. 54, pp. 189–201, Jan. 1975.

[65] ——, "On the independence theory of equalizer convergence," *Bell Syst. Tech. J.*, vol. 58, pp. 963–993, May–June 1979.

[66] ——, "Analysis of decision-directed equalizer convergence," *Bell Syst. Tech. J.*, vol. 59, pp. 1857–1876, Dec. 1980.

[67] D. G. Messerschmitt, "A geometric theory of intersymbol interference: Part I," *Bell Syst. Tech. J.*, vol. 52, pp. 1483–1519, Nov. 1973.

[68] ——, "Design of a finite impulse response for the Viterbi algorithm and decision feedback equalizer," in *Proc. IEEE Int. Conf. Communications, ICC-74* (Minneapolis, MN, June 17–19, 1974).

[69] ——, "Echo cancellation in speech and data transmission," *IEEE J. Selected Areas Commun.*, vol. SAC-2, pp. 283–296, Mar. 1984.

[70] A. Milewski, "Periodic sequences with optimal properties for channel estimation and fast start-up equalization," *IBM J. Res. Develop.*, vol. 27, pp. 426–431, Sept. 1983.

[71] P. Monsen, "Feedback equalization for fading dispersive channels," *IEEE Trans. Inform. Theory*, vol. IT-17, pp. 56–64, Jan. 1971.

[72] ——, "MMSE equalization of interference on fading diversity channels," *IEEE Trans. Commun.*, vol. COM-32, pp. 5–12, Jan. 1984.

[73] M. Morf, T. Kailath, and L. Ljung, "Fast algorithms for recursive identification," in *Proc. 1976 IEEE Conf. Decision Contr.* (Clearwater Beach, FL, Dec. 1976), pp. 916–921.

[74] M. Morf, A. Vieira, and D. T. Lee, "Ladder forms for identification and speech processing," in *Proc. 1977 IEEE Conf. Decision Contr.* (New Orleans, LA, Dec. 1977), pp. 1074–1078.

[75] K. H. Mueller, "A new, fast-converging mean-square algorithm for adaptive equalizers with partial-response signaling," *Bell Syst. Tech. J.*, vol. 54, pp. 143–153, Jan. 1975.

[76] K. H. Mueller and D. A. Spaulding, "Cyclic equalization—A new rapidly converging equalization technique for synchronous data communication," *Bell Syst. Tech. J.*, vol. 54, pp. 369–406, Feb. 1975.

[77] K. H. Mueller and J. J. Werner, "A hardware efficient passband equalizer structure for data transmission," *IEEE Trans. Commun.*, vol. COM-30, pp. 538–541, Mar. 1982.

[78] M. S. Mueller, "Least-squares algorithms for adaptive equalizers," *Bell Syst. Tech. J.*, vol. 60, pp. 1905–1925, Oct. 1981.

[79] ——, "On the rapid initial convergence of least-squares equalizer adjustment algorithms," *Bell Syst. Tech. J.*, vol. 60, pp. 2345–2358, Dec. 1981.

[80] M. S. Mueller and J. Salz, "A unified theory of data-aided equalization," *Bell Syst. Tech. J.*, vol. 6, pp. 2023–2038, Nov. 1981.

[81] K. Murano, Y. Mochida, F. Amano, and T. Kinoshita, "Multiprocessor architecture for voiceband data processing (application to 9600 bps modem)," in *Proc. IEEE Int. Conf. Commun.*, pp. 37.3.1–37.3.5, June 1979.

[82] T. Murase, K. Morita, and S. Komaki, "200 Mb/s 16-QAM digital radio system with new countermeasure techniques for multipath fading," in *Proc. IEEE Int. Conf. Commun.*, pp. 46.1.1–46.1.5, June 1981.

[83] R. Price, "Nonlinearly feedback-equalized PAM vs. capacity for noisy filter channels," in *Proc. 1972 IEEE Int. Conf. Commun.*, pp. 22-12 to 22-17, June 1972.

[84] J. G. Proakis and J. H. Miller, "An adaptive receiver for digital signaling through channels with intersymbol interference," *IEEE Trans. Inform. Theory*, vol. IT-15, pp. 484–497, July 1969.

[85] J. G. Proakis, "Adaptive nonlinear filtering techniques for data transmission," in *IEEE Symp. on Adaptive Processes, Decision and Control*, pp. XV.2.1-5, 1970.

[86] ——, "Advances in equalization for intersymbol inter-

ference," in *Advances in Communication Systems*, vol. 4, A. J. Viterbi, Ed. New York: Academic Press, 1975, pp. 123–198.

[87] ——, *Digital Communications.* New York: McGraw-Hill, 1983.

[88] S. U. H. Qureshi, "New approaches in adaptive reception of digital signals in the presence of intersymbol interference," Ph.D. dissertation, Univ. of Toronto, Toronto, Ont., Canada, May 1973.

[89] S. U. H. Qureshi and E. E. Newhall, "An adaptive receiver for data transmission over time-dispersive channels," *IEEE Trans. Inform. Theory*, vol. IT-19, pp. 448–457, July 1973.

[90] S. U. H. Qureshi, "Fast start-up equalization with periodic training sequences," *IEEE Trans. Inform. Theory*, vol. IT-23, pp. 553–563, Sept. 1977.

[91] S. U. H. Qureshi and G. D. Forney, Jr., "Performance and properties of a T/2 equalizer," in *Nat. Telecomm. Conf. Rec.*, Dec. 1977.

[92] W. Rudin, *Real and Complex Analysis.* New York: McGraw-Hill, 1966.

[93] J. Salz, "Optimum mean-square decision feedback equalization," *Bell Syst. Tech. J.*, vol. 52, pp. 1341–1373, Oct. 1973.

[94] ——, "On mean-square decision feedback equalization and timing phase," *IEEE Trans. Commun. Technol.*, vol. COM-25, pp. 1471–1476, Dec. 1977.

[95] Y. Sato, "A method of self-recovering equalization for multilevel amplitude modulation," *IEEE Trans. Commun.*, vol. COM-23, pp. 679–682, June 1975.

[96] E. H. Satorius and S. T. Alexander, "Channel equalization using adaptive lattice algorithms," *IEEE Trans. Commun.* (Concise Paper), vol. COM-27, pp. 899–905, June 1979.

[97] E. H. Satorius and J. D. Pack, "Application of least squares lattice algorithms to adaptive equalization," *IEEE Trans. Commun.*, vol. COM-29, pp. 136–142, Feb. 1981.

[98] C. A. Siller, Jr., "Multipath propagation," *IEEE Commun. Mag.*, vol. 22, pp. 6–15, Feb. 1984.

[99] T. A. Schonhoff and R. Price, "Some bandwidth efficient modulations for digital magnetic recording," in *Proc. IEEE Int. Conf. Communications, ICC-81* (Denver, CO, June 15–18, 1981).

[100] S. Y. Tong, "Dataphone II service: Data set architecture," in *Nat. Telecomm. Conf. Rec.* (New Orleans, LA, Dec. 1981), pp. B.3.2.1–B.3.2.5.

[101] T. Tsuda, Y. Mochida, K. Murano, S. Unagami, H. Gambe, T. Ikezawa, H. Kikuchi, and S. Fujii, "A high performance LSI digital signal processor for communication," in *Proc. 1983 IEEE Int. Conf. Commun.*, pp. A5.6.1–A5.6.5, June 1983.

[102] G. Ungerboeck, "Theory on the speed of convergence in adaptive equalizers for digital communication," *IBM J. Res. Devel.*, vol. 16, pp. 546–555, Nov. 1972.

[103] ——, "Adaptive maximum-likelihood receiver for carrier-modulated data-transmission systems," *IEEE Trans. Commun.*, vol. COM-22, pp. 624–636, May 1974.

[104] ——, "Fractional tap-spacing equalizer and consequences for clock recovery in data modems," *IEEE Trans. Commun.*, vol. COM-24, pp. 856–864, Aug. 1976.

[105] ——, "Channel coding with multilevel/phase signals," *IEEE Trans. Inform. Theory*, vol. IT-28, pp. 55–67, Jan. 1982.

[106] P. J. van Gerwen, N. A. M. Verhoeckx, and T. A. C. M. Claasen, "Design considerations for a 144 kbit/s digital unit for the local telephone network," *IEEE J. Selected Areas Commun.*, vol. SAC-2, pp. 314–323, Mar. 1984.

[107] L.-F. Wei, "Rotationally invariant convolutional channel coding with expanded signal space—Part II: Nonlinear codes," *IEEE J. Selected Areas Commun.*, vol. SAC-2, pp. 672–686, Sept. 1984.

[108] S. B. Weinstein, "A baseband data-driven echo canceller for full-duplex transmission on two-wire circuits," *IEEE Trans. Commun.*, vol. COM-25, vol. 654–666, July 1977.

[109] B. Widrow and M. E. Hoff, Jr., "Adaptive switching circuits," in *IRE WESCON Conv. Rec.*, pt. 4, pp. 96–104, Aug. 1960.

[110] B. Widrow, J. R. Glover, Jr., J. M. McCool, J. Kaunitz, C. S. Williams, R. H. Hearn, J. R. Zeidler, E. Dong, Jr., and R. C. Goodlin, "Adaptive noise cancelling: Principles and applications," *Proc. IEEE*, vol. 63, pp. 1692–1716, Dec. 1975.

[111] B. Widrow, J. M. McCool, M. G. Larimore, and C. R. Johnson,

Jr., "Stationary and nonstationary learning characteristics of the LMS adaptive filter," *Proc. IEEE*, vol. 64, pp. 1151–1162, Aug. 1976.

[112] M. Yasumoto, T. Enomoto, K. Watanabe, and T. Ishihara, "Single-chip adaptive transversal filter IC employing switched capacitor technology," *IEEE J. Selected Areas Commun.*, vol. SAC-2, pp. 324–333, Mar. 1984.

[113] K. Feher, *Digital Communications: Satellite/Earth Station Engineering.* Englewood Cliffs, NJ: Prentice-Hall, 1983.

[114] P. Monsen, "Theoretical and measured performance of a DFE modem on a fading multipath channel," *IEEE Trans. Commun.*, vol. COM-25, pp. 1144–1153, Oct. 1977.

[115] H. Nyquist, "Certain topics in telegraph transmission theory," *Trans. AIEE*, vol. 47, pp. 617–644, Apr. 1928.

[116] J. W. Smith, "The joint optimization of transmitted signal and receiving filter for data transmission systems," *Bell Syst. Tech. J.*, vol. 44, pp. 2363–2392, Dec. 1965.

[117] D. W. Tufts, "Nyquist's problem—The joint optimization of transmitter and receiver in pulse amplitude modulation," *Proc. IEEE*, vol. 53, pp. 248–260, Mar. 1965.

[118] CCITT Recommendation V.27 bis, "4800/2400 bits per second modem with automatic equalizer standardized for use on leased telephone-type circuits," Int. Telegraph and Telephone Consultative Committee, Geneva, Switzerland, 1980.

[119] CCITT Recommendation V.29, "9600 bits per second modem standardized for use on point-to-point 4-wire leased telephone-type circuits," Int. Telegraph and Telephone Consultative Committee, Geneva, Switzerland, 1980.

[120] S. U. H. Qureshi and H. M. Ahmed, "A custom ship set for digital signal processing," in *VLSI Signal Processing*, P. R. Cappello *et al.*, Eds. New York: IEEE PRESS, 1984.

[121] K. J. Wouda, S. J. M. Tol, and W. J. M. Reiknkjens, "An ISDN transmission system with adaptive echo cancelling and decision feedback equalization—A two-chip realization," in *Proc. IEEE Int. Conf. Commun.* (Amsterdam, The Netherlands), pp. 685–690, 1984.

[122] J. Makhoul, "Linear prediction: A tutorial review," *Proc. IEEE*, vol. 63, pp. 561–580, Apr. 1975.

[123] O. Agazzi, D. G. Messerschmitt, and D. A. Hodges, "Nonlinear echo cancellation of data signals," *IEEE Trans. Commun.*, vol. COM-30, pp. 2421–2433, Nov. 1982.

[124] M. Holte and S. Stueflotten, "A new digital echo canceller for two-wire subscriber lines," *IEEE Trans. Commun.*, vol. COM-29, pp. 1573–1581, Nov. 1981.

[125] V. B. Lawrence and J. J. Werner, "Low-speed data transmission by using high-speed modems," in *Proc. IEEE Globecom* (Atlanta, GA), pp. 677–682, Nov. 1984.

[126] E. Eleftheriou and D. D. Falconer, "Restart methods for stabilizing FRLS adaptive equalizers in digital HF transmission," in *Proc. IEEE Globecom* (Atlanta, GA), pp. 1558–1562, Nov. 1984.

[127] B. Friedlander, "Lattice filters for adaptive processing," *Proc. IEEE*, vol. 70, pp. 829–867, Aug. 1982.

[128] D. L. Lyon, "Envelope-derived timing recovery in QAM and SQAM systems," *IEEE Trans. Commun.*, vol. COM-23, pp. 1327–1331, Nov. 1975.

[129] S. U. H. Qureshi and G. D. Forney, Jr., "A 9.6/16 kb/s speech digitizer," in *Conf. Rec. IEEE Int. Conf. Commun.*, pp. 30–31 to 30–36, June 1975, also in *Waveform Quantization and Coding*, N. S. Jayant, Ed. New York: IEEE PRESS, 1976, pp. 214–219.

[130] G. Picchi and G. Prati, "Self-orthogonalizing adaptive equalization in the discrete frequency domain," *IEEE Trans. Commun.*, vol. COM-32, pp. 371–379, Apr. 1984.

Adaptive Equalization for TDMA Digital Mobile Radio

John G. Proakis, *Fellow, IEEE*

Abstract—Adaptive equalization for a time division multiple access (TDMA) digital cellular system is treated in this paper. First, a survey of adaptive equalization techniques is presented, including their performance characteristics and limitations, and their implementation complexity. Then, the design of adaptive equalization algorithms for a narrow-band TDMA system is considered.

I. INTRODUCTION

THE demand for mobile radio telephone service throughout the United States, Canada, Europe, and Japan has accelerated the development spectrally efficient digital modulation/demodulation techniques for replacing the existing spectrally inefficient systems based on analog modulation. The proposed narrow-band and wide-band time division multiple access (TDMA) digital cellular systems require adaptive equalization at the demodulator to combat the intersymbol interference resulting from the time-variant multipath propagation of the signal through the channel.

Adaptive equalization techniques that have been developed during the last two decades for high speed, single carrier serial transmission over telephone channels and radio channels such as microwave line-of-sight, troposcatter and HF, may also be applied to digital transmission over mobile radio channels in the VHF band. However, the TDMA signal structure and the rapidly varying channel characteristics (fading) impose some stringent conditions on the design of the adaptive equalizer.

In this paper we first present a survey of adaptive equalization techniques, including their performance characteristics and limitations, and their implementation complexity. Then, we consider the design of an effective adaptive equalizer structure and associated adaptation algorithm for a narrow-band TDMA system.

II. SURVEY OF ADAPTIVE EQUALIZATION TECHNIQUES

Equalization techniques for combatting intersymbol interference (ISI) on bandlimited time dispersive channels may be subdivided into two general types—linear and nonlinear equalization. Associated with each type of equalizer is one or more structures for implementing the equalizer. Furthermore, for each structure there is a class of algorithms that may be employed to adaptively adjust the equalizer parame-

Manuscript received May 11, 1990; revised September 1, 1990.

The author is with the Department of Electrical and Computer Engineering, Communications and Digital Signal Processing Research Center, Northeastern University, Boston, MA 02115.

IEEE Log Number 9144472.

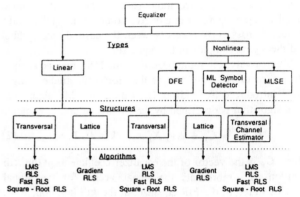

Fig. 1. Equalizer types, structures, and algorithms [2].

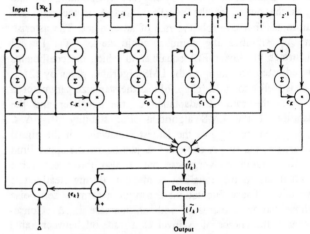

Fig. 2. Adaptive linear FIR equalizer with LMS algorithm [2].

ters according to some specified performance criterion. Fig. 1 provides an overall categorization of adaptive equalization techniques into types, structures, and algorithms.

A. Linear Equalization Techniques

A linear equalizer may be implemented as a finite-duration impulse response (FIR) filter (also called a transversal filter) with adjustable coefficients. The adjustment of the equalizer coefficients is usually performed adaptively during the transmission of information by using the decisions at the output of the detector in forming the error signal for the adaptation, as shown in Fig. 2. For symbol error rates below 10^{-2}, the occasional errors made by the detector have a negligible effect on the performance of the equalizer. During the start-up period, a short known sequence of symbols is transmitted for the purpose of initial adjustment of the equalizer coefficients.

0018-9545/91/0500-0333$01.00 © 1991 IEEE

The criterion most commonly used in the optimization of the equalizer coefficients is the minimization of the mean square error (MSE) between the desired equalizer output and the actual equalizer output. The minimization of the MSE results in the optimum Wiener filter solution for the coefficient vector, which may be expressed

$$C_{\text{opt}} = \Gamma^{-1}\xi \qquad (1)$$

where Γ is the autocorrelation matrix of the vector of signal samples in the equalizer at any given time instant and ξ is the vector of cross correlations between the desired data symbol and the signal samples in the equalizer.

Alternatively, the minimization of the MSE may be accomplished recursively by use of the stochastic gradient algorithm introduced by Widrow [1], called the LMS algorithm. This algorithm is described by the coefficient update equation

$$C_{k+1} = C_k + \Delta e_k X_k^*, \qquad k = 0, 1, \cdots \qquad (2)$$

where C_k is the vector of the equalizer coefficients at the kth iteration X_k represents the signal vector for the signal samples stored in the FIR equalizer at the kth iteration, e_k is the error signal, which is defined as the difference between the kth transmitted symbol I_k and its corresponding estimate $\hat{I}_k$ at the output of the equalizer, and Δ is the step size parameter that controls the rate of adjustment. The asterisk on X_k^* signifies the complex conjugate of X_k. Fig. 2 illustrates the linear FIR equalizer in which the coefficients are adjusted according to the LMS algorithm given by (2).

It is well known [2, chap. 6] that the step size parameter Δ controls the rate of adaptation of the equalizer and the stability of the LMS algorithm. For stability, $0 < \Delta < 2/\lambda_{\max}$, where $\lambda_{\max}$ is the largest eigenvalue of the signal covariance matrix. A choice of Δ just below the upper limit provides rapid convergence, but it also introduces large fluctuations in the equalizer coefficients during steady-state operation. These fluctuations constitute a form of self-noise whose variance increases with an increase in Δ. Consequently, the choice of Δ involves a trade-off between rapid convergence and the desire to keep the variance of the self-noise small.

The convergence rate of the LMS algorithm is slow due to the fact that there is only a single parameter, namely Δ, that controls the rate of adaptation. A faster converging algorithm is obtained if we employ a recursive least squares (RLS) criterion for adjustment of the equalizer coefficients. For the linear FIR equalizer, the RLS algorithm that is obtained for the minimization of the sum of exponentially weighted squared errors, i.e.,

$$\mathscr{E} = \sum_{n=0}^{k} w^{k-n} | I_n - \hat{I}_n |^2$$

$$= \sum_{n=0}^{k} w^{k-n} | I_n - C_k' X_n^* |^2$$

may be expressed as [2, chap. 6]

$$C_{k+1} = C_k + P_k X_k^* e_k \qquad (3)$$

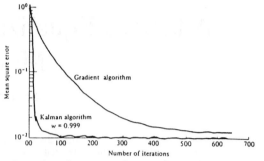

Fig. 3. Comparison of convergence between LMS and RLS algorithms [2].

where $\hat{I}_k$ is the estimate of the kth symbol I_k at the output of the equalizer, C_k' denotes the transpose of C_k, and

$$e_k = I_k - \hat{I}_k \qquad (4)$$

$$P_k = \frac{1}{w}\left[P_{k-1} - \frac{P_{k-1} X_k^* X_k' P_{k-1}}{w + X_k' P_{k-1} X_k^*} \right]. \qquad (5)$$

The exponential weighting factor w is selected to be in the range $0 < w < 1$. It provides a fading memory in the estimation of the optimum equalizer coefficient. P_k is an $(N \times N)$ square matrix which is the inverse of the data autocorrelation matrix:

$$R_k = \sum_{n=0}^{k} w^{k-n} X_n^* X_n'. \qquad (6)$$

Initially, P_0 may be selected to be proportional to the identity matrix. Fig. 3 illustrates a comparison of the convergence rate of the RLS and the LMS algorithms for an equalizer of length the $N = 11$ and a channel with a small amount of ISI. We note that the difference in convergence rate is very significant.

The recursive update equation for the matrix P_k given by (5) has poor numerical properties. For this reason, other algorithms with better numerical properties have been derived which are based on a square-root factorization of P_k as $P_k = S_k S_k'$, where S_k is a lower triangular matrix. Such algorithms are called *square-root RLS algorithms* [3]. These algorithms update the matrix S_k directly without computing P_k explicitly, and have a computational complexity proportional to N^2. Another type of RLS algorithm appropriate for a transversal FIR equalizer has been devised with a computational complexity proportional to N [2, chap. 6]. This algorithm is called a *fast RLS algorithm* [4]–[7].

The linear equalizer based on the RLS criterion may also be implemented as a lattice structure [8]. This structure is illustrated in Fig. 4. The equations for updating the parameters of this structure are given in [2, chap. 6]. The convergence rate is identical to that of the RLS algorithm given above for the adaptation of the linear transversal (FIR) structure. However, the computational complexity for this RLS lattice structure is proportional to N, but with a larger proportionality constant compared with the fast RLS algorithm for the transversal FIR equalizer structure [2, chap. 6]. A computationally simpler, albeit slower converging, algo-

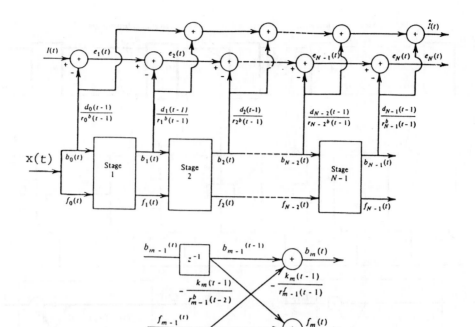

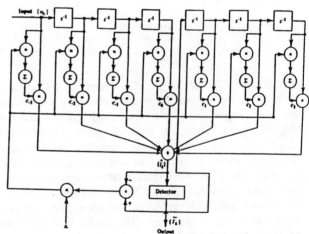

Fig. 4.   RLS lattice equalizer [2].

rithm for the lattice structure is the gradient lattice algorithm [2, chap. 6].

Both the linear transversal and lattice equalizers are all-zero filters. The implementation of a linear equalizer as an infinite duration impulse response (IIR) filter structure (direct form or lattice) can be easily accomplished by adding a filter section that contains poles. However, the addition of poles in an adaptive filter entails the risk that one or more poles may move outside the unit circle during adaptation and render the equalizer unstable. Usually, the risk of instability far outweighs the small benefits in reduced complexity (fewer filter coefficients) that the IIR may yield. Consequently, adaptive IIR equalizers are seldom used in practice.

B. Nonlinear Equalization Techniques

Nonlinear equalizers find use in applications where the channel distortion is too severe for a linear equalizer to handle. In particular, the linear equalizer does not perform well on channels with spectral nulls in their frequency response characteristics. In an attempt to compensate for the channel distortion, the linear equalizer places a large gain in the vicinity of the spectral null and, as a consequence, significantly enhances the additive noise present in the received signal.

Three very effective nonlinear equalization methods have been developed over the past three decades. One is decision feedback equalization. The second is a symbol-by-symbol detection algorithm based on the maximum *a posteriori* probability (MAP) criterion proposed by Abend and Fritchman [9]. The third is a sequence detection algorithm, based on the maximum-likelihood sequence estimation (MLSE) criterion, which is efficiently implemented by means of the Viterbi algorithm (VA) [10]. We briefly describe the key features of these equalization methods.

Fig. 5.   Decision-feedback equalizer [2].

1) Decision Feedback Equalizer (DFE): The basic idea in the DFE is that once an information symbol has been detected, the ISI that it causes on future symbols may be estimated and subtracted out prior to symbol detection. The DFE may be realized either in the direct form or as a lattice. The direct-form structure of the DFE is illustrated in Fig. 5. It consists of a feedforward filter (FFF) and a feedback filter (FBF). The latter is driven by decisions of the output of the detector and its coefficients are adjusted to cancel the ISI on the current symbol that results from past detected symbols (postcursors). The coefficient adjustment may be performed as in the linear equalizer by the relatively simple gradient LMS algorithm or by the faster converging RLS algorithm, e.g., the square-root RLS or the fast RLS algorithm.

An alternative form of the DFE is the RLS lattice structure which is illustrated in Fig. 6. This structure is equivalent to a direct form (transversal-type) DFE having an FFF of length

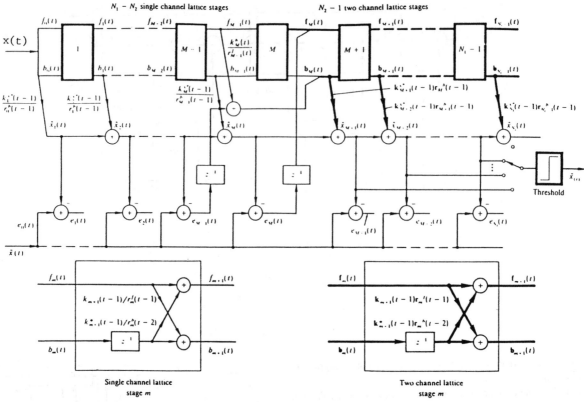

Fig. 6.   RLS lattice DFE [2].

N_1, and an FBF of length N_2, where we assume that $N_1 \geq N_2$. In the lattice structure shown in Fig. 6 we note that the lattice stages are of two types, single channel and two channel lattices. There are $N_1 - N_2$ single channel lattice stages, which have basically the same form as the lattice stages in a linear FIR lattice filter and one transitional stage. There are also $N_2 - 1$ two-channel lattice stages whose input and output consists of two two-dimensional vectors, and the lattice gain factors (the reflection coefficients) are also two-dimensional vector quantities. The ladder part (the joint process estimator) which forms estimates of the information symbols is similar in form to the ladder section in the linear RLS lattice–ladder equalizer. The RLS lattice algorithm may be replaced by the simpler gradient lattice algorithm to adjust the coefficients of the lattice DFE, as described in [2, chap. 6].

Another form of a DFE, proposed by Belfiore and Park [11], is called a predictive DFE. The basic structure for the predictive DFE is illustrated in Fig. 7. It also consists of a FFF as in the conventional DFE. However, the FBF in the predictive DFE is driven by an input sequence formed by the difference of the output of the detector and the output of the FFF. (In the conventional DFE, the input to the FBF is the output of the detector.) As a consequence, the FBF is called a noise predictor, because it forms an estimate (or a prediction) of the noise and residual ISI contained in the signal at the output of the FFF and subtracts from it the detector output after some feedback delay. It can be shown [2], [11] that the predictive DFE performs as well as the conventional DFE in

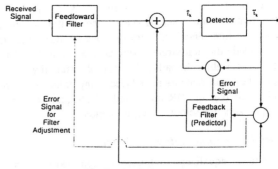

Fig. 7.   Predictive DFE [2].

the limit as the number of taps in the FFF and FBF approach infinity.

The FFF of the predictive DFE may also be realized as a lattice structure as demonstrated by Zhou et al. [12]. Thus the RLS lattice algorithm may be used in the FFF to yield fast convergence if necessary. The FBF may be implemented either as a direct form (transversal) FIR filter structure or as a lattice structure and either the gradient LMS or an RLS algorithm may be used to adapt the filter weights depending on the speed of convergence required to track the channel variations.

In general, the class of RLS algorithms provide faster convergence and better tracking of time-variant channel characteristics than the LMS algorithm. The convergence rate of the LMS algorithm is especially slow in channels which

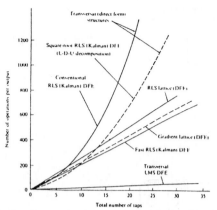

Fig. 8. Computational complexity of DFE algorithms [2].

TABLE I
COMPUTATIONAL COMPLEXITY OF ADAPTIVE DFE ALGORITHMS [2]

Algorithm	Total number of complex operations	Number of divisions
LMS DFE	$2N + 1$	0
Fast RLS (Kalman) DFE	$20N + 5$	3
Conventional RLS Kalman) DFE	$2.5N^2 + 4.5N$	2
Square-root RLS (Kalman) DFE	$1.5N^2 + 6.5N$	N
Gradient lattice DFE	$13N_1 + 33N_2 - 36$	$2N_1$
RLS lattice DFE	$18N_1 + 39N_2 - 39$	$2N_1$

contain spectral nulls, whereas the convergence rate of the RLS algorithm is unaffected by the channel characteristics.

The price to be paid for faster convergence is an increase in computational complexity. Fig. 8 illustrates the computational complexity of the LMS and RLS algorithms when applied to a DFE. We observe that the LMS algorithm is by far the simplest to implement. Its computational complexity is proportional to $2N$, where N is the total (FFF and FBF) equalizer length. The fast RLS algorithm for a transversal structure equalizer is the most computationally efficient among the RLS algorithms. Its computational complexity is approximately proportional to $20N$. The gradient lattice algorithm, which is a computationally simpler version of the RLS algorithm, albeit suboptimum in terms of rate of convergence, also has a computational complexity proportional to N, but the proportionality constant is smaller that than for the RLS algorithm. Finally, the square-root RLS algorithm has a computational complexity proportional to N^2, which renders the algorithm computationally inefficient for large equalizer lengths, as can be observed from Fig. 8. Table I summarizes the computational complexities of the various DFE algorithms. In this table, N_1 denotes the number of coefficients in the FFF, N_2 denotes the number of coefficients in the FBF, and $N = N_1 + N_2$. The graphs shown in Fig. 8 are based on the assumption that $N_1 = N_2 = N/2$.

2) Probabilistic Detection Algorithm: The MAP algorithm and the MLSE algorithm are optimal in the sense that they minimize the probability of the error. In the MAP algorithm, it is the symbol error rate that is minimized. In MLSE, the VA minimizes the probability of a sequence error. In practice, these two probabilistic detection algorithms provide comparable performance. A description of the

algorithms and their performance characteristics are given in [9], [10], [2].

The symbol-by-symbol MAP and the MLSE algorithms require knowledge of the channel characteristics in order to compute the metrics (probabilities) for making decisions. In the absence of such knowledge the channel must be estimated. Channel estimation can be accomplished adaptively as illustrated in Fig. 9. The channel estimator is usually an FIR transversal filter with adjustable coefficients. Either the gradient LMS algorithm or one of the class of the faster converging RLS algorithms may be used to adjust the coefficients of the channel estimator. These estimated coefficients are fed to the probabilistic symbol-by-symbol MAP algorithm or the MLSE-based VA for use in the metric computations.

In addition to knowledge of the channel characteristics, the MAP and MLSE algorithms also require knowledge of the statistical distribution of the noise corrupting the signal. Thus the probability distribution of the additive noise determines the form of the metric for optimum demodulation of the received signal.

For time-dispersive channels in which the ISI spans many symbols, these probabilistic algorithms become impractical due to an exponentially growing computational complexity with ISI span. Nevertheless, they serve as benchmarks against which we can compare the performance of suboptimal algorithms such as the DFE and the linear equalizer.

C. Symbol Versus Fractionally Spaced Equalizers

It is well known [2] that the optimum receiver for a digital communication signal corrupted by additive white Gaussian noise consists of a matched filter which is sampled periodically at the symbol rate. If the received signal samples are corrupted by intersymbol interference, the symbol-spaced samples are further processed by either a linear or nonlinear equalizer.

In the presence of channel distortion, the matched filter prior to the equalizer must be matched to the channel corrupted signal. However, in practice, the channel impulse response is usually unknown and consequently, the optimum matched filter to the received signal must be adaptively estimated. A suboptimum solution in which the matched filter is matched to the transmitted signal pulse may result in a significant degradation in performance. In addition, such a suboptimum filter is extremely sensitive to any timing error in the sampling of its output [13].

A fractionally spaced equalizer (FSE) is based on sampling the incoming signal at least as fast as the Nyquist rate. For example, if the transmitted signal consists of pulses having a raised cosine spectrum with rolloff factor β, its spectrum extends to $F_{max} = (1 + \beta)/2T$. This signal may be sampled at the receiver at the minimum rate of

$$2F_{max} = \frac{1 + \beta}{T} \qquad (7)$$

and then passed through an equalizer with tap spacing of $T/(1 + \beta)$. For example, if $\beta = 1$, we require a T/2-spaced equalizer. If $\beta = 1/2$, we may use a $2T/3$-spaced equalizer and so forth. In general, a digitally implemented FSE has tap

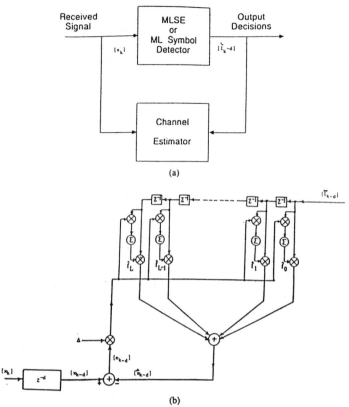

Fig. 9.   Probabilistic symbol detection with adaptive channel estimation [2].

spacings of KT/L, where K and L are integers and $K < L$. Often, a $T/2$-spaced equalizer is used in many applications, even in cases where a larger tap spacing is possible.

The frequency response of a FSE is [2]

$$C_{T'}(f) = \sum_{k=0}^{N-1} c_k e^{-j2\pi f k T'} \qquad (8)$$

where $T' = KT/L$. Hence, $C_{T'}(f)$ can equalize the received signal spectrum beyond the Nyquist frequency up to $f - L/KT$. The equalized spectrum is

$$C_{T'}(f)Y_{T'}(f) = C_{T'}(f)\sum_{n} X\left(f - \frac{n}{T'}\right)e^{j2\pi(f-n/T')\tau_0}$$
$$= C_{T'}(f)\sum_{n} X\left(f - \frac{nL}{KT}\right)e^{j2\pi(f-nL/KT)\tau_0} \qquad (9)$$

where $X(f)$ is the input analog signal spectrum $Y_{T'}(f)$ is the spectrum of the sampled signal and τ_0 is a timing delay. Since $X(f) = 0$ for $|f| > L/KT$, the above expression reduces to

$$C_{T'}(f)Y_{T'}(f) = C_{T'}(f)X(f)e^{j2\pi f\tau_0}, \qquad |f| \le \frac{1}{2T'}. \qquad (10)$$

Thus the FSE compensates for the channel distortion in the received signal before aliasing effects occur due to symbol rate sampling. In addition, the equalizer with transfer function $C_{T'}(f)$ can compensate for any timing delay τ_0, i.e., for any arbitrary timing phase. In effect, the fractionally spaced equalizer incorporates the functions of matched filtering and equalization into a single filter structure.

The FSE output is sampled at the symbol rate $1/T$ and has a spectrum

$$\sum_{k} C_{T'}\left(f - \frac{k}{T}\right)X\left(f - \frac{k}{T}\right)e^{-j2\pi(f-k/T)\tau_0}. \qquad (11)$$

Its tap coefficients may be adaptively adjusted once per symbol as in a T-spaced equalizer. There is no improvement in convergence rate by making adjustments at the input sampling rate of the FSE.

Simulation results demonstrating the effectiveness of the FSE over a symbol rate equalizer have been given in the papers by Qureshi and Forney [13] and Gitlin and Weinstein [14]. We cite one example from the first paper. Fig. 10 illustrates the performance of the symbol rate equalizer and $T/2$-spaced equalizer for a channel with high frequency amplitude distortion, whose characteristics are also shown in this figure. In this example, the symbol spaced equalizer was preceded by a filter matched to the transmitted pulse which had a (square-root) raised cosine spectrum with a 20 % rolloff factor ($\beta = 0.2$). The FSE did not have any filter preceding it. The symbol rate was 2400 Bd and the modulation was QAM. The received SNR was 30 dB. Both equalizers had 31 taps; hence the $T/2$ FSE spanned one-half of the time interval of the symbol rate equalizer. Nevertheless, the FSE outperformed the symbol rate equalizer even when the latter was optimized at the best sampling time. Furthermore, the FSE did not exhibit any sensitivity to timing phase, as shown in Fig. 10. Similar results were later obtained by Gitlin and Weinstein [14].

The above results clearly demonstrate the advantages of

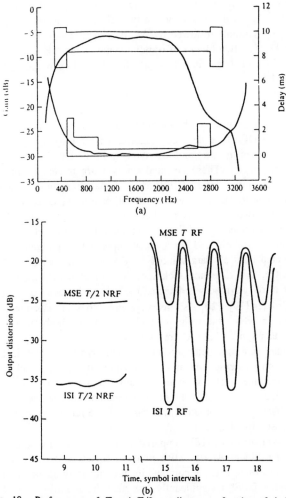

Fig. 10. Performance of T and $T/2$ equalizer as a function of timing phase for 2400 symbols/s. (From paper by Qureshi and Forney [13].) (NRF-no receive filter; RF-receive filter.) (a) Channel with high-end amplitude distortion (HA). (b) Equalizer performance [2].

FSE's over symbol rate equalizers. FSE's are currently in use in nearly all commercially available high speed modems over voice frequency channels.

In the implementation of the DFE, the FFF should be fractionally spaced, e.g., $T/2$-spaced taps, where $1/T$ is the symbol rate, and its length should span the total anticipated channel dispersion. The FBF has T-spaced taps and its length should also span the anticipated channel dispersion.

III. EQUALIZATION OF DIGITAL MOBILE RADIO CHANNELS

Mobile radio channels are generally characterized as fading multipath channels with time dispersion (multipath spreads) ranging from a few microseconds (μs) up to as much as 100 μs. Such large multipath spreads result in ISI which necessitates the use of an equalizer.

The frequency selective channel characteristics in mobile radio generally result in channel spectral nulls. As a consequence, linear equalizers must be ruled out since it is generally recognized that their performance is poor on such channel characteristics [2]. Hence, our choice is limited to nonlinear equalization techniques namely, DFE, MLSE, and MAP.

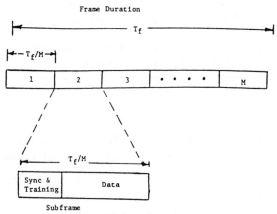

Fig. 11. TDMA signal frame format [2].

Of the two probabilistic algorithms, the MLSE is generally simpler to implement. For the anticipated multipath spreads in the channel (≤ 100 μs), both the DFE and the MLSE algorithms are relatively simple to implement for narrow-band TDMA. The DFE may prove to be more robust than MLSE, since the former does not require knowledge of the statistical characteristics of the additive noise whereas the latter does require this information in the computation of distance metrics among sequences of information symbols.

We assume that the access scheme is TDMA with some nominal channel bandwidth B. The modulation scheme may be four-phase PSK (QPSK) or, more generally, multiphase PSK. The symbols are differentially encoded.

The basic transmission frame for TDMA is denoted as T_f. Each frame is equally subdivided into M subframes of duration $T_{sf} = T_f/M$, where M is the number of users per frame. In each subframe we transmit a packet of digitized voice preceded by synchronization and equalizer training bits. Fig. 11 illustrates the signalizing format.

Below, we present an example of an equalizer design based on the use of a DFE for narrow-band TDMA.

A. A Design Example Based on a DFE

As a specific example, we shall take a TDMA system with a nominal 30-kHz channel bandwidth. The frame duration is selected as $T_f = 20$ ms and, consistent with the 30-kHz nominal channel bandwidth and QPSK modulation, we select the symbol interval as $T_s = 40$ μs. Consequently, there are 500 symbols per frame. Let us assume that the frame is subdivided into $M = 3$ subframes and that we allocate 165 symbols per user. Then, the remaining time interval ($5T_s$) may be used to provide guard times between subframes.

Let us assume that a DFE will be used to suppress the ISI. In the implementation of the DFE, the FFF is assumed to be fractionally spaced, e.g., $T_s/2$-spaced taps, where T_s is the signaling interval, and its length is assumed to span the total anticipated channel dispersion. The FBF has T_s-spaced taps and its length is assumed to also span the total anticipated channel dispersion.

At the transmitter the QPSK signal spectrum is shaped by a filter having a square root raised-cosine frequency response with a specified excess bandwidth (rolloff factor β). For the nominal 30-kHz channel and a symbol duration of $T_s = 40$

μs, an excess bandwidth of approximately 25 % is feasible. Hence, a fractionally spaced FFF must have tap spacing of $4T_s/5$ or less.

If we allow for channel dispersion up to 100 μs and $T_s = 40$ μs, the FFF of the DFE must span 2.5–3 symbols in duration. A fractionally spaced FFF with tap spacing of $T_s/2$ would require six taps while a tap spacing of $4T_s/5$ would require four taps. The FBF is assumed to have three T_s-spaced taps, in order to span the duration of the ISI. Hence, a DFE (6, 3) having a 6-tap FFF with $T_s/2$ tap spacing and a three-tap FBF is sufficient to handle the total channel dispersion.

B. Equalizer Training

An important issue to be considered is the algorithm for training the equalizer. The choice of a short frame duration and our desire to keep the overhead of training symbols to data symbols low dictates the use of a rapidly converging algorithm for training the equalizer. To be specific, in the 20 ms frames for the 30-kHz mobile radio system, we allocated 165 symbols per frame for each user. A 10 % overhead would allow a maximum of 16 symbols for training. Such a short sequence rules out any gradient-type algorithm, such as the LMS algorithm, for equalizer adaptation. One of the faster converging RLS algorithms is required to accommodate for such a short training sequence.

A very important consideration in the use of an RLS algorithm is computational complexity. As previously indicated, the RLS lattice and the fast RLS algorithms have a computational complexity proportional to N, where N is the total number of equalizer tap coefficients. In contrast, the RLS square-root algorithms have a computational complexity proportional to N^2. In Table I, we listed the computational complexity of these algorithms for the DFE (N_1, N_2) in terms of complex multiplications and divisions, and the results were graphed in Figure 8 for the special case where $N_1 = N_2 = N/2$. From these graphs we note that for a DFE with fewer than 10 taps, the square-root RLS algorithm is the most efficient. For the DFE (6, 3) the total number of complex operations per symbol, obtained from Table I, is 180. This number, which is about 10 times larger than the computational complexity of the LMS algorithm, is well within computational capabilities of modern programmable digital signal processors (DSP) for the signal bandwidths under consideration.

The square-root RLS algorithm is known to have excellent numerical properties and has been shown to be very robust to round off noise in the computations [2], [16], [17]. For example, results from simulation [16] illustrate that the algorithm performs well with fixed point arithmetic down to about 12 b.

C. Block Processing Algorithms

An alternative to a sequential adaptation algorithm, such as RLS, is to use a block processing algorithm, where we first estimate the channel impulse (or frequency) response from the training sequence and, then, use the channel estimate in the detection of the data symbols. Such an approach has been investigated by Hsu [15], Davidson et al. [18], and Crozier [19], [20]. In these schemes, a typical block consists of M_t training symbols followed by M_d data symbols as shown in Fig 12. The two blocks of training symbols on either side of the block of data symbols is used for channel estimation and data detection.

Hsu [15] investigated a nonlinear decision-feedback symbol detection technique in which the training symbol blocks on each side of a data block are used for channel estimation. Then, a DFE-type detection algorithm is employed to make decisions on the first and last symbols in the data block, say $\tilde{b}_1$ and $\tilde{b}_{M_d-1}$. These detected symbols are treated as additional training symbols to estimate the channel response again and, then, to use the new channel estimate in the detection of the next two outermost symbols b_1 and b_{M_d-2}. These detected symbols $\tilde{b}_2$ and $\tilde{b}_{M_d-1}$ are again treated as additional training symbols and the estimation/detection procedure is repeated until all data symbols in the block are detected. In essence, this detection technique in a block DFE, where a pair of data symbols is detected recursively in each iteration.

Crozier [19], [20] also investigated a block detection technique in which the training-symbol blocks on both sides of the data-symbol block are used in the estimation of the channel and the detection of the data symbols. The LSE criterion was employed in the estimation of the data symbols for both linear estimation and decision-feedback estimation of the block of symbols. The detection method for the decision-feedback block estimation was basically the recursive scheme proposed by Hsu [15].

Davidson et al. [18] compared the computational complexity of a block-adaptive DFE which a square-root RLS-DFE and illustrated that the computational complexity of the block adaptive DFE is significantly less than the square-root RLS for equalizer lengths of 10 or more.

D. Effect of Channel Fading

Another important consideration in the design of the adaptation scheme is channel fading. The fade rate for the received multipath signal depends on the vehicle speed. For example, at a speed of 80 km/h and a carrier frequency of 900 MHz, the peak Doppler shift that may occur is 67 Hz. Hence, the coherence time is about 15 ms. With such a small coherence time, a channel measurement made at the beginning of a block from the training symbols may be inadequate for reliably detecting symbols that occur later in the block. In such a case, the block-adaptive DFE scheme of Hsu [15], which augments the channel measurement by the inclusion of data symbols that have been detected, should prove very effective. Other schemes based on block adaptation with channel tracking via interpolation between successive blocks of training symbols have been proposed by Davidson et al. [18] and by Lo et al. [21]. Alternatively, in an RLS implementation, the DFE may be continuously (sequentially) adapted during symbol detection, based on decision-directed adaptation. A comparison of the performance of these adaptation schemes under various channel conditions is required before a decision is made on which scheme is selected for the TDMA system.

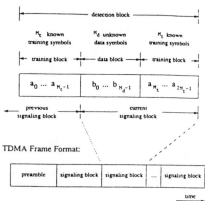

Fig. 12.   Block signal format in TDMA frame [2].

IV. Concluding Remarks

We have provided a brief survey of adaptive equalization techniques and described their performance characteristics, their limitations, and their computational complexity. Based on this background of available adaptive equalization algorithms and their characteristics, we considered the high-level design of adaptive equalization for a narrow-band TDMA system. On the basis of implementation complexity and performance in the presence of multipath distortion and signal fading, both MLSE and DFE are viable equalization methods for mobile radio. The issue of whether the DFE or MLSE should be designed to operate and adapt either sequentially or on the entire block of symbols is still an open question and requires further investigation.

Many papers have been published over the last few years on the performance of linear, decision-feedback, and maximum-likelihood equalization techniques for the digital mobile radio channel. The majority of these papers are concerned with the design and performance of equalizers for the pan-European GSM system. Relatively few results are available for the proposed narrow-band (30-kHz) TDMA digital mobile radio system for the U.S. Several investigations are currently underway in the design and performance of equalizers for this system.

References

[1] B. Widrow, "Adaptive filter, I: Fundamentals," Stanford Electronics Lab., Stanford Univ., Stanford, CA, Tech. Rep. 6764-6 Dec. 1966.

[2] J. G. Proakis, *Digital Communications.* New York: McGraw-Hill, 1989, 2nd ed.

[3] G. J. Bierman, *Factorization Methods for Discrete Sequential Estimation.* New York: Academic, 1977.

[4] D. D. Falconer and L. Ljung, "Application of fast Kalman estimation to adaptive equalization," *IEEE Trans. Commun.*, vol. COM-26, pp. 1439–1446, Oct. 1978.

[5] J. M. Cioffi and T. Kailath, "Fast recursive-least-squares transversal filters for adaptive filtering," *IEEE Trans. Acoust., Speech, Signal Processing*, vol. ASSP-32, pp. 304–337, Apr. 1984.

[6] D. T. M. Slock and T. Kailath, "Numerically stable fast recursive least-squares transversal filters," in *Proc. Int. Conf. Acoust., Speech, Signal Processing*, Apr. 1988, pp. 1365–1368.

[7] G. Carayannis, D. G. Manolakis, and N. Kalouptsidis, "A fast sequential algorithm for least-squares filtering and prediction," *IEEE Trans. Acoust., Speech, Signal Processing*, vol. ASSP-31 pp. 1394–1402, Dec. 1983.

[8] E. H. Satorius and J. D. Pack, "Application of least squares lattice algorithms to adaptive equalization," *IEEE Trans. Commun.*, vol. COM-29, pp. 136–142, Feb. 1981

[9] K. Abend and B. D. Fritchman, "Statistical detection for communication channels with intersymbol interference," *Proc. IEEE*, pp. 779–785, May 1970.

[10] G. D. Forney, Jr., "Maximum-likelihood sequence estimation of digital sequences in the presence of intersymbol interferences," *IEEE Trans. Inform. Theory*, vol IT-18, pp. 363–378, May 1972.

[11] C. A. Belfiori and J. H. Park Jr., "Decision-feedback equalization," *Proc. IEEE*, vol. 67, pp. 1143–1156, Aug. 1979.

[12] K. Zhou, J. G. Proakis, and F. Ling, "Decision-feedback equalization of time dispersive channels with coded modulation," *IEEE Trans. Commun.*, vol. 38, pp 18–24, Jan. 1990.

[13] S. U. H. Qureshi and G. D. Forney Jr., "Performance properties of a T/2 equalizer," pp. 11.1.1–11.1.14, Los Angeles, CA, Dec. 1977.

[14] R. D. Gitlin and S. B. Weinstein, "Fractionally spaced equalization: An improved digital transversal equalizer," *Bell Syst. Tech. J.*, vol. 60, pp. 275–296, Feb. 1981.

[15] F. Hsu, "Data directed estimation techniques for single-tone HF modems," in *Proc. MILCOM '85*, Boston, MA, Oct. 1985, pp. 12.4.1–12.4.10.

[16] F. Ling and J. G. Proakis, "Numerical accuracy and stability: Two problems of adaptive estimation algorithms caused by round-off error," in *Proc. ICASSP '84*, San Diego, CA, Mar. 1984, pp. 30.3.1–30.3.4.

[17] F. Hsu, "Square-root Kalman filtering for high-speed data received over fading dispersive HF channels," *IEEE Trans. Inform. Theory*, vol. IT-28, pp. 753–763, Sept. 1982.

[18] G. W. Davidson, D. D. Falconer, and A. U. H Sheikh, "An investigation of block-adaptive decision feedback equalization for frequency selective fading channels," *Can. J. Elec. Comp. Eng.*, vol. 13, no. 3-4, pp. 106–111, Mar. 1988.

[19] S. N. Crozier, D. D. Falconer, and S. Mahmoud, S. "Short-block equalization techniques employing channel estimation for fading time-dispersive channels," *Proc. IEEE Veh. Technol. Conf.*, San Francisco, CA, May 1989, pp. 142–146.

[20] S. N. Crozier, "Short-block data detection techniques employing channel estimation for fading, time-dispersive channels," Ph.D. dissertation, Dept. of Systems and Computer Eng., Carleton Univ., Ottawa, Ont., Canada, Apr. 1990.

[21] N. K. W. Lo, D. D. Falconer, and A. U. H. Sheikh, "Adaptive equalization and diversity combining for a mobile radio channel," in *Proc. GLOBECOM '90*, San Diego, CA.

John G. Proakis (S'58–M'62–SM'82–F'84) received the E.E. degree from the University of Cincinnati, Cincinnati, OH, in 1959, the S.M. degree in electrical engineering from the Massachusetts Institute of Technology, Cambridge, in 1961, and the Ph.D., degree in engineering from Harvard University, Cambridge, MA, in 1966.

From June 1959 to September 1963 he was associated with MIT, first as a Research Assistant and later as a Staff Member at the Lincoln Laboratory, Lexington, MA. During the period 1963–1966 he was engaged in graduate studies at Harvard University, where he was a Research Assistant in the Division of Engineering and Applied Physics. In December 1966 he joined the Staff of the Communication Systems Laboratories of Sylvania electronic Systems and later transferred to the Waltham Research Center of General Telephone and Electronics Laboratories, Inc., Waltham, MA. Since September 1969 he has been with Northeastern University, Boston, MA, where he holds the rank of Professor of Electrical Engineering. From July 1982 to June 1984 he held the position of Associate Dean of the College of Engineering and Director of the Graduate School of Engineering. Since July 1984 he has been Chairman of the Department of Electrical and Computer Engineering. His interests have centered on digital communications, spread spectrum systems, system modeling and simulation, adaptive filtering and digital signal processing. He is the author of the book *Digital Communication* (New York: McGraw-Hill, 1983; 1989, second edition) and the coauthor of the book *Introduction to Digital Signal Processing* (New York: Macmillan, 1988).

Dr. Proakis has served as an Associate Editor for the IEEE Transactions on Information Theory (1974–1977), and the IEEE Transactions on Communications (1973–1974). He has also served on the Board of Governors of the Information Theory Group (1977–1983), is a Past Chairman of the Boston Chapter of the Information Theory Group, a Registered Professional Engineer on the State of Ohio, and a member of Eta Kappa Nu, Tau Beta Pi, and Sigma Xi.

Diversity Techniques for Mobile Radio Reception

J. D. PARSONS, MIGUEL HENZE, P. A. RATLIFF, AND MICHAEL J. WITHERS

Abstract—After discussing the nature of the electromagnetic field in urban environments, and the use of diversity techniques in fading media, the paper presents a survey of these techniques as applied to mobile radio, with emphasis on systems designed for analogue communication. It includes a unified approach to systems based on the double-heterodyne phase-stripper principle. The discussion is not restricted to present-day receivers only, but includes considerations which may apply in future designs.

I. INTRODUCTION

DURING the last 30 years, there has been a rapid expansion in the use of mobile radio, for both civilian and military purposes. Since the portions of the radio spectrum available for this type of communication have not been expanded in proportion to the number of users, channel separation has been progressively reduced. The users have also come to expect a higher quality of service, regions of poor reception well within the nominal coverage area of a base station not being readily tolerated; such regions are most likely to occur in urban and wooded areas. Because of interference and the mounting concern about electromagnetic pollution, this problem cannot be solved by indiscriminate increases in transmitter power. Although the majority of users still requires analogue communications, the demand for data transmission is likely to increase in the future, and this will place even more stringent requirements on the quality of the channel.

Diversity reception is one way to improve the reliability of communication without increasing either the transmitted power, or bandwidth.

This paper reviews the nature of the electromagnetic field in urban environments, and the use of diversity in fading media; it also provides a survey of techniques by which diversity reception systems can be implemented for mobile radio use. The stress is on analogue communication, but most of the systems can, in principle, be used for the reception of data.

II. THE FIELD IN URBAN ENVIRONMENTS

A situation which often occurs in land-based mobile radio links is that an elevated base station on a good site attempts to control a number of vehicles located some distance away. Due to natural and man-made obstacles, there is often no direct line of sight path between transmitter and receiver and as the vehicle moves, the received signal fluctuates or fades. There are two approximately separable effects known as fast and slow fading. Fast fading is characterized by deep fades which occur within fractions of a wavelength and is caused by multiple reflexions from buildings and other obstacles in close proximity to the vehicle. It is most severe in heavily built-up areas where the number of waves arriving from different directions with different amplitudes and phases is often sufficient to cause the signal amplitude to follow a Rayleigh distribution over relatively small distances. However, it also occurs to various degrees in suburban and wooded areas.

When the fast fading is removed by averaging over distances of the order of tens of wavelengths, a so-called slow fading is still apparent. This is caused by variations in both the terrain profile and the general nature of the environment. The averaged signal strength in urban areas can be as much as 50 dB below the value predicted by computing the free-space path loss and is found to have a log-normal distribution with a standard deviation between 5 and 10 dB.

The signal received in an urban area is therefore locally Rayleigh with the mean of this process varying slowly with position. A vehicle moving through this spatially varying field experiences a fading rate proportional to its speed and the transmission frequency. At 48 km/h (30 miles/h) the fading rate is about 10 Hz per 100 MHz transmission frequency and is associated with a Doppler spread in the received spectrum.

Although there is no satisfactory mathematical model to explain the behaviour of the slow-fading, the log-normal distribution being the best-fit to experimental data, several models [1], [2] have been proposed for the local fast fading, the most useful probably being that of Clarke [3]. He assumes that for vertically polarized transmissions the field at any receiving point is also vertically polarized, and is composed of N equal-amplitude plane waves, the rth wave arriving at an angle α_r with respect to an arbitrary spatial axis in the horizontal plane, with a phase angle ϕ_r relative to an arbitrary reference. With the further assumption that the phase angles ϕ_r are independent of each other and of the spatial angles α_r,

Manuscript received July 1975. J. D. Parsons and M. Henze wish to thank the UK Science Research Council for a grant in support of work on diversity reception methods for mobile radio.

J. D. Parsons is with the Department of Electronic and Electrical Engineering, University of Birmingham, Birmingham, England.

M. Henze is with the Department of Electronic and Electrical Engineering, University of Birmingham, Birmingham, England, on leave from the Instituto Tecnologico de Aeronautica, Sao Paulo, Brazil.

P. A. Ratliff was with the Department of Electronic and Electrical Engineering, University of Birmingham, Birmingham, England. He is now with the BBC Research Department, Kingswood, Surrey, England.

M. J. Withers was with the Department of Electronic and Electrical Engineering, University of Birmingham, Birmingham, England. He is now with the British Aircraft Corporation, Stevenage, Herts., England.

Editor's Note: This paper is reprinted by permission from *The Radio and Electronic Engineer*, vol. 45, no. 7, pp. 357-367, July 1975, the journal of the Institution of Electronic and Radio Engineers in London, England. It is an expansion of work described by Prof. Parsons at the Symposium on Microwave Mobile Communications, Boulder, CO, in September 1973. Since no record was produced for that symposium, and since the use of diversity techniques for mobile communications has received additional emphasis as a result of frequency allocations at 900 MHz for land mobile radio, it is appropriate to bring this survey paper to the attention of our readers.

Clarke is able to predict the Rayleigh distribution of the received signal strength, the base-band spectrum received as a result of motion through the fading field, and the spatial correlation functions of the envelopes of the various field components. The predictions fit well with experimental data collected in areas where there is no direct wave from the transmitter. Based on this work, Gans has produced a power spectral theory [4] and more general work has been reported by Lin [5].

A further physical insight into the nature of the field has been provided by the work of Cox [6]-[9]. He conducted experiments at 910 MHz, which yielded statistical descriptions of the time delays and Doppler shifts associated with propagation in urban and suburban environments. The results show that the typical delay spread is about 0.25 μs and the coherence bandwidth at 0.9 correlation coefficient is greater than 250 kHz. However, in some extreme cases significant signal amplitudes with excess delays up to 7 μs were recorded, and in many instances it was possible to identify the actual buildings from which reflected signals were received. This work also yielded various other statistical parameters of the propagation link and showed that whereas the signal amplitude at fixed delays often had a Rayleigh distribution, large departures from the Rayleigh also occurred.

III. DIVERSITY RECEPTION

In a situation where the received signal is subject to fading, some kind of diversity reception can be employed to reduce the depth and duration of fades and hence improve the quality of communication. For example, if two signals with uncorrelated envelopes can be obtained, then in Rayleigh fading conditions the probability of occurrence of simultaneous fades in excess of 20 dB is 0.01% whereas for a single signal it is 1%. It is thus clear that a signal composed of a suitable combination of the two independently fading components will have much less severe fading properties than either signal alone.

Spatial correlation functions based on expressions derived by Clarke are shown in Fig. 1, and give an indication of how almost independently fading signals can be obtained on a vehicle. The autocorrelation coefficient of both the electric and magnetic fields is less than 0.2 at displacements greater than $\lambda/4$, and thus it is reasonable to suppose that omnidirectional electric- or magnetic-field sensing aerials spaced at least $\lambda/4$ apart will receive signals suitable for a diversity system. It is worth noting that completely uncorrelated signals are not necessary; it has been shown [10] that the efficiency of a two-branch diversity receiver is negligibly impaired if the correlation coefficients between the signals on the branches are less than 0.3, and much of the diversity advantage is still apparent with correlation coefficeints as high as 0.7. Fig. 1 also shows that at zero separation the electric and magnetic fields are uncorrelated, and this indicates the possibility of a field-diversity system using an aerial capable of independently receiving the electric and magnetic components of the field. Lee [11] has built and tested such an aerial and demonstrated its usefulness in a type of diversity receiver which will be des-

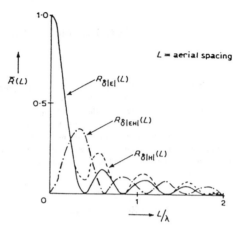

Fig. 1.   Normalized autocorrelation functions for the envelopes of isotropically scattered waves.

cribed later. Frequency diversity is also theoretically possible since Cox has shown that the coherence bandwidth is not high. However, because of the congested spectrum this is not really a practicable proposition. Neither is it likely that there is sufficient depolarization along the various transmission paths to make polarization diversity a viable solution, although experiments using two transmissions with different polarizations have shown promise [12].

Space diversity transmission from the base station may be used as an alternative to diversity reception at the mobile and the economic implications of this are discussed in Section V. For the return link, it is possible to use diversity reception at the base station. Since the major reflecting surfaces are close to the mobile, the aerial separation required at the base station is quite large, probably tens of wavelengths [13], but there is no reason in principle why the diversity reception techniques described in this paper cannot be applied at the base station.

Diversity reception methods are usually classified in the literature in three categories as selection diversity, equal-gain combining, and maximal-ratio combining; all three methods are relatively easy to realize in a practical form. As far as the combining methods are concerned, it is irrelevant in principle whether combining takes place before or after demodulation when the demodulation process is linear, but of vital concern, for example, in systems employing frequency modulation, where the demodulation process is inherently nonlinear [14]. In this case predetection combining yields a higher baseband signal/noise ratio (SNR) than postdetection combining, and since most of the world's mobile radio systems use frequency modulation it is predetection combining which is of major interest.

In ideal selection diversity, the signal with the highest instantaneous SNR is used and so the output SNR is equal to that of the best incoming signal. In practice the system cannot function on a truly instantaneous basis, and so to be successful it is essential that the internal time-constants of a selection combiner are substantially shorter than the reciprocal of the signal fading rate. Whether this can be achieved depends upon the bandwidth available in the receiving system. Practical systems of this type usually select the branch with the highest

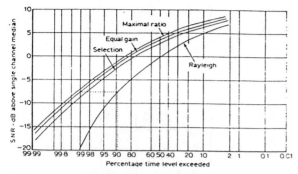

Fig. 2. Theoretical cumulative probability distributions for two-branch diversity systems.

signal plus noise or utilize a simpler technique known as scanning diversity. In scanning diversity a selector scans the various branches in a fixed sequence until a signal above a predetermined threshold is found. This signal is then used until it drops below the threshold, when the scanning process restarts and continues until another acceptable signal is found. There is no attempt to find the best signal, merely one which is acceptable.

In maximal ratio combining, first proposed by Kahn [15], each signal is weighted in proportion to its SNR before summation. When this takes place before demodulation it is necessary to cophase the individual signals before combining, and various techniques for doing this will be described. Ideally, maximal ratio combining produces an output SNR equal to the sum of the individual SNR's and is the best that can be achieved. Although the technique involves some complexity in the form of the weighting, cophasing, and summing circuits, it has the advantage that it is possible to produce an output with an acceptable SNR even when none of the signals on the individual branches are themselves acceptable. This, of course, is not possible with selection diversity.

Equal gain combining is similar to maximal ratio, but there is no attempt at weighting the signals before addition. The possibility of producing an acceptable signal from a number of unacceptable inputs is retained, and the performance is only marginally inferior to that of maximal ratio combining.

The effectiveness of diversity can be assessed in a number of ways [10]. Most important as far as mobile radio is concerned is its ability to increase the reliability of communication by reducing the depth and rate of fading. Fig. 2 shows theoretical cumulative probability distribution functions for two-branch diversity systems, assuming equal, locally-incoherent noise powers on each branch, with the Rayleigh fading characteristic shown for comparison. Clearly diversity has most to offer at low signal levels, and taking the equal-gain case as an example, it can be seen that the predicted reliability of a two-branch system is 99% in circumstances in which a single-aerial system would be only about 88% reliable. This means that for a fixed transmitter power, the coverage of the service area is far more "solid", and there are far fewer areas in which signal flutter causes poor reception. This may be very significant in radio links designed to carry data transmissions, especially since it can also be shown that if all branches are

statistically alike, the average fade duration below any particular reference level is reduced by a factor equal to the number of diversity branches used [16]. To achieve the same result by altering the transmitter power would involve an increase of 12 dB, which would be both costly and socially undesirable. Although strictly speaking the above figures are only valid in areas where there is Rayleigh fading, in practice the detailed nature of the fading is relatively unimportant and it is only when there is a strong direct wave from the transmitter (in which case diversity has little to offer) that there is any serious error.

Equal-gain and maximal ratio combiners attempt to add the noises incoherently, whilst providing coherent addition of the signals. We have already stated that little degradation of performance results if the inputs have low correlations. However, in unfavourable conditions where the noise inputs may be highly correlated (e.g., ignition interference) maximal-ratio combining, far from being the ideal diversity combiner may in fact turn out to provide the lowest SNR with respect to the wanted output, since correlated noise appears as signal to this kind of receiving system. Selection diversity is the optimum technique in this case [16]. Because field conditions of various kinds can occur in mobile radio, and since the main advantage of diversity arises from making the best use of the strongest signal branch, the relatively minor differences between the various techniques are not important. No one technique can be said to be superior to the others in all conditions, even if the system realization is ideal.

IV. DIVERSITY RECEPTION SYSTEMS

Having obtained signals which are sufficiently decorrelated, the task of the diversity receiver is to use them in a way which will provide the optimum benefit. There are many ways in which the signals from the various branches can be processed, and practical systems based on the techniques mentioned above will be described in the Section.

A. Systems using Single Conventional Receivers

1) Switched diversity: Selection diversity and its derivative scanning diversity are both switched systems in the sense that one of a number of possible inputs is allowed into the receiver. A major disadvantage of implementing true selection diversity is the expense of continuously monitoring the signals on all the branches, and this makes it impracticable for mobile use. On the other hand, scanning diversity is inherently cheap to build, since irrespective of the number of branches it requires only one circuit to measure the short-term average power of the signal actually being used. Switching occurs when the output of this circuit falls below a threshold. In this context "short-term" refers to a period which is small compared with the time taken by the vehicle to travel a significant fraction of a wavelength. A basic form of scanning diversity is shown in Fig. 3(a), although of course it is not essential for the averaging circuit to be connected to the front-end of the receiver. In its simplest form only two aerials are used, and switching from one to the other occurs whenever the signal level on the one in use falls sufficiently to activate the changeover switch.

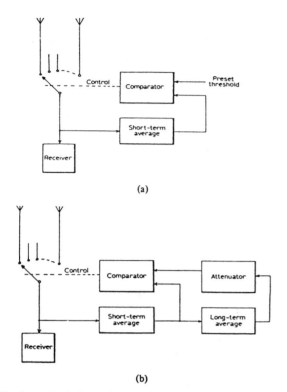

(a)

(b)

Fig. 3.   (a) Basic form of scanning diversity. (b) Scanning diversity
system with variable threshold.

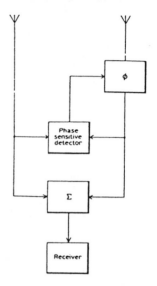

Fig. 4.   Equal gain combiner with cophasing control information
derived from a direct phase comparison.

The matter of threshold setting is worthy of mention because although it should clearly be low enough to prevent unnecessary switching, a setting which is optimum in one area is not necessarily the best when the vehicle has moved some distance away. Fig. 3(b) shows a modified system which incorporates a variable threshold derived from the mean signal level in the vicinity of the vehicle [17] and this takes some account of the slow fading encountered by the vehicle as it moves. The long-term average is computed over a period comparable with the time the vehicle takes to travel tens of wavelengths, and the setting of the attenuator determines how far below the long-term average the signal is allowed to fall before switching occurs. A practical form of switched diversity for FM mobile radio has been described [18] that also includes circuits to ensure that the receiver is always connected to the aerial with the stronger signal when the vehicle is stationary or moving very slowly. However, it is not directly compatible with present-generation receivers.

There is modification of switched diversity which allows the mobile to carry only one receiving aerial. It involves using the mobile transmitter to inform the base station that the vehicle is entering a fade, and is known as feedback diversity [19]. In this system the spaced aerials are at the base station, and whenever the signal at the mobile falls below the threshold, the mobile transmitter sends a message (in the form of a tone-burst) back to the base station which then switches to the other aerial. Due to the inherent delay in this system, its performance is marginally inferior.

2) Combining methods using perturbation techniques: A number of single-receiver systems which have continuous

rather than switched inputs have been devised, and usually form combiners of the equal gain type. In a two-branch combiner, perhaps the simplest method is to place a phase shifter in one of the aerial leads, and to adjust it in a "hill-climbing" manner using a suitable dc voltage, e.g., the receiver AGC line, as an indication of the input signal level. The disadvantages are that the receiver time-constants may be such that the system cannot follow the fast fades at higher frequencies, and   there is no simple method of extending the system to three or more branches.

A more flexible technique is to adjust the phase-shifter by means of a control voltage derived from the relative phase between the RF signals, as illustrated in Fig. 4, and in AM systems a convenient scheme for detecting whether a phase lag or phase lead condition exists has been described by Lewin [20]. A low-frequency sinusoidal oscillator drives a small-deviation phase-perturbing network which is connected in series with the phase-shifter, and as shown in Fig. 5, this produces amplitude perturbation of the sum signal. The phase of this perturbation changes by 180° as the perturbed signal changes from lagging to leading the sum signal. When the RF signals are cophased only a small second harmonic is present, and a "sense-of-correction" indication can therefore be obtained by low-frequency phase detection of the amplitude perturbations detected by the receiver with the original perturbing oscillator. This procedure produces a positive or negative dc control signal that is used to determine the direction in which the phase-shifter must be adjusted in order to approach the optimum position. The method is readily extended to N branches, provided $(N - 1)$ control circuits are used each with its own perturbing frequency. For FM receivers a dual technique has been proposed in which small amplitude perturbations are applied to the RF signals [21]. The resulting phase perturbations in the sum signal can be detected by the receiver and similar signal processing can be used to provide the control information. Various practical embodiments of the

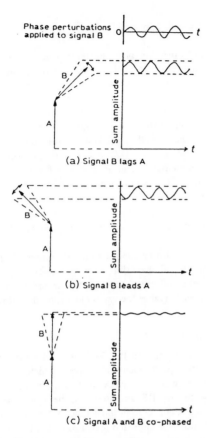

Fig. 5. Method of detecting phase relationship using phase perturbations.

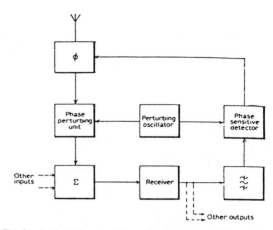

Fig. 6. Predetection combiner using phase perturbations to derive control signals for cophasing.

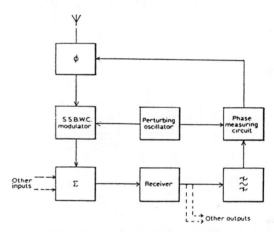

Fig. 7. Predetection combiner using phase measurements to derive control signals for cophasing.

technique have been described for AM systems [22]-[24] and a typical system is shown in Fig. 6.

The major disadvantage in the practical systems is that since it is not easy to build an analogue 360° phase shifter for insertion in the RF path, it is usual to use a unit that is quantized into a number of steps. However, the association of a quantized phase shifter with a "sense-of-correction" detector leads to a situation in which the phase-shifters oscillate about their optimum position. This is due to the inability of such systems to detect the optimum position, the polarity of the control signal being reversed only after that position has been passed. This oscillation can be eliminated by using a true phase measurement scheme, and a new compatible direct-reading method has been proposed [25]. In this method, the RF signals are single-sideband modulated without carrier suppression by different low-frequency signals. The phases of the introduced sidebands are transposed on to the sum signal, and thus the phase of the detected modulation differs from that applied by exactly the phase angle between the appropriate RF signal and the sum. It is thus possible to generate control signals which can be used to set the phase shifters. A block diagram of such a system is shown in Fig. 7, and due to the nature of the SSB signal, an FM or AM receiver can be used.

There are several ways in which the cophased signals can be added together. An active summing unit can be used, but unless very carefully designed, it is likely to cause a degrada-

tion in receiver intermodulation performance. A simple hybrid is commonly used, but this has the disadvantage that some power is always dissipated at the difference port except in the special case of equal-amplitude input signals. One possible arrangement using hybrids to provide total power transfer [26] is shown in Fig. 8. Phase-shifter *A* is adjusted to give quadrature signals at the inputs of the first hybrid and so it provides equal-amplitude signals at its output ports. Phase-shifter *B* is than adjusted to cophase these signals so that the second hybrid produces an output only at its sum port. However, in the mobile radio environment, the improvement in performance compared with that of a single hybrid is unlikely to justify the cost and complexity involved in implementing such a system.

Systems of this kind are mainly limited by the maximum rate of phase compensation attainable. This is determined by the number of input branches and the total processing bandwidth available for the perturbing signals, since these factors determine the response times of the filters used to isolate these signals after detection. The perturbing frequencies must be such that the control information passes through the receiver

293

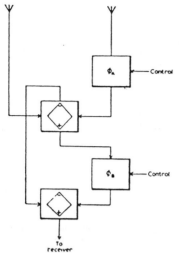

Fig. 8.   Total power combiner (after Sidwell [26]).

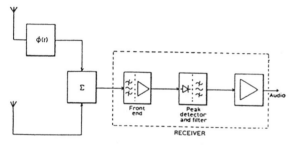

Fig. 9.   Basic phase-sweeping diversity receiver.

transmitter is located, a lobe of the receiving pattern will s(across it.

The way in which the phase-shifter sweeps out the radians is of importance to avoid serious spectrum spread: and hence simplify signal retrieval while maintaining SNR. appears that the best choice is to cause the phase to char linearly with time. This can be achieved by using a repetiti sawtooth phase modulation, and is equivalent to frequen translating the whole of the spectrum passing through t phase-shifter by an amount $1/T$, where T is the repetiti period. An experimental system as shown in Fig. 9 has be evaluated, using a staircase approximation to give the requir phase shift [28].

The majority of work on the technique has been direct(towards AM systems although, in theory, it appears applicab to FM systems [30].

The method can also be applied to higher orders of dive sity but each additional phase-shifter has then to be swept at progressively higher rate in proportion to the order. In th case it seems easier to realize the system in a form described i the next section.

B. Receivers with Multiple Front-Ends

The increase in losses, and the degradation in noise figur and intermodulation performance of receivers, which is cause by the use of units such as RF phase-shifters and perturbatio networks prior to the RF amplifier can be overcome b: receivers designed along conventional lines but having mor than one front-end. This approach may also render feasible a otherwise unattractive solution, or simplify previously con sidered alternatives.

With this type of receiver, for example, a direct detectio of the relative phase between the signals is possible at an inter mediate frequency, provided this frequency is sufficiently low Each diversity branch then requires its own RF amplifier anc mixer, but one local oscillator, with various outputs whose phases can be independently controlled, is sufficient. The out puts of the phase comparators are used to control the phase: of the local oscillator outputs before mixing, and cophasing of the signals takes place at the intermediate frequency [31].

The perturbation methods can also be implemented in a similar way. The perturbations themselves can be introduced either in the signal path after the RF amplifiers, or added to the local oscillator outputs. However, if the perturbation con tains amplitude modulation components, true multipliers must be used as mixers.

The phase-sweeping method becomes particularly attractive in association with a multiple front-end receiver. Since the linear phase sweep is equivalent to a frequency shift, the desired effect can be achieved by connecting the aerials to front-ends which have local oscillators at different frequencies. To obtain good adjacent channel rejection, a suitably-tuned IF amplifier is used in each branch, and the signals are summed before demodulation and filtering. This system has been proposed for HF communication [29], [32] and a block diagram is shown in Fig. 10. It is interesting to note that it is possible to apply AGC around each individual IF amplifier and thereby produce a maximal-ratio diversity combiner.

without causing interference to the wanted information sidebands.

3) Phase-sweeping method: Another technique for use with a single standard receiver employs a method of continuously sweeping the phase in one of the signal paths. This has been considered, both theoretically and practically, by a number of authors [27]-[29].

The method of operation is relatively simple and can be explained by considering the part of Fig. 9 prior to the receiver. If the phase-shifter continuously changes its value, between 0 and 2π radians, there are instants in time when the two signals come into phase and a peak occurs. If the rate at which the 2π phase excursion is repeated is at least twice the highest modulation frequency expected in the received signal, then peaks occur at a rate to satisfy the sampling theorem. By using a peak detector and a suitable filter, demodulation of AM-signals is possible giving the improvement expected from a diversity technique.

The operation of the system can be considered from another viewpoint [27]. The two spaced aerials form a two-element array which has a multilobed directional pattern. This directional pattern is electronically scanned by the phase-shifter and a 2π radian change causes it to move round in angle by one lobe spacing. Therefore, no matter where the

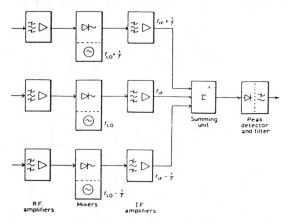

Fig. 10. Three-branch diversity receiver with continuous phase-sweeping produced by offset local oscillators.

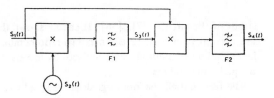

Fig. 11. Double-heterodyne phase-stripper.

C. Special Receivers

A whole class of predetection combiners can be derived from a circuit sometimes known as the "heterodyne phase-stripper." This circuit, shown in the block diagram of Fig. 11 has the property that the phase of the output signal is independent of the phase of the input. To illustrate the action of the circuit let us assume an input signal of the form

$$S_1(t) = A(t) \cos [\omega_c t + \phi(t) + \phi_0] + A_p \cos (\omega_p t + \phi_0)$$

in which the first term represents an amplitude and phase modulated carrier, and the second is a pilot tone at a frequency sufficiently close to the carrier that it is valid to assume that the propagation medium affects the phase of both in the same way. The signal is multiplied by a local oscillator signal

$$S_2(t) = B \cos (\omega_0 t + \theta)$$

at a frequency lower than that of the carrier, and the resultant signal is passed through a very narrow bandpass filter $F1$ which is centred at a frequency equal to the difference between the local oscillator and either the pilot-tone or the carrier. If the center frequency corresponds to the latter case, $F1$ should be narrow enough to reject all modulation associated with the carrier. The output of $F1$ will then be of the form

$$S_3(t) = A_F \cos (\omega_F t - \theta + \phi_0)$$

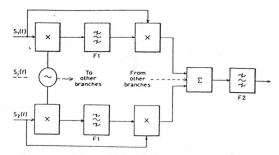

Fig. 12. Predetection combiner using phase-stripper principle.

with ω_F equal to either $(\omega_c - \omega_0)$ or $(\omega_p - \omega_0)$, and this is now multiplied by the fed-forward signal $S_1(t)$. Filter $F2$, which is wide enough to pass the modulation sidebands of the output, has a center frequency given by the difference between the centre frequency of $F1$ and either the pilot-tone or the carrier. If the pilot-tone frequency is chosen to determine the center frequencies of the filters then the output of the circuit is

$$S_4(t) = k\{A(t) \cos [(\omega_c - \omega_p + \omega_0)t + \phi(t) + \theta]$$
$$+ A_p \cos (\omega_0 t + \theta)\}$$

and if the carrier frequency is chosen, the output is

$$S_4(t) = k\{A(t) \cos [\omega_0 t + \phi(t) + \theta]$$
$$+ A_p \cos [(\omega_p - \omega_c + \omega_0)t + \theta]\}.$$

It can be seen that in both cases the static input phase ϕ_0 has been cancelled at the output, the phase of the local oscillator has been acquired, and the modulation has been preserved.

There are various alternative choices which can be made for the local oscillator and filter frequencies in order to obtain this cancellation of ϕ_0, the output being always at the local oscillator frequency. For example, if the frequency of the local oscillator is higher than that of the carrier and $\omega_F = \omega_0 - \omega_c$, then $F2$ is centered at $\omega_c + \omega_F$ to obtain cancellation, and the local oscillator frequency is regenerated at the output. However, in most practical embodiments of this technique it is usual to perform an initial frequency down-conversion, and then to use the circuit described with a local oscillator at another lower frequency. In this way phase-stripping and conversion to a second intermediate frequency are both accomplished at the same time.

Since the phase of the output signal is that of the local oscillator, applying the same local oscillator to the various branches of a diversity reception system will automatically cophase the various signals provided that the "static" phases vary slowly, or in other words that the fading rate is slow compared with the bandwidth of filter $F1$. In this case the signals can be added after the second multiplier, with filter $F2$ following the summing unit, and a predetection combiner results. Fig. 12 shows this arrangement with only two diversity branches fully drawn.

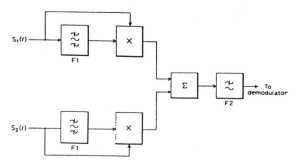

Fig. 13.   Phase-stripping combiner using transmitted pilot-tone in place of local oscillator.

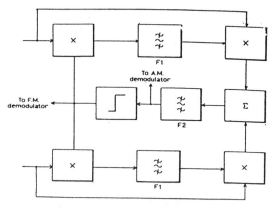

Fig. 15.   Closed-loop phase-stripping combiner using limiter (Granlund receiver [39]).

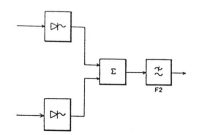

Fig. 14.   Phase-stripping combiner using transmitted pilot-tone with nonlinear device as mixer.

Systems using this principle can be broadly classified into two categories: systems with and without pilot-tones. The following sections describe different methods of implementation.

1) Pilot-tone systems: The idea of the "phase-stripper" was originally applied to a diversity system that used a pilot-tone located in the middle of the baseband to compensate for phase changes associated with the information modulation [33]. The filters $F1$ in Fig. 12 selected a frequency determined by the pilot-tone. However, if the pilot-tone is located outside the baseband, and can be recovered by $F1$, the local oscillator is not needed since the recovered pilot-tone can be used in its place. In this case $F2$ can be a low-pass filter and its output is

$$S_4(t) = k\{A(t) \cos [(\omega_c - \omega_p)t + \phi(t)] + A_p\}.$$

A block diagram of such a system [34] is shown in Fig. 13.

In other embodiments of this principle, illustrated in simplified form by the block diagram of Fig. 14, the filters $F1$ are eliminated altogether, and the remaining multipliers are replaced by either true squarers [35] or by devices whose characteristic exhibits a second-order nonlinearity [34]. Filter $F2$ then selects the difference frequency $(\omega_c - \omega_p)$ which is subsequently demodulated.

2) Systems without pilot-tone: It has been pointed out [36] that the principle of the phase-stripper can be applied to signals which contain no pilot-tone, provided a carrier is present in the received signal, and this is also apparent from the equations at the beginning of Section IV-C. However, for the system to function correctly as a predetection combiner, the filters $F1$ must reject all the information sidebands and care must be taken to isolate the local oscillator from the

output of the circuit [37], [38]. It can be shown that in this case the circuit of Fig. 12 is a maximal-ratio combiner but if the second set of multipliers are replaced by mixers it becomes an equal-gain combiner.

The fact that all the input signals are reduced to a common frequency and phase equal to those of the local oscillator points the way to an interesting solution [39], [40] which historically preceded those already mentioned. The combined output signal, suitably amplified, can be used as the local oscillator, forming a regenerative closed loop. Fig. 15 shows such a system with outputs indicated for AM and FM demodulators. This solution has two main advantages over systems using a local oscillator. Firstly, good isolation between the local oscillator and the output is no longer necessary, and secondly, for FM signals the filters $F1$ although narrow-band, do not need to be very sharp. The latter advantage results from the presence of modulation at both inputs to the first multipliers, and since $F1$ selects the difference frequency, modulation components in the range of interest will cancel. A theoretical analysis of this type of combiner has recently been published [41].

An alternative way of solving the problem of the gain of each loop being a function of the amplitude of its input signal is to use an AGC amplifier [42] and this is illustrated in Fig. 16. As with the local oscillator system, the closed-loop approach can produce maximal ratio or equal-gain combining depending on whether multipliers or mixers are used prior to the summation point.

D. Engergy-Density Reception

Diversity reception systems so far considered have been designed to operate with any set of fading signals which are sufficiently uncorrelated. However, Pierce [2] has devised a technique for mobile radio which specifically includes a composite aerial. He observed that in simple fading fields the total electromagnetic energy density remains constant although the electric and magnetic field components both exhibit deep fades, and proposed the use of co-sited but separate electric- and magnetic-field sensing aerials with their outputs combined so as to produce a signal proportional to the energy density $D = \frac{1}{2}[\epsilon |E|^2 + \mu |H|^2]$. A suitable aerial

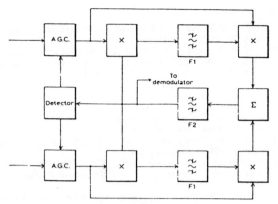

Fig. 16.   Closed-loop phase-stripping combiner using automatic gain control (after Cease [42]).

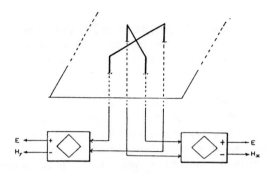

Fig. 17.   Composite aerial with hybrids to resolve components of electromagnetic field (energy-density antenna).

has been devised [43] and consists of two small mutually orthogonal semi-loops above a ground plane as shown in Fig. 17. The loops are taken to hybrid sum and difference networks, the sum ports giving outputs proportional to the electric field, and the difference ports outputs proportional to the resolved components of the magnetic field.

A problem arising with the use of this method is that the squaring process causes distortion with conventional modulation methods and it is clearly undesirable to resort to taking the square root of the receiver output. Since Clarke's model predicts that the electric and magnetic components of the field at any point are uncorrelated, thus making "field-diversity" a possibility, a more promising approach would appear to be to use the composite aerial to provide the uncorrelated signals, and then to combine these signals using one of the methods described above. In this connexion other versions of the composite aerial have been proposed, in which the crossed loops are replaced by slots, and a separate vertical monopole is used [44], [45]. These aerials provide three outputs of comparable amplitude without the matching difficulties which arise from the use of loops. A signal proportional to the magnetic field can be obtained by combining the outputs of the two loops or slots via a $\pi/2$ phase shifter (since $H = H_x + jH_y$), and thus it would seem possible to construct a simple two-branch field-diversity system in this manner.

V. DISCUSSION

Mobile radio in this country is an established system and therefore any changes have to be introduced by evolution rather than revolution. The capital investment in both base-station and mobile equipment is large and therefore the introduction of diversity techniques restricts the choice, in general, to methods compatible with the existing system.

The ideal advantages offered by diversity are that improved communication performance can be obtained without an increase in transmitter power or bandwidth. The ideal cannot be achieved with all the methods described, in that an increase in bandwidth is sometimes required and in other cases the full diversity gain cannot be achieved. However, the improvement that can be obtained, in practice, even with only a two-branch

system, is of the order of 10 dB at the 99% reliability level. This is more easily found by using diversity, which only requires a moderate amount of equipment, than by increasing the transmitted power with the resulting increase in interference.

The usefulness of applying diversity depends very much upon environment, the reliability required of the mobile service, the format of the signals conveyed and other techniques which may be used to provide improved communication efficiency. Discussing these in turn, the environment may consist of geographical areas where screening of signals or multipath exist. Also there may be good coverage in the region from a cochannel transmitter (the same frequency utilized by another user), or there may be other local sources of interference. Provided the wanted signal is slightly larger than the unwanted ones, the diversity systems exploit the strong-signal capture effect and the interference is reduced [17]. The reliability of the service is very dependent upon the user's requirements. Reliability is being used here in the communication sense rather than with reference to the equipment. For emergency services, near 100% reliability needs to be attained. For other services, holes in coverage might be tolerable but are only accepted with reluctance by the user. The other important factor is the type of message to be conveyed. For example, if speech is used, then short fades and flutter generally do not matter since there is sufficient redundancy in the language to cope. The redundancy may also be effectively increased by the use of standard phrases, careful choice of words, by repeating the message, or by requesting a repetition.

Some users require facsimile facilities, and others "hard-copy" teleprinter type messages. These message formats are very susceptible to fades and diversity techniques, as described in this paper, appear well suited to reducing this problem. Obviously redundancy coding is a competitor, but this eats into the information bandwidth compared with diversity. However, in a number of cases it is not possible to use the ultimate channel spacing due to stability problems, and a wider bandwidth has to be provided. Therefore, although redundancy may well be the best choice, care is necessary since the fading-rate encountered in mobile radio is a function of vehicle speed.

Base station transmitter diversity [19], [46], [47] has been mentioned earlier and it offers certain advantages and disadvantages over mobile receiver diversity. There is obviously a strong argument for keeping the complexity at the base station whilst keeping the mobile units simple. However, since the diversity spacing is large, several aerial towers are necessary and so more land is needed. The cost of the structures, etc., will not be insignificant, but may compare favourably with the cost of modifying all the mobile units. For some users, it may be impossible to increase the area occupied by the base-station if, for example, they operate from the roof of a building.

One possible diversity technique which may be used at base-station transmitters is similar to the phase sweeping method described for the mobile receiver. The disadvantage is that it occupies more radiated bandwidth than a non-diversity channel. This increase in bandwidth, when using the same technqiue for diversity reception in the mobile is of no importance since it occurs only in the receiver circuits and not in the medium, and the mutual interference problem is not made worse.

Diversity applied to the mobile receiver has the advantage that the receiver is usable with any base-station and therefore wide area-coverage is possible without having a diversity system at every base-station. As to the choice of diversity to be used in the mobile, it is impossible to say that one method is superior to all others. In order to choose a system it is first necessary to ensure that it will work with the type of modulation in use. Secondly, there is the choice of using either an existing range of receivers, possibly with minor modifications, or adopting a method integrated into a new design of receiver. The final choice must be an economic one as, of course, is the choice between using diversity as opposed to either redundancy or (if allowed by the licensing authority) a straight increase in transmitter power.

REFERENCES

[1] J. F. Ossana, "A model for mobile radio fading due to building reflections: theoretical and experimental fading waveform power spectra," *Bell Syst. Tech. J.,* vol. 43, no. 6, pp. 2935-71, November 1964.
[2] E. N. Gilbert, "Energy reception for mobile radio," *Bell Syst. Tech. J.,* vol. 44, no. 8, pp. 1779-1803, October 1965.
[3] R. H. Clarke, "A statistical theory of mobile radio reception," *Bell Syst. Tech. J.,* vol. 47, no. 6, pp. 957-1000, July-August 1968.
[4] M. J. Gans, "A power-spectral theory of propagation in the mobile-radio environment," *IEEE Trans. on Vehicular Technology,* vol. VT-21, no. 1, pp. 27-38, February 1972.
[5] S. H. Lin, "Statistical behaviour of a fading signal," *Bell Syst. Tech. J.,* vol. 50, no. 10, pp. 3211-70, December 1971.
[6] D. C. Cox, "Doppler spectrum measurements at 910 MHz over a suburban mobile radio path," *Proc. IEEE,* vol. 59, no. 6, pp. 1017-18, June 1971.
[7] —, "Delay-Doppler characteristics of multipath propagation at 910 MHz in a suburban mobile radio environment," *IEEE Trans. on Antennas and Propagation,* vol. AP-20, no. 9, pp. 625-35, September 1972.
[8] —, "Time and frequency domain characterizations of multipath propagation at 910 MHz in a suburban mobile-radio environment," *Radio Science,* vol. 7, no. 12, pp. 1069-77, December 1972.
[9] —, "A measured delay-Doppler scattering function for multipath propagation at 910 MHz in an urban mobile radio environment," *Proc. IEEE,* vol. 61, no. 4, pp. 479-80, April 1973.
[10] D. G. Brennan, "Linear diversity combining techniques," *Proc. IRE,* vol. 47, no. 6, pp. 1075-1102, June 1959.
[11] W. C.-Y. Lee, "Theoretical and Experimental study of the properties of the signal from an energy density mobile-radio antenna," *IEEE Trans. on Vehicular Technology,* vol. VT-16, no. 1, pp. 25-32, October 1967.
[12] W. C.-Y. Lee and Y. S. Yeh, "Polarization diversity system for mobile radio," *IEEE Trans. on Communication Technology,* vol. COM-20, no. 5, pp. 912-23, October 1972.
[13] W. C.-Y. Lee, "Antenna spacing requirement for mobile radio base-station diversity," *Bell Syst. Tech. J.,* vol. 50, no. 6, pp. 1859-76, July-August 1971.
[14] E. D. Sunde, *Communication Systems Engineering Theory.* Chap. 9 (Wiley: New York, 1969).
[15] L. R. Kahn, "Ratio squarer," *Proc. IRE,* vol. 42, p. 1704, November 1954 (Correspondence).
[16] M. Schwartz, W. R. Bennett, and S. Stein, *Communication Systems and Techniques.* Chap. 10 (McGraw-Hill: New York, 1966).
[17] L. Schiff, "Statistical suppression of interference with diversity in a mobile-radio environment," *IEEE Trans. on Vehicular Technology,* vol. VT-21, no. 4, pp. 121-8, November 1972.
[18] W. E. Shortall, "A switched diversity receiving system for mobile radio," *IEEE Trans. on Communication Technology,* vol. COM-21, no. 11, pp. 1269-75, November 1973.
[19] A. J. Rustako, Y. S. Yeh, and R. R. Murray, "Performance of feedback and switch space diversity 900 MHz f.m. mobile radio systems with Rayleigh fading," *IEEE Trans. on Communication Technology,* vol. COM-21, no. 11, pp. 1257-68, November 1973.
[20] L. Lewin, "Diversity reception and automatic phase correction," *Proc. Instn Elect. Engrs,* vol. 109B, 46, pp. 295-304, July 1962.
[21] J. D. Parsons and P. A. Ratliff, "Self-phasing aerial array for f.m. communication links," *Electronics Letters,* vol. 7, no. 13, pp. 380-1, 1st July 1971.
[22] P. A. Ratliff, "VHF Mobile Radio Communications: A Study of Multipath Fading," University of Birmingham, Department of Electronic and Electrical Engineering, Ph.D. thesis 1973.
[23] W. T. Blackband and D. E. Nichols, "Active aerials for aircraft communications," in W. T. Blackband, (Ed.), *Signal Processing Arrays.* p. 319 (Technivision Services, Wokingham, Berkshire, 1968).
[24] M. J. Withers, D. E. N. Davies, and R. H. Apperley, "Self-focusing aerial arrays for airborne communication," in W. T. Blackband, (Ed.), *loc. cit.,* p. 307.
[25] J. D. Parsons and M. Henze, "A proposed pre-detection combining system for mobile radio," *Proceedings of Conference on Radio Receivers and Associated Systems,* 1972. (IERE Conference Proceedings No. 24.)
[26] J. M. Sidwell, "A diversity combiner giving total power transfer," *Proc. Instn Elect. Engrs,* vol. 109B, No. 46, pp. 305-9, July 1962.
[27] S. Kazel, "Antenna pattern smoothing by phase modulation," *Proc. IRE,* vol. 52, no. 4, p. 435, April 1964 (Correspondence).
[28] M. J. Withers, "A diversity technique for reducing fast-fading," *Proceedings of Conference on Radio Receivers and Associated Systems,* 1972. (IERE Conference Proceedings No. 24.)
[29] O. G. Villard, Jr., J. M. Lomasney, and N. M. Kawachika, "A mode-averaging diversity combiner," *IEEE Trans. on Antennas and Propagation,* vol. AP-20, no. 4, pp. 463-9, July 1972.
[30] A. J. Rogers, "A double phase sweeping system for diversity reception in mobile radio," *The Radio and Electronic Engineer,* vol. 45, no. 4, pp. 183-91, April 1975.
[31] T. Inatomi, H. Kohno, K. Ikai, and B. Miyamoto, "I.f. combined space diversity reception," *Fujitsu Sci. Tech. J.,* vol. 8, no. 1, pp. 91-118, March 1972.
[32] N. M. Kawachika, and O. G. Villard, Jr., "Computer simulation of h.f. frequency-selective fading and performance of the mode-averaging diversity combiner," *Radio Science,* vol. 8, no. 3, pp. 203-12, March 1973.
[33] P. Bello and B. Nelin, "Predection combining with selectively fading channels," *IRE Trans. on Communication Systems,* vol. CS-10, no. 3, pp. 32-42, March 1962.
[34] M. J. Withers, D. E. N. Davies, A. H. Wright, and R. H. Apperley, "A self-focusing receiving array," *Proc. Instn Elect. Engrs,* vol. 112, no. 9, pp. 1683-8, September 1965.
[35] E. N. Gilbert, "Mobile radio diversity reception," *Bell Syst. Tech. J.,* vol. 48, pp. 2473-92, September 1969.
[36] C. C. Cutler, R. Kompfner and L. C. Tillotson, "A Selfsteering array repeater," *Bell Syst. Tech. J.,* vol. 42, no. 5, pp. 2013-

31, September 1963.

[37] D. M. Black, P. S. Kopel, and R. G. Novy, "An experimental u.h.f. dual-diversity receiver using a predetection combining system," *IEEE Trans. on Vehicular Communications*, vol. VC-15, no. 2, pp. 41-7, October 1966.

[38] J. W. Boyhan, "A new forward-acting predetection combiner," *IEEE Trans. on Communication Technology*, vol. COM-15, no. 5, pp. 689-94, October 1967.

[39] J. Granlund, "Topics in the design of antennas for scatter," MIT Lincoln Lab., Lexington, Mass., U.S.A. Tech. Report No. 135, pp. 105-113, November 1956.

[40] R. T. Adams and H. L. Smith, Reply to "Digital transmission capabilities of a transportable troposcatter system," *IEEE Trans. on Communication Technology*, vol. COM-16, no. 4, pp. 345-8, April 1968 (Correspondence).

[41] S. W. Halpern, "The theory of operation of an equal-gain predection regenerative diversity combiner with Rayleigh fading channels," *IEEE Trans. on Communication Technology*, vol. COM-22, no. 8, pp. 1099-1106, August 1974.

[42] R. G. Cease, "A predetection combiner for telemetry diversity applications," *Proceedings of National Telemetry Conference* (U.S.A.), 1969, pp. 26-30.

[43] W. C.-Y. Lee, "An energy-density antenna for independent measurement of the electric and magnetic field," *Bell Syst. Tech. J.*, vol. 46, no. 7, pp. 1587-99, September 1967.

[44] K. Itoh and D. K. Cheng, "A slot-unipole energy-density mobile antenna," *IEEE Trans. on Vehicular Technology*, vol. VT-21, no. 2, pp. 59-62, May 1972.

[45] J. D. Parsons, "Field-diversity antenna for u.h.f. mobile radio," *Electronics Letters*, vol. 10, no. 7, pp. 91-2, 4th April 1974.

[46] J. S. Bitler, H. H. Hoffman, and C. O. Stevens, "A mobile radio single-frequency 'two-way' diversity system using adaptive retransmission from the base," *IEEE Trans. on Communication Technology*, vol. COM-21, no. 11, pp. 1241-7, November 1973.

[47] W. Gosling and V. Petrovic, "Area coverage in mobile radio by quasisynchronous transmissions using double-sideband diminished-carrier modulation," *Proc. Instn. Elect. Engrs*, vol. 120, no. 12, pp. 1469-76, December 1973.

A Comparison of Specific Space Diversity Techniques for Reduction of Fast Fading in UHF Mobile Radio Systems

WILLIAM C. JAKES, JR., FELLOW, IEEE

Abstract—Over the past few years a variety of space diversity system techniques have been considered for the purpose of reducing the rapid fading encountered in microwave mobile radio systems. Basic diversity methods are first reviewed in the framework of mobile propagation effects, and then specific techniques are compared from the standpoint of transmitter power required to achieve a certain performance. Criteria of comparison used included baseband SNR while moving and reliability when the vehicle stops at random. System parameters are type and order of diversity and transmission bandwidth. Tradeoffs between performance properties and system parameters are indicated.

The calculations show that relatively modest use of diversity techniques can afford savings in transmitter power of 10–20 dB. For example, at a range of 2 mi, to obtain 30-dB baseband SNR while moving and 99.9-percent reliability when stopped requires a transmitted power of 8 W for a conventional FM system with no diversity. Two-branch selection diversity provides the same performance for a transmitter power of only 300 mW.

I. INTRODUCTION

IT IS WELL KNOWN [1] that the envelope of the field strength seen by a mobile antenna usually obeys a Rayleigh distribution[1] when examined for distances along the street of a few hundred wavelengths. In such a distribution, fades of 10 dB or more below the rms value of the envelope occur 10 percent of the time, 20 dB 1 percent of the time, 30 dB 0.1 percent of the time, and so on. At a transmission frequency of 840 MHz, a mobile receiver traveling 30 mi/h through such a signal would expect 10 dB or greater fades 30 times/s, 20 dB fades 10 times/s, and so on. In addition, when the vehicle stops for some reason, there is a finite chance of landing in a deep fade. With conventional transmission methods there is thus a tradeoff between the amount of power transmitted and the quality of service provided to the mobile user. It will be shown later that without the use of diversity, transmitter powers typically must be in the range of about 5–15 W to provide a good voice circuit at 2-mi range.

Random amplitude fading is also accompanied by random changes in phase. This introduces additional noise as the vehicle moves, with a baseband spectrum extending from zero to a value equal to approximately twice the Doppler shift. Above this value the spectral density falls at $1/f$. Thus, at UHF and 60 mi/h, the random FM spectrum extends to about 150 Hz. This effect imposes an upper limit on baseband signal-to-noise ratios (SNR) for FM systems, which is independent of the transmitter power. Fortunately, however, at UHF the random FM is not severe, and its effect decreases with increasing modulation index. The use of any kind of space diversity mitigates this effect, and some techniques eliminate it entirely, as will be seen later.

A. Space Diversity Concepts in Mobile Radio

The generally accepted model [1] of the mobile transmission path consists of an elevated base station antenna essentially removed from the influence of local (a few hundred feet or so) scattering objects, a line-of-sight path to the general location of the mobile in question, and scattering to the mobile antenna from many objects in its immediate vicinity. No direct path is assumed to exist between mobile and base antennas. This model produces most of the observed small-scale statistics of the received signal, including those already mentioned specifically.

From this model one can immediately deduce [2] that the signals received from spatially separated antennas on the mobile should have essentially uncorrelated envelopes for antenna separations on the order of one-half wavelength or more. Indeed, the probability of simultaneous fades of 20 dB or more from two such antennas is 0.01 percent, whereas it is 1 percent in each antenna individually, as mentioned earlier. Thus a signal made up of some combination of the two would have much less severe fading properties than a signal from one antenna alone.

These characteristics of independent signals also apply to separate antennas at the base station. However, since the important scatterers in the mobile-base transmission path are in the immediate vicinity of the mobile, the base antennas must be considerably farther apart to achieve decorrelation. Actually, a correlation coefficient as high as 0.7 can still provide almost as much diversity advantage as zero correlation [3]. Separations on the order of tens of wavelengths would probably be adequate at the base station.

Space diversity may thus be used either at the mobile or the base station, subject to the peculiar constraints of the mobile transmission path. However, mobile communications are two-way, requiring that the mobile station and the base station each have a transmitter and a receiver. Another classification of diversity systems thus appears which specifies whether the diversity array is located at the transmitter or receiver. This property has important implications in system design, as we

Manuscript received December 31, 1970; revised July 9, 1971.
The author is with Bell Telephone Laboratories, Inc., Merrimack Valley, Mass. 01860.
[1] Various statistical properties of the mobile radio signal are collected for reference in Appendix I.

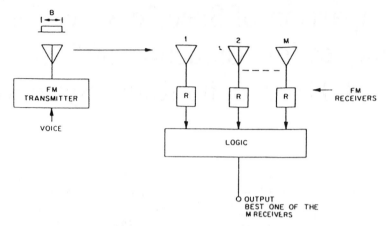

Fig. 1. Principles of selection diversity.

shall see, but is quite separate from the mobile path geometry. We can thus identify four types of diversity, which, in general, will have different design and performance properties:

1) mobile transmitter diversity
2) mobile receiver diversity
3) base transmitter diversity
4) base receiver diversity.

From the economic standpoint it would appear at the outset that only types 3) and 4) should be considered, i.e., keep all the diversity and its attendent complexity at the base station. However, the final system configuration will be influenced by many factors in addition to cost, such as the frequency spectrum required and the number of customers served. Thus it is worthwhile to explore all four possibilities to establish their performance characteristics, so that comparisons can be made later on an overall system basis.

B. Diversity Methods

Over the years a number of methods have evolved to take advantage of the uncorrelated fading from separate antennas. These will be described briefly in this section to illustrate their principles of operation. Specific embodiments appropriate to UHF mobile telephone systems will be detailed and their performance properties shown in Section II.

1) *Selection Diversity*: This is perhaps the simplest technique of all. Referring to Fig. 1, that one of the M receivers having the highest baseband SNR is connected to the output.[2] As far as the statistics of the output signal are concerned, it is immaterial where the selection is done. (If switching is done at IF or RF the problem of switching transients arises, although it does not appear to be a serious one.) The antenna signals themselves could be sampled, for example, and the best one sent to one receiver. The cumulative distribution curves for the output signals from selection diversity systems with 1, 2, 3, 4, and 6 branches are shown in Fig. 2. The theory for these curves follows Brennan [3]; it is summarized in Appendix II for reference. The potential power savings offered by

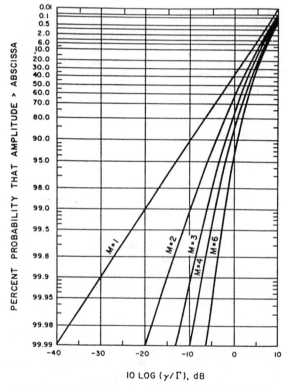

Fig. 2. Probability distribution of SNR γ for M-branch selection diversity system. Γ = SNR on 1 branch.

diversity are immediately obvious: 10 dB for 2-branch diversity at the 99-percent reliability level, for example, and 16 dB for 4 branches.

2) *Maximal Ratio Combining Diversity*: In this method, first proposed by Kahn [4], the M signals are weighted proportionately to their individual signal voltage-to-noise power ratios and then summed. Fig. 3 shows the essentials of the method. The individual signals must be cophased before combining in contrast to selection diversity; a technique to be described later does this very simply. The distribution curves are presented in Fig. 4, and the theory in Appendix II. This kind of combining gives the best statistical reduction of fading of any known linear diversity combiner. In comparison with

---

[2] In practice, the branch with the largest $(S + N)$ is usually used, since it is difficult to measure SNR.

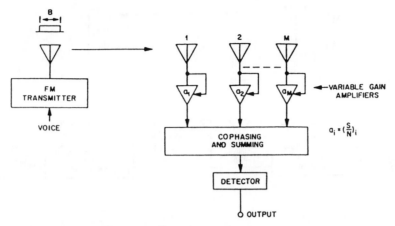

Fig. 3. Principles of maximal ratio combining.

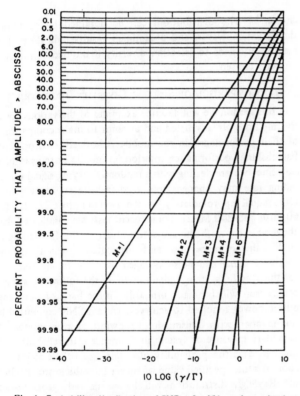

Fig. 4. Probability distribution of SNR γ for M-branch maximal
ratio diversity combiner. Γ = SNR on 1 branch.

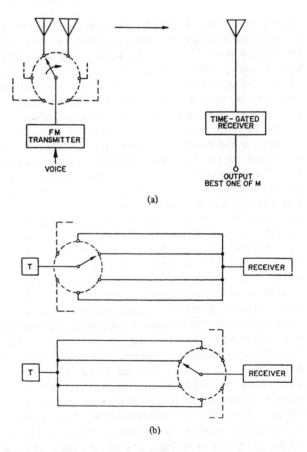

Fig. 5. Switched transmission path diversity. (a) Switched diversity at
base transmitter; M independent fading paths. (b) Equivalent configurations.

selection diversity, for example, 2 branches give 11.5-dB gain
at the 99-percent reliability level and 4 branches give 19-dB
gain, improvements of 1.5 and 3 dB, respectively. In some
cases it may not be convenient to provide the variable weighting capability required for true maximal ratio combining. The
weights are all then set equal to unity, and equal-gain combining results. The performance is only slightly inferior to
maximal ratio combining, however, and still better than selection diversity, particularly for larger values of M.

3) *Switched Transmission Paths*: In an attempt to provide
base transmitter diversity for mobile radio, the system shown
in Fig. 5(a) has been proposed. The transmitter continuously
switches between M spaced antennas at a suitably high rate; the

receiver then selects the best signal out of the appropriate time
slot. This is a selection diversity system and the signal distributions of Fig. 2 apply. The switching can be done either at the
transmitter or receiver, as shown in Fig. 5(b). The switching
process introduces out-of-band noise, however, which may
limit the applicability of this technique. Some means of
limiting the RF spectrum would also be needed.

4) *Feedback Diversity*: A very elementary type of diversity
receiver, called "scanning diversity" [3], is similar to selection

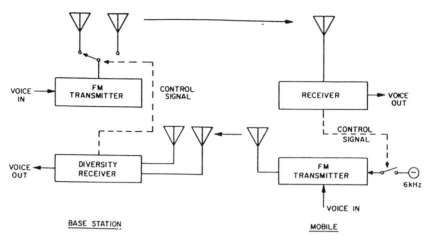

Fig. 6. Feedback diversity.

diversity except that instead of always using the best one of M signals, the M signals are scanned in a fixed sequence until one is found above threshold. This signal is used until it falls below threshold, when the scanning process starts again. The resulting fading statistics are somewhat inferior to those from other diversity systems; however, a modification of this technique appears promising for UHF mobile radio applications and is currently under investigation. The principles of operation are shown in Fig. 6 for 2-branch base transmitter diversity. In this system the fact that every base–mobile contact is a two-way affair is exploited by using the mobile-to-base path as a signaling channel in addition to carrying the voice modulation. It is assumed that the mobile-to-base path is reliable, using some sort of base receiver diversity. The base transmitter is connected to one of its two antennas by a switch, and remains there until the received signal at the mobile falls below a preset threshold level. It signals this fact over the mobile-to-base path and the transmitter then switches to the other antenna and remains there until the new signal again falls below the threshold. Since the chance of having both transmission paths poor simultaneously is smaller than either one being weak, there should be an average improvement in the signal received by the mobile. The performance is affected by the total time delay in actuating the switch, which is the sum of the roundtrip propagation time and the time delay corresponding to the bandwidth of the control channel. If this delay is too great the signal at the mobile could continue into a fade below the threshold before the transmitter switches and the new signal arrives. At UHF and vehicle speeds of 60 mi/h, however, the expected signal drop is only 1 or 2 dB; thus the technique appears promising for use in this frequency range. A variation on this scheme can provide very simple mobile receiver diversity, where the receiver switches between two antennas using the same logic principles previously described. In this case the time delay in the switching process can be made very small, with a consequent improvement in the signal fading statistics.

Although these systems are conceptually simple, obtaining the signal distribution curves analytically becomes very difficult because the condition of the output signal at any instant depends on the past history of the available signals from the transmitting antennas, and also on the particular decision logic

chosen.[3] The distributions can be obtained by computer simulation, however, and preliminary results indicate performance very near to selection diversity if the decision level is properly chosen.

II. Comparison of Specific Diversity Techniques

In this section we will describe examples of the various diversity methods already listed and attempt to make comparisons of their performance in the mobile radio UHF environment. Establishing a performance criterion for mobile systems is difficult since we are dealing with a randomly varying signal whose fading rate may change from zero (vehicle stationary) to a value equal approximately to the Doppler shift (75 Hz for 60 mi/h and 840 MHz). Some recent tests have indicated that above about 5 mi/h the subjective effect of the fast fading at UHF is that of noise, and can be evaluated by calculating the baseband signal and noise averaged over the fading statistics for the specific system under study. In the following, system comparison will be made primarily on the basis of the transmitter power required to achieve a certain SNR computed in this manner, with transmission bandwidth as a parameter. A secondary measure of system performance is its reliability, defined here as the probability of obtaining an SNR at least as good as some specified value whenever the vehicle stops. Only the Rayleigh distribution will be considered; gross terrain effects (shadows due to hills, etc.) will be neglected.

A. Reference System (No Diversity)

We assume the following parameters for a system in an urban environment:

frequency	840 MHz
base antenna height	30 m (100 ft)
mobile antenna height	1.5 m (5 ft)
range d	3.2 km (2 mi)
base antenna gain G_B	9 dB
mobile antenna gain G_M	3 dB.

[3]In fact, obtaining the distribution curves requires knowledge of the distribution function for the length of the interval between two successive level crossings of a randomly varying signal. This is still one of the unsolved problems in statistical communication theory.

The line-of-sight loss between these antennas is

$$L_0 = 20 \log \frac{4\pi d}{\lambda} - (G_B + G_M) = 101 - 12 = 89, \quad \text{dB}.$$

Applying an additional loss of 41 dB to account for the mobile path characteristics [5, fig. 41], the net loss is

$$L = 89 + 41 = 130, \quad \text{dB}.$$

We will assume that FM is used with Gaussian modulation flat over the band W extending from 0–3 kHz. The baseband SNR is a function of the parameters α and ρ, where $\alpha = B/2w$, B is the RF bandwidth, and ρ is the carrier-to-noise ratio (CNR). Now ρ represents the Rayleigh fading signal at the receiver input. To obtain the effect of the fading, Davis [6] has averaged the signal over the fading distribution and calculated the resulting baseband SNR as a function of ρ_0 (the mean value of ρ) and the parameter α. Using his results the required value of ρ_0 to obtain a given SNR was determined as a function of α, with the results shown in Table I (ρ_0 (dB) = $10 \log_{10} \rho_0$). In this and subsequent systems we will choose a baseband SNR of 30 dB as a reference, and random FM will be neglected.

Knowing ρ_0 and the receiver noise figure, assumed to be 3 dB, the transmitter power required may then be calculated:

$$\begin{aligned} P_T &= \rho_0 + \text{NF}_{\text{dB}} + 10 \log kTB + L \\ &= \rho_0 + 3 - 174 + 10 \log B + L \\ &= \rho_0 + 10 \log \alpha - 3.2, \quad \text{dBm} \end{aligned}$$

where

$$kT = (1.38 \times 10^{-20})(290°) = 4 \times 10^{-18}, \quad \text{mW/Hz}$$

NF_{dB} is the noise figure, and L is the net loss of 130 dB, previously computed. The variation of required power with bandwidth B is shown in Fig. 7, and we see that the power falls from 16 to 5.5 W as B increases from 12 to 40 kHz. For $B \geqq 40$ kHz, the power slowly increases, thus $B = 40$ kHz would probably be the maximum value considered.

The reliability of this system may be obtained from the Rayleigh distribution curve of Fig. 2. Suppose we require the probability of stopping in a location where $\rho \geqq \rho_t$, the FM click threshold given by

$$\rho_t \, \text{erfc} \sqrt{\rho_t} = \frac{0.75 \sqrt{2\pi}}{(10\alpha)^2}.$$

Values of ρ_t versus α are listed below.

α	2	3	4	6	8	10	15	20
ρ_t (dB)	7.5	8.2	8.6	9.1	9.4	9.6	10.0	10.3

The Rayleigh curve is then entered at a signal level of ρ_t/ρ_0 for the various values of α and ρ_0 already given, and the corresponding probability is then obtained. The results are plotted in Fig. 8 for $B \leqq 400$ kHz.

In order to select the system design parameters, the relative importance of the SNR experienced while moving compared with the chance of stopping at an unfavorable signal level will have to be examined from a subjective viewpoint. In this paper we will arbitrarily assume that enjoying a 30-dB SNR while

TABLE I

α	B (kHz)	ρ_0 (dB)
2	12	42.0
3	18	37.4
4	24	35.0
6	36	32.7
8	48	31.5
10	60	30.6
15	90	29.7
20	120	29.0

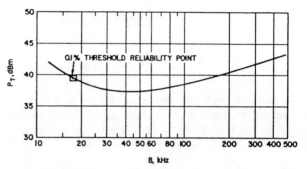

Fig. 7. Transmitter power required at 836 MHz with no diversity for 30-dB SNR at baseband with Gaussian FM modulation. B is RF bandwidth, vehicle speed is 60 mi/h, range is 2 mi in city; base antenna height is 30 m, total antenna gain is 12 dB.

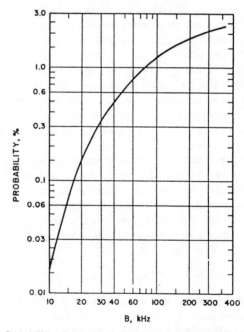

Fig. 8. Probability of stopping in location where signal level is at or below click threshold in reference system (no diversity).

moving is just as important as stopping at the click threshold only one time in a thousand, i.e., a reliability of 99.9 percent with respect to threshold. This point corresponds to a transmitted power of 39.4 dBm and a bandwidth of 17.5 kHz, as marked in Fig. 7. These numbers characterize the reference system, and will be compared with equivalent values for diversity systems in the sections following.

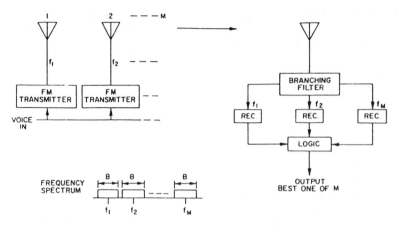

Fig. 9. Selection diversity scheme with diversity antenna array at transmitter.

The reader is reminded that random FM has been neglected in the analysis here. At 60 mi/h, requiring 30-dB SNR, this assumption is valid for $\alpha > 4$ ($B > 24$ kHz), or, alternatively, for $B > 12$ kHz if the speed is limited to 15 mi/h.

B. Selection Diversity

The selection diversity system shown in Fig. 1 is a receiver diversity type that can be used either at the base or mobile, the only difference being the larger antenna separation required at the base station. It is possible to conceive of a selection diversity scheme where the diversity antenna array is at the transmitter, as shown in Fig. 9 for 2-branch diversity. T_1 and T_2 are two FM transmitters operating on adjacent frequency bands centered at f_1 and f_2. The receiver separates these bands in a branching filter; each signal is then separately detected and the best one chosen as before. Although more frequency bandwidth is required, the transmitter antenna array spacing may be slightly reduced by taking advantage of a certain amount of frequency diversity.[4]

We will proceed as before to determine the required transmitter power and reliability for these two diversity schemes. Davis [6] gives the required value of ρ_0, the mean CNR per diversity branch, to obtain a desired baseband SNR in the presence of fading for various numbers of diversity branches M. We again assume 30-dB baseband SNR. The transmitted power for the system of Fig. 1 is then

$$P_T = \rho_0 + 10 \log \alpha - 3.2, \quad \text{dBm}.$$

Table II lists the values of ρ_0 in dB and P_T in dBm. The values of required power are plotted in Fig. 10, and behavior similar to the no-diversity case is seen except for a substantial reduction in required power.

For the transmitter diversity arrangement of Fig. 9 the required total power and RF bandwidth are simply M times the values for the receiver diversity scheme, that is,

$$P_T = \rho_0 + 10 \log \alpha - 3.2 + 10 \log M, \quad \text{dBm}.$$

These values are shown in Fig. 10 as dotted curves.

<hr/>

[4] If the transmitted bands were separated widely enough one could completely exchange frequency diversity with space diversity, i.e., only one antenna would be required at the transmitter.

TABLE II

α	B(kHz)	$M=2$		$M=3$		$M=4$	
		ρ_0	P_T	ρ_0	P_T	ρ_0	P_T
2	12	30.0	29.8	28.1	27.9	27.2	27.0
3	18	23.3	24.9	20.6	22.2	19.6	21.2
4	24	20.0	22.8	16.4	19.2	15.0	17.8
6	36	17.7	22.3	13.2	17.8	11.1	15.7
8	48	16.5	22.3	12.2	18.0	10.0	15.8
10	60	16.0	22.8	11.6	18.4	9.6	16.4
15	90	15.1	23.7	10.9	19.5	8.9	17.5
20	120	14.8	24.6	10.5	20.3	8.5	18.3

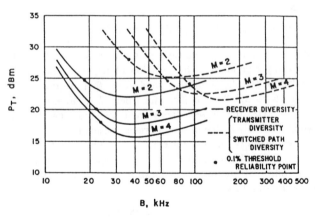

Fig. 10. Variation in required transmitter power for selection diversity with 2, 3, and 4 branches; 2-mi path, 30-dB SNR.

The 99.9-percent threshold reliability points are also marked on the curves of Fig. 10 and tabulated in Table III. They are obtained as previously described for the no-diversity case: enter the distribution curves of Fig. 2 for signal levels of ρ_t/ρ_0 for the various values of α and ρ_0 already given and the corresponding probability is obtained. As an example, let $M = 2$ and $\alpha = 4$; then $\rho_0 = 20.0$ dB, $\rho_t(4) = 8.6$ dB, and $\rho_t/\rho_0 = -11.4$ dB. The curve for $M = 2$ in Fig. 2 gives a probability of 99.5 percent. Plotting the probability versus α enables us to find the point for 99.9 percent. The results are given in Table III.

In the foregoing treatment the effects of random FM have again been neglected. It was stated earlier that any kind of diversity helps to reduce random FM; this is due to the fact

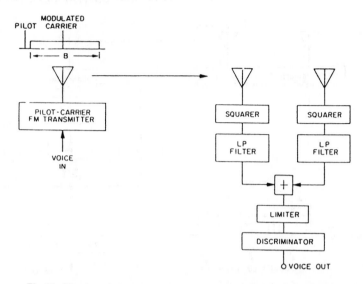

Fig. 11. Two-branch maximal ratio receiver diversity of pilot-carrier type.

TABLE III

M	Receiver Diversity		Transmitter Diversity	
	P_T (dBm)	B(kHz)	P_T (dBm)	B(kHz)
2	24.8	18.5	27.8	37.0
3	20.0	22.0	24.8	66.0
4	18.0	23.5	24.0	94.0

that the phase of the RF signal changes more and more rapidly as it drops deeper into a fade. Thus the worst random FM excursions occur during the fading instants of the signal. Selection diversity utilizes the signal during times when the fading is less, thus it discriminates against the more severe bursts of random FM. For this reason it is felt that the preceding results may be applied for vehicle speeds up to 60 mi/h without restrictions on bandwidth, contrary to the case with no diversity.

C. Maximal Ratio Combining

A diversity scheme that has been the subject of recent tests [7] uses a separate unmodulated carrier (the "pilot") transmitted along with a modulated carrier to provide predetection combining. The scheme is illustrated in Fig. 11, and it is shown in Appendix III that true maximal ratio combining action is achieved.

Referring to Fig. 11, the principles of operation are as follows: the pilot and carrier, separated by one-half the RF modulation bandwidth, are multiplied together and the difference frequency is selected by a low-pass filter. Any phase modulation (random FM) imparted by the transmission path is thus canceled and the resulting signal is at a phase determined by the initial phase difference between pilot and carrier at the transmitter. All diversity branches are at the same phase, thus may be simply summed to obtain predetection combining. The price paid for these features is an overall penalty of 6 dB in CNR: 3 dB because half the power is in the pilot and 3 dB because both pilot and carrier heterodyne noise into the baseband, as shown in Appendix III.

As in the case of selection diversity, the maximal ratio technique can also be applied to transmitter diversity, as shown in

Fig. 12 for 2-branch diversity. Two transmitters each radiate a pilot-carrier signal on adjacent frequencies from individual antennas of the diversity array. The receiver has one antenna followed by a branching filter to separate the two signals. Operation then proceeds as in the receiver diversity case.

For maximal ratio receiver diversity the total required transmitter power for our 2-mi path is given by

$$P_T = \rho_0 + 10 \log \alpha + 2.8, \quad \text{dBm}$$

where ρ_0 is the CNR per branch to obtain a 30-dB SNR while moving, given by Davis [6], or maximal ratio combining, as shown in Table IV. The 6-dB penalty has also been applied. The values of required power are plotted in Fig. 13. The curves are similar to selection diversity except for an additional 4–6 dB of transmitter power.

For the transmitter diversity scheme of Fig. 12 the required total power and RF bandwidth are M times the values for receiver diversity, and the performance is shown by the dotted curves of Fig. 13. Following the method previously described, we can obtain the 99.9-percent threshold reliability points. They are marked on the curves shown in Fig. 13 and are listed in Table V.

D. Switched Transmission Paths

The principles of this technique are shown in Fig. 5 and were briefly described earlier. A considerable number of embodiments of the time-gated receiver were tried in the laboratory in an attempt to reduce or eliminate the effect of noise foldover due to the switching process. Although the techniques used differed substantially in detail, their performance was found to be equivalent to that of the following model.

Assume M antennas are used with a transmitter of bandwidth B, and the receiver selects the best of the M time samples out of each switching cycle. Then this is a selection diversity system of order M, and the procedure described earlier for selection diversity applies. The effective bandwidth per diversity branch is reduced, however, to B/M, thus $\alpha = B/(2WM)$. The effective transmitter power per branch is also reduced by the factor $1/M$. We can then calculate the required transmitter

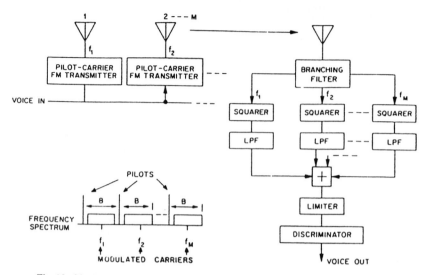

Fig. 12. Maximal ratio diversity scheme with diversity antenna array at transmitter.

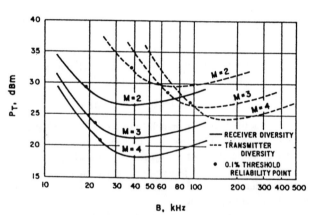

Fig. 13. Variation in required transmitter power for pilot-carrier maximal ratio diversity with 2, 3, and 4 branches; 2-mi path, 30-dB SNR.

TABLE IV

α	B(kHz)	$M = 2$		$M = 3$		$M = 4$	
		ρ_0	P_T	ρ_0	P_T	ρ_0	P_T
2	12	28.6	34.4	25.6	31.4	23.9	29.9
3	18	21.9	29.5	18.1	25.7	16.3	23.9
4	24	18.6	27.4	13.9	22.7	11.7	20.5
6	36	16.3	26.9	10.7	21.3	7.8	18.4
8	48	15.1	26.9	9.7	21.5	6.7	18.5
10	60	14.6	27.4	9.1	21.9	6.3	19.1
15	90	13.7	28.3	8.4	23.0	5.6	20.2
20	120	13.4	29.2	8.0	23.8	5.2	21.0

TABLE V

M	Receiver Diversity		Transmitter Diversity	
	P_T (dBm)	B(kHz)	P_T (dBm)	B(kHz)
2	29.3	19.0	32.3	38.0
3	23.5	21.8	28.3	65.4
4	20.8	23.0	26.8	92.0

power as before:

$$P_T = \rho_0 + 3 - 174 + 10 \log \frac{B}{M} + L + 10 \log M$$

$$= \rho_0 + 10 \log B - 41$$

$$= \rho_0 + 10 \log \alpha - 3.2 + 10 \log M, \quad \text{dBm}$$

where ρ_0 is the CNR given by Davis [6] to realize a given baseband SNR, again chosen to be 30 dB. We note that this expression is exactly the same as that for transmitter diversity given before in the selection diversity case; thus the dotted curves of Fig. 10 apply in all respects, where B is the transmitter bandwidth and P_T its power.

E. Feedback Diversity

The feedback diversity scheme, outlined in Fig. 6, was described earlier. The switching time must be short to realize a statistical advantage; on the other hand, abrupt switching introduces phase transients which produce noise in the baseband output. A method of reducing this switching noise somewhat is currently under study. Assuming the existence of this technique, Davis has calculated the power required under the same assumptions used here, and his results are reproduced in Fig. 14.

III. SUMMARY AND CONCLUSIONS

The performance characteristics of the various diversity systems have been presented in terms of total power required at the transmitter versus bandwidth for a typical mobile transmission path in an urban environment. This gives one a feel for the order of magnitude of the powers involved. To make comparisons between systems the ratio of the total required transmitter power for the various diversity systems to that of the reference system with no diversity is of interest. The curves of Figs. 15 and 16 show these ratios expressed in dB as a function of total RF bandwidth for selection diversity and maximal ratio combining, respectively. The solid curves compare receiver diversity with the no-diversity case, and the dotted curves are for transmitter diversity. Considering the receiver

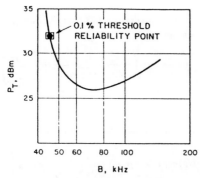

Fig. 14. Variation in required transmitter power for 2-branch feedback diversity; 2-mi path, 30-dB SNR ratio; switching delay is 266 μs, vehicle speed is 60 mi/h.

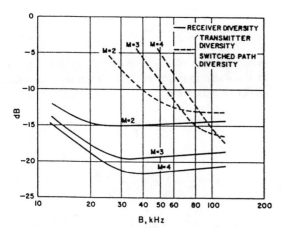

Fig. 15. Ratio of transmitter power required in selection diversity system to that of no-diversity system; 30-dB SNR.

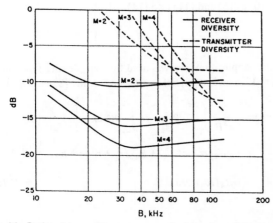

Fig. 16. Ratio of transmitter power required in pilot-carrier maximal ratio diversity combiner to that of no-diversity system; 30-dB SNR.

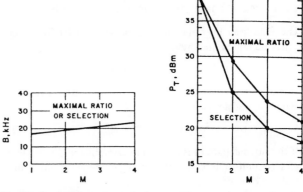

Fig. 17. Bandwidth and required transmitter power versus diversity order for 30-dB SNR in motion and 99.9-percent threshold reliability when stopped.

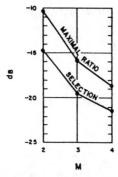

Fig. 18. Ratio of transmitter power required in receiver diversity systems to that of no-diversity system versus diversity order for 30-dB SNR in motion and 99.9-percent threshold reliability when stopped.

diversity cases, substantial reductions of transmitter power of 10–20 dB result from either selection or maximal ratio combining of 2–4 branches, with selection somewhat better by about one order of diversity. That is, 2-branch selection diversity is about equivalent to 3-branch maximal ratio combining. This is due mainly to the 6-dB noise penalty incurred by the pilot-carrier scheme. The advantages of transmitter diversity appear for larger bandwidths and generally are not as great.

Another interesting comparison uses the design criteria that the transmitted power and bandwidth are chosen to provide 30-dB SNR while moving and 99.9-percent threshold reliability when stopped. These values have been given earlier for the various techniques and are plotted in Fig. 17 as a function of diversity order. Fig. 18 shows the transmitter power saving for these design points, which again is in the range of 10–20 dB. These savings are substantial; for example, with no diversity the design point requires 8 W of transmitter power, whereas with 2-branch selection receiver diversity only 300 mW are needed.

It should be pointed out that in this paper the large-scale shadowing caused by gross terrain features, such as large hills or especially high buildings, have been neglected. This shadowing causes variations in mean signal level which may be described by a log-normal distribution [5]. Thus there may be times when the mean signal is so low that diversity schemes addressed to the problem of reducing multipath interference of the Rayleigh type will still not provide a usable signal. The mean signal must be increased in these cases, but the use of diversity makes possible much smaller an increase by the aforementioned amounts. The effects of possible cochannel interference arising from reuse of the transmission frequency in a distant area are also neglected, but the change in fading statistics brought about by the use of diversity can be shown to be beneficial in this case.

APPENDIX I
STATISTICAL PROPERTIES OF MOBILE RADIO SIGNAL [1]

The field seen by a mobile antenna at a carrier frequency ω_c may be represented by

$$E = x \cos \omega_c t + y \sin \omega_c t \qquad (1)$$

where x and y are independent zero-mean Gaussian random variables of equal variance:

$$\langle x^2 \rangle = \langle y^2 \rangle = \langle |E|^2 \rangle = E_{\text{rms}}^2. \qquad (2)$$

If the antenna pattern is uniform in azimuth and the incident waves are uniformly distributed in angle, the spectrum of this signal can be expressed

$$S(f) = \frac{2E_{\text{rms}}^2}{\omega_m} \left[1 - \left(\frac{f - f_c}{f_m} \right)^2 \right]^{-1/2}, \qquad |f - f_c| \leq f_m \qquad (3)$$

where $f_m = \omega_m/2\pi = V/\lambda$, the Doppler shift corresponding to vehicle speed V and carrier wavelength λ. The moments of this process are [8]

$$b_n = (2\pi)^n \int_0^\infty S(f)(f - f_c)^n \, df$$

$$= \begin{cases} \dfrac{2E_{\text{rms}}^2}{\pi} \omega_m^n \displaystyle\int_0^{\pi/2} \sin^n \theta \, d\theta, & n \text{ even} \\ 0, & n \text{ odd} \end{cases}$$

$$= \begin{cases} E_{\text{rms}}^2 \, \omega_m^n \dfrac{1 \cdot 3 \cdot 5 \cdots (n-1)}{2 \cdot 4 \cdot 6 \cdots n}, & n \text{ even} \\ 0, & n \text{ odd.} \end{cases} \qquad (4)$$

Thus

$$b_0 = E_{\text{rms}}^2 = \text{total power in the field}$$

$$b_2 = \tfrac{1}{2} \omega_m^2 \, b_0$$

$$b_4 = \tfrac{3}{8} \omega_m^4 \, b_0.$$

The probability densities of x and y are

$$p_x(\alpha) = p_y(\alpha) = \frac{1}{\sqrt{2\pi b_0}} \exp\left(-\frac{\alpha^2}{2b_0} \right). \qquad (5)$$

Since x and y are independent, $p(x,y) = p(x)p(y)$. The density function of the envelope of E, $r = (x^2 + y^2)^{1/2}$, is

$$p(r) = \frac{r}{b_0} \exp\left(-\frac{r^2}{2b_0} \right), \qquad r \geq 0 \quad \text{(Rayleigh)} \qquad (6)$$

$$\langle r^2 \rangle = 2b_0 = R_{\text{rms}}^2. \qquad (7)$$

The rate at which the envelope makes crossings of the level $r = R$ with positive slope (the level-crossing rate) is

$$N_R = \int_0^\infty \dot{r} p(R, \dot{r}) \, d\dot{r} \qquad (8)$$

where $p(R, \dot{r})$ is the joint density function of r and $\dot{r}$ at $r = R$. Rice gives the joint density function [8, eq. 4.5]

$$p(r, \dot{r}, \theta, \dot{\theta}) = \frac{r^2}{4\pi^2 b_0 b_2} \exp\left[-\frac{1}{2} \left(\frac{r^2}{b_0} + \frac{\dot{r}^2}{b_2} + \frac{r^2 \dot{\theta}^2}{b_2} \right) \right] \qquad (9)$$

where $\tan \theta = y/x$. Integrating (9) over θ from 0 to 2π and $\dot{\theta}$ from $-\infty$ to $+\infty$, we get

$$p(r, \dot{r}) = \underbrace{\frac{r}{b_0} \exp\left(-\frac{r^2}{2b_0} \right)}_{p(r)} \underbrace{\frac{1}{\sqrt{2\pi b_2}} \exp\left(-\frac{\dot{r}^2}{2b_2} \right)}_{p(\dot{r})}. \qquad (10)$$

Thus r and $\dot{r}$ are independent. Substituting in (8),

$$N_R = \int_0^\infty \dot{r} p(R) p(\dot{r}) \, d\dot{r} = p(R) \frac{1}{\sqrt{2\pi b_2}} \int_0^\infty x \exp\left(-\frac{x^2}{2b_2} \right) dx$$

or

$$N_R = f_m \sqrt{2\pi} \, \rho e^{-\rho^2} \qquad (11)$$

where

$$\rho = \frac{R}{R_{\text{rms}}} = \frac{R}{\sqrt{2b_0}}. \qquad (12)$$

The average duration of fades below $r = R$ is also of interest. Let τ_i be the duration of the ith fade. Then the probability that $r \leq R$ for time duration T is

$$P\{r \leq R\} = \frac{1}{T} \sum \tau_i. \qquad (13)$$

Now

$$\bar{\tau} = \frac{1}{TN_R} \sum \tau_i = \frac{P\{r \leq R\}}{N_R}. \qquad (14)$$

But

$$P\{r \leq R\} = \int_0^R p(r) \, dr = 1 - e^{-\rho^2}. \qquad (15)$$

Substituting (15) and (11) into (14),

$$\bar{\tau} = \frac{1}{\rho f_m \sqrt{2\pi}} \left(e^{\rho^2} - 1 \right). \qquad (16)$$

In other applications we will need the joint density function of $r, \dot{r}, \ddot{r}$. Rice gives a general expression for this [8, eq. 3.8-5]:

$$p(r, \dot{r}, \ddot{r}) = \frac{r^3}{8\pi^3 B} \int_0^{2\pi} d\theta \int_{-\infty}^{\infty} d\dot{\theta} \int_{-\infty}^{\infty} d\ddot{\theta} \exp\left\{ -\frac{1}{2B^2} [B_0 r^2 \right.$$

$$- 2B_2(r\ddot{r} - r^2\dot{\theta}^2) + B_{22}(\dot{r}^2 + r^2\dot{\theta}) + B_4(\ddot{r} - r\dot{\theta}^2)$$

$$\left. + B_4(r\ddot{\theta} + 2\dot{r}\dot{\theta})^2] \right\} \qquad (17)$$

where

$$B = b_2(b_0 b_4 - b_2^2), \qquad B_0 = b_2 b_4 B, \qquad B_2 = -b_2^2 B$$

$$B_{22} = (b_0 b_4 - b_2^2)B, \qquad B_4 = b_0 b_2 B.$$

It is assumed here that the spectrum is symmetrical so that the odd moments are zero. Equation (17) can be integrated immediately on θ and $\ddot{\theta}$:

$$p(r, \dot{r}, \ddot{r}) = \frac{r}{\pi} \sqrt{\frac{b_0}{B}} p(r) p(\dot{r}) \int_0^\infty \exp\left\{ -\frac{1}{2} \left[\frac{r^2 \dot{\theta}^2}{b_2} \right. \right.$$

$$\left. \left. + \frac{B_4}{B^2} \left(\ddot{r} - \frac{B_2 + B_4 \dot{\theta}^2}{B_4} r \right)^2 \right] \right\} d\dot{\theta}. \qquad (18)$$

APPENDIX II
DERIVATION OF DIVERSITY SIGNAL DISTRIBUTION

We will use the approach given by Brennan [3], with some shortcuts that have recently been pointed out. We assume the signals in each diversity branch are independent and Rayleigh distributed with equal mean power b_0. The density function of the signal envelope has been given in Appendix I:

$$p(r_i) = \frac{r_i}{b_0} \exp\left(-\frac{r_i^2}{2b_0}\right). \qquad (19)$$

We will be interested in SNR, thus it is convenient to introduce new variables. The local (averaged over one RF cycle) mean signal power per branch is $r_i^2/2$. Let the mean noise power per branch n_i^2 be the same for all branches $\overline{n_i^2} = N$, and let

$$\gamma_i = \frac{\text{local mean signal power per branch}}{\text{mean noise power per branch}} = \frac{r_i^2}{2N} \qquad (20)$$

$$\Gamma = \frac{\text{mean signal power per branch}}{\text{mean noise power per branch}} = \overline{\gamma_i} = \frac{b_0}{N}. \qquad (21)$$

Then

$$p(\gamma_i) = \frac{1}{\Gamma} e^{-\gamma_i/\Gamma}. \qquad (22)$$

A. Selection Diversity

The probability that the SNR in one branch γ_i is less than or equal to some specified value γ is

$$P\{\gamma_i \leq \gamma\} = \int_0^{\gamma_i} p(\gamma_i)\, d\gamma_i = 1 - e^{-\gamma/\Gamma} \qquad (23)$$

The probability that the γ_i in all M branches are simultaneously less than or equal to γ is then

$$P\{\gamma_1 \cdots \gamma_M \leq \gamma\} = (1 - e^{-\gamma/\Gamma})^M = P_M(\gamma). \qquad (24)$$

This is the distribution of the best signal, i.e., largest SNR, selected out of the M branches, and is plotted in Fig. 2.

To get the mean signal-to-mean noise ratio of the selected signal we need the density

$$p_M(\gamma) = \frac{d}{d\gamma} P_M(\gamma) = \frac{M}{\Gamma}(1 - e^{-\gamma/\Gamma})^{M-1} e^{-\gamma/\Gamma} \qquad (25)$$

$$\overline{\gamma} = \int_0^\infty \gamma p_M(\gamma)\, d\gamma = \Gamma \int_0^\infty Mx(1 - e^{-x})^{M-1} e^{-x}\, dx \qquad (26)$$

where $x = \gamma/\Gamma$. This integral may be evaluated to get

$$\frac{\overline{\gamma}}{\Gamma} = \sum_{k=1}^M \frac{1}{k}. \qquad (27)$$

The value of $\overline{\gamma}/\Gamma$ increases slowly with M, as shown below.

M	$\overline{\gamma}/\Gamma$	$10 \log(\overline{\gamma}/\Gamma)$, (dB)
1	1.00	0.0
2	1.50	1.77
3	1.833	2.64
4	2.083	3.20
5	2.283	3.60
6	2.450	3.90

B. Maximal Ratio Combining

In this case the signals in the M branches are cophased and added with appropriate branch weighting factors a_i. The resulting signal envelope is

$$r_M = \sum_{i=1}^M a_i r_i. \qquad (28)$$

Likewise, the total noise power is the weighted sum of the branch noises, here assumed to be

$$N_T = N \sum_{i=1}^M a_i^2. \qquad (29)$$

The resulting SNR is then

$$\gamma_M = \frac{r_m^2}{2N_T}. \qquad (30)$$

Brennan [3] has shown that if $a_i = r_i/N$, then γ_M will be maximized:

$$\gamma_M = \frac{1}{2} \frac{\sum (r_i^2/N)^2}{N \sum (r_i^2/N^2)} = \frac{1}{2} \sum_{i=1}^M \frac{r_i^2}{N} = \sum_{i=1}^M \gamma_i. \qquad (31)$$

That is, the SNR out of the combiner equals the sum of the branch SNR. Now

$$\gamma_i = \frac{1}{2N} r_i^2 = \frac{1}{2N}(x_i^2 + y_i^2) \qquad (32)$$

where x_i and y_i are independent Gaussian random variables of equal variance b_0 and zero mean. Thus γ_M is a chi-square distribution of $2M$ Gaussian random variables with variance $b_0/2N = \frac{1}{2}\Gamma$. The density function is thus

$$p(\gamma_M) = \frac{\gamma_M^{M-1} e^{-\gamma_M/\Gamma}}{\Gamma^M (M-1)!}, \qquad \gamma_M \geq 0. \qquad (33)$$

The probability that $\gamma_M \leq \gamma$ is

$$P\{\gamma_M \leq \gamma\} = \int_0^\gamma p(\gamma_M)\, d\gamma_M = 1 - e^{-\gamma/\Gamma} \sum_{k=1}^M \frac{(\gamma/\Gamma)^{k-1}}{(k-1)!}. \qquad (34)$$

Equation (34) is the distribution function for maximal ratio combining, and is plotted in Fig. 4. The mean signal-to-mean noise ratio is obtained immediately from (31):

$$\overline{\gamma}_M = \sum_{i=1}^M \gamma_i = \sum_{i=1}^M \Gamma = M\Gamma. \qquad (35)$$

Thus γ_M goes linearly with M, whereas for selection diversity it increases much more slowly.

It is interesting to note that, although the fluctuations in SNR have been reduced by this combining process, the variations in signal power have increased. In fact, the signal power is $\frac{1}{2} r_M^2 = 2\gamma_M^2$, thus varies as the square of the output SNR.

311

Chapter 6

SPEECH CODING

Speech Codecs for Personal Communications

For a standard to be accepted, it must meet the needs of its users, and these needs should be reflected in the requirements promulgated by the standards body.

Raymond Steele

Reprinted from *IEEE Communications Magazine,* pp. 76-83, November 1993.

e've got it Percy, we ve got it!" The chief executive officer Ray Dundant burst into Percy's office, "All the financing we need to construct and run a personal communications network. And you, Percy," he said as he lit up with a this-one's-in-your-lap grin, "are in charge of the project."

Percy Comms' posture sagged toward a fetal position. "That's terrific news, but no one agrees what personal communications is."

"Yes they do, it's that old Spencer Tracy communicating wristwatch stuff. Let's you talk to anyone, anywhere, at any time."

"Dick."

The chief executive looked offended.

"Dick Tracy," Percy mumbled.

"Of course. Well, can't stop. I know you mobile radio engineers like to have a good challenge."

A Sad Tale

Half the air in the room departed as the CEO strode through the door leaving Percy pondering what to do. A wristwatch communicator. That sounded bad enough, but the thought of constructing a network to support mass personal communications was an order far more daunting. Involuntarily he put on his old raincoat and absent-mindedly wandered out of the building.

Whenever a crisis occurred Percy Comms felt hungry, and the brightly colored FISH & CHIPS sign cleared his mind in an instant. Without hesitating he entered the shop, his nose seeking odors that never came.

"Can I help you?" The man in the sports jacket looked out of place.

"A large haddock and a small p-portion of chips," Percy stammered.

"A what? This is a speech codec shop."

"Says FISH and CHIPS outside."

"I know, that is an abbreviation. We sell fast integrated silicon hybrids (FISH) and codecs for high integrity perceptual speech (CHIPS). I'm J. N. D. Codit at your service."

Percy's tummy rumbled. Could his error be prov-

idential? "Would any of your FISH and CHIPS be of use in a personal communications network?"

"For a Spencer Tracy communicator?" He gave Percy a superior-marketing-director smile.

Dick, thought Percy, "Yes, I've got to put a personal communications network together."

"There is no doubt I can help you. We carry a complete range of speech codecs. These are our waveform speech codecs." A tray of pulse code modulation chips, half hidden under a half-price summer sale sign, was placed on the counter. "This," said the young man, "is the model-T Ford of speech codecs. A miracle in its day, low cost, low delay, and although easily damaged by transmission errors, these errors are not that difficult to identify and correct — it's an international standard, a pet of the fixed network designers. It's a bit heavy on the gas, but if you've got the bandwidth you might seriously consider it."

Percy, the future personal communication system (PCS) designer sighed, "That sort of bandwidth is beyond me, strictly for optical guys."

J.N.D. Codit continued with his well rehearsed patter, "There are these traditional waveform codecs with feedback. Tremendous step forward, once they made them adaptive." He held up an adaptive delta modulation (ADM) codec chip in the style of a famous Dutch engineer, "The military got really excited with this. One bit words, easy to encrypt, remarkably robust to transmission errors, low cost, low delay, low power consumption. Tried it in the British Cordless Telecommunication (CT) scheme called CT2. Worked rather well, but lost, some say on a technicality, to this fellow." The salesman fingered an adaptive differential pulse code modulation (ADPCM) codec chip. "International standard you see, sir. Takes more power than ADM and is more vulnerable to transmission errors, although it has marginally better performance in error-free conditions." Percy murmured that he had read about error-free conditions in technical fairy stories.

"Don't misunderstand me, ADPCM works quite well with moderate error rates. It's just not as

RAYMOND STEELE is
Professor of Communications
in the Department of
Electronics and Computer
Science at the University of
Southampton.

robust as ADM. They're using 32 kb/s ADPCM in the Digital European Cordless Telecommunications (DECT) system, at multiple access rates of 1152 kb/s without, I am told, either channel coding or channel equalization."

A few photons of amusement reflected from Percy's lugubrious eyes, "Let's hope their excess path delays don't exceed half a microsecond with that bit rate."

The loquacious J. N. D. Codit moved on to the speech codecs he understood. "This works at only 16k b/s. Smarter than ADPCM, it's a subband codec (SBC), the poor relative of the more complex adaptive transform codec (ATC)."

"And a lot more expensive."

"Of course, you don't get the bits per Nyquist sample down for nothing. I shudder to think how many million dollars and how many years it takes to reduce the number of bits per sample by only one bit while maintaining speech quality. You see, SBC is really a number of codecs. The speech is split into subbands, and the signal in each subband is adaptively coded, with its bits per sample depending on the actual block of speech being coded." He smiled, "However, you might not want SBC. It seems to have gone out of fashion. Came after ADPCM, was respected in the 1980s. Of the six contending codecs for the pan-European GSM system, four were SBC codecs. As the codecs were required to operate at 16 kb/s, SBC looked a good bet, particularly as its coding delays were relatively short. But it lost the beauty contest." Percy noticed the flicker of sadness on J. N. D Codit's face in the multiple mirror quadrature filters.

"What beat it?"

"The candidate codecs that offered more. There was multi-pulse excited linear predictive coding, and regular pulse excited with long term prediction (RPE-LTP) coding. The opinions of the judges — we call it a mean opinion score (MOS) — was highest for RPE-LTP at 3.54."

"Is that high?"

"It could have been 5.0, but who's counting. It won. Here it is. A good performer with drawbacks." He dropped the codec into the palm of Percy's hand.

"Which are?" Part of Percy's mind continued to dwell on haddock and french fries while his tummy rumbled in sympathy.

"Well it's complex, and relatively power hungry. Redundancy has been squeezed from it so it needs FEC coding. The speech sequence is first pre-emphasised, ordered into blocks of 20 ms duration and then Hamming windowed. STP analysis filtering follows, which involves calculating the logarithmic area ratios (LARs). Then comes the LTP analysis where the LTP prediction error is minimized. This complex process," he paused, put off his stride by ambient noise interference as Percy's tummy made another plea, "involves finding the LTP delay D and gain factor G. The LTP residual is weighted and decomposed into three candidate excitation sequences. The energies of these sequences are identified and the one with the highest energy is selected to represent the LTP residual. The pulses in the excitation sequence are normalized to the highest amplitude and quantized." He paused, detecting that his customer's interest was experiencing Rayleigh fading.

"You want to know about the multiplexing of the bits for transmission and RPE-LTP decoding?"

Percy's eyes glazed over, "No thank you. Tell me what else you have on the shelf."

"I do a nice range in code excited linear prediction (CELP) codecs. Would you like to know how they work?" The salesman, Codit, had the insensitivity of a poor receiver.

"Not today. Just tell me the rates."

"This little beauty works at 6.5 kb/s. We call it a half rate codec, as the GSM codec eventually went at 13 kb/s. Then there are these at 8 kb/s, and if you don't mind a little less quality we have this 4 kb/s one. Of course, if you are really desperate we could go right down to vocoder rates with their somewhat synthetic speech quality." His voice petered to a whisper. "Perhaps," his speech increased by 20 dBs, "sir does not require a bespoke speech codec. Would he like to visit our tailored codec department where we can fit you up with vector quantization, tree coding, trellis coding, wavelet filtering, entropy coding, just noticeable distortion (JND) coding — we even have a line in fractals, you know, snow flakes and all that?"

Percy staggered out of the speech codec shop, rumbling, dazed and baffled, into Sub-System Street. He wandered past the FEC codec shop with its range of BCH, Reed Solomon, and convolutional code chips, and passed the House of Multilevel Modems, not even sparing a glance into the brightly lit Equalizer Emporium. A spent man, he left mobile radio communications and went into silicon real estate.

This true story took place in the East-End of London. The shops on Sub-System Street are now closed due to the recession, and speech coding engineers are selling cut price FISH and CHIPS on the streets of London. There is not much to salvage from our tale, save a moral. When shopping for a speech codec, be sure you know what you need.

We will now consider some of the many factors that go into the design of a PCS, and how these factors affect the choice of the speech codec. The "Percy Comms," after reading this discourse will, we hope, be better placed to ask their speech coding colleagues the right questions when they select the codec(s) for their PCS.

Setting the Scene

We are concerned here with the role of speech codecs in PCS, and how the speech codec designer might be advised to view PCS. There are many difficulties in designing a PCS, and most can be traced to radio propagation. A complex subject, we will make just a few pertinent statements on mobile radio propagation and advise those who wish to know more to consult references 1 to 3. At their best, mobile radio channels can approach an additive Gaussian white noise channel. Generally the received signal level is subjected to fades, which may be up to 40 dBs in flat fading channels. These deep fades often result in error bursts. The data is subject to dispersion in frequency selective channels, resulting in inter-symbol interference and the need for channel equalization. Worse, the fading conditions may change rapidly if the mobile is moving or objects near to the mobile are in motion, necessitating frequent estimations

A tray of pulse code modulation chips, half hidden under a half-price summer sale sign, was placed on the counter. "This," said the young man, "is the model-T Ford of speech codecs."

of the channel impulse response. There are sources of interference, such as from other mobiles using the same channel in different locations, adjacent channel noise, ignition noise, and so forth, as well as receiver noise. The mobile radio channels do have one good characteristic, and that is that they are linear.

Therefore, transmitting digital speech over mobile radio channels is not a trivial task. A PCS designer is always conscious of the time variant nature of the mobile channel and the channel signal-to-noise ratios (SNRs) and signal-to-interference ratios (SIRs) that his speech transmission will face. Transmission errors are inevitable. If the speech codec is not robust, then methods to prevent transmission errors causing devastation of the recovered speech is high on the designer's agenda.

By contrast, many speech codec research workers conceive codecs to put in the "codec store." Their latest algorithms are often solutions waiting for applications. The PCS engineer shops around for the best codec to solve his need. Fundamentally, the speech codec designer seeks ways to change the speech to a form that requires fewer bits to code. At the receiver the reverse process is performed to recover the original speech, as in waveform coding, or to generate signals that can deceive the brain into believing the speech is undistorted, as in analysis-by-synthesis coding.

Does this mean that the processed signal being coded with fewer bits is more vulnerable to transmission errors? If we inflict 56 kb/s log-PCM and 40 kb/s ADM with the same bit error rate, we find that the ADM sounds much better. We can use other lower bit rate codecs having toll quality in error-free conditions, and find that they also perform better in the presence of errors than log-PCM. So, how far may we decrease the speech coding rate and simultaneously improve the robustness to transmission errors? It appears not too far. As we exploit the parameters of the glottis, the vocal tract, and the auditory mechanism, the bit rate decreases, but some bits are now carrying vital information and if corrupted during the transmission process, the reversing algorithm will produce unacceptable distortion. To transmit low bit rate encoded speech we must do a systematic search to find the degree of perceptual degradation in the recovered speech, due to each bit in every frame being erroneously regenerated at the receiver. It is a common practice to group bits according to their sensitivity to errors, and to use different channel coders on each group in order to achieve toll quality speech for transmission over the hostile mobile radio channels.

The speech coding engineer removes as much redundancy as possible by a myriad of techniques [4, 5], while the channel coding engineer puts back redundancy in a controlled way [6]. The combined low bit rate codec and the forward error correction (FEC) codec yield an acceptably low bit rate, while the majority of bit errors incurred due to transmission over the radio channel can be corrected at the receiver to yield toll quality speech. Notice that the FEC is chosen to combat the BER, but the BER depends on the PCS design. More specifically, having opted for a method of modulation, the BER depends on the signal-to-noise ratio (SNR) at the receiver, the in-band interference, the fading characteristic of the

channel, etc., while these factors are dependent on the higher order organization of the PCS.

If the fading rate is sufficiently fast, as in vehicular mobiles, a long FEC block code can be used to bridge one or more fades. The duration between fades depends on the propagation frequency and speed of the mobile. Provided the burst errors resulting from the fades are a small fraction of the bits used in the FEC coding, they will be corrected. Long FEC block codes are complex to implement, cause delay, and decrease the digital speech throughput. An alternative approach is to use a shorter FEC coding length, and interleave the FEC coded digital speech prior to transmission. Provided the interleaving depth covers the duration of a few fades, bursts of errors will be randomised by the de-interleaving process. The shorter FEC codec only needs to correct random errors. By this means a short FEC code can have a similar performance to a longer FEC code, have similar delays, and be significantly more simple to implement [3].

The specification of the speech codec for a PCS in terms of its delay, coding rate, BER and speech quality will only be made when many issues have been addressed. These will depend on the type of PCS, the network protocols, the number of users to be supported per base station (BS), cell size, multiple access method, modulation, FEC coding, diversity method, and so on. Some aspects in the design of a PCS are more important than others, and spectral efficiency is of prime importance. We now consider the factor that is the most influential on spectral efficiency, namely, the cell size.

Cell Size

The regulators allocate a specific bandwidth of radio spectrum to an operator. In cellular radio this bandwidth is divided up into subbands, and each subband is used by a base station (BS). The BSs communicate by radio with mobiles in their cell, where a cell is the coverage area in which the received signal strength is sufficiently high to support two-way communications. A group of BSs that collectively use all the allocated bandwidth is referred to as a cluster. Clusters of base stations are now tessellated with each cluster, reusing the same frequency band. Mobiles in different clusters have the potential to interfere with each other. This cochannel interference is maintained at an acceptably low level by making the distances between interfering mobiles and BSs sufficiently large. All cellular systems are essentially cochannel interference limited.

If the bandwidth allocated is capable of supporting say, 100 channels, and if a country had 1000 clusters, the total number of channels nationwide is 10^5. It is apparent that if the cells are reduced in size with a concomitant reduction in radiated power levels, then more clusters can be supported. For 10^6 clusters the same allocated bandwidth now supports 10^8 users nationwide. In practice we have a mixture of cell sizes ranging from large cells in rural areas to small cells in urban areas.

Second generation mobile systems include digital cellular networks and cordless telecommunication (CT) networks. The former will offer wide area coverage, e.g., all of Western Europe, while

317

CT networks are conceived to handle dense tele-traffic levels in office environments. The terminology relating to cells in mobile radio is confused, with new names being coined all the time. Let us crudely define cell size in terms of the maximum dimension from the BS to the cell boundary. Starting with the smallest cell, which we will call a picocell (a few meters), we proceed to the nanocell (up to 10 m), the nodalcell (up to 300 m, an isolated cell acting as a high capacity network node), the microcell (10 to 400 m for pedestrians, 300 m to 2 km for vehicles), the minicell (500 m to 3 km), the macrocell (1 to 5 km), the large cell (5 to 35 km), the megacell (20 to 100 km), and the largest cell of all, the satellite cell (> 500 km). In first generation mobile networks we use macrocells and large cells, with some satellite cells. Second generation systems will add microcells and minicells, while the third generation networks for the turn of the century will have all of the above cells, often in a tier arrangement, for example, clusters of street microcells, with each cluster overlaid by a macrocell. The macrocells will in turn be clustered, and will essentially act as a support for the microcells when used in this tier arrangement [3].

As the cell size changes, so does the radio propagation phenomena. A microcell is not just smaller than a macrocell, it has better propagation properties for the frequencies of interest, 800 MHz to 3 GHz. With a macrocell, wide area coverage is obtained by siting antennas high above the urban skyline. Down in the street canyons where the mobiles travel, the propagation can take many forms, from exhibiting flat Rayleigh fading to frequency selective fading. In microcells the antennas are located below the urban skyline, e.g., on lamp posts or on the side of buildings; the transmitted power is much lower and the fading is less severe. If flat fading occurs it will be Rician, and often the channel will be close to Gaussian. If the mobile radio channel becomes dispersive, the average delay between the reception of the first path and the last path will be much smaller than for the macrocells. These favorable propagation phenomena in microcells contribute towards a greater density of bits/sec per unit area compared to macrocells.

Armed with these brief remarks concerning cell size in PCS, we can now discuss why cell size has a dramatic effect on the choice of speech codec. We have argued that the smaller the cell the greater the bandwidth per unit area that is available. This means the higher the teletraffic, measured in Erlangs, that can be accommodated. The unit of spectral efficiency in mobile radio communications is Erlangs/Hz/m^2 and this goes up as the cell size goes down. We emphasize that spectral efficiency is not measured in bits/sec/Hz, although this is an important system parameter.

When the cell size is sufficiently small such that the bearer rate can be high, a mobile may communicate to its BS using a simple high rate speech codec. Indeed, even PCM can be used [7, 8]. In CTs [9] with their picocells, nanocells and microcells, 32 kb/s ADPCM codecs are used, and although the multiplexed transmitted bit rate may exceed a megabit per second, the speech performance may be acceptable, even without channel coding and channel equalization. Cellular systems on the other hand operate with much larger cells

which have lower quality radio channels. FEC coding is necessary, and often channel equalization is required. The consequence is that the speech codecs are required to operate at lower bit rates, currently 6.7 to 13 kb/s, and there is pressure to halve these rates.

In mobile satellite communications the cell sizes are vast and the assigned Hz/m^2 relatively tiny compared to other cells. To accommodate realistic numbers of users the speech rate must be of the order of 3 kb/s, requiring the use of vocoder techniques. FEC coding will increase this rate to 4 to 10 kb/s.

Multiple Access Methods

There are three basic multiple access (MA) methods: frequency division multiple access (FDMA), time division multiple access (TDMA), and spread spectrum multiple access (SSMA). The latter has two versions, known as frequency hopping SSMA and direct sequence SSMA. The first generation mobile radio systems used FDMA, where each user is given a unique frequency subband for the duration of the call. The second generation systems opted for TDMA, where N users are organized in a time frame and transmitted on one radio carrier. FDMA of the TDMA carriers ensues. Direct sequence SSMA, usually referred to as CDMA, is also a second generation system in which all the users making a call use the total bandwidth all of the time. The users are separated from each other at the receiver by virtue of their unique codes.

The MA method allows many mobile users to access the network via the radio medium. However, the choice of MA has powerful ramifications on the selection of the speech codec because after cell size, MA is the next most important factor in determining spectral efficiency (η). By operating with few cells per cluster, η is increased. The best you can do is to operate with single cell clusters. This means that every cell uses the same frequency band, i.e., the entire allocated bandwidth. Cochannel interference will occur, and the values of signal-to-interference ratios (SIRs) required for satisfactory operation using FDMA and TDMA cannot be achieved. However, CDMA was conceived to combat jamming, and can operate in single cell clusters, albeit with some degradation in its performance due to intercellular interference.

The TDMA system originally known as IS-54 and now called D-AMPS was conceived to increase the capacity of the first generation analog system AMPS. In D-AMPS three digital radio channels in a TDMA mode are housed in the same 30 kHz that accommodates one AMPS analog speech FM channel. This ability of D-AMPS to support these extra channels in the same channel spacing was accomplished by using an 8 kb/s VSELP speech codec and a $\pi/4$-DQPSK modem supporting two bits per baud. The bit rate (after the addition of system overheads) is 48.6 kb/s transmitted at 24.3 kBd.

The European GSM system is much more complex than D-AMPS, and was introduced as a new system unrelated to any first generation system. Conceived for large cells with BS to MS separation distances up to 35 km, it is nevertheless able to operate in street microcells where it is

The choice of MA has powerful ramifications on the selection of the speech codec because after cell size, MA is the next most important factor in determining spectral efficiency.

undoubtedly over-engineered. With 8 channels per carrier in a TDMA format, the transmission rate is 271 kb/s. In large cells at this rate dispersion occurs. GSM combats this effect by means of FEC coding and channel equalization. Adding FEC coding to the digital speech data, while trying to contain the transmission rate and obtain a reasonable number of users per carrier, forced the GSM designers to select the lowest bit rate speech codec that offered acceptable quality. At the time the selection of the speech codec was made the RPE-LTP codec operating at 13 kb/s was considered to be the best choice.

The first so-called PCN system is the DCS 1800, which is currently being deployed in the United Kingdom. It is a close relative of GSM, having most of its parameters. Two important exceptions are that it operates in the 1800 MHz band, and it uses lower transmission power to work in smaller cells.

The second-generation Qualcomm CDMA cellular system, called IS-95, overcomes the traditional difficulties associated with CDMA, namely, that of finding sufficient codes for millions of users, the provision of accurate synchronization, and the ability to ensure that all the users' signals are received at the BS with essentially the same power. Each user's data symbol of duration T is converted into a binary sequence where the duration of each pulse, called a chip, is T_c, and $T >> T_c$. The binary sequence is a code unique to the user, and the ratio of T/T_c is called the processing gain G. Binary phase shift keying (BPSK) is used, and $G = W/R$, where W is the bandwidth required to transmit the chip sequence and R is the data rate. The received signal at a base station may have U users, and each user experiences jamming from (U-1) mobiles within the cell, and interference from other mobiles in adjacent cells. A particular user is able to suppress this jamming because of its processing gain.

To the PCS designer CDMA is philosophically attractive. As eloquently described by Viterbi [10], the speech codec should be designed to yield the lowest bit rate for toll quality speech without regard to its innate robustness to transmission errors. Knowing the modulation to be used, G, and U, the signal-to-jamming noise ratio (SJR) can be found to ensure that the BER is below a level that provides speech with imperceptible degradation. The coding rate of the FEC codec can then be determined to achieve the target SJR. The bandwidth expansion of the data due to the FEC is small compared to that resulting from the CDMA coding, enabling the designer to use powerful error correcting codes. The CDMA waveform is noise-like due to the pseudo-random nature of the codes. Consequently in CDMA the communications are conducted in the presence of cochannel interference having Gaussian-like noise properties. So not only is CDMA spectrally efficient, it encourages the speech coding engineer to forget (nearly) the hostile mobile channel and concentrate on achieving the best low bit rate speech codec design in the knowledge that the CDMA engineer will FEC code it with sufficient power to achieve toll quality speech. The speech encoder designer need not bother about ordering the coded bits in the order of their bit sensitivity. With FEC correcting power to spare, all the speech bits can be coded with the same power.

Although the speech encoding engineer need not be too concerned regarding the innate robustness of his speech codec if CDMA is to be used, there is a requirement that this codec only transmits at the minimum rate to yield toll quality speech. This is also necessary with other forms of multiple access because, if the transmissions are curtailed so is the interference to other mobile users. Voice activity detection (VAD) is therefore a vital sub-system in PCS. The VAD can be used crudely to prevent transmissions during silences, but it is better to use the VAD to ratchet the encoding rate to suit the activity in the encoding process. However, the effect of VAD on cochannel interference is different in TDMA and FDMA where it benefits particular individual users, compared to CDMA where any lowering of interference is shared among all users.

A particular form of TDMA, known as packet reservation multiple access (PRMA) uses the VAD in an on-off capacity, avoiding transmissions in silences and even in intra-word silences [11]. PRMA supports significantly more users than there are channels. Provision has to be made to convey the essential features of background ambient noise so that the received speech sounds natural.

While CDMA has significant advantages in cellular networks within a PCS, it is not necessarily the best in all situations. For example, high capacity nodes which convey very high rates to users within a microcell, or nodalcell, would probably be better off employing FDMA. TDMA works well in cellular, as witnessed by GSM and D-AMPS, although this writer prefers CDMA. However, in microcellular networks and in CTs, where the cochannel interference problems are radically different, it is unclear whether TDMA with dynamic channel allocation (DCA) [12] will be out-performed by CDMA when operating in the same bandwidth. Ideally CDMA should increase its bandwidth in city street microcells and in office environments to exploit the multipath while offering an impressive high bit rate service.

After the analog CT1 service the digital CT2 service was introduced. Operating over short distances, typically 100 m, a CT2 user can cordlessly access a CT2 BS and hence the PSTN using the telepoint service. The CT2 telepoint portable is restricted to making calls, although facilities are provided by some operators for a user to log on to a CT2 BS in order to receive calls. In a PBX environment, a CT2 portable has all the facilities associated with a fixed phone. There is also the DCT 900 system which supports 8 channels/carrier compared to CT2's one channel/carrier. However, DECT supports 12 channels/carrier, and is likely to be widely deployed in the future. All CTs use the G.721 ADPCM codec. This codec was preferred because it is an international standard and the small cell sizes can support its bit rate. Table 1 lists parameters of some of the cordless and cellular systems, and the reader is referred to [13] for further details.

Trade-offs

Determining the bit rate used in a PCS is a complex task. In making the cells smaller and/or using fewer cells per cluster, the spectral efficien-

System	GSM	DSC1800	Qualcomm CDMA	D-AMPS	CT2 CAI	DECT	Units
Forward band	935-960	1805-1880	869-894	869-894	864.15-868.05	1880-1900	MHz
Reverse band	890-915	1710-1785	824-849	824-849	864.15-868.05	1880-1900	MHz
Multiple access	TDMA	TDMA	CDMA	TDMA	FDMA	TDMA	
Duplex	FDD	FDD	FDD	FDD	TDD	TDD	
Carrier spacing	200	200	1250	30	100	1728	kHz
Channels/carrier	8	8	55-62	3	1	12	
Bandwidth/channel	25*	25*	20*	10*	100	144	kHz
Modulation	GMSK	GMSK	QPSK	π/4-DQPSK	GFSK	GMSK	
Modulation rate	271	271	1228	48.6	72	1152	kb/s
Voice rate	22.8	22.8	19.2/28.8	13	32	32	kb/s
Speech codec	RPE-LTP	RPE-LTP	CELP	VSELP	ADPCM(G721)	ADPCM (G721)	
Uncoded voice rate	13	13	1.2 to 9.6	8	32	32	kb/s
Control channel name	SACCH	SACCH	FACCH	SACCH	D	C	
Control channel rate	967	967	800	600	1000/2000	6400	b/s
Control message size	184	184	1	65	64	64	b
Control delay	480	480	1.25	240	32/16	10	ms
Peak power (mobile)	2-20	0.25-2	0.6-3	0.6-3	0.01	0.25	W
Mean power (mobile)	0.25-2.5	0.03-0.25	0.6-3	0.6-3	0.005	0.01	W
Power control	Yes	Yes	Yes	Yes	Yes	No	
VAD	Yes	Yes	Yes	Optional	No	No	
Handover	Yes	Yes	Yes	Yes	No	Yes	
DCA	No	No	N/A	No	Yes	Yes	
Minimum cluster size	3	3	1	7	N/A	N/A	

* Per duplex channel

■ Table 1. *Comparison of second generation cellular and cordless systems.*

cy η is increased. By decreasing the bit rate of the speech codec we get more users per bandwidth. It is desirable to use channel coding, even in flat fading microcellular environments. This is because the FEC codec can be used like the general on the battlefield, directing operations. For example, the FEC can decide when to use post-enhancement procedures on speech that is badly corrupted due to the FEC's inability to correct sufficient transmission errors. Handovers, i.e., when the mobile is switched from one BS to another, or communicates with two BSs simultaneously, can be initiated by the FEC codec, and the FEC can also be part of the decision process of adapting the transceiver to the changing channel conditions.

In street microcells having dimensions of some 100 m from their microcellular BSs the transmitted bit rate supported without channel equalization is approximately 1 Mb/s. As the multiplexed transmission bit rate increases, dispersion will increase and, to contain the BER, channel equalization will be used in TDMA, and path diversity reception will be employed in CDMA. FDMA mobiles will not be in trouble because their individual communication rates are relative-

ly low. When there is a need to increase the FEC coding power either the same speech codecs are used resulting in a higher transmission rate per user, or lower bit rate speech codecs are deployed to balance the extra bit rate due to the FEC codec.

If we repeated the above arguments for much larger cells the only difference would be that the effects happen at lower transmission rates. Even at rates as low as 25 ksymbols/s, dispersion can arise if the cell extends for many tens of kms. Thus large cells have relatively little bandwidth per unit area, have hostile propagation conditions and can only support a limited number of mobiles by using low bit rate speech codecs, powerful FEC coding and must exploit the large dispersion that occurs. On the other hand, by citing microcellular BSs in offices, the opposite situation occurs, with transmissions in excess of 4 Mb/s not experiencing dispersion. With a limited number of users, office workers will not only be offered roaming within the building while making telephone calls, but may also be able to have video mobile communications.

Modulation also has a considerable impact on the speech codec. By changing the modulation to support more bits per baud, the bandwidth of the modulated signal decreases. For the same transmitted power, increasing the number of bits per baud increases the BER for the same SNR and SIR. This has a knock-on effect in that the requirement for high SNR effects the cell size, while the SIR influences the choice of the number of cells per cluster and the method of modulation and FEC to be used. Consequently if the PCS designer needs to accommodate more mobile users, he will seek to trade the complexity of his sub-systems in the transceivers, with the higher features of cell size, cluster size, multiple access, system protocols, etc. [12]. These latter features affect infrastructure costs and capacity, while changes to the transceivers influence the costs of the mobile equipment. The

users will pay either way. If the portable is inexpensive to buy, the user charges will be higher to pay for the higher network costs, and vice versa.

The speech codec is a small but vital component in PCS. Even in third generation PCS to be deployed at the turn of the century, speech transmissions are likely to be the most used form of communications. For a given transmitted symbol rate the designer of the transceiver can decide which sub-system, or sub-systems, are given the "hot potato," the difficult part or parts to design. If the complexity falls on the modulator by arranging for it to go multilevel, such as 16-level star QAM [14], then for the same bandwidth the FEC coded digital speech rate can increase some 2.4 times when consideration is given to the smoother filtering requirements. FEC coding would still be necessary as QAM is more prone to errors than, say, GMSK. However, the FEC and speech codec could share the spoils, allowing the speech rate to rise, making it easier for the speech coding engineer to design a toll quality codec.

We have noted that by using microcells, CT operates with ADPCM, and by deploying large cells GSM uses a RPE-LTP codec, while D-AMPS has a VSELP codec, and Qualcomm CDMA uses a CELP codec. While digital cellular systems use macrocells, they must also use low bit rate coders and FEC coding.

The CTs operate in small cells and CT transceivers are simple. The CT is a consumer product, designed to be small, light, and inexpensive. Transceivers used in cellular environments tend to be a professional product, for use over very large areas. The complexity of the systems originates in their need to use large cells. When they operate in microcells they are over-equipped.

The Future

We may anticipate that the first intelligent multimode terminal will not be mobile. It will be a domestic high performance personal computer that will be both a communication and entertainment center in the home. It will accept B-ISDN from the fixed optical fiber network, and will also be able to emulate via its radio interfaces the different existing mobile radio transceivers. This would mean that the first versions of this terminal would be able to perform as stationary CT and cellular transceivers [12].

The advent of the third generation PCS is expected at the turn of the century. It will have a number of different radio interfaces, e.g., satellite mobile and microcellular. If these interfaces are to be optimally designed, the transceiver will need to be an intelligent multimode portable. The portable will need to reconfigure itself according to the radio interface being used, the geographical environment, the rapid variations of the radio channel, the local teletraffic loading, and the service required.

The implications for the speech codec designer when the transceiver is required to adapt to radically dissimilar situations is profound. In effect, the transceiver requires a number of different speech codecs, or a speech codec conceived in a modular form that can be rapidly reconfigured under software control using fast decisions from the portable and strategic deci-

List of Abbreviations

ADM	Adaptive delta modulation
ADPCM	Adaptive differential pulse code modulation
BCH	Bose Chandhuri Hocquenghem code
BER	Bit error rate
BPSK	Binary phase shift keying
BS	Base station
CDMA	Code division multiple access
CELP	Code excited linear prediction
CT	Cordless telecommunications
CT2	CT system two
D-AMPS	Digital AMPS (formerly IS54)
DCA	Dynamic channel allocation
DECT	Digital European cordless telecommunications
DQPSK	Differential quadrature phase shift keying
FACCH	Fast associated control channel
FDD	Frequency division duplex
FDMA	Frequency division multiple access
FEC	Forward error correction coding
G	Processing gain in CDMA
GFSK	Gaussian frequency shift keying
GMSK	Gaussian minimum shift keying
GSM	Global system of mobile communications
ISDN	Integrate service digital network
JND	Just noticeable distortion
J. N. D. Codit	A salesman
LAN	Local area network
LARs	Logarithmic area ratios
log-PCM	Logarithmic pulse code modulation
LTP	Long term prediction
MA	Multiple access
PBX	Private branch exchange
PCN	Personal communication networks
PCS	Personal communication system
Percy Comms	An engineer named after personal communications
PRMA	Packet reservation multiple access
PSTN	Public switched telephone network
QAM	Quadrature amplitude modulation
RPE-LTP	Regular pulse excited LPC with long term prediction
SACCH	Slow associated control channel
SBC	Subband coding
SIR	Signal-to-interference ratio
SJR	Signal-to-jamming ratio
SNR	Signal-to-noise ratio
SSMA	Spread spectrum multiple access
STP	Short term prediction
TDD	Time division duplex
TDMA	Time division multiple access
VAD	Voice activity detection
VSELP	Vector sum excited linear predication

sions from the PCS network conveyed via the communicating BS. The transceiver will be programmed to opt for the codec that is sufficient to do the function required, and with the minimum power consumption. Most of the codec chip will be in hibernation until activated.

The design of a codec that can adapt to a wide range of ambient noise, audio bandwidths up to 15 kHz, while yielding the lowest bit rate to obtain the required service quality and the minimum power consumption from the battery, is indeed a challenge. It pales to insignificance, however, compared to making the remainder of the transceiver fully adaptive. Fortunately microelectronics is developing at a rate that continually allows more complexity to be introduced. Considering how far we have come from a PCM codec to a CELP codec shows that we are not unduly concerned about adding massive complexity, with the proviso that it does not cost a great deal more and the enhancement in performance is significant.

Each year the latest speech codec model is proudly displayed in the speech codec show room. Other Percy Comms will visit the shop, not looking for fish and chips, but for the right codec to suit a variety of speech and high fidelity audio services. Let us hope that J. N. D. Codit will have the right model by the turn of the century when the third generation PCN should be deployed.

References

[1] W. C. Jakes, Microwave Mobile Communications, (John Wiley and Sons, New York 1974).
[2] J. D. Parsons, The Mobile Radio Propagation Channel, (Pentech Press, London, 1992).
[3] R. Steele, Mobile Radio Communications, (Pentech Press, London, 1992).
[4] N. S. Jayant and P. Noll, Digital Coding of Waveforms, (Prentice-Hall, 1984).
[5] N. Jayant, "Signal Compression: Technology Targets and Research Directions," IEEE JSAC, vol. 10, no. 5, June 1992, pp. 796-818.
[6] L. Hanzo, R. Steele, and P. M. Fortune, "A subband coding, BCH coding and 16-QAM system for mobile radio speech communica-

tions," IEEE Trans. on Veh. Tech., vol. 39, no. 4, Nov. 1990, pp. 327-339.
[7] R. Steele, C-E. Sundberg, and W. C. Wong, "Transmission Errors in Companded PCM over Gaussian and Rayleigh Fading Channels," BSTJ., vol. 63, no. 6, July/Aug. 1984, pp. 955-990.
[8] C-E. Sundberg, W. C. Wong, and R. Steele, "Logarithmic PCM Weighted QAM Transmissions over Gaussian and Rayleigh Fading Channels," IEE Proc., vol. 134, pt. F, no. 6, Oct. 1987, pp. 557-570.
[9] W. H. W. Tuttlebee, Cordless Telecommunications in Europe, (Springer-Verlag, 1990).
[10] A. J. Viterbi, "Wireless Digital Communication: A View Based on Three Lessons Learned," IEEE Comm. Mag., Sept. 1991, pp. 33-36.
[11] D. J. Goodman and S. X. Wei, "Efficiency of Packet Reservation Multiple Access," IEEE Trans. on Veh. Tech., vol. 40, no. 1, Feb. 1991, pp. 170-176.
[12] R. Steele and J. E. B. Williams, "The Intelligent Multimode Mobile Portable for Third Generation Networks," IEE Electronics & Communications J., vol. 5, no. 3, June 1993, pp. 147-156.
[13] Special Issue, "PCS: The Second Generation," IEEE Commun. Mag., vol. 30, no. 12, Dec. 1992.
[14] W. T. Webb, "QAM, The Modulation Scheme for Future Mobile Radio Communications?" IEE Electronics & Communications J., vol. 4, no. 4, Aug. 1992, pp. 1167-1176.

Biography

RAYMOND STEELE [SM '80] received a B.Sc.in electrical engineering from Durham University, England in 1959 and Ph.D. and D.Sc. degrees from Loughborough University of Technology, England in 1975 and 1983, respectively. Before attaining his B.Sc., he was an indentured Apprentice Radio Engineer. After R & D posts with E. K. Cole, Cossor Radar and Electronics, and Marconi, he joined the lecturing staff at the Royal Naval College. He move to Loughborough University in 1968 where he lectured and directed a research group in digital encoding of speech and picture signals. During the summers of 1975, 1977, and 1978 he was a consultant to the Acoustics Research Department at Bell Laboratories in the U.S.A., and in 1979 he joined the company's Communications Methods Research Department, Crawford Hill Laboratory. He returned to England in 1983 to become professor of communications in the Department of Electronics and Computer Science at the University of Southampton, a post he retains. From 1983 to 1986, he was a non-executive director of Plessey Research and Technology and from 1983 to 1989, a consultant to British Telecom Research Laboratories. In 1986, he formed Multiple Access Communications Ltd, a company concerned with digital mobile radio systems. He is the author of the book Delta Modulation Systems, editor of the book Mobile Radio Communications, and the editor of a series of books on Digital Mobile Communications. He and his co-authors were awarded the Marconi Premium in 1979 and in 1989, and the Bell System Technical Journal's Best Mathematics, Communications, Techniques, Computing and Software, and Social Sciences Paper in 1981. He has been a conference and session organizer of numerous international conferences and a keynote speaker at many international meetings. He is also a senior technical editor of the IEEE Communications Magazine, a Fellow of The Royal Academy of Engineering, and a Fellow of the IEE in the U.K.

The advent of the third generation PCS is expected at the turn of the century.

Coding speech at low bit rates

Advanced algorithms and hardware for voice telecommunications are paring bit rates by at least a factor of four, without losing intelligibility

New digital coding techniques are emerging that promise to dramatically enhance the applicability of voice communications and storage. The techniques allow more speech to be represented with a given number of binary digits, without losing natural voice quality.

The advanced coding techniques just becoming available will yield natural-sounding telephone speech at digital transmission rates of 16, 8, and eventually 4 kilobits per second, not just the 64 and 32 kb/s that have become international standards.

Although high-bandwidth channels and networks are becoming more viable, coding speech at low bit rates has retained its importance. One reason is the growing need to transmit speech messages with a high level of security over low data-rate channels such as radio links. Another factor is the desire for memory-efficient systems for voice storage, voice response, and "voice mail."

Low bit-rate coding of voice is critical for accommodating more users on channels that have inherent limitations of bandwidth or power—like cellular radio or satellite links. It can lend flexibility in the design of the evolving integrated-services digital network (ISDN), which will reduce communication signals—voice, graphics, video, or computer data—to the common denominator of binary digit sequences. In particular, low bit-rate voice coding can ease the transition to shared channels for voice and data. Further, the low bit rates can readily adapt voice messages for packet switching and help to make voice mail practical and popular. Encryption of sensitive messages can become more readily available to business, as well as to the military, and the capacity of recording devices like answering machines can rise dramatically.

With lucrative applications like these in the offing, speech processing laboratories are in a state of ferment. Developers are trying a wide variety of coding schemes. The industry is still immature, preoccupied with planning, standardization, and research. In fact, totally new coding concepts may still emerge from the laboratories to add to those that now look so promising.

Although the technology is in a state of flux, marketable low bit-rate products have already begun to appear; over the next decade, digital speech coding hardware will proliferate [see "Low bit-rate hardware: boxes, boards, and chips," p. 60]. It is likely that 16 kb/s will be accepted internationally by 1988 as a standard coding frequency, joining 64 and 32 kb/s. Bit rates around 8 and 4 kb/s may also eventually achieve the same status [see "Standardizing low bit rates," p. 63.]

Developers of digital speech coders are striving to optimize the interplay of four parameters: bit rate, quality, complexity, and delay time. As bit rate is reduced, quality naturally drops off, unless the complexity of the coding scheme (and the very large-scale integration chips embodying it) is increased. But complexity raises the cost, and in many types of coding it increases

Nuggehally S. Jayant   AT&T Laboratories

processing delay. (Delay, of course, is not a problem in voice storage applications, like voice mail.)

Developers can find encouragement in the fact that in the interplay between voice quality and bit rate, the fundamental limits on bit rate suggested by speech perception and information theory are fairly low. Some researchers believe that high-quality speech coding may eventually be practical at rates as low as 2 kb/s.

At the high rate of 64 kb/s, quality is not a problem. Straightforward pulse-code modulation (PCM) yields highly acceptable quality; in fact, few people can tell whether the voice at the other end of the telephone line has been transmitted digitally. The analog speech waveform is sampled, quantized, digitally encoded, and transmitted at 64 kb/s. At the receiving end, the process is reversed by a decoder.

The only special algorithm in 64-kb/s PCM is nonuniform quantization: using a nonlinear quantizer in which the fineness of quantization increases as the magnitude of the speech input decreases. This nonlinearity favors low amplitudes, which predominate in speech; it quantizes them with imperceptible quantization error. It also exploits a characteristic of human hearing—that large amplitudes mask quantization errors to some extent.

New coding strategies

At lower bit rates, however, more elaborate strategies are needed. The general function of these strategies is to scrutinize the

---

Defining terms

Linear predictive filter: a circuit that performs speech predictions based on linear operations.
Pitch prediction: a process in which the quasiperiodic nature of a voiced speech waveform is used to predict speech amplitudes during a given period, based on samples taken during the preceding period.
Predictive coding: a speech coding scheme based on the tendency of speech waveforms to follow predictable patterns.
Pulse-code modulation: a process in which binary codes are used to represent quantized values of instantaneous samples of a signal.
Quantization: a process in which the continuous range of values of an input signal is divided into subranges, each of which is assigned a discrete value.
Quantization error: error caused by conversion of a variable having a continuous range of values into a quantized form having discrete values.
Vector quantization: quantization of a group, or block, of samples instead of individual samples.
Vocoder: a speech digitizing system that reduces voice signals to a compact set of parameters that can be transmitted over communications systems of limited bandwidth.

0018-9235 86 0800-0058$1.00 1986 IEEE

Reprinted from *IEEE Spectrum Magazine*, pp. 58-63, Aug., 1986.

speech signal more carefully, to eliminate the redundancies in the signal more completely, and to use the available bits to code the nonredundant parts of the signal in a perceptually efficient manner. As the available bit rate is reduced to 32, 16, 8, and 4 kb/s, the strategies for redundancy removal and bit allocation need to be ever more sophisticated.

One method is linear prediction, which compresses a speech signal by estimating it as a linear function of past outputs of the speech quantizing system. The prediction error tends to be lower in energy than the original speech, and hence can be coded using fewer bits, for a given level of reconstruction error. Another approach is to use adaptive subband coding, which separates speech into frequency bands and allocates the available bits to suit the input speech spectrum and the properties of hearing.

These approaches can be combined with each other in a complementary fashion. Further, they can be combined with yet another strategy, vector quantization, in which a block of input is quantized all at once rather than sample by sample. The result is again a considerable savings in bits for a given level of reconstructed speech quality.

Systems using these techniques are waveform coders, as distinguished from vocoders. Waveform coders employ algorithms to produce at the system output an approximation of the input speech waveform.

Vocoders, in contrast, distill a compact description of the input and digitize only the parameters of this compact description. (This description is based in general on the notion of an excitation signal feeding a linear filter—a model that attempts to simulate the processes of excitation and modulation in the human vocal mechanism.) The standard result is an artificial-sounding output in which the words may be clearly understandable but the speaker is not always identifiable.

Vocoder quality is tolerated for secure military communications that must be carried at very low bit rates—4 kb/s and less. The Government standard LPC-10 has been a workhorse of U.S. secure voice communications for years. The coding algorithms for vocoders are not robust; they may break down when the background is noisy and when several speakers use the encoder simultaneously.

A promising application of vocoder speech is in voice-response services, in which the user encodes a message for later delivery. The encoding can be carried out by a careful process that takes into account vocoder sensitivities. Delivery of the message involves merely decoding at a predetermined quality.

Some of the newer approaches to low bit-rate waveform coders, however, borrow from vocoder technology. These hybrid coders—or "soft" vocoders—are more robust than true vocoders; the goal is to provide communication quality that is good enough for routine commercial service—for both transmission and storage—at bit rates of 8 kb/s and eventually 4 kb/s.

Quantifying quality

Subjective tests of experimental low bit-rate systems have demonstrated that users find waveform coder quality more than adequate. As might be expected, a 64-kb/s rate scores highly in such tests. But substantially lower rates, when implemented by advanced, high-complexity coders, get surprisingly favorable ratings [Fig. 1].

Subjectively, quality is measured by a mean opinion score (MOS) obtained by formal tests with human subjects. An MOS of 5 indicates perfect quality; a score of 4 or more represents high quality. (Telephone engineers call this level "toll-quality" in standard waveform coders when they also meet certain transmission specifications.)

An MOS exceeding 4 means that test subjects find the speech as intelligible as the original and free of distortion. MOS values of 3 to 4 are known as "communication-quality"; at these values distortion is present but not obvious, and intelligibility is still very high. MOS values between 3.5 and 4.0 represent a very useful level of communication quality. At the bottom of the scale is the "syn-

thetic quality" typical of vocoders; words are mostly intelligible, but the voice cannot always be identified. Also characteristic of synthetic-quality devices is a low level of robustness under conditions such as noisy inputs or multiple speakers.

Subjective MOS ratings supplement objective measurements like signal-to-noise ratio and in fact are often more revealing. For example, on the basis of signal-to-noise measurements, 16-kb/s coders, no matter how complex, are far inferior to 64-kb/s PCM coders. But in subjective measurements, the best 16-kb/s coders approach the higher bit-rate PCM coders in quality, and get MOS ratings very close to 4.

Even sophisticated low bit-rate schemes fall short of 64- and 32-kb/s coding in one respect: their quality drops sharply with successive coding-decoding stages, like those encountered in a combined analog-digital transmission path. The higher bit rates still yield high quality after as many as eight coding-decoding stages. Of course, in all-digital links, coded signals are decoded into analog form only once, and the multistage advantage of higher bit rates is less significant.

Clearly, reducing bit rates for coding requires greater scrutiny of the speech process. This means greater processing complexity, as well as longer processing delays.

Implementing algorithms on chips

Algorithms for digital coding are usually implemented in digital signal processors—combinations of general-purpose and special-purpose processing chips. The complexity of these processors is measured by the number of multiply-add operations required to code speech, usually expressed in millions of instructions per second (MIPS). As a rule of thumb, in the 64- to 8-kb/s range the number of MIPS increases by an order of magnitude when the coding rate is halved, for approximately equal speech quality [see table]. Encoders are typically much more complex than decoders.

An algorithm is generally considered practical if it can be implemented on a single chip. Low bit-rate coders are headed for wide use—perhaps one for every telephone—so they must be cheap. By this standard, most of the coding schemes listed in the table are currently practical, since a single general-purpose signal processor can handle up to about 10 MIPS. Even the exception—the stochastically excited linear prediction coder, with a complexity of 50 to 100 MIPS—may be implemented on a set of a few specially designed chips.

The table compares tradeoffs for representative types of speech coding algorithms. It shows the best overall match between complexity, bit rate, and quality. A coder type is not necessarily limited to the bit rate stated. For example, the medium-complexity adaptive differential pulse-code modulation coder can be redesigned to give communication-quality speech at 16 kb/s instead of high-quality speech at 32 kb/s. In fact, a highly complex ver-

Low bit-rate speech coding schemes compared

Coder type	Bit rate, kb/s	Complexity, MIPS	Delay, ms	Quality
Pulse-code modulation	64	0.01	0	High
Adaptive differential pulse-code modulation	32	0.1	0	High
Adaptive subband coding	16	1	25	High
Multipulse linear predictive coding	8	10	35	Communication
Stochastically excited linear predictive coding	4	100	35	Communication
LPC vocoder	2		35	Synthetic

sion can provide high-quality speech at the lower bit rate. Similarly, lower-complexity multipulse linear predictive coding can yield high-quality coding at 16 kb/s, and a lower-complexity stochastically excited linear predictive coder (LPC) can be designed if the bit rate can be 8 kb/s instead of 4 kb/s.

In specific cases, the values of complexity can differ substantially from the order-of-magnitude estimates in the table. For example, the stochastically excited linear predictive coder can be simplified to 50 MIPS with only a slight loss of speech quality.

The values for delay are rounded values and generally reflect the minimum for the various coders. A widely used version of the LPC vocoder, for instance, has much higher delay than that given in the table. This is because speech segments are much longer than the 10-millisecond value assumed in the table, and because it uses additional subsystems for parameter interpolation and error protection.

Cost, of course, is a tradeoff factor too, but it is hard to quantify in a table. The cost of coding hardware generally increases with complexity. However, advances in signal processor technology tend to decrease cost for a given level of complexity and, more significantly, to reduce the cost difference between low-complexity and high-complexity techniques.

Of course, as encoding and decoding algorithms become more complex, they take longer to perform. Complex algorithms introduce delays between the time the speaker utters a sound and the time a coded version of it enters the transmission system. These coding delays can be objectionable in two-way telephone conversations, especially when they are added to delays in the transmission network and combined with uncanceled echoes. Coding delay is not a problem if the coder is used in only one stage of coding and decoding, such as in voice storage. If the delay is objectionable because of uncanceled echoes, the addition of an echo canceler to the voice coder can eliminate or mitigate the problems. Finally, coding delay is not a concern if the speech is merely stored in digital form for later delivery.

Algorithms for prediction and bit allocation

Digital speech compression assumes that a portion of a waveform that starts out in a certain way is likely to continue in a known way for a while. This predictability makes it unnecessary to quantize the entire wave; instead the encoder and decoder use a prediction algorithm based on the statistical properties of the wave and quantize only the prediction error signal.

Exploiting this principle makes it possible to reduce the quantization to 4 bits per sample at an 8-kilohertz sampling rate in 32-kb/s coding, compared with 8 bits per sample at the same sampling rate in 64-kb/s coding. In the 32-kb/s scheme, the subjective quality of the transmitted voice is comparable to that at the higher rate, and there is no increase in processing delay. The scheme is called differential pulse-code modulation, because the input to the quantizer is the difference between a speech sample and the prediction of it. The prediction is linear; it is an estimate of the present speech sample from a weighted linear combination of past quantized samples.

The decoder in this scheme performs an inverse operation similar to integration. It adds the quantized difference signal to its own prediction of the current speech sample.

Although the principle of differential pulse-code modulation has been known for 30 years, it was not possible to standardize such a 32-kb/s coder until 1983, after efficient and robust algorithms become available. These adaptive algorithms are efficient in the sense that they adapt quantization and prediction synchronously at the encoder and decoder without transmitting explicit adaptation information. They are robust in the sense that they

Low bit-rate hardware: boxes, boards, and chips

At the IEEE International Conference on Acoustics, Speech, and Signal Processing in Tokyo last April, low bit-rate coding was a prime topic. A sampling of hardware exhibited at the conference clearly demonstrated the mature state of coding at 32 kb/s and the formative, rapidly changing nature of the technology at lower rates. Exhibitors showed hardware in varying stages of integration, ranging from prototype equipment boxes to circuit boards to very large-scale integration.

Japanese manufacturers were well represented at ICASSP, as might be expected from the site of the conference. A handful of U.S. companies also displayed hardware, but European manufacturers were not conspicuous, even though companies like Siemens AG of West Germany and NV Philips Gloeilampenfabrieken of the Netherlands are key players.

A digital signal processor chip was featured in a 16-kb/s adaptive subband coding system developed by Nippon Telegraph & Telephone Corp. The NTT system employs three subbands. The subband coding algorithm performs 12 million operations per second. This system has somewhat greater delay but slightly higher quality than NTT's 16-kb/s adaptive differential pulse-code modulation, also demonstrated.

Fujitsu Ltd. showed a 16-kb/s adaptive differential pulse-code modulation system based on an encoder and decoder on single chips. Fujitsu used its own digital signal processor chip in a 9.6- or 8-kb/s residually excited linear predictive coding scheme (which, like multiphase linear predictive coding, is a hybrid algorithm).

Hitachi Ltd. also showed hybrid coders based on single-chip digital signal processors for 8- and 9.6-kb/s coding rates. Hitachi also uses residually excited linear predictive coding.

Texas Instruments Inc. demonstrated voice coding algorithms using its TMS320 family of digital signal processors at rates from the conventional 64-kb/s PCM to 2.4-kb/s vocoding.

NEC Corp. showed a 16-kb/s multipulse linear predictive coder unit with a modem interface for easy connection to domestic and international networks and to multipoint links. Other NEC hardware included a 9.6-kb/s multichannel adaptive differential pulse-code modulation unit and 2.4- and 4.8-kb/s linear predictive coders for secure network communications.

Among other companies demonstrating hardware were Kokusai Denshin Denwa Co. Ltd. with a 16-kb/s noise-shaping adaptive differential pulse-code modulation system and a 4.8-kb/s hybrid coder, and Oki, with a 32- and 24-kb/s adaptive pulse-code modulation system that accommodates data at 9.6 and 1.2 kb/s as well as digital voice. The Oki system features single-chip encoders and decoders.

GTE Corp. announced in a technical session a 16-kb/s system using adaptive transform coding, a technique like adaptive subband coding.

In voice storage services, equipment is even more diverse. Small entrepreneurial companies abound, and the technologies are almost as varied as the players. Bit rates vary from 2.4 to 16 kb/s. Among the big telecommunications companies, Northern Telecom has developed prototype equipment using 16-kb/s multipulse linear predictive coding for voice storage. Signetics Corp. and Texas Instruments are prominent suppliers of vocoder chips, particularly for voice response applications.

AT&T Technologies Inc. is developing voice storage and forwarding equipment as well as voice response equipment. The latter hardware uses precoded speech messages, such as responses for automated travel reservations, and operates readily with more complex algorithms and lower bit rates since only a decoder is needed. AT&T uses a 16-kb/s adaptive subband coding system for voice storage and forwarding. For voice response, AT&T uses a 9.6-kb/s multipulse linear predictive coder. An AT&T affiliate, Conversant Systems, offers these voice storage and forwarding systems.
—Spectrum Staff

function reasonably well even in moderate bit-error environments—up to about one error per hundred bits.

Capitalizing on how people hear

Going from 8 bits per sample to 4 involves a fairly simple combination of adaptive quantization and adaptive prediction [Fig. 2]. Going from 4 bits per sample to 2 (16 kb/s coding) is much harder. It makes use of the periodic, repetitive nature of voiced speech and a characteristic of the way people hear called noise masking.

Periodicity of speech—the fact that people speak with a characteristic pitch frequency—allows pitch prediction and a further reduction in the level of the prediction error signal that needs to be quantized in differential pulse-code modulation. The amount of information that must be transmitted can therefore be greatly reduced—again without seriously decreasing speech quality. In general, all the complex forms of lower bit-rate coding can build on the hierarchy of strategies established in higher-rate coding by making the coder respond to the fine pitch structure of the amplitude spectrum.

The number of bits can be further reduced by noise masking. As far as a listener is concerned, a strong formant (vowel resonance) tends to mask the noise in its frequency locality as long as the noise is about 15 decibels below the signal. This means that a relatively larger coding error—the equivalent of noise—can be tolerated near formants, and that the coding rate can be correspondingly reduced.

For example, the more complex schemes for adaptive subband coding and adaptive differential pulse-code modulation use pitch prediction and noise-spectrum shaping. They use information about formants so the frequency of the quantization-error noise can be allowed to rise and fall according to the formant frequency. Noise shaping is achieved by means of error feedback or postfiltering techniques in differential pulse-code modulation. In adaptive subband coding, noise shaping is done by adaptive bit assignment. More bits are assigned to the perceptually more important frequencies, while an average of 2 bits per sample is maintained.

In adaptive subband coding, the speech band is divided into four or more contiguous bands by a bank of bandpass filters. The scheme employs a specific coding strategy for each band. At the receiving end, the subband signals are decoded and summed into a close reproduction of the original speech signal.

In the example of a four-band system with equal-width subbands, assuming that the sampling rate of each subband is one-fourth of 8 kHz, or 2 kHz, an average rate of 2 bits per sample implies a total rate of 8 bits per sample. In this case, the bit allocation appropriate to a speech segment rich in low frequencies could be, for instance, 5, 2, 1, 0 bits for the four subbands, in increasing order of frequency; the bit allocation appropriate for a segment with a predominance of high-frequency components might be 1, 1, 3, 3 bits per subband sample.

If the allocation of the bits to different bands is varied, the number of quantization levels can be controlled separately in each band, and the shape of the overall quantization noise spectrum can be controlled as a function of frequency. In the lower-frequency bands, where pitch and formant information have to be preserved, more quantizer levels are used on the average. However, if high-frequency energy predominates in the input speech segment, the adaptive algorithm will automatically assign a greater number of quantization levels for higher frequencies. Furthermore, in a subband coding system, the quantization noise of a

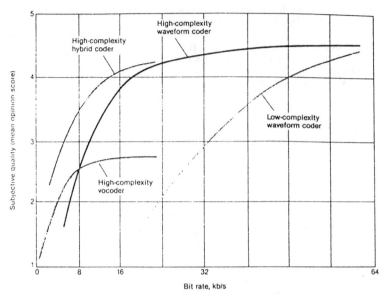

[1] The perceived quality of speech produced by waveform coders at varying digital coding rates (red and purple curves) was measured in subjective tests at AT&T Bell Laboratories with many listeners. The speech quality for vocoders and hybrid coders (green and blue curves) is estimated based on informal listening tests at several laboratories. A mean opinion score (MOS) of 5 is considered excellent, 4 is good, 3 fair, 2 poor, and 1 bad. In all tests, a standard telephone bandwidth of 3.2 kilohertz was maintained. A lower bandwidth, like 2.7 kHz, may be adequate for some uses. The bit rates will then be proportionally lower, but the MOS values (and more significantly, the intelligibility levels) will also be somewhat lower because of the lower bandwidth. If silences in speech are eliminated, the effective average bit rates will be as much as 30 percent lower than those shown for the same subjective quality.

band is kept within that band; a low-level speech input cannot be hidden by quantization noise in another band.

Hybrid coding allows lower data rates

Still lower rates of coding can be achieved by a kind of coding that feeds a carefully optimized excitation signal to a linear predictive filter—a hybrid approach that adopts much of the efficiency of traditional vocoding, yet is flexible enough to follow the subtle properties of the speech waveform. The approach uses high-quality waveform coding principles to optimize the excitation signal, instead of using the rigid two-state excitation of vocoding. With the most advanced hybrid techniques, only 1 to 0.5 bit per sample may be needed to code the speech signal, resulting in bit rates from 8 to 4 kb/s.

Optimizing the excitation and filter parameters is a big challenge to developers. Both types of parameters must vary with time to achieve coding quality and naturalness.

A good candidate for coding at 8 kb/s is multipulse linear predictive coding, in which a suitable number of pulses are supplied as the excitation sequence for a speech segment—perhaps 10 pulses for a 10-ms segment. The amplitudes and locations of the pulses are optimized, pulse by pulse, in a closed-loop search. The bit rate reserved for the excitation information is more than half the total bit rate of 8 kb/s. This does not leave much for the linear predictive filter information, but with the sophistication of vector quantization, the coding of the prediction parameters can be made accurate enough.

In vector quantization, the quantizer looks up in its memory the set that most closely matches a sequence of samples—a sequence 40 samples long, for instance. It chooses the address code stored in the memory's codebook for the matching set of sequences. The quantizer then transmits that address code to the receiver, rather than a sequence of 40 quantized samples.

If the codebook contains 1024 candidate output sequences, only

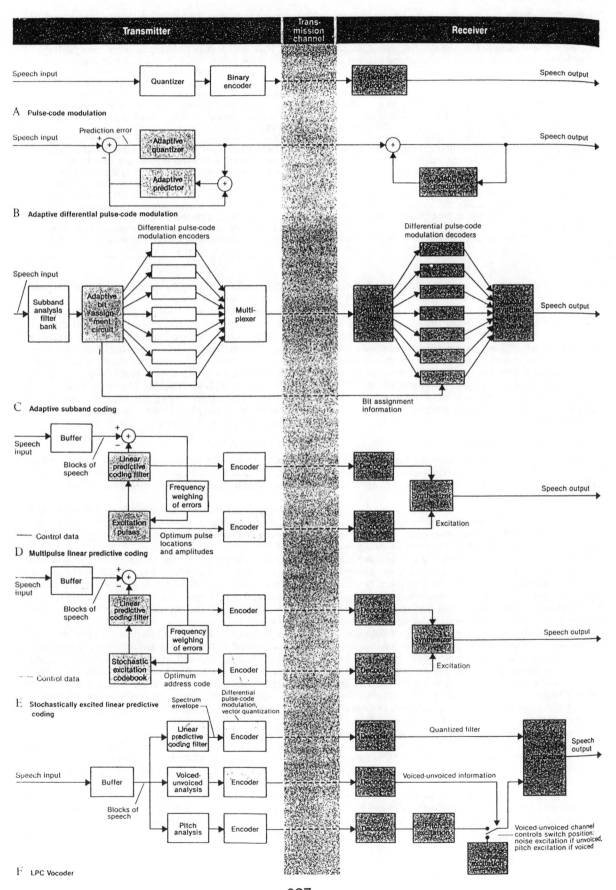

Speech input → Quantizer → Binary encoder → [Binary decoder] → Speech output

A Pulse-code modulation

Speech input → + / − → Adaptive quantizer → ... → + → Speech output
Prediction error
Adaptive predictor
Adaptive predictor

B Adaptive differential pulse-code modulation

Differential pulse-code modulation encoders

Speech input → Subband analysis filter bank → Adaptive bit assignment circuit → Multiplexer → [Demultiplexer] → Differential pulse-code modulation decoders → [Adder] → Speech output

Bit assignment information

C Adaptive subband coding

Speech input → Buffer → + / − → Linear predictive coding filter → Encoder → [Decoder] → Synthesizer → Speech output
Blocks of speech
Frequency weighing of errors
Excitation pulses → Encoder → [Decoder] → Excitation
Optimum pulse locations and amplitudes
—— Control data

D Multipulse linear predictive coding

Speech input → Buffer → + / − → Linear predictive coding filter → Encoder → [Decoder] → Synthesizer → Speech output
Blocks of speech
Frequency weighing of errors
Stochastic excitation codebook → Encoder → [Decoder] → Excitation
Optimum address code
---- Control data

E Stochastically excited linear predictive coding

Spectrum envelope
Differential pulse-code modulation, vector quantization

Linear predictive coding filter → Encoder → [Decoder] → Quantized filter → [Synthesizer] → Speech output
Speech input → Buffer → Voiced-unvoiced analysis → Encoder → [Decoder] → Voiced-unvoiced information
Blocks of speech
Pitch analysis → Encoder → [Decoder] → [Pitch excitation]

Voiced-unvoiced channel controls switch position: noise excitation if unvoiced, pitch excitation if voiced

F LPC Vocoder

10 bits are needed to transmit the output address code for the 40 samples, and the quantization rate is therefore only 0.25 bit per sample.

For 4-kb/s coding, code-excited or stochastically excited linear predictive coding is promising. The coder stores a repertory of candidate excitations, each a stochastic or random, sequence of pulses. The best sequence is selected by a closed-loop search. Vector quantization in the linear predictive filter is almost a necessity here to guarantee that enough bits are available for the excitation and prediction parameters. Vector quantization ensures good quality by allowing enough candidates in the excitation and filter codebooks.

Two key trends at work

Two complementary trends are at work in digital telecommunications: as speech coding developers drive down the bit rate for a given quality level, developers of modulation and demodula-

[2] Block diagrams of representative coding schemes illustrate the growth in complexity as coding rates decrease from 64 kb/s for a pulse-code modulation system (A) (not shown are the binary encoder and decoder on the transmitting and receiving sides, respectively, of the more complex systems). The complexity increases because more coding strategies are needed to exploit properties of speech and hearing. Each sublevel of coding builds on the techniques of the preceding level, adding one or two techniques of its own. The nonuniform quantization in (A) turns to advantage the greater probability of lower amplitudes in speech. The nonuniform quantization of pulse-code modulation is replaced by adaptive quantization and prediction in differential pulse-code modulation (B)—adaptive meaning responsive to the changing level and spectrum of the input speech wave. This technique makes good use of the statistically predictable nature of speech. Adaptive subband coding adds subbands and adaptive assignment of bits to the subbands (C). Multipulse linear predictive coding adds an analysis-synthesis approach and adaptive pulse excitation (D). Continuing the hierarchy of strategies, stochastically excited linear predictive coding (LPC) adds adaptive codebook excitation (E). The LPC vocoder (F) is not part of the hierarchy, but some of its principles are adapted in the hybrid MP-LPC and SE-LPC systems; specifically, for the rigid two-state vocoder excitation, the multipulse (MP) and stochastically excited (SE) coders substitute continuously optimized, time-varying excitation.

tion techniques are driving up the bit rate that a channel of a given bandwidth can accommodate.

For example, it may soon be practical to send high-quality digital speech signals at about 8 kb/s over a wide range of channels, instead of the wider-band digital channels to which they are now restricted. Robust, high-quality coding algorithms will cut the bit rate, and new modulators and demodulators will transmit the lower bit rate, with a low bit-error probability, over an analog channel having a bandwidth of about 3 kilohertz. Analog voice links, now used for transmitting high-quality analog speech, will therefore be able to carry high-quality digital speech, with added benefits such as voice security.

To probe further

Details on advanced digital speech coding abound in papers presented at the annual IEEE International Conference on Acoustics, Speech, and Signal Processing, most recently held in Tokyo (April 8–11).

A comprehensive reference is Nuggehally S. Jayant and Peter Noll's *Digital Coding of Waveforms—Principles and Applications to Speech and Video*, Prentice-Hall Inc., Englewood Cliffs, N.J., 1984.

James L. Flanagan et al review basic principles of traditional coding technology in "Speech Coding," *IEEE Transactions on Communications*, April 1979. Ronald E. Crochiere et al discuss issues of real-time implementation for waveform coders in "Real-time Speech Coding, *IEEE Transactions on Communications*, April 1982.

John Makhoul, S. Roucos, and Herbert Gish provide an up-to-date review of vector quantization in "Vector Quantization in Speech Coding," *Proceedings of the IEEE*, November 1985.

Descriptions of new approaches to speech coding are covered regularly in the *IEEE Transactions on Acoustics, Speech, and Signal Processing* and in the yearly proceedings of the International Conference on Acoustics, Speech, and Signal Processing.

About the author

Nuggehally S. Jayant (F) is head of the Signal Processing Research Department at AT&T Bell Laboratories in Murray Hill, N.J. He has worked on the coding and transmission of waveforms, particularly in digital communications. He received a B.S. in physics and mathematics from Mysore University in India in 1962 and a B.E. and Ph.D. in electrical communications engineering from the Indian Institute of Science, Bangalore, in 1965 and 1970, respectively.   ♦

63

328

Reprinted from *IEEE Transactions on Communications*, Vol. COM-27, No. 4, pp. 710-735, April, 1979.

Speech Coding

JAMES L. FLANAGAN, FELLOW, IEEE, MANFRED R. SCHROEDER, FELLOW, IEEE, BISHNU S. ATAL, SENIOR MEMBER, IEE
RONALD E. CROCHIERE, SENIOR MEMBER, IEEE, NUGGEHALLY S. JAYANT, SENIOR MEMBER, IEEE, AND
JOSE M. TRIBOLET, MEMBER, IEEE

(Invited Paper)

Author's Note—This paper is a camel.* Short of a book-length text (which, incidentally, one of us (NSJ) is writing), our task to discuss the field of digital speech coding is well-nigh impossible. Nevertheless, we have attempted to respond to this charge in a circumscribed, if not disjoint, way. Our hope here is to expose the reader to a selection of critical topics and current techniques in digital voice. Our choice is highly parochial—a collection of "islands"—and many good colleagues will feel slighted. We have tried to minimize the latter effect through extensive references. At the same time have concentrated on those topics with which we have first-hand familiarity and one which we can comment confidently. We hope the reader will appreciate this stance and make allowances for it.

* A camel is a horse designed by a committee.

Paper approved by the Editor-in-Chief of the IEEE Communications Society for publication without oral presentation. Manuscript received September 15, 1978; revised January 3, 1979.

J. L. Flanagan is with Bell Laboratories, Murray Hill, NJ 07974.
M. R. Schroeder is with Bell Laboratories, Murray Hill, NJ 07974 and Drittes Physikalisches Institut Universitat Göttingen, Göttingen, Germany.
B. S. Atal is with Bell Laboratories, Murray Hill, NJ 07974.
R. E. Crochiere is with Bell Laboratories, Murray Hill, NJ 07974.
N. S. Jayant is with Bell Laboratories, Murray Hill, NJ 07974.
J. M. Tribolet was on leave at Bell Laboratories, Murray Hill, NJ 07974. He is with the Instituto Superior Técnico, Lisbon, Portugal.

I. DIGITAL CODING OF SPEECH

Scope of the Paper

A S digital technologies evolve, and as the economies o large-scale integration begin to be achieved, renewe interest focuses on efficient methods for digitally encodin; and transmitting speech. The underlying goal is not new. It i to transmit speech—with the highest possible quality, over th least possible channel capacity, and with the least cost. But th intent to accomplish this goal through novel and sophisticatec digital methods is new—triggered by the promise of digita. hardware economies.

Typically the cost of speech encoding is positively correlated with coder complexity—and complexity, in turn, is positively correlated with code efficiency and channel utilization. Heretofore studies of complex (and potentially efficient) digital codes have often been deterred by the spectre of high costs. But advances in device integration are dramatically changing this climate. It is therefore timely to inquire further into properties of the speech signal that make it particulary amenable to digital representation. Our aim in this paper, therefore, is to outline current understanding and capabilities in "digital voice". and to opine on the near-time outlook for new advances. Our viewpoint will necessarily be parochial, because it is easier for us to talk about those results with which we have first-hand acquaintance. We hope the reader will excuse this bias.

The complete design of any transmission system requires optimal selection (in some sense) of a combination of factors—such as signal quality, transmission bit-rate, and coder cost. The proper selection depends very much upon the transmission environment (for example, terrestrial wire, glass fiber, or radio). Issues of total system optimization are largely outside the scope of this paper. But, we will strive to indicate the implications that code designs have for certain inevitable system issues—such as transmission error rate, multiple coding (tandeming), variable-rate coding, packet transmission and encryption.

Fidelity Criteria

Any assessment of signal quality implies a fidelity measure. For most communication systems this measure is difficult to specify quantitatively—because it involves human perception. Speech quality is traditionally assessed by the criterion that a listener understands *what* is being said, and *who* said it. Objective measures that accurately reflect these factors are intensively sought, but are difficult to establish with generality.

Despite this incomplete state of knowledge, several implements are available for quantifying speech quality. They derive largely from word intelligibility tests, speaker recognition tests and signal masking experiments—all conducted with human listeners. One can often use these data as guideposts in speech coder designs. Furthermore, short-time spectral measures and weighted signal-to-noise ratios, prudently interpreted, represent steps toward the elusive, objective quantification of perception. These points will enter for significant comment as the discussion unfolds.

Waveform Coders and Source Coders (Vocoders)

A broad class of speech coders is termed *waveform coders*. As the name implies, these coders essentially strive for facsimile reproduction of the signal waveform. In principle, they are designed to be signal-independent, hence they can code equally well a variety of signals—speech, music, tones, voiceband data. They also tend to be robust for a wide range of talker characteristics and for noisy environments. To preserve these advantages with minimal complexity, waveform coders typically aim for moderate economies in transmission bit rate.

Waveform coders can be optimized and made more signal-specific for greater coding efficiency. This typically is done by observing statistics of a given signal set, so that the waveform coder yields minimal encoding error for this signal class, (i.e., speech). The tailoring of these coders is thus based on a statistical characterization of speech waveforms, as distinct from parameterization of speech information according to some physical model of the signal.

A second class of speech coders depends upon a parsimonious description of speech using *a priori* knowledge about how the signal was generated at the source. The idea is that certain physical constraints of the signal generation can be quantified, and turned to advantage in efficiently describing the signal. This implies that the signal must be fitted into a specific (in our case, speech-specific) mold and parameterized accordingly. We refer to coding techniques that exploit constraints of signal generation as *"source coders."* Source coders for speech are generically referred to as vocoders (a contraction of the words *voice coders*).

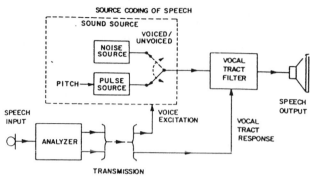

Fig. 1.   Speech generation source-system model.

The traditional speech generation model, that dates from so-called "channel vocoder" days, is the source-system model shown in Fig. 1. The sound-generating mechanism (the source) is assumed to be linearly-separable from the intelligence-modulating, vocal-tract filter (the system). Other assumptions are that speech sounds are either voiced or unvoiced, and that they are generated either from quasi-periodic vocal-cord sound, or from random sound produced by turbulent air flow. The parameters of this model (voice pitch, the pole frequencies of the modulating filters, and corresponding amplitude parameters) will be discussed subsequently. Here, it suffices to note that by exceedingly meticulous adjustments of parameters, one can demonstrate speech reproduction with good quality. More generally, however, in actual one-pass analysis/synthesis transmission, vocoders tend to be fragile (in terms of parameters such as voiced/unvoiced decision and pitch values), the performance is often talker-dependent, and the output speech has a synthetic (less than natural) quality. These characteristics constitute a ceiling on the performance that vocoders can achieve. But by virtue of their signal parameterization, vocoders can achieve very high economies in transmission bandwidth.

The boundary between waveform coders and vocoders might be thought of as a sort of middle ground, where the design criterion is neither waveform preservation nor signal modeling. Rather, the guiding principle is the preservation of the short-time amplitude spectrum of the speech signal in an auditorily palatable way. This middle ground offers opportunities to combine some advantages of both waveform and source coders.

Transmission Rates in Speech Coding

Figure 2 shows a spectrum of speech coding transmission rates currently of interest. The figure highlights the dichotomy between nonspeech-specific waveform coders that need relatively higher transmission rates, and speech-specific vocoders for digitization at relatively lower bit rates. The figure also indicates the quality of speech reproduction that can presently be attained at a prescribed bit rate. The quality characterizations are denoted as *commentary*, *toll*, *communications* and *synthetic*.

We use the term toll quality somewhat loosely, but typically to imply quality comparable to that of an analog speech signal having approximately the properties: [Frequency range = 200 to 3200 Hz; Signal-to-noise ratio $\geq$ 30 dB; Harmonic distortion $\leq$2-3%]. As Fig. 2 suggests, we presently know how to make digital coders that achieve telephone toll quality for speech

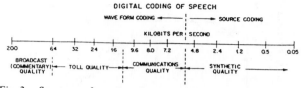

Fig. 2. Spectrum of speech coding transmission rates (nonlinear scale) and associated quality.

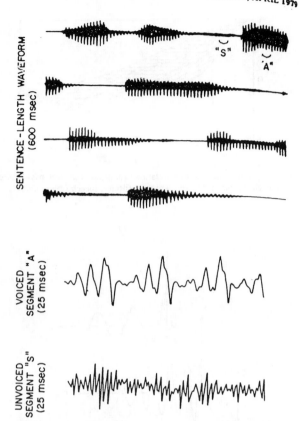

Fig. 3. Time waveforms of speech for a long (sentence-length) segment and for short-time (25 ms) segments.

signals at coding rates of 16 kbits/s and above. At the present time we know of no way, with any amount of complexity, to achieve toll quality at rates much below 16 kbits/s.

At bit rates exceeding 64 kbits/s, it is possible to obtain the signal-to-noise ratios and harmonic distortion characteristics of toll quality with input signal bandwidths significantly wider than normal telephone (e.g., 0 to 7 kHz or better). This grade of quality is loosely referred to as commentary quality. It is appropriate for digitizing some varieties of radio broadcast material.

At rates below 16 kbits/s and specifically the data-speed range 9.6 to 7.2 kbits/s, we know how to make waveform coders that provide communications quality speech. The signal is highly intelligible, but has noticeable quality reduction, some detectable distortion, and perhaps lessened talker recognition. Again, circuit complexity is a function of transmission rate.

Coders in the source-coding range (vocoders), 4.8 kbits/s and below, provide synthetic quality, where the signal usually has lost substantial naturalness, typically sounds reedy, and sometimes automaton-like. Talker recognition is substantially degraded, and coder performance is talker dependent.

In the remainder of this paper we propose to outline the properties of waveform coders and vocoders. In a nonexhaustive way, we further propose to discuss specific current examples of both categories of coders. When possible, we will strive to point to earlier work and "cousin" coders through appropriate references.

II. WAVEFORM CODING

Before describing waveform coders for speech, it is useful to mention several speech properties that can be utilized in an efficient waveform coder design. These properties include distributions for waveform amplitude and power, the nonflat characteristics of speech spectra (and equivalent autocorrelation functions), the quasi-periodicity of voiced speech, and the presence of silent intervals in the signal. As we progress through this enumeration, we deal with properties whose utilization calls for increasing coder memory. Thus, for example, amplitude and power distributions can be used in the design of zero-memory quantizers; while coders that remove speech redundancy, as manifested in speech spectra and speech periodicity, may require anywhere from 0.2 to 32 ms of coder memory; and coding systems that utilize the silent intervals in speech waveforms may need seconds of memory or encoding delay. Let us speak briefly about these fundamental properties.

A. Characterization of Speech Waveforms

Perhaps the most basic property of speech waveforms is that they are bandlimited. The bandlimitation begins in the speech production process, but an additional contribution is the finite bandwidth of typical speech transmission systems— for example, the 200 to 3200 Hz bandwidth associated with conventional voice circuits. In any case, the finite bandwidth of the speech waveform means that it can be time-sampled at a finite rate (the Nyquist rate which for a low-pass signal is twice the highest frequency therein; 8000 Hz is a conservative sampling frequency used in commercial telephony). Waveform coding systems are based not only on time-discretization, but also on different strategies for amplitude discretization. These strategies exploit a hierarchy of waveform properties that are best described in a formal statistical framework, although the underlying notions can be qualitatively perceived by merely examining an adequately long plot of speech waveform amplitudes versus time as in Fig. 3.

The waveform oscillogram immediately puts in evidence the differences between the relatively intense, quasi-periodic voiced sounds of speech (produced by periodic vibration of the vocal cords) and the lower amplitude unvoiced sounds (produced by the random noise of turbulent air flow through a constriction). Short-time signal statistics reflect the properties of these two vocal sound sources.

Amplitude Distributions—The probability density function (PDF) of speech amplitudes is characterized in general by a very high probability of zero and near-zero amplitudes (related to pauses and low-energy segments of the speech waveform), by a significant probability of 'very high' amplitudes, and by a

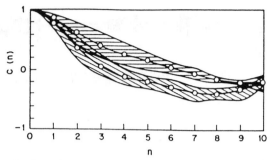

Fig. 4.   Long-time autocorrelation function for low-pass-filtered speech (upper curve) and bandpass-filtered speech (lower curve).

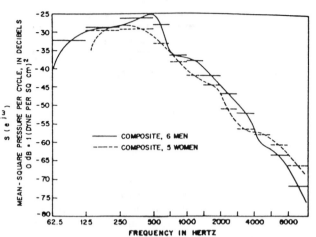

Fig. 6.   Long-time-averaged spectral density of speech.

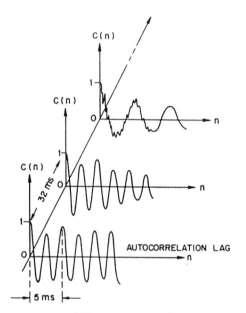

Fig. 5.   Short-time autocorrelation at different times of a speech waveform.

monotonically decreasing function of amplitude in between these extremes. An analytical fit to the long-time-averaged PDF of speech amplitudes is provided by a sum of Gaussian and Laplacian (two-sided exponential) functions, or by a Gamma function. A simple Gaussian model is usually adequate for a short-time (20 ms-based) PDF.

Autocorrelation Function (ACF)—The correlations that exist among the amplitude samples of a speech waveform are described by an autocorrelation function $C(n)$ where C, the correlation measure, is the expected (time-averaged) product of (zero-mean) amplitudes at times t and $t + n$, normalized to the expected (time-averaged) value of squared amplitudes. By definition, $C(n)$ is constrained in the range $[-1, 1]$, and $C(0) = 1$. The condition $[C(n) = 0; n \neq 0]$ represents a segment of uncorrelated (white-noise-like) amplitudes. Positive and negative values of $C(1)$ indicate, respectively, slowly changing and rapidly changing amplitude sequences.

A long-time (55 s)-averaged $C(n)$ function is illustrated in Fig. 4 for the example of 8 kHz speech. The upper and lower sets of curves refer respectively to low-pass filtered (0 to 3400 Hz) and bandpass filtered (200 to 3400 Hz) speech. In

each set, upper and lower limits represent maxima and minima, respectively, over four speakers (two male and two female); while the middle curves give mean values (averages over four speakers). For $n = 1$, this average value is $C(1) \sim .9$, suggesting a very high adjacent sample correlation in Nyquist-sampled (8 kHz) speech.

Short-time correlations (based on observations averaged over about 20 ms) are sketched in Fig. 5. Unlike the previous figure, the maximum value of n (the autocorrelation lag in Nyquist samples) here is large enough to show secondary peaks (related to periodicities that often exist in the speech waveform). In fact, Fig. 5 also shows the evolution of the short-time ACF as one proceeds from a highly periodic, slowly varying, speech segment (lower) to an aperiodic, more rapidly changing segment (upper).

Power Spectral Density (PSD)—The average probabilities of different frequency components in speech are illustrated by the long-time-averaged spectral density $S(e^{j\omega})$ plot of Fig. 6. It is clear that high-frequency waveform components contribute very little to the total speech energy. Nevertheless, these high-frequency components are very important carriers of speech information, and as such, they must be adequately represented in coding systems. It is useful to seek analytical models for the average spectrum of Fig. 6. A model derived from an integrated white noise spectrum, with an asymptotic 6 dB per octave decay characteristic, is often employed. This model is related, in discrete-time description, to a first-order Markov process. Further, this model is characterized by an exponentially decreasing $C(n)$ function. More realistic models for average speech spectra are obtained by seeking 12-dB per octave decay characteristics (for example, second-order Markov processes), and by truncating model spectra at an appropriate frequency.

As in correlation descriptions, short-time statistics (as measured over a typical 20 ms) are also of interest. Fig. 7 shows that short-time speech spectra do not always permit simple low-pass descriptions. Unvoiced waveform segments can have high-pass spectra; while voiced segments, although globally low-pass, also display local resonances, called *formants*. In fact, the time course of the formant frequencies is extremely significant to speech intelligibility.

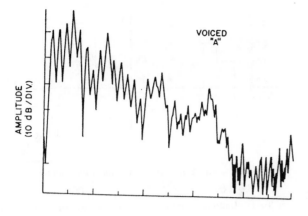

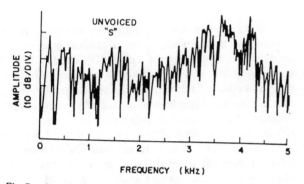

Fig. 7.   Short-time spectra for typical voiced and unvoiced segments.

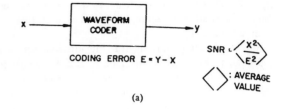

(a)

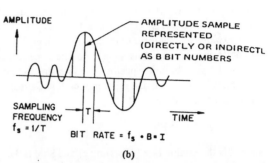

(b)

Fig. 8.   (a) SNR measurement of a waveform coder and (b) represen
tion of uniform sampling and B bit quantization per sample.

Spectral Flatness: Prediction or Transform-Coding Gain–
The ACF and PSD descriptions of a waveform $x(n)$ are equivalent in that the functions $C(n)$ and $S(e^{j\omega})$ form a Fourier
transform pair:

$$S(e^{j\omega}) = |X(e^{j\omega})|^2 = \sum_{n=-\infty}^{\infty} C(n)e^{-jn\omega T} \qquad (1)$$

where $T = 1/f_s$ = the sampling interval in seconds. The redundancy in a speech waveform can be exploited, for realizing
coding economies, either in the time-domain (using ACF
properties) or in the frequency-domain (using PSD properties).
In fact, the maximum theoretical performance of redundancy
removing coders (the so-called prediction gain, or an equivalent
transform coding gain) is related to the nonflatness of the
PSD, as described by a certain arithmetic-to-geometric-mean
ratio. If samples of the PSD, taken at uniform intervals in frequency, are denoted by S_k (abbreviation for $S(e^{j\omega_k})$), where
$k_{max} = N$ corresponds to the highest frequency, and if N is
sufficiently large, then a spectral flatness measure (SFM) can
be expressed by the ratio

$$\text{SFM} = \left[1/N \sum_{k=1}^{N} S_k^2 \right] \bigg/ \left[\prod_{k=1}^{N} S_k^2 \right]^{1/N} \qquad (2)$$

A long-time average value of SFM, for speech, is typically on
the order of 8, while short-time values can vary widely and fall
in the range $2 < \text{SFM} < 500$.

*Speech-Specific Waveform Properties–*Last in our hierarc
of waveform descriptions are characterizations that are qui
specific to the speech signal. One important characterizatio
relates to the quasi-periodicity of the voiced sounds of speec
Another relates to the "activity factor" of the signal, in oth
words, to the silent intervals in the speech stream. The form
will be appropriately characterized in our subsequent discussio
of adaptive prediction. The latter will be treated in terms (
silence statistics and interval modification as part of commen
on packet transmission of speech.

Fidelity Criteria for Waveforms: Signal-to-Noise Rati
(SNR)–The difference between samples of a coded wave
form and the original input waveform is traditionally define
as the coding error. The square of this quantity averaged ove
an appropriate interval is termed the coding noise. The rati
of the average value of the square of the input signal (average
over the same interval) to the coding noise is defined as th
signal-to-noise ratio (SNR). The quantity is often expressed in
dB as $10 \log_{10}$ SNR. A general objective in waveform coding is
to attain, for a given coding bit rate I, the most preceptuall
advantageous value of SNR. Frequently, but not always, this
is the maximum possible SNR.

Figure 8 shows amplitude samples quantized, the latter as
B-bit numbers, directly or indirectly. The meaning of this
qualification will be clear as we proceed from a simple amplitude quantizer (as in PCM) to a differential quantizer (as in
differential PCM or predictive coding), and then to a block
quantizer (as in transform coding). The uniform time discretization implied in Fig. 8 is tacitly assumed for the rest of
this paper. We shall regard speech as a band-limited waveform
so that we can invoke the sampling theorem which says that
the spectral informaton in speech can be preserved by sampling
the waveform at a frequency f_s equal to, or in practice slightly
higher than, the Nyquist rate f_{nyq}. We shall note later, in the
context of delta modulation, that there are certain benefits to
be gained by oversampling ($f_s > f_{nyq}$). Undersampling

$(f_s < f_{nyq})$, on the other hand, results in an unacceptable form of degradation referred to as aliasing distortion [Recording number 1].

We have found it convenient to categorize speech waveform coder algorithms into time-domain and frequency-domain classes. But it is important to stress that coders in different classes can be equivalent in terms of the properties of speech that they exploit. For example, both adaptive predictive coding (APC) (which is a time-domain technique) and adaptive transform coding (ATC) (which is a frequency-domain technique) exploit the same redundancy in the speech signal and their performances are thus subject to the same information theoretical upper bound: specifically the distortion-rate function (derived for a Gaussian input model; non-Gaussian signals such as speech can do better)

$$D(R) = 2^{-2R} \cdot \text{SFM} \qquad (3)$$

where R is the coding rate (bits/sample) and SFM is the spectral flatness measure (2).

B. Time-Domain Coders

Pulse Code Modulation (PCM)—*General Quantizer Designs*— Waveform coders quantize amplitude samples by rounding off each sample value to one of a set of several discrete values. In a B-bit quantizer, the number of these discrete amplitude levels is 2^B. A fundamental result in quantization theory is that the quantization error power is proportional to the square of the quantizing step-size; and since the step-size is inversely proportional to the total number of levels for a given total amplitude range, a signal-to-quantization-error ratio SNR can be defined that is proportional to 2^{2B}. In logarithmic units, SNR increases linearly with B. A Pulse Code Modulation (PCM) coder, which is basically a quantizer of sampled amplitudes, can be associated, for example, with a performance characteristic of the form

$$\text{SNR}_{\text{PCM}} \text{ (in dB)} = 10 \log_{10} \text{SNR}_{\text{PCM}} = (6B - \theta) \qquad (4)$$

where θ is a step-size dependent parameter.

An SNR formulation implies that quantization error samples can be modeled as additive noise samples provided the quantization is sufficiently fine. It is good to bear in mind that for coarse quantization (say, $B < 5$) the error waveform has too much structure and too much correlation with the input speech itself, to be regarded as additive noise.

Implicit in the derivation of (4) is the assumption that the range of the quantizer is aligned with that of the speech amplitudes at its input. This requirement is realized if speech amplitudes do not exceed the overload points of the quantizer ($+4.0$ and -4.0 in the uniform 3-bit quantizing characteristic of Fig. 9a) with any significant probability; and if all quantizer ranges are utilized in some equitable fashion. In practice, such quantizer-input alignment is realized by one of two techniques, nonuniform quantization or adaptive quantization.

Nonuniform quantization (Fig. 9b) is characterized by fine quantizing steps (and hence, a relatively small noise variance) for the very frequently occurring low amplitudes in speech; while much coarser quantizing steps take care of the occasional

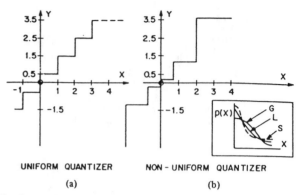

Fig. 9. (a) Uniform quantizer characteristic and (b) a non-uniform quantizer characteristic.

large amplitude excursions in the speech waveform. Average distributions of speech amplitudes are decreasing functions of amplitude (as typified by the dashed PDF labeled "S" in the inset of Fig. 9b), and nonuniform quantization constitutes a direct utilization of this speech property. An amplitude quantizer in commerical telephony uses, for example, an exponential quantizer characteristic that is equivalent to the uniform quantization of a logarithmically compressed input. Logarithmic quantization is in fact more robust than PDF-optimized (say Gamma-optimized) quantization in terms of dynamic range and idle-channel noise performance. The amplitude compression characteristics used in log-quantization follow either the so-called μ-law or the A-law. For a signal input $x(n)$, both characteristics are symmetrical about $x(n) = x = 0$. For $x > 0$ and $x_{\max} = 1$ compressed signals $x_c(n)$ are defined as follows:

$$\mu\text{-law: } x_c = [\ln(1 + \mu x)/\ln(1 + \mu)]; \qquad \mu > 0$$

$$A\text{-law: } x_c = [Ax/1 + \ln A], \qquad 0 \leqslant x \leqslant A^{-1}$$

$$= [1 + \ln Ax/1 + \ln A], \qquad A^{-1} \leqslant x \leqslant 1 \qquad (6)$$

or, in practice, piecewise linear versions thereof. A 7-bit (128-level) log-quantizer for speech indeed realizes the 35 dB SNR predicted by a formula such as (4). A uniform quantizer as in Fig. 9a will need, on the other hand, about 12 bits (4096 levels) to assure such an SNR with speech inputs, because of the large (although infrequent) amplitude excursions. (Recording number 2).

In typical voice communication systems, the dynamic range of speech signals (range of syllabically averaged speech volumes or powers, considering inter-talker as well as intra-talker variations), can be as much as 40 dB. While time-invariant nonuniform quantization has been a traditional solution to this dynamic range problem, better results can be obtained by recognizing that the large dynamic range of speech signals is a result of a nonstationary or time-varying process at the coder input; so that a truly optimal quantization strategy is one that is also time-variable, or adaptive to the input signal.

Adaptive quantization utilizes a quantizer characteristic (uniform or nonuniform) that shrinks or expands in time like an accordion. Snapshots of such an adaptive quantizer at two

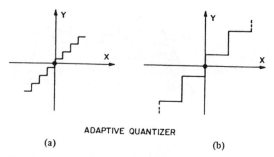

Fig. 10. Adaptive quantizer characteristic for (a) low signal power and (b) large signal power.

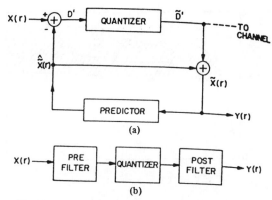

Fig. 11. (a) DPCM quantizer and (b) D*PCM quantizer.

instants of time may therefore look like the pictures in Fig. 10 (a) and (b), indicating adaptation to low and high speech powers, respectively. Although speech signals have a large dynamic range over a long period of time, input power-levels vary slowly enough to facilitate the design of simple adaptation algorithms to keep track of these power variations. For example, in adaptive quantization with a one-word memory, the adaptation of step size Δ_{r+1} at time sample $r + 1$ is characterized by

$$\Delta_{r+1} = \Delta_r \cdot M(\,|\,\text{Quantizer Output}_r\,|\,) \qquad (7)$$

where the step-size multiplier M is a function only of the latest quantizer output. Since adaptations follow quantizer output rather than input, step-size information in this scheme does not have to be explicitly communicated but, in the case of error-free transmission, can be recreated exactly by the receiver.

Differential Pulse Code Modulation (DPCM)–We now consider techniques that encode speech more intelligently than in PCM, by utilizing the redundancies present in the speech. The redundancy, or predictability, can be utilized either in the time-domain or in the frequency domain. There is also a large class of redundancy reducing codes that operate on coder output symbols, rather than on the input speech. These "noiseless" entropy coding schemes have been well documented in general coding literature.

Adjacent amplitudes in speech waveforms are often highly correlated (Fig. 4). This means, for example, that the variance of the difference D between speech amplitudes $x(r)$ and $x(r - 1)$ is much smaller than the variance of $x(r)$. This immediately suggests a scheme where speech is represented not in terms of waveform amplitudes, but in terms of the differences between waveform amplitudes. Crudely speaking, in differential coding, one can take these differences, quantize them, and recover an approximation to $x(r)$ by essentially integrating the quantized difference samples. Quantization error variance tends to be proportional to quantizer input variance, for a given number of bits B. Therefore, by reducing the variance at the quantizer input by a factor G, one reduces the variance of the coding errors by a factor G as well, and thus increases the signal-to-noise ratio SNR by a factor G. If the correlation $C(1)$ between adjacent samples of the input signal is c_1 (note that by definition $-1 < c_1 < 1$), the value of G for the first-difference coding scheme can be shown to be $[2(1 - c_1)]^{-1}$, which can

be significant if $c_1 > 0.5$. More generally, if the quantizer input is $x(r) - a_1 x(r - 1)$, it can be shown that the variance of this quantity is minimum for $a_1 = c_1$ and the prediction gain in this case is $(1 - c_1^2)^{-1}$, a gain which is greater than 1 for all values of c_1. Formally the quantity $a_1 x(r - 1)$ is called a first-order prediction of $x(r)$; and the differential coding scheme is called predictive coding. An even more general (linear) predictor would be one of order p, where the estimate of the input signal sample $x(r)$ is

$$\dot{x}(r) = \sum_{n=1}^{p} a_n x(r - n). \qquad (8)$$

The best values of the prediction coefficients a_n are calculated in terms of the first p samples of the autocorrelation of $x(n)$.

The decoder for the predictive scheme involves an integrating process, and this can mean a corresponding integration, or accumulation of quantizing errors. Accumulation of quantizing noise is controlled by the feedback arrangement in Fig. 11, wherein predictions are based on quantized values $\tilde{x}(r - n)$:

$$\dot{x}(r) = \sum_{n=1}^{p} a_n \tilde{x}(r - n) = \sum_{n=1}^{p} a_n y(r - n). \qquad (9)$$

Figure 11a thus represents a practical predictive quantizer known as differential PCM (DPCM). For any number of bits B, the SNR for DPCM shows a gain over that of PCM. This gain, except for very coarse quantization (say, $B = 1$) is very close to the prediction gain discussed earlier. The design and performance of the pth order DPCM can be characterized under the assumption of negligible quantization noise (say, $B > 2$) by

$$A_{\text{opt}} = \Gamma^{-1} \Sigma, \qquad (10)$$

$$G_{\text{opt}} = [1 - A_{\text{opt}} \cdot \Sigma]^{-1}, \qquad (11)$$

$$\text{SNR}_{\text{DPCM}} \text{ (in dB)} = \text{SNR}_{\text{PCM}} \text{ (in dB)} + 10 \log G \qquad (12)$$

where A^T is the pth order predictor vector $[a_1, a_2 \cdots a_p]$, G is the SNR gain over PCM (prediction gain), Σ^T is the correlation vector $[c_1, c_2 \cdots c_p]$ and Γ is the correlation matrix $[c_{ij}]$ with

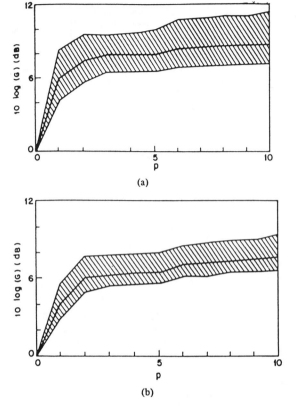

Fig. 12. Plots of DPCM prediction gain (in dB) as a function of predictor order for (a) low-pass-filtered and (b) bandpass-filtered speech.

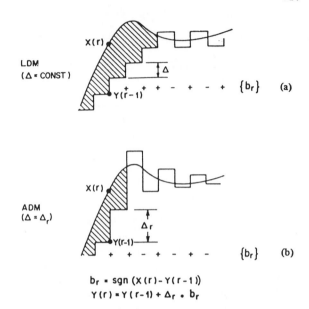

$$b_r = \operatorname{sgn}(x(r) - y(r-1))$$
$$y(r) = y(r-1) + \Delta_r \cdot b_r$$

Fig. 13. (a) Linear delta modulation (LDM) and (b) adaptive delta modulation (ADM).

$i, j = 0, 1, 2 \cdots p - 1$. (As before, we have simplified the notation in that $c_{ij} = c_{|i-j|} = C(|i - j|)$.) For example, with 8 kHz samples and typical (long-time averaged) correlations of [$c_0 = 1$, $c_1 = 0.85$, $c_2 = 0.562$, $c_3 = 0.308$], (time-invariant) predictor vectors of order 1 through 3 are described respectively by [0.85], [1.13, −.38] and [1.10, −0.28, −.08]; while time-invariant first-order predictors for 16 and 24 kHz samples would use typical a_1 values of 0.95 and 0.98, in veiw of corresponding higher values of c_1.

Figure 12 shows plots of DPCM gain G (gain over PCM) as a function of predictor order p, for the case of low-pass-filtered and bandpass-filtered speech inputs. These curves correspond to the conditions of Figure 4. As in that case, each G-characteristic includes a maximum, a minimum and an average over four speakers. Notice that the prediction gain tends to saturate, in general, around a predictor order p of about 2 or 3. The saturation point can be pushed to larger values of p in the case of time-varying, or adaptive prediction which exploits the short-time ACF statistics, such as those in Figure 5. In fact, the greatest achievements in differential encoding of speech have depended on predictions that are both spectrum adaptive (for near-sample-based predictions, say with $p \leq 12$) and pitch-adaptive (for distant-sample-based predictions, say with $20 \leq p \leq 120$). This is discussed further in the following section on adaptive predictive coding (APC).

Refer now to Fig. 11b. With proper design of a prefilter and a postfilter, the quantization error spectrum in the decoded

speech can be shaped to provide a perceptually palatable characteristic. This arrangement, called D*PCM in recent formalizations, can provide a mean-square error performance that lies between that of PCM and DPCM. For example, if the prefilter is constrained to be a first-order predictor, the optimum predictor coefficient would be $c_1(1 - \sqrt{1 - c_1^2})$ (instead of c_1 in the DPCM system); and the corresponding gain over PCM would be $(1 - c_1^2)^{-1/2}$, in place of $(1 - c_1^2)^{-1}$ for DPCM. This D*PCM filter can be characterized as a partially whitening filter (while the DPCM predictor would attempt full-whitening), and the post-filter would be the inverse of the prefilter. Although DPCM out performs D*PCM in the sense discussed above, the latter scheme can be more attractive if one desires to minimize the mean square value of errors with certain types of frequency weighting. An example is an error weighting that is the inverse of the input PSD: $S^{-1}(e^{j\omega})$. For speech, such error-weighting would lead to designs that suppress high frequency errors much more than in DPCM.

Delta Modulation (DM)—An important special case of DPCM is the one bit ($B = 1$) version known as Delta Modulation (DM). In this one bit (2-level) quantizer, the prediction error is quantized to one of only 2 values, a positive one or a negative one. This is shown in Fig. 13 where $y(r - 1)$ represents the quantized prediction of $x(r)$. Figure 13a represents the simplest form of DM in that it assumes a perfect integrator, which provides a special case of first-order prediction.

The operation of this DM coder can be characterized very simply. At each sample time r, one notes the polarity b_r of the difference between the present input $x(r)$ and the latest staircase approximation to it, $y(r - 1)$; and one then updates the staircase function in the direction of this difference. The update magnitude Δ_r is time-invariant in nonadaptive (linear) delta modulation (LDM), while Δ_r tracks changes in input slope in case of adaptive delta modulation (ADM). The hatched areas in Fig. 13 represent bursts of "slope-overload" distortion,

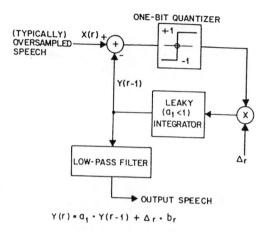

Fig. 14.   Block diagram of delta modulation.

in contrast with granular noise that occurs when the DM coder is tracking an input with rapidly alternating step polarities. The distinction is very significant (both statistically and perceptually) in the special case of $B = 1$, because of the coarse quantization involved.

The block diagram in Fig. 14 shows that DM retains the feedback-around-quantizer feature of DPCM. (It also notes that DM can use a leaky integrator—a general form of first-order predictor, $a_1 \leqslant 1$, rather than $a_1 = 1$ as in Fig. 13.) The figure also indicates that in DM, the input speech is generally oversampled. Clearly, this is to make up for the coarseness of one bit quantization: oversampling increases adjacent sample correlation, increases prediction gain, and makes the job of the quantizer easier. As noted earlier, the prediction gain is related to c_1. In DM, this correlation is more appropriately written as a function of sampling frequency in the form $c_1(f_s)$. It turns out that the prediction gain $(1 - c_1^2(f_s))^{-1}$, for many important characterizations of input spectrum, is proportional to f_s^2. Furthermore, the quantization noise in DM coding has out-of-band components because of input over-sampling, and an output low pass filter (as shown in Fig. 14) is invariably employed to eliminate these out-of-band noise components. The gain due to filtering is typically proportional to f_s, so that the resultant SNR in DM (including contributions of prediction gain and filtering gain) is proportional to the cube of sampling frequency (also to the cube of bit rate, which in DM is numerically equal to f_s). Formally, for single-integration (first-order prediction) DM coding,

$$\mathrm{SNR_{DM}} \text{ (in dB)} = 10 \log \left[k_1 f_s^3 \right]$$

$$= 10 \log k_1 + 30 \log f_s \qquad (13)$$

where k_1 is a constant related to the performance of the 1-bit quantizer and to the input spectral shape (PSD). The best performance of the system is obviously realized with a quantizer step-size Δ that is designed to provide the best value of speech-to-in-band noise ratio. As with multi-bit quantizers, the best way of realizing the most useful quantizer loading is by an adaptive procedure which lets the step-size follow the input slope

statistics. In a very simple one-bit-memory procedure, one pands the step-size Δ_r by a factor $P(>1)$ when the DM se slope-overload ($b_r = b_{r-1}$), and reduces the step-size by a fa $P^{-1}(<1)$ when it senses granularity ($b_r \neq b_{r-1}$). A m general adaptation rule is of the form

$$\Delta_r = f(\Delta_{r-1}, b_r, b_{r-1}, b_{r-2} \cdots)$$

and the utility of this more general algorithm is discussed l: in the context of transmission errors.

Finally, one can design "double-integration" DM cod which attain a 15 (rather than 9) dB per-octave performan in terms of sampling frequency. These encoders, as the na implies, encode the speech input in terms of its second, well as the first derivative. The advantage over single-integrati decreases as the sampling frequency decreases. Further, stabili considerations are more important in double-integration desig (as in higher-order predictor designs, in general), especia when certain types of adaptive quantization are involved.

In concluding our discussion of DPCM and DM, we no that multi-bit DPCM provides an SNR gain over PCM for : bit rates B, as given by $10 \log G$ dB (see (12)), while D provides a gain over PCM if the bit rate is low enough so th. the linear-with-bit-rate function in (4) lags behind the log-bi rate function in (13). Typically, at bit rates such as 32 kbits or less, both DPCM and DM provide SNR as well as perceptu advantages over PCM (Recording number 3). At 16 kbits/s, th performances of DPCM and DM are not significantly differer from one another. (Recording number 4). Also, at bit rate such as 16 kbits/s or less, differential coder performance canno be adequately described by a simple SNR value; but depend instead on a variety of parameters that characterize DPCM- o: DM-noise structure (Recording number 5), as in the case o: low-bit-rate PCM.

Finally, we mention that there are numerous, sophisticated "relatives" of DPCM and DM, based upon similar underlying principles of waveform representation. Among them are Delayed (Tree) Encoding, Aperture Coding and Gradient-Search Coding. Space precludes their detailed discussion here. Suffice it to say that their various advantages depend upon greater memory and more extensive processing power in the encoder.

Adaptive Predictive Coding (APC)–Predictive coding systems discussed so far have been limited to linear predictors with fixed coefficients. However, due to the nonstationary nature of the speech signals, a fixed predictor cannot predict the signal values efficiently at all times. Thus, the predictor must vary with time to cope with the changing spectral envelope of the speech signal as well as with the changing periodicities in voiced speech.

Adaptive prediction of speech is done most conveniently in two separate stages: a prediction exploiting the correlations between successive speech samples (or, equivalently, the non-uniform nature of the short-time spectral envelope of speech signals), and another prediction exploiting the quasi-periodic nature of voiced speech (or, the spectral fine structure representing the harmonic lines of the voice pitch).

Prediction Based on Spectral Fine Structure–A simple method of predicting the present value of a periodic signal is equate it to the value of the signal one or more periods earlier. For speech, the predictor has to provide some gain adjustment as well as to account for amplitude variations from one period to another. The predictor can be characterized in the z-transform notation by

$$P_d(z) = \beta z^{-M} \tag{15}$$

where M represents a relatively long delay in the range 2 to 20 ms and β is scaling factor. In most cases, the delay would correspond to a pitch period (or possibly, an integral number of pitch periods). The amount of prediction depends on the correlation between adjacent pitch periods. Such correlations typically vary considerably across speech sounds and speakers. For voiced speech, the average prediction gain is about 13 dB. For unvoiced speech which is noise-like in nature, the factor β is a small number and the value of M is unimportant.

The method of determining unknown parameters β and M of (15) is relatively straightforward. The delay M is chosen so that the correlation between speech samples delayed M samples apart is highest. The parameter β is then given by

$$\beta = \langle s_n s_{n-M} \rangle_{\text{av}} / \langle s^2_{n-M} \rangle_{\text{av}} \tag{16}$$

where s_n is the nth speech sample and the $\langle \ \rangle_{\text{av}}$ indicates the averaging over all the samples in a given time segment during which the predictor is to be optimum.

The prediction gain can be increased by using additional samples on both sides of the pitch delay for prediction of the present sample. The predictor is then represented in the z-transform notation by

$$P_d(z) = \beta_1 z^{-M+1} + \beta_2 z^{-M} + \beta_3 z^{-M-1}. \tag{17}$$

The three amplitude coefficients provide a frequency-dependent gain factor which is very desirable since voice periodicity does not show up as strongly at higher frequencies in speech. The sampling of the speech waveform also tends to reduce the correlation between adjacent pitch periods at higher frequencies. In computer simulation tests, the third-order pitch predictor provides an additional average prediction gain of approximately 3 dB over a first-order pitch predictor.

Prediction Based on Short-Time Spectral Envelope—The short-time spectral envelope of speech is determined by the frequency response of the vocal tract and for voiced speech also by the spectrum of the vocal-cord sound pulses. The predictor can be characterized in the z-transform notation as

$$P_s(z) = \sum_{k=1}^{p} a_k z^{-k} \tag{18}$$

where z^{-1} represents a delay of one sample interval and a_1, $a_2, \cdots a_p$ are p predictor coefficients. The value of p typically is 10 for speech sampled at 8 kHz. The predictor coefficients a_k are determined such that the power of the prediction residual is minimized in an adaptive manner over frame lengths whose duration may vary from 10 to 30 ms (a time interval within which the vocal-tract shape can be assumed to be nearly sta-

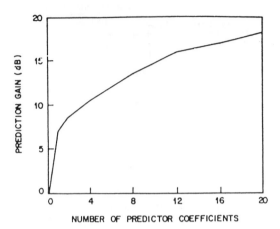

Fig. 15. Variation of the prediction gain (expressed in dB) with the number of predictor coefficients for voiced speech sampled at 10 kHz.

tionary). The methods for determining time-varying predictor coefficients are exactly identical to ones used for the fixed predictors in the DPCM section, except that the covariance matrix is updated periodically (once every 10 to 30 ms).

The asymptotic value of the prediction in the limit as p becomes large is solely determined by the short-time spectral envelope of the signal, specifically by the flatness measure SFM of (2). The variation of the average prediction gain (expressed in dB) with the number of predictor coefficients for voiced speech sampled at 10 kHz is shown in Fig. 15. For $p = 10$, the average prediction gain for voiced speech is 14 dB. It is important to remember here that, in predictive coding, the input signal samples are predicted from the past reconstructed signal samples as obtained by the receiver, and not from the past samples of the actual input signal. Since the output signal contains quantizing noise, the prediction gain is decreased when the quantization is coarse.

Combining the Two Types of Adaptive Prediction—The two types of prediction can be combined serially, in either order, to produce a predictor operator $[1 - P(z)]$ that is essentially the product of $[1 - P_d(z)]$ and $[1 - P_s(z)]$. The prediction with $P(z)$ can realize a higher prediction gain than is possible with either P_d or P_s acting alone. However, the total prediction gain (in dB) is not the sum of the gains for P_d and P_s acting singly on the speech signal. Used in combination, the first predictor achieves the bulk of the gain (typically 13-14 dB), and the second predictor (now operating on a signal that is less predictable than the original speech) achieves the balance (typically another 3 dB).

Perceptual Criteria for Optimizing Predictive Coders—So far our emphasis has been on minimizing the power of the quantization noise. However, to ensure that the distortion in the speech signal is perceptually small, it is necessary to consider the spectrum of the quantization noise and its relation to the speech spectrum. The theory of auditory masking suggests that noise in the frequency regions where speech energy is concentrated (such as formant regions) would be partially or totally masked by the speech signal. Thus, a large part of the perceived noise in a coder comes from those frequency regions where the signal level is low. Furthermore, what needs to be minimized is

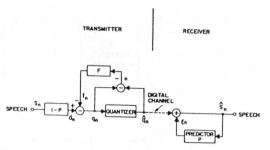

Fig. 16. Block diagram of a generalized predictive coder capable of producing any desired spectrum of the quantizing noise.

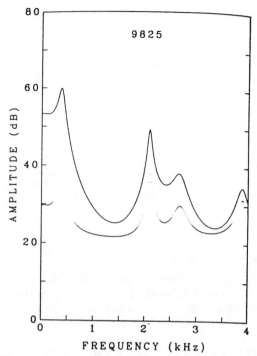

Fig. 17. An example showing the envelope of the output noise spectrum (dotted curve) shaped to reduce perceived distortion (F as in Eq. 21) and the corresponding speech spectrum (solid curve).

perhaps not the power of the quantization noise, but its subjective loudness. Methods for minimizing the loudness of quantizing noise are outside the scope of this paper. But, we will present a simple method of modifying the spectrum of quantizing noise in predictive coders to make it perceptually less audible. Fig. 16 shows a predictive coder capable of producing any desired spectrum of the quantizing noise. It can be shown that the ratio of the spectrum of the output noise to the spectrum of the quantizer noise in the feedback coder of Fig. 16 is given by

$$\lambda(\omega) = |\,[1 - F(e^{j\omega T})]\,/\,[1 - P(e^{j\omega T})]\,|^2 \qquad (19)$$

where T is the sampling interval. Eq. 19 implies an important constraint on the average value of log $\lambda(\omega)$, that is

$$\frac{1}{\pi f_s} \int_0^{\pi f_s} \log \lambda(\omega)\, d\omega = 0 \qquad (20)$$

where f_s is the sampling frequency. Expressed on a dB scale, the average value of log $\lambda(\omega)$ is 0 dB. The filter F, however, redistributes the noise power from one frequency to another. Thus, reduction in noise at one frequency can be obtained only at the expense of increasing the noise at another frequency. Since a large part of perceived noise in a coder comes from the regions where the signal level is low, the filter F can be used to reduce the noise in such regions while increasing the noise in the formant regions where the noise could be effectively masked by the signal.

Several arbitrary but illustrative choices for the feedback filter F can be considered: (i) $F = P$: This results in minimum unweighted noise power. The spectrum of the output noise is white, producing a very high SNR at the formants, but a poor one in between the formants (where the signal spectrum is low in magnitude). (ii) $F = 0$: This implies no feedback; the output noise has the same spectrum as the original speech, but shifted downward—a good choice if our ears were equally sensitive to quantizing distortion for all frequencies. (iii) an intermediate design exemplified by

$$F = P(\alpha z^{-1}). \qquad (21)$$

where α is an additional parameter introduced to increase the bandwidths of the zeros of $1 - F$. The increased bandwidth

causes the noise to peak up in the formant regions accompanied by a reduction in noise in regions where the signal level is low. An increase of 400 Hz in bandwidth has been found to be satisfactory. An example of the envelope of the output noise spectrum together with the corresponding speech spectrum is shown in Fig. 17.

Generalized Predictive Coder for Speech—The block diagram of a generalized predictive coder for speech signals is shown in Fig. 18. The high frequencies in the speech signal are pre-emphasized using a filter with the transfer function $[1 - 0.4z^{-1}]$ prior to analysis. The spectral envelope predictor is P_s while the second predictor P_d is based on pitch periodicity. The quantizer error is filtered and peak-limited to produce the sample f_n. The composite signal $q_n = d_n - d_n' - f_n$ is quantized by an adaptive 3-level quantizer. The parameters of the quantizer are selected to be optimum for a Gaussian PDF. (For a uniformly spaced 3-level quantizer, the optimum spacing between the quantizer output levels is 1.22 times the rms value of the prediction error.) Predictors and quantizer parameters are reset once every 10 ms. Proper choice of F provides an average SNR of approximately 20 dB (lower than for $F = P$, but higher than for $F = 0$) and an output speech quality that is subjectively close to that of 7-bit log-PCM, which has an SNR of 33 dB. The nonuniform noise spectrum together with the predictive coding thus produces a gain of about 13 dB in the subjective SNR.

Two examples of speech with the prediction error quantized by a 3-level and a 2-level quantizer are included in the record (Recording no. 6). The predictor parameters have not been quantized in these recordings.

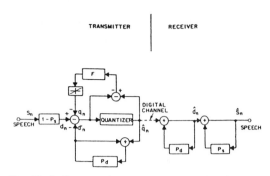

Fig. 18. Block diagram of a generalized predictive coder with two stages of prediction: a predictor P_s based on the short-time spectral envelope and another predictor P_d based on the pitch periodicity. The filter F is chosen to reduce the perceived distortion of quantizing noise.

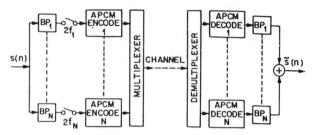

Fig. 19. Block diagram of a sub-band coder.

C. Frequency Domain Coders

In the class of "time domain" coding algorithms, speech is treated as a single full-band signal; and in the predictive coders, the speech redundancy is removed prior to encoding by prediction and inverse filtering. The main differences in the various algorithms are determined by the degree of prediction or interpolation that is exploited, and by whether schemes are adaptive or not.

In this section another class of encoding algorithms is discussed in which the approach is to divide the speech signal into a number of separate frequency components and to encode each of these components separately. These "frequency domain" coding techniques have the additional advantage that the number of bits used to encode each frequency component can be varied dynamically and shared with other bands, so that the encoding accuracy is always placed where it is needed in the frequency domain. In fact, bands with little or no energy may not be encoded at all.

As in the time domain techniques, a large variety of frequency domain algorithms, from simple to complex, are available and the main differences are usually determined by the degree of "prediction" that is employed in the technique. We will begin by describing one technique of lower complexity and then proceed to one of higher complexity.

Sub-Band Coding (SBC)—In the sub-band coder the speech band is divided into typically four to eight sub-bands by a bank of bandpass filters. Each sub-band is, in effect, low-pass translated to zero frequency by a modulation process equivalent to single-side-band modulation. It is then sampled (or resampled) at its Nyquist rate (twice the width of the band) and digitally encoded with an adaptive-step-size PCM (APCM) encoder (using techniques discussed in Section II-B). In this process, each sub-band can be encoded according to perceptual criteria that are specific to that band. On reconstruction, the sub-band signals are decoded and modulated back to their original locations. They are then summed to give a close replica of the original speech signal.

Encoding in sub-bands offers several advantages. Quantization noise can be contained in bands to prevent masking of one frequency range by quantizing noise in another frequency range. Separate adaptive quantizer step-sizes can be used in

each band. Therefore, bands with lower signal energy will have lower quantizer step-sizes and contribute less quantization noise. By appropriately allocating the bits in different bands, the shape of the quantization noise can be controlled in frequency (see also, bit allocation in adaptive transform coding in the next section). This is effectively a frequency-domain implementation of short-time prediction. In the lower frequency bands, where pitch and formant structure must be accurately preserved, a larger number of bits/sample can be used; whereas in upper frequency bands, where fricative and noise-like sounds occur in speech, fewer bits/sample can be used.

Figure 19 illustrates a basic block diagram of the sub-band coder. Essentially it consists of a bank of band-pass filters, APCM encoders, and a multiplexer. The modulation is obtained essentially "for free" by using the technique of integer-band sampling. The most complex part of the coder is the filter bank. However, with newer filter technologies such as CCD filters and digital filters, this complexity is rapidly being reduced to a chip level. Also the design technique of quadrature mirror filters (QMF) affords distinct advantages in digital implementation of this coder.

The perceptual advantages of sub-band coding are well put in evidence by the perceptual data of Fig. 20. Fig. 20a shows the relative preference of a 16 kbit/s sub-band coder versus that of an ADPCM coder at various ADPCM coder bit rates (a 50-50 percent preference implies that the two qualities are about equal). Fig. 20b shows similar results for a 9.6 kbit/s sub-band coder compared against an ADM coder at various ADM bit rates. In comparison with ADM and adaptive DPCM (ADPCM) at transmission bit rates in the order of 16 kbits/s, one sees that the effect of SBC is worth the equivalent of about 10 kbits/s of transmission. (Recording no. 7 demonstrates examples of sub-band coding performance for bit rates in the range from 32 to 9.6 kbits/s.)

Adaptive Transform Coding (ATC)—A more complex technique, transform coding, involves block transformation of windowed input segments of the speech waveform. Each segment is represented by a set of transform coefficients, which are separately quantized and transmitted. At the receiver, the quantized coefficients are inverse transformed to produce a replica of the original input segment. Successive segments, when joined, represent the input speech signal. Figure 21 illustrates the process.

The Block Transformation—The class of block transformations of interest for speech processing are time-to-frequency transformations. Since a primary goal is to generate the least audible coding noise, it is natural to control the quantization

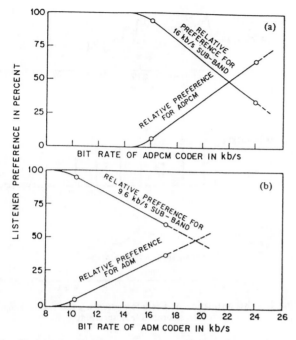

Fig. 20. (a) Relative comparison of quality of 16-kbit/s sub-band coding against ADPCM coding (based on listener preference) for different ADPCM coder bit rates. (b) Relative comparison of quality of 9.6-kbit/s sub-band coding against ADM coding for different ADM coder bit rates.

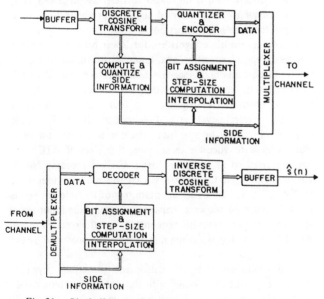

Fig. 21. Block diagram of an adaptive transform coder (ATC).

noise by controlling its characteristics in the frequency domain. Also, speech production can be modeled, on a short-time basis, in terms of the linear, time-invariant filtering model of Fig. 1. This model provides a framework for adapting the transform coder to the time-varying properties of speech.

A particular time-to-frequency transform, referred to as the Discrete Cosine Transform (DCT), has been found to be well suited for speech coding. One of the attractions of the DCT is

that, in a long-time average sense, it happens to be a good fit to the optimally orthogonalizing Karhunen-Lòeve Transform (KLT) for speech waveforms. Also, the even symmetry inherent ir. the structure of the DCT framework helps to minimize end effects. The DCT of a N-point sequence $x(n)$ is formally defined as:

$$X_c(k) = \sum_{n=0}^{N-1} x(n) g(k) \cos\left[\frac{(2n+1)k\pi}{2N}\right]$$
$$k = 0, 1, 2, \cdots N-1 \qquad (22)$$

where $g(0) = 1$ and $g(k) = \sqrt{2}$, $k = 1, \cdots, N-1$. The inverse DCT (IDCT) is defined as:

$$x(n) = \frac{1}{N} \sum_{k=0}^{N-1} X_c(k) g(k) \cos\left[\frac{(2n+1)k\pi}{2N}\right]$$
$$n = 0, 1, 2, \cdots N-1. \qquad (23)$$

Fast algorithms have been derived for implementing the DCT with great computational efficiency.

The choice of the analysis window for the DCT is important in controlling block boundary effects. Trapezoidal shaped windows have been found very useful for low-bit rate transform coding of speech. By allowing a small overlap between the successive blocks, such that the sum of the overlapped windows is always unity, an additional reduction of end effect noise is achieved without significantly lowering the number of bits available for encoding each block.

Quantization Strategy—The transform coefficients are usually quantized individually using a uniform step size Δ_i and a number of levels 2^{b_i}. The choice of Δ_i and the number of bits b_i for a given transform coefficient, i, is of fundamental importance. Ideally, Δ_i must be just large enough so that overloading of the quantizer does not occur. If this is assumed, the number of quantization levels as a function coefficient number and the amplitude spectrum determines the coarseness of the quantization, and thus the spectral characteristics of the quantization noise.

If the transform coefficients were stationary Gaussian variables with variances σ_i^2, $i = 0, \cdots, N-1$, and if a flat noise spectral distribution is desired, then the optimal bit assignment for the ith coefficient is

$$b_i = \delta + \frac{1}{2} \log_2 \frac{\sigma_i^2}{D^*} \qquad i = 0, \cdots, N-1, \qquad (24)$$

where δ is a correction term that reflects the performance of practical quantizers, and D^* denotes the noise power:

$$D^* = \frac{1}{N} \sum_{i=0}^{N-1} E_i^2 \qquad (25)$$

where E_i^2 is the noise power incurred in quantizing the i-th transform coefficient. These bit assignments have to satisfy the constraint:

$$B = \sum_{i=1}^{N} b_i \qquad (26)$$

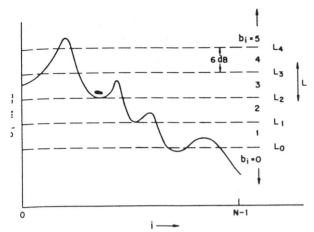

Fig. 22.   Interpretation of bit assignment rule.

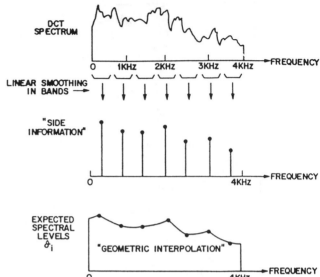

Fig. 23.   Representation of side information as equal spaced samples of the smoothed DCT spectrum.

where B is the number of bits/block available for transmission over the binary channel.

An interpretation of this bit assignment rule can be seen in Fig. 22. for given positions of the horizontal dashed thresholds, i.e., for given values of spectral levels L_K, an assignment is made so that every coefficient with variance below L_0 will be assigned zero bits (i.e., it is not transmitted), those with variances between L_1 and L_0 will be assigned 1 bit, and so on. If the resulting total number of bits assigned in this way is less than the number of bits available for transmission, B, the values of L_K are decreased; if it is greater than B, the values of L_K are increased. This process continues until (26) is satisfied. Given the number of levels per quantizer and the variance of the transform coefficients, the choice of the optimal step sizes for the uniform quantizers depends only on the PDF of the transform coefficients (a Gaussian PDF is typically assumed).

Adaptation Strategy—Application of the bit assignment scheme takes account of the spectral non-flatness of the signal characteristics as described by the SFM in Section II-A. The expected spectral levels $\hat{\sigma}_i$ of the transform coefficients for any particular speech block are not known *a priori* and must be estimated. This information, which reflects the dynamical properties of speech in the transform domain, is commonly referred to as "side information". Transform coders that operate within this framework are called Adaptive Transform Coders (ATC).

Two basic adaptation techniques for ATC of speech have been proposed. In one, illustrated in Fig. 23, the side information consists of a small number of samples computed by averaging (smoothing) the DCT spectral magnitudes, and represents the expected spectral levels at specified frequencies. These samples are then geometrically interpolated (i.e., linearly interpolated in log amplitude) to yield the expected spectral levels at all frequencies. This simple algorithm has been referred to as "non-speech specific" since it does not take into account the dynamical properties of speech production. This adaptation technique is however quite appropriate for speech transmission at or above 16 kbits/s, since there are enough bits to allow accurate representation of the fine structure of the DCT

spectrum—in particular, the voice pitch information. As one attempts to operate this approach at rates below 16 kbits/s, however, it becomes increasingly more difficult to accurately encode the fine structure. In fact, at 8 kbits/s, for example, the pitch information is no longer sufficiently preserved, and, as a consequence, the received signal is degraded by a very perceptible "burbling" distortion.

A more appropriate algorithm for lower bit rates is a more complex, "speech specific," adaptation algorithm which utilizes the traditional model of speech production to predict the DCT spectral levels. The prediction therefore involves two components, as illustrated in Fig. 24. The first is associated with a (formant) spectral envelope, and the second with the harmonic (fine) structure of the spectrum. Because this spectrum modeling is basic to vocoder techniques, this class of ATC's has been referred to as Vocoder-Driven ATC's Recording No. 8. An important point, however, is that this model is used only to "steer" the ATC in its adaptation—not to code the signal per se! Figure 25 shows a snapshot in time of the input signal spectrum, the dynamic bit assignment, and the receiver reconstruction for a digital coding rate of 8 kbits/s produced by the ATC.

Noise Shaping—The bit assignment prescribed by (24) produces quantization noise with flat spectral characteristics. Such characteristics are however known to be perceptually suboptimal (see the section on APC). Referring to the case where the input to the coder is a stationary random process with transform coefficient variances σ_i^2, the distribution of quantization noise across frequency can be controlled by a modified bit assignment rule:

$$b_i = \delta + \frac{1}{2} \log_2 \frac{W_i \sigma_i^2}{D^*} \qquad i = 0, \cdots, N-1 \qquad (27)$$

where W_i represents a positive weighting.

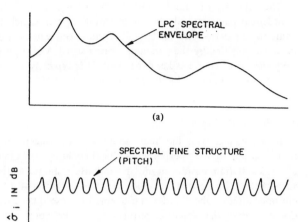

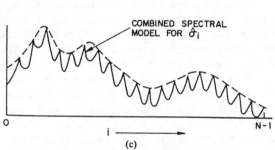

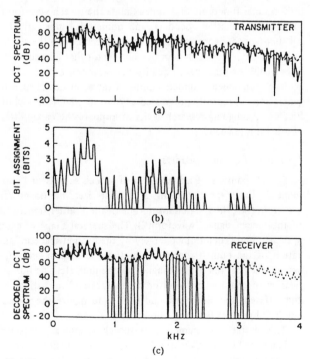

Fig. 24. Components of the speech spectrum model, (a) formant structure, (b) pitch structure, and (c) combined model.

Fig. 25. Illustration of operation of the "vocoder-driven" ATC algorithm, (a) DCT spectrum and spectral estimate, (b) bit assignment and (c) reconstructed signal.

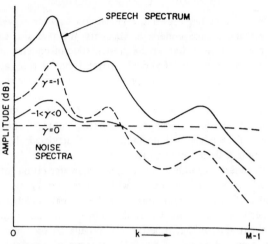

Fig. 26. Control of noise shaping in ATC by parameter γ.

This bit assignment minimizes the frequency-weighted distortion measure

$$D^* = \frac{1}{N} \sum_{i=0}^{N-1} W_i E_i^2, \qquad (28)$$

where E_i^2 represents the error variance associated with the quantization of the ith transform coefficient. The resulting noise spectrum is given by:

$$E_i^2 = CW_i^{-1} \qquad i = 0, \cdots, N-1 \qquad (29)$$

where C is constant.

Consider the class of weighting functions given by

$$W_i = \hat{\sigma}_i^{2\gamma} \qquad (30)$$

where γ is a constant and $\hat{\sigma}_i$ denotes the spectral envelope as denoted by the top curve in Fig. 24. For the case $\gamma = 0$ (uniform weighting), discussed previously, the noise spectrum is flat and the bit assignment will follow the log spectrum of the input, with appropriate bias. The case $\gamma = -1$ (inverse spectral weighting) leads to a constant bit assignment. Here, the noise spectrum will follow the input spectrum, and the SNR is constant as a function of frequency. As the value of γ varies between these two extremes $(-1 < \gamma < 0)$, the noise spectrum likewise changes from a flat distribution to one that follows the input signal spectrum. This variation is depicted in Fig. 26.

Notice that the weighting function W_i is really the spectral envelope of segment i raised to the 2γ power. The weighting is thus not associated with the harmonic structure of the spectrum, but captures only, in a controllable way, the spectral envelope of each block. For any particular transmission rate, the optimal value of γ can be determined by means of listening tests or, in principle, by an appropriate audibility criterion.

III. SOURCE CODERS; VOCODERS

We have earlier drawn the distinction between waveform coders and source coders, or vocoders. We have also earlier exposed our view that, at the present state of understanding, the synthetic quality of vocoder speech is not broadly appropriate for commercial telephone application. Nevertheless, vocoder techniques are valuable for special communications purposes, especially where digital speech must be conveyed over very low bit-rate channels. Reflecting this view, our treatment of vocoders will be even more abbreviated than our selection of topics in waveform coding.

Vocoders depend upon a rigid parameterization of the speech signal in accordance with a linear, quasi-stationary model of speech production. The traditional model, shown earlier in Fig. 1, is one where the *source* of the sound and the resonant *system* that modifies (or modulates) the sound are separable and do not interact. Let us summarize the elements of this model.

Sound Source Characterization

Sound can be generated by the vocal apparatus in three ways.

(i) Voiced sounds. The vocal cords, acting as an aerodynamic oscillator, "valve" the flow of air from the lungs into the pharynx. The resulting flow is a nearly periodic sequence of pulses of air. The acoustic impedance of the valve, relative to the driving point impedance of the tract, is moderately high, so that (to first order) the vocal cords can be thought of as a current (acoustic volume velocity) source which excites a linear, passive, slowly time-varying network. The spectral envelope of this periodic excitation is typically monotonously falling at about 12 dB/octave.

(ii) Fricatives: Sustained unvoiced (voiceless) sounds are produced from random sound pressure resulting from turbulent air flow at a constricted point in the vocal system. From a network standpoint, (and also to first order), this source appears as a serial voltage (pressure) source, whose internal impedance is essentially that of the constriction and is typically of large value. Actually this impedance is flow-dependent, as well as time varying. The spectrum of the pressure source is typically broad in bandwidth with gentle attenuation at the band edges.

(iii) Stops: Transient, unvoiced sounds are produced from an air pressure build-up behind a complete occlusion, and an abrupt release of this pressure. The result is a transient of pressure, much like a step function and hence having a spectrum falling as $1/f$, applied as a serial voltage (pressure) at the constricted place. The source impedance is dictated by the constriction size.

For vocoder purposes, the source for voiced sounds is represented by a periodic pulse generator. The source for unvoiced sounds is represented by a random noise generator (Fig. 1). The sources are normally considered mutually exclusive with a parametric signal indicating switching between voiced and unvoiced sources.**

** Voiced fricative sounds such as /v, z/ are normally classified as voiced.

The intensity of sound excitation for each source is also represented parametrically by an amplitude or gain signal. In addition, the periodicity or "pitch" of the voiced pulse source must be specified by the parametric pitch signal. Voice pitch frequency is of course talker-dependent. It is typically a two-octave range, about 50-200 Hz for men and about 100-400 Hz for women.

System Characterization

In accordance with the linear source-system concept, the sound output of the vocal tract is the convolution (in time) of the excitation waveform and the impulse response of the vocal system. In the frequency domain, this is equivalent to a multiplication of the Fourier transforms of these quantities. Consequently, the acoustic resonances of the vocal tract modulate or shape the broader spectre of the sources. Different speech sounds correspond uniquely to different spectral shapes. In a sense, the sound sources serve as an acoustic carrier for the speech intelligence so that it can be radiated into the air medium. The intelligence appears largely as time variations in the spectral *envelope* of the radiated signal.*** Vocoders depend upon a parametric description of the vocal-tract transfer function. This parametric description can take a variety of forms: for example, values of the short-time amplitude spectrum of the speech signal evaluated at specific frequencies (as in the channel vocoder), linear prediction coefficients that describe the spectral envelope (LPC vocoder), frequency values of major spectral resonances (formant vocoder), specified samples of the short-time autocorrelation function of the speech signal (autocorrelation vocoder), coefficients of a set of orthonormal functions that approximate the speech waveform (orthogonal function vocoder), and numerous other variants. All such vocoder approaches, however, depend upon the signal model of Fig. 1. And therein lies one factor that immediately fixes a ceiling to the speech quality that vocoders can presently achieve. Fundamental understanding of the acoustics of speech generation has recently progressed beyond the simple model of Fig. 1 and on-going research seeks to improve the capabilities of low bit-rate vocoders.

Frequency Domain Vocoders

Speech consists of a succession of speech sounds or "phonemes." Each speech sound is characterized by its power spectrum (i.e., the squared magnitude of the Fourier transform of the speech signal "waveform"). The spectral shape of each speech sound is determined largely by the geometrical configuration of the human speech production mechanism—in other words, the position of the tongue, lip opening, etc. The human hearing mechanism has evolved to analyze and to distinguish the different speech spectra and thus to recognize the succession of phonemes that make up speech.

This simple fact, namely that speech sounds are encoded and analyzed by their spectra, is exploited directly in a so-called channel vocoder. Clearly, instead of transmitting the telephone

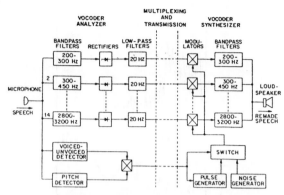

Fig. 27.   Block diagram of a channel vocoder.

Fig. 28.   Sound spectrograms of the utterance "Speech Communications" for a 15 channel vocoder and the original speech.

speech signal directly (with a bandwidth of nearly 4000 Hz), one can transmit only the spectrum of each speech signal, which can be reasonably well described by 16 values along the frequency axis, taken every 20 ms. According to the sampling theorem, each spectral channel requires a bandwidth of $1/(2 \times 20$ ms$) = 25$ Hz. The total bandwidth required for transmitting the speech information equals $16 \times 25 = 400$ Hz or about one tenth of the bandwidth of the speech signal itself. (In digital transmission of the channel signals, further savings can be realized by using DPCM in combination with maximum-entropy coding because the spectrum signals are relatively constant during steady-state portions of speech.)

A block diagram of a spectrum channel vocoder is shown in Fig. 27. The speech signal is separated into contiguous spectral bands. The output of each filter is connected to a rectifier and a low-pass filter whose output represents the time-varying average signal amplitude for each frequency band.

Also shown in Fig. 27 are a voiced-unvoiced detector and a pitch detector which determine the fine structure of the speech signal and produce a corresponding narrow-band signal. These narrow-band signals are combined into a single signal with a total bandwidth of about 300 Hz. Thus, the transmission bandwidth is only one-tenth of that required for the original speech signal.

At the synthesizer, the original channel signals are utilized to control the frequency response of a time-varying filter (consisting of modulators and narrow band-pass filters), to correspond to the spectral envelope measured at the analyzer. The input of this time-varying filter is supplied with a flat spectrum excitation signal of the proper spectral fine structure (quasi-periodic pulses for voiced speech sounds, or "white" noise for unvoiced sounds). The performance of a 15-channel vocoder is illustrated by the spectrograms in Fig. 28.

An even more efficient description of the speech information is obtained by specifying only the frequencies of spectral peaks (or formants) and their amplitudes. The resulting vocoder is called a formant vocoder. This technique requires much more sophisticated analysis to derive the formant data.

Time Domain Vocoders

Instead of the spectral information, samples of the autocorrelation function can be used to characterize speech sounds.

For telephone-bandwidth speech such samples would have to be spaced $1/(2 \times 4000$ Hz$) = 0.125$ ms apart. But the number of samples would have to be quite large (perhaps 30) to ensure sufficient spectral resolution at low frequencies. (By contrast, for a channel vocoder, the spectral resolution can be tailored to the bandwidths of speech formants and to characteristics of auditory perception: higher resolution at low frequencies and lower resolution at intermediate and higher frequencies.) Thus, an autocorrelation vocoder would require somewhat greater bandwidth for transmitting the spectral information than a channel vocoder. Another problem is the synthesis of speech whose autocorrelation function has been specified. Simply generating a quasi-periodic sequence of replicas of the autocorrelation function results in a speech signal whose spectrum is the squared spectrum of the original speech signal. Thus, spectral square-rooting is required.

The spectral square-rooting is avoided by cross-correlation vocoders in which samples of the cross-correlation functions between the speech signal and a "spectrally-flattened" coherent signal are transmitted. As in autocorrelation vocoders, the number of samples, and thus the total transmission bandwidth, are higher than in channel vocoders of comparable output speech quality. Also, for digital transmission, the number of bits per correlation sample needs to be about twice as high as for spectral samples (7-8 bits/sample instead of 3-4 bits/sample for good speech fidelity).

Thus, "time-domain" vocoders, as auto- and cross-correlation vocoders have been called collectively, appear to suffer from several defects compared to "frequency-domain" vocoders such as channel and formant vocoders. Yet, one of the most successful innovations in speech analysis/synthesis, linear predictive coding, is based on autocorrelation analysis (also called autoregression or maximum entropy analysis). The difference lies in the way in which the autocorrelation information is used for synthesis (see following comments on LPC vocoders).

Orthogonal Expansion Vocoders

The autocorrelation function can be viewed as an expansion of the power spectrum into an harmonic cosine-series. Similarly, the power or amplitude spectrum, or the autocorrelation function, or even the signal waveform itself, can be expanded into a series of orthogonal functions. For waveform expansion, the Laguerre polynomials are particularly suited because of the weighting function e^{-t} in conformity with the time envelope of a speech formant.

For the power spectrum (or, in fact, any other suitable representation of the signal), the eigenfunctions of the auto-covariance matrix are an optimum set of expansion functions for minimum r.m.s. error under quantization. Such expansions are called Karhunen-Lòeve ("KL") expansions.

Instead of expanding the power spectrum, its logarithm can be expanded into a cosine series. The sequence of expansion coefficients is called "cepstrum" and a vocoder based on the cepstral coefficients is called a cepstral or "homomorphic" vocoder. The advantage of cepstral vocoders is that the cepstrum is often available already for measuring voice pitch and no additional analysis is needed.

LPC Vocoders

Another method of representing the spectrum of a speech sound is by means of the linear predictive coefficients (see the earlier discussion of APC). In essence, an LPC vocoder is an APC system in which the prediction residual has been replaced by the pulse and noise sources as in Fig. 1. Figure 29 shows the LPC approximation of speech spectrum envelope as a function of p, the number of predictor coefficients. For the telephone band, the spectral resolution provided by $p = 8$ is as good as, or better than that in a 15-channel vocoder. This superiority of predictor coefficients stems from the fact that they define an "all-pole" spectrum—precisely the kind of spectrum that vowels possess. They can also cope with formants of narrow bandwidths. Movements of formant frequencies are followed continuously—not stepwise as in channel vocoders.

The predictor coefficients, as well as so-called "partial correlations" (PARCOR), pseudo-areas, or log-area coefficients—like vocoder channel signals—show a high degree of correlation that can be exploited by Karhunen-Lòeve or similar transformations.

Our remarks here have centered largely on the "spectral envelope" of speech sounds—as distinguished from the spectral fine structure exhibited by voiced speech sounds. Voiced speech sounds are quasi-periodic and thus their spectra show spectral "lines" lying equidistantly ("harmonically") along the frequency axis. These spectral lines, or harmonics, impart a comb-like fine structure to the spectrum. (In the topmost spectrum in Fig. 29, this fine structure has a fundamental fre-

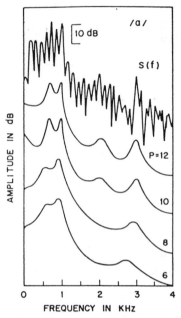

Fig. 29.   Comparison of the actual speech spectrum with the spectral envelope determined by the linear prediction method for different values of p, the number of predictor coefficients.

quency of 140 Hz.) To a listener, these spectral lines represent intonation, or pitch, of a spoken phrase. For synthesis of natural speech, this pitch information must be accurately rendered and, in analysis-synthesis speech coding, has to be extracted from the input speech signal. Thus measurement of the pitch parameter is a fundamental requisite for all vocoders.

Pitch Detection

The extraction of the fundamental frequency from a running speech signal, traditionally called "pitch detection", is one of the more difficult tasks in speech analysis—especially for telephone signals in which the fundamental frequency is physically absent (but still fully perceived, owing to the "residue pitch" effect of hearing).

A remarkably extensive literature exists on the problem of voice pitch detection—reflecting, perhaps, the challenge that it represents. As yet, no completely satisfactory solution exists, owing at least partially to the ill-posed nature of the parameter. The basic fault lies in forcing the speech signal into the simplistic model of voiced and unvoiced sounds. There simply are instances where the excitation for voiced sounds is not well represented by a periodic pulse train, or that of unvoiced sounds by a noise generator. Nevertheless, sophisticated techniques exist for approximating the vagaries of voice pitch. Among the foremost are techniques such as the cepstrum, the autocorrelation waveform periodicity measures (using diversity and majority vote) and others. Traditionally, also, the positive detection of periodicity is used to provide the voiced/unvoiced switching of the sources.

Related Vocoder Devices: Hybrid Waveform Coders/Vocoders

The sheer difficulty of pitch detection has spawned a number of vocoder devices where the objective is to settle for

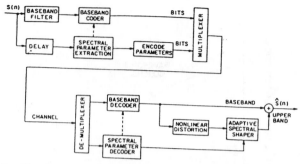

Fig. 30. Hybrid waveform-coder/vocoder in which the baseband is coded by waveform coding and the remaining upper frequency range is vocoded by a voice-excited vocoder.

less bandwidth conservation to avoid the problem and complexity of pitch detection. The family of voice-excited vocoders (VEV) and phase vocoders (ϕV) are representative of earlier efforts. More recently, hybrid arrangements of SBC, APC and LPC are coming into vogue, where a portion of the coding (generally a low frequency baseband) is accomplished by waveform techniques and a portion (upper frequency band) by voice-excited vocoder techniques (see Fig. 30). These combinations aim particularly for the transmission bit-rate range of about 4.8 to 9.6 kbits/s. This is an especially interesting range for reasons to be mentioned shortly.

Examples of speech output from some of the vocoders discussed above are presented on the sound record. The performance of vocoders varies considerably from one speaker to another and from one speaking environment to another. Thus, it would not be very meaningful to compare the performance of these vocoders on the basis of speech samples presented in the record. The intent here is to provide a (more or less random) sample of the possible differences between the sound outputs of various vocoders. The recordings include a channel vocoder (no. 9), a linear predictive vocoder (no. 10), a phase vocoder (no. 11), and a voice-excited vocoder (no. 12). The listening environment has a considerable influence also on the preceived quality of vocoded speech.

IV. TRANSMISSION ISSUES

Transmission issues are largely orthogonal to the issues of speech coding *per se*, but sometimes transmission factors are critical to the choice or design of a coding strategy. Some considerations that impact coder design include transmission error environment; tandem (or multiple) codings; encryption; and packet formats.

Channel Errors

Subject to some qualifications and exceptions, one can say that "tolerable" bit error rates in most speech coding procedures (waveform coders as well as vocoders) are in the order of 10^{-3}. One typically gains order-of-magnitude advantages (tolerable error rate = 10^{-2} or more) by using so-called robust versions of coding algorithms, and by using explicit methods of bit protection (error-correction/coding), or by speech smoothing operations at the receiver. The numbers given above are only rules of thumb and they can vary depending on coder

type and speech-usability criterion. Also, the perceptual eff of transmission errors can be very different with diffe: coding strategies. The only issues that we can address ourse to in the present paper are the notions of 'robust' quanti: and coders. The following are some of the procedures use(realize robust versions of speech coders. These procedures also relevant to vocoder-parameter transmission.

(a) Subdued Quantizer-Adaptation Algorithm—The c word memory algorithm (7) can be replaced by a 'lea adaptation logic

$$\Delta_r = \Delta_{r-1}^{1-\beta} \cdot M(\,|\, \text{quantizer output}\,|\,)$$

$$\beta : \text{positive}, \Rightarrow 0 \text{ e.g., } \beta = 1/64, 1/32 \qquad (3$$

which 'leaks' out the effects of a wrong step-size caused transmission error (and a consequent erroneous quanti: output). This modification has been employed successfully time-domain coding (ADPCM) as well as for APCM qu: tizers used in frequency domain (sub-band) coding.

Similarly, a robust version of ADM can be realized in adaptation procedure that is characterized by a longer-memo: M (typically $M = 2$ or 3) for step-size increases, and an eve longer (syllabic, several ms long) time constant (determined b the typical number .996 below) for step-size decreases:

$$\Delta_r = 0.996\,\Delta_{r-1} + \Delta_0 \cdot [\text{ADAPT}]_r$$

$$[\text{ADAPT}]_r = 1 \text{ if } \left| \sum_{s=0}^{M} b_{r-s-p} \right| = 4 \text{ for } p = 0 \text{ or } 1 \text{ or } 2$$

positive,
$[\text{ADAPT}]_r = 0$ otherwise; Δ_0: input-proportional   (32
scaling factor

where

$$b_u = \pm 1 \text{ (DM bits).}$$

Known sometimes as CVSD (continuously variable slope deltamod), the above type of syllabically companded DM is well-known for its transmission error performance. Differential coders are in fact more robust, as a class, than non-differential PCM.

(b) Explicit Transmission of Code Parameters—Quantizer and predictor parameters are often inherently contained in the mainstream of a speech coder output. This is true for example of the step-size information in adaptive quantization with a one-word memory; and of the predictor coefficient information in adaptive prediction procedures that use so-called Gradient Search or Stochastic Approximation methods. When the channel error rates are very high it may be useful to dedicate a fraction (typically 10% or less) of the coder bit rate for explicit transmission of adaptation information in a special error-protected format. Clearly, the procedure works best with a block strategy, with the adaptation information being updated once every block. For this reason, explicit transmission of side information is naturally appropriate for the block quantizers in Transform Coding.

(c) Subdued Prediction—A simple illustration of subdued prediction is the use, in first-order DPCM, of a predictor value a_1 that is lower than the autocorrelation-based value of c_1. The notion can be extended to higher-order prediction, and presumably even to Transform Coding.

Recordings no. 13 to 16 demonstrate effects of channel errors for log-PCM (recording no. 13), syllabically companded deltamod (recording no. 14), sub-band coding without and with channel error protection (recording no. 15) and linear predictive vocoding (recording no. 16).

Tandem Coding

Tandem codings of speech in communication links may involve identical or nonidentical coder stages. For identical coders in tandem, one is concerned with speech quality as a function of the number of identical encodings. If the encoding stages are separated by intermediate stages of digital-to-analog conversion, the distortions introduced by different coding stages tend to be statistically independent, identical and therefore additive and predictable in some sense. On the other hand, if one is concerned with multiple encoding/decoding and code conversions in all-digital format, there is generally less reason to expect identical, statistically independent quality losses at each stage. In any case, with both hybrid and all-digital networks, there is a tendency for most of the quality loss to occur in the first stage. Thus in DPCM or SBC most of the slope or band-limiting effects occur in the first stages of encoding; and the second stage faces a simpler task because its input (due to slope or band-limiting) is simpler to handle than the original speech. Nevertheless, quality losses in the second and subsequent stages are significant enough to set a limit (typically 2 to 4) on the tolerable number of encoding stages.

The category of nonidentical coders is exemplified by the tandeming of 'wideband' (e.g., 16 kbit/s DM/DPCM/SBC) coders with 'narrowband' (e.g., 2.4 kbit/s LPC) vocoders, not necessarily in that order. The above situation motivates two interesting problems: (1) Redesign of the wideband coder to improve speech analysis (e.g., voiced/unvoiced decision, pitch extraction) in the narrowband coder; and (2) Redesign of the narrowband coder (e.g., use of all-pass filtering and non-impulsive excitation in LPC) to provide a simpler input to the waveform coder. The gains from the above reoptimizations have typically been of second-order; and in an experiment with the three 16 kbit/s coders mentioned above, the only coder with a fair degree of compatibility with 2.4 kbit/s LPC was sub-band coding.

Encryption

One of the attractive properties of digitized speech is the ease with which it can be encrypted. Digital encryption can be accomplished either by masking speech-carrying bits (with a pseudorandom binary noise sequence known to the intended receiver), or by permuting their positions within a block of length N (Recording number 17). In general, the residual intelligibility from permutation is always higher than in masking. But the residual intelligibility does decrease with increasing N. In experiments with ADM, APCM and ADPCM, bits permutations were found to be least effective (for a given N) in the case of ADM, and most effecitve in the case of ADPCM.

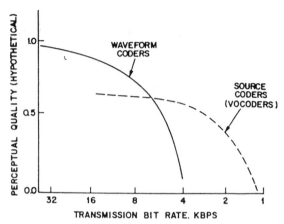

Fig. 31.   Variable rate coding configuration.

With 24 kbit/s or 16 kbit/s ADPCM, $N = 16$ would be adequate, implying a simple encryption procedure with an attractively short encoding delay of about 1 ms. The procedure could be very appropriate for providing casual privacy in applications like mobile radio.

Speech can also be encrypted without digitization. In fact, many conventional and commercially available speech scramblers use analog encryption techniques such as frequency inversion and time segment permutation. Most analog techniques have the nice property of not expanding the signal bandwidth as a result of encryption. But, as a class, analog encryption tends to be less secure than a digital procedure (Recording number 18).

Variable Rate Coding: TASI and Packet Systems

It is often assumed in the design of speech coders that the coder and channel operate at fixed bit rates. In reality, however, speech is an intermittent, nonstationary process and user demand on a communication system is a variable process. The first property, that of an intermittent source is exploited in systems such as TASI (Time Assignment Speech Interpolation) systems. The second property, that of a variable demand on the system, is being explored for use in packet transmission systems.

In both of the above systems an important element is a variable rate coder. In its simplest form it may amount to a trivial transmit/no-transmit decision as is used in initial TASI systems. In such systems a large number of users (say 100) share a smaller number of channels at any instant. Channels are allotted only to users who are actively speaking. Since, on the average, individuals are actively talking less than 50 percent of the time, a factor of at least 2 in efficiency is gained in the system.

In the most general case, variable rate coding may be characterized according to Fig. 31 where both the source activity and the channel rate are assumed to be variable. The buffer is used to "take up the slack" between the source and the channel and smooth out the fluctuations. A number of variations on this approach are possible. Perhaps the simplest is block processing, where a block of N samples is encoded with a total of B bits such that B/N is the average transmission rate at that time. The allocation of bits across the block can be made according to rate-distortion relations. The SNR gain over that of a fixed bit/sample allocation over the block can be shown to be related to the arithmetic/geometric mean ratio of the expected signal variance across the block. (This relation is similar to that between the Spectral Flatness Measure [SFM] and prediction gain in the frequency domain.) If the speech is highly nonstationary across the block (i.e., the signal variance fluctuates greatly), a large gain can be expected. For a single speaker, however, speech is locally stationary over periods of 30-50 ms and block sizes much greater than this would have to be used before nonstationarity in time can be exploited. For the case of multiple users, as in TASI, P speakers can share a single channel by assigning each user a sub-block of N/P samples, and concatenating the sub-blocks into one large block.

Voice Storage

In voice storage applications, the objective is not to transmit encoded bits, but to simply store them in memory in an efficient manner. Here again, concepts similar to variable rate coding apply. Ideally, the input signal is encoded and stored as bits only when the speaker is talking and the encoder bit rate slows down or stops during silent intervals. The problem is analogous to a voice-activated tape recorder that records during speech and stops during silence. (Included in the bit stream must be a sufficient amount of information about silence-durations, so that one can reproduce the original timing of the speech during playback.)

Transmission of Non-Speech Signals

An important problem in certain speech communication links is the need for transmitting non-speech signals such as voice-band data. Data signals have waveforms and spectra that are statistically very different from speech signals. Consequently encoders designed for efficient digitization of speech are, in general, very sub-optimal for the encoding of data signals. One approach to this problem is the development of compromise designs that will not be severely sub-optimal for either class of signal (for example, compromise predictors for DPCM). Perhaps a more useful, if not simpler, approach is one that first identifies the coder input and then switches to an appropriate strategy.

V. ASSESSMENT OF CODER PERFORMANCE

Objective performance measures are not well established and are typically used only as guideposts in coder design. Formal judgments on coded speech quality almost inevitably depend on subjective testing. The methodology of these testing procedures, as well as the statistical techniques used to interpret the results, are subjects beyond the scope of this paper. What we shall attempt here is a simplified, qualitative outline of techniques for assessing coder performance.

A. Subjective Assessment

Intelligibility—Speech intelligibility usually is not a problem in waveform-coded speech unless the bit rate is low and, in addition, there are demanding transmission requirements (such as very high bit error rates, or tandem encoding involving vocoders or multiple stages of the waveform coding. Intelligibility is more of an issue with vocoders. The intelligibility of vocoded speech depends—in some cases strongly the nature of the input speech, such as sex and age of speaker; additive noise, peak limiting and other amplitude nonlinearities, frequency shift (suffered in carrier transmission systems) and prior speech processing (e.g., DM, PCM, adaptive quantizing, or even vocoding itself). Intelligibility tests under difficult listening conditions are sometimes used to assess quality of synthetic speech. Although the intelligibility may high in a quiet environment, it may drop appreciably (in manner that depends on speech quality), in a noisy environment; in the presence of distracting signals; or during the performance of a distracting task.

The literature on intelligibility tests is extensive and several paradigms are used for the tests. All, however, depend upon human listener indicating his understanding of *what* was said whether it be single syllables, words, phrases or sentence Tests can therefore differ in their contextual redundancies Nonredundant stimuli lead to the smallest intelligibility scores in view of the redundancies in real-life speech stimuli, these scores tend to constitute conservative lower bounds for "acceptability."

Speaker Recognizability—A particularly important aspect of coder performance is speaker recognizability. This is important not only in telephone conversations among relatives and friends, but even more so in many business and government transactions by voice. In fact, speaker recognition by human listeners is a significant paradigm in its own right, as well as being an additional subjective measure of speech quality. Speaker recognizability, like intelligibility, tends to be a minimal issue in waveform coding. Many vocoders, on the other hand, have a tendency to make everybody sound alike (namely like a vocoder) and thus have poor speaker recognizability.

B. Objective Assessment

Most auditory experiments support one basic point—that acceptable voice communication is achieved by the preservation of the short-time amplitude spectrum of the signal. For a real signal $x(t)$, the short-time Fourier transform is the complex quantity

$$X(\omega, t) = \int_{-\infty}^{t} x(\lambda) h(t - \lambda) e^{-j\omega\lambda} \, d\lambda \qquad (33)$$

where $h(t)$ is a realizable low-pass function that essentially specifies the requisite resolution in frequency ω and time t. For the auditory mechanism, the $h(t)$ analysis window corresponds roughly to the low-pass transformation of contiguous "critical-band" filters. The short-time amplitude spectrum is the absolute value $|X(\omega, t)|$ with a discrete-time equivalent $|X(e^{j\omega})|$ as in (1)

One of the first things that the engineering student learns is that the ear makes a crude Fourier analysis of signals, and that the ear does not pay much attention to phase—that is, the ear is "phase deaf." Preservation of the quantity $|X(\omega, t)|$ is essentially consistent with both these statements, but there are finer points. (White noise and a single impulse have the same Fourier amplitude spectrum, but they sound quite different because of different phase spectra. They also have distinctly different short-time spectra.) What is significant for speech coding is that some loss of phase information is indeed permissible; and this implies also that exact facsimile reproduction of speech waveforms (while being completely sufficient for a high perceived quality) is not necessary for acceptable voice communication. If an important class of speech coders, viz., waveform coders, are indeed phase-preserving, the property is an incidental by-product of their simplicity, rather than a result of intentional design!

Objective Quality Measures That Apply Only to Waveform Coders—Waveform coders, at least in their simpler versions, preserve the amplitude spectrum by the simple process of mimicking the speech waveform. This means that simple signal-to-error ("noise") measures, where an error sample is defined as the difference between corresponding samples of input (x) and output (y) speech, have the potential of characterizing waveform coder quality. (Refer to Fig. 8).

The simplest of these measures is the long-time-averaged signal-to-noise (s/n) ratio

$$\text{SNR} = 10 \log \frac{\sum_n x^2(n)}{\sum_n [y(n) - x(n)]^2}, \tag{34}$$

where summations are typically over the duration of a sentence—length utterance. This measure is strongly influenced by the high energy components of a speech waveform, and a waveform coder design based on this SNR measure tends to be unfair to low energy segments (such as fricatives) whose preservation is perceptually very important.

An improved measure, referred to as a segmental ratio SNRSEG, computes SNR in short-time (typically 20 ms) segments and only for active portions of the signal, i.e., silence is discarded. A long-time-average of the "local" SNR values then tends to assign more equitable weightings to loud and soft parts of a sentence:

SNRSEG = Long-time averaged value of (short-time SNR in dB).

(35)

Despite these refinements, measures based solely on waveform differences tend to be inadequate, especially for low bit rate encodings and for cross-comparisons between widely

different types of coders. Part of this inadequacy is due to the fact that a low value of SNR (as derived from the waveform) does not necessarily mean perceptually unacceptable spectrum preservation.

Generalized concepts of objective quality measurement— Let us pursue briefly the notion of the short-time amplitude spectrum as a basis for objective quality measurement. To quantify the fidelity required in preservation of this spectrum, explicit account must be taken of the properties of human audition.

From the previous discussions we recall the short-time spectrum exhibits two attributes that are highly speech-specific— the spectral envelope, which primarily reflects the vocal-system transfer function and, in the case of voiced speech sounds, a fine structure which reflects the harmonic lines of the fundamental frequency, or voice pitch. Several psychoacoustic experiments have established human auditory acuity for perceiving changes in spectral envelope and voice pitch under conditions that are the most favorable for detecting the change, that is, in so-called differential discriminations, or close comparisons. These data constitute limits to audition perception, and can be used as guide posts in coder designs.

Because speech intelligibility is carried largely in the spectral envelope, this attribute provides a logical basis for an objective measure. This view, in fact, spawned the earlier concept of density of interfering noise can be used, together with the so-called auditory bands of equal contribution to intelligibility (similar to critical bands), to compute word and sentence intelligibility. This technique, which has served for a long time, is applicable for linear frequency distortions and additive interfering noise (which quantizing noise can be considered to be under some circumstances). It is not accurate for non-linear processing, such as utilized in vocoders.

More recent work has aimed to use the short-time spectral envelope to develop perceptually meaningful objective spectral distance measures that can be accumulated over the running speech signal. Alternatively, the approach can be used to calculate a spectral density signal-to-noise ratio for short-time segments. The fundamental ingredients are the evaluation of the short-time amplitude spectrum on a frequency-warped scale (corresponding to the equal articulation bands, or to the auditory critical bands), and a non-linear transformation of spectral magnitudes to approximate the relationship between subjective loudness and amplitude. Typically, a 20-30 ms duration of signal is used. The technique has already proved useful for shaping the quantizing noise of waveform coders (see sections on APC and ATC), and appears to have utility for assessing vocoder quality.

VI. CONCLUSION

In nailing the final hump on this "committee camel," some comment is appropriate about the trades that are possible among quality, complexity and bit-rate—across a broad spectrum of coders. Comprehensive, quantitative data on these interplays are not available, but some speculation and hypoth-

350

esizing can be done on the basis of experience. In particular, some comparisons can be drawn between waveform coders and vocoders.

Complexity

In the foregoing we have touched upon principles involved in a variety of speech coders. The coders range widely in complexity. We can put this point into perspective by estimating their complexity relative to a simple adaptive delta modulator (ADM).

Relative Complexity †		Coder
1	ADM:	adaptive delta modulator
1	ADPCM:	adaptive differential PCM
5	SUB-BAND:	sub-band coder (with CCD filters)
5	P-P ADPCM:	pitch-predictive ADPCM
50	APC:	adaptive predictive coder
50	ATC:	adaptive transform coder
50	ΦV:	phase vocoder
50	VEV:	voice-excited vocoder
100	LPC:	linear-predictive coefficient (vocoder)
100	CV:	channel vocoder
200	ORTHOG:	LPC vocoder with orthogonalized coefficients
500	FORMANT:	formant vocoder
1000	ARTICULATORY:	vocal-tract synthesizer; synthesis from printed English text.

† Essentially a relative count of logic gates. These numbers are very approximate, and depend upon circuit architecture. By way of comparison, Log PCM falls in the range 1-5.

Quality—Our experience to date suggests that toll quality (say, 56 kbit/s log PCM-equivalent) coding of speech can be obtained with the following coders running at, or above, the indicated transmission bit rates:

Toll-Quality Transmission

Coder	kbits/s
Log PCM	56
ADM	40
ADPCM	32
SUB-BAND	24
Pitch Predictive ADPCM	24
APC, ATC, ΦV, VEV	16

Similarly, our experience suggests that communications quality can be achieved by the following combinations of coders and *minimal* bit rates:

Communications-Quality Transmission

Coder	kbits/s
Log PCM	36
ADM	24
ADPCM	16
SUB-BAND	9.6
APC, ATC, ΦV, VEV	7.2

Crossing down into the synthetic-quality range, consistent. with source-coding techniques (vocoders), characterizations become even more difficult, but performance is typified by:

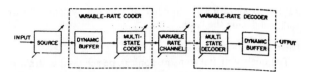

Fig. 32. Impressionistic assessment of performance of current waveform coders and vocoders.

Synthetic-Quality Transmission

Coder	kbits/s
CV, LPC	2.4
ORTHOG	1.2
FORMANT	0.5

Trades Among Quality, Complexity and Coding Rate

If one compares—in a loose, impressionistic way—the performance of current waveform coders and vocoders, the relationship to transmission bit rate might be sketched as in Fig. 32. The quality of waveform coders diminishes significantly as the bit rate is reduced below 16 kbits/s. Their quality tends to diminish precipitously below about 8 kbits/s. Vocoders achieve a maximum quality at about 4.8 kbits/s and cannot rise much further, no matter how much the bit-rate is increased. Reduction of their bit rate to the order of 2.4 kbits/s is accompanied by relatively small reduction in quality. Additional processing to orthogonalize the parametric description of the speech signal—or even to select from an ensemble of spectral patterns—results in similar synthetic quality (down to about 1.2 kbits/s). Further bit rate reduction usually imposes significant quality reductions. Presently, the quality crossover point between waveform coders and vocoders possibly lies in the vicinity of 7.2 kbits/s, with a difference in coder complexity of the order of 2. In the middle range of about 4 to 10 kbits/s, the hybrid vocoder methods mentioned in Section III serve to provide a useful compromise between the two techniques, with the possibility of a better overall quality than provided by either method individually.

The present limitations on the quality of vocoded speech seems to be a result, at least in part, of the simple model used to parameterize the speech information. A more sophisticated signal model, that better duplicates the short-time amplitude spectrum of the signal and the natural acoustic interactions between source and system, might elevate this ceiling on vocoder quality. What are the possibilities in this direction?

One quite parochial effort focuses on a much more sophisticated acoustic model of the signal. The sound source and resonant system are not considered independent in this model, but are allowed to interact (i.e., acoustically load one another). The model incorporates some deeper understanding—of the acoustics of speech production. Voiced sounds are produced by aerodynamic vocal cord vibration, and voiceless sounds by turbulent air flow in vocal constrictions. The formulation accounts for sound radiation from the yielding side walls of the vocal tract, and includes detailed loss factors for viscosity, heat conduction, and radiation resistance. Quite significantly, this model requires no greater information rate for its parameters than does the simpler source-system model, but it uses the information more effectively. This more sophisticated "mold" for the speech signal may therefore provide part of the key to elevating the quality of vocoder systems. Future work will decide the issue.

Where the Action Is

Toll quality digital transmission can be achieved with simple coders at 40 kbits/s (ADM), 32 kbits/s (ADPCM) and 24 kbits/s (SBC). Mobile radio telephone quality at 24 kbits/s with the same relatively simple coders also seems feasible. With increased complexity (APC, ATC), toll quality at 16 kbits/s can be attained. The necessary complexity for this is becoming much less frightening—with circuit complexities and densities comparable to complete fast Fourier transforms being planned for single-chip fabrication. New designs for digital transmission and for customer systems will benefit substantially from these hardware advances. The understanding is essentially in hand. Development is needed.

Research efforts might press in two directions. Look again at Figure 2. The boundary between toll quality and communications quality probably is a "soft" one, and susceptible to new fundamental understanding. This boundary is, in fact, the wall that we're "leaning" on at present, with new studies in APC and ATC. Pushing this boundary down to the region of 14 to 9.6 kbits/s would constitute a valuable advance. Why? Private lines, adaptively equalized, can transmit 9.6 kbits/s with 10^{-5} or better bit-error-rate (ber). At 16 kbits/s, however, current expectations are closer to 10^{-2} ber, not good enough for toll quality speech transmission. Toll quality speech transmission at bit rates of 10 kbits/s to 14 kbits/s might therefore be interesting both for digital radio and for wire transmission, using modems which could achieve 10^{-3} ber, or better.

By the same token conventional, dialed-up terrestrial lines can be expected to support 7.2 to 9.6 kbits/s with ber of 10^{-3} to 10^{-2}. These rates for speech coding and bit error presently correspond to less-than-desirable communications quality. Again, finding sophisticated ways to elevate speech quality at data-coding speeds, and to moderate the effects of transmission errors on conventional channels, is of substantial interest. Even relatively small advances here might make the difference between achieving good useful systems and unacceptable performance.

In one direction, speech research in progress aims to push the toll-quality/bit-rate boundary down toward the neighborhood of 9.6 kbits/s, while at the same time exploring possibilities for error detection and partial correction that would allow the coder to perform in a more severe error environment. In another direction, reducing the ber on the 9.6 kbits/s channel by sophisticated processing, is a research challenge that properly belongs to data transmission. In still another, but more speculative direction, a more accurate and comprehensive model of the speech signal may allow the full potential of source coding to be realized without inordinate sacrifice in speech quality. This is a research frontier that acoustic analysis and perceptual quantification will continue to address.

APPENDIX A
DESCRIPTION OF SPEECH CODING RECORD

(All speech files, except three files in [12] have been filtered to a bandwidth of 200-3200 Hz)

[1] (12-bit quantized) PCM Performance as a function of sampling frequency. 4 examples: 10, 5, 2.5 and 1.25 kHz.

[2] (10 kHz-sampled) PCM Performance as a function of the number of quantization bits. 5 examples: 12, 9, 4, 2 and 1 bits.

[3] 32 kbit/s coders. 3 examples: Log PCM, adaptively quantized DPCM, adaptively quantized Deltamod.

[4] 16 kbit/s coders. 2 examples: adaptively quantized DPCM, adaptively quantized Deltamod.

[5] Subjective attributes in differential coding. 4 examples: Crackling Noise, Hoarseness, Background Noise and Muffling.

[6] Adaptive Predictive Coding. 2 examples: 3-level quantization and 2-level quantization.

[7] Sub-band Coding. 5 examples: Uncoded original followed by encodings at 32, 24, 16 and 9.6 kbits/s.

[8] Vocoder-driven Transform Coding. 5 examples: Uncoded original followed by encodings at 16, 9.6, 8 and 4.8 kbits/s.

[9] Channel vocoder. 14 channels. 2.4 kbits/s.

[10] Linear Predictive Vocoder. Uncoded original followed by encoding at 2.4 kbits/s.

[11] Phase vocoder. 7.2 kbits/s followed by unquantized version.

[12] Voice-Excited Vocoder. 3 examples: Each example is a word-pair. The first word of each pair is a 3.2 kHz version. The second is a 10 kHz-bandwidth version. This Voice Excited Vocoder version uses a 2 kHz baseband and a 6-channel representation of the higher frequencies.

[13] Effects of Transmission Errors on 56 kbit/s log-PCM. 2 examples: Random error probabilities of 0.1 and 1.0 percent.

[14] Effects of Transmission Errors on syllabically companded deltamod at 32 kbits/s. 3 examples: Random error probabilities of 0.1 and 10%.

[15] Effects of Transmission Errors on Sub-band Coding at 16 kbits/s and random error probability of 2%. 3 examples: [Coder with Adaptive (PCM) Quantizer], [Coder with Robust Adaptive Quantizer] and [Coder with part of speech band dedicated for error protection].

[16] Effects of Transmission Error on Linear Predictive Vocoder at 3.6 kbits/s. 3 examples: zero errors, 1% random errors and 1% errors followed by error corrections, provided by a 50% expansion of channel capacity.

[17] Encryption of 3-bit ADPCM speech. 3 examples: [Unscrambled original], [bit-masking] and [bit permutation in 16-bit blocks].

[18] Encryption of analog speech. 4 examples: Frequency Inversion, Temporal Permutation, Gain Permutation and unscrambled original.

[19] Silence Elimination in Speech Transmission. 5 examples: Fractional duration of eliminated signal equals 0, 12, 18, 40 and 50%.

[20] Manipulation of Silence Duration. 6 examples: Durations of synthetic silences, relative to original silences, are 100, 0, 50, 150, 200 and 400%.

REFERENCES

1. N. Ahmed and K. R. Rao, *Orthogonal Transforms for Digital Signal Processing*, Springer-Verlag, NY 1975.
2. J. B. Anderson and J. B. Bodie, "Tree encoding of speech," *IEEE Trans. Inform. Theory*, vol. IT-21, pp. 379–387, July 1975.
3. B. S. Atal and S. L. Hanauer, "Speech analysis and synthesis by linear prediction of the speech wave," *J. Acoust. Soc. Amer.*, vol. 50, no. 2, pp. 637-655, August 1971.
4. B. S. Atal and L. R. Rabiner, "A pattern recognition approach to voiced-unvoiced-silence classification with applications to speech recognition," *IEEE Trans. Acoust., Speech, Signal Processing*, vol. ASSP-24, no. 3, pp. 201-212, June 1976.
5. B. S. Atal and M. R. Schroeder, "Predictive coding of speech signals," *Bell Syst. Tech. J.*, vol. 49, pp. 1973-1986, October 1970.
6. B. S. Atal and M. R. Schroeder, "Predictive coding of speech signals and subjective error criteria," *Proc. Int. Conf. on Acoustics, Speech and Signal Processing*. Tulsa, 1978, pp. 573-576.
7. B. S. Atal, M. R. Schroeder, and V. Stover, "Voice-excited predictive coding system for low-bit rate transmission of speech," *Proc. Int. Congress on Commun.*, San Francisco, pp. 30-37 to 30-40, June 1975.
8. J. W. Bayless, S. J. Campanella, and A. J. Goldberg, "Voice Signals: Bit by bit," *IEEE Spectrum*, October 1973.
9. T. Berger, *Rate Distortion Theory, A Mathematical Basis for Data Compression*, Prentice-Hall, Englewood Cliffs, NJ, 1971.
10. E. R. Berlekamp (ed.), *Key Papers in the Development of Coding Theory*, IEEE Press, 1974.
11. K. W. Cattermole, *Principles of Pulse Code Modulation*, Iliffe Books Ltd., London, 1969 and Elsevier Publishing Co., New York.
12. W. Chen and S. C. Fralick, "Image enhancement using cosine transform filtering," *Proc. of the Symposium on Current Mathematical Problems in Image Science*, Monterey, CA, pp. 186-192, November 1976.
13. D. L. Cohn and J. L. Melsa, "The residual encoder—An improved ADPCM system for speech digitization," *Proc. Int. Commun. Conf.*, pp. 30-26 to 30-31, San Francisco, June 1975.
14. R. E. Crochiere, "On the design of sub-band coders for low-bit-rate speech communication," *Bell Syst. Tech. J.*, vol. 56, pp. 747-770, May-June 1977.
15. R. E. Crochiere, 'An analysis of 16 kb/s sub-band coder performance: Dynamic range, tandem connections and channel errors," *Bell Syst. Tech. J.*, vol. 57, pp. 2927-2952, October 1978.
16. R. E. Crochiere, "A mid-rise/mid-tread quantizer switch for improved idle channel performance in adaptive coders," *Bell Syst. Tech. J.*, vol. 57, no. 8, pp. 2953-2955, October 1978.
17. R. E. Crochiere, L. R. Rabiner, N. S. Jayant, and J. M. Tribolet, "A study of objective measures for speech waveform coders," *Proc. Int. Zurich Seminar on Digital Communications*, Zurich, Switzerland, pp. H1.1-H1.7, March 1978.
18. R. E. Crochiere, S. A. Webber, and J. L. Flanagan, "Digital coding of speech in sub-bands," *Bell Syst. Tech. J.*, vol. 55, pp. 1069-1085, October 1976.
19. M. G. Croll, M. E. B. Moffat, and D. W. Osborne, "Nearly instantaneous compander for transmitting six sound program signals in a 2.048 MB/s multiplex," *Electron. Letters*, July 1973.
20. P. Cummiskey, N. S. Jayant, and J. L. Flanagan, "Adaptive quantization in differential PCM coding of speech," *Bell Syst. Tech. J.*, vol. 52, pp. 1105-1118, September 1973.
21. L. D. Davisson, "Rate-distortion theory and application," *Proc. IEEE*, vol. 60, pp. 800-808, July 1972.
22. L. D. Davisson and R. M. Gray (ed.), *Data Compression*, Halsted Press, 1976.
23. M. Dietrich, "Coding of speech signals using a switched quantizer," *Proc. IEEE Zurich Seminar on Digital Commun.*, pp. A4(1)-A4(4), March 1974.
24. J. J. Dubnowski and R. E. Crochiere, "Variable rate coding of speech," *Bell Syst. Tech. J.*, Mar. 1979.
25. H. Dudley, "The vocoder," *Bell Labs Record*, vol. 17, pp. 122-126, 1939.
26. D. Esteban and C. Galand, "Application of quadrature mirror filters to split band voice coding schemes," *Proc. of the 1977 Int. Conf. on Acoustics, Speech and Signal Processing*, Hartford, Conn., pp. 191-195, May 1977.
27. D. Esteban and C. Galand, "32 KBPS CCITT compatible split band coding scheme," *Proc. of the 1979 Int. Conf. on Acoustic Speech and Signal Processing*, Tulsa, OK, pp. 320-325, Apr 1978.
28. D. Esteban, C. Galand, D. Maudvit, and J. Menez, "9.6/7.2 KBP voice excited predictive coder (VEPC)," *Proc. 1979 IEEE Int. Conf. on Acoustics, Speech and Signal Processing*, Tulsa, OK, pp 307-311, April 1978.
29. J. L. Flanagan, "Bandwidth and channel capacity necessary t transmit the formant information of speech," *J. Acoust. Soc. Amer.*, vol. 28, pp. 592-596, 1956.
30. J. L. Flanagan and R. M. Golden, "Phase vocoder," *Bell Syst. Tech. J.*, vol. 45, pp. 1493-1509, 1966.
31. J. L. Flanagan, "Focal points in speech communication research," *IEEE Trans. Commun. Tech.*, vol. COM-19, pp. 1006-1015, Dec. 1971.
32. J. L. Flanagan, *Speech Analysis, Synthesis and Perception*. Second Edition, Springer-Verlag, New York, 1972.
33. J. L. Flanagan, "Digital techniques for speech communication some new dimensions," IEEE Int. Conf. on Cybernetics and Society, Washington, D.C., September 1977.
34. J. L. Flanagan, "Opportunities and issues in digitized voice," IEEE Eascon '78, Arlington, Virginia, September 1978.
35. J. L. Flanagan, "Computational models of speech sound generation," *J. Acoust. Soc. Am.*, 64, S40(A), November 1978.
36. B. Gold, "Digital speech networks," *Proc. IEEE*, vol. 65, no. 12 pp. 1636-1658, December 1977.
37. B. Gold and L. R. Rabiner, "Parallel processing techniques for estimating pitch periods of speech in the time domain," *J. Acoust. Soc. Amer.*, vol. 46, pp. 442-448, August 1969.
38. B. Gold and C. M. Rader, "Systems for compressing the band width of speech," *IEEE Trans. Audio Electroacoust.*, vol AU-15, no. 3, pp. 131-135, September 1967.
39. B. Gold and C. M. Rader, "The channel vocoder," *IEEE Trans Audio Electroacoust.*, vol. AU-15, no. 4, pp. 148-160, December 1967.
40. B. Gold and J. Tierney, "Digitalized voice-excited vocoder for telephone quality inputs using bandpass sampling of the base band signal," *J. Acoust. Soc. Amer.*, vol. 73, pp. 753-754, 1965.
41. A. J. Goldberg, "2400/16000 bps multirate voice processor," *Proc. Int. Conf. on Acoustics, Speech and Signal Processing*, Tulsa, Oklahoma, pp. 299-302, 1978.
42. A. J. Goldberg, R. L. Freudberg, and R. S. Cheung, "High quality 16 kb/s voice transmission," *Proc. Int. Conf. on Acoustics, Speech and Signal Processing*, pp. 244-246, Hartford, Conn., 1976.
43. A. J. Goldberg and H. L. Shaffer, "A real-time adaptive predictive coder using small computers," *IEEE Trans. Commun.*, vol. COM-23, no. 12, pp. 1443-1451, December 1975.
44. R. M. Golden, "Digital computer simulation of a sampled-data voice-excited vocoder," *J. Acoust. Soc. Amer.*, vol. 35, pp. 1358-1366, 1963.
45. R. M. Golden, "Vocoder filter design: practical considerations," *J. Acoust. Soc. Amer.*, vol. 43, pp. 803-810, 1968.
46. D. J. Goodman and J. L. Flanagan, "Direct digital conversion between linear and adaptive delta modulation formats," presented at the IEEE Int. Communications Conf., Montreal, June 1971.
47. D. J. Goodman, C. Scagliola, R. E. Crochiere, L. R. Rabiner, and J. Goodman, "Objective and subjective performance of tandem connections of waveform coders with an LPC vocoder," *Bell Syst. Tech. J.*, March 1979.
48. D. J. Goodman and R. M. Wilkinson, "A robust adaptive quantizer," *IEEE Trans. on Commun.*, vol. COM-23, pp. 1362-1365, November 1975.
49. J. A. Greefkes, "A digitally controlled delta codec for speech transmission," *Proc. IEEE Int. Conf. on Commun.*, pp. 7-33-7-48, 1970.
50. J. R. Haskew, J. M. Kelly, R. M. Kelly, Jr., and T. H. McKinney, "Results of a study of the linear prediction vocoder," *IEEE Trans. Commun.*, vol. COM-21, pp. 1008-1015, September 1973.
51. E. M. Hofstetter, J. Tierney, and O. Wheeler, "Microprocessor realization of a linear predictive vocoder," *IEEE Trans. Acoust. Speech, Signal Processing*, vol. ASSP-25, no. 5, pp. 379-387, October 1977.
52. J. Huang and P. Schultheiss, "Block quantization of correlated Gaussian random variables," *IEEE Trans. Commun. Syst.*, vol. CS-11, pp. 289-296, September 1963.

53. F. Itakura and S. Saito, "Analysis-synthesis telephony based upon the maximum-likelihood method," *Proc. 6th Int. Congress on Acoustics,* pp. C17-20, Tokyo, 1968.

54. F. deJager, "Delta modulation, a method of PCM transmission using the 1-unit code," *Philips Res. Rept.,* pp. 442-466, 1952.

55. N. S. Jayant, "Digital coding of speech waveforms: PCM, DPCM and DM quantizers," *Proc. IEEE,* vol. 62, pp. 611-632, May 1974.

56. N. S. Jayant (ed.), *Waveform Quantization and Coding,* IEEE Press, New York, 1976.

57. N. S. Jayant and S. A. Christensen, "Tree encoding of speech using the (M, L)-algorithm and adaptive quantization," *IEEE Trans. Commun.,* vol. COM-26, pp. 1376-1379, September 1978.

58. D. Kahn, *The Code Breakers,* New York: Macmillan, 1967.

59. S. C. Kak and N. S. Jayant, "On speech encryption using waveform scrambling," *Bell Syst. Tech. J.,* vol. 56, pp. 781-808, May-June, 1977.

60. C. M. Kortman, "Redundancy reduction—a practical method of data compression," *Proc. IEEE,* vol. 55, pp. 253-263, March 1967.

61. J. Makhoul, "Linear prediction: A tutorial review," *Proc. IEEE,* vol. 63, pp. 561-580, April 1975.

62. J. Makhoul, "Stable and efficient lattice methods for linear prediction," *IEEE Trans. Acoust., Speech, Signal Processing,* vol. ASSP-25, no. 5, pp. 423-428, October, 1977.

63. J. Makhoul and M. Berouti, "High quality adaptive predictive coding of speech," *Proc. Int. Conf. on Acoustics, Speech and Signal Processing,* Tulsa, Oklahoma, 1978, pp. 303-306.

64. J. D. Markel, "Application of a digital inverse filter for automatic formant and F_0 analysis," *IEEE Trans. Audio Electroacoust.,* vol. AU-21, no. 3, pp. 149-153, June 1973.

65. J. D. Markel and A. H. Gray, Jr., "A linear prediction vocoder simulation based upon the autocorrelation method," *IEEE Trans. Acoust., Speech and Signal Processing,* vol. ASSP-22, no. 2, pp. 124-134, April 1974.

66. J. D. Markel and A. H. Gray, Jr., *Linear Prediction of Speech,* New York: Springer-Verlag, 1976.

67. J. Max, "Quantizing for minimum distortion," *IRE Trans. Information Theory,* vol. 6, pp. 16-21, March 1960.

68. R. A. McDonald, "Signal-to-noise and idle channel performance of DPCM systems—Particular application to voice signals," *Bell Syst. Tech. J.,* vol. 45, pp. 1123-1151, September 1966.

69. C. A. McGonegal, L. R. Rabiner, and A. E. Rosenberg, "A subjective evaluation of pitch detection methods using LPC synthesized speech," *IEEE Trans. Acoust., Speech and Signal Processing,* vol. ASSP-25, no. 3, pp. 221-229, June 1977.

70. E. McLarnon, "A method for reducing the transmission rate of a channel vocoder by using frame interpolation," *Proc. Int. Conf. on Acoustics, Speech and Signal Processing,* Tulsa, Oklahoma, pp. 458-461, 1978.

71. D. R. Morgan and S. E. Craig, "Real-time adaptive linear prediction using the least mean square gradient algorithm," *IEEE Trans. Acoustics, Speech and Signal Processing,* vol. ASSP-24, no. 6, pp. 494-507, December 1976.

72. K. Murakami, K. Tachibana, H. Fujishita, and K. Omura, "Variable sampling rate coder," *Tech. Report,* vol. 26, Univ. of Osaka, Japan, pp. 499-505, October 1974.

73. A. M. Noll, "Cepstrum pitch determination," *J. Acoust. Soc. Amer.,* vol. 41, pp. 293-309, February 1967.

74. P. Noll, "A comparative study of various quantization schemes for speech encoding," *Bell Syst. Tech. J.,* vol. 541, pp. 1597-1614, November 1975.

75. P. Noll, "On predictive quantizing schemes," *Bell Syst. Tech. J.,* vol. 57, pp. 1499-1532, May-June 1978.

76. P. Noll, "Digital coding of speech and broadcast signals: design approaches and developments," *Proc. Conf. on Information and System Theory in Digital Communications,* Berlin, pp. 30-41, September, 1978.

77. J. B. O'Neal, Jr., "A bound on signal-to-quantizing noise ratios for digital encoding systems," *Proc. IEEE,* vol. 55, pp. 287-292, March 1967.

78. J. B. O'Neal, Jr., "Bounds on subjective performance measures for source encoding systems," *IEEE Trans. Inform. Theory,* vol. IT-17, pp. 224-231, 1971.

79. J. B. O'Neal, Jr. and R. W. Stroh, "Differential PCM for speech and data signals," *IEEE Trans. Commun.,* vol. COM-20, pp. 900-912, October 1972.

80. A. V. Oppenheim, "A speech analysis-synthesis system based on homomorphic filtering," *J. Acoust. Soc. Amer.,* vol. 45, no. 3, pp. 243-248, June 1976.

81. A. V. Oppenheim and R. W. Schafer, *Digital Signal Processing,* Englewood Cliffs, NJ: Prentice-Hall, 1975.

82. P. F. Panter and W. Dite, "Quantizing distortion in pulse-count modulation with nonuniform spacing of levels," *Proc. IRE,* vol. 39, pp. 44-48, January 1951.

83. M. R. Portnoff, "Implementation of the digital phase vocoder using the fast Fourier transform," *IEEE Trans. Acoust., Speech and Signal Processing,* vol. ASSP-24, no. 3, pp. 243-248, June 1976.

84. S. U. H. Qureshi and G. D. Forney, Jr., "A 9.6/16 kbit/s speech digitizer," *Proc. IEEE Int. Conf. on Commun.,* pp. 30-31 to 30-36, June 1975.

85. L. R. Rabiner and R. W. Schafer, *Digital Processing of Speech Signals,* Englewood Cliffs, NJ: Prentice-Hall, 1978.

86. L. R. Rabiner, M. J. Cheng, A. E. Rosenberg, and C. A. McGonegal, "A comparative performance study of several pitch detection algorithms," *IEEE Trans. Acoustics, Speech and Signal Processing,* vol. ASSP-24, pp. 399-418, October 1976.

87. D. L. Richards, *Telecommunication by Speech: The Transmission Performance of Telephone Networks,* London: Butterworths, 1973.

88. M. J. Ross, H. L. Shaffer, A. Cohen, R. Freudberg and H. J. Manley, "Average magnitude difference function pitch extractor," *IEEE Trans. Acoustics, Speech and Signal Processing,* vol. ASSP-22, pp. 353-362, October 1974.

89. M. R. Sambur, "A efficient linear prediction vocoder," *Bell Syst. Tech. J.,* December 1975.

90. R. W. Schafer and L. R. Rabiner, "System for automatic formant analysis of voiced speech," *J. Acoust. Soc. Amer.,* vol. 47, part 2, pp. 634-648, February 1970.

91. H. R. Schindler, "Delta modulation," *IEEE Spectrum,* pp. 69-78, October 1970.

92. M. R. Schroeder, "Vocoders: Analysis and synthesis," *Proc. IEEE,* vol. 54, pp. 720-734, May 1966.

93. M. R. Schroeder, "Period histogram and product spectrum: New methods for fundamental frequency measurement," *J. Acoust. Soc. Amer.,* vol. 43, no. 4, pp. 829-834, April 1968.

94. M. R. Schroeder, J. L. Flanagan, and E. A. Lundry, "Bandwidth compression of speech by analytic signal rooting," *Proc. IEEE,* vol. 55, pp. 396-401, 1967.

95. M. R. Schroeder and E. E. David, Jr., "A vocoder for transmitting 10 kc/s speech over a 3.5 kc/s channel," *Acustica,* vol. 10, pp. 35-43, 1960.

96. B. Smith, "Instantaneous companding of quantized signals," *Bell Syst. Tech. J.,* vol. 36, pp. 653-709, May 1957.

97. M. M. Sondhi, "New methods of pitch extraction," *IEEE Trans. Audio Electroacoust.,* vol. AU-16, pp. 262-266, June 1968.

98. R. Steele, *Delta Modulation Systems,* Pentech Press, London, 1975.

99. R. Steele and D. J. Goodman, "Detection and selective smoothing of transmission errors in linear PCM," *Bell Syst. Tech. J.,* vol. 51, pp. 399-409, March, 1977.

100. C. E. W. Sundberg, "Soft decision demodulation for PCM encoded speech signals," *IEEE Trans. on Communications,* vol. COM-26, pp. 854-859, June 1978.

101. J. M. Tribolet and R. E. Crochiere, "A vocoder-driven adaptation strategy for low-bit-rate adaptive transform coding of speech," *Proc. 1978 Int. Conf. on Digital Signal Processing,* Florence, Italy, pp. 638-642, September 1978.

102. J. M. Tribolet, P. Noll, B. J. McDermott, and R. E. Crochiere, "A comparison of the performance of four low bit rate speech waveform coders," *Bell Syst. Tech. J.,* (to be published). See also (same authors, "A study of complexity and quality of speech waveform coders," *Proc. of the 1978 IEEE Int. Conf. on Acoustics, Speech and Signal Processing,* Tulsa, OK, pp. 1586-1590, April 1978).

103. S. A. Webber, C. J. Harris, and J. L. Flanagan "Use of variable-quality coding and time-interval modification in packet transmission of speech," *Bell Syst. Tech. J.,* vol. 56, pp. 1569-1573, October 1977.

104. C. J. Weinstein, "A linear prediction vocoder with voice excitation," presented at Eascon, September 1975.

105. P. A. Wintz, "Transform picture coding," *Proc. IEEE,* vol. 60, pp. 809-820, July 1972.

106. R. Zelinski and P. Noll, "Adaptive block quantization of speech signals," Heinrich-Hertz-Institute, Berlin, Germany, Technical Report 181, 1975 (in German).
107. R. Zelinski and P. Noll, "Adaptive transform coding of speech signals," *IEEE Trans. Acoustics, Speech and Signal Processing,* vol. ASSP-25, pp. 299-309, August 1977.
108. L. H. Zetterberg and J. Uddenfeldt, "Adaptive delta modulation with delayed decision," *IEEE Trans. Commun.,* vol. COM-22, pp. 1195-1198, September 1974.

James L. Flanagan (A'51-M'57-SM'67-F'69) received the Sc.D. degree in electrical engineering from the Massachusetts Institute of Technology in 1955. He joined Bell Laboratories in 1957, where he has specialized in voice communications, digital techniques and communications acoustics. He holds patents in the fields of speech-coding, digital processing, and underwater acoustics, and has published technical papers in these and related fields. He is Head of the Acoustics Research Department at Bell Laboratories.

Dr. Flanagan is Past President of the IEEE Acoustics, Speech and Signal Processing Society. He is currently President of the Acoustical Society of America. He is a member of the National Academy of Engineering.

Manfred R. Schroeder (SM'67-F'71) was born in Ahlen, Germany, on July 12, 1926. He studied physics and mathematics at the University of Göttingen from which he received the degree of Diplom-Physiker in 1951 and that of Dr.rer.nat. in physics in 1954. The same year he came to the United States to join the research staff of Bell Telephone Laboratories at Murray Hill, New Jersey.

Dr. Schroeder's work at Bell has encompassed a variety of phases in the field of acoustics. He has been engaged in fundamental studies of architectural acoustics, electroacoustics, underwater sound, speech and hearing. He is also noted for his contributions in the communications sciences, for his statistical theory of random wave fields and for new measurement methods in acoustics and other disciplines. He holds over 40 United States patents for inventions in these fields. In 1958 Dr. Schroeder was appointed Head of the Acoustics Research Department at Bell Laboratories. In 1963 he became Director of the Acoustics and Speech Research Laboratory and in 1964 he assumed responsibility for all areas of acoustics and ultrasonics research at Bell as Director of the Acoustics, Speech and Mechanics Research Laboratory. He was appointed Professor of Physics and Director of the Drittes Physikalisches Institut of the University of Göttingen in 1969. In 1972 he served as Chairman of the Physics Department at Göttingen. Dr. Schroeder served on the "National Stereophonic Radio Committee", an advisory group to the Federal Communication Commission which formulated the standards for stereophonic broadcasting in the United States. He is a member of the Committee on Hearing and Bio-Acoustics of the National Research Council. In 1966 he served on the New Technologies Panel of the National Advisory Commission on Health Manpower. In 1963, he was a guest of the Soviet Government to consult on the acoustics of the Palace of Congresses in the Kremlin.

Prof. Schroeder is a past Member of the Executive Council of the Acoustical Society of America and a former Associate Editor of the *Journal of the Acoustical Society of America.* He is also a Fellow of the Audio Engineering Society. In 1973 he was elected to the Academy of Sciences at Göttingen. In 1975 he was elected a Foreign Scientific Member of the Max-Planck-Institut for Biophysical Chemistry. In 1972 Prof. Schroeder was awarded the Gold Medal of the Audio Engineering Society for "the successful merging of theory and practice in the analysis of sound transmission and reproduction, and for pioneering efforts to apply computer technology to the problems of architectural acous In 1976 he received the Baker Prize Award of the Institute of Electrical and Electronics Engineers for his work on models in hearing.

Bishnu S. Atal (M'76-SM'78) was born in Kanpur, India, on May 10, 1933. He received the B.Sc. (Honors) degree in physics from the University of Lucknow (India) in 1952, the Diploma in electrical communication engineering from the Indian Institute of Science, Bangalore (India) in 1955, and the Ph.D. degree in electrical engineering from the Polytechnic Institute of Brooklyn (New York) in 1968.

From 1957 to 1960 he was a Lecturer in Acoustics at the Department of Electrical Communication Engineering, Indian Institute of Science, Bangalore. In 1961 he came to the United States to join the Research Staff of Bell Telephone Laboratories, Murray Hill, New Jersey. At Bell Laboratories his work has covered a wide range of topics in acoustics such as computer simulation of sound transmission in rooms, new measurement techniques for concert halls, fading in mobile radio, automatic speaker recognition, and speech coding. More recently his research interests have centered on new methods for analysis and synthesis of speech signals. He is the author of a number of technical papers in architectural acoustics and speech communication.

Dr. Atal is a Fellow of the Acoustical Society of America.

Ronald E. Crochiere (S'66-M'67-SM'78) was born in Wausau, Wisconsin, on September 28, 1945. He received the B.S. degree in 1967 from the Milwaukee School of Engineering, Milwaukee, Wisconsin, and the M.S. and Ph.D. degrees in 1968 and 1974, respectively from the Massachusetts Institute of Technology, Cambridge, Massachusetts, all in electrical engineering.

From 1968 to 1970 he was employed with Raytheon Co. In 1970 he returned to M.I.T. and joined the Research Laboratory of Electronics where he was engaged in graduate studies for the Ph.D. degree.

In 1974 Dr. Crochiere joined the Acoustics Research Department of Bell Laboratories where he has been involved in research activities in concepts of decimation and interpolation, sub-band and transform coding of speech, and the measurement of digital speech quality. In 1976 he received the IEEE ASSP paper award for his paper on decimation and interpolation of digital signals.

Dr. Crochiere is an active member of the ASSP AdCom Committee and the ASSP Technical Committee on Digital Signal Processing. He has served for two years as a Technical Editor on digital signal processing for the *IEEE Trans. Acoust., Speech, Signal Processing* and is presently Secretary-Treasurer for the ASSP Society.

Nuggehally S. Jayant (M'68-SM'77) received the B.Sc. degree in Physics and Mathematics from the Mysore University (India) in 1962, and the B.E. and Ph.D. degrees in Electrical Communication Engineering from the Indian Institute of Science, Bangalore, in 1965 and 1970, respectively.

From 1967 to 1968, he was a Research Associate at Stanford University, California. From 1968, he has been on the Technical Staff of Bell Laboratories, Murray Hill, New Jersey, as a member of the Acoustics Research Department. During January-March 1972 and August-October 1975, he was a Visiting Professor at the Indian Institute of Science, Bangalore.

His research interests have included communication over burst-noisy and fading channels, statistical pattern discrimination, spectral analysis, and, most recently, digital coding and transmission of waveforms with special reference to speech signals.

He has several technical papers and patents on the above subjects, and is also the editor of an IEEE Press reprint book, *Waveform Quantization and Coding.*

Dr. Jayant is the 1974 winner of the Browder J. Thompson Prize for the best IEEE publication by an author or authors under thirty years of age.

José M. Tribolet (S'74–M'77) was born in Tancos, Portugal, on December 20, 1949. He received the Engenheiro Electrotécnico degree from the Instituto Superior Técnico, Lisbon, Portugal, in 1972, and the M.S., E.E., and Sc.D. degrees in electrical engineering from the Massachusetts Institute of Technology, Cambridge, in 1974, 1975, and 1977, respectively.

From 1970 to 1972 he was an Assistente Eventual of the Instituto Superior Técnico and a Researcher of the Centro de Estudos de Electrónica of the Instituto de Alta Cultura. From 1972 to 1977 he was a member of the Massachusetts Institute of Technology Research Laboratory of Electronics, with an Instituto Nacional de Investigação Científica Fellowship. During this period his research activities involved the application of homomorphic signal processing to speech and seismic data analysis. He is currently with the Department of Electrical Engineering, Instituto Superior Técnico, Lisbon, Portugal, where he is a Professor. He was on leave until September 1978 at the Acoustics Research Department, Bell Laboratories, Murray Hill, NJ, where he was working on adaptive transform coding of speech.

Dr. Tribolet is a member of Sigma Xi.

Signal Compression: Technology Targets and Research Directions

Nikil Jayant, *Fellow. IEEE*

(*Invited Paper*)

Abstract—Recent years have witnessed significant progress in the compression of speech, audio, image, and video signals. This has been the result of an interworking of coding theory, signal processing, and psychophysics. Advances in compression have been accompanied by numerous standards for digital coding and communication, both national and international. Emerging technology goals require even greater levels of signal compression. As we address these goals, and as we seek to define and approach fundamental limits in coding, we are guided by several promising trends in compression research, some of which represent cross-disciplinary evolution in the seemingly very different domains of acoustic and visual signals. Important aspects of new work in the field are the creation of refined methodologies for the measurement of signal distortion and coded signal quality, and increased interaction of source coding technology with other communications disciplines such as channel coding and networking.

I. INTRODUCTION

SIGNAL coding is the process of representing an information signal in a way that realizes a desired communications objective such as analog-to-digital conversion, low bit-rate transmission, or message encryption. In the literature, the terms *source coding, digital coding, data compression, bandwidth compression*, and *signal compression* are all used to connote techniques used for achieving a compact digital representation of a signal, including the important subclasses of analog signals such as speech, audio, and image. The subject of this paper is the art and science of signal compression. When the terms *coding, encoding*, and *decoding* are used in this paper, they will all refer to the specific common objective of compression. An important theme of our discussion is the human receiver at the end of the communication process (Fig. 1).

Fig. 2 defines the role of signal compression (source coding) in digital communication. While the source coder attempts to minimize the necessary bit rate for faithfully representing the input signal, the *modulator–demodulator* (modem) seeks to maximize the bit rate that can be supported in a given channel or storage medium without causing an unacceptable level p_e of bit error probability. The bit rate in source coding is measured in bits per sam-

ple or bits per second (b/s). In modulation, it is measured in bits per second per Hertz (b/s/Hz). The channel coding boxes add redundancy to the encoder bit stream for the purpose of error protection. In so-called coded modulation systems, the operations of channel coding and modulation are integrated for greater overall efficiency. The processes of source and channel coding can also be integrated in ways that will be illustrated in the last section of this paper.

The capability of signal compression has been central to the technologies of robust long-distance communication, high-quality signal storage, and message encryption. Compression continues to be a key technology in communications in spite of the promise of optical transmission media of relatively unlimited bandwidth. This is because of our continued and, in fact, increasing need to use bandlimited media such as radio and satellite links, and bit-rate-limited storage media such as CD-ROM's and miniaturized memory modules.

A. Background

The information-theoretical foundations of signal compression date back to the seminal work of Shannon [129], [130]. His mathematical exposition defined the information content or *entropy* of a source and showed that the source could be coded with zero error if the encoder used a transmission rate equal to or greater than the entropy and further, if it used a long processing delay, tending in general to infinity. In the special case of the infinite-alphabet or analog source, the encoding error tends to approach zero only at an infinite bit rate. However, in practice, the error is close enough to zero at finite rates. In the case of a finite-alphabet or discrete-amplitude source, the entropy is finite, and the bit rate needed for zero encoding error is finite as well. An important example of a finite-entropy source is an analog signal stored in a computer as a sequence of discrete amplitudes. The raw (uncompressed) bit rate of such a signal is typically 8, 16, or 24 bits per sample, respectively, for a grey-level image, high-quality speech (or audio), and a color image with three 8-bit components. The entropy, or the minimum bit rate for zero encoding error, will be smaller because of the statistical redundancy in the input sequence.

The inadequacies of the classical source coding theory are twofold. First, the theory is nonconstructive, offering

Manuscript received October 15, 1991; revised January 20, 1992.

The author is with the Signal Processing Research Department. AT&T Bell Laboratories. Murray Hill. NJ 07974.

IEEE Log Number 9107597.

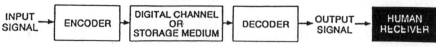

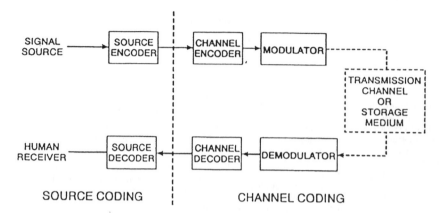

Fig. 1. Digital coding for signal compression.

Fig. 2. Block diagram of a digital communication system.

bounds on distortion-rate performance rather than techniques for achieving these targets. We note, however, that the theory does anticipate important qualitative recipes such as the benefits of delayed encoding algorithms, as in vector quantization with a large vector dimension or block length. Second, the source model used in the classical theory does not capture what are now recognized as fundamental nuances in speech, audio and visual signals. These include the facts that the input signal is non-Gaussian, nonstationary, in general has a complex and intractable power spectrum, and the human receiver does not employ a mean-squared-error criterion in judging the similarity of a coded signal to the uncoded signal. As a result of the above complications, some of the observations of classical source coding do not carry over in an obvious way to signal compression as discussed in this paper. One such result is that the source entropy measured with a perceptual distortion criterion is different from, and generally much lower than, the classical entropy measured with a mean-squared-error criterion for coding distortion. Another classical result that needs to be reexamined is the thesis that, in principle, the processes of source and channel coding can be separated without loss of optimality. This result would hold very nicely for the digital communication of data sequences, but does not necessarily suggest an optimal solution to the complex problem of communicating audio or visual signals over a noisy channel with high perceptual fidelity, robustness, and bit-rate efficiency.

The technology and literature of signal compression have therefore evolved somewhat independently—with some valuable inspiration from Shannon's theory and the rate-distortion theory that followed it [5], [23], [36], [43], [54], [156], but really with a great deal of innovative en-gineering on the part of scientists closely familiar with the signals in question [1], [3], [4], [18], [19], [30], [32], [35], [42], [44], [47], [49], [53], [54], [59], [60], [62], [78], [80], [89], [90], [98], [100], [101], [102], [108], [113], [116], [121], [127], [132], [133], [135], [139], [151], [152], [154], [157]. In particular, work on speech compression has benefited greatly from studies of speech production and speech perception by humans, and research on visual perception has similarly impacted the parallel field of image compression. Although the mathematical theory of source coding is a common denominator, the fields of speech and image coding have been generally discussed by different schools, with a few recent exceptions. One of the purposes of this paper is to point out that as we address future technology targets in the disciplines of speech and image compression, common threads continue to exist. One such commonality is the increasing importance of matching the compression algorithm to the human perceptual mechanism—the auditory process in one case and the visual process in the other, leading to newly emerging generic techniques for quantization and time-frequency analysis in support of perceptually tuned coding, or *perceptual coding* for short.

B. The Dimensions of Performance in Signal Compression

The generic problem in signal compression is to minimize the bit rate in the digital representation of the signal while maintaining required levels of signal quality, complexity of implementation, and communication delay. We will now provide brief descriptions of the above parameters of performance.

Signal Quality: Perceived signal quality is often measured on a five-point scale that is well known as the *mean opinion score* or *mos* scale in speech quality testing: an average over a large number of speech inputs, speakers, and listeners evaluating the signal quality [22], [46], [62], [73], [74]. The five points of quality are associated with a set of standardized adjectival descriptions: *bad*, *poor*, *fair*, *good*, and *excellent*, and every example of an input being evaluated is assigned one of these levels in the course of a subjective test. Five-point scales of quality have also been used in image and audio testing, sometimes in the form of an inverted scale that categorizes levels of impairment [8] (*very annoying*, *annoying*, *slightly annoying*, *perceptible but not annoying*, and *imperceptible*) rather than quality. Other variations include the notion of averaging measurements over a selected *difficult* subset of input signals [136], in order to provide conservative scores of coder performance. In this paper, we use the original notion stated in this paragraph, a quality (rather than an impairment) scale, and an average over a large set of typical inputs, in each of the four generic categories: telephone-bandwidth speech, 20 kHz-bandwidth audio, still images, and video.

Our quantitative discussion of image and video quality will be impressionistic at best, given the multidimensionality of the problem (dependence of subjective quality on input scene, picture resolution, image size, and viewing distance) and the general lack of formal quality assessments in recent image-coding literature. In the field of speech coding, *mos* evaluations are well accepted and sometimes supplemented with measurements of *intelligibility* [146] and *acceptability* [145].

Bit Rate: We measure the bit rate of the digital representation in *bits per sample*, *bits per pixel* (b/p), or *bits per second* (b/s) depending on context, where *pixel* (sometimes shortened to *pel*) refers to a picture element or an image sample. The rate in bits per second is merely the product of the sampling rate and the number of bits per sample. The sampling rate is typically slightly higher than about twice the respective signal bandwidth, as required by the Nyquist sampling theorem [62].

Table I defines four commonly used grades of audio bandwidth. Typical sampling rates are 8 kHz for telephone speech, 16 kHz for AM-radio-grade audio, 32 kHz for FM-audio, and 44.1 or 48 kHz for CD (compact-disk) audio or DAT (digital audio tape) audio, both of which are signals of 20 kHz bandwidth. Respective bandwidths are strictly lower than half the corresponding sampling rates, following the principle of Nyquist sampling.

Table II defines commonly used grades of video in terms of sampling rate in pixels per second (p/s) or Hertz (Hz). The sampling rates for the CIF, CCIR, and HDTV formats defined in the table are 3, 12, and 60 MHz. Respective Nyquist bandwidths are approximately 1.5, 6, and 30 MHz, although bandwidth-limiting of image and video signals is in general less formal than the bandlimiting operations used for speech and audio signals. The HDTV format in the table is merely a specific example, one of several alternative formats. The sampling rates in

TABLE I
DIGITAL AUDIO FORMATS

Format	Sampling Rate (kHz)	Bandwidth (kHz)	Frequency Band
Telephony	8	3.2	(200–3400 Hz)
Teleconferencing	16	7	(50–7000 Hz)
Compact Disk (CD)	44.1	20	(20–20000 Hz)
Digital Audio Tape (DAT)	48	20	(20–20000 Hz)

TABLE II
DIGITAL TELEVISION FORMATS

Format	Spatio-Temporal Resolution	Sampling Rate
CIF	$360 \times 288 \times 30 =$	3 MHz
CCIR	$720 \times 576 \times 30 =$	12 MHz
HDTV	$1280 \times 720 \times 60 =$	60 MHz

CIF: Common Intermediate Format
CCIR: International Consultative Committee for Radio
HDTV: One Example of a High Definition Television Format

the table refer to luminance information. Overheads for including color information are system-dependent.

In the CIF format, the color overhead is 50% in sampling rate, corresponding to a 50% subsampling relative to luminance in each of the horizontal and vertical directions, and a total of two chrominance components. Higher degrees of subsampling are sometimes used, leading to overall color overheads lower than 50%. In the line-interlaced CCIR format, the subsampling of color is performed only in the horizontal direction, and the final overhead in sampling rate is 100%.

Complexity: The complexity of a coding algorithm is the computational effort required to implement the encoding and decoding processes in signal processing hardware, and it is typically measured in terms of arithmetic capability and memory requirement. Coding algorithms of significant complexity are currently being implemented in real time, some of them on single-chip processors. Other related measures of coding complexity are the physical size of the encoder, decoder, or codec (encoder plus decoder), their cost (in dollars), and the power consumption (in watts or milliwatts, mW), a particularly important criterion for portable systems [15], [134].

Fig. 3 defines the evolution of digital signal processing (DSP) technology in terms of the number of instructions per second (mi/s) on a single general-purpose processor, as a function of time. The evolution is exponential, with no evidence of saturation in the near-term [86]. In the five-year period from 1990 to 1995, the typical per-chip capability is expected to increase tenfold, from about 25 mi/s to about 250 mi/s per chip. Supporting this evolution in arithmetic capability is a parallel advance in memory capability. The significance of the above advances is that sophisticated compression algorithms that demand increasing levels of complexity will be supported by DSP technology in the form of single-chip, multichannel im-

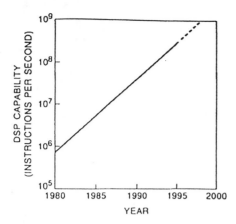

Fig. 3. Evolution of arithmetic capabilities in digital signal processing.

Digital Communication

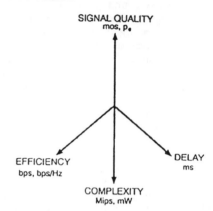

Fig. 4. The dimensions of coder performance.

plementations of the relatively less complex algorithms and realistic parallel-processing machines for the more complex techniques. Power dissipation and cost are also expected to decrease steadily, making DSP technology increasingly useful for personal portable devices and for other high-volume consumer applications.

Communication Delay: Increasing complexity in a coding algorithm is often associated with increased processing delays in the encoder and decoder. Although improved DSP capability can be used as an argument in favor of more sophisticated algorithms, the need to constrain communication delay should not be underemphasized. This need places important practical restrictions on the permissible sophistication of a signal compression algorithm. Depending on the communication environment, the permissible total delay for one-way communication (coding plus decoding delay) can be as low as about 1 ms (as in network telephony under conditions of no echo control) and as high as about 500 ms (as in very low bit-rate videotelephony (or *videophony*) where the delay performance is severely compromised in the interest of obtaining a received picture good enough for communication). Communication delay is largely irrelevant for applications involving one-way communication (as in television broadcasting) or storage and message-forwarding (as in voice mail).

C. Coding and Digital Communication

Fig. 4 describes performance criteria in digital communication by recapitulating the four dimensions of coder performance, explained specifically for source coding in Section I-B. These dimensions of performance apply to channel coding and modulation as well, although the units of quality and bit rate are different. Respective units, defined either in Section I-B or in the second paragraph of the paper, appear along the *quality* and *efficiency* axes in Fig. 4. Along each axis, the left and right entries refer to source and channel coding, respectively. The units of *complexity* and *delay* are identical for source and channel

coding, although those parameters are used for different reasons in the two cases. Processing delay is used in source coding to remove signal redundancy. In channel coding, delay may be used for adding error protection bits and for processes such as interleaving for the randomization of burst errors.

The axes in Fig. 4 define a four-dimensional space in which some regions are theoretically allowable, and some regions are desirable for specific communication applications. Researchers in source and channel coding attempt to describe the allowable regions and tradeoffs as quantitatively as possible. The focus of this paper is on the domain of source coding. In particular, we shall comment extensively on the quantitative relationship between compressed signal quality and bit rate.

D. Road Map

In the remainder of this paper, we discuss applications and technology goals (Section II), the current tools of the trade (Section III), and emerging research directions (Section IV). The last section includes a brief discussion of efficient transmission, channel error protection, combined source and channel coding, and networking.

II. APPLICATIONS, STANDARDS, AND TECHNOLOGY GOALS

A. Applications

Fig. 5 depicts various applications of signal compression. The vertical axis in the figure does not have any special meaning. The numbers on the horizontal axis are bit rates *after* compression. The labels in the figure represent, in an approximate fashion, current capabilities. The bit rates spanned by these labels and, in some cases, the bit rates on which the labels are centered, represent the rates at which compressed signals render the corresponding application practical. As our capabilities in compression improve, the labels in the figure tend to drift

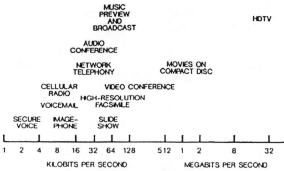

Fig. 5. Applications of signal compression.

to the left. Signals covered in Fig. 5 include telephone speech, wideband audio (speech and music), and a wide range of image signals, including still pictures and motion video. In the following paragraphs, we provide very brief summaries of current capabilities in the compression of speech (Fig. 6), audio (Fig. 7), and image (Fig. 8) signals.

Telephone Speech: Speech compressed to 2.4 kb/s provides a high level of intelligibility. However, speech quality, naturalness, and speaker recognizability are all poor at this bit rate. The need for digital encryption over a very wide range of transmission media is the main reason why 2.4 kb/s speech is widely used, particularly in government and defense communications [139].

A bit rate of 4.8 kb/s is sufficient to provide measurable gains in naturalness and speaker recognition. This bit rate is also of interest to government and defense applications [71]. With the increased demands of mobile telephony over bandlimited channels, 4.8 kb/s speech coding is also becoming very important for commercial communications using digital cellular radio. At 8 kb/s, which is the bit rate chosen for first-generation digital cellular telephony in North America [41], speech quality is high although significantly lower than that of the uncoded telephone-band signal.

At 16 kb/s and beyond, the speech quality is extremely close to that of the original, especially after a single stage of encoding and decoding. We use the term *network quality* to signify a performance level at which there is sufficient margin for additional functions such as multiple stages of encoding and decoding for speech, as well as high-accuracy transmission of nonspeech voiceband signals such as modem waveforms. The bit rate required for network-quality telephony, which was historically 64 kb/s and later reduced to 32 kb/s [9], is now coming down to 16 kb/s [13].

The application of *voicemail* involves speech storage as well as speech transmission for forwarding the voice message. Depending on the network environment used for this service and the desired speech quality in the received message, the bit rate can range from 4 to 32 kb/s, with increased focus expected in the 4 to 16 kb/s range in the future.

Wideband Speech: The 7 kHz signal has a higher voice

quality than traditional telephony. This is partly due to increases in speaker presence and the naturalness of speech, as provided by the low-frequency enhancement (the added band from 200 to 50 Hz; see Table I), and partly due to increased intelligibility and crispness provided by high-frequency enhancement (the added band from 3400 to 7000 Hz). The higher quality of wideband speech is desirable for the extended communication task of a long audioconference call. It is also appropriate for other applications of loudspeaker telephony and for systems that include a high-quality speakerphone. It is also known that low-cost electret microphones can, in principle, support an incoming bandwidth of 7 kHz.

The standardized bit rate for high-quality coding of 7 kHz speech is 64 kb/s [10], [91], typically for an audioconference application using the Integrated Services Digital Network (ISDN). Recent algorithms have provided 7 kHz capability at 32 kb/s [115], permitting stereo-teleconferencing or dual-language programming over basic-rate ISDN. The projected capability for high-quality coding of 7 kHz speech is at least as low as 16 kb/s [34], [63], [64], [122].

The lower bit rates for wideband speech are also central to high-quality conferencing with combined audio and video. Current practice, at low values of total bit rate such as 64 kb/s, is to limit speech to the traditional telephone bandwidth of 3.2 kHz and to use 8 kb/s, or at the most, 16 kb/s for the coding of the audio channel. With advances in wideband speech compression, 16 kb/s coding of 7 kHz audio is expected to be an important component of conferencing, especially at higher ISDN rates such as 128 and 384 kb/s.

Wideband Audio: On a compact disk (CD), 20 kHz audio is sampled at 44.1 kHz and stored at 16 bits per sample or 706 kb/s per sound channel. Current algorithms for audio compression [6], [66], [97], [136], [137] provide CD-quality at 128 kb/s per channel for nearly all tested inputs, and CD-*like* quality at 64 kb/s per channel. These capabilities are important for emerging digital systems for audio broadcast and music preview. The capabilities are also central to applications that combine audio and visual functions, such as CD-ROM multimedia with a total bit rate of about 1.5 Mb/s and digital HDTV with a total bit rate of about 20 Mb/s (see Fig. 5).

Still Images: A 500 × 500 pixel color image, with the uncompressed format of 24 bits per pixel (b/p), will require about 100 seconds of transmission time over a 64 kb/s link. With 0.25 b/p coding, the transmission time is about 1 s, a number that would be deemed excellent for an interactive "slide show" [114]. Current technology for coding a 500 × 500 image is capable of providing good picture quality at 0.25 b/p for a wide class of color images, assuming a viewing distance of about six times the picture height [124], [147]. For most images, increasing the bit rate to 1 b/p provides excellent and, in some cases, perfect image quality. The corresponding transmission time over a 64 kb/s link is 4 s. Likewise, high-resolution facsimile typically takes several seconds of transmission

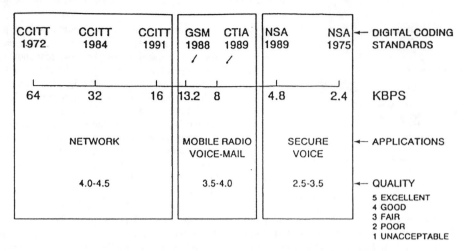

Fig. 6. Applications of telephone-speech compression, grades of digital
speech quality, and standards for digital coding.

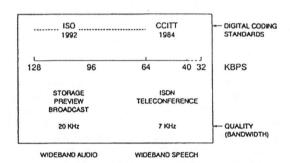

Fig. 7. Applications of wideband speech and audio, grades of bandwidth,
and standards for digital coding.

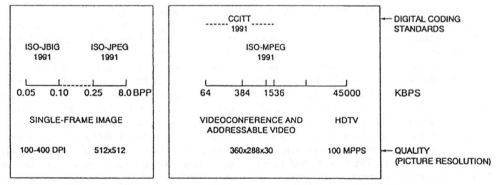

Fig. 8. Applications of image compression, grades of signal quality, and
standards for digital coding.

time over a 64 kb/s channel even after the use of powerful techniques for fax compression [55], [56].

Techniques for *progressive transmission* [55], [147], [149] involves a first stage of coding characterized by a low bit rate and rapid picture access followed, if needed, by additional stages of transmission that upgrade the picture quality. Progressive transmission is ideal for applications such as telebrowsing. It is also appropriate for applications where one expects display modalities (terminals and printers) of varying resolutions.

The image-phone application in Fig. 5, which assumes the use of a telephone line and a 9.6 kb/s modem, involves pictures of very low spatial and temporal resolution. A typical resolution would be 100 × 100 pixels per frame, and about 3–6 frames per second. With an even lower temporal resolution, such as about 1 frame per second, the system degenerates to a sequence-of-stills service, sometimes referred to as *freeze-frame video*.

Digital Video: Serious videoconferencing requires CIF resolution (360 × 288 pixels per frame, Table II) or at least quarter-CIF resolution (180 × 144 pixels per frame). Input temporal resolutions are usually submultiples of 30 frames per second, say 15 or even 10, for bit rates lower than about 1.5 Mb/s. With CIF resolution and a bit rate of 1.5 Mb/s, the communications quality of the service is generally agreed to be high. With quarter-CIF or somewhat lower resolution and correspondingly lower values of temporal resolution, it is possible to achieve lower bit rates such as 48 or 112 kb/s. But the video quality is useful only if one accepts low levels of sharpness in the output picture and very low levels of motion activity in the input scene, as in the head-and-shoulders view of a single person—an environment sometimes referred to as video*telephony*, rather than video*conferencing*. The bit rates of 48 and 112 kb/s are appropriate for ISDN systems with total bit rates of 64 and 128 kb/s, respectively, and a bit rate of 16 kb/s for voice transmission. The bit rate of 384 kb/s is a very interesting number in the current state of technology. At this bit rate, it is possible to provide a fair, if not high, level of picture quality in the coding of a videoconference scene.

CD-ROM media have a net throughput rate of about 1.5 Mb/s for source data and a total bit capacity of a few gigabits. If video can be compressed to about 1 Mb/s, a CD-ROM device could store and play out about an hour or more of the video signal together with compressed stereo sound. This capability is central to various emerging applications of CD-ROM multimedia, including the specific example of a movie on an audio compact disk [81]. The additional capability of selecting a still-image snapshot of a desired part of the image sequence leads to the concept of *addressable video*. This is an important feature in emerging systems for video storage.

Uncompressed high-definition television (HDTV) has a bit rate of over a gigabit per second (the product of a sampling rate on the order of 60 MHz, as in Table II, and the representation of three color components with a total of 24 bits per sample). Compression of the HDTV signal to a bit rate on the order of a few tens of Mb/s will create several important opportunities for HDTV broadcasting. In particular, a bit rate in the range of 20 Mb/s will bring the service into the realm of a 6 MHz transmission channel [103], implying the capability of simulcasting the HDTV version of a program in vacant slots of an NTSC channel set. Transmission rates higher than 20 Mb/s are appropriate for higher-quality transmissions over satellite and broadband ISDN channels and applications of HDTV for movie production.

B. Compression Standards

The need to interoperate different realizations of signal encoding devices (transmitters) and signal decoding devices (receivers) has led to the formulation of several international and national standards for compression algorithms. Figs. 6–8 provide a nonexhaustive summary of compression standards for speech [9], [13], [41], [71], [78], [85], [139], audio [91], [97], [136], image [55], [56], [147], and video [81], [85], [103], [106]. Additional information provided in these illustrations includes typical applications, typical levels of signal quality, and the approximate date of formulating the standard.

The recent explosion in standards activity has had an important impact on research and development in the field. Standards have led to an increased focus in applied research. They have sometimes stimulated highly productive new research as well. They have also elevated the threshold of performance that a novel research algorithm needs to exceed before it is widely accepted, given that the supplanting of a recently endorsed standard is generally difficult and expensive.

Several applications of signal compression are decoder-intensive in the sense that users need access only to a decoder, the encoding being a one-time operation by the provider of the service. Examples are multimedia and HDTV decoders. In recognition of this, corresponding standards have specified the decoder algorithms and bit stream syntax rather than the encoder. In these cases, compatible enhancements to the standard are possible in the encoding module, as well as in optional modules of pre- and postprocessing—prefiltering at the encoder and postfiltering at the decoder.

C. Quality of the Compressed Signal

We have noted earlier that there are several dimensions defining the performance of a coding system. If we ignore the dimensions of algorithmic complexity and communication delay for the moment, coder improvements can be demonstrated in two ways: by measuring signal quality improvement at a specified bit rate, or by realizing a specified level of signal quality at a lower bit rate. Depending on the application, one of the above approaches would be more relevant than the other. For example, in the problems of coding telephone speech at 4 kb/s and HDTV at 15 to 20 Mb/s, the bit rates are defined by important generic applications and the goal of coding research is to enhance signal quality at those rates. On the other hand, in the field of digital audio broadcasting (DAB) where the signal quality needs to be transparent to the coding algorithm and equivalent, say, to that of CD-audio, the goal is to demonstrate such performance at progressively lower rates (say, at 48 or 64 kb/s per channel rather than at 96 or 128 kb/s per channel as in currently proven algorithms).

So far, our discussions of bit rate have been in terms of kilobits per second (kb/s) for speech and audio, and both kilobits per second (kb/s) and megabits per second (Mb/s) in the case of video. All of these numbers can obviously be converted to equivalent numbers in bits per sample based on sampling rates such as the illustrative numbers in Tables I and II. For example, the 4 kb/s, 48 kb/s, and 15 Mb/s rates for 8 kHz-sampled speech, 48 kHz-sampled audio, and 60 MHz-sampled HDTV correspond to 0.5, 1, and 0.25 bits per sample, respectively. In our ensuing definition of technology targets in signal compression, we shall use the normalized unit of bits per sample in the interest of a unified perspective for audio and visual signals.

D. Technology Targets

Fig. 9 is a simplified and impressionistic summary of current capabilities in signal coding, expressed in terms of subjective quality as a function of bit rate. The results are derived from a combination of published work [9], [10], [22], [91], unpublished reports [136], and by collective impressions of experts, especially in the case of image and video signals where formal evaluations of quality are not generally available. Signal quality is measured on a subjective five-point scale ranging from *bad* to *excellent*, as in our earlier description of the *mos* scale.

One of the implications in Fig. 9 is that video signals are the easiest to compress on a bit-per-sample basis. This is attributable to the well-known redundancy in video information in both the spatial and temporal domains of the signal, a property that is also reflected in the extreme lowpass nature of the power spectral density of typical video. By contrast, it is not unusual to encounter relatively flat power spectra in 20 kHz audio. This, combined with the universal expectation of very high levels of quality in entertainment audio, leads to generically lower subjective scores in compressed audio at a given number of bits per sample.

In the category of still images, facsimile documents constitute a special subclass if we agree to regard text and line graphics, rather than grey-level photographs, as typical fax documents. A half-toned (black–white) document is generally highly compressible. The bit rate for the lossless coding of a fax document can be typically on the order of 0.1 bit per sample.

As we seek to advance the state of the art as depicted earlier in Fig. 5, it is useful to talk about bit rate targets at which one expects the four signals in Fig. 9 to be digitized with a quality rating such as 4 or higher. Without loss of generality or realism, all of these targets can be collectively described by the bit-rate-independet horizontal broken line of 4.5–quality in Fig. 9. Clearly, we are closer to this goal in some signal domains than others. It is also possible that the 4.5–quality goal at rates down to 0.25 bit per sample is impossible to achieve in some cases, regardless of coder complexity or processing delay, because of fundamental limits imposed by information theory and the acuity of the human perceptual system. But it is fair to ask the question: as we seek to approach these (sometimes unattainable) levels of high quality at low bit rates, what are the techniques most likely to succeed?

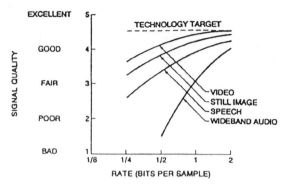

Fig. 9. Current capabilities in the coding of audiovisual information.

III. TOOLS OF THE TRADE

There are three fundamental operations that are common to low bit-rate signal coding: reduction of signal redundancy in the input signal, removal of irrelevant information in the operation of quantization, and signal enhancement by postfiltering. Of these, postprocessing is generally considered to be a process outside of the coding operations *per se*, although the benefits of performing the process can be very significant, as in low bit rate speech coding [12], [117]. Prefiltering can, likewise, increase the performance of a compression algorithm. This is accomplished in video coding, for example, by the reduction of camera noise in the input image or by the insertion of an explicit bandlimiting filter. The remainder of this paper will focus on the two operations that are intrinsic to signal coding: *removal of redundancy* and *reduction of irrelevancy*.

Almost all sampled signals in coding are redundant because Nyquist sampling typically tends to preserve some degree of intersample correlation. This is reflected in the form of a nonflat power spectrum. Greater degrees of nonflatness, as resulting from a lowpass function for signal energy versus frequency or from resonances (in audio) and periodicities (in audio and video), lead to greater gains from redundancy removal. These gains are also referred to as prediction gains or transform coding gains, depending on whether the redundancy is processed in the time or frequency (or transform) domain.

In a signal compression algorithm, the inputs to the quantizing system are typically sequences of prediction error or transform coefficients. The idea is to quantize the time components of the prediction error or the transform coefficients just finely enough to render the resulting distortion imperceptible, although not mathematically zero. If the available bit rate is not sufficient to realize this kind of perceptual transparency, the intent is to minimize the perceptibility of the distortion by shaping it advantageously in time or frequency, so that as many of its components as possible are masked by the input signal itself. We use the term *perceptual coding* to signify the matching of the quantizer to the human auditory or visual system, with the goal of either minimizing perceived distortion or driving it to zero where possible.

The parts of a coder that process redundancy and irrelevancy are sometimes separate, as in the above explanation. On the other hand, there are examples where the two functions cannot be easily separated. One example is a vector quantizer that combines intersample processing and quantization in a single stage of processing.

Almost all coding systems depend on the complementary interworking of the two basic operations defined above. A notable exception is a pulse code modulation (PCM) system, based on memoryless coding and quantizing algorithms, where there is no attempt to remove signal redundancy. This simple procedure is adequate for high-quality coding at bit rates in the range of 8–16 bits per sample, depending on the input signal. On the other hand, low bit rate coders, such as those evaluated in Fig. 9, depend heavily on more sophisticated signal analysis, processing delay, and redundancy removal prior to perceptually tuned quantization.

Speech signals have a universal production model which provides a very powerful framework for redundancy removal. By contrast, audio and image signals, although often very structured and redundant, lack a universal model of signal production. In this sense, the role of perceptually-efficient quantization becomes even greater in the coding of such signals, especially at lower bit rates.

We now discuss some generic examples of coding algorithms, drawing from the fields of speech, audio, and image signals. Our attempt here is not to provide an exhaustive overview, but to provide a suitable background against which to portray some evolutionary trends in current coding literature. This discussion, in turn, will set the stage for the research directions described in Section IV.

A. Closed-Loop LPC Coding of Speech

The universal model of speech production, consisting of an excitation followed by a linear filter [Fig. 10(a)], has made possible the ubiquitous use of the linear predictive coding (LPC) model in various applications of low bit rate speech, prominently in the form of the LPC vocoder for the extremely low bit rate of 2.4 kb/s [139] and code-excited linear prediction (CELP) and related algorithms for bit rates such as 4.8, 6.4, 8, and 16 kb/s [3], [13], [71], [79], [131]. The lower bit rates of CELP are typical of cellular radio and satellite applications. The Regular Pulse Excitation algorithm (RPE) [78] used in the European digital cellular radio system at the bit rate of 13.2 kb/s, as well as algorithms such as multipulse speech coding [4], vector-adaptive predictive coding [12], and vector-sum-excited linear predictive coding (VSELP), are similar to CELP in that they are *closed-loop* LPC systems unlike the *open-loop* algorithm used in the LPC vocoder. The 8 kb/s VSELP system [41] is the basis of the first-generation North American digital cellular telephone standard. The 16 kb/s CELP coder, with a backward-

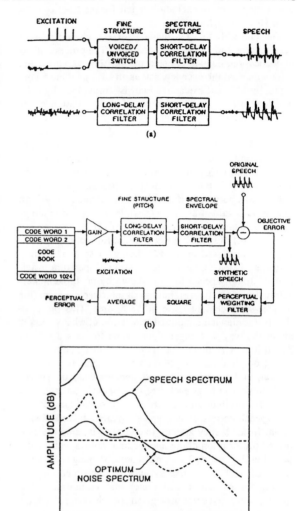

Fig. 10. (a) The excitation-filter model of speech production. (b) A block diagram of a code-book excited linear predictive coder (CELP). (c) Generic form of the optimally shaped spectrum of coding distortion (after [4]).

adaptive LPC predictor, is the basis for a network-quality speech coding system [13].

In the closed-loop systems, the basic model for speech synthesis is the traditional LPC scenario of an excitation signal driving an all-pole filter. However, the nature of the excitation is not decided by a simple binary categorization of speech into *voiced* and *unvoiced* segments (as in vocoding), but by an exhaustive search procedure that is reminiscent of waveform coding. This procedure picks, for each all-pole filter model, the best possible excitation signal for each speech segment (with a typical duration in the range of 5–20 ms) based on a frequency-weighted mean square error criterion. The closing of the *analysis-by-synthesis* loop provides the mechanism whereby a frequency-weighted mean squared error criterion is used to select the optimum excitation signal for the LPC synthesizer [Fig. 10(b)].

The closed-loop LPC method for speech coding is a very mature and successful paradigm. The process of redundancy removal is effected by the time-varying LPC filter which, at the encoder, acts as an LPC predictor. Perceptually-efficient quantization is provided by the weighted-error-steered selection of the excitation codevector (which is equivalent to vector quantization of the LPC residual). The voiced–unvoiced binary excitation codebook in LPC vocoding can be regarded as an extremely low bit rate version of that vector quantizer.

CELP and CELP-like methods are also very appropriate for the low bit rate coding of wideband speech. In currently reported work, with 7 kHz inputs, very high quality is obtained at 32 kb/s while very good communications quality is possible at 16 kb/s [63].

Fig. 10(c) illustrates the qualitative form of noise-shaping used in closed-loop LPC coding of speech. Coding distortion is least audible when its power spectrum is neither white nor speech-like (as shown by the broken characteristics in the figure), but has an intermediate form (the thinner solid characteristic in the figure). This prescription is a somewhat simplified translation of what we know about the way a stronger signal tends to mask a weaker signal in its frequency vicinity in the human auditory system. Complete masking of the weaker signal (coding distortion, in the current discussion) occurs when the signal-to-distortion ratio at each frequency equals or exceeds a critical threshold. In the current state of the art, low bit rate speech coders do not typically realize this ideal situation. Instead, they attempt to approach it using the simplified prescription of the intermediate noise spectrum depicted in Fig. 10(c). This kind of noise shaping is achieved in CELP and CELP-like systems by the process of selecting the excitation codevector that minimizes a frequency-weighted error for a given speech segment and the corresponding LPC filter.

B. Perceptual Transform Coding of Wideband Audio

Fig. 11(a) illustrates a key concept of perceptual coding that we shall call *just noticeable distortion* (JND, which stands for just noticeable *difference* in psychophysical literature). If the coding distortion is at or below the level of the staircase JND function at all frequencies, the input signal masks the distortion completely and transparent coding results, meaning that the compressed audio is indistinguishable from the original uncompressed audio. Important consequences of the above result are that: a) most signal components can be quantized fairly coarsely; and in particular, b) signal components below the JND can be completely discarded without causing any perceivable distortion. This is because the objective distortion at the discarded frequencies is equal to the (unsent) input magnitude and, if this is below the corresponding JND, the resulting degradation is perceptually irrelevant. A good example of discardable frequencies in Fig. 11(a) is the 5-to-6 kHz band. The combined result of discarding

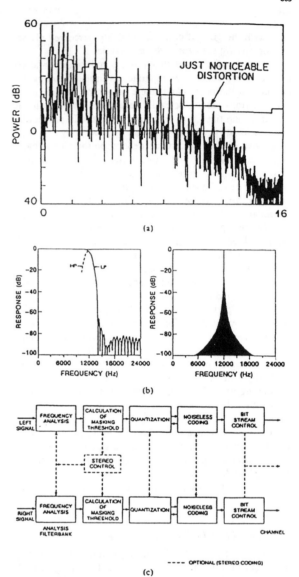

Fig. 11. (a) The just-noticeable distortion as a function of frequency. (b) Frequency response of QMF and MDCT filterbanks. (c) Block diagram of a perceptual coder capable of transparent coding (after [66]).

several frequencies and quantizing many others quite coarsely is that the average bit rate is very low, on the order of 2–3 bits per sample. These rates provide perceptually transparent coding of the 16 bit-per-sample input because of the use of the JND principle.

The JND profile is calculated from a short-term frequency description of the signal to be coded, using principles of human audition—in particular, detectability criteria and properties of the cochlea, the frequency-separation mechanism of the human ear. Quantitative descriptions of JND data are well understood for simple and idealized descriptions of masking and masked signals (for example, tones and noise). But in the general case of an arbitrary audio signal masking a complicated type of coding distortion, the JND description is by necessity approximate and empirical.

The result of Fig. 11(a) is one such empirical description, for the example of a trumpet signal with a bandwidth of 16 kHz. In this example, the staircase description of the JND profile is the result of interpolating between known results of noise-making-tone and tone-masking-noise [66]. The interpolation algorithm is based on a measure of how tone-like or noise-like the input signal is. The staircase treads in Fig. 11(a) correspond to the 25 *critical bands* in the human auditory system, narrow at low frequencies and wide at the higher frequencies. The phenomenon of masking is largely localized in each critical band, although there is some degree of interband masking (masking of distortion in a band by a signal in a different critical band). The interband masking effect falls at about 15 dB per critical band for the higher masked frequencies, and at about 25 dB per critical band for the lower masked frequencies. This behavior is related to the frequency shape of the cochlear filter.

The JND is a function of a (finely described) input spectrum and the ear model, rather than a simple transformation of the LPC spectrum of the input signal. While it is conceivable that a very high-order LPC analysis can describe the input spectrum finely enough to permit a useful JND model, current methodologies for transparent or near-transparent audio coding have depended on high-resolution frequency analysis using either a subband coding or a transform-coding framework.

Fig. 11(b) illustrates the typical characteristics of a 96-tap *quadrature-mirror filterbank* (QMF) used for subband analysis of audio, and the alternative framework of a 512-line *modified discrete cosine transform* (MDCT).

The QMF system is based on the division of a frequency band into contiguous but overlapping subbands. The partially-completed broken line shows the characteristic of a highpass filter that is the mirror image of the solid-line lowpass filter characteristic. The extent of overlap is a decreasing function of the number of filter taps. But the allowing of a nonzero overlap simplifies filter design. Frequency aliasing is caused in QMF analysis by sampling each of the two bands in the QMF split at twice the *nominal* bandwidth (rather than twice the actual, say 90 dB, bandwidth). However, with a special design of QMF filters, the process of QMF synthesis provides cancellation of this aliasing if the quantization noise inserted in the system is zero [16], [26], [62], [65], [140].

The modified DCT system [111] is a dual of the QMF approach in that it permits an overlap between successive transform blocks in the time-domain but decimates the resulting sequence to maintain the original sampling rate. In place of the frequency aliasing in the QMF system, the MDCT exhibits time-domain aliasing. But this is canceled by the inverse MDCT process in the receiver due to the design of the DCT basis vectors and the analysis window. The overlap in the MDCT is typically 50% and the decimation rate is 2:1 [97],[111]. While a 50% fre-

quency-domain overlap is admissible in QMF design in principle, it is untypical.

Although QMF and MDCT systems are essentially dual, they exhibit different properties in the context of a specific overall coder. For example, operations such as signal anticipation and temporal bit allocation can be used to some extent to control quantizing distortion and the consequent phenomenon of uncanceled time-domain aliasing in the MDCT system. This, in turn, permits the use of a 50% time overlap which provides a smooth handling of the time process as well as a simple decimator design. In QMF design, on the other hand, one still prefers, in the current state of the art, to employ a small spectral overlap within the constraints on filter complexity. This, in turn, implies a relatively discontinuous handling of the frequency process. Finally, in both MDCT and QMF, the available time-frequency behavior is a compromise between a match to sustained stationary inputs and the ability to handle true input discontinuities in time or frequency.

Fig. 11(c) is a schematic of a perceptual audio coder based on subband or transform coding. The block diagram of Fig. 11(c) includes the possibility of jointly coding the two channels in a stereo-pair to maximize the gains due to redundancy removal and perceptual tuning [6], [66], [137].

C. DCT and Motion-Compensated DCT Coding of Image and Video

As in the case of audio, the absence of a strong autoregressive (LPC) model for the source signal has led to the well-accepted use of subband and transform coding methods for image and video compression. Source redundancy is addressed by decomposing the input signal into components of differing variance in the frequency or transform domain and following this by variable bit allocation in the quantization of the transform coefficients. A greater number of bits is allocated to the components of higher variance, and the overall mean squared error is minimized for a given constraint on total bit rate. Perceptual matching can be realized, at least partly, in the bit allocation process if a good model is used for the best possible profile of distortion versus frequency.

Of particular interest to image processing and image coding standards is the two-dimensional discrete cosine transform (2-D DCT) [1], [62], [100], [147]. The DCT provides a good match to the optimum (covariance-diagonalizing or Karhunen–Loeve) transform [62] for most image signals, and fast algorithms exist for computing the DCT.

A ubiquitously used frequency-selection or bit-allocation rule for 2-D DCT is one based on a lowpass process described by the *zig-zag* scan of Fig. 12(a), which refers to an 8 × 8 DCT operation with 64 samples of 2-D frequency ranging from DC (0, 0) to the highest frequency (7, 7). The low frequencies in the initial parts of the scan are given higher priority for retention or bit allocation,

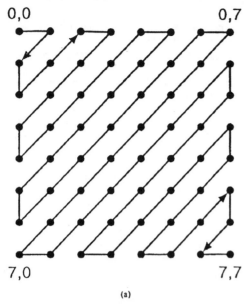

0,0   0,7

7,0   7,7

(a)

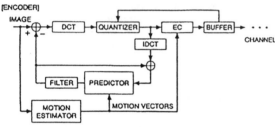

[ENCODER]

IMAGE

DCT | QUANTIZER | EC | BUFFER · · ·

CHANNEL

IDCT

FILTER | PREDICTOR

MOTION ESTIMATOR | MOTION VECTORS

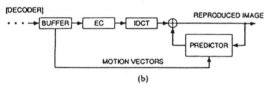

[DECODER]

· · · → BUFFER → EC → IDCT → REPRODUCED IMAGE

PREDICTOR

MOTION VECTORS

(b)

Fig. 12. (a) The zig-zag scan for frequency selection and bit allocation processes in DCT coding. (b) Block diagram of video coding based on motion compensation followed by DCT coding of the interframe residual (IDCT: Inverse DCT. EC: Entropy Coding) (after [81], [85], [100]).

and the zig-zag scan represents a monotonic deemphasis of the higher 2-D frequencies. More sophisticated algorithms (Section IV-B) can provide for a nonmonotonic bit allocation process for those inputs that demand it.

Fig. 12(b) is a block diagram of a video coding system based on interframe motion compensation followed by 2D-DCT coding of the residual signal after motion compensation. Motion vectors are derived from the input process and are used to define a space-varying interframe predictor. Abrupt changes in hypothesized motion are prevented by means of a filter before computing the motion-compensated interframe residual that is quantized by the 2-D DCT module. The transmitted information consists of the quantized residual and the motion vector. An

entropy coder (EC) is used to reduce the redundancies in the DCT and motion vector sequences. Important sources of these redundancies are the unequal probability distributions in the possible ranges of quantized DCT components and the quantized motion vectors.

The system of Fig. 12(b) is widely used in video coding. It forms the basis of two international coding standards: the $p * 64$ kb/s standard of the CCITT ($p = 1$ to 24) [85] and the 1.1 Mb/s standard of ISO-MPEG [81]. The CCITT standard is intended for videotelephony and videoconferencing applications. The MPEG standard is intended for addressable video in multimedia applications. As such, the MPEG standard also allows for high-quality *intraframe* coding of selected frames in the video sequence. The MPEG algorithm also permits greater processing delays than the CCITT system, given the storage focus in MPEG.

Removal of the motion compensation loop in Fig. 12(b) results in a still-image coder based on 2-D DCT. The resulting system forms the basis of the international coding standard known as ISO-JPEG [147].

D. The Next Generation of Signal Coders

The techniques of Sections III A–C are low bit-rate systems with several important features. With the exception of the LPC vocoder, these systems are phase-preserving waveform coders with an emphasis on naturalness in output signal quality. These coders use generic techniques for reducing redundancy and for matching the quantizing system to the human perceptual mechanism. They generally include means for variable bit rate coding, provide for some degree of inherent resistance to transmission errors, and permit efficient implementation in existing signal processing technology. They form the bases for various international standards.

As we look toward the next generation of coding algorithms, we seek to decrease the bit rate even further for specified levels of signal quality and, in some applications, we need to increase reproduced signal quality at a specified bit rate. We need to develop the possibility of increasing the signal bandwidth that can be realized at a given bit rate and at a given level of quantizing distortion. Finally, we need to drive several technologies towards the ideal of (perceptually) distortion-free coding, especially speech and video technologies that currently provide only *communications quality* rather than *high* or *transparent* quality.

IV. RESEARCH DIRECTIONS

In discussing directions of research, it is impossible to be exhaustive; and in predicting what the successful directions may be, we do not necessarily expect to be accurate. Nevertheless, it may be useful to set down some broad research directions, with a range that covers the obvious as well as the speculative. The remaining parts of Section IV are addressed to this task.

A. Subsampling, Interpolation, and Multiresolution Processing

Signal distortion in digital coding is a combination of prefiltering, aliasing, and quantizing components. The tradeoffs among these components are not rigorously understood. Typically, at high and moderate bit rates, prefiltering distortion is effectively a nonissue in that users accept or get used to a predefined and well-accepted bandwidth, as in telephone-grade speech. Aliasing distortion can be very noticeable even when it is small in a mean squared error sense. Typically, aliasing components are either avoided by proper prefiltering or, in the case of relatively complex arrangements such as subband coding, these components are carefully minimized. Quantizing distortion is, therefore, the typically most relevant part of reconstruction error; it is the component that one typically seeks to minimize in designing a coding algorithm at a specified bit rate.

As we consider lower bit rates and higher quality, all the components of coding distortion can become significant, and issues of their interaction and of relative tradeoffs become important as well.

One example of such a tradeoff is that between bandwidth (or sampling rate) and the bits per sample for a given total bit rate. With the possible exception of speech coding, where the telephone bandwidth of 3.2 kHz can be regarded as an essential minimum, the notion of optimizing bandwidth for a given bit rate can be quite an interesting problem. A good example is the definition of optimum bandwidth in audio given an overall low bit rate such as 16 or 32 kb/s. Another example is the definition of optimum spatio-temporal resolution in the coding of video signals at low bit rates such as 64 or 128 kb/s. Specific low resolutions such as CIF and quarter-CIF are selected using reasonably good criteria in low bit rate coding, and displayed signal resolution is enhanced by means of interpolation especially in the time domain [81], [85]. However, fundamental unanswered problems remain, such as the selection of the best fixed set of spatial and temporal resolutions for a given bit rate and the definition of efficient multiresolution algorithms [7], [28], [88], [141] for dynamic adaptation of spatial and/or spatio-temporal frequency resolutions. Rigorous solutions of these problems are difficult even with the simplifying assumption of zero aliasing, but refined models of coding and psychophysics may be able to translate what is currently no more than an art to a science with a reasonable potential for generalization.

Although there is little flexibility in the bandwidth or sampling rate of telephone speech, as mentioned earlier, the techniques of subsampling and interpolation are still extremely valuable in the spectral domain. For example, the interpolation of LPC parameters is a widely practiced tool for low bit-rate speech coding [41], [71], [79]. More recently, an interpolation technique for the compact description of the excitation waveform of voiced speech has been proposed. The technique, called *prototype waveform interpolation* [76], is based on the transmission of a prototype excitation waveform and its pitch once every update time, on the order of 20 to 30 ms. A complete excitation signal is obtained by means of interpolative techniques in the Fourier series domain. The process preserves a high level of periodicity, and is flexible enough to realize low levels of reverberation and tonal artifacts. Combined with CELP for the coding of unvoiced speech, the technique provides a promising framework for high-quality coding at rates on the order of 2 to 4 kb/s.

B. Techniques for Time-Frequency Analysis

The assumption of zero aliasing is a conspicuous simplification in almost all subband coding literature. Techniques such as quadrature-mirror filtering and the modified discrete cosine transform can provide perfect cancelation of aliasing in the absence of quantization errors. This ideal solution never occurs in a practical coding situation, and the assumption of the ideal case becomes increasingly inappropriate at lower bit rates because of a corresponding increase of quantization error. Recent work has given us a fairly good understanding of filterbanks that provide zero or near-zero reconstruction error in the absence of quantization. However, the design of a filterbank that minimizes the combined effect of quantizing and aliasing errors is an entirely unsolved problem. Here again, a rigorous optimization is extremely intractable in our current state of knowledge but we sorely need at least partial solutions.

A somewhat orthogonal, though not unrelated, research area is that of efficient time-frequency analysis. Considerations of signal nonstationarity and perceptual distortion criteria have resulted in increasingly sophisticated frameworks for signal analysis. In particular, techniques that provide flexible combinations of time-support and bandwidth represent a powerful generic tool for efficient coding. The discrete Fourier transform and a uniform-bandwidth quadrature mirror filterbank are well-understood and widely used analysis tools. But in their simplest forms, they lack the flexibility for time-frequency analysis mentioned above. QMF trees with unequal-bandwidth branches, as well as subband-DFT hybrids, are relatively newer structures with more flexible features. This is also true of *wavelet* filters [21], [88], [120].

Wavelets: Unlike the basis vectors of a DFT (sinusoids and cosinusoids of various frequencies and constant time-support), the wavelet filter structure is characterized by a shorter time support at higher frequencies and a longer time support at lower frequencies, a direct result of a dilating operation that is a basic component of wavelet design (Fig. 13). The time-frequency characteristic of a wavelet filterbank is a natural match to some of the properties of audio-visual information: high-frequency events often occur for a short time and stand to benefit from a finer resolution in time analysis, while low-frequency events are often sustained in time and require less frequent sampling in time. The wavelet approach, especially if used in a time-varying framework, may therefore offer

FT

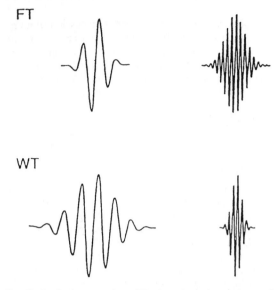

WT

Fig. 13. Qualitative comparison of Fourier and wavelet transforms (after [120]).

powerful forms of adaptive analysis in a more basic sense than a nonuniform frequency-band QMF system or a variable window-length MDCT system. Wavelet transforms provide the additional feature of perfect reconstruction, a property not generally offered in conventional methods of analysis. Conventional methods, however, do offer the property of *almost-perfect* reconstruction which is adequate for many low bit rate applications.

Wavelet filtering is a promising analytical tool and applications of it are also beginning to emerge. What is still lacking, however, is a thorough understanding of what wavelets can do for coding that the more sophisticated examples of conventional analyses cannot. As we seek to apply wavelet (or nonwavelet) tools to low bit rate coding, attention must necessarily shift to the yet-untouched problems of uncanceled aliasing and the computationally-intensive but naturally appealing notion of a signal-adaptive filterbank.

Multidimensional Subband Coding: The discussion in Section III-C implied a two-dimensional DCT (2-D DCT) for still-image coding and the hybrid approach of motion compensation combined with 2-D DCT for the coding of video. Alternative approaches in current research include 2-D subband coding for still-image coding [25], [124], [150], [154] and 3-D subband coding for video [69], [112], [144]. The subband filters in each case may belong to any desired class, such as QMF's or wavelets. As in our discussion of (1-D) QMF and (1-D) MDCT techniques for audio, we note that (2-D and 3-D) transform and subband techniques are, in principle, dual operations. However, in the context of an overall coding system, differences can exist in terms of implementation, matching to the human perceptual system and robustness to transmission errors.

Fig. 14 shows the analysis filterbanks for 2-D and 3-D subband coding using separable (but not necessarily identical) filters in the different dimensions. In the illustrated examples, the 2-D filterbank provides a 16-band partition of the spatial 2-D frequency space and the 3-D filterbank provides an 8-subband partitioning of the spatio-temporal 3-D frequency space. We shall comment again on the 2-D subband coder in the context of perceptual coding (Section IV-D-2). The 3-D subband coder represents a significant departure from the current practice of motion compensation followed by a spatial transform. Rather than comparing two adjacent frames to realize good models of local motion, the 3-D approach processes two or more adjacent frames in order to capture spatial and temporal activity in a more integrated fashion. Most of the signal energy tends to be in the subband with low temporal and horizontal frequencies. Motion detection occurs prominently in the subband with high temporal frequency and low spatial frequencies. Low bit rate coding results from a variable bit allocation algorithm that matches not only the energies in the spatio-temporal subbands but also the respective measures of perceptual significance (ideally, a spatial-temporal 3-D JND profile).

C. Vector Quantization

It is generally recognized that in the compression of audio and visual signals, a suitably global distortion metric is more meaningful than a local single-sample-oriented metric. Simple forms of waveform coding, such as PCM and differential PCM (DPCM), are based on a local distortion metric. More complex coders, such as delayed-decision DPCM and block-oriented coders of various kinds, are characterized by the use of a more global distortion criterion. Examples of block coders are CELP systems for speech coding and 1-D and 2-D transform coders for audio and image signals. The general mechanism that permits the use of a global distortion criterion is called *delayed decision coding* [20], [62].

Vector quantization (VQ), tree coding, and trellis coding are all techniques for delayed-decision coding [62]. Unlike the block-oriented approach in VQ, tree and trellis structures use a sequential procedure to minimize a suitably global distortion measure. The trellis structure, which can be realized by using a finite-impulse-response code generator, has the advantage that the total number of possible output values is a finite number by definition, permitting efficient searches for the best path through the trellis (the coded output sequence) [33]. The VQ paradigm also has a finite output alphabet, by definition.

Several examples of vector, tree, and trellis coders have appeared in coding literature [62], together with some examples of hybrid algorithms using both block and sequential approaches [107]. The block and sequential approaches are not fundamentally different from a rate-distortion viewpoint. The block approach of VQ seems, however, to be somewhat better understood and more widely practiced, with a broad repertoire of special techniques for codebook design, fast codevector search, in-

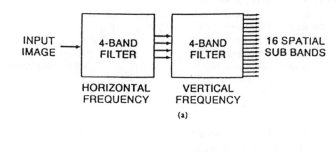

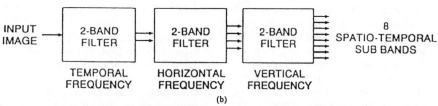

Fig. 14. Examples of analysis filterbanks for: (a) two-dimensional and (b) three-dimensional subband coding (after [124] and [69], [112]).

terblock coding, and optimization for noisy transmission channels [11], [14], [24], [27], [28], [38], [39], [40], [48], [49], [51], [72], [82], [83], [84], [92], [105], [107], [112].

Vector quantization (VQ) is known to be most efficient at very low bit rates (on the order of $R = 0.5$ bit per sample and less). This is because when R is small, one can afford to use a large vector dimension N and yet have a reasonable size 2^{RN} of the VQ codebook. Use of a large dimension N tends to bring out the inherent capability of VQ to address linear as well as nonlinear redundancies [87] in the components of the vector being quantized.

In speech coding, the use of VQ has been most successful in the coding of parameters describing the speech spectrum (LPC-VQ) [37], [67], [87], [109], [123] and speech excitation [4], [12], [13], [41]. Refinements such as gain-shape VQ [125] and multistage VQ [67], [83] have been routinely applied in the quantization of the spectrum and excitation parameters in speech. Direct VQ coding of the waveform has been less successful to date. This is due partly to intervector discontinuities resulting from a block-quantization process, and partly to the lack of a perceptually good model for minimizing intravector distortion. With LPC-VQ and CELP-coding, the distortion in the VQ process is conveniently frequency-shaped. This is accomplished by a spectral distortion metric in LPC quantization, and by the subsequent LPC filter itself in the case of excitation VQ. Direct VQ coding of the speech waveform, while not very successful to date, is a very interesting research problem and it offers an excellent challenge to the notion of adaptive and perceptually-tuned vector quantization.

Scalar quantization, followed by entropy coding [54], [55], [81], [85], [153], has some equivalences with vector quantization, at least in terms of attaining high values of signal-to-noise ratio at relatively low bit rates, especially if the probability density function of the input signal is Laplacian or gamma (rather than, say, uniform or Gaussian). This equivalence has perhaps diminished the application of VQ. However, powerful algorithms result when selected vectors from a DCT process are identified for low bit-rate vector quantization. Such techniques for DCT-VQ coding have been used successfully in image and audio coding [11].

Direct VQ coding of the signal waveform has been relatively more successful in the coding of intensities in 2-D images [49], [51], [96], [99], [118], [125], [141], [149], [150], [155] and in the coding of interframe 2-D residuals in motion-compensated coding of video. In the usual case of iteratively trained codebooks, the need for a typical database for training and the possibility of a significant mismatch between the input sequence and the trained VQ codebook are both important, if not problematic, issues. The importance of finite-state vector quantization to provide 2-D adaptivity of the codebook has been well demonstrated, resulting for instance in high-quality coding of a still image at 0.5 bit per sample. In the case of high-frequency image subbands dominated by sparse intensity profiles such as edges, an untrained system called *geometric vector quantization* [112] has been shown to provide very efficient coding at extremely low bit rates, on the order of 0.1 bit per sample and less.

As we move toward the next generation of low bit-rate algorithms for image and video coding, more ubiquitous use of vector quantization is very likely. Further research is needed to support these applications. We need a better understanding of tradeoffs and interactions between vector quantization and entropy coding, better experience with VQ-type structures such as pruned trees, better algorithms for adaptive VQ in 1, 2, and 3 dimensions, and more powerful perceptual models for characterizing and minimizing the vector-distortion process.

In expanding the frontiers of VQ research, a potential source of cross-fertilization is the field of *fractal block coding* [57]. Recent work on the subject has shown image quality similar to that of 2-D VQ in the coding of still images at rates on the order of 0.5 bit per sample. The fractal method utilizes a subtle form of self-similarity in a scene, in particular, similarities between selected pairs of blocks in 2-D images in the presence of a powerful set of transformations. In each such pair, the block to be encoded is called a child and the potential matching reference is called a parent (Fig. 15). Allowed transformations include changes of scale, orientation and grey level, or color. An image is encoded by means of a fractal code that consists of a sequence of child–parent maps and a corresponding sequence of transformations chosen from a transformation codebook. The child–parent map and transformation, for any given input block, are results of a joint optimization that minimizes the error in child–parent matching. Since there is no insistence on a zero-matching error (perfect child–parent similarity), the method can be called a *soft-fractal* technique. Decoding consists of iterative calls of the fractal code on an arbitrary initial image. The block-fractal algorithm can be viewed as an interblock coding algorithm with nonlinear block prediction. In this sense, it is reminiscent of predictive and finite-state vector quantization algorithms [48], [51], [72], [107]. It appears that new research on adaptive quantization may enrich the fields of fractal coding and VQ alike, and that these disciplines, as we know them today, may also have mutually orthogonal strengths.

D. Models of Signal Production and Perception

Central to the success of the techniques discussed in Sections IV A–C are models of signal production and perception. These models need to be physically realistic and computationally feasible. In some problems, as in filter-bank design, it is desirable to seek architectures that are matched reasonably both to the production and perception models. On the other hand, it is important to realize that entirely different requirements may exist in the matching processes mentioned above. In order to address the complex and not identical phenomenon in signal production and signal perception, it is important that the signal processing architectures used in signal coding are as general and flexible as possible.

1) Models of Signal Production: As mentioned earlier, the speech waveform is unique in that it enjoys a reasonably well-understood and universal source model that is efficiently approximated as the result of an excitation signal modulated by a linear transfer function. In the search for high-quality LPC coders, recent algorithms such as CELP have gotten away from the restrictive voiced-unvoiced binary model for excitation. However, as we increase the focus on very low bit rates such as 2–4 kb/s, the voiced-unvoiced model is receiving renewed attention because of its compactness and the resulting economies in bit rate. These include the tech-

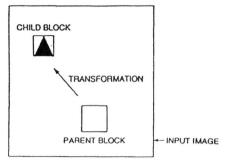

Fig. 15. Image coding based on soft-fractal analysis (after [57]).

niques of multiband excitation coding [50], [52], sinusoidal coding [90], and prototype-waveform interpolation [76]. In the new generation of coders based on the voiced-unvoiced classification, the attempt is to make the classification soft and robust, so that the effect of a wrong classification is imperceptible. A generalization of the binary voiced-unvoiced classification is a system where classification is based on a larger number of states and on a phonetic criterion for state-switching [148].

Indirect analogies to the voiced-unvoiced model exist in audio and image coding as well. In audio, the classification of an input signal as a tone-like or a noise-like signal provides a good basis for adaptive perceptual coding. More powerful adaptation results if the signal can be associated with degrees of tonality and noisiness along a continuum.

In image processing, an analogous classification is into categories dominated by edges, textures, and flat regions of grey [118]. Another kind of classification results from segmenting an image into background and moving areas, a classification that is particularly meaningful in a typical teleconferencing scene. Resulting image classes call for different algorithms, redundancy removal, and perceptually-matched quantization.

As we seek to extend the capabilities of the above models, especially the less mature models of audio and image, we need to optimize prediction, transform, and quantization processes for the various signal classes. We also need to address the problem of maintaining perceptual continuity in a signal in the context of the breaking up of the signal into several sets of homogeneous parts in model-based classification processes [68], [80], [111].

In principle, the ultimate results in coding are those obtained when the models reflect the very earliest stages in the signal production processes. Examples are the articulatory vocal cord-vocal tract model of human speech production and the wire-frame image model of a human face (Fig. 16).

The articulatory model [128] of Fig. 16(a) extends the focus from LPC analysis to the analysis of vocal tract areas and, in principle, provides a much stronger domain for very low bit-rate vector quantization. The model also permits a better handle on the interaction between vocal cords and the vocal tract, a phenomenon that is conspic-

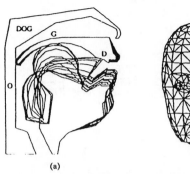

Fig. 16. (a) Articulatory model for speech coding (after [128]). (b) Wire-frame model for coding images of the human face (after [2], [94]).

uously ignored in the simplified traditional model of excitation-followed-by-filter.

The wire-frame model of the human face [2], [29], [94] in Fig. 16(b) is potentially a very powerful domain for compressing scenes dominated by the face image, as in a videotelephone scene with a closeup of the human face. The general wire-frame model is a 3-D lattice with a very large number of polygonal sections. Using affine transformations and the actual image of a given face, a personal facial model is created. Vector-quantized versions of dynamically changing mouth features driven by textual or speech cues, together with algorithms for representing text- or speech-independent facial features, are used to create a talking-head image.

The above models are a natural basis for an intelligent human–machine interface. If they are intended as a basis for human telecommunication, an important obvious challenge for articulatory and wire-frame models is the realization of natural, rather than synthetic, signal quality at the very low bit rates such models are intended for.

2) Models of Signal Perception: Perceptual criteria have been addressed since the very beginnings of speech and image coding, and as coding algorithms have matured, criteria for optimizing these algorithms have become increasingly complex [3], [6], [30], [58], [66], [98], [102]–[104], [110], [121], [124], [126], [127], [135], [137]. An interesting early example of perceptual coding is the use of dithering to improve the quality of low bitrate PCM by breaking up structured and, hence, highly visible patterns in the distortion process [62], [121]. Examples of complex perceptual coders are the time- and space-varying algorithms discussed in the remainder of this section.

Models of human hearing and vision can steer a coding algorithm in two related but distinct ways. If one can define, for each part of a signal being coded, a just-noticeable-distortion (JND) level below which reconstruction errors are rendered imperceptible because of masking, the model will be the basis of perceptually lossless coding, a process in which the perceived distortion D_p is zero even if a mathematically measurable distortion such as the mean-squared-error D_{mse} is nonzero (in fact, quite significantly so in typical examples). If, on the other hand, the average bit rate for coding is not sufficient to realize the JND profile in all parts of the signal, the perceptual model can still suggest a good match of the quantizing system to the perceptual model, in the sense of minimizing D_p, rather than D_{mse}. In this case, the input to the quantizing system is not a JND profile as in Fig. 11(a) but a *minimally-noticeable-distortion* or MND profile, as in Fig. 10.

A generic block diagram of a perceptual coder driven by JND (or MND) cues is shown in Fig. 17. The JND and MND profiles are meant to be dynamic, being functions of local signal properties such as dominant frequency, background intensity or texture, and local temporal activity. The mapping from these properties to the JND or MND profile is performed in real time, although the function defining the mapping is established prior to coding, based typically on extensive subjective experimentation. In the case of the JND coder, the system is a constant-quality, variable-bit-rate system by definition. However, feedback from a bit rate buffer can be used to realize a constant-bit-rate variable-quality system whose distortion profile approximates the JND. In a conservative design, the actual distortion would be less than the JND most of the time and greater than the JND on occasion.

The JND method has been extremely successful in the transparent and near-transparent coding of wideband audio [Fig. 11(a)] and, more recently, in the transparent coding of still images based on 2-D subband analysis at extremely low bit rates. Fig. 18 illustrates the JND as a function of space for the example of a subband image signal. Parts (a)–(c) of the figure display: (a) the input image; (b) the lowest frequency subband of it (low horizontal and low vertical frequencies in the 16 subband system of Fig. 14(a); and (c) the JND image for the coding of that subband. Following the coding of the 16 subbands based on respective JND profiles, the coded subbands are combined in the synthesis filter to obtain a low bit-rate coded image. Part (d) of the figure describes, as a function of space, the number of subbands (out of 16) that are retained (with nonzero bit allocation) in the JND-driven system. White, light grey, dark grey, and black blocks retain 4, 3, 2, and 1 subbands out of 16, respectively, indicating a high degree of compression.

The perceptual subband coder results in excellent image quality at extremely low bit rates, generally lower than the rates realized by the well-known ISO-JPEG DCT algorithm. This capability has also led to a new proposal for the coding of high-resolution facsimile—as a grey-level input compressed to extremely low rates and half-toned at the receiver rather than at the transmitter [104]. Retention of the grey-level domain in most of the system has been shown to result not only in lower transmission time on a given digital channel but also in a significantly sharper end-result with a printer of given resolution, provided that the half-toning algorithm is optimized using models of the printer and the human visual system [110].

In the coding of 3-D images, a major research challenge is the definition of temporal models of masking and

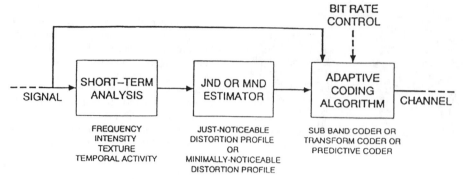

Fig. 17. Generic block diagram of perceptual coding.

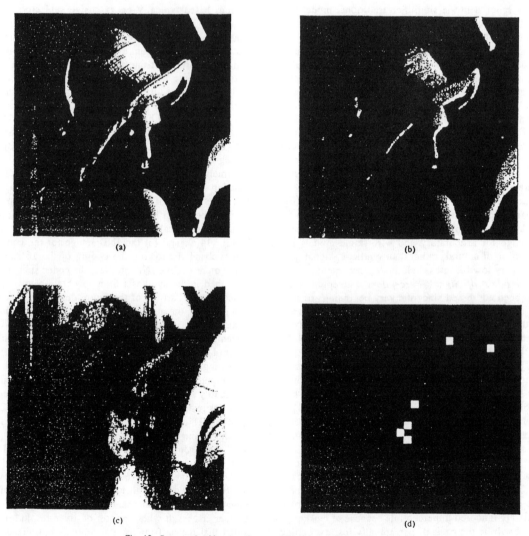

Fig. 18. Perceptual subband coding. (a) 512 × 512 image of *Lena*. (b) Low-frequency subband of *Lena* in a 16-band 2-D subband coder. (c) The JND image for (b). (d) The spatial bit-allocation profile (after [124]).

374

the definition of a suitable coding framework for exploiting these physical models. Recent experiments in motion-compensated coding for high-definition television have shown the potential of spatial masking models as in Fig. 18, as well as temporal masking models that describe the masking of distortion by certain kinds of motion activity [103]. It is possible that 3-D subband analysis [Fig. 14(b)] will provide an even better description of the motion process and, therefore, a better framework for perceptual coding [112].

Temporal models of noise masking are also of great interest in the coding of audio and speech. More powerful masking models are crucial to the goal of high-quality coding of speech at very low bit rates. The focus here is on MND functions that provide increasingly more efficient formulas for noise shaping. JND formulas resulting in transparent coding (at slightly higher bit rates) are expected to be byproducts rather than prime targets in low bit-rate telephony.

An MND-based coder has an easier function, in principle, than a JND-based coder because a nonzero value of the perceptual distortion D_p is allowed. However, the methodology leading to a good MND design can actually be more complex than that leading to a JND design at a higher bit rate. This has to do with the subjective experiments that are the bases for the JND and MND formulas. In such experiments, it is easier to identify a situation where D_p is zero (unnoticed distortion by a specified percentage, say 50 or 95, of the subject population) than to associate a perceptually meaningful value, along a continuum, to the distortion (given that it is clearly noticeable). For the same reason, if one were to seek a distortion-rate description of the signal, as in information theory, it is easier to identify the bit rate $R_p(0)$ required for $D_p = 0$ than to calculate a perceptually valid shape for the entire $D_p(R)$ curve. We expect, of course, that this curve is a significant left-shift of the traditional curve of D_{mse} versus R (Fig. 19).

The mean squared distortion at zero bit rate is equal to the signal power itself. The bit rate $R_{mse}(0)$ at which the mean squared error is zero is infinite for a continuous-amplitude source, and finite for a discrete-amplitude source such as a computer-stored file of 8-bit image pixels ($R_{mse}(0) = 8$ in this case; in fact less, typically in the neighborhood of five bits per sample if mathematically lossless coding is performed). The bit rate $R_p(0)$ at which the perceived distortion can be designed to be zero is a fundamental limit in coding, and we will call it the *perceptual entropy* of the signal. More generally, the $D_p(R)$ curve defines the fundamental limit of signal compression for a specified level of output signal quality.

E. Source and Channel Coding

Traditionally, source and channel coding have had complementary roles in digital communication. The source coder has tried to minimize the bits-per-sample for high-quality signal representation, while the channel coder

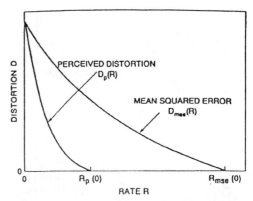

Fig. 19. Distortion-rate function for mean squared error and perceived distortion.

(modulator and error protection system) has attempted to maximize the bits-per-second-per Hertz that can be used on a transmission or storage medium to represent the digitally coded signal. While these complementary roles will continue in future technology, it will become increasingly important to define interactive and joint designs of the source and channel coding algorithms.

For example, if the output of a source coder can be categorized into parts of varying sensitivity to bit errors in transmission, a given total overhead for error protection can be used in an unequal error protection scheme that will have the final effect of extending the range of channel quality over which a specified quality of signal communication can be maintained [Fig. 20(a)] [17]. In situations where the transmitter has information about channel quality, a joint source-and-channel coding algorithm can make a suitable allocation of the total bit rate for source coding and error protection [93]. This, again, has the effect of utilizing transmission media at low levels of channel quality: a slight undercoding of the signal in a quantization-noise sense, together with a stronger focus on error protection, can realize a specified level of total (quantization-plus-channel) noise over a wider range of channel quality [Fig. 20(b)]. Finally, by refining our methods for source coding, channel coding, and cooperative source-and-channel coding, we can generally enhance the gains over analog communication [Fig. 20(c)]. Recent examples of this appear in digital cellular technology for mobile radio telephony and digital transmission proposals for high-definition television. Ideas of perceptual optimization, discussed in earlier sections of this paper for signal coding, carry over to signal communication as well, as in optimizing algorithms for unequal error protection and joint source and channel coding.

F. Signal Compression and Packet Networks

Earlier work on packet speech and video was concerned with the effects of packet losses and means for interpolating the signal in the presence of packet losses [45], [61], [70], [119], [142]. This line of research is being reacti-

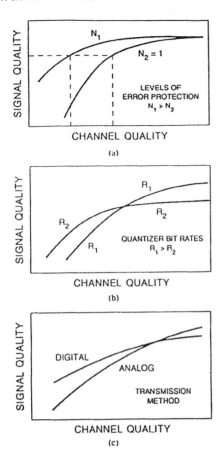

Fig. 20. Signal quality as a function of transmission channel quality, with: (a) digital transmission using unequal error protection; (b) digital transmission using joint source and channel coding; and (c) analog and digital transmission.

vated today because of the increasing importance of the asynchronous transfer mode (ATM) packet technology. An important concept here is that of *layered coding*. The output of the source encoder is divided into cells of varying significance, typically with two layers, or levels of it. When the packet network is congested, the idea is to drop cells of lower priority [75], [95], [138], [143]. In the case of uncompressed PCM data, the prioritization of encoder bits is straightforward. But in low bit rate coders, as in the unequal error protection scheme of Fig. 20(a), optimization of the communication network will depend on perceptual cues for cell layering and subjective methods for measuring the user acceptance of layered coding.

V. CONCLUSION

This paper has presented a description of technology targets in signal compression and a nonexhaustive, and very possibly biased, account of research directions that may lead us toward these targets. As we pursue these and other directions, some of which are undescribed here and some of which are quite unknown at this time, one broad

observation would perhaps stay uncontested: the new generation of algorithms will reflect, better than ever before, perceptual cues as integral parts of the coding process. In order to calibrate and steer our research progress in a discipline where the performance criteria are mathematically hard to model, we will also depend increasingly on subjective evaluations of signal quality, as in Fig. 9.

Quality evaluations have been invaluable in speech and audio coding and in the early history of television, but conspicuously lacking in contemporary work on digital image coding. Experiments to define specific distortion models for optimizing a coder, and experiments to measure the overall quality of the coded signal, are both very time-consuming and intricate. However, both of these endeavors will be necessary investments if we seek to advance signal coding technology to the fundamental limits defined by information theory and psychophysics.

This paper has also pointed out opportunities for integrating source coding and channel coding technologies. Such integration, which has hitherto been an informal exercise, will become increasingly essential as we stretch our communication capabilities with capacity-limited channels such as wireless media. In parallel, as we seek greater sophistication in the integration of speech and data with broadband signals such as CD-audio and high-resolution video, we will witness an increased interaction of signal compression technology with the field of communication networking.

ACKNOWLEDGMENT

The author is indebted to three anonymous reviewers whose extensive and constructive criticism has added significantly to the value and credibility of this article. Also, for their invaluable inputs, many thanks to his esteemed colleagues P. Noll, L. Rabiner, B. Atal, J. Johnston, N. Seshadri, S. Quackenbush, and J. Schroeter. Thanks also to T. Pappas for providing the pictures in Fig. 18.

REFERENCES

[1] N. Ahmed, T. Natarajan, and K. Rao, "Discrete cosine transform," *IEEE Trans. Comput.*, pp. 90–93, Jan. 1974.
[2] K. Aizawa, H. Harashima, and T. Saito, "Model-based analysis-synthesis image coding (MBASIC) system for a person's face," *Signal Processing: Image Communication*. New York: Elsevier Science, Oct. 1989, pp. 139–152.
[3] B. S. Atal and M. R. Schroeder, "Predictive coding of speech signals and subjective error criteria," *IEEE Trans. Acoust., Speech and Signal Proces.*, pp. 247–254, June 1979.
[4] B. S. Atal, "High-quality speech at low bit rates: Multi-pulse and stochastically excited linear predictive coders," in *Proc. ICASSP*, 1986, pp. 1681–1684.
[5] T. Berger, *Rate Distortion Theory*. Englewood Cliffs, NJ: Prentice Hall, 1971.
[6] K. H. Brandenburg, "OCF—A new coding algorithm for high quality sound signals," in *Proc. ICASSP*, Apr. 1987, pp. 141–144.
[7] P. J. Burt and E. H. Adelson, "The Laplacian pyramid as a compact image code," *IEEE Trans. Commun.*, vol. COM-31, pp. 532–540, Apr. 1983.
[8] CCIR, "Method for the subjective assessment of the quality of television pictures," Rec. 500-1, 1978.
[9] CCITT Study Group XVIII, "32 kb/s adaptive differential pulse code modulation (ADPCM)," Working Party 8, Draft Revision of

Recommendation G.721, Temporary Document No. D.723/XVIII, Source: USA; Geneva, Switzerland.

[10] CCITT Study Group XVIII, "7 kHz audio coding within 64 kb/s," CCITT Draft Recommendation G.722, Report of Working Party XVIII/8, July 1986.

[11] W.-Y. Chan and A. Gersho, "Constrained-storage quantization of multiple vector sources by codebook sharing," IEEE Trans. Commun., vol. 39, pp. 11-13, Jan. 1991.

[12] J.-H. Chen and A. Gersho, "Real-time vector APC speech coding at 4800 b/s with adaptive postfiltering," in Proc. ICASSP, Apr. 1987, pp. 2185-2188.

[13] J.-H. Chen, R. V. Cox, Y-C. Lin, N. S. Jayant, and M. J. Melchner, "A low-delay CELP coder for the CCITT 16 kb/s speech coding standard," this issue, pp. 830-849.

[14] P. A. Chou, T. Lookabaugh, and R. M. Gray, "Entropy-constrained vector quantization," IEEE Trans. Signal Proces., vol. 37, pp. 31-42, 1989.

[15] D. C. Cox, "Portable digital radio communications—An approach to tetherless access," IEEE Commun. Mag., pp. 30-40, July 1989.

[16] R. V. Cox, "The design of uniformly and nonuniformly spaced pseudoquadrature mirror filters," IEEE Trans. Signal Proces., vol. 34, pp. 1090-1096, 1986.

[17] R. V. Cox, J. Hagenauer, N. Seshadri, and C-E. W. Sundberg, "Subband speech coding and matched convolutional channel coding for mobile radio channels," IEEE Trans. Signal Proces., vol. 39, no. 8, pp. 1717-1731, Aug. 1991.

[18] R. E. Crochiere, S. A. Webber, and J. L. Flanagan, "Digital coding of speech in subbands," Bell Syst. Tech. J., pp. 1069-1085, 1976.

[19] C. C. Cutler, "Differential quantization for communication signals," U.S. Patent 2 605 361, July 29, 1952.

[20] —, "Delayed encoding: Stabilizer for adaptive coders," IEEE Trans. Commun., vol. 19, pp. 898-904, Dec. 1971.

[21] I. Daubechies, "Orthonormal bases on compactly supported wavelets," Comm. Pure Appl. Math., pp. 909-996, 1988.

[22] W. R. Daumer, "Subjective evaluation of several efficient speech coders," IEEE Trans. Commun., pp. 655-662, Apr. 1982.

[23] L. D. Davisson, "Rate distortion theory and application," Proc. IEEE, pp. 800-808, July 1972.

[24] J. R. B. deMarca and N. S. Jayant, "An algorithm for assigning binary indices to the codevectors of a multidimensional quantizer," in Proc. ICC, June 1987, pp. 1128-1132.

[25] C. Diab, R. Prost, and R. Goutte, "Block-adaptive subband coding of images," in Proc. ICASSP, Apr. 1990, pp. 2093-2096.

[26] D. Esteban and C. Galand, "Application of quadrature mirror filters to spit band voice coding schemes," in Proc. ICASSP, 1987, pp. 191-195.

[27] N. Farvardin, "A study of vector quantization for noisy channels," IEEE Trans. Inform. Theory, pp. 799-809, July 1990.

[28] T. R. Fischer, "A pyramid vector quantizer," IEEE Trans. Inform. Theory, pp. 568-583, July 1986.

[29] R. Forscheimer and T. Kronander, "Image coding—from waveforms to animation," IEEE Trans. Signal Proces., pp. 2008-2023, Dec. 1989.

[30] J. L. Flanagan et al., "Speech coding," IEEE Trans. Commun., vol. 27, pp. 710-737, Apr. 1979.

[31] J. L. Flanagan, D. A. Berkley, G. Elko, J. E. West, and M. M. Sondhi, "Autodirective microphone systems," Acustica, vol. 73, pp. 58-71, 1991.

[32] J. L. Flanagan, Speech Analysis, Synthesis and Perception. New York: Springer-Verlag, 1972.

[33] G. D. Forney, Jr., "The Viterbi algorithm," Proc. IEEE, pp. 268-278, Mar. 1973.

[34] A. Fuldseth, E. Harborg, F. T. Johansen, and J. E. Knudsen, "A real time implementable 7 kHz speech coder at 16 kbps," presented at Proc. Eurospeech 91, Genoa, 1991.

[35] S. Furui and M. M. Sondhi, Eds., Advances in Speech Signal Processing. New York: Marcell-Dekker, 1992.

[36] R. G. Gallager, Information Theory and Reliable Communication. New York: McGraw Hill, 1965.

[37] A. Gersho and V. Cuperman, "Vector quantization: A pattern matching technique for speech coding," IEEE Commun. Mag., pp. 15-21, 1983.

[38] A. Gersho, "Asymptotically optimum block quantization," IEEE Trans. Inform. Theory, pp. 373-380, July 1979.

[39] A. Gersho, "On the structure of vector quantizers," IEEE Trans. Inform. Theory, pp. 157-166, Mar. 1982.

[40] A. Gersho and R. M. Gray, Vector Quantization and Signal Compression. Kluwer Academic, 1992.

[41] I. A. Gerson and M. A. Jasiuk, "Vector sum excited linear prediction (VSELP)," presented at IEEE Workshop on Speech Coding for Telecommun., Sept. 5-8, 1989.

[42] J. D. Gibson, "Sequentially adaptive backward prediction in ADPCM speech coders," IEEE Trans. Commun., pp. 145-150, Jan. 1978.

[43] T. J. Goblick Jr. and J. L. Holsinger, "Analog source digitization: A comparison of theory and practice," IEEE Trans. Inform. Theory, pp. 323-326, 1967.

[44] R. C. Gonzalez and P. Wintz, Digital Image Processing. Reading, MA: Addison-Wesley, 1977.

[45] D. J. Goodman, G. B. Lockhart, O. J. Wasem, and W-C. Wong, "Waveform substitution techniques for recovering missing speech segments in packet voice communication," IEEE Trans. Signal Proces., pp. 1440-1448, Dec. 1986.

[46] D. J. Goodman, "Speech quality of the same speech transmission conditions in seven different countries," IEEE Trans. Commun., pp. 642-654, Apr. 1982.

[47] D. J. Goodman, "Embedded DPCM for variable bit rate transmission," IEEE Trans. Commun., pp. 1040-1066, July 1980.

[48] R. M. Gray, J. Foster, and M. O. Dunham, "Finite-state vector quantization for waveform coding," IEEE Trans. Inform. Theory, pp. 348-359, 1985.

[49] R. M. Gray, "Vector quantization," IEEE ASSP Mag., pp. 4-29, 1984.

[50] D. W. Griffin and J. S. Lim, "Multiband excitation vocoder," IEEE Trans. Signal Proces., pp. 1223-1235, 1988.

[51] H.-M. Hang and J. W. Woods, "Predictive vector quantization of images," IEEE Trans. Commun., vol. 37, pp. 1208-1219, 1989.

[52] J. C. Hardwick and J. S. Lim, "The application of the IMBE speech coder to mobile communication," in Proc. ICASSP, May 1991, pp. 249-252.

[53] J. J. Y. Huang and P. M. Schultheiss, "Block quantization of correlated Gaussian random variables," Trans. IRE, vol. CS-11, pp. 289-296, 1963.

[54] D. Huffman, "A method for the construction of minimum redundancy codes," in Proc. IRE, Sept. 1952, pp. 1098-1101.

[55] ISO, "Coded representation of picture and audio information—Progressive bi-level image compression standard," ISO/IEC Draft, Dec. 1990.

[56] International Telephone and Telegraph Consultative Committee, "Facsimile coding schemes and coding control functions for group 4 facsimile apparatus," Red Book, Fascicle VII.3 Rec. T.6, 1980.

[57] A. E. Jacquin, "A novel fractal block-coding technique for digital images," in Proc. ICASSP, pp. 2225-2228, 1990.

[58] J. Jang and S. A. Rajala, "Segmentation-based image coding using fractals and the human visual system," in Proc. ICASSP '90, Apr. 1990, pp. 1957-1960.

[59] N. S. Jayant, "Adaptive quantization with a one-word memory," Bell Syst. Tech. J., pp. 1119-1144, Sept. 1973.

[60] N. S. Jayant, Waveform Quantization and Coding. New York: IEEE Press, 1976.

[61] N. S. Jayant and S. W. Christensen, "Effects of packet losses on waveform coded speech and improvements due to an odd-even interpolation procedure," IEEE Trans. Commun., vol. 29, pp. 101-109, Feb. 1981.

[62] N. S. Jayant and P. Noll, Digital Coding of Waveforms: Principles and Applications to Speech and Video. Englewood Cliffs, NJ: Prentice Hall, 1984.

[63] N. S. Jayant, J. D. Johnston, and Y. Shoham, "Coding of wideband speech," presented Proc. 2nd Europ. Conf. Speech Commun. Techol., Sept. 1991.

[64] N. S. Jayant, "High-quality coding of telephone speech and wideband audio," IEEE Commun. Mag., pp. 10-19, Jan. 1990.

[65] J. D. Johnston, "A filter family designed for use in quadrature mirror filter banks," presented at Proc. ICASSP, 1980.

[66] —, "Transform coding of audio signals using perceptual noise criteria," IEEE J. Select. Areas Commun., pp. 314-323, Feb. 1988.

[67] B. H. Juang and A. H. Gray, "Multiple stage vector quantization for speech coding," in Proc. ICASSP, Apr. 1982, pp. 597-600.

[68] M. Kaneko, A. Koike, and Y. Hatori, "Coding with knowledge-

based analysis of motion pictures," in *Proc. Picture Coding Symp.*, June 1987, vol. 12-3, pp. 167-168.

[69] G. Karlsson and M. Vetterli, "Three dimensional subband coding of video," presented at Proc. ICASSP, 1988.

[70] ——, "Packet video and its integration into the network architecture," *IEEE J. Select. Areas Commun.*, pp. 739-751, June 1989.

[71] D. P. Kemp, R. A. Sueda, and T. E. Tremain, "An evaluation of 4800 b/s voice coders," presented at Proc. ICASSP, May 1989.

[72] T. Kim, "New finite state vector quantizers for images," in Proc. ICASSP, 1988, pp. 1180-1183.

[73] N. Kitawaki, M. Honda, and K. Itoh, "Speech-quality assessment methods for speech-coding systems," *IEEE Commun. Mag.*, pp. 26-32, Oct. 1984.

[74] N. Kitawaki and H. Nagabuchi, "Quality assessment of speech coding and speech synthesis systems," *IEEE Commun. Mag.*, pp. 36-44, Oct. 1988.

[75] N. Kitawaki, H. Nagabuchi, M. Taka, and K. Takahashi, "Speech coding technology for ATM networks," *IEEE Commun. Mag.*, pp. 21-27, Jan. 1990.

[76] W. B. Kleijn and W. Granzow, "Methods for waveform interpolation in speech coding," *Digit. Signal Proces.*, 1991.

[77] P. Kroon and B. S. Atal, "On improving the performance of pitch predictors in speech coding systems," in *Proc. ICASSP*, 1990, pp. 661-664.

[78] P. Kroon, E. F. Deprettere, and R. J. Sluyter, "Regular-pulse excitation—A novel approach to effective and efficient multipulse coding of speech," *IEEE Trans. Signal Proces.*, vol. ASSP-34, no. 5, pp. 1054-1063, Oct. 1986.

[79] P. Kroon and K. Swaminathan, "A high quality multi-rate real-time CELP coder," this issue, pp. 850-857.

[80] M. Kunt, A. Ikonmopoulos, and M. Kocher, "Second-generation image coding technique," *Proc. IEEE*, pp. 549-574, Apr. 1985.

[81] D. LeGall, "MPEG: A video compression standard for multimedia applications," *Commun. ACM*, pp. 47-58, Apr. 1991.

[82] D. H. Lee and D. L. Neuhoff, "Conditionally corrected two-stage vector quantization," in *Proc. 1990 Conf. Inform. Sci. Syst.*, Princeton, NJ, Mar. 1990, pp. 802-806.

[83] ——, "An asymptotic analysis of two-stage vector quantization," presented at Proc. ISIT, 1991.

[84] Y. Linde, A. Buzo, and R. M. Gray, "An algorithm for vector quantization design," *IEEE Trans. Commun.*, vol. 28, pp. 84-95, Jan. 1980.

[85] M. Liou, "Overview of the $p * 64$ kbits/s video coding standard," *Commun. ACM*, pp. 60-63, Apr. 1991.

[86] S. S. Magan, "Trends in DSP system design," in short course on *Digital Signal Processing*, IEEE-Int. Electron. Device Meet., Dec. 1989.

[87] J. Makhoul, S. Roucos, and H. Gish, "Vector quantization in speech coding," *Proc. IEEE*, pp. 1551-1588, Nov. 1985.

[88] S. Mallat, "A theory for multiresolution signal decomposition: The wavelet representation," *IEEE Trans. Patt. Anal. Mach. Intel.*, July 1989.

[89] J. Max, "Quantizing for minimum distortion," *IRE Trans. Inform. Theory*, pp. 7-12, Mar. 1960.

[90] R. J. McAulay and T. F. Quatieri, "Speech analysis and synthesis based on a sinusoidal model," *IEEE Trans. Signal Proces.*, pp. 744-754, Aug. 1986.

[91] P. Mermelstein, "G.722: A new CCITT coding standard for digital transmission of wideband audio signals," *IEEE Commun. Mag.*, pp. 8-15, Jan. 1988.

[92] N. Moayeri, D. L. Neuhoff, and W. E. Stark, "Fine-coarse vector quantization," *IEEE Trans. Inform. Theory*, pp. 1072-1084, July 1991.

[93] J. W. Modestino and D. G. Dant, "Combined source-channel coding of images," *IEEE Trans. Commun.*, pp. 1644-1659, Nov. 1979.

[94] S. Morishima, K. Aizawa, and H. Harashima, "A real-time facial action image synthesis system driven by speech and text," in *Proc. SPIE Vis. Commun. Image Proces.*, 1990, pp. 1151-1158.

[95] G. Morrison and D. Beaument, "Two-level video coding for ATM networks," *Sign. Proces.: Image Commun.*, pp. 179-195, June 1991.

[96] T. Murakami, K. Asai, and A. Itoh, "Vector quantization of color images," in *Proc. IEEE-ICASSP*, 1986, pp. 133-135.

[97] H. G. Musmann, "The ISO audio coding standard," presented at Proc. IEEE GLOBECOM, Dec. 1990.

[98] H. G. Musmann, P. Pirsch, and H.-J. Grallert, "Advances in picture coding," *Proc. IEEE*, pp. 523-548, Apr. 1985.

[99] N. M. Nasrabadi and R. A. King, "Image coding using vector quantization: A review," *IEEE Trans. Commun.*, pp. 957-971, Aug. 1988.

[100] A. N. Netravali and B. G. Haskell, *Digital Pictures*. New York: Plenum Press, 1988.

[101] A. N. Netravali and J. A. Stuller, "Motion compensation transform coding," *Bell Syst. Tech. J.*, pp. 1703-1718, Sept. 1974.

[102] A. N. Netravali and J. O. Limb, "Picture coding: A review," *Proc. IEEE*, pp. 366-406, Mar. 1980.

[103] A. N. Netravali, E. Petajan, S. Knauer, K. Mathews, R. J. Safranek, and P. Westerink, "A high quality digital HDTV codec," *IEEE Trans. Consum. Electron.*, pp. 320-330, Aug. 1991.

[104] D. L. Neuhoff and T. N. Pappas, "Perceptual coding of images for halftone display," presented at Proc. ICASSP, May 1991.

[105] D. L. Neuhoff and D. H. Lee, "On the performance of tree-structured vector quantization," presented at Proc. ICASSP, 1991.

[106] Y. Ninomiya, "HDTV broadcasting systems," *IEEE Commun. Mag.*, pp. 15-23, Aug. 1991.

[107] J.-R. Ohm and P. Noll, "Predictive tree encoding of still images with vector quantization," presented at Proc. ISSSE89, Nurmberg, Sept. 1989.

[108] J. B. O'Neal, "Predictive quantizing systems for the transmission of television signals," *Bell Syst. Tech. J.*, pp. 689-719, May/June 1966.

[109] K. S. Paliwal and B. S. Atal, "Efficient vector quantization of LPC parameters at 24 bits per frame," presented at Proc. ICASSP, 1991.

[110] T. N. Pappas and D. L. Neuhoff, "Model-based halftoning," presented at Proc. SPIE/IS&T Symp. Electron. Imag. Sci. Technol., Feb./Mar. 1991.

[111] J. Princen, A. Johnson, and A. Bradley, "Sub-band transform coding using filterbank designs based on time-domain aliasing cancellation," in *Proc. ICASSP*, 1987, pp. 2161-2164.

[112] C. I. Podilchuk, N. S. Jayant, and P. Noll, "Sparse codebooks for the quantization of non-dominant sub-bands in image coding," presented at Proc. ICASSP, 1990.

[113] W. Pratt, *Digital Image Processing*. New York: Wiley, 1978.

[114] S. R. Quackenbush, "Hardware implementation of a color image decoder for remote database access," presented at Proc. ICASSP, 1990.

[115] ——, "A 7 kHz bandwidth, 32 kbps speech coder for ISDN," presented at Proc. ICASSP, 1991.

[116] L. R. Rabiner and R. W. Schafer, *Digital Speech Processing*. Englewood Cliffs, NJ: Prentice Hall, 1978.

[117] V. Ramamoorthy, N. S. Jayant, R. V. Cox, and M. M. Sondhi, "Enhancement of ADPCM speech coding with backward-adaptive algorithms for postfiltering and noise feedback," *IEEE J. Select. Areas Commun.*, pp. 364-382, Feb. 1988.

[118] B. Ramamurthi and A. Gersho, "Classified vector quantization of images," *IEEE Trans. Commun.*, pp. 1105-1115, Nov. 1986.

[119] A. R. Reibman, "DCT-based embedded coding for packet video," *Signal Proces.: Image Commun.*, pp. 333-343, Sept. 1991.

[120] P. Rioul and M. Vetterli, "Wavelets and signal processing," *IEEE Signal Proces. Mag.*, pp. 14-38, Oct. 1991.

[121] L. G. Roberts, "Picture coding using pseudo-random noise," *IRE Trans. Inform. Theory*, pp. 145-154, Feb. 1962.

[122] G. Roy and P. Kabal, "Wideband CELP speech coding at 16 kbps," in *Proc. ICASSP*, 1991, pp. 17-20.

[123] M. J. Sabin and R. M. Gray, "Product vector quantizers for waveform and voice coding," *IEEE Trans. Signal Proces.*, pp. 474-488, June 1984.

[124] R. J. Safranek and J. D. Johnston, "A perceptually tuned subband image coder with image-dependent quantization and post-quantization," in Proc. ICASSP, 1989.

[125] T. Saito, H. Takeo, K. Aizawa, H. Harashima, and H. Miyakawa, "Adaptive image coding using gain-shape vector quantization," in *Proc. IEEE-ICASSP*, Apr. 1986, pp. 129-132.

[126] D. J. Sakrison, "Image coding applications of vision models," in *Image Transmission Techniques*, W. K. Pratt, Ed. New York: Academic, May 1979, pp. 21-51.

[127] W. F. Schreiber, "Psychophysics and the improvement of television picture quality," *SMPTE J.*, pp. 717-725, Aug. 1984.

[128] J. Schroeter and M. M. Sondhi, "Speech coding based on physiological models of speech production," in *Advances in Speech Signal*

Processing, S. Furui and M. M. Sondhi. Ed.   New York: Marcel Dekker, 1991.

[129] C. E. Shannon, "A mathematical theory of communications," *Bell Syst. Tech. J.*, vol. 27, 1948, pp. 379–423, and 623–656.

[130] ——, "Coding theorems for a discrete source with a fidelity criterion," *IRE Nat. Conv. Rec.*, part 4, pp. 142–163, 1959.

[131] Y. Shoham, "Constrained excitation coding of speech at 4.8 kbps," in *Advances in Speech Coding*.   New York: Kluwer Academic, 1991, pp. 339–348.

[132] B. Smith, "Instantaneous companding of quantized signals," *Bell Syst. Tech. J.*, pp. 653–709, May 1957.

[133] R. Steele, *Delta Modulation Systems*.   New York: Halsted, 1975.

[134] ——, "The cellular environment of lightweight handheld portables," *IEEE Commun. Mag.*, pp. 20–29, July 1989.

[135] T. G. Stockham, "Image processing in the context of a visual model," *Proc. IEEE*, pp. 828–842, July 1972.

[136] Swedish Radio, unpublished report on the quality of low rate audio algorithms submitted for the ISO-MPEG standard, Aug. 1990.

[137] G. Theile, G. Stoll, and M. Link, "Low bit-rate coding of high-quality audio signals," *EBU Tech. Rev.*, no. 230, pp. 71–94, Aug. 1988.

[138] H. Tominaga, H. Jozawa, M. Kawashima, and T. Hanamura, "A video coding method considering cell losses in ATM networks," *Signal Process.: Image Commun.*, pp. 291–300, Sept. 1991.

[139] T. E. Tremain, "The government standard linear predictive coding algorithm: LPC-10," *Speech Techno.*, vol. 1, no. 2, pp. 40–49, Apr. 1982.

[140] P. P. Vaidyanathan, "Quadrature mirror filter banks, M-band extensions and perfect-reconstruction techniques," *IEEE ASSP Mag.*, pp. 4–20, 1987.

[141] J. Vaisey and A. Gersho, "Variable block-size image coding," in *Proc. ICASSP*, Apr. 1987, pp. 1051–1054.

[142] R. Valenzuela and C. N. Animalu, "A new voice-packet reconstruction technique," in *Proc. ICASSP*, 1989, pp. 1334–1337.

[143] W. Verbiest and L. Pinnoo, "A variable bit rate video coder for asynchronous transfer mode networks," *IEEE J. Select. Areas Commun.*, pp. 761–770, June 1989.

[144] M. Vetterli, "Multidimensional sub-band coding: Some theory and algorithms," *Signal Proces.*, pp. 97–112, Apr. 1984.

[145] W. D. Voiers, "Diagnostic acceptability measure for speech communication systems," in *Proc. ICASSP*, May 1977, pp. 204–207.

[146] ——, "Diagnostic evaluation of speech intelligibility," in *Speech Intelligibility and Speaker Recognition*, M. Hawley, Ed.   Stroudsburg, PA: Dowden Hutchinson Ross, 1977.

[147] G. K. Wallace, "The JPEG still picture compression standard," *Commun. ACM*, pp. 31–43, Apr. 1991.

[148] S. Wang and A. Gersho, "Phonetically-based vector excitation coding of speech at 3.6 kbps," in *Proc. ICASSP*, May 1989, pp. 49–52.

[149] L. Wang and M. Goldberg, "Progressive image transmission using vector quantization on images in pyramid form," *IEEE Trans. Commun.*, pp. 1339–1349, Dec. 1989.

[150] P. H. Westerink, D. E. Boekee, J. Biemond, and J. H. Woods, "Subband coding of images using vector quantization," *IEEE Trans. Commun.*, pp. 713–719, June 1988.

[151] R. Wilson, H. E. Knutsson, and G. H. Granlund, "Anisotropic nonstationary image estimation and its applications: Part II—Predictive image coding," *IEEE Trans. Commun.*, pp. 398–406, Mar. 1983.

[152] P. A. Wintz, "Transform picture coding," *Proc. IEEE*, pp. 809–820, July 1972.

[153] I. H. Witten, R. M. Neal, and J. G. Cleary, "Arithmetic coding for data compression," *Commun. ACM*, pp. 520–540, June 1987.

[154] J. W. Woods and S. D. O'Neil, "Subband coding of images," *IEEE Trans. Acoust., Speech, Signal Proces.*, pp. 1278–1288, Oct. 1986.

[155] S.-W. Wu and A. Gersho, "Optimal block-adaptive image coding with constrained bit rate," presented at Proc. 24th Asilomar Conf. Signals, Syst. Comput., Nov. 1990.

[156] A. D. Wyner, "Fundamental limits in information theory," *Proc. IEEE*, pp. 239–251, Feb. 1981.

[157] R. Zelinski and P. Noll, "Adaptive transform coding of speech signals," *IEEE Trans. Acoust., Speech, Signal Proces.*, pp. 299–309, Aug. 1977.

Nikil Jayant (M'69–SM'77–F'82) received the Ph.D. degree in electrical communication engineering from the Indian Institute of Science, Bangalore, India.

He is Head of the Signal Processing Research Department at AT&T Bell Laboratories in Murray Hill, NJ. He is responsible for research in speech and image processing with applications to coding, communications, and recognition. He joined AT&T Bell Laboratories in 1968. He is the Editor of *Waveform Quantization and Coding* (New York: IEEE Press, 1976) and coauthor of *Digital Coding of Waveforms—Principles and Applications to Speech and Video* (Englewood Cliffs, NJ: Prentice-Hall, 1984).

Dr. Jayant was the first Editor-in-Chief of the IEEE ACOUSTICS, SPEECH, AND SIGNAL PROCESSING MAGAZINE.

Chapter 7

MULTIPLE ACCESS TECHNIQUES

IEEE TRANSACTIONS ON VEHICULAR TECHNOLOGY, VOL. 38, NO. 2, MAY 1989

69

Spectrum Efficiency in Cellular

WILLIAM C. Y. LEE, FELLOW, IEEE

Reprinted from *IEEE Transactions on Vehicular Technology*, Vol. 38, No. 2, pp. 69-75, May, 1989.

Abstract—The spectrum efficiency in cellular can be measured by a parameter called here radio capacity. Using radio capacity, it can be shown that splitting analog channels does not increase the spectrum efficiency in cellular radio systems, but using digital cellular channels does. A simple evaluation method of judging spectrum efficiency in different digital systems is introduced. Also, several new concepts of spectrum efficiency in cellular are described.

I. INTRODUCTION

SINCE THE FREQUENCY spectrum is a limited resource, we should utilize it very effectively. In order to approach this goal, we have to clearly define spectrum efficiency from either a total system point of view or a simple fixed point-to-point link perspective. For most radio systems, spectrum efficiency is the same as channel efficiency, the maximum number of channels that can be provided in a given frequency band. This is true for a point-to-point system that does not reuse frequency channels, but cellular mobile radio does. Therefore, in cellular mobile radio systems:

$$\text{spectrum efficiency} \neq \text{channel efficiency.}$$

The system capacity is directly related to spectrum efficiency but not to channel efficiency [1].

Evaluation of any communication system is based on its voice quality. The specified voice quality can be determined by a subjective test, usually based on a subjective mean opinion score (MOS) [2], [3]. For example, a voice quality is deemed acceptable based on 75 percent of total listeners' judgment that the voice quality is good or excellent (the top two scores among five merit scores). A specified voice quality then can be tested on any communication systems with their given channel bandwidth B_c to determine a required carrier-to-interference ratio $(C/I)_s$. Since the different modulation schemes will affect $(C/I)_s$ and the transmitted power can be set based on the $(C/I)_s$ at the receiving end. Therefore, $(C/I)_s$ and B_c are the only parameters to be used to evaluate spectrum efficiency of each system as described in this paper. A simple but very effective formula for evaluating the spectrum efficiency of each system called radio capacity has been derived in this paper.

II. RADIO CAPACITY IN CELLULAR SYSTEMS

A parameter named radio capacity by Lee [4] is derived in this section and used to measure the spectrum efficiency.

In cellular systems, although the frequency reuse scheme

Manuscript received December 28, 1987; revised December 12, 1988. A part of this work was presented at the 1988 IEEE Vehicular Technology Society Conference, Philadelphia, PA, June 15-17, 1988.

The author is with PacTel Cellular, Inc., 2355 Main Street, Ervine, CA 92714.

IEEE Log Number 8929439.

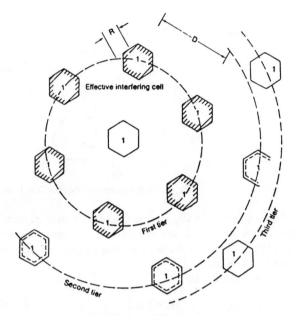

Fig. 1. Six effective interfering cells of cell 1.

proves increasing the system capacity, it also creates the co-channel interference from co-channel cells. The co-channel cells are the cells within them when the same set of frequency channels is used. We may show the co-channel cells in Fig. 1. The number of co-channel cells in the worst case is six at the first tier surrounding the desired cell, then another six at the second tier, and another six at the third tier. We concentrate only on the first tier of six co-channel since the co-channel cells at the first tier dominate the interference [5]. Let the distance between the two adjacent co-channel cells be called D. The cell radius is R. A required D/R ratio for avoiding co-channel interference is called co-channel interference reduction factor q_s

$$q_s = (D/R)_s.$$

q_s can be found from the required $(C/I)_s$ as shown in the following derivation. Since the interference in a worst case comes from six dominate interferers, then the C/I can be expressed as

$$C/I = \frac{C}{\sum_{k=1}^{6} I_k + n} \qquad (1)$$

where C is the received carrier power in a desired cell, n is the local noise, and I_k is the interference from one of six co-channel cells at the first tier. There are two conditions to be

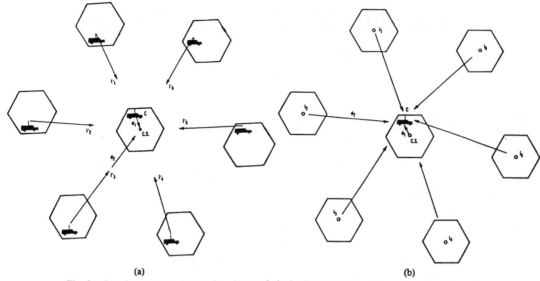

Fig. 2. Interference scenario. (a) Case 1: at a desired cell site. (b) Case 2: at a desired mobile unit.

disclosed as shown in Fig. 2. One is all the co-channel mobile units interference with the desired cell site, and the other condition is all the co-channel cell sites interference with the desired mobile unit.

For our analysis, let

$$C = \mathcal{L} R^{-\gamma} \qquad (2)$$

and

$$I_k = \mathcal{L} D_k^{-\gamma} \qquad (3)$$

where D_k is the distance from a transmitter (at cell site or mobile unit) of a kth co-channel cell to a receiver (at cell site or mobile unit) located in the center cell as shown in Fig. 1. γ is the path loss slope in a mobile radio environment, $\mathcal{L}$ is a constant. The local noise n can be neglected. Substituting (2) and (3) into (1) yields

$$C/I = \frac{R^{-\gamma}}{\sum_{k=1}^{6} D_k^{-\gamma}}. \qquad (4)$$

Since all D_k are not much different from D, which is the separation between two co-channel cell sites, we let $D_k = D$ for simplifying the derivation in (4) be

$$C/I = \frac{R^{-\gamma}}{6 \cdot D^{-\gamma}}. \qquad (5)$$

Assume that all C/I values have to be greater than $(C/I)_s$, then

$$\frac{1}{6} \left(\frac{R}{D} \right)^{-\gamma} \geq \left(\frac{C}{I} \right)_s \qquad (6)$$

or

$$\left(\frac{R}{D} \right)_s^{-\gamma} = 6 \left(\frac{C}{I} \right)_s. \qquad (7)$$

The co-channel interference reduction factor q_s is obtained from

$$q_s = \left(\frac{D}{R} \right)_s = \left(6 \left(\frac{C}{I} \right)_s \right)^{1/\gamma}. \qquad (8)$$

Let K be a number of cells in a frequency reuse pattern. Then the relationship between q_s and K can be found from the hexagon-cell configuration as [3]

$$q_s \triangleq \sqrt{3K}. \qquad (9)$$

The radio capacity is defined by [4]

$$m = \frac{B_t}{B_c \cdot K} \quad \text{number of channels /cell} \qquad (10)$$

where B_t is the total allocated spectrum for the system. In the present cellular, $B_t = 12.5$ MHz for each carrier operator and B_c is the channel bandwidth. Substituting (8) and (9) into (10) results in

$$m = \frac{B_t}{B_c \frac{q_s^2}{3}} = \frac{B_t}{B_c \left(\frac{6}{3^{\gamma/2}} \cdot \left(\frac{C}{I} \right)_s \right)^{2/\gamma}}. \qquad (11)$$

In the mobile radio environment, we may assume a fourth power rule, i.e., $\gamma = 4$ [6], then (11) becomes

$$m = \frac{B_t}{B_c \sqrt{\frac{2}{3} \left(\frac{C}{I} \right)_s}} \quad \text{number of channels/cell.} \qquad (12)$$

Equation (12) is called radio capacity by Lee [4]. It is the most general equation. From this equation, we can convert to a different representation of radio capacity m from a dimension generally known as the number of channels per cell.

III. DIFFERENT UNITS REPRESENTING RADIO CAPACITY

The radio capacity m, in general, is quantified by the number of channels per cell. With different given conditions, the radio capacity also can be represented in different units.

Case 1   Assume that a given blocking probability (grade of service) is P_B in the system. The radio capacity m_1 will be obtained from Erlang B model by inserting the number of channels per cell m and the blocking probability P_B as follows:

$$m_1 = f(m, P_B) \quad [\text{Erlang/cell}].$$

The notation $f(x)$ means a function of x.

Case 2   Assume that a cell area A_1 is in square miles, the radio capacity m_2 will be represented by

$$m_2 = f(m_1, A_1) = f(m, P_B, A_1)$$

$$[\text{Erlang/m}^2 \text{ or Erlang/km}^2].$$

Case 3   Assume that an average holding time per each call is T, then the radio capacity can be represented by

$$m_3 = f(m_2, T) = f(m, P_B, A_1, T)$$

$$[\text{number of calls/h/mi}^2].$$

Case 4   Assume that the average calls per user in a busy hour is k, the radio capacity can be represented by

$$m_4 = f(m_3, k) = f(m, P_B, A_1, T, k)$$

$$[\text{number of users/mi}^2].$$

Case 5   Assume that the total area in a system is A_t in square miles. The radio capacity can be represented by

$$m_5 = f(m_4, A_t) = f(m, P_B, A_1, T, k, A_t)$$

$$[\text{number of users in a system}].$$

Example: Let $m = 45$ channels/cell, $P_B = 0.02$, $A_1 = 12.5$ mi^2 (i.e., cell radius $R = 2$mi), $T = 100$ s, $k = 0.8$ calls/user in a busy hour, and $A_t = 7000$ mi^2 wide covered by a system, then using Erlang B model, we obtain $m_1 = 35.6$ Erlang/cell in an omnidirectional-antenna cell. If a three-sectorized cell uses three 120° directional antennas, one antenna per sector, each sector operates 15 channels, then the Erlang/sector is 9 and $m_1 = 3 \times 9 = 27$ Erlang/cell.

kind of cells	m	m_1	m_2	m_3	m_4	m_5
omnicell	45	35.6	2.86	103	128.75	901250
directional cell	45	27	2.16	77.8	97.25	680750

The other representation of radio capacity such as m_6 (number of calls/h/cell), m_7 (number of users/cell in a busy hour) and m_8 (number of uses/channel in a busy hour) can also be derived from the general radio capacity m

$$m_6 = f(m_1, T) = f(m, P_B, T)$$

$$m_7 = f(m_6, k) = f(m, P_B, T, k)$$

$$m_8 = f(m_6, m).$$

Using the same values shown in the example above, the results of m_6, m_7, and m_8 are

kind of cells	m	m_6	m_7	m_8
omnicell	45	1281.6	1602	35.6
directional cell	45	972	1215	27

The above example illustrates all the different presentation units that can be generated from the number of channels per cell.

IV. RADIO CAPACITIES OF FM CHANNELS VERSUS NARROWBANDING CHANNELS [1], [7]

A. Radio Capacity of 30 kHz FM Channels

Assume that a 30 kHz FM receiver has a two-branch diversity with a preemphasis/deemphasis device. The required C/I of 18 dB of this receiver can be obtained through a subjective test with a criterion that 75 percent of the listeners evaluate the voice quality as good and excellent while driving in a mobile radio fading environment with various vehicle speeds. From a required C/I, $(C/I)_s$, of 18 dB, we can determine the required signal-to-noise (S/N) ratio at the baseband by the following calculation.

In the fading environment, a 30 kHz FM receiver without a diversity scheme and a preemphasis-deemphasis device, the C/I = 18 dB is converted to the $S/N = 15$ dB [8] shown in Fig. 3. The improvement of FM with preemphasis/deemphasis is [9], [10]

$$P_{\text{FM}} = \frac{(W_1/W_2)^2}{3} = \frac{(3000 \text{ Hz}/300 \text{ Hz})^2}{3} = 33.3 (=)15 \text{ dB}. \quad (13)$$

W_1 and W_2 are the two ends of a voice band. The notation ($=$) means a conversion between a decibel and a linear ration. The advantage of two-branch diversity receiver is that the baseband signal-to-noise ratio of a two-branch FM shows an 8 dB improvement over the S/N ratio of a single FM channel [3, p. 311] based on a 90% signal level. Then the baseband signal-to-noise ratio of a two-branch FM receiver with a preemphasis/deemphasis device is

$$(S/N)_{2Br \text{ FM}} = -3 + P_{\text{FM}} + \text{diversity gain} + (C/I)_{\text{FM}}$$

$$= 38.2 \text{ dB}. \quad (14)$$

This is an expected level for a reasonable voice quality [11, p. 371]. For introducing a parameter $\eta = (C/N) \cdot (B_c/W_1)$ then the baseband $(S/N)_{\text{FM}}$ in a nonfading case can be expressed as [11]

$$(S/N)_{\text{FM}} = \frac{3}{8} \left(\frac{B_c}{W_1} \right)^2 \cdot \eta = \frac{3}{8} \left(\frac{B_c}{W_1} \right)^3 \cdot \left(\frac{C}{N} \right)_{\text{FM}} \quad (15)$$

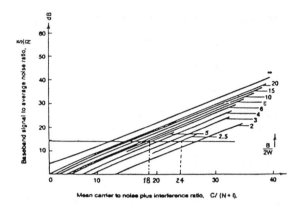

Fig. 3. C/I versus S/N of FM channels fading environment.

where $W_1 = 3000$ Hz. Equation (15) is plotted in Fig. 4. For a 30 kHz FM channel, $B_c/W_1 = 10$ and $S/N = 38$ dB corresponding to $C/N = 12.5$ dB in a nonfading case.

In a fading environment, as we have mentioned, a requirement of $(C/N)_s$ or $(C/I)_s$ has to be 18 dB in a two-branch diversity receiver equipped with a preemphasis/deemphasis device to obtain the same voice quality as one requiring only a $(C/N)_s = 12.5$ dB in a nondiversity receiver with no preemphasis/deemphasis but operating in a nonfading environment.

To find a radio capacity of this 30-kHz FM receiver, only $(C/I)_s$ and B_c are needed. But keep in mind that $(C/I)_s = 18$ dB at the RF band is equivalent to $S/N = 38$ dB at the baseband. Substituting $(C/I)_s = 63$ (which is 18 dB), $B_c = 30$ kHz, and $B_t = 12.5$ MHz into (12), then

$$m = \frac{B_t}{B_c\sqrt{\frac{2}{3}\left(\frac{C}{I}\right)_s}} = 64.3 \text{ channels/cell.} \quad (16)$$

B. Radio Capacity of 3-kHz SSB Channels

When a 30-kHz FM channel is replaced by six 3-kHz single sideband (SSB) channels, the channel efficiency counted by the number of channels increases six times. However, the spectrum efficiency of these two systems remains the same as shown below.

The SSB receiver is a linear modulation receiver [12]. It means that the required $(C/I)_s$ value at the RF will be the same value of S/N at the baseband. For calculating the $(C/I)_s$ of a SSB receiver for the same voice quality as the $(C/I)_s = 18$ dB of a 30-kHz FM receiver, we simply use

$$(S/N)_{\text{FM}} = (S/N)_{(\text{SSB})} = 38 \text{ dB.} \quad (17)$$

Since $B_c = 3$ kHz, $W_1 = 3$ kHz then the parameter η can be simplified as $\eta = (C/I)\cdot(B_c/W_1) = C/N$. In a nonfading case [11]

$$(S/N)_{\text{SSB}} = (C/N)_{\text{SSB}} = 38 \text{ dB.} \quad (18)$$

Equation (18) is plotted in Fig. 4. $(C/I)_{\text{SSB}} = 38$ dB is needed for a qualified voice quality in a nonfading case [12]. Normally $(C/I)_{\text{SSB}}$ in a fading case should be higher than 38

dB just as the same $(C/I)_{\text{FM}}$ in a fading case is higher than that in a nonfading case. We may apply $(C/I)_{\text{SSB}} = 38$ dB in a fading case if the best fading-remove scheme is used, i.e., $(C/I)_{\text{SSB}} = 38$ dB can be used as if in a nonfading case. Substituting $B_c = 3$ kHz, $(C/I)_s = 38$ dB into (12). The radio capacity m becomes

$$m = \frac{B_t}{B_c\sqrt{\frac{2}{3}\left(\frac{C}{I}\right)_s}} = 64.25 \text{ channels/cell.} \quad (19)$$

Comparing the radio capacities of both 30-kHz FM channels shown in (16) and 3-kHz SSB channels shown in (19), we found that they are almost the same. Therefore, the spectrum efficiencies of the two systems are the same.

C. Radio Capacity of 15 kHz FM Channels

When a 30-kHz FM channel is split into two 15-kHz FM channels, the channel efficiency is increasing by two. However, the spectrum efficiencies of using two different channel bandwidths in an FM system are the same. The explanation is shown as follows.

The FM is an exponential modulation, C/I at the RF band is not linearly proportional to S/N at the baseband as shown in Fig. 3. The $(C/I)_s = 24$ dB of a 15 kHz channel is equivalent to $(C/I)_s = 18$ dB of a 30-kHz channel to obtain the same S/N as depicted in Fig. 3. Now assume that the 15-kHz FM channel uses the same diversity scheme and preemphasis/deemphasis device, then let $C/I = 24$ dB, $B_c = 15$ kHz and $B_t = 12.5$ MHz, the radio capacity m of this 15-kHz FM system can be found from (12) as

$$m = \frac{B_t}{B_c\sqrt{\frac{2}{3}\left(\frac{C}{I}\right)_s}} = 64.3 \text{ channels/cell.} \quad (20)$$

Comparing (20) with (16), we find that the number of channels per cell does not increase by splitting every 30-kHz channel in half.

From (16), (19), and (20), we may conclude that in analog cellular systems, splitting channels does not increase the spectrum efficiency. Moreover, when splitting channels, due the increase of $(C/I)_s$, the size of cells is reduced. With a given transmitted power at each cell site, more cells are needed in a given area [1]. Therefore, no advantages of splitting analog channels in an analog cellular system.

V. Radio Capacity in Digital Cellular Channels

In digital cellular systems because the voice waveforms can be treated digitally and protected by adding channel coding which can protect different levels of important bits for a low bit-rate transmission. Since the nature of digital signals are not susceptible to the interference, the required $(C/I)_s$ in digital cellular can be lower than that in analog systems for a same voice quality as we have found [13]–[15]. In digital cellular systems, the $(C/I)_s$ is about 10 to 12 dB for a 30 kHz bandwidth and about 16 to 18 dB for a 10-kHz bandwidth. Based on these figures, Table I can be generated.

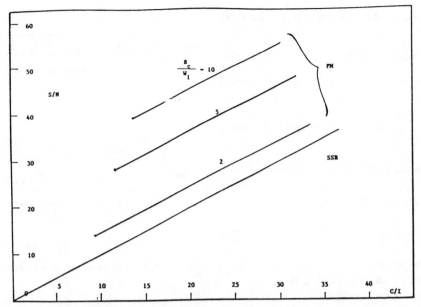

Fig. 4. *C/I* versus *S/N* in nonfading case.

TABLE I
DIGITAL CELLULAR CHANNELS

Channel Bandwidth (kHz)	(C/I)ₛ (dB)	K	D/R	m
30	10	2.58	2.78	161.5
	12	3.27	3.13	127.42
10	16	5.16	3.93	242.25
	18	6.48	4.41	192.9

The increase in spectrum efficiency by using digital cellular channels can be shown in the above table by comparing the radio capacities m with the value shown in (16). Also the concept of increasing spectrum efficiency can be expressed by either keeping a channel bandwidth of 30 kHz and lowering a value of required C/I; or keeping a required C/I of 18 dB and narrowing a channel bandwidth.

VI. EVALUATION OF DIFFERENT DIGITAL SYSTEMS

Since the different digital techniques have been applied on the speech codings, channel codings, and modulations with different channel bandwidths specified in individual systems, $(C/I)_s$ of each proposed digital system should be determined by a subjective test [15] through a standard voice-quality evaluation center. After the $(C/I)_s$ is found then a fair evaluation method can be used.

A. Conversion of (C/I)ₛ

Since each digital cellular system has different $(C/I)_s$ and B_c, we can convert $(C/I)_s$ into equivalent $(C/I)'_s$ if new B'_c is different from B_c by the following equation:

$$(C/I)' = (C/I) \left(\frac{B_c}{B'_c} \right)^2 \tag{21}$$

Equation (21) is derived from (12).

B. Evaluation Method

The evaluation method can be illustrated as follows:

Input: manufacture A $B_c = 18$ kHz $(C/I)_s = 17$ dB
        manufacture B $B_c = 25$ Hz $(C/I)_s = 13$ dB
        manufacture C $B_c = 15$ kHz $(C/I)_s = 20$ dB.

Normalization and Evaluation

Converting all B_c to 15 kHz, then three $(C/I)_s$ change to

manufacture A $B_c = 15$ $(C/I)_s = 1.58 + 17 + 18.58$ dB
manufacture B $B_c = 15$ $(C/I)_s = 4.4 + 13 = 17.3$ dB
manufacture C $B_c = 15$ $(C/I)_s = 20$ dB.

Since the smallest value of $(C/I)_s$ provides the largest radio capacity (see (12)), manufacture B gives the highest spectrum efficiency.

VII. CONCEPTS OF SPECTRUM EFFICIENCY IN CELLULAR

A. Relation Between (C/I) and Bc

We have seen from (21) that by maintaining the same voice quality the channel bandwidth reduces by half, and $(C/I)_s$ increases by four times.

B. Similarity Between Channel Capacity and Radio Capacity

Shannon's well-known channel capacity is expressed as

$$\hat{C} = B_c \log_2 (1 + C/I). \tag{22}$$

Interestingly enough, both the radio capacity m shown in (12) and the channel capacity C shown in (22) involve two parameters B_c and C/I.

The relationship between two different sets of B_c and C/I in channel capacity is (from (22))

$$\frac{\log_2 \left(\frac{C}{I} \right)}{\log_2 \left(\frac{C}{I} \right)'} = \frac{B'_c}{B_c}. \tag{23}$$

Equation (23) shows that B_c reduces by half, (C/I) increases by squaring itself, i.e.,

$$(C/I)' = (C/I)^2. \qquad (24)$$

Since the radio capacity shown in (21) and the channel capacity shown in (22) both involve C/I and B_c, the relationship of C/I and B_c shown in radio capacity (see (21)) and that in channel capacity (see (23)) are similar. Therefore the radio capacity is named.

C. Relationship Between E_b/N_o and B_c in Digital Cellular

The C/I in digital systems can be expressed as

$$C/I = \frac{E_b \cdot R_b}{I} = \frac{E_c \cdot R_c}{I} \qquad (25)$$

R_b rate of speech coding
E_b energy per bit of each speech coding bit
R_c rate of channel coding
E_c energy per bit of each channel coding bit.

Then

$$C/I \quad \Leftarrow \quad \frac{E_c \cdot R_c}{I} \quad \Leftarrow \quad \frac{E_b \cdot R_b}{I}$$

determined from   applied in   applied in
the subjective test   fading case   nonfading case.

$$(26)$$

Substituting (26) into (21) yields

$$C/I \, (C/I)' = \frac{E_c R_c / I}{E_c' R_c' / I'} = \left(\frac{B_c'}{B_c}\right)^2. \qquad (27)$$

Assume that the interference level I is the same in the environment, then (27) becomes

$$\frac{E_c R_c}{E_c' R_c'} = \left(\frac{B_c'}{B_c}\right)^2. \qquad (28)$$

Since the linear relationship between R_c and B_c always holds

$$R_c = kB_c, \quad R_c' = kB_c'.$$

Then (28) becomes

$$E_c E_c' = (B_c'/B_c)^3. \qquad (29)$$

Equation (29) shows that if B_c is reduced by half, the energy per bit has to increase eight times.

VIII. ANALYSIS ON SPECTRUM EFFICIENCY FOR FDMA AND TDMA IN AN IDEAL CASE

We may compare the spectrum efficiency between frequency division multiple access (FDMA) and time division multiple access (TDMA) in theory as follows.
A.

In FDMA, the total allocated bandwidth B_t can be divided

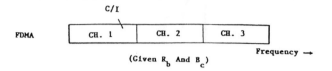

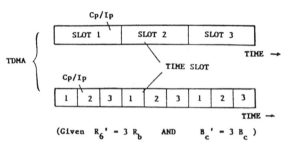

Fig. 5.   Multiple access schemes, FDMA versus TDMA.

by M channels. The radio capacity m in (12) is

$$m = \frac{B_t}{\dfrac{B_t}{M}\sqrt{\dfrac{2}{3}\left(\dfrac{C}{I}\right)_s}} = \frac{M}{\sqrt{\dfrac{2}{3}\left(\dfrac{C}{I}\right)_s}}. \qquad (30)$$

A three-channel FDMA with $B_t = 30$ kHz and $M = 3$ is illustrated in Fig. 5. Now both the carrier and the interference can be expressed, respectively, as

$$C = E_b \cdot R_b \qquad (31)$$

and

$$I = I_0 \cdot B_c \qquad (32)$$

where R_b is the transmission rate, E_b is the energy per bit, and I_0 is the interference (power) per hertz. For comparing the spectrum efficiency between FDMA and TDMA in a simple manner, we may use 10-kHz FDMA channels and 30-kHz TDMA channels. Let the notations used in (31) and (32) represent the parameters for a 10 kHz FM channel. The following notations are for the 30-kHz TDMA channels.

$$C' = E_b \cdot R_b' \qquad (33)$$

$$I' = I_0 \cdot B_c'. \qquad (34)$$

Note that E_b and I_0 are the same for both channel bandwidths.
B.

For easier understanding in TDMA, we may also have to show the calculation of radio capacity from the example depicted in Fig. 5. The radio capacities between the following two multiple access schemes need to be compared:

1) (M = three channels, $B_c = 10$ kHz, and $R_b = 10$ kbps)
2) TDMA (S = three time slots, $B_c' = 30$ kHz, and $R_b' = 30$ kbps).

The received carrier-to-interference ratio C_p/I_p at a TDMA 30-kHz channel is only measured in its 333 ms time in a second.[1] It can be equivalent to the C/I of a 10-kHz FDMA channel as

$$C_p = \frac{10 \text{ Kb} \cdot E_b}{\frac{1}{3} \text{ s}} = 3 \ (10 \text{ kbps}) \ E_b = 3R_b E_b = 3C \quad (35)$$

and

$$I_p = I_0 \frac{10 \text{ K cycles}}{\frac{1}{3} \text{ s}} = I_0 \cdot B_c' = I' = 3I. \quad (36)$$

The carrier-to-interference ratio $(C/I)_p$ in 333 ms of a second is the same as the (C/I) of an FDMA 10-kHz channel.

$$\frac{C_p}{I_p} = \frac{C'}{I'} = \frac{C}{I}. \quad (37)$$

Substituting (37) into (30) indicates that FDMA and TDMA have the same radio capacity, and therefore, the same spectrum efficiency. However, the power of a TDMA signal has to be three times (or 5 dB) higher than that of a FDMA signal. If a TDMA signal has k time slots and a bandwidth of kB_c, where B_c is the FDMA channel bandwidth, then the required power for TDMA is

$$C_p = C' = 10 \log k + C \quad \text{(in dB)}. \quad (38)$$

It means that the required power of a TDMA is $10 \log k$ higher than that of a FDMA signal.

IX. SUMMARY AND CONCLUSION

A general radio capacity equation for cellular system was derived. It involves two parameters; channel bandwidth and the required carrier-to-interference ratio. Its dimension is number of channels per cell. The spectrum efficiency can be measured from the radio capacity formula. The radio capacity can be also represented in various units with additional given information. Using the radio capacity, we have proved that splitting analog channels does not increase the spectrum efficiency.

The concept of spectrum efficiency in cellular is also described. Use of digital cellular channels shows a higher spectral efficient than use of analog cellular channels. A simple evaluation method of spectrum efficiency is introduced in judging different digital systems. The similar nature of channel capacity and radio capacity has been mentioned. From radio capacity, we may conclude that maintaining the same voice quality reduces the channel bandwidth by half, increases $(C/I)_s$ four times, or increases the energy per bit eight times.

---

[1] A three time-slot TDMA system means that three slots serve three different users. The number of time slots in one second can be a number of multiples of three. However, the total time spent for each user in one second is always 333 ms.

REFERENCES

[1] W. C. Y. Lee, "Spectrum efficiency: A comparison between FM and SSB in cellular mobile systems," presented at OST/FCC, August 2, 1985. (Also "New concept redefines spectrum efficiency of cellular mobile systems," *Telephony*, p. 82, Nov. 11, 1985.)
[2] V. H. MacDonald, "The cellular concept," *Bell Syst. Tech. J.*, vol. 58. pp. 15–42, Jan. 1979.
[3] W. C. Y. Lee, *Mobile Communications Engineering*. New York: McGraw-Hill, 1982, p. 429.
[4] ——, "Spectrum efficiency and digital cellular," in *38th IEEE Veh. Technol. Conf. Rec.*, June 1988, p. 643. (Also W. C. Y. Lee, "New concept redefines spectrum efficiency of cellular mobile systems," *Telephony*, p. 82, Nov. 11, 1985.)
[5] ——, *Mobile Communications Design Fundamentals*. New York: McGraw-Hill, 1988, p. 58.
[6] ——, *Mobile Communications Design Fundamentals*. New York: Howard W. Sams Co., 1986, p. 65.
[7] ——, "Narrowbanding in cellular mobile systems," *Telephony*, p. 44, Dec. 1, 1986.
[8] ——, "Mobile communications engineering," *Telephony*, p. 248, Dec. 1, 1986.
[9] P. F. Panter, *Modulation Noise, and Spectral Analysis*. New York: McGraw-Hill, 1965, p. 447.
[10] A. B. Carlson, *Communications Systems*. New York: McGraw-Hill, 1968, p. 279.
[11] ——, *Communications Systems*. New York: McGraw-Hill, 1975, p. 366.
[12] ——, *Communications Systems*. New York: McGraw-Hill, 1975, p. 267. (Also H. Taub and D. L. Schilling, *Principles of Communication Systems*. New York: McGraw-Hill, p. 290.)
[13] AT&T Demonstration of a Proposed Digital Cellular System, Chicago, IL, Mar. 1988.
[14] J. Swernp and J. Uddenfeldt, "Digital cellular," *Personal Commun. Technol.*, pp. 6-12, May 1986.
[15] W. C. Y. Lee, "How to Evaluate Digital Cellular Systems," presented at FCC Commission Meet., Washington, DC, Sept. 3, 1987. (Also W. C. Y. Lee, "How to evaluate digital cellular systems," *Telecommun.*, p. 45, Dec. 1987.)

William C. Y. Lee (M'64–SM'80–F'82) received the B.Sc. degree from the Chinese Naval Academy, Taiwan, and the M.S. and Ph.D. degrees from The Ohio State University, Columbus, in 1954, 1960, and 1963, respectively.

From 1959 to 1963 he was a Research Assistant at the Electroscience Laboratory, The Ohio State University. He was associated with Bell Laboratories from 1964 to 1979 where he was concerned with the study of wave propagation and systems, millimeter and optical wave propagation, switching systems, and satellite communications. He developed a UHF propagation model for use in planning the Bell System's new Advanced Mobile Phone Service and was a pioneer in mobile radio communication studies. He applied the field component diversity scheme over mobile radio communication links. While working in satellite communications, he discovered a method of calculating the rain rate statistics which would affect the signal attenuation at 10 GHz and above. He successfully designed a 4 × 4 element printed circuit antenna for tryout use. He studied and set a 3-mm-wave link between the Empire State Building and Pan American Building in New York City, experimentally using the newly developed IMPATT diode. He also studied the scanning spot beam concept for satellite communication using the adaptive array scheme. From April 1979 until April 1985 he worked for ITT Defense Communications Division and was involved with advanced programs for wiring military communications systems. He developed several simulation programs for the multipath fading medium and applied them to ground mobile communication systems. In 1982 he was Manager of the Advanced Development Department, responsible for the pursuit of new technologies for future communication systems. He developed an artificial intelligence application in the networking area and filed a patent application before leaving ITT. He joined PacTel Mobile Companies in 1985, where he is engaged in the improvement of system performance and capacity.

Capacity of Digital Cellular TDMA Systems

Krister Raith, *Member, IEEE,* and Jan Uddenfeldt, *Member, IEEE*

Abstract—The market for cellular radio telephony is expected to increase dramatically during the 1990's. Service may be needed for 50% of the population. This is beyond what can be achieved with the present generation analog cellular systems. The evolving digital time division multiple access (TDMA) cellular standards in Europe, North America, and Japan will give important capacity improvements and may satisfy much of the improvements needed for personal communication. The capacity of digital TDMA systems is addressed in this paper. Capacity improvement will be of the order 5–10 times that of analog FM without adding any cell sites. For example, the North American TIA standard offers around 50 Erlang/km^2 with a 3-km site-to-site distance. However, in addition, the TDMA principle allows a faster handoff mechanism (mobile assisted handoff, MAHO), which makes it easier to introduce microcells with cell radius of, say, 200 m. This gives substantial additional capacity gain beyond the 5–10 factor given above. Furthermore, TDMA makes it possible to introduce adaptive channel allocation (ACA) methods. ACA is a vital mechanism to provide efficient microcellular capacity. ACA also eliminates the need to plan frequencies for cells. A conclusion is that the air-interface of digital TDMA cellular may be used to build personal communication networks. The TDMA technology is the key to providing efficient handover and channel allocation methods.

I. Introduction

HIGH capacity radio access technology is vital for cellular radio. Digital time division multiple access (TDMA) is becoming a standard in the major geographical areas (Europe, North America, and Japan). Digital technology is capable of giving higher capacity than in the present analog FM systems. For example, the demonstration performed by Ericsson in 1988 [1] showed that multiple conversations could be carried out on a 30-kHz radio channel without degradation in radio range or carrier-to-interference (C/I) radio performance.

This paper gives capacity estimates for all digital TDMA standards and compares this with analog FM. First the different standards are described in Section II. Section III deals with a capacity comparison between different systems. In Section IV the benefits of digital for microcellular operation is discussed.

II. Digital Cellular Systems

Digital cellular technology will be introduced around 1991. There are three emerging standards, the pan-European GSM system specified by European Telecommunications Standards Institute (ETSI), the American Digital Cellular (ADC) specified by Telecommunications Industry Association (TIA), and

the Japanese Digital Cellular (JDC) specified by the Ministry of Post and Telegraph (MPT).

The standardization bodies have had different driving forces, time plans and scopes of work but all three have had in common that they address the lack of capacity in existing analog systems and that the new systems are digital and use TDMA as access method.

The GSM and ADC systems will be described in more detail in the following text. The JDC standardization work started recently, and detailed decisions have not yet been made. In short, the JDC system can be described as being very similar to the ADC system regarding the digital traffic channels whereas it has more in common with the GSM system in the lack of backward compatibility and in that the scope of the work is a single phase standardization process. Table I describes some of the characteristics regarding the air-interface for all the three systems.

A. The GSM System

The GSM system specifies many interfaces but only a part of the air-interface will be considered here. The frame and slot structure is shown in Fig. 1. There are also a super- and hyperframe (not shown in the figure) for various purposes, e.g., synchronization of crypto and provision for mobiles to identify surrounding base stations.

There are eight voice channels on one carrier (full rate) with the capability to introduce half rate speech codecs in the future. The carrier spacing is 200 kHz. Thus 25 kHz (200/8) is allocated to a full rate user. In all, there is a bandwidth of 25 MHz giving 125 radio channels i.e., 1000 traffic channels.

The gross bit rate is 270.8 kb/s. The modulation scheme is GMSK with the normalized pre-Gaussian filter bandwidth equal to 0.30 e.g., constant envelope allowing a class-C amplifier. The 33.85 kb/s per user are divided into

speech codec	13.0 kb/s
error protection of speech	9.8 kb/s
SACCH (gross rate)	0.95 kb/s
guard time, ramp up, synch.	10.1 kb/s

The overhead part could be defined to be 10.1/33.85 = 30%.

The bits in a speech block (20 ms) consist of two main classes according to sensitivity to bit errors. The most sensitive bits (class 1) are protected by a cyclic redundant check (CRC) code and a rate = 1/2 convolutional code with constraint length equal to 5. The coded speech block is interleaved over eight TDMA frames to combat burst errors. To further enhance the performance, frequency hopping, where each slot is transmitted on different carriers, can be used by

Manuscript received June 30, 1990; revised October 10, 1990.

K. Raith is with Ericsson GE Mobile Communication Inc., Box 13969, Research Triangle Park, NC 27709.

J. Uddenfeldt is with Ericsson Radio Systems AB, S-164 80 Stockholm, Sweden.

IEEE Log Number 9144475.

Reprinted from *IEEE Transactions on Vehicular Technology*, Vol. 40, No. 2, pp. 323-331, May, 1991.

391

TABLE 1
AIR-INTERFACE CHARACTERISTICS OF THREE DIGITAL CELLULAR STANDARDS

	Europe (ETSI)	North America (TIA)	Japan (MPT)
Access method	TDMA	TDMA	TDMA
Carrier spacing	200 kHz	30 kHz	25 kHz
Users per carrier	8 (16)	3 (6)	3 (tbd)
Modulation	GMSK	π/4 DQPSK	π/4 DQPSK
Voice codec	RPE 13 kb/s	VSELP 8 kb/s	tbd
Voice frame	20 ms	20 ms	20 ms*
Channel code	convolutional	convolutional	convolutional*
Coded bit rate	22.8 kb/s	13 kb/s	11.2 kb/s
TDMA frame duration	4.6 ms	20 ms	20 ms*
Interleaving	$\approx$ 40 ms	27 ms	27 ms*
Associated control channel	extra slot	in slot	in slot*
Handoff method	MAHO	MAHO	MAHO*

*Ericcson proposal

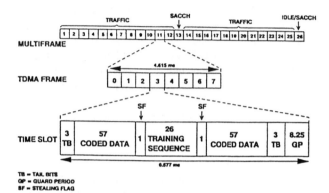

TB = TAIL BITS
GP = GUARD PERIOD
SF = STEALING FLAG

Fig. 1. GSM slot and frame structure showing 130.25 b per time slot (0.577 ms), eight time slots/TDMA frame (full rate), and 13 TDMA frames/multiframe.

the system. This is a mandatory function for the mobile but is optional for the system operator to use.

There are two control channels associated with the traffic channels, the slow and the fast ACCH. The FACCH is a blank-and-burst channel and replaces a speech block whenever it is to be used. Two frames in the multiframe (see Fig. 1) are allocated for the slow associated control channel (SACCH). With full rate users the second SACCH frame is idle. In a SACCH frame the slots are assigned in the same way as for traffic frames. The gross bit rate on this channel is 950 b/s and the net rate is 383 b/s. A SACCH message is interleaved over four multiframes.

With the fast growing number of subscribers anticipated in conjunction with smaller cell sizes it becomes increasingly important that the locating of mobiles can be done more accurately and faster than in the present analog systems. The method chosen by GSM is that the mobiles shall measure the signal strengths on channels from neighboring base stations and report the measurements to their current base station (Mobile Assisted Handoff (MAHO)). The land system evaluates these measurements and determines to which base station the mobile shall be transferred (handoff) if the mobile is about to leave its present cell or for other reasons would gain in radiolink quality by a handoff. The number of handoffs increases with the amount of traffic carried in a cell and the reduction of cell size. In analog systems where neighboring base stations measure the signal transmitted from a mobile, a

very high signaling load is introduced on the links between base stations and the switch and also higher processing requirement in the switch. Thus a decentralized location procedure where each mobile is a measurement point will reduce the burden on the network.

Of the eight time slots in a TDMA frame, two are used on different frequencies for transmission and reception. In the remaining time the mobile can measure the received signal strength on a broadcast control channel (BCCH) from its own and surrounding base stations. These measurements are averaged over a SACCH block (480 ms) before they are transmitted to the base station using the SACCH. The maximum number of surrounding base stations contained in the measurement list is 32 but only the result from the six strongest ones is reported back to the land system. Thus the mobiles preprocess the measurements and the reports contain results from different base stations for every SACCH block. Since there is a possibility that the signal strength measurements can be affected by a strong cochannel, and thereby be highly unreliable, the mobile is required to identify the associated base stations on a regular time basis. Therefore, it is necessary for the mobile to synchronize to and demodulate data on the BCCH in order to extract the base station identity code. This code is included in the measurement report informing the land system which base station is measured.

The mobile performs this identification process in its idle TDMA frame. There is one of these per multiframe, see Fig. 1. For half-rate, this idle frame is used for SACCH for the new traffic channels created. The mobile measurement report also contains an estimate of the bit error rate on the traffic channel used. This additional information is useful to determine the radio link quality since the received signal strength measurement cannot indicate a cochannel interferer or severe time dispersion.

B. The ADC System

This standard covers only the air-interface. Another subgroup of TIA is currently dealing with the intersystem connection. Since there is a single analog standard in North America and roaming is already possible, it has been decided that the first mobiles shall be dual mode, i.e., they should be capable of operating on both analog and digital voice channels. This makes it possible for the operators to introduce

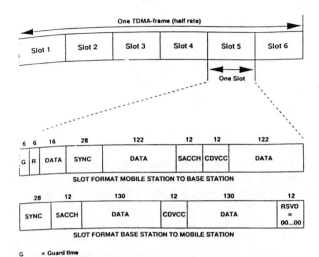

Fig. 2. ADC slot and frame structure for down- and uplink with 324 bits per time slot (6.67 ms) and 3(6) time slots/TDMA frame for full rate (half-rate).

digital radio channels according to capacity needs. In this first phase of digital technology the current analog control channels are used. Later on, provision for digital mode only mobiles will be made by introducing digital control channels.

With the dual mode requirement, it was natural to select a 30-kHz TDMA radio format. Each burst is 6.7 ms and for full rate users the TDMA frame length is 20 ms, see Fig. 2. Thus 10 kHz are allocated to a full rate user. In all, this gives 2500 traffic channels over a 25-MHz bandwidth.

The gross bit rate is 48.6 kb/s. The modulation scheme is differentially encoded $\pi/4$ QPSK with root-raised cosine pulse shaping and a roll off factor equal to 0.35. The 16.2 kb/s per user are divided into

speech codec	7.95 kb/s
error protection of speech	5.05 kb/s
SACCH (gross rate)	0.6 kb/s
guard time, ramp up, synch., color code	2.6 kb/s.

The overhead part could be defined to be 2.6/16.2 = 16% (compare corresponding calculation for the GSM system). The color code is an 8-bit signature to provide the capability to distinguish between connections using the same physical channel i.e., cochannels. This signature is transmitted in each burst and is protected by a shortened Hamming code to form the 12-bit field CDVCC.

The 20-ms speech block consisting of 159 b has two classes of bits with different sensitivity to bit errors. The most sensitive class of bits is protected by a CRC code and then coded with rate = 1/2. The other part (class 2 b) is not protected at all. The channel coding method used for speech and signaling is a convolutional code with constraint length equal to six. The coding rate for speech and SACCH is 1/2 and for FACCH 1/4. The interleaving for speech (and FACCH) is diagonal over two slots. A SACCH message is distributed over 22 slots by means of a self-synchronized interleaving process. The net rate on SACCH is 227 b/s.

Mobile Assisted Handoff is also used in the ADC system.

Perhaps the major difference in comparison to the GSM system is that the mobiles are not required to extract the base station identity code. In the dual mode phase of the ADC system there are no digital control channels on which to perform this task. There are only three time slots in a TDMA frame, and there is no idle frame as for GSM. Thus there is not enough remaining time to synchronize and demodulate data on another carrier without introducing high complexity in the mobile. Instead, there is the capability for a neighboring base station to identify a mobile, using the unique synch. word to identify a time slot and the CDVCC to distinguish the intended user from a cochannel.

Thus an implementation of the handoff process in an ADC system is that the land system evaluates the measurements from the mobile and lets the candidate base station verify that it can take over the call, before ordering the intended handoff. The MAHO is a mandatory function in the mobile but can optionally be turned on or off by the system. Thus, a traditional handoff implementation is also a possible method in which only information related to the traffic channel in use is considered.

The measurement reports contain the same information as in GSM (signal strength and estimated bit error rate) with the difference that in ADC the measurements from all base stations are reported, rather than only the six strongest. The list may contain up to 12 channels including the current traffic channel. For the same number of channels in the channel list, the GSM measurement reports are somewhat more accurate because of better averaging out the Rayleigh fading. The total number of samples with a certain time period is dependent on the number of TDMA frames within that time. There are 50 TDMA frames per second in the ADC system and approximately 216 per second in the GSM system. The reporting interval is once every second in the ADC system and once every 0.48 s in the GSM system.

C. The JDC System

As stated earlier, the JDC system is very similar to the ADC system i.e., it has a three-split TDMA air-interface. The main difference lies in the narrower channel bandwidth of 25 kHz compared to the 30-kHz bandwidth selected for the ADC system. The same type of modulation, $\pi/4$ DQPSK, as for the ADC system has been selected. To avoid extreme complexity in the power amplifier the gross bit rate has to be lower than in the ADC system (48.6 kb/s) and has been chosen to be 42.0 kb/s. The pulse shaping in the modulation scheme is root-raised cosine with a roll off factor equal to 0.5.

As was the case in North America, the speech and channel coding algorithm will be selected by testing candidates implemented in hardware. 11.2 kb/s has been selected for the total bit rate of the test. The difference between the gross bit rate per user (14 kb/s) and the protected speech rate (11.2 kb/s) is 2.8 kb/s and it will be allocated to the same functions as in the ADC system (see Section II-B) but the details will be different. Since the JDC system does not have any backward compatibility, all the control channels have to be specified within the first specification.

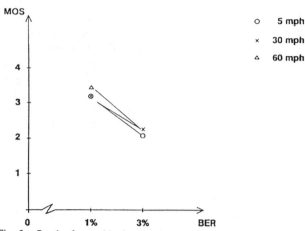

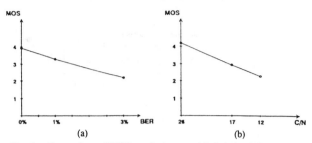

Fig. 3. Results from subjective listening tests (formal TIA tests) with a DCELP 8.7 kbs voice codec at different vehicle speeds.

Fig. 4. Comparison of MOS results between (a) digital and (b) analog.

III. CAPACITY OF SECOND GENERATION CELLULAR SYSTEMS

To accurately evaluate the capacity for the emerging digital systems, field trials are needed. Until these systems are implemented and operated in a multicell environment under high traffic conditions, the capacity can only be estimated. Nevertheless, it is possible to evaluate the performance of the system since there is experience from speech codec evaluations, computer simulations and limited field trials e.g., [1]–[4]. This is also necessary because the system operators have to make cell planning assumptions and system suppliers have to design handoff algorithms and do traffic analysis.

A. C/I Performance of the ADC System

In the selection process of the speech codec for the ADC system, subjective listening tests of hardware implemented candidate speech and channel-coding algorithms under a channel imperfection environment was used. The result from this test can be used as a starting point to estimate the capacity of the ADC system.

Fig. 3 shows the result with the Ericsson candidate 8.7 kb/s DCELP speech codec [4]. This is a deterministic code excited linear predictive codec running at 8.7 kb/s and error protected with a convolution code at a total of 13 kb/s. The mean opinion score (MOS) is the relative subjective quality perceived by the listening group. In the channel between the coder and the decoder bit errors were introduced. The bit error sequence was generated with a flat Rayleigh fading model using three-split TDMA and QPSK modulation. Soft information was not included in the test procedure. The channel model was tuned to give a bit error rate (BER) of 1% and 3% at a vehicular speed of 5, 30, and 60 mi/h. The 0% BER case is the basic voice quality, i.e., with an undistorted channel. The conclusion which can be drawn from the figure is that the speech quality is not sensitive to vehicular speed. At high vehicular speeds the channel coding and interleaving can combat the burst errors, whereas the interleaving (27 ms) is not long enough at low vehicular speeds. On the other hand, the speech frame repetition mechanism is more effective at lower speeds because incorrect speech blocks generally are followed by a number of correct ones.

In Fig. 4 the effect of vehicular speed has been taken out by averaging over the three different vehicular speeds. In the listening test there was an analog FM 30-kHz reference so that the results for the candidates could be evaluated against the quality perceived in the present analog system.

A design criterion used in cell planning of present analog FM systems is to achieve a C/I power ratio not less than 18 dB over most of the cell area (e.g., 90%). The speech quality at 18 dB C/I is about equal to a carrier-to-noise power ratio (C/N) measured over 30 kHz equal to 17 dB. As seen in Fig. 4(b) this subjective speech quality for analog is an MOS of 3.0. The BER before channel decoding giving the same subjective quality for digital is about 1.5% (see Fig. 4(a)).

We can now translate the BER figure to derive the minimum required C/N and C/I for digital. To this end, a computer implemented ADC radio format for performance evaluation is used, see Figs. 5 and 6. From Fig. 5, a BER at the demodulator output (class 2) of about 1.5% corresponds to about 18-dB C/I. From Fig. 6 we derive the peak E_b/N_b to be 15 dB. The peak E_b/N_o is converted to peak C/N over 30 kHz by 48/30 (+2 dB), e.g., the required peak C/N for digital is about 17 dB. (Here we define C/N as the carrier-to-noise power ratio measured in a 30-kHz bandwidth and E_bN_o as the bit energy divided by the noise spectral density. E_bN_o can also be viewed as the ratio between carrier power and noise power measured in a bandwidth equal to the bit rate.) This means that a 0.6-W peak power digital mobile will have the same coverage as a 0.6-W analog mobile. Thus the power consumption needed for the same coverage is only a third for the digital compared to the analog because of the three-split TDMA. In summary, the speech quality perceived in analog with 18-dB C/I is about the same as for digital at 18-dB C/I. Defining coverage with the same quality, both systems need about 17-dB C/N.

So far, the enhancement of using soft information in the channel decoding process has not been considered because the speech test did not provide any soft information. At a C/I of 18 dB the gain of using soft information is about 3 dB, see Fig. 5. Since only the performance of the coded bits (class 1) is affected by the use of soft information, we assume here for further calculation that the overall gain is around 2 dB. This means that the minimum required C/I for digital is 16 dB.

B. C/I Performance of the GSM System

We can now make use of the information derived for the ADC system to determine the necessary C/I for the GSM system by comparing BER curves for the two systems. Fig. 7 shows simulated BER performance for the GSM system for

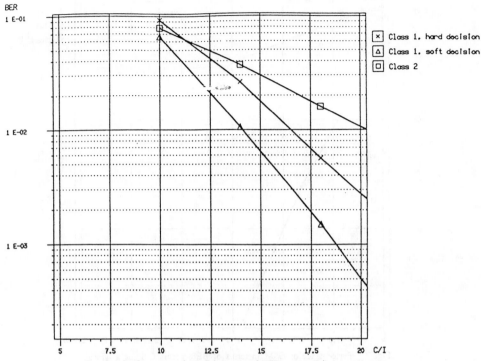

Fig. 5.   Performance of ADC in terms of BER versus C/I for a Rayleigh
fading channel at 55 mi/h.

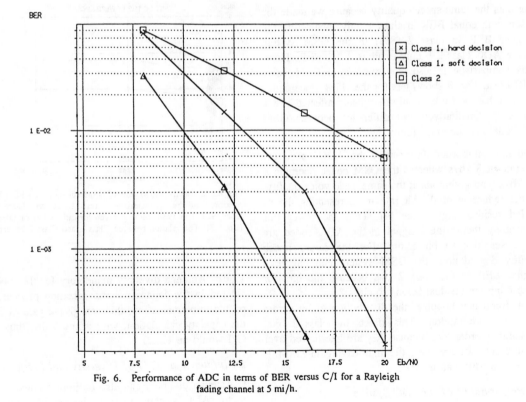

Fig. 6.   Performance of ADC in terms of BER versus C/I for a Rayleigh
fading channel at 5 mi/h.

class 1 and class 2, respectively. The vehicle speed is 25 mi/h and frequency hopping is used. The channel contains a small amount of time dispersion with a delay spread of about 1 μs which represent a typical case in urban environment. All paths are within the required capability of the equalizer.

One method to compare C/I performance is to use the BER of class 1. At the required C/I = 16 dB the BER in class 1 is 0.4% for ADC, see Fig. 5 (soft information). From Fig. 7 it is seen that the C/I value for GSM corresponding to a 0.4% BER in class 1 is C/I = 9 dB. It can be questioned if

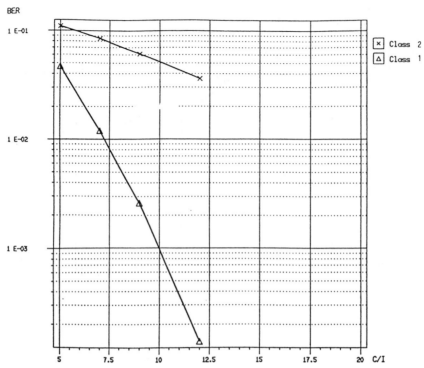

Fig. 7. Performance of GSM in terms of BER versus C/I for a typical urban fading channel at 25 mi/h.

this represents the same speech quality because we made the comparison with equal BER in class 1 only, neglecting the much higher BER in class 2 (8%) in the GSM system whereas the ADC class 2 b only exhibits 2.3% BER. However, this comparison is also confirmed by a GSM voice codec MOS-test. Fig. 8 shows test for the GSM voice codec [6]. It is seen that the MOS is only marginally affected at 8% BER (class 2). Qualitatively this difference between GSM and ADC codecs is due to the following.

1) The most important factor why the ADC codec can operate at 8 kb/s whereas the GSM codec operates at 13 kb/s, and giving about the same basic voice quality, is the reduction of the bit rate in representing the so called residual signal, see [4]. Thus the fewer bits describing the residual signal in the ADC codec are more sensitive to bit errors. For the same speech quality degradation, the GSM system can handle a higher BER in the class 2 bits because the bits for describing the residual belong to class 2.

2) With frequency hopping, the distribution of bit errors will be more random. This affects the class 2 bits, although unprotected, because they are interleaved over 8 different carriers, i.e., the interleaving in GSM is more powerful than in ADC.

C. C/I Performance of the JDC System

At the time of writing this paper, the selection for the JDC system has not yet taken place. Because of the similarities with the ADC system, at least the same minimum C/I is required. Since the total bit rate (11.2 kb/s) is less than for the ADC system we add, somewhat arbitrarily, 1 dB to give

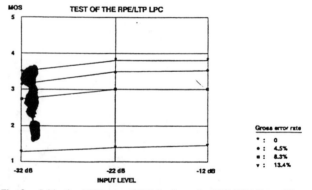

Fig. 8. Subjective MOS tests of GSM voice codec RPE/LTP (from [6]) as measured in the characterization tests based on channel simulation by GSM/SCEG in 1988. MOS is given for different error rates on the channel (Class 2). The relation between Class 1 and Class 2 bit errors is shown in Fig. 7.

17 dB. Antenna diversity mandatory for the mobiles is considered in the Japanese standardization process. Depending on the implementation, this can give a gain of 3–7 dB. For the capacity calculation, we assume 5 dB, thus the required C/I would be 12 dB.

D. Influence on C/I of Diversity and Frequency Hopping

Antenna diversity can also be implemented in the mobiles in the ADC system. In fact, the specification of the radio format has been derived so that a simple antenna diversity method, only using one RF receiver, can be implemented. With this simple antenna selection scheme, the gain is about 4 dB, thus an ADC system with mobiles using this type of diversity the minimum required C/I is also about 12 dB.

TABLE II
MINIMUM REQUIRED C/I IN DIFFERENT DIGITAL SYSTEMS (IN FADING CHANNEL CONDITIONS)

GSM		ADC		JDC	
With frequency hopping	Without frequency hopping	With antenna diversity	Without antenna diversity	With antenna diversity	Without antenna diversity
9 dB	11 dB	12 dB	16 dB	13 dB	17 dB

TABLE III
APPROXIMATE CAPACITY IN ERLANG PER KM^2 FOR ONE FIRST GENERATION ANALOG SYSTEM (AMPS) AND SECOND GENERATION DIGITAL SYSTEMS IN EUROPE, NORTH AMERICA, AND JAPAN ASSUMING A CELL RADIUS OF 1 KM (SITE DISTANCE OF 3 KM) IN ALL CASES AND 3 SECTOR PER SITE. THE LEE-MERIT IS THE NUMBER OF CHANNELS PER SITE ASSUMING AN OPTIMAL REUSE PLAN

	ANALOG FM	GSM pessimistic	optimistic	ADC pessimistic	optimistic	JDC pessimistic	optimistic
Bandwidth	25 MHz	25 MHz		25 MHz		25 MHz	
Number of voice channels	833	1000		2500		3000	
Reuse plan	7	4	3	7	4	7	4
Channels per site	119	250	333	357	625	429	750
Erlang/km^2	11.9	27.7	40.0	41.0	74.8	50.0	90.8
Capacity gain	1.0	2.3	3.4	3.5	6.3	4.2	7.6
(Lee merit gain)	(1.0)	(2.7)	(3.4)	(3.8)	(6.0)	(4.0)	(7.2)

In some countries in Europe the frequency band allocated for the GSM system is presently used for current analog systems (NMT 900 and TACS). When introducing the GSM system in those countries, not all of the band can be used. The remaining band has to be divided by the number of operators. In most countries the number of operators will be at least two. This will make the use of frequency hopping more difficult, or the available band is not large enough to provide carriers with independent fading resulting in performance degradation to ideal conditions. The degradation if no frequency hopping is used is about 2dB. If frequency hopping is used but the available band is limited, the degradation will be in the range from almost no degradation to 2 dB.

E. Importance of Handoff Algorithm

A handoff margin could be defined as the difference between the C/I used for frequency planning and the C/I where the speech quality is just about acceptable. A tight frequency plan will reduce the handoff margin. The more diversity used to lower the required C/I, the less the handoff margin will be. This effect arises because the BER curves for the coded bits will be more steep. Only a little reduction of the C/I will result in unintelligible speech. This is also valid for the signaling system, e.g., the mobile measurement reports.

A system designed for a very low C/I and operated near the C/I limit needs a fast and reliable handoff procedure and adequate error protection of the signaling messages. Thus, it is not only the static C/I figure that determines the possible frequency plan but also the dynamics (rate of C/I changes) in the system.

F. Capacity Comparison

We can now summarize the minimum required C/I, see Table II. Before a capacity comparison can be made we need to assume a channel plan for each system. Based on [7] we can convert the C/I figures in Table II to channel plans. We assume a sectorized plan with three sectors/site. The values in Table II will correspond to the following reuse plan:

GSM: 3 or 4 site reuse plan corresponding to optimistic or pessimistic case;

ADC: 4 or 7 site reuse plan corresponding to optimistic or pessimistic case;

JDC: as ADC above.

The capacity result is shown in Table III where 25 MHz has been used as the total bandwidth for all systems. A blocking figure of 2% has been used. The capacity gain figure is obtained by dividing the Erlang per square kilometer figure relative to the analog FM. In parenthesis an additional row is showing the capacity gain obtained by using the Lee formula given in [5].

As seen in Table III the digital systems can achieve a capacity increase of up to 7.6 times that of analog FM for the same number of sites in the system. If antenna diversity is used, 90.8 Erlang/km^2 can be achieved with a 3-km site to site distance. Even without diversity substantial capacity improvement can be obtained. For ADC a pessimistic capacity gain factor is 3.5. For GSM a similar pessimistic capacity gain factor is 2.3. This indicates that the GSM system has lower capacity than the ADC system. However, on the other hand, the GSM system will be the first system to make it possible to introduce the half-rate channels. The voice-coding technology already exists, although not mature enough to meet all the requirements of the GSM, because the gross bit rate for the GSM half-rate is equal to the JDC full-rate. This will improve the GSM system regarding capacity by almost a factor of two.

The next step would be the introduction of the half-rate in the ADC system giving a capacity gain factor in the range of 7–12 (depending on whether an optimistic or pessimistic frequency plan is used). In summary, we can conclude that digital TDMA can achieve a capacity gain of up to around 10 times as compared to analog systems. This is without introducing microcells. On top of this the TDMA principal makes it more attractive to introduce microcells which will be addressed in the next section.

397

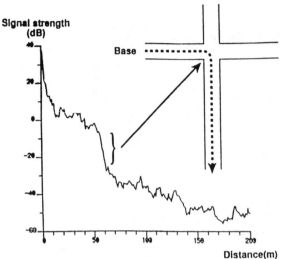

Fig. 9.   Example of radio field strength drop in a street corner.

TABLE IV
ILLUSTRATIVE (PESSIMISTIC) EXAMPLE OF CHANNEL PLAN AND CAPACITY WITH MICROCELLS AND FIXED CHANNEL ALLOCATION
GROUPS (WITH d_0 ASSUMED TO BE 400 M).

	With Microcells			Without Microcells
	10 MHz Umbrellas	15 MHz Street Microcells		25 MHz Umbrellas only
Site–site distance	3 km	0.8 km	0.4 km	0.2 km	3 km
Reuse plan	7-cell sector hex	25 cell omni (rect)	49 cell omni (rect)	100 cell omni (rect)	7-cell sector hex
Channels per site	140	60	30	15	357
Channels per km^2	20	97	195	390	45
Total channels/km^2		117	215	410	45
Total Erlang/km^2		107	174	280	41

IV. MICROCELLULAR OPERATION

A. Microcells and Handoff

The digital TDMA systems are using the MAHO method, which is a more rapid handover method enabling the call to be handed over between base stations at much faster rate than in first generation analog cellular. This is important since it makes the introduction of microcells much easier.

For street microcells the street corners will be the dimensioning case since radio field strength may drop 20 dB in a street corner over a short distance, see Fig. 9. A total reaction time of 1–2 s is needed.

MAHO means that the mobile stations (MS) makes measurements of the radio field strength from surrounding base stations (BS) and report these to the BS in use. It is possible to do these measurements in the MS due to the TDMA structure since there is idle time between received and transmitted time slots as explained in Section II.

The cochannel interference situation in microcells is difficult to predict due to very different radio propagation situation at different sites. This makes traditional frequency planning with fixed channel allocation to every site difficult to pursue. A very conservative reuse frequency plan must be employed. In addition, street microcells will not provide full coverage, i.e., overlaid umbrella cells must also be used. For

both these reasons capacity gain will not be proportional to the inverse of the squared cell radius.

An example may illustrate this. Street microcell measurements show propagation results of the following kind. Close to the BS propagation will be close to free space, i.e., a loss of 6 dB per octave in distance. Beyond a certain breakpoint d_0 the propagation loss will be higher. We assume a 12-dB per octave loss beyond a breakpoint $d_0 = 400$ m which is a typically figure encountered at 900 MHz. When the cell radius is decreased below the breakpoint d_0 there will be a C/I penalty for a given distance ratio between wanted and unwanted signals. This loss is of the order 6 dB per octave in cell radius for this example when the cell radius is below d_0. This 6-dB penalty roughly means that the number of channel groups N has to be doubled. In other words decreasing the cell radius by a factor of 2 will increase the capacity per square kilometer by a factor of 2 (instead of the expected factor 4).

In practice there will be some additional loss both due to less trunking efficiency and due to the need to use umbrella cells. Table IV gives an example where it is seen that the total capacity is not increased beyond 280 Erlang per km^2 even with very small street microcells (100-m radius). Even if 280 Erlang per km^2 is a large figure it must be noticed that

TABLE V
COMPARISON OF ADAPTIVE AND FIXED CHANNEL ALLOCATION IN INDOOR PROPAGATION ENVIRONMENT

With adaptive channel allocation		With fixed channel allocation		
Total number of traffic channels in the system	Simulated capacity per site	Required number of channels per site	Required reuse cell plan	Total number of required traffic channels in the system
16	1.1E	5	27	135
32	2.5 E	8	27	216

with a 100-m radius there will be a substantial number of handoffs per call, i.e., the handoff failure rate per call may be too high.

The above indicates that microcells alone do not solve all problems. Capacity may not increase dramatically when cell radius is decreased unless the investment in new cell sites is dramatically increased. Table IV is only an example but it illustrates the phenomenon with microcells. In practice it may be possible to plan frequencies in a more clever way and thus obtain higher capacity than indicated in Table IV.

B. Adaptive Channel Allocation

With adaptive channel allocation (ACA) there is no fixed frequency plan. Instead each BS is allowed to use any channel in the system. The ACA procedure is very attractive in a TDMA frame structure. The basic idea is to allocate channels depending on 1) the actual traffic situation and 2) the actual interference situation.

Early work only took 1) into account with relatively poor capacity gain as a result. In more recent proposals [8]–[10] it has been proposed to dynamically assign channels to every *call* instead of assigning channels to every *cell*.

Channel allocation is done both at call setup and at handoff. Due to the TDMA structure it is possible for the MS to continuously search for new better channels also during the call. An efficient ACA procedure requires that interference measurements of unused channels are performed both in the BS and in the MS. The TDMA format allows that interference measurements are done in the MS during conversation by measuring in idle time slot(s).

The ACA method is built into the standard for the Digital European Cordless Telecommunication (DECT) system specified by ETSI. DECT is a 24-slot TDMA system operating with a carrier spacing of 1.73 MHz in the 1880–1900-MHz band [11]. Similar ACA methods can be built into modification of both the GSM and the ADC standards. Simulations of ACA have revealed large capacity gain. This is particularly true when the number of available channels is relatively few. See Table V.

In a large system the capacity gain is 50% or more [8], [10]. In small systems capacity gains of 3–4 times has been found in [9]. An example is given in Table V from [9]. This shows that separate channel sets can be set aside for use by ACA for indoor picocells and outdoor microcell.

This means that ACA is an extremely powerful method for the following reasons: 1) Microcells and picocells can be deployed without frequency planning and 2) capacity gain in micro- and picocells operation is significant.

The advantages of ACA are from a capacity point of view:

i) Trunking efficiency loss is reduced almost entirely since there is no fixed allocation of channels to each BS.

ii) It is not necessary to take worst case propagation condition into account. With ACA one can design against the average interference situation as opposed to the case with fixed allocation where the worst possible interference situation must be used as a design criteria to choose the size of the channel group N.

ACA gives significant capacity improvement varying from a factor 1.5–2 in large systems to a factor of 4–6 in small systems. The large capacity gain in small systems (i.e., small allocated bandwidth) is due to an extra trunking efficiency.

V. CONCLUSION

The evolving digital TDMA cellular standards will provide a substantial capacity increase in short term perspective. In the medium term perspective half rate voice coding technology will provide a further improvement yielding a capacity gain of around 10 times that of the present analog cellular technology. In the long range, microcellular technology combined with ACA will move this into the personal communication capacity range.

REFERENCES

[1] K. Raith *et al.*, "Performance of a digital cellular experimental testbed," presented at VTC'89, San Francisco, CA, May 1–3, 1989.

[2] J.-E. Stjernvall and J. Uddenfeldt, "Performance of a cellular TDMA system in severe time dispersion," presented at Globecom '87, Tokyo, Japan.

[3] M. Nilsson, S. Johansson, and L. Borg, "A GSM validation system —Configuration cell lay-out, technical details, field measurements," presented at the Fourth Nordic Seminar on Digital Radio Communications DMR IV," Oslow, Norway, June 26–28, 1990.

[4] T. B. Minde and U. Wahlberg, "An implementation of a 8.7 kbit/s speech coder and a study on the channel error robustness for the combination of the speech and channel coding," presented at the IEEE workshop on Speech Coding, Vancouver, Canada, 1989.

[5] W. C. Y. Lee, "Spectrum efficiency in cellular," *IEEE Trans. Veh. Technol.*, vol. 38, pp. 69–75, 1989.

[6] J. Natvig *et al.*, "European DMR—The standardization procedure on the way from full-rate to the half-rate system," presented at the 2nd EURASIP Workshop on Medium- to Low-Rate Speech Coding, Hersbruck, Sept. 1989.

[7] J.-E. Stjernvall *et al.*, "Calculation of capacity and co-channel interference in a cellular system," presented at the Nordic Seminar on Digital Radio Communication, DMRI, Espoo, Finland, Feb. 5–7, 1985.

[8] H. Eriksson, "Capacity improvement by adaptive channel allocation," presented at the *Globecom*, Hollywood, FL, Nov. 28–Dec. 1, 1988.

[9] H. Ochsner, "DECT—Digital European cordless telecommunications," presented at the VTC'89, San Francisco, CA, May 1–3, 1989.

[10] R. Beck and H. Panzer, "Strategies for handover and dynamic channel allocation in micro-cellular mobile radio," presented at VTC'89, San Francisco, CA, May 1–3, 1989.

[11] D. Åkerberg, "Properties of a TDMA pico cellular orifice communication system," presented at VTC'89, San Francisco, CA, May 1–3, 1989.

Krister Raith (M'89) was born in Hofors, Sweden, in 1958. He received the M.S. degree in electronic engineering from the Royal Institute of Technology, Stockholm, Sweden.

In 1983 he joined Ericsson Radio Systems. From 1983 to 1990 he was a member of the System Research Department where he has been involved in signal processing and system design for digital cellular. Since 1990 he has been working at Ericsson GE Mobile Communication, Inc., Research Triangle Park, NC, a research and development facility for wireless communication applications. During the last two years he has been active in the TIA 45.3 air-interface standardization committee.

Jan Uddenfeldt (M'79) received the M.Sc. degree in electrical engineering in 1973 and the Ph.D. degree in telecommunications from the Royal Institute Technology, Stockholm, Sweden, in 1978.

He is currently the Vice President of Research and Technology at Ericsson Radio Systems, Stockholm. He is responsible for the company's worldwide research for cellular radio, cordless telephony, and mobile radio. He has been with Ericsson Radio Systems since 1978. During the years 1981–1987 he was also a part time Industry Professor at the Royal Institute of Technology in Teletransmission theory. His recent research has been in digital cellular. He has been active in developing the digital TDMA cellular technology for the pan-European GSM standard and the North-American TIA standard. He has also been involved in establishing the technology for the digital European cordless telecommunication (DECT) standard.

IEEE TRANSACTIONS ON VEHICULAR TECHNOLOGY, VOL. 40, NO. 2, MAY 1991

On the Capacity of a Cellular CDMA System

Klein S. Gilhousen, *Senior Member, IEEE*, Irwin M. Jacobs, *Fellow, IEEE*, Roberto Padovani, *Senior Member, IEEE*, Andrew J. Viterbi, *Fellow, IEEE*, Lindsay A. Weaver, Jr., and Charles E. Wheatley III, *Senior Member, IEEE*

Abstract—The use of spread spectrum or code division techniques for multiple access (CDMA) has long been debated. Certain advantages, such as multipath mitigation and interference suppression are generally accepted, but past comparisons of capacity with other multiple access techniques were not as favorable. This paper shows that, particularly for terrestrial cellular telephony, the interference suppression feature of CDMA can result in a many-fold increase in capacity over analog and even over competing digital techniques.

I. Introduction

SPREAD-SPECTRUM techniques, long established for antijam and multipath rejection applications as well as for accurate ranging and tracking, have also been proposed for code division multiple access (CDMA) to support simultaneous digital communication among a large community of relatively uncoordinated users. Yet, as recently as 1985 a straightforward comparison [1] of the capacity of CDMA to that of conventional time division multiple access (TDMA) and frequency division multiple access (FDMA) for satellite applications suggested a reasonable edge in capacity for the latter two more conventional techniques. This edge was shown to be illusory shortly thereafter [2] when it was recognized that since CDMA capacity is only interference limited (unlike FDMA and TDMA capacities which are primarily bandwidth limited), any reduction in interference converts directly and linearly into an increase in capacity. Thus, since voice signals are intermittent with a duty factor of approximately 3/8 [3], capacity can be increased by an amount inversely proportional to this factor by suppressing (or squelching) transmission during the quiet periods of each speaker. Similarly, any spatial isolation through use of multibeamed or multisectored antennas, which reduces interference, also provides a proportional increase in capacity. These two factors, voice activity and spatial isolation, were shown to be sufficient to render CDMA capacity at least double that of FDMA and TDMA under similar assumptions for a mobile satellite application [2].

While previous comparisons primarily applied to satellite systems, CDMA exhibits its greatest advantage over TDMA and FDMA in terrestrial digital cellular systems, for here isolation among cells is provided by path loss, which in terrestrial UHF propagation typically increases with the fourth power of the distance. Consequently, while conventional techniques must provide for different frequency allocation for contiguous cells (only reusing the same channel in one of every 7 cells in present systems), CDMA can reuse the same (entire) spectrum for all cells, thereby increasing capacity by a large percentage of the normal frequency reuse factor. The net improvement in capacity, due to all the above features, of CDMA over digital TDMA or FDMA is on the order of 4 to 6 and over current analog FM/FDMA it is nearly a factor of 20.

The next section deals with a single cell system, such as a hubbed satellite network, and develops the basic expression for capacity. The subsequent two sections derive the corresponding expressions for a multiple cell system and determine the distribution on the number of users supportable per cell. The last section presents conclusions and system comparisons.[1]

II. Single Cell CDMA Capacity

The network to be considered throughout consists of numerous mobile (or personal) subscribers communicating with one or multiple cell sites (or base stations) which are interconnected with a mobile telephony switching office (MTSO), which also serves as a gateway to the public switched telephone network. We begin by considering a single cell system, which can also serve as a model for a satellite system whose "cell site" is a single hub.

Each user of a CDMA system occupies the entire allocated spectrum, employing a direct sequence spread spectrum waveform. Without elaborating on the modulation and spreading waveform, we assume generic CDMA modems at both subscriber units and the cell site with digital baseband processing units as shown in Fig. 1 for the transmitter sides of each. These consist of (digital) forward-error correction (FEC), modulation and (direct sequence) spreading functions, preceding the (analog) amplification and transmission functions. Each of the digital functions can be performed using binary sequences in the subscriber modulator.

At the cell-site transmitter, the spread signals directed to the individual subscribers are added linearly and phase randomness is assured by modulating each signal with independent pseudorandom sequences on each of the two quadrature phases. The weighting factors $\emptyset_1, \emptyset_2, \cdots, \emptyset_N$ can be taken to be equal for the time being, but for the multiple cell case they will provide power control based on considerations to be

Manuscript received May 1, 1990; revised September 14, 1990. This paper was presented at the 1990 IEEE GLOBECOM Conference, San Diego, CA.

The authors are with QUALCOMM, Inc., 10555 Sorrento Valley Road, San Diego, CA 92121-1617.

IEEE Log Number 9144470.

[1]It should be noted that our purpose is not to evaluate or optimize modem performance for the channels under consideration. Rather, assuming an efficient modulation and FEC code for the given channels, we shall establish conditions under which the modems will achieve an acceptable level of performance, particularly in terms of the maximum number of users supportable per cell.

Reprinted from *IEEE Transactions on Vehicular Technology*, Vol. 40, No. 2, pp. 303-311, May, 1991.

401

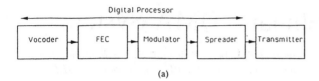

(a)

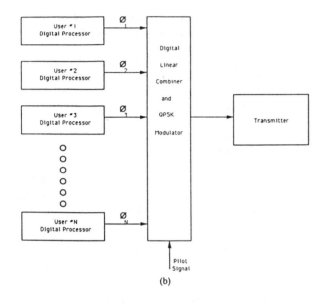

(b)

Fig. 1. Cellular system simplified block diagram. (a) Reverse link subscriber processor/transmitter. (b) Forward link cell-site processor/transmitter.

described later. The receiver processors in both subscriber and cell-site receivers provide the inverse baseband functions, which are of course considerably more complex than the transmitter baseband functions.

One other key feature of the cell-site transmitter is the inclusion of a pilot signal in the forward (cell-site-to-subscriber) direction. This provides for acquisition by the mobile terminals, including initial power control by the mobile, which adjusts its output power inversely to the total signal power it receives. Power control is a basic requirement in CDMA and will be expanded on in a later section.

We note also that the pilot signal is used by the subscriber demodulator to provide a coherent reference which is effective even in a fading environment since the desired signal and the pilot fade together. In the subscriber-to-cell-site (reverse) direction, no pilot is used for power efficiency considerations, since unlike the forward case, an independent pilot would be needed for each signal. A modulation consistent with, and relatively efficient for, noncoherent reception is, therefore, used for the reverse direction.

Without elaborating further on the system implementation details, we note that for a single cell site with power control, all reverse link signals (subscribers-to-cell site) are received at the same power level. For N users, each cell-site demodulator processes a composite received waveform containing the desired signal having power S and $(N - 1)$ interfering signals each also of power S. Thus the signal-to-noise (inter-

ference) power is

$$\text{SNR} = \frac{S}{(N - 1)S} = \frac{1}{N - 1}.$$

Of greater importance for reliable system operation is the bit energy-to-noise density ratio, whose numerator is obtained by dividing the desired signal power by the information bit rate, R, and dividing the noise (or interference) by the total bandwidth, W. This results in

$$E_b/N_0 = \frac{S/R}{(N - 1)S/W} = \frac{W/R}{N - 1}. \tag{1}$$

This paper does not explicitly address modulation techniques and their performance. Rather, an E_b/N_0 level is assumed which ensures operation at the level of bit error performance required for digital voice transmission. Among the factors to be considered in establishing the modulation and the resulting required E_b/N_0 level are phase coherence, amplitude fading characteristics and power control techniques and their effectiveness, particularly for the reverse link. One of the lesser considerations, albeit one of the most cited, is the probability distribution of the interfering signals. While Gaussian noise is often assumed, this is not strictly necessary to establish the E_b/N_0 requirements. Nonetheless, the assumption is quite reasonable when powerful forward error-correcting codes are employed, particularly at low code rates, because in such cases decisions are based on long code sequence lengths over which the interfering signal sequence contributions are effectively the sums of a large number of binomial variables, which closely approximate Gaussian random variables.

Equation (1) ignores background noise, η, due to spurious interference as well as thermal noise contained in the total spread bandwidth, W. Including this additive term in the denominator of (1) results in a required

$$E_b/N_0 = \frac{W/R}{(N - 1) + (\eta/S)}. \tag{2}$$

This implies that the capacity in terms of number of users supported is

$$N = 1 + \frac{W/R}{E_b/N_0} - \frac{\eta}{S} \tag{3}$$

where W/R is generally referred to as the "processing gain" and E_b/N_0 is the value required for adequate performance of the modem and decoder, which for digital voice transmission implies a BER of 10^{-3} or better. In words, the number of users is reduced by the inverse of the per user signal-to-noise ratio (SNR) in the total system spread bandwidth, W. In a terrestrial system, the per user SNR is limited only by the transmitter's power level. As will be justified below, we shall assume SNR just below unity corresponding to a reduction in capacity equivalent to removing one user. The background noise, therefore establishes the required received signal power at the cell site, which in turn fixes the subscriber's power or the cell radius for a given maximum transmitter power.

For the reverse (subscriber-to-cell-site) direction, noncoherent reception and independent fading of all users is assumed. With dual antenna diversity, the required $E_b/N_0 = 7$ dB for a relatively powerful (constraint length 9, rate 1/3) convolutional code. Since the forward link employs coherent demodulation by the pilot carrier which is being tracked, and since its multiple transmitted signals are synchronously combined, its performance in a single cell system will be much superior to that of the reverse link. For a multiple cell system however, other cell interference will tend to equalize performance in the two directions, as will be described below.

All this leaves us at the point of our previous conclusions [1], only worse because of the Rayleigh fading encountered in terrestrial mobile applications. In the next section we begin to remedy the situation.

III. AUGMENTED PERFORMANCE THROUGH SECTORIZATION AND VOICE-ACTIVITY MONITORING

Short of reducing E_b/N_0 through improved coding or possibly modulation, which rapidly reaches the point of diminishing returns for increasing complexity (and ultimately the unsurmountable Shannon limit), we can only increase capacity by reducing other user interference and hence the denominator of (1) or (2). This can be achieved in two ways.

The first is the common technique of sectorization, which refers to using directional antennas at the cell site both for receiving and transmitting. For example, with three antennas per cell site, each having 120° effective beamwidths, the interference sources seen by any antenna are approximately one-third of those seen by an omnidirectional antenna. This reduces the $(N - 1)$ term in the denominator of (2) by a factor of 3 and consequently, in (3) N is increased by nearly this factor. Henceforth, we shall take N_s to be the number of users per sector and the interference to be that received by one sector's antenna. Using three sectors, the number of users per cell $N = 3N_s$.

Secondly, voice activity can be monitored, a function which virtually already exists in most digital vocoders, and transmission can be suppressed for that user when no voice is present. Extensive studies show that either speaker is active only 35 % to 40 % of the time [3]. We shall assume for this the "voice activity factor," $\alpha = 3/8$ throughout. On the average, this reduces the interference term (in the denominator of (2)) from $(N - 1)$ to $(N - 1)\alpha$. Below, we will find through a more careful analysis that the net improvement in capacity due to voice activity is reduced from 8/3 to about 2 due to the fact that with a limited number of calls per sector, there is a nonnegligible probability that an above average number of users are talking at once. We ignore this in this preliminary discussion but include it in the results described below. Thus with sectorization and voice activity monitoring, the average $\overline{E_b}/N_0$, is increased relative to (2) to become[2]

$$\frac{\overline{E_b}}{N_0} = \frac{W/R}{(N_s - 1)\alpha + (\eta/S)} . \quad (4)$$

[2] These arguments leading to (4) were first advanced by Cooper and Nettleton [4].

This suggests that the average number of users per cell is increased by almost a factor of 8. In fact, because of variability in E_b/N_0 this increase will need to be backed off to a factor of 5 or 6. We shall return to this variability issue and other more precise results after we consider multiple cell interference in the next section. For now, note from (3) and (4) that this is enough to bring the number of users/cell up to the processing gain, $N \approx W/R$ users/cell which makes CDMA at least competitive with other multiple access techniques (FDMA or TDMA) on a single cell basis. As we will presently show, in multiple-cell systems additional advantages accrue through frequency reuse of the same spectrum in all cells. To assess this advantage, we must first consider the power control techniques and their effect on multicell interference.

IV. REVERSE LINK POWER CONTROL IN MULTIPLE-CELL SYSTEMS

As should be clear by now, power control is the single most important system requirement for CDMA, since only by control of the power of each user accessing a cell can resources be shared equitably among users and capacity maximized. In a single cell system, the principle is straightforward, though the implementation may not be. Prior to any transmission, each of the subscribers monitors the total received signal power from the cell site. According to the power level it detects, it transmits at an initial level which is as much below (above) a nominal level in decibels as the received pilot power level is above (below) its nominal level. Experience has shown that this may require a dynamic range of control on the order of 80 dB. Further refinements in power level in each subscriber can be commanded by the cell site depending on the power level it receives from the subscriber.

The relatively fast variations associated with Rayleigh fading may at times be too rapid to be tracked by the closed-loop power control but variations in relative path losses and shadowing effects, which are modelled as an attenuation with log-normal distribution, will generally be slow enough to be controlled. Also, while Rayleigh fading may not be the same for forward and reverse links, log-normal shadowing normally will exhibit reciprocity. For the forward link, no power control is required in a single cell system, since for each subscriber any interference caused by other subscriber signals remains at the same level relative to the desired signal; inasmuch as all signals are transmitted together and hence vary together, there are no resulting degradations due to fading assuming the background noise may be neglected.

In multiple-cell CDMA systems, the situation becomes more complicated in both directions. First, for the reverse link, subscribers are power controlled by the base station of their own cell. Even the question of cell membership is not simple. For it is not minimum distance which determines which base station (cell site) the subscriber joins, but rather the maximum pilot power among the cell sites the subscriber receives. In any case the interference level from subscribers in the other cells varies not only according to the attenuation in the path to the subscriber's cell site, but also inversely to the attenuation from the interfering user to his own cell site, which through power control by that cell site may increase, or decrease, the interference to the desired cell site. These issues will be treated in the next section.

As for the forward link for a multiple cell system, interference from neighboring cell sites fade independently of the given cell site and thereby degrade performance for any level of interference. This becomes a particularly serious problem in the region where two or even three cell transmissions are received at nearly equal strengths. Techniques for mitigating this condition are treated in Section VI.

V. Reverse Link Capacity for Multiple Cell CDMA

Recalling that power control to a given mobile is exercised by the cell whose pilot signal power is maximum to that mobile, it follows that if the path loss were only a function of distance from the cell site, then the mobile would be power controlled by the nearest cell site, which is situated at the center of the hexagon in which it lies, as shown in Fig. 2(a) for an idealized placement of cell sites. In fact, the loss is proportional to other effects as well, the most significant being shadowing. The generally accepted model is an attenuation which is the product of the fourth power of the distance and a log-normal random variable whose standard deviation is 8 dB. That is, the path loss between the subscriber and the cell site is proportional to $10^{(\xi/10)}r^{-4}$ where r is distance from subscriber to cell site and ξ is a Gaussian random variable with standard deviation $\sigma = 8$ and zero mean. Fast fading (due largely to multipath) is assumed not to affect the (average) power level.

We note that other propagation exponents can be found in different environments. In fact, within a single cell the propagation may vary from inverse square law very close to the cell antenna to as great as the inverse 5.5 power far from the cell in a very dense urban environment such as Manhattan. The present analysis is primarily concerned with interference from neighboring and distant cells so the assumption of inverse fourth law propagation is a reasonable one.

The interference from transmitter within the given subscriber's cell is treated as before; that is, since each user is power controlled by the same cell site, it arrives with the same power S, when active. Thus given N subscribers per cell, the total interference is never greater than $(N - 1)S$, but on the average it is reduced by the voice activity factor, $\propto$. Subscribers in other cells, however, are power controlled by other cell sites (Fig. 2(a)). Consequently, if the interfering subscriber is in another cell and at a distance r_m from its cell site and r_0 from the cell site of the desired user, the other user, when active, produces an interference in the desired user's cell site equal to

$$\frac{I(r_0, r_m)}{S} = \left(\frac{10^{(\xi_0/10)}}{r_0^4}\right)\left(\frac{r_m^4}{10^{(\xi_m/10)}}\right)$$

$$= \left(\frac{r_m}{r_0}\right)^4 10^{(\xi_0 - \xi_m)/10} \gtrless 1 \qquad (5)$$

where the first term is due to the attenuation caused by

distance and blockage to the given cell site, while the second term is the effect of power control to compensate for the corresponding attenuation to the cell site of the out-of-cell interferer.[3] Of course ξ_0 and ξ_m are independent so that the difference has zero mean and variance $2\sigma^2$. For all values of the above parameters, the expression is less than unity, for otherwise the subscriber would switch to the cell site which makes it less than unity (i.e., for which the attenuation is minimized).

Then, assuming a uniform density of subscribers, and normalizing the hexagonal cell radius to unity, and since the average number of subscribers/cell is $N = 3N_s$, the density of users is

$$\rho = \frac{2N}{3\sqrt{3}} = \frac{2N_s}{\sqrt{3}} \text{ per unit area.} \qquad (6)$$

Consequently, the total other-cell user interference-to-signal ratio is

$$I/S = \iint \psi\left(\frac{r_m}{r_0}\right)^4 \{10^{(\xi_0 - \xi_m)/10}\}$$

$$\cdot \varnothing(\xi_0 - \xi_m, r_0/r_m)\rho \, dA \qquad (7)$$

where m is the cell-site index for which

$$r_m^4 10^{-\xi_m} = \min_{k \neq 0} r_k^4 10^{-\xi_k} \qquad (8)$$

and

$$\varnothing(\xi_0 - \xi_m, r_0/r_m)$$

$$= \begin{cases} 1, & \text{if } (r_m/r_0)^4 10^{(\xi_0 - \xi_m)/10} \leq 1 \\ & \quad \text{or } \xi_0 - \xi_m \leq 40 \log_{10}(r_0/r_m) \qquad (9) \\ 0, & \text{otherwise} \end{cases}$$

and ψ is the voice activity variable, which equals 1 with probability $\propto$ and 0 with probability $(1 - \propto)$. To determine the moment statistics of the random variable I, the calculation is much simplified and the results only slightly increased if for m we use the smallest distance rather than the smallest attenuation. Thus (7), with (9), holds as an upper bound if in place of (8) we use that value of m for which

$$r_m = \min_{k \neq 0} r_k. \qquad (8')$$

In Appendix I, it is shown that the mean or first moment, of the random variable I/S is upper bounded (using (8') rather than (8) for m) by the expression

$$E(I/S) = \propto \iint \frac{r_m^4}{r_0} f\left(\frac{r_m}{r_0}\right)\rho \, dA$$

where

$$f\left(\frac{r_m}{r_0}\right) = \exp\left[(\sigma \ln 10/10)^2\right]\left\{1 - Q\left[\frac{40}{\sqrt{2\sigma^2}}\right.\right.$$

$$\left.\left. \cdot \log_{10}\left(\frac{r_0}{r_m}\right) - \sqrt{2\sigma^2}\frac{\ln 10}{10}\right]\right\} \qquad (10)$$

---

[3]Cooper and Nettleton [4] employed similar geometric arguments to compute interference, but did not consider log-normal statistical variations due to blockage.

and

$$Q(x) = \int_x^\infty e^{-y^2/2} \, dy / \sqrt{2\Pi} \; .$$

This integral is over the two-dimensional area comprising the totality of all sites in the sector (Fig. 2(a)). The integration, which needs to be evaluated numerically, involves finding for each point in the space the value of r_0, the distance to the desired cell site and r_m, which according to (8'), is the distance to the closest cell site, prior to evaluating at the given point the function (10). The result for $\sigma = 8$ dB is

$$E(I/S) \le 0.247 N_s.$$

Calculation of the second moment, var (I/S) of the random variable requires an additional assumption on the second-order statistics of ξ_0 and ξ_m. While it is clear that the relative attenuations are independent of each other, and that both are identically distributed (i.e., have constant first-order distributions) over the areas, their second-order statistics (spatial correlation functions) are also needed to compute var (I). Based on experimental evidence that blockage statistics vary quite rapidly with spatial displacement in any direction, we shall take the spatial autocorrelation functions of ξ_0 and ξ_m to be extremely narrow in all directions, the two-dimensional spatial equivalent of white noise. With this assumption, we obtain in Appendix I that

$$\text{var}\,(I/S) \le \iint \left(\frac{r_m}{r_0}\right)^8 \left[\propto g\left(\frac{r_m}{r_0}\right) - \propto^2 f\left(\frac{r_m}{r_0}\right)\right] \rho \, dA$$

where

$$g\left(\frac{r_m}{r_0}\right) = \exp\left[\left(\frac{\sigma \ln 10}{5}\right)^2\right] \left\{1 - Q\left[\frac{40}{\sqrt{2\sigma^2}} \right.\right.$$
$$\left.\left. \cdot \log_{10}\left(\frac{r_0}{r_m}\right) - \sqrt{2\sigma^2}\left(\frac{\ln 10}{5}\right)\right]\right\}. \quad (11)$$

This integral is also evaluated numerically over the area of Fig. 2(a), with r_m defined at any given point by condition (8'). The result for $\sigma = 8$ dB is var $(I/S) \le 0.078 N_s$. The above argument also suggests that I, as defined by (7), being a linear functional on a two-dimensional white random process, is well modelled as a Gaussian random variable.[4]

We may now proceed to obtain a distribution on the total interference, both from other users in the given cell, and from other-cell users on the desired user's reverse link transmission. With sectorization, variable voice activity and the other-cell interference statistics just determined, the received E_b/N_0 on the reverse link of any desired user becomes the random variable

$$E_b/N_0 = \frac{W/R}{\displaystyle\sum_{i=1}^{N_s-1} \chi_i + (I/S) + (\eta/S)} \quad (12)$$

where N_s is the users/sector and I is the total interference from users outside the desired user's cell. This follows easily

[4]Of course, it can never be negative, but since the ratio of mean-to-standard deviation is approximately $\sqrt{N_s}$, with typical values of $N_s > 30$ the approximating Gaussian distribution is nearly zero for negative values.

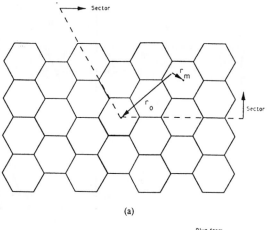

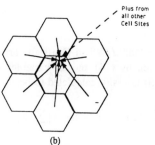

Fig. 2. Capacity calculation geometries. (a) Reverse link geometry. (b) Forward link allocation geometry.

from (2) with the recognition that the $N_s - 1$ same sector normalized power users, instead of being unity all the time, now are random variables χ_i with distribution

$$\chi_i = \begin{cases} 1, & \text{with probability } \propto \\ 0, & \text{with probability } 1 - \propto . \end{cases} \quad (13)$$

The additional term I represents the other (multiple) cell user interference for which we have evaluated mean and variance,

$$E(I/S) \le 0.247 N_s \quad \text{and} \quad \text{var}\,(I/S) \le 0.078 N_s \quad (14)$$

and have justified taking it to be a Gaussian random variable. The remaining terms in (12), W/R and S/η, are constants.

As previously stated, with an efficient modem and a powerful convolutional code and two-antenna diversity, adequate performance (BER $< 10^{-3}$) is achievable on the reverse link with $E_b/N_0 \ge 5$ (7 dB). Consequently, the required performance is achieved with probability $P = \text{Pr(BER} < 10^{-3}) = \text{Pr}(E_b/N_0 \ge 5)$. We may *lower bound* the probability of achieving this level of performance for any desired fraction of users at any given time (e.g., $P = 0.99$) by obtaining an upper bound on its complement, which according to (12), depends on the distribution of χ_i and I, as follows

$$1 - P = \text{Pr(BER} > 10^{-3}) = \text{Pr}\left(\sum_{i=1}^{N_s} \chi_i + I/S > \delta\right) \quad (15)$$

where

$$\delta = \frac{W/R}{E_b/N_0} - \frac{\eta}{S}, \qquad E_b/N_0 = 5.$$

Since the random variable χ_i has the binomial distribution

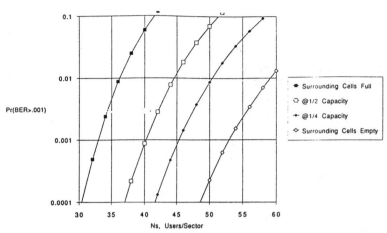

Fig. 3.   Reverse link capacity/sector. ($W = 1.25$ MHz, $R = 8$ kb/s, voice activity $= 3/8$).

given by (13) and I/S is a Gaussian variable with mean and variance given by (14) and all variables are mutually independent, (15) is easily calculated to be

$$1 - P = \sum_{k=0}^{N_s-1} \Pr\left(I/S > \delta - k \,\Big|\, \sum x_i = k\right) \Pr\left(\sum x_i = k\right)$$

$$= \sum_{k=0}^{N_s-1} \binom{N_s - 1}{k} \propto^k (1 - \propto)^{N_s-1-k}$$

$$\cdot\, Q\left(\frac{\delta - k - 0.247 N_s}{\sqrt{0.078 N_s}}\right). \quad (16)$$

This expression is plotted for $\delta = 30$ (a value chosen as discussed in the conclusion) and $\propto = 3/8$, as the leftmost curve of Fig. 3. The rightmost curve applies to a single cell without other cell interference ($I = 0$), while the other intermediate curves assume that all cells other than the desired user's cells are on the average loaded less heavily (with averages of $1/2$ and $1/4$ of the desired user's cell).

We shall discuss these results further in the concluding section, and now concern ourselves with forward link performance.

VI. MULTIPLE-CELL FORWARD LINK CAPACITY WITH POWER ALLOCATION

As noted earlier, although with a single cell no power control is required, with multiple cells it becomes important, because near the boundaries of cells considerable interference can be received from other cell-site transmitters fading independently.

For the forward link, power control takes the form of power allocation at the cell-site transmitter according to the needs of individual subscribers in the given cell. This requires measurement by the mobile of its relative SNR, defined as the ratio of the power from its own cell site transmitter to the total power received. Practically, this is done by acquiring (correlating to) the highest power pilot and measuring its energy, and also measuring the total energy received by the mobile's omnidirectional antenna from all cell site transmitters. Both measurements can be transmitted to the

selected (largest power) cell site when the mobile starts to transmit. Suppose then that based on these two measurements, the cell site has reasonably accurate estimates of S_{T_1} and $\sum_{i=1}^{K} S_{Ti}$, where

$$S_{T_1} > S_{T_2} > \cdots > S_{T_K} > 0 \quad (17)$$

are the powers received by the given mobile from the cell site sector facing it, assuming all but K (total) received powers are negligible. (We shall assume hereafter that all sites beyond the second ring around a cell contribute negligible received power, so that $K \leq 19$.) Note that the ranking indicated in (17) is not required of the mobile—just the determination of which cell site is largest and hence which is to be designated T_1.

The ith subscriber served by a particular cell site will receive a fraction of S_{T_1} the total power transmitted by its cell site, which by choice and definition (17) is the greatest of all the cell site powers it receives, and all the remainder of S_{T_1} as well as the other cell site powers are received as noise. Thus its received E_b/N_0 can be lower bounded by

$$\left(\frac{E_b}{N_0}\right)_i \geq \frac{\beta \varnothing_i S_{T_1}/R}{\left[\left(\sum_{j=1}^{K} S_{T_j}\right)_i + \eta\right]/W} \quad (18)$$

where S_{T_j} is defined in (17), β is the fraction of the total cell site power devoted to subscribers ($1 - \beta$ is devoted to the pilot) and $\varnothing_i$ is the fraction of this devoted to subscriber i. Because of the importance of the pilot in acquisition and tracking, we shall take $\beta = 0.8$. It is clear that the greater the sum of other cell-site powers relative to S_{T_1}, the larger the fraction $\varnothing_i$ which must be allocated to the ith subscriber to achieve its required E_b/N_0. In fact, from (18) we obtain

$$\varnothing_i \leq \frac{(E_b/N_0)_i}{\beta W/R}\left[1 + \left(\frac{\sum_{j=2}^{K} S_{T_j}}{S_{T_1}}\right)_i + \frac{\eta}{(S_{T_1})_i}\right] \quad (19)$$

where

$$\sum_{i=1}^{N_s} \emptyset_i \leq 1 \qquad (20)$$

since βS_{T_1} is the maximum total power allocated to the sector containing the given subscriber and N_s is the total number of subscribers in the sector. If we define the relative received cell-site power measurements as

$$f_i \triangleq \left(1 + \sum_{j=2}^{K} S_{T_j}/S_{T_1}\right)_i, \qquad i = 1, \cdots, N_s \quad (21)$$

then from (19) and (20) it follows that their sum over all subscribers of the given cell site sector is constrained by

$$\sum_{i=1}^{N_s} f_i \leq \frac{\beta W/R}{E_b/N_0} - \sum_{i=1}^{N_s} \frac{\eta}{S_{T_{1_i}}} \triangleq \delta'. \qquad (22)$$

Generally, the background noise is well below the total largest received cell site signal power, so the second sum is almost negligible. Note the similarity to δ in (15) for the reverse link. We shall take $\beta = 0.8$ as noted above to provide 20 % of the transmitted power in the sector to the pilot signal, and the required $E_b/N_0 = 5$ dB to ensure BER $\leq 10^{-3}$. This reduction of 2-dB relative to the reverse link is justified by the coherent reception using the pilot as reference, as compared to the noncoherent modem in the reverse link. Note that this is partly offset by the 1-dB loss of power due to the pilot.

Since the desired performance (BER $\leq 10^{-3}$) can be achieved with N_s subscribers per sector provided (22) is satisfied with $E_b/N_0 = 5$ dB, capacity is again a random variable whose distribution is obtained from the distribution of variable f_i. That is, the BER can not be achieved for all N_s users/sector if the N_s subscribers combined exceed the total allocation constraint of (22). Then following (15),

$$1 - P = \Pr\left(\text{BER} > 10^{-3}\right) = \Pr\left(\sum_{i=1}^{N_s} f_i > \delta'\right). \qquad (23)$$

But unlike the reverse link, the distribution of the f_i, which depends on the sum of ratios of ranked log-normal random variables, does not lend itself to analysis. Thus we resorted to Monte Carlo simulation, as follows.

For each of a set of points equally spaced on the triangle shown in Fig. 2(b), the attenuation relative to its own cell center and the 18 other cell centers comprising the first three neighboring rings was simulated. This consisted of the product of the fourth power of the distance and the log-normally distributed attenuation

$$10^{(\xi_k/10)}r_k^{-4}, \qquad k = 0, 1, 2 \cdots, 18.$$

Note that by symmetry, the relative position of users and cell sites is the same throughout as for the triangle of Fig. 2(b). For each sample, the 19 values were ranked to determine the maximum (S_{T_1}), after which the ratio of the sum of all other 18 values to the maximum was computed to obtain $f_i - 1$. This was repeated 10 000 times per point for each of 65 equally spaced points on the triangle of Fig. 2(b). From this, the histogram of $f_i - 1$ was constructed, as shown in Fig. 4.

From this histogram the Chernoff upper bound on (23) is obtained as

$$1 - P = < \min_{s>0} E \exp\left[s \sum_{i=1}^{N_s} f_i - s\delta'\right]$$

$$= \min_{s>0} \left[(1 - \propto) + \propto \sum_k P_k \exp\left(sf_k\right)\right]^{N_s} e^{-s\delta'} \quad (24)$$

where P_k is the probability (histogram value) that f_i falls in the kth interval. The result of the minimization over s based on the histogram of Fig. 4, is shown in Fig. 5.

VII. CONCLUSIONS AND COMPARISONS

Figs. 3 and 5 summarize performance of reverse and forward links. Both are theoretically pessimistic (upper bounds on probability). Practically, both models assume only moderately accurate power control.

The parameters for both links were chosen for the following reasons. The allocated total spread bandwidth $W = 1.25$ MHz represents ten percent of the total spectral allocation, 12.5 MHz, for cellular telephone service of each service provider. which as will be discussed below, is a reasonable fraction of the band to devote initially to CDMA and also for a gradual incremental transition from analog FM/FDMA to digital CDMA. The bit rate $R = 8$ kb/s is that of an acceptable nearly toll quality vocoder. The voice activity factor, 3/8, and the standard sectorization factor of 3 are used. For the reverse channel, the received SNR per user $S/\eta = -1$ dB reflects a reasonable subscriber transmitter power level. In the forward link, 20 % of each site's power is devoted to the pilot signal for a reduction of 1 dB ($\beta = 0.8$) in the effective processing gain. This ensures each pilot signal (per sector) is at least 5 dB above the maximum subscriber signal power. The role of the pilot, as noted above, is critical to acquisition, power control in both directions and phase tracking as well as for power allocation in the forward link. Hence, the investment of 20 % of total cell site power is well justified. These choices of parameters imply the choices $\delta = 30$ and $\delta' \approx 38$ in (16) and (24) for reverse and forward links, respectively.

With these parameters, according to Fig. 3, the reverse link can support over 36 users/sector or 108 users/cell, with 10^{-3} bit error rates better than 99 % of the time. This number becomes 44 users/sector or 132 users/cell if the neighboring cells are kept to half of this loading. The forward link according to Fig. 5, can do the same or better for 38 users/sector or 114 users/cell.

Clearly, if the entire cellular allocation is devoted to CDMA, these numbers are increased tenfold. Similarly, if a lower bit rate vocoder algorithm is developed, or if narrower sectors are employed, the number of users may be increased further.

Remaining with the parameters assumed, interesting comparisons can be drawn to existing analog FM/FDMA cellular systems as well as other proposed digital systems. First, the former employs 30-kHz channel allocation, and assuming 3

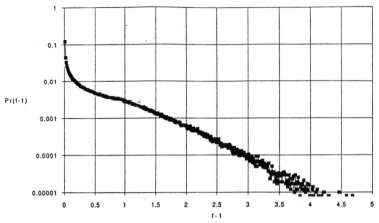

Fig. 4. Histogram of forward power allocation.

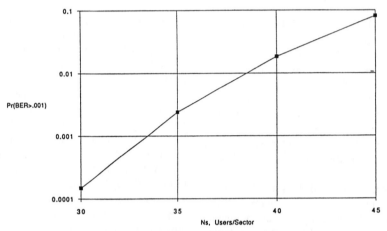

Fig. 5. Forward link capacity/sector. ($W = 1.25$ MHz, $R = 8$ kb/s,
voice activity $= 3/8$, pilot power $= 20$ %).

sectors/cell, requires each of the six contiguous cells in the first ring about a given cell to use a different frequency band. This results in a "frequency reuse factor" of $1/7$. Hence, given the above parameters, the number of channels in a 1.25-MHz band is slightly less than 42, and with a frequency reuse factor of $1/7$, this results in slightly less than 6 users/cell for a 1.25-MHz band. Thus, CDMA offers at least an eighteenfold increase in capacity. Note further that use of CDMA over just ten percent of the band supports over 108 users/cell whereas analog FM/FDMA supports only 60 users/cell using the entire 12.5 MHz band. Thus by converting *only 10 % of the band* from analog FDMA to digital CDMA, overall capacity is increased *almost threefold*.

Comparisons of CDMA with other digital systems are more speculative. However, straightforward approaches such as narrower frequency channelization with FDMA or multiple time slotting with TDMA can be readily compared to the analog system. The proposed TDMA standard for the U.S. is based on the current 30-kHz channelization but with sharing of channels by three users each of whom is provided one of three TDMA slots. Obviously, this triples the analog capacity but falls over a factor of 6 short of CDMA capacity.

In summary, properly augmented and power-controlled multiple-cell CDMA promises a quantum increase in current cellular capacity. No other proposed scheme appears to even approach this performance. Other advantages of CDMA not treated here include inherent privacy, flexibility in supporting multiple services and multiple voice and data rates, lower average transmit power requirements and soft limit on capacity, since if the bit error rate requirement is relaxed more users can be supported. With all these inherent advantages, CDMA appears to be the logical choice henceforth for all cellular telephone applications.

APPENDIX I
REVERSE LINK OUTER-CELL INTERFERENCE

Outer-cell normalized interference, I/S, is a random variable defined by (7), (8), and (9), and upper bounded by replacing (8) by (8'). Then the upper bound on its first moment, taking into account also the voice activity factor of the outer-cell subscribers, $\propto$, becomes

$$E(I/S) \leq \iint_{\text{sector}} \left(\frac{r_m}{r_0}\right)^4 E(\psi \cdot 10^{x/10} \emptyset(\chi, r_0/r_m)) \rho \, dA$$

where r_m is defined by (8') for every point in the sector, $\psi = 1$ with probability $\propto$ and 0 with probability $(1 - \propto)$, and $\chi = \xi_0 - \xi_m$ is a Gaussian random variable of zero mean and variance $2\sigma^2$ with $\oslash(\chi, r_0/r_m)$ defined by (9),

$$\oslash(\chi, r_0/r_m) = \begin{cases} 1, & \text{if } \chi \le 40 \log(r_0/r_m) \\ 0, & \text{otherwise.} \end{cases}$$

The expectation is readily evaluated as

$$\propto f\left(\frac{r_m}{r_0}\right) \triangleq E(\psi)E\left(e^{\chi \ln 10/10}\oslash(\chi, r_0/r_m)\right)$$

$$= \propto \int_{-\infty}^{40 \log r_0/r_m} e^{\chi \ln 10/10}\frac{e^{-x^2/4\sigma^2}}{\sqrt{4\Pi\sigma^2}}\,dx$$

$$= \propto e^{(\sigma \ln 10/10)^2} \int_{-\infty}^{40 \log(r_0/r_m)}$$

$$\cdot \frac{\exp\left[-\frac{1}{2}\left(x/\sqrt{2\sigma^2} - \sqrt{2\sigma^2}\ln 10/10\right)^2\right]}{\sqrt{2\Pi(2\sigma^2)}}\,dx$$

$$= \propto e^{(\sigma \ln 10/10)^2}\left\{1 - Q\left[\frac{40 \log(r_0/r_m)}{\sqrt{2\sigma^2}}\right.\right.$$

$$\left.\left. - \sqrt{2\sigma^2}\,\frac{\ln 10}{10}\right]\right\}$$

which yields (10).

To evaluate $\text{var}(I/S)$, assuming the "spatial whiteness" of the blockage variable, we have

$$\text{var}(I/S) \le \iint_{\text{sector}} \left(\frac{r_m}{r_0}\right)^8$$

$$\cdot \text{var}\left(\psi \cdot 10^{\chi/10}\oslash(\chi, r_0/r_m)\right)\rho\,dA.$$

Rewriting the variance in the integral as

$$E(\psi^2)E\left[10^{2\chi/10}\oslash^2(\chi, r_0/r_m)\right]$$

$$- \left\{E(\psi)E\left[10^{\chi/10}\oslash(\chi, r_0/r_m)\right]\right\}^2$$

$$= \propto g\left(\frac{r_m}{r_0}\right) - \propto^2 f^2\left(\frac{r_m}{r_0}\right)$$

where $f(r_m/r_0)$ was derived above and

$$\propto g\left(\frac{r_m}{r_0}\right) = E(\psi^2)E\left[e^{\chi \ln 10/5}\oslash^2(\chi, r_0/r_m)\right]$$

$$= \propto e^{(\sigma \ln 10/5)^2}\left\{1 - Q\left[\frac{40 \log(r_0/r_m)}{\sqrt{2\sigma^2}}\right.\right.$$

$$\left.\left. - \frac{\sqrt{2\sigma^2}\ln 10}{5}\right]\right\}$$

which yields (11).

ACKNOWLEDGMENT

The authors gratefully acknowledge the contribution of Dr. Audrey Viterbi and Dr. Jack Wolf.

REFERENCES

[1] A. J. Viterbi, "When not to spread spectrum—A sequel," *IEEE Communications Mag.* vol. 23, pp. 12–17, Apr. 1985.
[2] K. S. Gilhousen, I. M. Jacobs, R. Padovani, and L. A. Weaver, "Increased capacity using CDMA for mobile satellite communications," *IEEE Trans. Select. Areas Commun.*, vol. 8, pp. 503–514, May 1990.
[3] P. T. Brady, "A statistical analysis of on–off patterns in 16 conversations," *Bell Syst. Tech. J.*, vol. 47, pp. 73–91, Jan. 1968.
[4] G. R. Cooper and R. W. Nettleton, "A spread spectrum technique for high capacity mobile communications," *IEEE Trans. Veh. Technol.*, vol. VT-27, pp. 264–275, Nov. 1978.

Klein S. Gilhousen (M'86–SM'91) was born in Coshocton, OH, in 1942. He received the B.S. degree in electrical engineering from the University of California, Los Angeles, in 1969.

In 1985, he became a cofounder and Vice President for Systems Engineering for QUALCOMM, Inc., San Diego, CA. his professional interests include satellite communications, cellular telephone systems, spread spectrum systems, communications privacy, communications networks, video transmission systems, error correcting codes and modem design. He holds six patents in these areas with five more applied for. Prior to joining QUALCOMM, he was Vice President for Advanced Technology at M/A-COM LINKABIT San Diego, CA, from 1970 to 1985 and Senior Engineer at Magnavox Advanced Products Division, Torrance, CA from 1966 to 1970.

Irwin M. Jacobs (S'55–M'60–F'74) received the B.E.E. degree in 1956 from Cornell University, Ithaca, NY, and the M.S. and Sc.D. degrees in electrical engineering from the Massachusetts Institute of Technology, Cambridge, in 1957 and 1959, respectively.

On July 1, 1985, he became a founder and the Chairman and President of QUALCOMM, Inc. From 1959 to 1966, he was an Assistant/Associate Professor of Electrical Engineering at M.I.T. and a staff member of the Research Laboratory of Electronics. During the academic year 1964–1965, he was a NASA Resident Research Fellow at the Jet Propulsion Laboratory. In 1966, he joined the newly formed Department of Applied Electrophysics, now the Department of Electrical Engineering and Computer Science, at the University of California, San Diego (UCSD). IN 1972, he resigned as Professor of Information and Computer Science to devote full time to LINKABIT Corporation. While at M.I.T., he coauthored a basic textbook in digital communications, *Principles of Communication Engineering*, published first in 1965, and still in active use. He retains academic ties through memberships on the Cornell University Engineering council, the visiting committees of the M.I.T. Laboratory for Information and Decision Systems, as Academic/Scientific member of the Technion International Board of Governors, and as a Board Member of the UCSD Green Foundation for Earth Sciences. He is a past Chairman of the Scientific Advisory Group for the Defense Communications Agency, and of the Engineering Advisory Council for the University of California. He has served on the governing boards of the IEEE Communications Society, the IEEE Group on Information Theory, and as General Chairman of NTC'74. In 1980, he and Dr. A. Viterbi were jointly honored by the American Institute of Aeronautics and Astronautics (AIAA) with their biannual award "for an outstanding contribution to aerospace communications." In 1984, he received the Distinguished Community Service Award for the Anti-Defamation League of B'nai B'rith. The local American Electronics Association's First Annual ExcEL Award was presented to Dr. Jacobs in 1989 for excellence in electronics and his "dedication and innovation, which have set the highest standards in the local electronics industry."

Dr. Jacobs was elected a member of the National Academy of Engineering for "Contributions to communication theory and practice, and leadership in high-technology product development." He is a member of Sigma Xi, Phi Kappa Phi, Eta Kappa Nu, and the Association for Computing Machinery (ACM).

Roberto Padovani (S'83–M'84–SM'91) received the Laurea degree from the University of Padova, Italy, and the M.S. and Ph.D. degrees from the University of Massachusetts, Amherst, in 1978, 1983, and 1985, respectively, all in electrical and computer engineering.

In 1986 he joined QUALCOMM, Inc. and he is now Director of System Engineering in the Engineering Department. As a member of the engineering department of QUALCOMM, Inc., he has been involved in the design and development of CDMA modems for the mobile satellite channel, various satllite modems, and VLSI Viterbi decoders. He is currently involved in the development of the CDMA digital cellular telephone system. In 1984 he joined M/A-COM Linkabit, San Diego where he was involved in the design and development of satellite communication systems, secure video systems, and error-correcting coding equipment.

Andrew J. Viterbi (S'54–M'58–SM'63–F'73) received the S.B. and S.M. degrees in electrical engineering from the Massachusetts Institute of Technology, Cambridge, in 1957, and the Ph.D. degree in electrical engineering from the University of Southern California, Los Angeles, in 1962.

He has devoted approximately equal segments of his career to academic research, industrial development, and entrepreneurial activities. In 1985, he became a founder and Vice Chairman and Chief Technical Officer of QUALCOMM, Inc., a company concentrating on mobile satellite communications for both commercial and military applications. In 1968, he cofounded LINKABIT Corporation. He was Executive Vice President of LINKABIT from 1974 to 1982. In 1982, he took over as President of M/A-COM LINKABIT, Inc. From 1984 to 1985, he was appointed Chief Scientist and Senior Vice President of M/A-COM, Inc. After graduating from M.I.T., he was a member of the project team at C.I.T. Jet Propulsion Laboratory which designed and implemented the telemetry equipment on the first successful U.S. satellite, Explorer I. From 1963–1973 he was a Professor with the UCLA School of Engineering and Applied Science. He did fundamental work in digital communication theory and wrote books on the subject, for which he received numerous professional society awards and international recognition. These include three paper awards, culminating in the 1968 IEEE Information

Theory Group Outstanding Paper Award. He has also received three major society awards: the 1975 Christopher Columbus International Award (from the Italian National Research Council sponsored by the City of Genoa); the 1980 Aerospace Communications Award jointly with Dr. I. Jacobs (from AIAA); and the 1984 Alexander Graham Bell Medal (from IEEE sponsored by AT&T) "for exceptional contributions to the advancement of telecommunications." He has a part-time appointment as Professor of Electrical and Computer Engineering at the University of California, San Diego.

Dr. Viterbi is a member of the National Academy of Engineering.

Lindsay A. Weaver, Jr., received the S.B. and S.M. degrees from the Massachusetts Institute of Technology, Cambridge, in 1976 and 1977, respectively.

He is Vice President of Engineering at QUALCOMM, Inc. He was a key member of the design teams at QUALCOMM for the Mobile Satellite CDMA voice system, the OmniTRACS mobile satellite messaging system (hybrid frequency hopping and direct sequence), and the CDMA cellular telephone system. He has also lead projects developing FDMA modems, Viterbi decoders, highspeed packet switches, and satellite video scrambling.

Charles E. Wheatley III (SM'91) received the B.S. degree in physics from the California Institute of Technology, Pasadena, in 1956, the M.S. degree in electrical engineering from the University of Southern California, Los Angeles, in 1958 and the Ph.D. degree in electrical engineering from the University of California, Los Angeles in 1972.

He joined QUALCOMM, Inc., in 1987 as Principal Engineer, and has worked on both government and commercial programs, concentrating on system performance issues. The last two years have been spent working on RF hardware and system design for CDMA cellular phone applications. He has over 30 years of experience in RF satellite-based communications systems. His areas of expertise include time/frequency, anti-jam and LPI, all of which he has applied to a wide variety of systems. Prior to joining QUALCOMM, he held the position of Technical Assistant Vice President at M/A COM LINKABIT, Inc. in San Diego, CA.

Advantages of CDMA and Spread Spectrum Techniques over FDMA and TDMA in Cellular Mobile Radio Applications

Peter Jung, *Member, IEEE*, Paul Walter Baier, *Senior Member, IEEE*, and Andreas Steil

Reprinted from *IEEE Transactions on Vehicular Technology*, Vol. 42, No. 3, pp. 357-364, Aug., 1993.

Abstract—In this paper, a unified theoretical method for the calculation of the radio capacity of multiple-access schemes such as FDMA (frequency-division multiple access), TDMA (time-division multiple access), CDMA (code-division multiple access) and SSMA (spread-spectrum multiple access) in noncellular and cellular mobile radio systems shall be presented for AWGN (additive white Gaussian noise) channels. The theoretical equivalence of all the considered multiple-access schemes is found.

However, in a fading multipath environment, which is typical for mobile radio applications, there are significant differences between these multiple-access schemes. These differences are discussed in an illustrative manner revealing several advantages of CDMA and SSMA over FDMA and TDMA. Furthermore, novel transmission and reception schemes called coherent multiple transmission (CMT) and coherent multiple reception (CMR) are briefly presented.

I. INTRODUCTION

IN cellular mobile radio systems, the problem of multiple access can be solved by the basic multiple-access schemes FDMA (frequency-division multiple access), TDMA (time-division multiple access), CDMA (code-division multiple access) and SSMA (spread-spectrum multiple access) or by combinations thereof [1], [2]. When selecting a multiple-access scheme, perhaps the most important question is the number of admissible users per cell for a given available total bandwidth, for given radio propagation conditions, and for a required transmission quality. This number is termed the cellular radio capacity. In several papers [3]–[7], the cellular radio capacity of cellular radio systems using special multiple-access schemes has been studied. However, a unified theory of cellular radio capacity which is applicable independently of the used multiple-access scheme has not yet been presented.

Independently of the used multiple-access scheme, all reasonably well-designed multiple access schemes are theoretically equivalent if AWGN (additive white Gaussian noise) channels are considered. This shall be demonstrated in Sections II and III, both for noncellular and for cellular systems by the calculation of the radio capacity. When pursuing such a principle aim, effects such as transmitter power control, receiver synchronization, and channel estimation, which are important in practical system designs, cannot be considered because such effects are beyond the scope

Manuscript received February 18, 1991; revised September 26, 1991.

The authors are with the Research Group for RF Communications, University of Kaiserslautern, D-6750, Kaiserslautern, Germany.

IEEE Log Number 9207169.

of a paper dealing with basic considerations. Nevertheless, the authors believe that this paper is helpful even for the practically oriented engineer when comparing the cellular radio capacity of competing multiple-access schemes.

If the channel exhibits time variance and frequency selectivity, which are typical for a mobile radio environment, the situation is different. In order to give an impression of this issue, Section IV shall present a brief introduction to fading multipath channels in mobile radio. In Section V, the diversity potential of the different multiple-access schemes shall be presented. In the case of such channels, multiple-access schemes having at lest a CDMA or SSMA component are superior to other multiple-access schemes because by CDMA and SSMA, the frequency selectivity of the radio channel, which severely impairs the system performance, can be averaged out.

In Section VI, further advantages of CDMA and SSMA over FDMA, and TDMA shall be discussed. Furthermore, novel transmission and reception schemes called coherent multiple transmission (CMT) and coherent multiple reception (CMR) are briefly presented.

II. BASIC MULTIPLE-ACCESS PROBLEM

Firstly, the basic multiple-access problem is considered for a noncellular system. This problem consists in dividing up the available frequency-time space among z users in such a way that there is no interference between the users. If a transmission interval of duration T is considered, and if the available total bandwidth of the noncellular system is B, a function $\Phi_\mu(t)$, $\mu = 1, 2, \cdots, z$, from a finite set of z orthonormal bandpass functions of ensemble duration T and ensemble bandwidth B can be exclusively assigned to each transmitter. The simultaneous limitation of both the duration and the bandwidth of these functions $\Phi_\mu(t)$ can be understood on the basis of the uncertainty principle, see e.g., [8]. Such an assignment is assumed in what follows. In this case, the number of admissible transmitters is

$$z \leq 2BT. \tag{1}$$

In the following, it is assumed that the time−bandwidth product BT is an integer, and therefore the equality in (1) holds:

$$z = 2BT. \tag{2}$$

Theoretically, an infinite number of sets of z orthonormal functions $\Phi_\mu(t)$ with ensemble duration T and ensemble bandwidth B exist. The most usual kinds of orthonormality are frequency orthonormality, time orthonormality and code orthonormality, which correspond to FDMA, TDMA, CDMA/SSMA, respectively. Also, combinations of FDMA, TDMA, CDMA, and SSMA are possible. Due to the close relationship between CDMA and SSMA, SSMA shall not be treated separately, although in contrast to CDMA in the case of SSMA, usually

$$z \ll 2BT \tag{3}$$

holds.

It is assumed that each transmitter uses its function $\Phi_\mu(t)$ as its individual carrier signal and that transmitter μ transmits information using the signal

$$s_\mu(t) = \sum_{\nu=-\infty}^{\infty} x_{\mu\nu}\Phi_\mu(t - \nu T), \qquad \mu = 1, 2, \cdots, z \tag{4}$$

where the factors $x_{\mu\nu}$ are samples of a Gaussian process $\{x_{\mu\nu}\}$ with zero mean. The signals $x_{\mu\nu}\Phi_\mu(t - \nu T)$ transmitted by transmitter μ generate the signals

$$a \cdot x_{\mu\nu}\Phi_\mu(t - \nu T - t_0), \qquad 0 < a < 1;$$
$$a, t_0 \in \mathbb{R} \tag{5}$$

which represent the attenuated and time-delayed versions of the transmitted signals at the corresponding receiver μ. The average energy E_μ per received signal $a \cdot x_{\mu\nu}\Phi_\mu(t - \nu T - t_0)$ is assumed to be the same for all μ, i.e., with $\mathrm{E}\{\cdot\}$ denoting the expectation

$$E = E_\mu = \mathrm{E}\left\{(a \cdot x_{\mu\nu})^2\right\}, \qquad \mu = 1, 2, \cdots, z. \tag{6}$$

In the case of FDMA, each function $\Phi_\mu(t)$ uses the total ensemble duration T; therefore, the duration T_u of the individual function $\Phi_\mu(t)$ is given by

$$T_\mathrm{u} = T. \tag{7}$$

In this case, the bandwidth B_u of the individual function $\Phi_\mu(t)$ assumes the value

$$B_\mathrm{u} = \frac{B}{z} = \frac{1}{2T}. \tag{8}$$

In the case of TDMA, each function $\Phi_\mu(t)$ uses the total ensemble bandwidth B, i.e., the bandwidth B_u of the individual function $\Phi_\mu(t)$ is

$$B_\mathrm{u} = B. \tag{9}$$

In this case, the duration T_u of the individual function $\Phi_\mu(t)$ is only the zth part of the ensemble duration T:

$$T_\mathrm{u} = \frac{T}{z} = \frac{1}{2B}. \tag{10}$$

With CDMA and SSMA, each function $\Phi_\mu(t)$ uses the total ensemble duration T as well as the total ensemble bandwidth B; therefore,

$$T_\mathrm{u} = T \tag{11}$$

and

$$B_\mathrm{u} = B \tag{12}$$

are valid.

On account of their orthogonality, the signals from the z transmitters can be perfectly separated in the corresponding receivers by correlation or matched filtering, if all transmitters and receivers are synchronized, which shall be the case as already mentioned above. After the separation, with the average energy E of the received signals and with the one-sided spectral power density N_0 of the thermal noise, the signal-to-noise ratio γ is given by

$$\gamma = \frac{E}{N_0/2}. \tag{13}$$

The average information transmitted per signal $x_{\mu\nu}\Phi_\mu(t - \nu T)$ is $0.5 \cdot \log_2(1 + \gamma)$ if the thermal noise is assumed to be Gaussian [9]. Because every T seconds a signal $x_{\mu\nu}\Phi_\mu(t - \nu T)$ is transmitted, the channel capacity per user assumes the value

$$C = \frac{1}{2T} \cdot \log_2(1 + \gamma). \tag{14}$$

Substituting $2T$ in (14) by z/B (see (2)), yields

$$z \cdot \frac{C}{B} = \log_2(1 + \gamma). \tag{15}$$

For a given available total bandwidth B, a required channel capacity C per user and a given signal-to-noise ratio γ, the number z of admissible users can be calculated from (15).

If the signals from the z transmitters are not perfectly separated at the corresponding receivers, the signal-to-noise ratio is given by

$$\gamma = \frac{E}{f \cdot E + N_0/2} \tag{16}$$

where the term $f \cdot E$ represents the interference from the $(z - 1)$ other users. Nonperfect signal separation is the case if e.g., the signals $\Phi_\mu(t)$ are not perfectly orthogonal and at the same time the signal separation is performed by conventional matched filtering instead of applying optimum unbiased estimation. In such cases, f may assume rather large values, which leads to a considerable decrease of γ.

III. CELLULAR SYSTEMS IN THE CASE OF IDEAL RADIO PROPAGATION

In the case of cellular systems, the set of $2BT$ orthonormal functions $\Phi_\mu(t)$ has to be subdivided into subsets of size

$$z = \frac{2BT}{r} \tag{17}$$

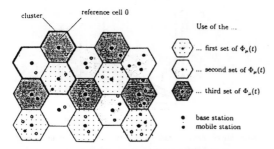

Fig. 1. Part of a cellular system with hexagonal cells for $r = 3$.

with r being the reuse factor and z being the number of users per cell. A fraction $1/r$ of the cells use the same subset of orthonormal functions $\Phi_\mu(t)$. Whereas it is possible in each cell to separate the transmitted signals $x_{\mu\nu}\Phi_\mu(t - \nu T)$ originating from this cell from one another, it is impossible to separate signals coming form the other cells using the same orthonormal functions $\Phi_\mu(t)$. Rather, these signals have to be treated as interference which has a similar effect as thermal noise [1].

Regular cellular systems are considered in which each cell contains a base station in the center of the cell and z mobile stations communicating with the base station of the cell. Conventionally, the shape of a cell in such a cellular scheme is assumed to be a regular hexagon [1], [10]. In Fig. 1, a part of a cellular system with such hexagonal cells is schematically shown for the case $r = 3$. The base stations and the mobile stations of those cells displayed with the same texture use the same set of orthonormal functions $\Phi_\mu(t)$. Each group of three neighboring cells using disjointed sets of $\Phi_\mu(t)$ is combined to form a cluster.

It is assumed that, by a suitable power control, all mobile stations belonging to a certain cell receive equal powers from their base station, and that all mobile stations generate equal powers in the receiver of their base station. Cell 0 with its base station BS_0 is taken as the reference cell. A worst-case situation is considered in which the interference power arriving at the receivers in cell 0 from extra-cell transmitters, i.e., from transmitters in cells other than cell 0, is maximum. In order to obtain this worst-case situation, both in the case of the mobile stations calling the base stations (uplink) and in the case of the base stations calling the mobile stations (downlink), all extra-cell transmitters must transmit maximum power, which means that the extra-cell mobile stations must have maximum distance R from their base stations (Condition I). In addition to this condition, the intra-cell and extra-cell mobile stations have to be arranged in such a way that the interference power received by the receivers in cell 0 is maximized (Condition II).

Following the discussion of the preceding section, the total interference power I in a cellular interference scenario as referred to here is given by

$$I = \frac{E}{T} \cdot f(r, \alpha) \tag{18}$$

with α being the attenuation coefficient [1]. In (18), the function $f(r, \alpha)$ is introduced, for which the term cellular

interference function is proposed. In order to determine the interference power I, $f(r, \alpha)$ must be calculated.

In what follows, only the uplinks in a hexagonal cellular system comparable to the one shown in Fig. 1 are considered. The cellular interference function $f(r, \alpha)$ is dependent on the distances between the extra-cell transmitters and the base station of the reference cell 0. Due to the regular structure of such a hexagonal cellular system, setting out from cell 0, the cellular system can be divided into six 60° segments. In the above-mentioned worst-case situation, these 60° segments are equivalent. Therefore, only one 60° segment must be evaluated. Now it is advantageous to introduce an affine coordinate system with the basis vectors

$$\vec{n}_0 = D \cdot \begin{pmatrix} \cos(0 \cdot 60°) \\ \sin(0 \cdot 60°) \end{pmatrix} = D \cdot \begin{pmatrix} 1 \\ 0 \end{pmatrix} \tag{19}$$

and

$$\vec{n}_2 = D \cdot \begin{pmatrix} \cos(2 \cdot 60°) \\ \sin(2 \cdot 60°) \end{pmatrix} = \frac{D}{2} \cdot \begin{pmatrix} -1 \\ \sqrt{3} \end{pmatrix} \tag{20}$$

where D denotes the distance between the two adjacent cells with base stations and mobile stations using the same set of $\Phi_\mu(t)$ [1]. According to [1], with the cell radius R_{cell}, D is given by

$$D = R_{\text{cell}} \cdot \sqrt{3r}. \tag{21}$$

These cells are called co-channel cells [1]. Fig. 2 gives a graphical representation of the considered situation. Each base station is associated with a unique vector $\vec{R}_{uv}$ with

$$\vec{R}_{uv} = u \cdot \vec{n}_0 + v \cdot \vec{n}_2,$$
$$u \in \mathsf{N}, \qquad v \in \mathsf{N}_0, \qquad v < u \tag{22}$$

of length

$$R_{uv} = R_{\text{cell}} \cdot \sqrt{3r(u^2 + v^2 - uv)},$$
$$u \in \mathsf{N}, \qquad v \in \mathsf{N}_0, \qquad v < u. \tag{23}$$

In order to simplify the mathematical solution, the hexagons are approximated by circular regions of radius R which have the same area as the hexagons. The cell radius R_{cell} can then be expressed in the following way:

$$R_{\text{cell}} = \frac{\sqrt{2\pi\sqrt{3}}}{3} \cdot R. \tag{24}$$

Substituting (24) into (23) yields

$$R_{uv} = R \cdot \sqrt{\frac{2\pi\sqrt{3}}{3}r(u^2 + v^2 - uv)},$$
$$u \in \mathsf{N}, \qquad v \in \mathsf{N}_0, \qquad v < u. \tag{25}$$

The distances $d_{uv}^{(r)}$ between the extra-cell transmitters in the co-channel cells and the base stations of the reference cell 0 are now given by

$$d_{uv}^{(r)} = R_{uv} - R = R \cdot \left(\sqrt{\frac{2\pi\sqrt{3}}{3}r(u^2 + v^2 - uv)} - 1 \right),$$
$$u \in \mathsf{N}, \qquad v \in \mathsf{N}_0, \qquad v < u. \tag{26}$$

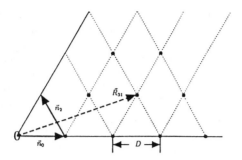

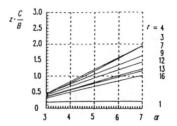

Fig. 4.   Normalized cellular radio capacity $z \cdot C/B$ versus α with r as a parameter for the case of vanishing N_0.

• base station of a co-channel cell

Fig. 2.   Affine coordinate system in a 60° segment of a cellular system similar to the one in Fig. 1, used for the determination of $\vec{R}_{uv}$, $d_{uv}^{(r)}$, and $f(r, \alpha)$.

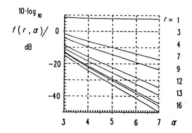

Fig. 3.   Cellular interference function $f(r, \alpha)$ versus α with r as a parameter.

Using (26), the cellular interference function $f(r, \alpha)$ of the whole cellular system yields

$$f(r, \alpha) = 6 \cdot \sum_{u=1}^{\infty} \sum_{v=0}^{u-1} \left(\frac{R}{d_{uv}^{(r)}} \right)$$

$$= 6 \cdot \sum_{u=1}^{\infty} \sum_{v=0}^{u-1} \left(\frac{R}{d_{uv}^{(r)}} \right) \left(\sqrt{\frac{2\pi\sqrt{3}}{3} r(u^2 + v^2 - uv)} - 1 \right)^{-\alpha},$$
$$r > 1 \qquad (27)$$

in the case of the uplink. In order to reduce the error arising form the circular approximation of the hexagons in the case of $r = 1$, the interference power equal to $6E/T$ resulting from the six neighboring cells of the reference cell 0 is considered separately. In this case, $f(1, \alpha)$ is given by

$$f(1, \alpha) = 6 \cdot \left\{ 1 + \sum_{u=2}^{\infty} \sum_{v=0}^{u-1} \left(\sqrt{\frac{2\pi\sqrt{3}}{3}(u^2 + v^2 - uv)} - 1 \right)^{-\alpha} \right\}.$$
$$(28)$$

For the considered cellular system, Fig. 3 shows $f(r, \alpha)$ according to (27) and (28) versus α with r as a parameter. As expected, $f(r, \alpha)$ decreases with increasing r and α.

If the extra-cell interference is modeled as white noise over the available total bandwidth B, the one-sided spectral interference power density assumes

$$I_0 = \frac{z \cdot I}{B/r} = \frac{f(r, \alpha) \cdot r \cdot z \cdot E/T}{B}$$
$$= 2E \cdot f(r, \alpha) \qquad (29)$$

by using (17) and (18). Now, instead of (13) the expression:

$$\gamma = \frac{E}{I_0/2 + N_0/2} \qquad (30)$$

is obtained for the signal-to-noise ratio after the signal separation. Substituting (29) into (30) yields

$$\gamma = \frac{E}{E \cdot f(r, \alpha) + N_0/2}. \qquad (31)$$

With this expression for γ, and considering the fact that the available time–bandwidth product per user is BT/r, instead of (15), the expression:

$$z \cdot \frac{C}{B} = \frac{1}{r} \cdot \log_2 \left(1 + \frac{E}{E \cdot f(r, \alpha) + N_0/2} \right) \qquad (32)$$

is obtained. For each quadruple r, α, N_0 and E, the expression $z \cdot C/B$ attains a certain value. It is recommended to term this quantity the normalized cellular radio capacity. If no thermal noise has to be considered, (32) reduces to

$$z \cdot \frac{C}{B} = \frac{1}{r} \cdot \log_2 \left(1 + \frac{1}{f(r, \alpha)} \right). \qquad (33)$$

By substituting $f(r, \alpha)$ according to (27) and (28) into (33), the normalized cellular radio capacity $z \cdot C/B$ for the cellular system shown in Fig. 1 is obtained. In Fig. 4, $z \cdot C/B$ is depicted versus α with r as a parameter for vanishing N_0. As expected, $z \cdot C/B$ increases with increasing α. With respect to the dependence on r, it can be stated by inspection of Fig. 4 that the maximum normalized cellular radio capacity $(z \cdot C/B)|_{\max}$ is obtained for r equal to four for the considered example.

The theoretical results for AWGN channels presented in this section are independent of the chosen multiple-access scheme. Nevertheless, the situation changes for fading multipath channels typical for mobile radio applications. This shall be discussed in the following two sections.

IV. FADING MULTIPATH RADIO CHANNELS

The mobile radio channel can be characterized by its time-variant impulse response $\underline{h}(\tau, t)$ [11], [12]. A short impulse sent into the channel results in a finely structured response of duration T_M. Typical experimental results are given in [13] and the references therein. The parameter T_M is called delay window. The spreading of the transmitted impulse is caused by the fact that the transmitted signal reaches the receiver via a number of different paths on account of reflections, diffractions and scattering [11], [12]. The order of the delay window T_M is

$$T_M \approx \begin{cases} 0.3 \ \mu s, & \text{for indoor channels,} \\ 10 \ \mu s, & \text{for outdoor channels} \end{cases} \tag{34}$$

which corresponds to maximum path differences of 100 m and 3 km, respectively.

In order to characterize the mean energy spread caused by the mobile radio channel, the delay spread S_D is introduced as the standard deviation of the delay time parameter τ (see below):

$$S_D = \sqrt{\frac{1}{P_m}\int_0^{T_M}\tau^2|\underline{h}(\tau,t)|^2\,d\tau - \left(\frac{1}{P_m}\int_0^{T_M}\tau|\underline{h}(\tau,t)|^2\,d\tau\right)}$$

$$P_M = \int_0^{T_M}|\underline{h}(\tau,t)|^2\,d\tau. \tag{35}$$

The order of the delay spread S_D is

$$S_D \approx \begin{cases} 10\cdots50 \ \text{ns}, & \text{for indoor channels} \\ 0.1\cdots5.0 \ \mu s, & \text{for outdoor channels}. \end{cases} \tag{36}$$

Due to the motion of the mobile transceiver and the scatterers in the surrounding environment, the dependence of $\underline{h}(\tau,t)$ on the time t results. The strength of the time dependence is closely related to the velocity v of the mobile stations, which also causes a Doppler shift of the transmitted frequency spectrum. Nevertheless, the impulse responses $\underline{h}(\tau,t+\Delta)$ and $\underline{h}(\tau,t)$ are similar for small increments Δ because $\underline{h}(\tau,t)$ is varying continuously versus time. However, the impulse response $\underline{h}(\tau,t)$ may be completely different after a certain minimum time T_{coh} has elapsed, which is typical for the channel. The parameter T_{coh} is called coherence time of the channel [14]. The time dependence of the impulse response results from the motion of the transmitter and/or the receiver [11], [12]. With the velocity v of the mobile station and the wavelength λ_0 at center frequency, T_{coh} can be approximated by

$$T_{coh} \approx \frac{\lambda_0}{2\cdot v}. \tag{37}$$

The approximation (37) is based on the fact that the impulse response $\underline{h}(\tau,t)$ may look entirely different when the mobile station has changed its position by half the wavelength λ_0. With λ_0 equal to 0.3 m and v equal to 5 km/h for the indoor channel and v equal to 50 km/h for the outdoor channel, one obtains from (37)

$$T_{coh} \approx \begin{cases} 110 \ \text{ms}, & \text{for the indoor channel} \\ 11 \ \text{ms}, & \text{for the outdoor channel}. \end{cases} \tag{38}$$

The typical symbol durations T in mobile speech communication are smaller than T_{coh} according to (38) [1]. Therefore, for the duration T of one symbol the channel can be considered as time invariant.

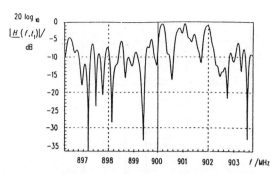

Fig. 5. Part of the absolute value of a sample $\underline{H}(f,t_1)$ of the transfer function $\underline{H}(f,t)$ for a typical urban channel.

By Fourier transform of the time-variant impulse response $\underline{h}(\tau,t)$, the time-variant transfer function

$$\underline{H}(f,t) = \int_{-\infty}^{\infty}\underline{h}(\tau,t)\cdot\exp[-j2\pi f\tau]\,d\tau \tag{39}$$

of the channel is obtained. In Fig. 5, part of a sample of the transfer function $\underline{H}(f,t)$ around a center frequency of 900 MHz for a typical urban channel at a time instant t_1 is displayed. The fine structure of $\underline{H}(f,t)$ along the frequency axis is determined by the delay spread S_D and the delay window T_M and has the approximate structure:

$$\begin{aligned} B_{coh} &\leq 1/T_M, \\ B_{coh} &\approx 1/[8S_D] \end{aligned} \tag{40}$$

[12], [14]. For a typical urban channel the parameters S_D and T_M assume the following values:

$$\begin{aligned} T_M &\approx 3\cdots5 \ \mu s \\ S_D &\approx 1 \ \mu s \end{aligned} \tag{41}$$

thus resulting in

$$125 \ \text{kHz} \approx B_{coh} \leq 333 \ \text{kHz}. \tag{42}$$

The width B_{coh} is termed coherence bandwidth [14]. With T_M and S_D according to (34) and (36), respectively, one obtains typical values for the coherence bandwidth as follows:

$$B_{coh} \approx \begin{cases} 3 \ \text{MHz}, & \text{for indoor channels} \\ 0.1 \ \text{MHz}, & \text{for outdoor channels}. \end{cases} \tag{43}$$

The transfer function $\underline{H}(f,t)$ can be considered as a sample function of a two-dimensional stationary and ergodic complex process. The width of the main autocorrelation peak of this process can be approximated by B_{coh} in the f-direction and by T_{coh} in the t-direction. Consequently, if a grid of widths B_{coh} and T_{coh}, respectively, is imposed on the frequency–time plane (see Fig. 6) the values of $\underline{H}(f,t)$ in different time–frequency elements of this grid are mutually uncorrelated.

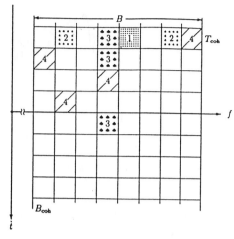

Diversity principles:
1 none
2 spectral
3 temporal
4 spectral and temporal

Fig. 6.   Uncorrelated time-frequency elements.

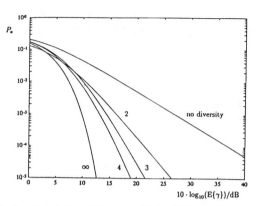

Fig. 7.   Typical plot of P_e versus $E\{\gamma\}$ with L as a parameter.

V. COMBATTING DEGRADATION CAUSED BY FADING MULTIPATH RADIO CHANNELS

The time and frequency dependences of the channel tend to degrade the system performance. This will be explained by the use of Fig. 6. In a fading multipath environment, with the average value of the received energy E per signal $x_{\mu\nu}\Phi_\mu(t - \nu T)$, the signal-to-noise ratio with the constant average value

$$E\{\gamma\} = \frac{E}{I_0/2} \qquad (44)$$

and with variance var$\{\gamma\}$ prevails at the receiver for an interference-limited cellular system (cf. Section III).

If the energy E is entirely concentrated within a single time–frequency element, (see case 1 in Fig. 6), for a constant $E\{\gamma\}$ the actual γ assumes quite different values, depending on the transfer function $\underline{H}(f, t)$ in the considered time–frequency element. Therefore, var$\{\gamma\}$ will be maximum. If E is distributed over two, three, or four time-frequency elements, the diversity parameter L being equal to two, three, and four, respectively (see cases 2–4 in Fig. 6) for still-constant average $E\{\gamma\}$, the variance var$\{\gamma\}$ will be reduced. The distribution of E over several time-frequency elements is termed diversity [15]. Diversity can be achieved by distributing the transmitted symbol energy along the frequency axis (see case 2 in Fig. 6) along the time axis (see case 3 in Fig. 6) or along both axes (see case 4 in Fig. 6).

In order to give a quantitative discussion, spectral diversity in the case of a multipath channel consisting of L paths that can be resolved in the receiver, with L depending on B_u, is considered. It can be shown that

$$var\{\gamma\} = \left(\frac{E}{I_0/2}\right)^2 \cdot \frac{1}{L}. \qquad (45)$$

[16]. Obviously, diversity reduces the dependence of the actual γ on the actual channel state.

The error probability P_e in the receiver is a strongly nonlinear convex function of γ. Consequently, for constant $E\{\gamma\}$, P_e decreases rapidly with decreasing var$\{\gamma\}$ which is equivalent to increasing diversity. Fig. 7 shows a typical plot of P_e as a function of $E\{\gamma\}$ with the diversity as a parameter for the case of coherently-detected binary orthogonal frequency shift-keying (FSK) [16], [17]. For infinite diversity, the minimum error probability P_e is obtained. As a conclusion it can be stated that the variance var$\{\gamma\}$ is reduced by diversity, which results in a decreasing error probability P_e.

In contrast to pure FDMA, the schemes of pure TDMA and of pure CDMA and SSMA are approaches to keep down the variance var$\{\gamma\}$ of γ at receivers operating in time-variant, frequency-selective channels, as long as B_u is considerably larger than the coherence bandwidth B_{coh} of the channel. However, CDMA and SSMA have a number of additional advantages which are not encountered in TDMA. Those advantages are:

- The Euclidean distances between symbols are virtually invariant to time displacements of the symbols, i.e., the distances do not decrease rapidly when time-displaced versions of the symbols are faced. Therefore, problems of intersymbol interference (ISI) and co-channel interference are less severe in CDMA and SSMA than in TDMA [2], [17].

- In order to maintain the required temporal order among the symbols, a complicated system organization is necessary in TDMA, but not in CDMA and SSMA [2], [17].

- CDMA and SSMA permit a CW-like operation of the transmitter power stages which leads to favorable circuitry [2], [17].

VI. FURTHER ADVANTAGES OF CDMA AND SSMA

The invariance of the Euclidean distances between symbols to time displacements entails a number of further advantages of CDMA and SSMA. One main advantage is that coherent multiple transmission and reception can be realized (cf. [17]).

In Fig. 8, three base stations $BS_{1,2,3}$ and one mobile station MS are depicted. In conventional systems, the mobile station

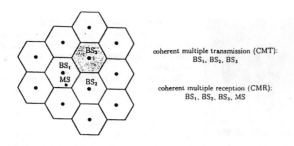

coherent multiple transmission (CMT):
BS₁, BS₂, BS₃

coherent multiple reception (CMR):
BS₁, BS₂, BS₃, MS

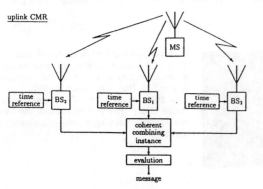

uplink CMR

Fig. 8. Coherent multiple transmission (CMT) and reception (CMR).

MS communicates with one of the base stations $BS_{1,2,3}$ and is handed over to another base station, if, by doing so, the communication quality can be improved.

In coherent multiple transmission (CMT), the base stations $BS_{1,2,3}$ surrounding the mobile station MS simultaneously transmit to the mobile station MS. All signals arriving at the mobile station MS are coherently combined by coherent multiple reception (CMR). CMR in the uplink is obtained if the signal transmitted by the mobile station MS is simultaneously received by several base stations and if the received signals are coherently combined to obtain the message (see Fig. 8). As a presupposition for CMT and CMR, reliable and fast digital communication between base stations is required, e.g., via optical-fiber links. However, it should be emphasized that this digital communication between base stations does not have to fulfill exact analog timing conditions if exact time standards, e.g., Rubidium clocks, are available at the base stations as shown in Fig. 8. In this case, the signals transmitted to the coherent combining instance can be supplied with the information of their absolute time of arrival at the base stations. This information can be used to perform the coherent combination digitally.

CMT and CMR are used to improve the exploitation of the transmitted power and therefore reduce the necessary transmitted power and the electromagnetic load of the air. The achievable gains by CMT and CMR are shown in Table I. With CMT there is also a reduction of the carrier-to-interference ratio C/I by approximately 3 dB due to the diminution of the transmission power in the base stations $BS_{1,2,3}$ by factor 3 [17]. With CMT, only three base stations in the first tier [17] contribute to the interference, whereas without CMT there are six interferes. The latter situation is considered in [5].

TABLE I
MINIMUM GAINS IN dB BY CMT AND CMR

	Antenna at $BS_{1,2,3}$	
	Omnidirectional	Directional (60°)
Downlink	0	7.8
Uplink	4.8	12.6

Especially when directional antennas having an angular beamwidth of 60° are used by the base stations $BS_{1,2,3}$, which is feasible in the configuration shown in Fig. 8, the gains are considerable. Additional favorable features of CMT and CMR are reduced shadowing and the possibility to locate the mobile station MS.

VII. CONCLUSIONS

In the present paper, a unified theoretical approach to the calculation of the normalized cellular radio capacity for multiple-access schemes in cellular mobile radio applications has been introduced in the case of AWGN channels. The considered multiple-access schemes FDMA, TDMA, CDMA, and SSMA are theoretically equivalent for AWGN channels. However, there are significant differences among these multiple-access schemes for fading multipath radio channels. These differences have been discussed in an illustrative way, revealing several advantages of CDMA and SSMA over FDMA and TDMA. In addition to the already presented advantages of CDMA and SSMA, there are further benefits:

* graceful degradation;
* less timing organization than TDMA;
* reduction of ISI and self-interference;
* additional gain by CMT and CMR;
* possibility of position location of MS;
* less bandwidth expansion due to Doppler spread than FDMA;
* less bandwidth expansion due to forward error correction than FDMA;
* independence of actual channel state; and
* potential exploitation of military research results.

ACKNOWLEDGMENT

The authors wish to thank Dipl.-Ing. Markus Naßhan and Karl-Heinz Eckfelder for their support during the preparation of this manuscript.

REFERENCES

[1] S. B. Rhee, (ED.): Special Issue on Digital Cellular Technologies, *IEEE Trans. Veh. Technol.*, vol. 40, 1991.
[2] M. K. Simon, J. K. Omura, R. A. Scholtz, and B. K. Levitt, *Spread Spectrum Communications, Volume III.* Rockville, MD: Computer Science, 1985, Ch. 5.
[3] D. N. Hatfield, "Measures of spectral efficiency in land mobile radio," *IEEE Trans. Electromagn. Compat.*, vol. EMC-19, pp. 266–268, 1977.
[4] G. R. Cooper and R. W. Nettleton, "A spread-spectrum technique for high-capacity mobile communications," *IEEE Trans. Veh. Technol.*, vol. VT-27, pp. 264–275, 1978.
[5] W. C. Y. Lee, "Spectrum efficiency in cellular," *IEEE Trans. Veh. Technol.*, vol. 38, pp. 69–75, 1989.
[6] W. C. Y. Lee, "Estimate of channel capacity in Rayleigh fading environment," *IEEE Trans. Veh. Technol.*, vol. 39, pp. 187–189, 1990.

[7] K. S. Gilhousen, I. M. Jacobs, R. Padovani, A. J. Viterbi, L. A. Weaver, and C. E. Wheatley, "On the capacity of a cellular CDMA system," *IEEE Trans. Veh. Technol.,* vol. 40 pp. 303–312, 1991.

[8] H. Baher, *Analog and Digital Signal Processing.* New York: Wiley, 1990.

[9] R. G. Gallager, *Information Theory and Reliable Communication.* New York: Wiley, 1968.

[10] W. C. Y. Lee, *Mobile Cellular Telecommunication Systems* New York: McGraw-Hill, 1989.

[11] J. G. Proakis, *Digital Communications.* 2nd Ed. New York: McGraw-Hill, 1989.

[12] S. Stein, J. J. Jones, *Modern Communication Principles.* New York: McGraw-Hill, 1967.

[13] T. S. Rappaport, S. Y. Seidel, R. Singh, "900-MHz multipath propagation meausrements for U.S. digital celluar radiotelephone," *IEEE Trans. Veh. Technol.,* vol. 39, pp. 132–139, 1990.

[14] W. C. Y. Lee, *Mobile Communications Engineering.* New York: McGraw-Hill, 1982.

[15] R. S. Kennedy, *Fading Dispersive Communication Channels.* New York: Wiley-Interscience, 1969.

[16] P. W. Baier and W. Kleinhempel, "Wide band systems," *AGARD EPP Lecture Series,* no. 172, on "Propagation limitations for systems using band-spreading," Boston, Paris, Rome, June 1990.

[17] P. Jung and P. W. Baier, "CDMA and spread spectrum techniques versus FDMA and TDMA in cellular mobile radio applications," in *Conf. Proc. 21st EuMC'91,* Stuttgart, Germany, Sept. 9–12, pp. 404–409, 1991.

Paul Walter Baier (M'82–SM'87) was born in 1938 in Germany. He received the Dipl.-Ing. degree in 1963 and the Dr.-Ing. degree in 1965 from the Munich Institute of Technology, Germany.

In 1965, he joined the Telecommunications Laboratories of Siemens AG, Munich, where he was engaged in various topics of communications engineering, including spread-spectrum techniques. Since 1973, he has been a Professor for Electrical Communications at the University of Kaiserslautern, Germany. His present research interests are spread-spectrum techniques, mobile radio systems, and digital radar-signal processing.

Dr. Baier is a member of VDE-ITG and of the German U.R.S.I. section.

Andreas Steil was born in 1964 in Germany. From 1985 until 1991, he studied electrical engineering at the University of Kaiserslautern, Germany. He received his Dipl.-Ing. degree in 1991.

In 1991, he joined the RF Comunications Research Group. His present research interests are novel signal-processing techniques for mobile radio systems.

Peter Jung (S'91–M'92) was born in 1964 in Germany. From 1983 until 1993, he studied physics and electrical engineering at the University of Kaiserslautern, Germany. He received the Dipl.-Phys. and Ph.D. (Dr.-Ing.) degrees in 1990 and 1993, respectively.

From 1990 until 1992, he was with the Microelectronics Centre (ZMK) of the University of Kaiserslautern, where he was engaged in the design and implementation of Viterbi equalizers for mobile radio applications. In 1992 he joined the RF Communications Research Group. His present research interests are signal processing, such as adaptive interference cancellation, and multiple-access techniques for mobile radio systems.

Dr. Jung is a student member of VDE-GME, VDE-OTG and AES.

Reprinted from *Proceedings of the IEEE Vehicular Technology Society 42nd VTS Conference*, Vol. 2, pp. 732-735, 1992

A Comparison of CDMA and TDMA Systems

Björn Gudmundson, Johan Sköld and Jon K. Ugland
Ericsson Radio Systems
S-164 80 Stockholm, Sweden

Abstract

In this reports two candidates for high capacity cellular systems are simulated and analysed, one CDMA and one TDMA system. Simulations of the CDMA example indicate a high sensitivity to variations in certain system parameters.

The TDMA example is a GSM system using random frequency hopping and operating without frequency planning. The simulations show that the TDMA system has at least the same capacity as the CDMA candidate. Soft capacity, efficient use of voice activity and diversity are features available in both systems.

1 Introduction

CDMA has been introduced as a candidate for high capacity cellular systems [1, 3]. Another atractive alternative is to refine todays TDMA systems already in operation into high capacity cellular TDMA or hybrid TDMA/CDMA systems for the future.

Introduction of dynamic channel allocation has been suggested as an evolution of TDMA [4, 5]. Another alternative is the frequency hopping TDMA system presented below. It can be seen as a TDMA/CDMA hybrid.

In this report, the issues of system capacity, flexibility and operational features are investigated for the CDMA system example and a comparison is made with the frequency hopping TDMA system.

2 The CDMA system

A description of the CDMA system is given in [3]. A direct sequence spreading technique is used, spreading the $R = 8$ kbps user data over the bandwidth $W = 1.25$ MHz trough a low-rate convolutional encoder and a Walsh-Hadamard transform.

Scrambling the data with PN-sequences provides decorrelation of the different users. The result is that the interference from all co-existing channels will add to a noise-like interference which is efficiently supressed in the despreading process.

The wide channel bandwidth allows a high resolution when extracing multipaths. Thus, a high degree of path diversity is availiable. For complexity reasons however only a few paths will be resolved. The remaining paths have to be supressed.

The analysed system is specified with a re-use factor of 1, i.e. the same frequency can be re-used in all base stations (BSs). This configuration is possible by the inherent interference diversity. Since the co-channel interference will be the average interference from several mobile stations (MSs), the worst case interferer will no longer determine the re-use factor as it does in conventional TDMA- and FDMA schemes.

For a CDMA system to work, the users sharing the same carrier must be received with equal power levels. Otherwise some users will jam the others. This makes heavy demands on the up-link design where the power control must have a wide dynamic range to compensate for the near/far effect. The power control must also be very fast since it has to compensate the multipath fading for slowly moving MSs. For fast moving MSs this is not necessary since the interleaving and coding probably provides sufficient quality in this case.

In [3] BS controls the MS power trough a closed loop. Every millisecond a command is transmitted to the MS demanding either an increase or a decrease by 0.5-1.0 dB of its power level. The algorithm is claimed able to track Rayleigh fading for vehicle speeds up to 25-100 mph.

Power control in the down-link is not as critical as in the up-link. Here all signals propagate along the same path, naturally giving balanced signal levels. It may however be used to increase capacity.

A study of the performance and capacity of the CDMA system is made in [1] and [2]. In both papers analytical expressions are used for the capacity calculations. Here, Monte-Carlo simulations are performed, allowing the introduction of hand-off margin, power control limitations etc. The evaluation is limited to the up-link since this is the most critical connection in the system.

A system comprising one local BS and three rings of interfering BSs is simulated. A hexagonal omni-cell pattern is assumed, and the MSs are randomly spread with uniform distribution in a circle having an equivalent area of the 37 hexagons. For each MS the power attenuation to each of the BSs are calculated. A distance dependent path-loss $r^{-\alpha}$ is used where α is the propagation exponent. The shadowing effects are modelled by independent, log-normally distributed random variables having a standard deviation of 8 dB. Hand-off is made to the BS resulting in the least attenuation.

In each simulation the C/I-level for the MSs con-

732

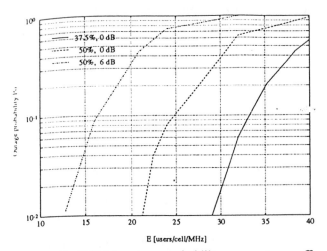

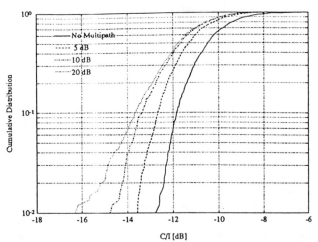

Figure 1: CDMA outage probability vs spectrum efficiency. *Right*: Ideal. *Middle*: canging channel activity to $\vartheta = 50\%$. *Left*: adding handoff margin $\Delta_{HO} = 6$ dB.

Figure 2: CDMA C/I-distributions in Rayleigh fading environment with max MS power limits $\nu = 5, 10$ and 20 dB.

nected to the local BS is calculated. Here C is the received power from the desired MS and I is the sum of undesired powers from all MSs sharing the same bandwidth. C/I-values for MSs not connected to the local BS are discarded due to boundary effects. Repeated simulations give a C/I-distribution.

The capacity is constrained by the minimum acceptable transmission quality $BER \leq 10^{-3}$ after decoding. In [1] this is claimed to be reached at $E_b/N_0 \geq 7$ dB. Assuming negligible background noise this is directly translateable to the minimum C/I-level

$$\gamma = \frac{R \cdot (E_b/N_0)}{W} \qquad (1)$$

which for the vocoder bitrate $R = 8$ kbps is $\gamma = -14.9$ dB. The probability of having a C/I-level below γ is called the outage probability P_O and is found from the C/I-distribution. Note that outage implies that *all* MSs connecetd a the BS have unacceptable transmission quality.

Simulated values of P_O vs different values of spectrum efficiency E (users/cell/MHz) are given in Figure 1. The rightmost curve yields the parameter values used in [1]. For $P_O = 1\%$ the capacity is found to be $E = 29$ users/cell/MHz which gives 36 users/cell according with the result in [1]. However, this capacity figure is achievable only under idealized conditions. Choosing slightly more pessimistic parameter values reduce performance significantly.

The voice activity factor of $\vartheta = 37.5\%$ claimed in [1] seems very optimistic. In [2] the value $\vartheta = 60\%$ was used instead. Here $\vartheta = 50\%$ has been used giving the expected capacity reduction of 30% shown in Figure 1.

The choice of propagation parameters is known to depend heavily on geography. Own measurements gave

path loss exponents mainly between $\alpha = 3$ and 4, but also values as low as 2 and as high as 5 occured. Changing the path loss exponent from 4 to 3 in the simulations gives a capacity reduction of 20%.

In deep fading dips the MSs must dramatically increase their output power levels in order to compensate the multipath fading. If a MS is located close to a neighbouring BS this may cause severe interference to the MSs at that BS. In the simulations a Rayleigh distributed fading is assumed. It is modeled by scaling the signals with independently Nakagami-2 distributed random variables corresponding to ideal, two-antenna diversity. All MSs will perfectly handle the fading, i.e. move at speeds less than 25 mph. However, parts of the deepest fading dips will remain due to an upper limiting of the MS output power. C/I-distributions for the upper limits $\nu = 5, 10$ and 20 dB are given in Figure 2. For the most realistic power limit, $\nu = 10$ dB, the C/I performance is reduced by 2 dB compared to a non-fading environment. Translated to capacity this is a reduction of 45%, implying that power control of the Rayleigh fading may cause severe problems to the system.

Perfect power control has been assumed in the capacity calculations in [1]. Here long-term power control er-

Table 1: Relative capacity degradation for variations in CDMA simulation parameters

Parameter:	Ideal[1]	Test value	Capacity change
Voice activity ϑ	37.5%	50%	−30%
Path-loss α	4	3	−20%
Multipath fading	No	Yes	−45%
HO-margin Δ_{HO}	0 dB	6 dB	−40%
Power control σ_{err}	0 dB	1 dB	−35%

733

420

rors are included by scaling the MS transmit-powers with an independent, zero mean, log-normally distributed random variable, having standard deviation σ_{err}. Simulations show that the degradation is large. For the moderate error value $\sigma_{\text{err}} = 1$ dB the capacity is reduced by 35%, demonstrating an extremely high accuracy needed in the power control to maintain operation.

To prohibit too frequent hand-offs, a hysteresis has commonly been used in cellular systems. This has been obtained by prohibiting the hand-off until a BS occures that gives more than Δ_{HO}dB reduction in path loss. Soft hand-off is an alternative to using hand-off margins, however, this has not been a prefered solution since connecting the MSs to more BSs simultaneously results is a dramatic increase in network complexity. The leftmost curve in Figure 1 shows P_O vs E for the CDMA system using hard hand-off with $\Delta_{\text{HO}} = 6$dB. A capacity reduction of 40% is observed compared to the case of ideal hand-off implying that soft hand-off is mandatory in this system.

Table 1 is a summary of the sensitivity to parameter variations. All capacity changes cannot be added, but the total picture is that the optimistic capacity claims in [1] are unrealistic. Accounting for the reduced voice activity of $\vartheta = 50\%$ and including the effect of multipath fading gives the capacity $E = 12$ users/cell/MHz. This can be compared to the AMPS analog 30 kHz system which with a 21 re-use frequency plan has the capacity 1.6 users/cell/MHz. Thus, a realistic CDMA capacity gain would probably be somewhat less than 10 compared to AMPS.

3 Frequency Hopping TDMA

Comparisons between CDMA and TDMA systems have been made by several authors [2, 4, 6].

Results reported by Baier et.al. [2] suggest that TDMA and CDMA system capacity is similar. They proposed a TDMA system using slow frequency hopping to provide interferer diversity [6]. The cluster size K was varied continously to simulate a continous set of spectrum efficiency values E. The interference level in [2] was calculated as the average interference over all frequencies used for hopping, i.e.

$$I = I_{\text{external}} = \frac{1}{N_{\text{f}}} \sum_i I_i \qquad (2)$$

where N_{f} is the number of frequencies used for frequency hopping and I_i is the interference contribution from the ith external user. The system parameters used were taken from the GSM-system with a half-rate speech codec, see Table 2.

Here we will simulate TDMA in the same way but with a fixed cluster size $K = 1$, i.e. all frequencies are re-used in all sites and no frequency planning is required. By varying the subscriber load, different values of E can be simulated. Random frequency hopping is performed over $N_f = 8$ frequencies. The frequency hopping patterns

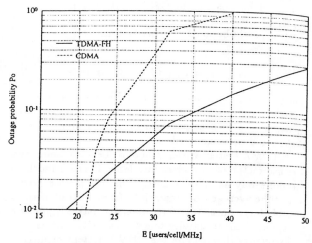

Figure 3: TDMA and CDMA outage probability vs. spectrum efficiency.

within the cell are strictly orthogonal, while the patterns are uncorrelated between the cells. Power control is used to achieve equal received powers at the site.

Each BS serves on the average N_s users per TDMA time slot, where $N_s \leq N_t$. N_s determines the spectrum efficiency E in users/cell/MHz:

$$E = \frac{N_s M}{N_t W} \qquad (3)$$

where M is the number of TDMA time slots and W is the bandwidth, see Table 2. The maximum available capacity of a BS is $E = 80$ users/cell/MHz.

The simulation model is in all other aspects equivalent to the one used for the CDMA simulations, including the ideal power control used. Up-link C/I-distributions are calculated and used to derive the outage probability. The result is plotted in Figure 3 together with the middle CDMA curve from Figure 1. The TDMA system capacity is 19 users/cell/MHz at 1% outage probability, while the corresponding CDMA figure is 21 users/cell/MHz. This shows that TDMA using 8 frequencies for interferer diversity has the same capacity as the CDMA system in [1]. The conclusion is confirmed in [2].

The GSM recommendation specifies a TDMA system with options for random frequency hopping and adaptive

Table 2: TDMA system parameters.

TDMA multiplex factor	M	16
No. of frequencies	N_{f}	8
User bit rate	R	6.5 kbit/s
Bandwidth	W	200 kHz
C/I limit [2]	$(C/I)_{\text{req}}$	6 dB
Channel activity	ϑ	50 %
HO-margin	Δ_{HO}	0 dB

734

power control. This indicates that one-cell frequency re-use could be possible in GSM. Further simulations of the GSM air interface operating in frequency hopping conditions will be needed to verify this.

4 System Comparison

The two system proposals presented above have approximately equal spectrum efficiency under the assumptions made. Below is a comparison of other system aspects that should be considered.

A TDMA system with random frequency hopping and a re-use factor of one gives interferer diversity much in the same way as CDMA does. The interference power will be the average from many interferers instead of a few possibly large ones. Interferer diversity makes it possible to exploit channel activity to get a capacity increase.

Frequency diversity in the CDMA system requires a multi-tap Rake receiver to exploit the wide bandwidth. In the TDMA system, the random frequency hopping inherently provides frequency diversity. Other diversity schemes such as antenna diversity, transmitter diversity and path diversity can be implemented in a TDMA system as well as in CDMA. The path diversity in TDMA can be provided by an equalizer.

In the CDMA system, a tight up-link power control is essential for operation. The TDMA system uses power control to reduce power comsumption in the mobiles and to gain capacity, but it is not as critical as in CDMA.

A CDMA system does not have a hard capacity limit. This is also true for the TDMA-proposal since it only uses 10-30% of the frequencies available at full capacity. Figure 3 shows that more soft capacity is available in TDMA since the outage probability increases more slowly.

The soft capacity also makes TDMA very flexible to unequal cell loading. A number of cells forming a line can be loaded up to 200% of the normal capacity if the surrounding cells have a 65% load. The high flexibility is due to the internal interference always being zero. In CDMA, the interference is mostly internal making it harder to increase capacity at peaks, since the cell will jam itself. A line of cells in the CDMA system can be loaded to 120% of nominal capacity, making it necessary to decrease the surrounding cells to 30%.

In the simulations presented in Section 3, 8 frequencies are used for frequency hopping. If the number is increased and averaging of interference can be performed, the system capacity will increase above that of the CDMA reference system. A TDMA system with a specific N_f will always have a higher spectrum efficiency than a similar CDMA system with the same interference diversity factor (processing gain). The reason is that the users within each cell are orthogonal i TDMA [6].

In a cellular system including both micro and macro-cells, MSs in the macro cells will use higher power levels and create high interference in the micro cells. The solution is to use different frequencies in macro and mi-cro cells. This is a disadvantage in CDMA if mobile assisted hand-off is used, since transmission is continous and no time slots are available for measurement such as in TDMA.

5 Conclusion

This report points out some strong and weak points in CDMA by simulation of a system example. The results indicate that performance can be severely degraded when assuming non-ideal conditions. A comparison is made with a frequency hopping TDMA system.

Simulations show that the capacity of the TDMA system is equal to or even better than in the CDMA example in ideal conditions. It also has most of the advantages that CDMA has, concerning e.g. frequency diversity and use of voice activity. The soft capacity properties of the TDMA system is superior to CDMA, making operation in inhomogeneous cell loads more efficient.

The conclusion is that when comparing CDMA and TDMA as multiple access schemes, capacity is not the only issue. The main point is how to exploit the capacity potential and achieve the system features wanted.

The overall picture is that a very careful system design is necessary to utilize the full potential of CDMA as well as TDMA systems. Further simulations are needed to verify the properties of both candidates.

References

[1] K. S. Gilhousen, I. M. Jacobs, R. Padovani, A. J. Viterbi, L .A. Weaver, Jr., C. E. Wheatly, "On the Capacity of a Cellular CDMA System". *IEEE Transactions on Vehicular Technology*, Vol. 40, No.2, May 1991, pp. 303-312 .

[2] A. Baier, W. Koch, "Potential and Limitations of CDMA for 3rd Generation Mobile Radio Systems", *MRC Mobile Radio Conference*, Nice, France, November 1991.

[3] A. Salmasi, K. S. Gilhousen, "On the System Design Aspects of Code Division Multiple Access (CDMA) Applied to Digital Cellular and Personal Communications Networks", Proceedings of *IEEE Vehicular Technology Conference*, VTC-91, St. Loius, USA, May 1991, pp. 57-63.

[4] B. Gudmundsson, "A Comparison of CDMA and TDMA Systems", *COST 231 Panel Meeting on Radio Subsystem Aspects of CDMA for Cellular Systems*, Berne, Switzerland, October 1991.

[5] K. Raith, J. Uddenfeldt, "Capacity of Digital Cellular TDMA Systems", *IEEE Transactions on Vehicular Technology*, Vol. VT-40, No. 2, May 1991, pp. 323-332.

[6] M. Mouly, "The Compared Capacity of CDMA and TDMA systems", *COST 231 Panel Meeting on Radio Subsystem Aspects of CDMA for Cellular Systems*, Florence, Italy, January 1991.

735

Chapter 8

CELLULAR NETWORKING

Reprinted from *Proceedings of the IEEE*, Vol. 80, No. 4, pp. 590-606, Apr. 1992.

An Overview of Signaling System No. 7

ABDI R. MODARRESSI, MEMBER, IEEE, AND RONALD A. SKOOG, MEMBER, IEEE

Invited Paper

In modern telecommunication networks, signaling constitutes the distinct control infrastructure that enables provision of ALL other services. The component of signaling systems that controls provision of services between the user and the network is the access signaling component, and the component that controls provision of services within the network, or between networks, is the network signaling component. There are international standards for both access signaling and network signaling protocols. From a network structure viewpoint, access signaling structures generally provide point-to-point connectivity between the user and a network node, while network signaling structures provide network-wide communication capability (directly or indirectly) between the nodes of the public network(s). Since the network signaling system acts as a traffic collector/distributor for many access signaling tributaries, its functions are more complex, its structure more involved, and its performance more stringent. This paper provides an overview of modern network signaling systems based on the Signaling System No. 7 international standard.

I. INTRODUCTION

In the context of modern telecommunications, signaling can be defined as the *system* that enables stored program control exchanges, network databases, and other 'intelligent' nodes of the network to exchange a) messages related to call setup, supervision, and tear-down (call/connection control); b) information needed for distributed application processing (inter-process query/response, or user-to-user data); and c) network management information. As such, signaling constitutes the control infrastructure of the modern telecommunication network.

Modern signaling systems are essentially data communication systems using layered protocols. What distinguishes them from other data communication systems are basically two things: their real time performance and their reliability requirements. No matter how complex the set of network interactions are for setting up a call, the call setup time should still not exceed a couple of seconds. This imposes quite a stringent end-to-end delay requirement on the signaling system. On the other hand, because of the absolute reliance of the telecommunication network on its signaling system, requirements for signaling network reliability (mes-

sage integrity, end-to-end availability, network robustness, recovery from failure, etc.) are extremely demanding. For example, current objectives require the down-time between any arbitrary pair of communicating nodes in the signaling network not to exceed 10 min/year. This is at least two orders of magnitude smaller than the corresponding requirement in a general-purpose data network. Requirements on real-time performance and reliability of signaling systems are likely to become even more stringent with advances in technology and new application needs.

Over the last century or so, signaling has evolved with the technology of telephony, although the pace of this evolution has never been faster than in the last two decades, a period characterized by the marriage of computer and switching technologies. The advent of the Integrated Services Digital Network (ISDN) has further accelerated the pace of development and deployment of signaling systems to support an ever increasing set of "intelligent network" services on a worldwide basis. When viewed as an end-to-end capability, signaling in ISDN has two distinct components: signaling between the user and the network (access signaling), and signaling within the network (network signaling). The current set of protocol standards for *access signaling* is known as the Digital Subscriber Signaling System No. 1 (DSS1). The current set of protocol standards for *network signaling* is known as the Signaling System No. 7 (SS7).

This paper provides an overview of Signaling System No. 7. It is a somewhat abridged and updated version of a tutorial on SS7 that was published in 1990 [1]. Following this introduction, the salient features of SS7's Network Services Part (NSP) are described in Section II. Functionally, NSP corresponds to the first three layers of the Open System Interconnection (OSI) Reference Model. This section also provides a discussion of signaling network structures that, in conjunction with the NSP, provide ISDN nodes with a highly reliable and efficient means of exchanging signaling messages. Once this reliable signaling message transport capability is realized, each network node has to be equipped with capabilities for processing of the transported messages in support of a useful function like setting up of a call (connection). In an increasingly large number of cases, call setup has to be preceded by invocation of some distributed

Manuscript received October 23, 1991; revised December 18, 1991.
A. R. Modarressi is with AT&T Bell Laboratories, Columbus, OH 43213.
R. A. Skoog is with AT&T Bell Laboratories, Holmdel, NJ 07733.
IEEE Log Number 9108075.

425

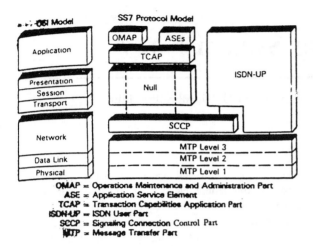

OMAP = Operations Maintenance and Administration Part
ASE = Application Service Element
TCAP = Transaction Capabilities Application Part
ISDN-UP = ISDN User Part
SCCP = Signaling Connection Control Part
MTP = Message Transfer Part

Fig. 1. SS7 protocol architecture.

application processes, the outcome of which determines the nature as well as the attributes of the subsequent call or connection control process. These nodal capabilities of call control and remote process invocation and management are part of the Signaling System No. 7 User Parts, which are described in Section III. In Section IV, we dwell on the very stringent performance requirements of signaling systems. These requirements reflect the critical nature of signaling functions and their real time exigencies. Finally, in Section V we sketch a broad outline of the likely evolution of network signaling in the remaining years of this century.

II. SIGNALING SYSTEM NO. 7 NETWORK SERVICES PART (NSP)

In this section, we describe the Signaling System No. 7 protocols that correspond to the first three layers (Physical, Data Link, and Network) of the OSI Reference Model. This component of the Signaling System No.7 protocol is called the Network Services Part (NSP), and it consists of the Message Transfer Part (MTP) and the Signaling Connection Control Part (SCCP). Figure 1 shows how these relate to each other and to the other components of the protocol. MTP consists of levels 1–3 of the Signaling System No. 7 protocol, which are called the Signaling Data Link, the Signaling Link, and the Signaling Network functions, respectively. SCCP is an MTP user, and therefore is in level 4 of Signaling System No. 7 protocol stack. MTP provides a connectionless message transfer system that enables signaling information to be transferred across the network to its desired destination. Functions are included in MTP that allow system failures to occur in the network without adversely affecting the transfer of signaling information. SCCP provides additional functions to MTP for both connectionless and connection-oriented network services.

MTP was developed before SCCP and it was tailored to the real time needs of telephony applications. Thus a connectionless (datagram) capability was called for which avoids the administration and overhead of virtual circuit

networks (one of the disadvantages of CCS6). Later, it became clear that there were other applications that would need additional network services (full OSI Network service capabilities) like an expanded addressing capability and connection-oriented message transfer. SCCP was developed to satisfy this need. The resulting structure, and specifically the splitting of the OSI Network functions into MTP level 3 and SCCP, has certain advantages in the sense that the higher overhead SCCP services can be used only when needed, allowing the more efficient MTP to serve the needs of those applications that can use a connectionless message transfer with limited addressing capability.

Sections II-A and II-B provide an overview of MTP and SCCP, respectively. Section II-C describes the signaling network structures that can be used to implement the Network Services Part.

A. The Message Transfer Part (MTP)

The overall purpose of MTP is to provide a reliable transfer and delivery of signaling information across the signaling network, and to react and take necessary actions in response to system and network failures to ensure that reliable transfer is maintained. Figure 2 illustrates the functions of MTP levels, and their relationship to one another and to the MTP users. These three levels are now described.

1) Signaling Data Link Functions (Level 1): A *Signaling Data Link* is a bidirectional transmission path for signaling, consisting of two data channels operating together in opposite directions at the same data rate. It fully complies with the OSI's definition of the physical layer (layer 1). Transmission channels can be either digital or analog, terrestrial or satellite.

For digital signaling data links, the recommended bit rate for the ANSI standard is 56 kb/s, and for the CCITT International Standard it is 64 kb/s. Lower bit rates may be used, but the message delay requirements of the User Parts must be taken into consideration. The minimum bit rate allowed for telephone call control applications is 4.8 kb/s. In the future, bit rates higher than 64 kb/s may be required (e.g., 1.544 Mb/s in North America and 2.048 Mb/s elsewhere), but further study is needed before these rates can be standardized.

2) Signaling Link Functions (Level 2): The Signaling Link functions correspond to the OSI's data link layer (layer 2). Together with a signaling data link, the signaling link functions provide a *signaling link* for the reliable transfer of signaling messages between two directly connected signaling points. Signaling messages are transferred over the signaling link in variable length messages called *signal units*. There are three types of signal units, differentiated by the length indicator field contained in each, and their formats are shown in Fig. 3. The Signaling Information Field (SIF) in a Message Signal Unit (MSU) must have a length less than or equal to 272 octets. This limitation is imposed to control the delay a message can impose on other messages due to its emission time (which is limited by the maximum standardized link speed of 64 kb/s).

426

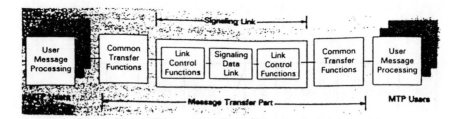

Fig. 2. MTP functional diagram.

The SS7 link functions show a strong similarity to typical data network bit-oriented link protocols (e.g., HDLC, SDLC, LAP-B), but there are some important differences. These differences arise from the performance needs of signaling (e.g., lost messages, excessive delays, out-of-sequence messages) that require the network to respond quickly to system or component failure events. The standard flag (01111110) is used to open and close signal units, and the standard CCITT 16-bit CRC checksum is used for error detection. However, when there is no message traffic, Fill-In Signal Units (FISU's) are sent rather than flags, as is done in other data link protocols. The reason for this is to allow for a consistent error monitoring method (described below) so that faulty links can be quickly detected and removed from service even when traffic is low.

a) Error correction: Two forms of error correction are specified in the signaling link procedures. They are the *Basic Method* and the *Preventive Cyclic Retransmission (PCR) Method.* In both methods only errored MSU's and Link Status Signal Units (LSSU's) are corrected, while errors in FISU's are detected but not corrected. Both methods are also designed to avoid out-of-sequence and duplicated messages when error correction takes place. The PCR method is used when the propagation delay is large (e.g., with satellite transmission).

The Basic Method of error correction is a non-compelled positive/negative acknowledgment retransmission error correction system. It uses the "go-back-N" technique of retransmission used in many other protocols. If a negative acknowledgment is received, the transmitting terminal stops sending new MSU's, rolls back to the MSU received in error, and retransmits everything from that point before resuming transmission of new MSU's. Positive acknowledgments are used to indicate correct reception of MSU's, and as an indication that the positively acknowledged buffered MSU's can be discarded at the transmitting end. For sequence control, each signal unit is assigned forward and backward sequence numbers and forward and backward indicator bits (see Fig. 3). The sequence numbers are seven bits long, which means at most 127 messages can be transmitted without receiving a positive acknowledgment.

The PCR method is a non-compelled positive acknowledgment cyclic retransmission, forward error correction system. A copy of a transmitted MSU is retained at the transmitting terminal until a positive acknowledgment for that MSU is received. When there are no new MSU's to be

sent, all MSU's not positively acknowledged are retransmitted cyclically. When the number of unacknowledged MSU's (either the number of messages or the number of octets) exceeds certain thresholds, it is an indication that error correction is not getting done by cyclic retransmission. This would occur, for example, if the traffic level was high, which causes the retransmission rate to be low. In this situation a *forced retransmission* procedure is invoked. In this procedure new MSU transmission is stopped and all unacknowledged MSU's are retransmitted. This forced retransmission continues until the unacknowledged message and octet counts are below specified threshold values. These threshold values must be chosen carefully, for if they are set too low, and the link utilization is large enough, the link will become unstable (i.e., once a forced retransmission starts, the link continues to cycle in and out of forced retransmission [2]).

b) Error monitoring: Two types of signaling link error rate monitoring are provided. A *signal unit error rate monitor* is used while a signaling link is in service, and it provides the criteria for taking a signaling link out of service due to an excessively high error rate. An *alignment error rate monitor* is used while a signaling link is in the proving state of the initial alignment procedure, and it provides the criteria for rejecting a signaling link for service during the initial alignment due to too high an error rate.

The signal unit error rate monitor is based on a signal unit (including FISU) error count, incremented and decremented using the "leaky bucket" algorithm. For each errored signal unit the count is increased by one, and for each 256 signal units received (errored or not), a positive count is decremented by one (a zero count is left at zero). When the count reaches 64, an excessive error rate indication is sent to level 3, and the signaling link is put in the out of service state. When loss of alignment occurs (a loss of alignment occurs when more than six consecutive 1s are received or the maximum length of a signal unit is exceeded), the error rate monitor changes to an octet counting mode. In this mode it increments the counter for every 16 octets received. Octet counting is stopped when the first correctly-checking signal unit is detected.

The alignment error rate monitor is a linear counter that is operated during alignment proving periods. The counter is started at zero at the start of a proving period, and the count is incremented by one for each signal unit received in error (or for each 16 octets received if in the octet counting mode). A proving period is aborted if the threshold for the

PROCEEDINGS OF THE IEEE, VOL. 80, NO. 4, APRIL 1992

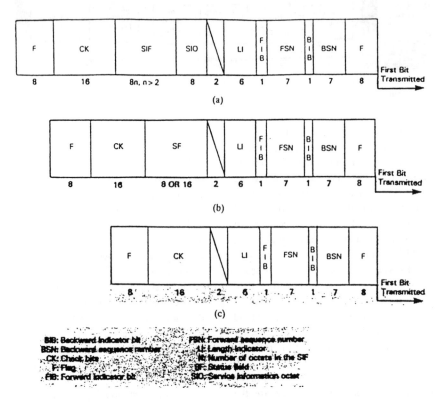

F	CK	SIF	SIO		LI	F I B	FSN	B I B	BSN	F	
8	16	8n, n > 2	8		2	6	1	7	1	7	8

First Bit Transmitted

(a)

F	CK	SF		LI	F I B	FSN	B I B	BSN	F	
8	16	8 OR 16		2	6	1	7	1	7	8

First Bit Transmitted

(b)

F	CK		LI	F I B	FSN	B I B	BSN	F	
8	16		2	6	1	7	1	7	8

First Bit Transmitted

(c)

BIB: Backward indicator bit   FSN: Forward sequence number
BSN: Backward sequence number   LI: Length indicator
CK: Check bits   N: Number of octets in the SIF
F: Flag   SF: Status field
FIB: Forward indicator bit   SIO: Service information octet

Fig. 3. Signal unit formats.

alignment error rate monitor count is exceeded before the proving period timer expires.

c) Flow control: The flow control procedure is initiated when congestion is detected at the receiving end of the signaling link. The congested receiving end notifies the transmitting end of its congestion with a link status signal unit (LSSU) indicating busy, and withholds acknowledgment of all incoming signal units. This action stops the transmitting end from failing the link due to a time-out on acknowledgment. However, if the congestion condition lasts too long (3–6 s), the transmitting end will fail the link.

A processor outage condition indication is sent by level 2, called signaling indication processor outage (SIPO), whenever an explicit indication is sent to level 2 from level 3 or when level 2 recognizes a failure of level 3. This indicates to the far end that signaling messages cannot be transferred to level 3 or above. The far-end level 2 responds by sending fill-in signal units and informing its level 3 of the SIPO condition. The far-end level 3 will reroute traffic in accordance with the signaling network management procedures described as follows.

3) Signaling Network Functions (Level 3): The signaling network functions correspond to the lower half of the OSI's Network layer, and they provide the functions and procedures for the transfer of messages between signaling points, which are the nodes of the signaling network. The signaling network functions can be divided into two basic categories: *signaling message handling* and *signaling network management.* The breakdown of these functions

and their interrelationship is illustrated in Fig. 4.

a) Signaling message handling: Signaling message handling consists of message routing, discrimination, and distribution functions. These functions are performed at each signaling point in a signaling network, and they are based on the part of the message called the *routing label*, and the Service Information Octet (SIO) shown in Fig. 3. The routing label is illustrated in Fig. 5 and consists of the Destination Point Code (DPC), the Origination Point Code (OPC), and the Signaling Link Selection (SLS) field. In the international standard the DPC and OPC are 14 bits each, while the SLS field is 4 bits long. For ANSI, the OPC and DPC are each 24 bits (to accommodate larger networks), while the SLS field has 5 bits, and there are 3 spare bits in the routing label. The routing label is placed at the beginning of the Signaling Information Field, and it is the common part of the label that is defined for each MTP user.

When a message comes from a level 3 user, or originates at level 3, the choice of the particular signaling link on which it is to be sent is made by the message routing function. When a message is received from level 2, the discrimination function is activated, and it determines if it is addressed to another signaling point or to itself based on the DPC in the message. If the received message is addressed to another signaling point, and the receiving signaling point has the transfer capability, i.e., the Signal Transfer Point (STP) function, the message is sent to the message routing function. If the received message is addressed to the

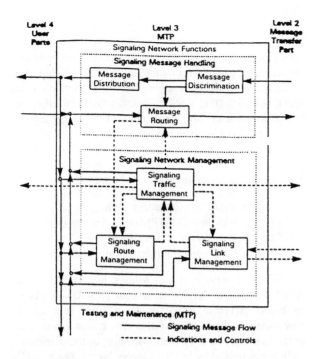

Fig. 4. Signaling network functions.

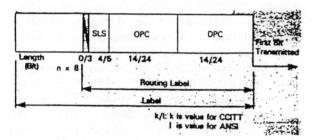

Fig. 5. Routing label structure.

receiving signaling point, the message distribution function is activated, and it delivers the message to the appropriate MTP user or MTP level 3 function based on the service indicator, a sub-field of the SIO field. Message routing is based on the DPC and the SLS in almost all cases. In some circumstances the SIO, or parts of it (the service indicator and network indicator), may need to be used.

Generally, more than one signaling link can be used to route a message to a particular DPC. The selection of the particular link to use is made using the SLS field. This is called load sharing. A set of links between two signaling points is called a *link set*, and load sharing can be done over links in the same link set or over links not belonging to the same link set. A load sharing collection of one or more link sets is called a *combined link set*.

The objective of load sharing is to keep the load as evenly balanced as possible on the signaling links within a combined link set. For messages that should be kept in sequence, the same SLS code is used so that such messages take the same path. For example, for trunk signaling with ISUP (see Section IV-A) the same SLS code is used for all

messages related to a particular trunk. In order to ensure proper load balance using SLS fields, it is critical that the SLS codes are assigned such that the load is shared evenly across all the SLS codes. Even then, the SLS load sharing method does not provide a fully balanced loading of signaling links in all cases. For example, if there are six signaling links in a combined link set, the 16 SLS codes would be assigned so that four signaling links would each carry three SLS codes and two of the signaling links would each carry only two SLS codes.

b) Signaling network management: The purpose of the signaling network management functions is to provide reconfiguration of the signaling network in the case of signaling link or signaling point failures, and to control traffic in the case of congestion or blockage. The objective is that, when a failure occurs, the reconfigurations be carried out so messages are not lost, duplicated, or put out of sequence, and that message delays do not become excessive. As shown in Fig. 4, signaling network management consists of three functions: signaling traffic management, signaling route management, and signaling link management. Whenever a change in the status of a signaling link, signaling route, or signaling point occurs, these three functions are activated as summarized below.

The *signaling traffic management* procedures are used to divert signaling traffic, without causing message loss, missequencing, or duplication, from unavailable signaling links or routes to one or more alternative signaling links or routes, and to reduce traffic in the case of congestion. When a signaling link becomes unavailable, a *changeover* procedure is used to divert signaling traffic to one or more alternative signaling links, as well as to retrieve for retransmission messages that have not been positively acknowledged. When a signaling link becomes available, a *changeback* procedure is used to reestablish signaling traffic on the signaling link made available. When signaling routes (succession of links from the origination to the destination signaling point) become unavailable or available, *forced rerouting* and *controlled rerouting* procedures are used, respectively, to divert the traffic to alternative routes or to the route made available. Controlled rerouting is also used to divert traffic to an alternate (more efficient) route when the original route becomes restricted (i.e., less efficient because of additional transfer points in the path). When a signaling point becomes available after having been down for some time, the *signaling point restart* procedure is used to update the network routing status and control when signaling traffic is diverted to (or through) the point made available.

The *signaling route management* procedures are used to distribute information about the signaling network status in order to block or unblock signaling routes. The following procedures are defined to take care of different situations. The *transfer-controlled* procedure is performed at a signaling transfer point in the case of signaling link congestion. In this procedure, for every message received having a congestion priority less than the congestion level of the signaling link, a control message is sent to the

OPC of the message asking it to stop sending traffic that has a congestion priority less than the congestion level of the signaling link to the DPC of the message. In ANSI Standards four congestion message priorities are used; in international networks only one is used. The *transfer-prohibited* procedure is performed at a Signal Transfer Point to inform adjacent signaling points that they must no longer route to a DPC via that STP. This procedure would be invoked, for example, if the STP had no available routes to a particular destination. The *transfer-restricted* procedure is performed at a Signal Transfer Point to inform adjacent signaling points that, if possible, they should no longer route messages to a DPC via that STP. The *transfer-allowed* procedure is used to inform adjacent signaling points that routing to a DPC through that STP is now normal. In the ANSI standards, the above procedures are also specified on a cluster basis (a cluster being a collection of signaling points), which significantly reduces the number of network management messages and related processing required when there is a cluster failure or recovery event. The *signaling-route-set-test* procedure is used by the signaling points receiving transfer prohibited and transfer restricted messages in order to recover the signaling route availability information that may not have been received due to some failure. Finally, in ANSI standards the *signaling-route-set-congestion-test* procedure is used to update the congestion status associated with a route toward a particular destination.

The *signaling link management* function is used to restore failed signaling links, to activate new signaling links, and to deactivate aligned signaling links. There is a basic set of signaling link management procedures, and this set of procedures are provided for any international or national signaling system. Two optional sets of signaling link management procedures are also provided, which allow for a more efficient use of signaling equipment when signaling terminal devices have switched access to signaling data links. The basic set of procedures are *signaling link activation* (used for signaling links that have never been put into service, or that have been taken out of service), *signaling link restoration* (used for active signaling links that have failed), *signaling link deactivation*, and *signaling link set activation*. The optional sets of procedures address automatic allocation of signaling terminals, and automatic allocation of data links and signaling terminals.

B. The Signaling Connection Control Part (SCCP)

SCCP enhances the services of the MTP to provide the functional equivalent of OSI's Network layer (layer 3). The addressing capability of MTP is limited to delivering a message to a node and using a four bit service indicator (a sub-field of the SIO) to distribute messages within the node. SCCP supplements this capability by providing an addressing capability that uses DPC's plus Subsystem Numbers (SSN's). The SSN is local addressing information used by SCCP to identify each of the SCCP users at a node. Another addressing enhancement to MTP provided by SCCP is the ability to address messages with global titles,

addresses (such as dialed 800 or free phone numbers) that are not directly usable for routing by MTP. For global titles a translation capability is required in SCCP to translate the global title to a DPC + SSN. This translation function can be performed at the originating point of the message, or at another signaling point in the network (e.g., at an STP).

In addition to enhanced addressing capability, SCCP provides four classes of service, two connectionless and two connection-oriented. The four classes are:

• Class 0: Basic connectionless class;
• Class 1: Sequenced (MTP) connectionless class;
• Class 2: Basic connection-oriented class;
• Class 3: Flow control connection-oriented class.

In Class 0 service, a user-to-user information block, called a Network Service Data Unit (NSDU), is passed by higher layers to SCCP in the node of origin; it is transported to the SCCP function in the destination node in the user field of a *Unitdata* message; at the destination node it is delivered by SCCP to higher layers. The NSDU's are transported independently and may be delivered out of sequence, so this class of service is purely connection-less.

In Class 1, the features of Class 0 are provided with an additional feature that allows the higher layer to indicate to SCCP that a particular stream of NSDU's should be delivered in sequence. SCCP does this by associating the stream members with a sequence control parameter and giving all messages in the stream the same SLS code.

In Class 2, a bidirectional transfer of NSDU's is performed by setting up a temporary or permanent signaling connection (a virtual channel through the signaling network). Messages belonging to the same signaling connection are given the same SLS code to ensure sequencing. In addition, this service class provides a segmentation and reassembly capability. With this capability, if an NSDU is longer than 255 octets, it is split into multiple segments at the originating node, each segment is transported to the destination node in the user field of a *Data* message, and at the destination node SCCP reassembles the original NSDU.

In Class 3, the capabilities of Class 2 are provided with the addition of flow control. Also the detection of message loss and missequencing is provided. In the event of lost or missequenced messages, the signaling connection is reset and notification is given to the higher layers.

The structure of SCCP is illustrated in Fig. 6, and consists of four functional blocks. The SCCP connection-oriented control block controls the establishment and release of signaling connections and provides for data transfer on signaling connections. The SCCP connectionless control block provides for the connectionless transfer of data units. The SCCP management block provides capabilities beyond those of MTP to handle the congestion or failure of either the SCCP user or the signaling route to the SCCP user. With this capability, SCCP can route messages to backup systems in the event failures prevent routing to the primary system. The SCCP routing block takes messages received from MTP or other SCCP functional blocks and performs the

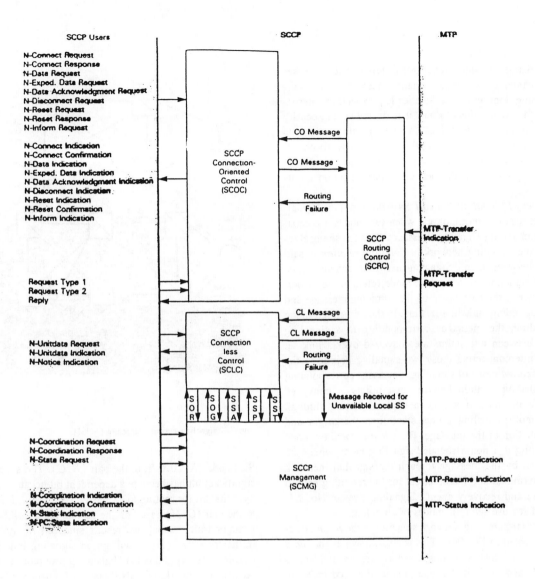

Fig. 6. SCCP functions.

necessary routing functions to either forward the message to MTP for transfer or pass the message to other SCCP functional blocks.

C. Signaling Network Structures

Signaling networks consist of signaling points and signaling links connecting the signaling points together. As alluded to earlier, a signaling point that transfers messages from one signaling link to another at level 3 is said to be a Signal Transfer Point (STP). Signaling points that are STP's can also provide functions higher than level 3, such as SCCP and other level 4 functions like ISUP (see Section IV-A). When a signaling point has an STP capability and also provides level 4 functions like ISUP, it is commonly said to have an *integrated* STP functionality. When the signaling point provides only STP capability, or STP and SCCP capabilities, it is commonly called a *standalone* STP. Signaling links, STP's (stand alone and integrated), and signaling points with level 4 protocol functionality can

be configured in many different ways to form a signaling network. The Signaling System No. 7 protocol is specified independent of the underlying signaling network structure. However, to meet the stringent availability requirements given in Section V-A (e.g., signaling relation unavailability not to exceed 10 min/year), it is clear that any network structure must provide redundancies for the signaling links, which by themselves have unavailabilities measured in many hours per year. In most cases the STP's must also have backups.

The worldwide signaling network is intended to be structured into two functionally independent levels: the national and the international levels. This allows numbering plans and network management of the international and the different national networks to be independent of one another. A signaling point can be either a national signaling point, an international signaling point, or both. If it serves as both, it is identified by a specific signaling point code in each of the signaling networks.

Administrations and Exchange Carriers can form agreements to interconnect their signaling networks, as is currently being done in North America [3], as well as internationally [4]. When this is done, it is desirable for security reasons to place restrictions on the signaling messages authorized to go from one network to another. To ensure these restrictions are complied with, screening procedures should be provided at the network interconnection points (gateways).

1) Types of Signaling Network Structures: In the Signaling System No. 7 terminology, when two nodes are capable of exchanging signaling messages between themselves through the signaling network, a *signaling relation* is said to exist between them. Signaling networks can use three different signaling *modes*, where mode refers to the association between the path taken by the signaling message and its corresponding signaling relation. In the *associated* mode of signaling, the messages corresponding to a signaling relation between two points are conveyed over a link set directly interconnecting those two signaling points. In the *non-associated* mode of signaling, a message corresponding to a signaling relation between two points is conveyed over two or more link sets in tandem passing through one or more signaling points other than the origin and the destination of the message. The *quasi-associated* mode of signaling is a non-associated signaling mode where the path taken by the message through the signaling network is predetermined and fixed, except for the rerouting caused by failure and recovery events. Signaling System No. 7 is specified for use with all modes of signaling.

A familiar signaling network structure is the *mesh* structure illustrated in Fig. 7(a). This is also known as the *quad* structure. The STP's in·this structure are "mated" on a pairwise basis. This is the type of structure used in North America [3].

The mesh network has 100% redundancy; that is, for any single point of failure, the traffic can be diverted to alternate paths that do not increase the number of transfer points. The network must be engineered so that each component under normal conditions can handle twice its peak load. Also, facility diversity requirements must be placed on the signaling links to meet availability requirements. For example, the two-way diversity rules state that the signaling links in each of the pairs $(AB, AC), (BD, BE), (CD, CE)$, etc., must be realized on physically diverse transmission facilities. For the signaling link quads (e.g., BD, BE, CE, CD), an additional requirement is usually required that either the pair (BD, CE) or the pair (CD, BE) be on diverse facilities (three-way diversity rule).

A routing example is shown in Fig. 7(b) that points out a difficulty that occurs with SLS code assignments when routing through multiple STP's (as in the interconnection of signaling networks in North America). In the example, the switching offices route to the STP's based on the least significant SLS bit and the STP's route to the next pair of STP's based on the second least significant bit. What happens is that the aggregate traffic to D from the pair (B, C) have SLS codes $XX1X$, and E has aggregate

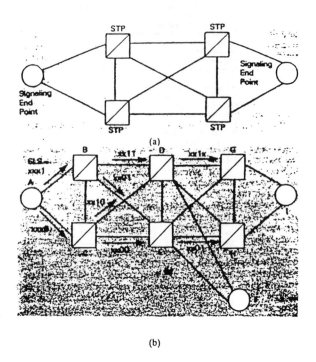

Fig. 7. Signaling network structure (ANSI).

SLS codes $XX0X$ from the pair (B, C). The second least significant bit has become a dependent bit because B and C used that bit for routing to D and E. As a result, if D routes to the pair (G, H) based on the second least significant bit, a major traffic imbalance results since all of the through traffic from pair (B, C) will go on signaling link DG, for example. To avoid this load balancing problem in the North American network, the ANSI standards specify a 5-bit SLS field, the least significant bit indicates the STP to route to, and the SLS code is "rotated" after routing so that the dependent bit is moved out of the least four significant bits, which are used for routing at the STP's.

Other signaling network structures using a mix of associated and quasi-associated signaling modes are possible. For example the associated signaling mode can be used as a first choice route between two signaling points, and a quasi-associated route through an STP as backup in case the associated path fails. This configuration is illustrated in Fig. 8(a). Here, associated mode signaling is used between nodes A and C, as well as nodes B and C, with quasi-associated mode through STP 1 as backup. Nodes A and B, however, load-share the routes through STP 1 and STP 2, thus using only a quasi-associated mode. When the STP's used in this structure are integrated with the switches at a relatively large percentage of nodes (perhaps more than 50%), the resulting *distributed* signaling transport architecture provides for more robustness and survivability due to the distribution of signaling transfer function throughout the network.

Another possibility is a generalization of the mesh network. In this case there is a backbone network of fully interconnected STP's, and different clusters of offices are

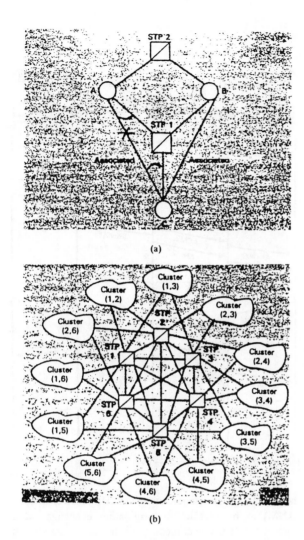

(a)

(b)

Fig. 8.   Alternate signaling network structures.

homed on different pairs of STP's. This configuration is illustrated in Fig. 8(b). An advantage of this structure is that when an STP fails its load is shed to a number of alternate STP's, and not just one mate as in the mesh network. Similar load shedding properties occur when backbone signaling links fail. The amount of reserve capacity that is needed to accommodate component failures is less in this type of network structure compared to the mated structure [5].

III.   SIGNALING SYSTEM NO. 7 USER PARTS

In this section we briefly describe three major Signaling System No. 7 User Parts that use the transport services provided by the MTP and the SCCP: the Integrated Services Digital Network User Part (ISUP), the Transaction Capabilities Application Part (TCAP), and the Operations, Maintenance, and Administration Part (OMAP). Other User Parts like the Telephone User Part (TUP) and the Data User Part (DUP) will not be covered as their functionalities are provided in the ISUP protocol.

A. The Integrated Services Digital Network User Part (ISUP)

The ISDN User Part of the Signaling System No. 7 protocol provides the signaling functions that are needed to support the basic bearer service, as well as supplementary services, for switched voice and non-voice (e.g., data) applications in an ISDN environment. Prior to ISUP, another user part called the Telephone User Part (TUP) was specified that provides the signaling functions to support control of telephone calls on national and/or international connections. ISUP, however, provides all the functions provided by TUP plus additional functions in support of non-voice calls and advanced ISDN and Intelligent Network (IN) services. The following summary is based on the 1988 Blue Book version of ISUP.

Services supported by the ISDN User Part include the basic bearer service, and a number of ISDN supplementary services. ISUP uses the services of the MTP for reliable in-sequence transport of signaling messages between exchanges. It can also use some services of SCCP as one method of end-to-end signaling. Figure 1 shows the relationship of ISUP with the other parts of the SS7 protocol. In accordance with the OSI model, the information exchange between ISUP and MTP (or SCCP) takes place through the use of parameters carried by inter-layer service primitives. The ISDN User Part message structure is shown in Fig. 9. As seen in this figure, ISUP messages have variable lengths (an ISUP message can consist of up to 272 octets including MTP level 3 headers). All ISUP messages include a routing label identifying the origin and destination of the message[1], a circuit identification code (CIC), and a message-type code that uniquely defines the function and format of each ISUP message.[2] In the mandatory fixed part of the message, the position, length, and order of parameters is uniquely determined by the message type. The mandatory fixed part of an ISUP message is followed by a series of pointers that point to mandatory (and possibly optional) variable length parameters. Mandatory variable-length parameters are specified by a length field and a parameter-value field, while optional variable-length parameters are specified by a parameter name field, a length field, and a parameter-value field.

1) Basic Bearer Service: The basic service offered by ISUP is the control of circuit-switched network connections between exchange terminations. Figure 10 shows a typical basic call setup and release procedure between an originating exchange and a destination exchange, through an intermediate (or transit) exchange, using ISUP messages. The user-to-network or access signaling in this figure is performed by use of the DSS1 protocol (Q.931) on the D-channel. Thus in response to a Q.931 setup message, the

---

[1] As discussed in Section III-A, the routing label is actually an MTP Level 3 header (see Fig. 5), and not an ISUP header. It is shown in Fig. 9 primarily to highlight the fact that i) the ISUP fields are preceded by the routing label and ii) for each individual circuit connection the same routing label must be used in all messages associated with that connection.

[2] It is the CIC that provides the identification that relates all ISUP messages for a given *call*.

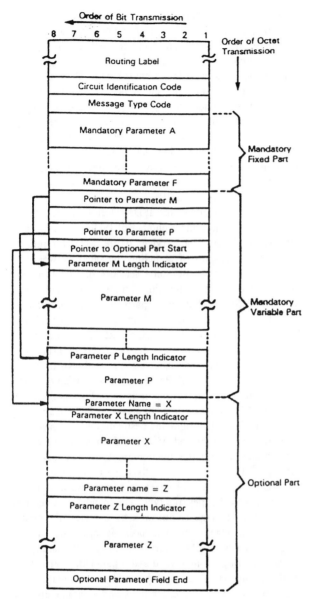

Fig. 9.   ISDN-UP message structure.

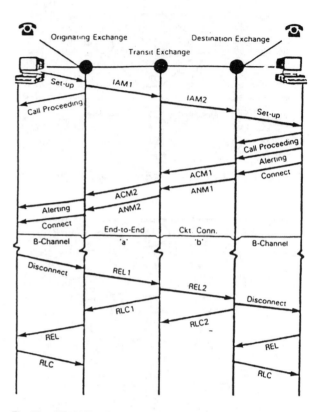

Fig. 10.   ISDN-UP call setup example.

originating exchange launches an Initial Address Message (IAM1) toward the transit exchange for the purpose of setting up trunk a. The transit exchange processes IAM1, sets up trunk a, and launches another Initial Address Message (IAM2) toward the destination exchange requesting use of trunk b. As the destination exchange sets up trunk b, an ISUP-Q.931 interworking takes place in this exchange, resulting in transmission of a Q.931 setup message on the D-channel toward the subscriber. After the subscriber has been alerted, an Address Complete Message (ACM) is sent by the destination exchange to the transit exchange, which processes it and generates and launches another ACM toward the originating exchange. A Q.931 alerting message is then generated by the originating exchange and sent to the calling station. When the called party answers, a Q.931 Connect message received on the D-channel at

the destination exchange causes an ISUP Answer Message (ANM) to be sent to the originating exchange, which sends a Q.931 Connect message to the calling station. The circuit-switched path between the calling and the called stations now consists of the cascade of a B-channel on the originating side, trunk a, trunk b, and a B-channel on the terminating side. Call tear-down is effected by use of ISUP messages Release (REL) and Release Complete (RLC), as shown in Fig. 10. Many other ISUP messages have been defined in support of the basic service for all contingencies that can arise on call setup, during the call, or on call tear-down, and for maintenance of associated circuits [6].

2) Supplementary Services: Supplementary services supported by ISUP include user-to-user signaling, closed-user group, calling line identification, call forwarding, etc. A short description of these services follows.

a) User-to-user signaling: The user-to-user signaling supplementary services provide a means of communication between two end users through the signaling network for the purpose of exchanging information of end-to-end significance. In order for these services to be possible, they need to be supported by the access protocol as well. Some of the end-to-end signaling methods use the services of SCCP while others use the services of MTP only. Three user-to-user signaling supplementary services have been defined. Service 1 allows the user-to-user signaling information to be included in the ISUP messages during call setup and clearing phases. Service 2 allows up to two ISUP User-to-User Information (UUI) messages to be transferred in each direction between end users during the

434

call setup phase. Service 3 provides for exchange of any number of ISUP-UUI messages between end users during the active phase of a call. All services can be implemented using the pass-along method. In this method, ISUP uses only the services of MTP. No processing of the pass-along information takes place at the transit exchanges. Services 2 and 3 can also be implemented using SCCP Connectionless (CL-SCCP) or Connection-Oriented (CO-SCCP) services.

b) Closed-user group: The closed user group (CUG) service allows a group of stations to communicate among each another only, with the option of some stations having incoming/outgoing access to users outside the group. A user can be a member of multiple CUG's. Each CUG is given an interlock code and this interlock code is assigned to all facilities associated with stations that belong to that CUG. When such a station initiates a call, a validation check is performed to verify that both the calling and the called stations belong to the CUG associated with the interlock code. The validation data can be in the exchange to which the station is connected (decentralized administration), or in a network database (centralized administration). In the centralized administration, the originating exchange sends a TCAP query to the appropriate network database and receives a response that determines the disposition of the call. Call setup can proceed normally from this point on if the call is permitted. Refer to Section III-B for more detail on TCAP interactions. This capability is the basis of a number of important virtual private network (VPN) services currently offered by a number of carriers.

In the decentralized administration of the CUG service, after the validation check is performed at the originating exchange, the interlock code of the selected CUG is transmitted in an IAM message to the transit exchange together with an indication on whether the calling station has outgoing access privileges. The transit exchange transmits this information to the succeeding exchanges. At the destination exchange, another validation check is performed to verify that the called party belongs to the CUG indicated by the interlock code. The call setup continues only if the information received checks with the information stored at the destination exchange.

c) Calling line identification: Another important supplementary service supported by ISUP is the calling line identity presentation and restriction. The calling line identity presentation (CLIP) service is used to present the calling party's number to the called party possibly with additional sub-address information. The calling party may have the option of activating the calling line identity restriction (CLIR) facility which would prevent the calling party's number from being presented to the called party. The transmission of the calling party's number to the destination exchange can be effected by either the originating exchange including it in the IAM, or by the destination exchange requesting it from the originating exchange through an ISUP Information Request Message. If CLIR is activated by the calling party, the originating exchange will provide the destination exchange with an indication in the IAM

that the calling party's number is not to be presented to the called station.

d) Call forwarding: The call forwarding service provides for the redirection of a call from the destination originally intended to a different destination. Three types of call forwarding service have been defined: call forwarding unconditional, call forwarding busy, and call forwarding no reply. By requesting the call forwarding unconditional service, a subscriber is able to have the network redirect all calls, or just calls associated with a basic service, originally intended for a user's number to another number. The call forwarding busy service allows the user to do the same but only if the original destination is busy. The call forwarding no reply works in a similar way to the call forwarding unconditional but only after allowing the original destination to be alerted for a specified length of time before redirecting the unanswered call.

Upon receipt of an IAM with a called party number for which call forwarding is in effect, the destination exchange determines if the redirection number is in the same exchange. If so, it alerts that station and sends back an Address Complete Message containing the redirection number to the originating exchange. If the redirection number is in another exchange, an IAM that contains the redirection number as the called party number is sent from the original destination exchange to the exchange with the redirection number. The latter exchange then sends an Address Complete Message containing the redirection number to the originating exchange.

B. The Transaction Capabilities Application Part (TCAP)

Transaction Capabilities (TC) refer to the set of protocols and functions used by a set of widely distributed applications in a network to communicate with one another. In the SS7 terminology, TC refers to the application-layer protocols, called Transaction Capabilities Application Part (TCAP), plus any Transport, Session, and Presentation layer services and protocols that support it. For all SS7 applications that have been designed thus far, TCAP directly uses the services of SCCP, which in turn uses the services of MTP, with Transport, Session, and Presentation layers being null-layers. In this context, then, the terms TC and TCAP are synonymous (see Fig. 1).

Essentially, TCAP provides a set of tools in a connectionless environment that can be used by an application at one node to invoke execution of a procedure at another node, and exchange the results of such invocation. As such, it includes protocols and services to perform remote operations. It is closely related and aligned (except for one extension[3]) with the OSI Remote Operation protocol (ROSE) specified in Recommendations X.219 and X.229 [7]. In the telecommunications network, the distributed applications that use TCAP can reside in exchanges and in network databases. The primary use of TCAP in these networks is for invoking remote procedures in support of

---

[3]This extension is the Return Result-Not Last (RR-NL) component whose purpose is to carry the segments of a result that would otherwise be longer than the maximum allowed message size.

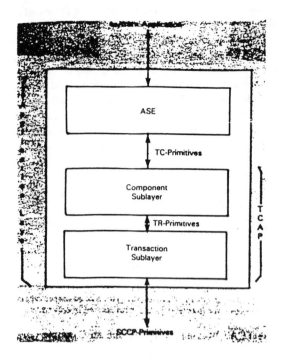

Fig. 11. Application layer structure.

Intelligent Network services like 800-service (free-phone). The application layer structure including TCAP is shown in Fig. 11. A TC-user Application Service Element (ASE) provides the *specific* information that a particular application needs (e.g., information for querying a remote database to convert an 800 number into a network-routable telephone number). TCAP provides the tools needed by *all* applications that require remote operation. TCAP itself is divided into two sub-layers: the component sub-layer and the transaction sub-layer. The component sub-layer involves exchange of "components" (the equivalent of Protocol Data Units or PDU's in ROSE) between TC-users. These components contain either requests for action at the remote end (e.g., invoking a process), or data indicating the response to the requested operation. The transaction sub-layer deals with exchange of messages that contain such components. This involves establishment and management of a dialogue (transaction) between the TC-users. A simplified discussion of the two TCAP sub-layers and an example are now given.

1) The Transaction Sub-Layer: A transaction (or dialogue)[4] defines the *context* within which a complete remote operation involving, for example, exchange of queries and responses between two TC-users, is executed. The transaction sub-layer is responsible for management of such a dialogue.[5] Two kinds of dialogues can take place between

[4] A dialogue refers to an explicit "association" between two TC-users, whereas a transaction refers to an explicit "association" between two peer transaction sub-layers. In TCAP, there is a one-to-one correspondence between dialogues and transactions, except in the case of an unstructured dialogue which does not use a transaction ID.

[5] The Transaction sub-layer is designed to provide an efficient *end-to-end* connection for exchange of "components" using the connectionless services of SCCP. In this role, it may be said to provide a very "skinny"

peer Transaction sub-layers: unstructured dialogue and structured dialogue. In the unstructured dialogue service, the Transaction sub-layer provides a means for a TC-user to send to its remote peer one or more components that do not require any responses. These components are received by the Transaction sub-layer from the TC-user (through the intervening Component sub-layer), and are packaged and sent to the remote Transaction sub-layer in a Unidirectional message. There is no explicit "association" established between peer Transaction sub-layers for this service.

The second kind of dialogue is the structured dialogue. Here, the TC-user issues a TC-BEGIN primitive containing a unique dialogue ID to the Component sub-layer. All the components that the TC-user sends within this dialogue would contain the same dialogue ID. The Component sub-layer maps this TC-BEGIN primitive into a TR-BEGIN primitive containing a *transaction* ID and issues it to the underlying Transaction sub-layer. There is a one-to-one correspondence between transaction ID's and dialogue ID's. Multiple components received by the Component sub-layer from the TC-user with the same dialogue ID can be grouped into a single TR-BEGIN message containing the appropriate transaction ID. The Transaction sub-layer manages each transaction (identified by its unique Transaction ID), groups components belonging to the same transaction into appropriate BEGIN, CONTINUE, END, and ABORT messages, and transmits them to its peer at the remote end (see Figs. 11 and 12).

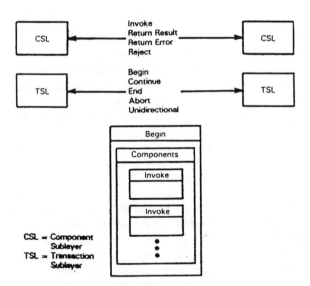

Fig. 12. TCAP sub-layer messages.

In short, the overall purpose of the Transaction sub-layer is to provide an efficient end-to-end connection between two TC-users over which they can exchange components related to one particular invocation of a distributed processing application. It avoids the OSI connection establishment and release overheads by packaging components in the 'connec-

and hybrid substitute for the missing layers (and specifically for the OSI Transport layer).

tion' setup and release messages. To reduce the number of signaling messages, it also supports a prearranged release facility where the peer Transaction sub-layers release their transaction resources related to a "dialogue" after a fixed period of time without the exchange of an END message.

2) The Component Sub-Layer: As alluded to earlier, a component consists of either a request to perform a remote operation, or a reply. Only one response may be sent to an operation request (which, however, could be segmented). The originating TC-user may send several components to the Component sub-layer before the Component sub-layer transmits them in a single message to its peer at the remote end. Components in a message are delivered individually to the TC-user at the remote end, and in the same order in which they were provided at the originating interface. Successive components exchanged between two TC-users for the purpose of executing an application constitute a dialogue. The Component sub-layer allows several dialogues to be run concurrently between two TC-users. Such dialogues can be unstructured or structured.

In the context of a structured dialogue, the Component sub-layer provides the function of associating replies with operations as well as handling abnormal situations. Associated with any invocation of an operation is a unique component ID. This allows several invocations of the same remote operation to be active simultaneously. The value of the invoke ID identifies an invocation of an operation unambiguously, and is returned in any reply to that operation. The Component sub-layer allows for four classes of remote operations. In class 1, both success and failure in performing the remote operation are reported. In class 2, only failure is reported, and in class 3 only success is reported. In class 4, neither failure nor success is reported. The replies to an operation could consist of one of the following components: Return Result (Last), Return Error, or Reject depending, respectively, on whether a result, error, or notification of syntax error in performing the remote operation is being provided (Fig. 12). Also, due to the signaling message size limitation, the segmentation of a successful result can be provided by the non-ROSE component Return Result-Not Last (RR-NL). In addition, any number of linked operations may be invoked prior to transmission of the reply to the original operation.

The reader is referred to CCITT Blue Book Recommendations .Q.771–Q.774 for more details [8]. The example in the next sub-section can help clarify the procedures involved.

3) The TCAP Message Structure and an Example: The overall TCAP message structure is shown in Fig. 13. The encoding is according to Recommendation X.209, where every information element is coded as Name (TAG), Length, and Value. The transaction portion of the message specifies, through the use of a message type identifier (TAG), whether the message is a Unidirectional message, a Begin message, a Continue message, an End message, or an Abort message. It also specifies the total length of the message. This is followed, except in the case of a

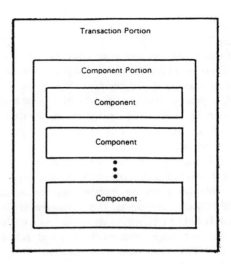

Fig. 13.   TCAP message structure.

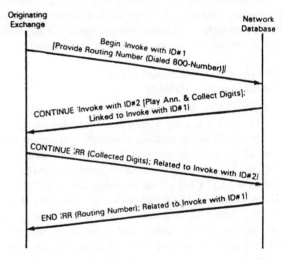

Fig. 14.   800 Service example using TCAP.

Unidirectional message, by the Name, Length, Value of the Transaction ID(s). The component portion of the message has a component portion Name and a component portion Length field, followed by the individual components. Each individual component in turn has a component type Name, a component Length field, and an information field specifying the required parameters for that component. The component type Name specifies whether the component is an Invoke, a Return Result, a Return Error, or a Reject component.

By way of a simplified example, consider an interactive 800 (free-phone) service and one way that it may be implemented using TCAP. Figure 14 shows the flow of TCAP messages between the originating exchange that has received the 800 call and the network database that contains the information for routing the call. In order for the database to provide the routing number, it is necessary for it to ask the exchange to play a certain announcement to the calling party, collect some more digits, and pass it on to the database.

437

The first TCAP message sent by the exchange is a BEGIN message that establishes a structured dialogue (and its associated transaction) between the exchange and the database in order to execute this application. Within the BEGIN message, a process that provides the routing number for the 800-service is invoked and given an invocation ID #1. The dialed 800 number is included as a parameter in the Invoke component. The database sends back a CONTINUE message as part of the structured dialogue in which it invokes a 'linked' operation that requires the exchange to play a certain announcement to the calling party and collects some digits. This invocation has Invoke ID #2 and is linked (related) to the invocation with Invoke ID #1. The exchange performs the required action and sends a CONTINUE message to the database. Within this CONTINUE, a Return Result component with invocation ID #2 is included containing the collected digits. Upon receipt and processing of this message, the database sends an END message to the originating exchange terminating the dialogue. Within the END message, however, a Return Result component is included with invocation ID #1, with the final routing number contained in it as a parameter.

C. The Operation, Maintenance, and Administration Part (OMAP)

The Operation, Maintenance, and Administration part (OMAP) of the Signaling System No. 7 provides the application protocols and procedures to monitor, coordinate, and control all the network resources that make communication based on Signaling System No. 7 possible. OMAP is specified in CCITT Blue Book Recommendation Q.791 [8].

The position of OMAP with respect to other parts of the Signaling System No. 7 is shown in Fig. 15. The collection of all the monitoring, control, and coordination functions above the application layer is known as the Systems Management Application Process (SMAP). All the management data that can be transferred or affected is contained in the Management Information Base (MIB). This data is gathered by the interaction of the MIB with the protocol entities at each layer through the Layer Management Interfaces (LMI). SMAP uses the services of TCAP through the OMAP Application Service Element (OMAP-ASE). OMAP-ASE sits on top of the TCAP Component sub-layer. An example of an OMAP-ASE is the ASE for MTP Routing Verification Test (MRVT) which uses the connectionless services of TCAP. MRVT is an important function of OMAP and is briefly described below and illustrated with an example.

The MRVT procedure tests MTP routing between any two signaling points in the network. A signaling point initiates an MRVT procedure for a given destination by sending MRVT messages to all its appropriate adjacent signaling points. Each node that receives this message processes it and "routes" it toward the given destination by sending an appropriate MRVT message to its adjacent nodes. This process is continued until the message reaches its destination. The destination node in response sends an acknowledgment message (MRVA) to the initiator of the MRVT in the same fashion. The example shown in Fig. 16

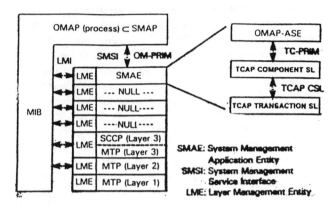

Fig. 15. SS7 management model.

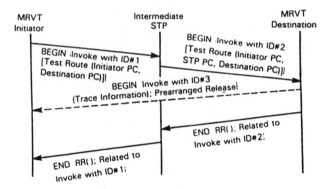

Fig. 16. MTP routing verification test with TCAP.

illustrates the procedure for a successful test of a message route that involves one intermediate STP. Notice that it is actually the BEGIN message here that has been *designated* as the MRVT message, and the END message that has been designated as the MRVA message. The MTP Routing Verification Result (MRVR) message is an optional message that the destination sends to the initiator of the MRVT procedure if the latter had asked for a trace in its original MRVT message.[6] This message is MTP-routed from the destination node to the initiator of the MRVT procedure. It is also an example of a prearranged release facility, alluded to in Section IV-B1), that would not require an END for release of transaction resources.

The MRVT procedure is designed to detect loops, excessive length of routes, excessive delays, inaccessibility of signaling points, and some other anomalies. In addition to MRVT, the OMAP procedures can be used for verifying SCCP routing and Global Title Translation process integrity (SRVT), management of routing data in different entities, circuit validation tests, link equipment failure management, link fault sectionalization, routine or on-occurrence measurement data collection, real time control and similar functions.

---

[6]The MRVR message is a particularly useful message when the MRVT test fails, as it can provide more detailed information that may be needed to trace the failure.

IV. Signaling System No. 7 Performance Objectives

There are three major categories for signaling network performance objectives: *Availability*, *Dependability*, and *Delay*. This section provides a summary of the current recommendations of CCITT, as well as those of ANSI. If an item is not directly identified as an ANSI or CCITT specification, it is common between both.

A. Availability Objectives

The unavailability of a signaling route set is determined by the unavailability of the components of the signaling network and the network structure. The *unavailability* of a system is defined as the probability that the system is in a failed state at a random point in time. This can be expressed in terms of the system mean-time-to-failure (MTTF) and mean-time-to-repair (MTTR) as $Unavailability = MTTR/(MTTF + MTTR)$. In applying this definition of unavailability to a pair of signaling network users, the system is defined to be the collection of all hardware and software in the user nodes, STP's, and the signaling links that are required to permit message transport between the pair of users.

MTP Unavailability Objective:
The unavailability of an MTP signaling route set should not exceed 1.9×10^{-5}, which equates to an expected downtime of less than 10 min/year.

From this objective and the network structure, availability objectives for the network components must be determined. Component objectives are not provided in the standards, since they depend on the network structure.

SCCP Relay Point Unavailability Objective:
The unavailability of an SCCP relay point should not exceed 10^{-4}, which equates to an expected downtime of less than 53 min/year.

B. Dependability Objectives

Dependability objectives relate to the ability of the network to reliably transport messages and not cause malfunctions. For the MTP there are four objectives:

Undetected Errors:
On each signaling link, not more than one in 10^{10} of all signal unit errors should be undetected by the MTP.

Lost Messages:
Not more than one in 10^7 messages should be lost due to failure of the MTP.

Messages Out-of-Sequence:
Not more than one in 10^{10} messages should be delivered out-of-sequence to the User Parts due to failure in the MTP. This includes duplicated messages.

Transmission Error Rate:
The signaling data link shall have a long term bit error rate that does not exceed 10^{-6}.

The ISUP dependability objectives are:

Probability of False Operation:
Not more than one in 10^8 of all signal units transmitted should be accepted and, due to errors, causes false operation.

Probability of Signaling Malfunction:
Unsuccessful calls can be caused by undetected errors, loss of messages, or messages delivered out-of-sequence. No more than 2 in 10^5 of all ISDN calls should be unsuccessful due to signaling malfunction. No more than 1 in 10^5 of all ISDN circuit connections should be unsuccessful due to signaling malfunction.

C. Delay Objectives

Delay performance is a very important attribute of signaling networks. There are no objectives given in the standards for MTP or User Part end-to-end signaling delays. Formulas are provided in Q.706 for signaling link queueing delays, which are based on M/G/1 queueing models. However, no recommendations are given in the CCITT standards on what delay criteria should be used in the traffic engineering and dimensioning of signaling networks. Work along these lines is currently being done in CCITT Study Group II. Reference [9] gives a good overview and list of references for performance modeling and delay considerations in Signaling System No. 7 networks.

At present, the standards provide some objectives on cross-office delays for STP's, SCCP message relay points, and ISUP switching exchanges. The cross-office transfer times that have been specified by CCITT and ANSI are given in Table 1. The cross-office transfer time is defined as the time from when the received message is delivered to level 2 to the time the corresponding outgoing message is delivered to level 1. Note that this delay includes the outgoing link queueing delay and emission time. No transmission system propagation times are included. Because queueing delays and emission times depend on the message length distribution, another useful office transit delay measure is *processor handling time*, which is the cross-office transit time less the outgoing link queueing and emission times. Work is in progress in the standards bodies to incorporate this measure into the recommendations.

V. Evolution of ISDN Signaling: Looking to the Future

In this section, we sketch a broad picture of the likely evolution of signaling systems in the remaining years of this decade based on trends that have emerged or are emerging in the global information age network. The discussion is divided into two parts: the first part is concerned with short-term trends (1992–1996), and the second part addresses the longer term trends (1996–2000 and beyond).

A. Evolution of Signaling Systems (1992–1996)

The short-term evolution of ISDN signaling systems is likely to be in the following directions.

- Widespread deployment of Signaling System No. 7 in public networks by various Administrations and Ex-

439

Table I Signaling Engineering and Performance Specifications

Signaling Point	Message Type	Load	Cross Office Transfer Time (ms)	
			Mean	95%
STP (ANSI)	All	Normal	45	80
		2×Normal	55	90
STP (CCITT)	All	Normal	20	40
		+15%	40	80
		+30%	100	200
ISDN Exchange	Simple	Normal	110	220
		+15%	165	330
		+30%	275	550
	Processing Intensive	Normal	180	360
		+15%	270	540
		+30%	450	900
SCCP Relay Point	Unit Data and Call Request	Normal	50–155	100–310
		+15%	100–233	200–465
		+30%	250–388	500–775
	Data and Conn. Conf.	Normal	30–110	60–220
		+15%	60–165	120–330
		+30%	150–275	300–550

Note: Assumes 64 kb/s emission and 15 octet average message length.

change Carriers. This will result in rapid proliferation of Intelligent Network capabilities and services within each carrier's SS7 network.

• Rapid growth of DSS1 signaling systems (both Basic and Primary Rate Interfaces) for user-to-network out-of-band signaling, and penetration of greater levels of intelligence to the edges of the network and to customer premise's equipment (e.g., very sophisticated PBX's as well as advanced ISDN workstations).

• Widespread interconnection of SS7 signaling networks between different Exchange Carriers (North America), and between different Administrations worldwide. This ushers in advanced ISDN and Intelligent Network services on an inter-network, international, end-to-end basis.

• Intensification and acceleration of work on signaling standards in support of multimedia and broadband applications as well as advanced Intelligent Network capabilities (see Section V-B).

B. Evolution of Signaling Systems (1996–2000)

In the closing years of this century, Broadband ISDN

(B-ISDN) capabilities are likely to emerge in the network. B-ISDN provides a cell-based[7] network infrastructure with extremely high-speed switching and transmission capabilities. It will provide a unified transport infrastructure that can be used for *all* kinds of traffic: voice, data, image, video, signaling, OA&M, etc. Although the extent of penetration of B-ISDN services in the telecommunication market place may be unclear, it is quite clear that broadband technologies will become commercially available and some new services using these capabilities will be offered. Given that the B-ISDN network will be based on Asynchronous Transfer Mode (ATM) switching/transmission principles, and implemented on a ubiquitous optical fiber facility infrastructure, use of enormous bandwidth on demand will become not only technologically possible, but also economically feasible. Truly integrated multimedia services involving voice, high-speed data, image, and video will emerge and penetrate the business and residential markets with a potential to profoundly impact and transform the very fabric of those markets and the nature of the "work place".

In the B-ISDN environment of tomorrow, signaling will play a crucial role. In the context of multimedia services, call control and connection control functions will have to be separated. A call is an end-to-end entity whereas a connection (bearer) may have only a link-by-link significance. In its duration, a multimedia multipoint "call" may require the capability to add/drop a number of connections and/or legs involving widely different bandwidths. Evolution of signaling to the broadband era has two major components, a transport component and a user-part component.

Work has been underway in CCITT for some time on signaling user part evolution to accommodate multimedia, broadband, and intelligent network needs. It has its origin in the effort to extend signaling capabilities for separation of call control from connection control. Initially, the call control and connection control functions of ISUP were identified and conceptually separated. This led to the concept of "separated ISUP" (in contrast to the monolithic ISUP). Subsequently, the term *ISDN Signaling Control Part (ISCP)* was coined to designate a new user part of the Signaling System No. 7 protocol in which such separation is built in from the outset. According to the ISCP Baseline Document [10], the principles and premises on which the ISCP effort is based include the following.

• The ISCP architecture should be based as much as possible on the OSI Application Layer Structure (OSI-ALS).

• ISCP should be viewed as both the network signaling protocol and the access signaling protocol for B-ISDN (in contrast to the current state of affairs where two different protocols with similar functionalities but different origins are used).

• ISCP functions and capabilities should be developed in the context of a B-ISDN environment to provide

---

[7] A cell is a fixed-size packet of 48 octets of information and 5 octets of control overhead, for a total of 53 octets.

for separation of bearer control from call control, to support supplementary services, and to be applicable to Intelligent Network services.

- To enhance portability, ISCP should be positioned to use an OSI Network Service for transport. If MTP/SCCP is to provide such a transport, work is needed to further align it with OSI.

ISCP is supposed to modularize the communication capabilities needed for its signaling applications into ASE's. Depending on the nature of the "call"/"connection" requested (the ISCP "Application Context"), appropriate ASE's will be dynamically combined to provide the required protocols. The nature of these ASE's and the various ISCP "Application Contexts" are the subject of current study in CCITT.

The signaling transport evolution work in CCITT, on the other hand, started quite some time later [11].

The issues to be addressed here relate to the kind of signaling transport architecture and protocol that can be used in an ATM environment to provide the reliability that signaling transport needs while making efficient use of the enormous broadband capabilities of ATM networks in support of new and vastly expanded signaling applications. Here, a number of alternatives, ranging from retention of MTP to fully associated signaling mode using signaling permanent virtual circuits and a skinny part of MTP Level 3, have been identified. Although it is generally agreed that MTP Levels 1 and 2 are going to be replaced by the Physical and ATM layers in the B-ISDN protocol model, questions related to the more complex evolution of MTP Level 3 and SCCP to the broadband environment are only beginning to be studied [12], [13].

The work currently underway on ISCP should be influenced appropriately with the work on signaling message transport issues.

REFERENCES

[1] A. R. Modarressi and R. A. Skoog, "Signaling System No. 7: A tutorial," *IEEE Commun. Mag.*, vol. 28, pp. 19–35, July 1990.
[2] R. A. Skoog, "Engineering common channel signaling networks for ISDN," *Proc. ITC-12*, Torino, Italy, June 1988.
[3] R. R. Goldberg and D. C. Shrader, "Common channel signaling interface for local exchange carrier to interexchange carrier Interconnection," *IEEE Commun. Mag.*, vol. 28, pp. 64–71, July 1990.
[4] J. J. Lawser, J. Matsumoto, and J. M. Pigott, " Common channel signaling for international service applications," *IEEE Commun. Mag.*, vol. 28, pp. 89–92, July 1990.
[5] R. A. Skoog, H. Ahmadi, and S. Boyles, "Network architecture planning for common channel signaling networks," in *Proc. 2nd Ann. Int. Symp. on Network Planning*, Brighton, UK, Mar. 1983.
[6] CCITT Study Group XI, "Specifications of Signaling System No. 7," Blue Book, vol. VI — Fascicle VI.8, Geneva 1989.
[7] CCITT Study Group VI, " Data communication networks," CCITT Blue Books , vol. VIII, Fascicles VIII.4 and VIII.5, Geneva, Switzerland, 1989.
[8] Study Group XI, "Specifications of Signaling System No. 7," Blue Book, vol.VI, Fascicle VI.9, Geneva, Switzerland, 1989.
[9] G. Willman and P. J. Kuhn, "Performance modeling of Signaling System No. 7," *IEEE Communications Mag.*, pp. 44–56, July 1990.
[10] ISCP Baseline Document, Temporary Document 699-E, CCITT Working Party XI/6 (Question 11/XI), Ottawa, Canada, Oct. 1989.
[11] R. A. Skoog and A. R. Modarressi, "Alternatives and issues for network signaling transport in a broadband environment," *Computer Networks and ISDN Systems*, vol. 20, pp 361–368, 1990.
[12] A. R. Modarressi and M. Veeraraghavan, "Network signaling architectures for broadband ISDN," in *Proc. Forum 91 Technical Symp.* pt. 2, vol. III, pp. 193–197, Geneva, Switzerland, Oct. 1991.
[13] G. I. Stassinopoulos and I. S. Venieris, "ATM adaptation layer protocols for signaling," *Computer Networks and ISDN Systems*, vol. 23, pp. 287–304, 1992.

Abdi R. Modarressi (Member, IEEE) received the B.E.E. degree from the American University of Beirut, Lebanon, in 1969, and the M.S. and Ph.D. degrees in electrical engineering from the University of Pittsburgh, Pittsburgh, PA, in 1973 and 1976, respectively.

In 1976, he joined AT&T Bell Laboratories and worked on new algorithms for optimization of traffic routing in the national toll network. Between 1978 and 1982 he taught control and communications courses at the National University of Iran. In 1983 he directed the Network Planning Group at the Telecommunications Company of Iran. He rejoined AT&T Bell Laboratories in 1984, where he has been involved in performance studies of common channel signaling networks, conformance testing in these networks, signaling architecture studies in narrowband and broadband environment, and applications of object-oriented methods to implementation of SS7 networks. He is currently a Distinguished Member of Technical Staff in the Signaling Platforms Department at AT&T. He has authored several journal and conference papers on multidimensional system theory and applications, congestion control performance in data networks, signaling technology, and signaling architecture in broadband networks. While at the National University of Iran, he also coauthored a two-volume text on the "Theory and Applications of Control Systems."

Ronald A. Skoog (Member, IEEE) received the B.S. degree from Oregon State University in 1964, and the M.S. and Ph.D. degrees from M.I.T., Cambridge, in 1965 and 1969, respectively, all in electrical engineering.

From 1969 to 1971 he taught in the Department of Electrical Engineering and Computer Sciences, University of California, Berkeley. He joined AT&T Bell Laboratories in 1971, where he worked in the area of transmission facility network planning and optimization. He became a supervisor in 1974. In 1981 he began working in the area of common channel signaling networks and supervised work on new signaling network architecture studies, performance, and simulation studies of signaling networks, establishing signaling network performance objectives, and developing signaling network engineering and dimensioning methodologies. He has written a number of journal and conference papers in the areas of non-linear systems, control theory, optimization, signaling network design, performance analysis of real-time systems, and performance of signaling networks.

Dr. Skoog is a member of Sigma Xi.

441

The European (R)evolution of Wireless Digital Networks

Wireless systems engineering is developing into a more conscious search for the best common culture for multiple users in a real environment.

Jens C. Arnbak

Reprinted from *IEEE Communications Magazine*, pp. 74-82, Sept. 1993.

JENS CHRISTIAN ARNBAK is professor of tele-information techniques at Delft University of Technology.

*I*n 1979, the author was invited to the chair of Radio Communications at Eindhoven University of Technology in the Netherlands. Several friends and professional associates questioned the sense of accepting this offer. "Radio is dead. Broadband optical fibre is the future," said the experts.

I wondered if I should trust their kind advice and found five reasons not to. First, during a decade spent in planning international satellite services, I had noted an emerging trend towards more flexible support of end-users by small maritime and land-mobile terminals. The most conspicuous example was probably the small dish immediately unfolding in all the foreign locations visited by the roving U.S. Secretary of State, Henry Kissinger. The new individual access links were different from the traditional multiplexed satellite trunk connections; indeed optical cables looked poised to replace in the 1980s. Second, a wireless access network could connect a new subscriber in most urban and rural areas to the public network at lower investment cost than any local loop requiring digging. Third, would the majority of subscribers really ask for the vast individual link capacity offered by installing optical subscriber loops, noting that standard video distribution could be supported by satellite (radio again!), possibly in tandem with existing local cable television (CATV) networks? Fourth, the Nordic Mobile Telephone (NMT) cellular standard was at that time being developed in close cooperation between different PTTs and competing manufacturers in Denmark, Finland, Iceland, Norway, and Sweden. NMT demonstrated a joint drive in European countries towards (inter-)national cellular networks. The divided and divesting United States looked less able to develop and follow a common strategy for mobile networking, even though Bell Laboratories had played a leading role in the initial development of the novel cellular technology in the early 1970s. And finally, the new mobile satellite and cellular networks would offer an opportunity to teach total communication systems engineering. This would

seem a dire necessity in most European electrical engineering departments, where the traditional academic separation of 'switching systems', 'transmission systems,' and 'enabling technologies' often leads to more focus on (sub-)system capacity than on system capabilities to meet users' needs.

Back to the Future

*T*hese — personally biased — observations were made almost 15 years ago. Many things look different now, including my own university affiliation and responsibilities. Still, the general emphasis of radio communications designs has shifted away from maximizing the capacity of single links limited by Gaussian noise and available bandwidth, towards optimizing the capabilities (including, but not limited to the capacity) of multi-user networks. The decisive interference now seldom comes from outside, but is produced by authorized users of the very same wireless network. Users thus share an interest in developing and adhering to the best possible protocols and standards for allocating the joint network resources. Accordingly, wireless systems engineering is developing into a more conscious search for the best common culture for multiple users in a real environment. The capacity, spectrum efficiency, and cost-effectiveness of modern radio networks can no longer be won from nature (or an adversary) in a classical pursuit of individual gain.

This development has been commercially reinforced by the recent business trend in the computer and defence sectors away from "selling the high-tech products that you can make" toward "making high-tech products that you can sell." It has become more important to consider the different requirements and "cultures" of new communities using wireless networks, including their needs for innovation of regulatory conditions and standards. In this article, some of the related European developments of two different personal communication services, digital cellular telephony and mobile data networks, are compared with each other and with developments in the United States. The related agendas for R & D of wireless technologies in the past decade are also reviewed and compared.

In 1992, the new standard for Pan-European digital cellular telephony known as GSM [1] saw its first operational successes. The name GSM originated early in the 1980s as the French acronym for Groupe Special Mobile. This international working group was tasked by most European PTT administrations to develop a standard for cellular networks allowing international roaming across the many European borders. A truly Pan-European standard also provides economies of scale in mass production of hand-held and car terminals. In the early 1980s, these two objectives were seen as the most critical success factors for achieving a much larger penetration of mobile telephone services in Europe. Below, we shall discuss a third factor, the introduction of competition in the monopolistic European markets — a cultural import from the United States.

Although the GSM standard provides digital circuits to mobile terminals, alternative specialized wireless data networks for mobile computer communications were put into operation much earlier in several European countries. Some essential technology differences between the circuit-switched and packet-switched radio systems, and their relative merits in the increasingly competitive European environment, are also reviewed in this article.

Present Status of European Cellular Telephony

As of April 1993, 32 operators in 22 countries are committed to GSM. In Europe, this is formally expressed by a Memorandum of Understanding (MoU). The Commission of the European Communities (CEC) ensured that common frequency allocations were made in the 900-MHz band in the 12 member states and in many other European countries. The GSM standard is also rapidly being adopted outside Europe, notably in the Asia-Pacific region in Singapore, Malaysia, India, Hong Kong, and Australia, where a Pan-Asian MoU is being considered, and in the Middle East and Africa. Moreover, Taiwan, Thailand, and New Zealand are considering the GSM standard, or have already imposed it for at least one of their competing cellular networks. Accordingly, the old acronym GSM is nowadays taken to mean Global System for Mobile Communication.

The general advantages of international standards, MoUs, and roaming agreements for public mobile telephony are increasingly appreciated by different operators and markets. Where the inception and implementation of such conventions lag behind, subscribers are denied some of the following mutual benefits:

- interoperability with the national public switched telephone network(s), including ISDN.
- connectivity to all mobile users in a carrier's service area, stimulating cooperation between adjacent carriers.
- limiting inflexible cable infrastructure to the backbone network.
- low cost of introducing service-area coverage, i.e., proportional to the initial peak network use (erlang/unit area).
- simple upgrading of network capacity, when and where economically justified, by reducing cell size ('cell splitting').

- by virtue of the third, fourth and fifth benefits above, less economies of scale than in hard-wired networks and, hence, less basis for 'natural monopoly' arguments and for regulation against competition between local network operators [2].
- ability to locate vehicles and roaming user terminals, allowing automatic billing of users away from their home location or own operator.
- international portability of a subscriber identity module (SIM) smart card, authorizing customized personal log-in from compatible foreign terminals and competing networks by inserting the SIM card [1].

Such benefits have caused the growth of mobile communications markets to exceed 60 percent p.a. in many countries. No other telecommunications sector can boast similar growth rates at present. The infrastructure supply market is dominated by the few international manufacturers who combine expertise in both radio transmission and national switching systems. One of these manufacturers, L. M. Ericsson of Sweden, has some 40 percent of the cellular world market, including a major proportion of the many networks using the American analog standard AMPS. The terminal market, on the other hand, is subject to the typical supply principles and economies of consumer electronics: short product development times and a "killing" competition, due to the eroding profit margins on micro-electronics commodity products. However, the resulting drop in terminal prices now assists in developing the service market much faster than in the past.

At the end of 1992, commercial GSM telephone service was offered in eight countries. The coverage area rolled out in Europe at that time is shown in Fig. 1. The major transit routes in Europe had been reasonably covered within one year after the first GSM service went on air, despite economic recession in most countries and delays in type approval of the first telephone handsets and their complicated software. So far, the service is restricted to voice-type digital circuits, but a simple service allowing transfer of short (i.e., paging-type) alphanumeric messages is also planned. A more complete set of data rates to be supported as bearer services in GSM at a later stage (beyond 1995?) has also been defined [3].

Strikingly, the fastest take-up of GSM has been in Germany, where the national economy has stagnated in the wake of the unification of East and West. After only six months of operation in December 1992, Mannesmann Mobilfunk announced the 100,000th subscriber to its GSM-network, known as D2. In May 1993, 220,000 users subscribed to D2. Mannesmann's competitor in Germany, Deutsche Bundespost Telekom, holds the national monopoly on both public analog mobile telephony and the fixed telephone network. Some 90% of all telephone calls to or from mobile users in Germany originate or terminate in the latter, and Telekom's analog cellular network, C1, is already virtually saturated by its 800,000 subscribers. Nevertheless, Telekom's own GSM-network, D1, is believed to have attracted fewer subscribers than the competing D2 network. The successful marketing of the D2 network services and the associated offer of the 'Handy' portable GSM-terminal at a price of less than $1400 (including 15 percent tax, deductible for business users) is clearly designed by D2's American shareholder, Pacific

Strikingly, the fastest take-up of GSM has been in Germany, where the national economy has stagnated in the wake of the unification of East and West.

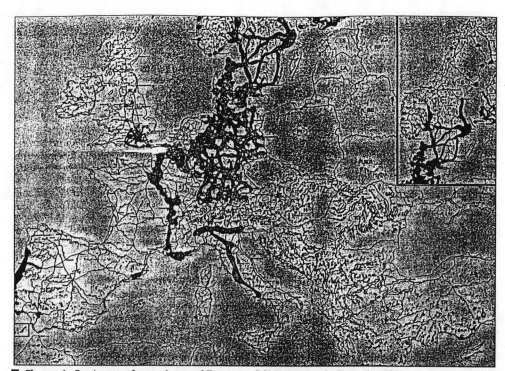

■ **Figure 1.** *Service area for carphones of European GSM-operators by December 1992.*

Telesis. Despite more advanced features consistent with the all-digital technology, the lightweight 'Handy' already costs less than a (heavier) handheld terminal for the analog C1 network did immediately before Deutsche Bundespost Telekom experienced competition. The explosion of the GSM market in Germany has caused considerable shortage of terminals, especially portables, resulting in waiting lists and loss of GSM-network revenues in other European countries.

The German case illustrates the general trend towards increased mobile network competition on the European continent, often spurred by experienced business partners from the less monopolistic Anglo-American shores of the Atlantic Ocean. Clearly, the international GSM-standard and MoU, with strict technical interface specifications and interworking requirements, have paved the way not only to international cooperation and roaming, but also to local rivalry between competing network operators. Introduction of network competition was in fact one of the GSM policy objectives of the CEC in Brussels. This regulatory change is the third key factor in the dynamics of European mobile communications, next to the more technological driving forces perceived by the founding fathers of GSM a decade ago. It also raises new technical issues, such as portability of numbers between competing operators. The GSM switches can readily be interconnected with Intelligent-Network (IN) nodes. Thus, new competitive service offerings can be extended into the fixed network, including storage of voice-mail messages to or from mobile subscribers who have temporarily logged out of the network, and call forwarding.

The European Cellular Radio Consortium (ECR-900—with Finnish Nokia, French Alcatel, and German AEG Telefunken as partners) and Ericsson of Sweden together have the lion's share of deliveries of GSM base stations and digital switching equipment. With a clear lead in digital and RF microelectronics, Japanese and U.S. manufacturers are poised to dominate the supply of the advanced digital terminals for GSM networks—a major consumer market with much competition, but only a peripheral system impact. In the core area of total systems engineering and standardization, US manufacturers and operators appear highly divided about the operational merits of various technologically advanced options for radio channel access, digital modulation, and networking, such as narrowband and broadband CDMA, different TDMA formats, and Low Earth-Orbiting Satellite (LEOS) systems. A single convincing first-generation digital alternative to GSM on the world market, similar to the US AMPS standard, which in the analog era captured about two-thirds of all mobile subscribers worldwide, has yet to emerge in the United States. As for the complementary Personal Communication Systems (PCS), the Digital European Cordless Telephone (DECT) standard in the 1800 MHz-band appears even further ahead of US efforts [4]. The DECT- standard includes most of the functionalities of the GSM architecture [1], but is intended to result in low-cost pocket phones suitable for both residential (cordless) use and wireless access to local (PBX) and wide-area (cellular) networks.

Arguably, the active presence of U.S. Bell Operating Companies (BOCs) in the operating consortia licensed by countries committed to the GSM standard can be taken as an American acknowledgment of the dominance of European manufacturers in the rapidly growing market for digital mobile network technologies. Conversely, the awards of such licenses in several EC countries can also be seen as a European acceptance of greater American operator experience and marketing skills. The remarkable role of the BOCs abroad suggests that innovative technologies imbedded in common standards are not enough to develop a modern mobile service market. This is confirmed by consideration of the evolution of another digital service in Europe, mobile data communications.

European telecommunication manufacturers have met more US competition in the supply of mobile data networks than in digital cellular telephone systems.

Mobile Data Networks in Europe

*I*n many European countries, the only possible way to conduct data transmission over public mobile networks is still to attach low-speed voice-band modems to analog radio telephone circuits. Generally, this results in a poor bit error rate due to the signal fading and shadowing, the extra-cellular interference and the handoffs between base stations inherent in mobile systems. Improvement by suitable error-correction codes or protocols is possible, but reduces the throughput of the narrowband radio channel further.

With the advent of the digital GSM, there is a popular belief that mobile computer communications will become easier, cheaper and better. The circuit quality and data rate do indeed improve, but a circuit-switched voice channel is not really suited to any bursty data source, even if using digital transmission. Dialing up (and paying for!) a real-time two-way circuit between end users is quite inefficient for the most frequent modes of mobile computer communications: electronic mail, interactive access to information services, EDI-type computer messaging, dispatch and other types of fleet management, and data "broadcasting." For such applications, the features of classical packet switching appear more desirable: non-blocking access for terminals and the ability to convert data rates and codes within the store-and-forward network. Wide-area flexibility and adaptability to serve computers and terminals with different functions and priorities point toward packet radio networks which can bill by traffic volume, rather than by connect time.

Such mobile data networks operate in several European countries. It is significant to note here that Deutsche Bundespost Telekom in the Spring of 1993 announced that its present experimental pilot mobile data network, Modacom, will be available for full operation in 80 percent of Germany by 1995. This wireless extension of Telekom's own public X.25 packet service to mobile computers such as Laptops, Notebooks, and Palmtops suggests that (at least) one of the major GSM operators in Europe has doubts about the competitiveness of the planned data transmission modes in the GSM network [3]. It is expected that the German government will license a second mobile data network later in 1993.

Well ahead of the German Modacom network, the Scandinavian PTT-administrations in Sweden, Norway, and Finland introduced the Mobitex Radio Data Standard in the last decade. At that time, Mobitex was supporting only a low-speed packet service at 1200 b/s plus an emergency voice service, reflecting the initial use by police patrols, fire brigades, and public utility services to exchange brief command-and-control messages with their headquarters. A decade later, the widespread acceptance of portable computers and the advent of more demanding commercial operators, including RAM Mobile Data in the United States, have pushed the Mobitex data rate up — and the hybrid voice channel out: the 19.2 kbit/s data rate defined in [7] is twice that foreseen in GSM [3]. Mobitex-based networks are also being introduced in the United Kingdom and the Netherlands by RAM and in France by France Telecom, assisted by Bell South. Again, we note the involvement of U.S. expertise in service provision, even where the network technology and standards are purely European.

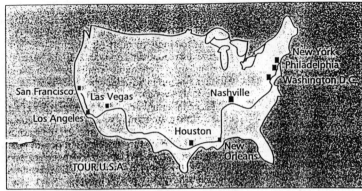

■ **Figure 2.** *EGA screen dump of route map encoded by differential chain coding into 1200-byte message [6].*

Nevertheless, European telecommunication manufacturers have met more U.S. competition in the supply of mobile data networks than in digital cellular telephone systems. Thus, Motorola's DataTAC technology, based on the RD LAP protocol for the logical radio link connection, was adopted by Deutsche Bundespost Telekom for Modacom in Germany, by Hutchison in the United Kingdom and Hong Kong, and by the ARDIS companies in the United States and Canada.

To illustrate the possibilities with narrowband mobile data throughputs, imagine the transfer of the simple American route map shown in Fig. 2 to (or from) a mobile terminal. It is possible to encode this image and the associated text as a message string of some 1200 bytes from a graphic tablet or a modern pen computer, using Differential Chain Coding [6]. This message occupies a nominal time of only half a second in a 19.2 kb/s packet data channel, considerably less than when dialling up a circuit-switched mobile telephone channel, especially if this were used to interconnect two facsimile terminals using slow run-length codes [6]. Even for the short 1200 byte message string, the time and the signalling overhead to build up the real-time circuit would likely result in a rather unattractive tariff. While a packet-radio protocol could also cause some extra delay, especially during heavy traffic loads, the billing would still correspond to an effective data transfer of only 1200 bytes.

Obviously, the need to link computers in a wireless mode does not only exist outdoor for public networks, but is often driven from the local-area networks at customer premises. The European DECT standard mentioned in the previous section will make data link throughputs up to 1 Mb/s available, but the limited access time (50 ms) makes it more suited for wireless linkage of multiple users with considerably lower individual throughputs to a PBX. In the office domain, the United States has more experience with wireless broadband networks than any other country. This gives microcellular wideband systems as Motorola's 18 GHz ALTAIR, meeting the IEEE 802.3 Ethernet standard, and NCR/AT&T's WaveLAN, using CDMA at 2.4 GHz, a competitive edge against the Europeans, when it comes to capacity. However, NCR has performed much of its related research in the Netherlands, where "Bell Labs Europe" is being created. This could result in closer coordination of the wireless-LAN standardization work in the European Telecommunications Standardisatin Institute (ETSI) and IEEE 802.11.

Finally, mention should be made of the present

446

work by ETSI on two Trans European Trunked Radio (TETRA) standards, one for pure packet data services to single or multiple destinations, and another also supporting additional circuit-switched data and speech channels (similarly to the initial Mobitex systems in Scandinavia mentioned above). As we shall see below, a hybrid system cannot be optimized in terms of capacity for both types of services; hybrid systems may nevertheless be required for certain applications, e.g., ones in the transport sector. The European frequency allocations for TETRA are likely to be in the UHF band (parts of 380-400 MHz, 410-430 MHz, 450-470 MHz and/or 870-890/915-933 MHz), and to adopt 25 kHz channels for co-existence with existing mobile services. TETRA is thus a typical narrowband data standard. European harmonization is deemed essential to achieve cross-border operation with this second-generation standard, in the interest of international courier services, railroads, and road and river transport companies.

Differences Between Packet Data and Digital Voice Systems

The ongoing introduction of separate digital networks for mobile telephony and mobile data applications in Europe described above will not come as a surprise to experienced network engineers. In the past, optimum use of classical hard-wired transmission resources motivated separate communications networks and signaling protocols for telephone and for computer traffic, adapted to the different statistical characteristics of the corresponding information sources. Real-time, blocking-type channels for telephone conversations pose other network requirements than does the delayed, but non-blocking exchange of bursty data in computer sessions, e.g., using the X.25 protocol. Where the physical transmission media contribute additional random fluctuations (as in mobile radio due to signal fading and mutual interference), or when the channel capacity is too precious to allow the overhead of ISDN-type service integration (as in mobile radio due to spectrum shortage), further distinctions arise between voice and data networks.

It is the instantaneous blocking probability, as determined by any unacceptable co-channel interference between cells, that determines the optimum design and real-time capacity of optimum cellular telephone networks. Although digital modulation schemes can be made more robust to interference than analog modulation, they still require considerable spacing between cells using the same radio frequencies, in order to avoid unacceptable degradation of voice circuits from time to time. A notable exception is spread-spectrum modulation as used in CDMA cellular systems, which trade carrier bandwidth for processing gain to tolerate a higher co-channel interference. The consequence is a stricter requirement for power control in mobile spread-spectrum telephone systems, in order to reap the theoretical network capacity and spectrum efficiency in the face of the fluctuating interference in the mobile channel [7].

Cellular structures and/or power control algorithms optimized for voice circuits are not optimum for a mobile network of virtual circuits between users who are prepared to accept delays of some data packets. Users of packet protocols are quite will-

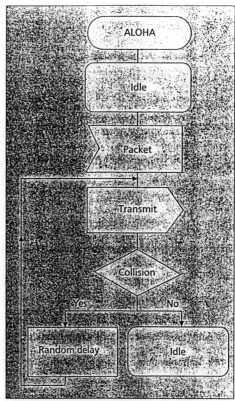

■ Figure 3. *Flow chart for network access by mobile data terminal using non-slotted ALOHA [11].*

ing to accept a significant risk of harmful collisions between coinciding packets, in return for not having to schedule their bursty accesses very strictly or to defer transmission for a long time. Packet protocols are designed to repair any harmful interference experienced between packets simply by retransmission. This contrasts with classical fixed and mobile telephone circuit requirements, which contain substantial *a priori* guarantees against occasional circuit outages, including those due to mutual harmful interference of radio signals.

When computer users are prepared to gamble against the risk of mutual conflicts in order to reap the benefit of lower average delay [7], they can also adopt a more tolerant attitude to the additional random vagaries of the mobile radio channel. Moreover, users prepared to accept occasional mutual conflicts between their accesses may find it counterproductive to invoke the strict power control assumed in mobile telephone systems. For if access powers are very unequal, there is a higher chance that at least one of the competitors wins the contest for the receiver than in the event of perfectly balanced signals, which will all annihilate each other in a collision.

Simple Multiple-Access Methods for Mobile Data Networks

Mobile transmitters sending bursty traffic in the form of data packets to a common base-station receiver can, in general, best use some kind of random access. The classical ALOHA protocol, according to which each mobile terminal is free to offer bursty packets to the channel in accordance with the simple flow diagram in Fig. 3, belongs in this category. It is well known that the (ideal) channel throughput can be doubled if active terminals are prepared to synchronize their packet transmissions into common time slots, such that the risk of partial packet overlap is avoided [7]. With high traffic loads, however, both unslotted and slotted ALOHA protocols become inefficient, since the free competition between all transmitters exposes most of the offered data traffic to collisions and, hence, multiple retransmissions and increasing delays.

To reduce this risk, a transmitter can follow a more cautious strategy. By first listening either to the common radio channel or to the "acknowledgement" return channel from the base station, a transmitter with a data packet can attempt to determine whether the shared radio facilities are already busy. The terminal approach based on the former listening method is known as carrier-sense multiple access (CSMA) [7]. In a realistic mobile channel, the various CSMA protocols may fail to detect ongoing radio transmissions of packets subject to deep fading on the listening path. Therefore, CSMA proves less efficient than in classical hard-wired and satellite networks, where contending terminals are not "hidden" from each other by individually different radio propagation effects. In such circumstances, mobile data terminals can better listen to the common base station, which broadcasts a "busy" signal to acknowledge an incoming transmission and inhibit prospective competitors and/or an "idle" signal to invite transmissions.

In principle, the simplest random-access protocols are inherently unstable, given the standard assumptions of infinitely many users and Poisson-distributed offered traffic. In practice, however, realistic ALOHA models based on finite populations of competing mobile terminals and proper propagation characteristics of the shared mobile channel have good stability properties [8]. Above all, this applies when the common base station receiver can be captured by a stronger packet in the presence of weaker competitors, and provides an incentive not to use too sophisticated random-access protocols in practical low-cost mobile networks.

Receiver Capture Effects in Mobile Data Networks

*I*n a mobile telephone network, every effort should be made to avoid propagation-induced outages of circuits, once these have been set up for real-time connections. It is a normal goal in both cellular engineering and adaptive power-control schemes to keep the capacity-limiting carrier-to-interference (C/I) ratios equally high at all receivers throughout a mobile telephone network, in order to secure the prescribed grade of service and voice circuit quality.

Equal C/I-ratios would not necessarily be a proper goal in multi-user packet communications. In 1976, Metzner showed that utilization of an ALOHA channel can be improved by deliberately introducing differences between the access powers of multiple users competing for a joint receiver [7]. In 1982, the author and one of his graduate students [9] demonstrated this effect in a particular case of a mobile packet radio system, where different propagation losses due to spatially distributed terminals and Rayleigh-distributed signal fluctuations due to multipath fading effects invariably introduce differences

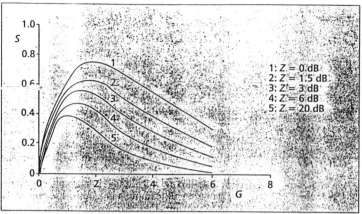

■ **Figure 4.** *Normalized throughput of single-cell mobile slotted ALOHA network, assuming uncorrelated Rayleigh fading of all packets and a base-station receiver with capture ratio Z [21].*

in C/I-ratios on the different links. To illustrate the effect of such differences, Fig. 4 shows the normalized throughput, S, of a slotted mobile ALOHA channel with offered normalized traffic, G, and uncorrelated multipath fading of the different packets. The formula for the multi-user throughput in this interference-limited network follows [10]:

$$S = G \exp \{ -G/(1 + 1/Z) \} \qquad (1)$$

The capture ratio of the base station, Z, indicates its ability to discriminate against a received signal in the presence of a co-channel signal which is Z times stronger. A thorough discussion of the validity of this simple receiver-capture model is found in the Ph.D. thesis of Linnartz [11]. For narrowband digital modulation, typical values for Z may be between 4 dB and 10 dB; an increase of S results, relative to the case without capture, i.e., when Z is infinite. In the latter case, the standard result

$$S = G \exp (-G) \qquad (2)$$

for slotted ALOHA in "ideal" (i.e., non-fading AWGN) channels [7] is recovered.

Note that the maximum throughput of (1)

$$S' = (1 + 1/Z) \exp(-1) \qquad (3)$$

occurs at increasingly higher offered packet traffic

$$G' = (1 + 1/Z), \qquad (4)$$

as the receiver capture ratio Z is decreased. Thus, both greater capacity and better stability result. With perfect receiver capture ($Z = 1$), equations 2 to 4 would suggest a doubled channel capacity due to Rayleigh fading and receiver capture capabilities. If Z would be allowed to tend to zero as in a spread-spectrum receiver, S would even remain close to the offered traffic G up to a level determined by the "processing gain" $1/Z$. Within the limitations [11] of this simple model, it does suggest that a combination of ALOHA and modulation with processing gain might support a very efficient random-access mobile data network. The absence of fast power control reaps the full contribution of random multipath fading to the random-access game played by the terminals!

*One of
the most
interesting
promises of
modern
wireless
communi-
cations
is a lower
initial cost
of connecting
each
subscriber
to the public
network,
largely
independent
of distance.*

Impact of the 'Poor' Mobile Radio Channel

*T*he foundations of packet radio were laid by U.S. researchers, mostly sponsored by military agencies. More emphasis was put on hostile interference and strategies for network survivability, than on optimum self-interfering systems and the random fluctuations of mobile radio channels. For this reason, historical terms like "packet radio" or "packet broadcasting" seldom refer to the typical propagation features of realistic wireless media, but reflect the purely architectural or information-theoretical notion of maximum connectivity among all terminals in a multi-user network. The experimental use of satellite links with their nearly perfect AWGN channels did not stimulate much consideration of real channel impairments, except hard-limiting satellite amplifiers and jamming by an adversary, where appropriate. When terrestrial networks were considered, these were often appropriate to a tactical battlefield scenario, with geographically distributed store-and-forward repeater nodes linked by random paths with fixed, but unknown losses. The desired packet communication modes were generally of the multi-hop type, designed to maximize the progress of packets in particular directions.

As a consequence of this strong research tradition, many researchers still intuitively expect the significant propagation impairments of typical terrestrial UHF/VHF mobile channels to reduce the moderate theoretical throughput of contention protocols ($S' = 0.38$ at $G' = 1$ for slotted ALOHA). However, as discussed above in the previous section, colliding packets with very different ground-wave losses or instantaneous fading levels do not necessarily all annihilate each other, given receiver capture capabilities. Indeed, throughput expressions such as equation 1 for "poor" mobile channels indicate a higher capacity than suggested by the classical studies of contention protocols in 'ideal' noiseless or AWGN channels.

The typical electromagnetic shadowing effects of large obstacles cause slow radio signal fluctuations in the mobile channel; the resulting slow-fading statistics is characterized by a log-normal distribution, with a r.m.s. spread σ between 6 dB and 12 dB. If individual data packets can be assumed to undergo uncorrelated shadowing, this gives rise to more significant capacity increases than does the Rayleigh fading due to multipath propagation. Figure 5 illustrates this for a receiver with moderate capture capability ($Z = 6$ dB). However, it should be noted that mutual correlation between the signal strengths of individual packets becomes much more likely in the presence of shadowing. Recent progress in the study of this very complicated phenomenon is reported in [12].

A related concern in radio propagation is that of the time constants of the fading fluctuations on links to or from mobile terminals. These time constants have a decisive influence on the throughput of packets. Generally, modelling is much easier if the channel state can be considered "frozen" during any one packet transmission, but completely decorrelated between any two packets [11]. The conditions for this obviously depend on carrier frequency, terminal velocities, packet lengths and the intensity of competing traffic. In general, short

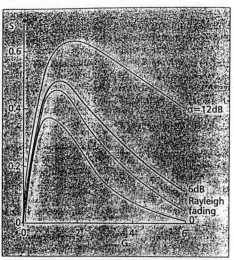

■ Figure 5. *Normalized throughput of single-cell mobile slotted ALOHA network, assuming log-normal shadow fading of all packets with spread σ, and a base station with capture ratio Z=6dB.*

packets profit most from the receiver capture effect described in the previous section, but on the other hand also suffer more from overhead losses in preambles, synchronization, or buffering. Optimal protocol design cannot ignore this trade-off between propagation-induced and higher OSI-level requirements posed by the particular data application.

Note that some of the statistical advantages reaped by packet networking to or from mobile terminals are due to the ergodic properties of the propagation: time and ensemble averages can be interchanged in many calculations, due to the movements of most terminals and the resulting fading fluctuations. Note also that this interchange does not apply to the near-far effect caused by the classical ground-wave path loss at UHF and VHF. This particular propagation effect creates a deterministic spatial discrimination around each base station [10] which, in effect, decides the effective frequency-reuse distance in a cellular system. The spatial bias can be studied very elegantly using a Laplace transform technique originating from the presence of Rayleigh-fading interferers [11], and can be repaired by a strategy restoring spatial fairness inside the cell by allocating higher retransmission rates for distant contenders.

Cellular Engineering Aspects

*T*here is an inherent tolerance against interferers built into all contention-oriented multiple-access schemes, since any packet failures will ultimately be repaired by protocol measures (retransmissions). This can be expected to lead to considerably smaller optimal cell re-use distances than in narrowband cellular telephone systems. Whether analog or digital, the latter require a considerable spacing between any two cells using the same frequency to ensure sufficient circuit quality. On the other hand, using contiguous "cells" with all base stations accepting the entire available bandwidth may prove better in many cases of packet networking, even in the absence of spread-spectrum processing gains in the base station receivers ($Z > 1$). A considerable chance of capturing a base station from outside its proper cell exists, especially if competition is not too heavy [11]. While this "site diversity" can be exploited in mobile data communications to reduce retransmissions of packets and so enhance throughput, it would be indicative of an unacceptable probability of harmful interference between cells in a mobile telephone system! It can thus be postulated that frequency re-use distances should be made much smaller in packet-switched mobile networks than in cellular circuit-switched systems.

Hence, hybrid or integrated cellular voice/data systems may be sub-optimal from a viewpoint of spectrum efficiency and capacity. If voice must be integrated in a computer-oriented cellular network for more than occasional use, it would seem necessary to develop more robust and delay-tolerant ways of packetizing it. Asynchronous transfer mode (ATM) experience may become relevant in this context. Conversely, if data packets are supported in a circuit-switched cellular network, inefficiencies in the use of transmission and signalling resources would appear to result. Where cost-based tariff regulations or plain competition apply, specialized mobile data networks therefore look more attractive to all users who attach only little value to the availability of voice transmission in the same network.

Economic and Regulatory Issues

One of the most interesting promises of modern wireless communications is a lower initial cost of connecting each subscriber to the public network, largely independent of distance. The access technologies employed in digital cellular radio such as GSM, Personal Communication Systems (PCS) such as first-generation Telepoint systems, the next-generation Cordless Telephone (DECT) and more full-fledged wireless office systems (WOS) using broadband radio LANs, all offer interconnection with public networks at an investment which is mainly determined by the capacity required. This contrasts with the high initial investment of a hard-wired local loop (whether optical or not), the cost of which increases dramatically in rural areas, roughly in inverse proportion to the population density and required traffic capacity! Fig. 6 illustrates these differences schematically [2].

It is the perverse cost structure of the traditional hard-wired access to public networks which caused 'market failure' to occur in many low-traffic rural regions until now and so justified a local operating monopoly for public telephony in most European countries and the United States. However, the novel wireless technologies may help avoiding any need to cross-subsidize the local access from monopoly profits made elsewhere, because costs directly proportional to area capacity remove the classical market-failure problem. In Fig. 6, wireless access to the public infrastructure would appear preferable to (installing new) twisted-pair loops left of the point 1, and to (installing) optical fibre to the home left of the point 2. In such circumstances, the economies of scale of the access network might become so marginal that local competition could be allowed under regulatory conditions to ensure fair spectrum allocation, consistent number planning and interconnectivity among the competitors. This is the case for the pan-European GSM-system, thanks to the policy directives from the CEC. Arguably, the recent acquisition of McCaw Cellular Communications by AT&T can be seen as the re-entry of the divested long-distance carrier into local telephone operations — and hence, as a US example of introduction of competition in a classical monopoly field, using wireless technology.

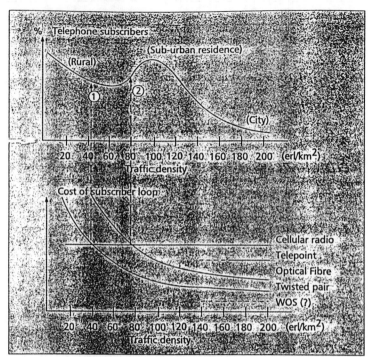

■ **Figure 6.** *The economy of various access techniques to public telephone network [2]. Above: Subscriber distribution in typical service area, as a function of traffic density. Below: Corresponding costs of different subscriber access techniques. Crossover points indicate the transition where cellular access techniques are cheaper than: 1) twisted-pair subscriber loop; 2) optical fiber to the home.*

It is not yet evident on which basis equitable access by competitors to scarce frequency and number resources can best be granted. In the United States, the FCC has replaced cumbersome administrative hearings by simple lotteries of frequency assignments. This resulted in rapid taking of "windfall profits" by fortunate winners, who simply sold their successful lots immediately after award. This proves that frequencies have a substantial "market" value. Where the Public Purse wishes to enjoy the profit of this, the government must design an auction system and the associated property rights very carefully to avoid being outsmarted by collusions of bidders.

Despite such practical difficulties with auctions of frequencies, it would at least seem desirable to assign some fee for spectrum occupancy. Fairer and more efficient use of scarce resources implies avoidance of extended "free parking" by inactive holders of frequency assignments. Even in the absence of competition, an incumbent holder of frequencies should be given sufficient incentive to vacate or share them for alternative use when not exploiting them fully. Broadcasters outside active operation hours occupy spectrum with considerable value for public and private mobile communications. In the successful European teletext standard, inactive video lines of a TV signal can be used to broadcast or download a substantial amount of data to arbitrary locations inside the coverage area of each TV transmitter network. This is commercially used in several European countries and by Luxembourg's broadcasting satellites ASTRA. Narrowband public datacasting of

The development of novel wireless technologies in the past 15 years has spurred a fascinating drive towards new multi-user systems and applications on either side of the Atlantic ocean.

traffic information to automobile radios is offered in several European countries, using residual transmit capacity on standard FM-radio carriers in the broadcast band 88 MHz to 108 MHz and the European Radio Data System (RDS) standard [13].

Conclusion: Different Cultural Paradigms.

The development of novel wireless technologies in the past 15 years has spurred a fascinating drive towards new multi-user systems and applications on either side of the Atlantic ocean. But while the enabling technologies are largely the same, some clear differences in the approach and involvement of European and American regulatory standardization bodies can be noted. The stronger European tradition of involvement of public authorities in telecommunications leads to firmer and more widely accepted standards for new public wireless systems, such as GSM, DECT, teletext and RDS. These common standards are proving of great commercial value not only to European manufacturers and system integrators of mobile infrastructure, but also to terminal equipment manufacturers, service operators and users world-wide.

In the United States, the government's role in steering communications system and technology R & D has traditionally been limited to the defense and aerospace sectors. The increasing competition between telecommunication operators now seems to result in an overwhelming 'smorgasbord' of alternative wireless technologies [4]. Frequently, these are based on military spin-offs such as CDMA and packet radio, with as yet no clear market winner in terms of successful public standards. On the other hand, the more competitive U.S. attitude to the use of computer and information technology, and the considerable financial resources of the BOCs, prove very powerful in the marketing of the new wireless services in the crumbling European telecommunication monopolies.

Most foundations of packet radio and CDMA were certainly laid in the United States, as evidenced by [7]. However, the shift from military to public multi-user networks required paradigm shifts in research focus which may perhaps have come easier in Europe. Thus, an earlier engineering attention appears in European system studies to the — often counter-intuitive — influence of physical channel impairments on random-access methods, cellular engineering and power control, as reviewed and further developed in [11] and [14]. Implications of other major paradigm shifts of information theory in the event of multiple self-interfering users [15], such as logarithmically unbounded traffic throughput with increased channel load, similar to the 'cock-tail-party' shouting effect, and non-applicability of the classical source/channel separation theorem in radio networks, have yet to be fully realized.

In more than one sense, all these changes and differences in economic, regulatory and technological research paradigms illustrate the fact that communication systems engineering is not merely a discipline related to the sciences of nature, but also to the protocols of culture.

References

[1] Moe Rahnema, "Overview of the GSM System and Protocol Architecture", *IEEE Commun. Mag.*, vol. 31, no. 4, pp. 92-100, April 1993.
[2] J. C. Arnbak, "Economic and Policy issues in the Regulation of Conditions for Subscriber Access and Market Entry to Telecommunications," in W. F. Korthals Altes *et al.*, eds., Information Law Towards the 21st Century, (Kluwer, Boston, 1992).
[3] European Telecommunication Standards Institute (ETSI), "Bearer Services Supported by GSM PLMN," Rec. GSM 02.02, Jan. 1990.
[4] K. Lynch, "U.S. Seen Losing Cellular Advantage," *Commun. Weekly*, March 22, 1993.
[5] K. Parsa, " The Mobitex Packet-Switched Radio Data System," *Proc. 3rd IEEE PIMRC Symposium*, Boston, MA, pp 534-538, 1992
[6] J. C. Arnbak, J. H. Bons, and J. W. Vieveen, " Graphical Correspondence in Electronic-Mail Networks Using Personal Computers," *IEEE Journ. Select. Areas. Comm.*, vol. J-SAC7, no. 2, pp. 257-267, Feb. 1989.
[7] N. Abramson, ed., *Multiple Access Communications — Foundations for Emerging Technologies*, (Selected Reprint Volume, IEEE Press, New York, 1993).
[8] C. van der Plas and J.P.M.G. Linnartz, "Stability of Mobile Slotted ALOHA Network with Rayleigh Fading, Shadowing and Near-far Effects," *IEEE Trans. Vehicular Technol.*, vol. VT-39, pp. 359-366, Nov. 1990.
[9] F. Kuperus and J. Arnbak, "Packet radio in a Rayleigh channel," *Electronics Lett.*, vol. 18, pp. 506-507, June 10, 1982.
[10] J.C. Arnbak and W. van Blitterswijk, "Capacity of slotted ALOHA in Rayleigh fading channel," *IEEE Journ. Select. Areas Commun.*, vol. JSAC-5, pp. 261-269, Feb. 1987
[11] J.P.M.G. Linnartz, "Effects of Fading and Interference in Narrow-band Land-mobile Networks," Ph.D. thesis, Delft University, 1991. (Also published as: J.-P. Linnartz, *Narrowband Land-Mobile Radio Networks*, Artech House, Boston and London, 1993.)
[12] A. Safak and R. Prasad, "Effects of correlated shadowing signals on channel reuse in mobile radio systems," *IEEE Trans. Vehicular Technol.*, vol. 40, no. 4, pp. 708-713, Nov. 1991.
[13] D.S. Chadwick *et al.*, "Communications Architecture for Early Implementation of Intelligent Vehicle Highway System," *IEEE Vehicular Technol. Society News*, vol 40, no 2, pp 63-70.
[14] J. Zander, "Distributed Cochannel Interference Control in Cellular Radio Systems," *IEEE Trans. Vehicular Technol.*, vol. 41, no. 3, pp. 305-311, 1992.
[15] T. M. Cover and J. A. Thomas, *Elements of Information Theory*, (Wiley, New York, 1991). See in particular Chapter 14: Network Information Theory.

Biography

JENS CHRISTIAN ARNBAK received Master's and Doctor's degrees from the Technical University of Denmark. From 1972 to 1980 he planned and designed international integrated digital communications networks and satellite systems for NATO at STC in The Netherlands. Between 1980 and 1986 he held a chair of radiocommunications at Eindhoven University of Technology, The Netherlands. Since May 1986, he has been a professor of tele-information techniques at Delft University of Technology, The Netherlands. He has published numerous scientific papers on electromagnetic wave propagation, satellite communication, and packet radio systems and was program chairman of the ICCC/IFIP-TC6 conference "ISDN in Europe" in 1989. Since 1982, he has participated in various telecommunications policy studies commissioned by the Dutch Government, *inter alia* recommending the new structure and status of the Netherlands PTT. He was a member of the Government Committee reviewing the Dutch Penal Code with respect to computer crime, and of the Council for Post and Telecommunications advising the Dutch Government. He is now the chairman of the Dutch Prime Minister's Coordinating Committee for Information and Communication Policy.

Cellular Packet Communications

DAVID J. GOODMAN, FELLOW, IEEE

(*Invited Paper*)

Abstract—Cellular mobile radio and residential cordless telephones are two new communication techniques with rapidly growing public acceptance. Present products and services use first-generation technology based on analog voice transmission. Second-generation equipment, conforming to at least five different sets of standards, is on the drawing boards. Based on digital speech transmission, second-generation systems will be introduced over the next three years.

This paper looks further into the future at third-generation wireless networking. The vision of the third-generation is a single set of standards that can meet a wide range of wireless access applications. Third-generation systems, in harmony with broad-band integrated services digital networks, will use shared resources to convey many information types. A single network architecture will serve its users efficiently in many environments including moving vehicles, indoor and outdoor public areas, residences, offices, and factories.

We are studying a switching architecture, referred to as a cellular packet switch, and a packet transmission technique referred to as packet reservation multiple access. This paper introduces both techniques and, by means of a design example, shows how they can work together to meet some of the demands of third-generation wireless networking.

I. INTRODUCTION

A. Fixed and Wireless Networking

INTEGRATED networking is the unifying theme of telecommunications systems evolution in the final quarter of the twentieth century. The purpose of an integrated network is to collect information from diverse sources and deliver it to designated destinations. In networks with wide area (national and international) coverage, advances in technology are coordinated under the heading ISDN (integrated services digital network). ISDN provides a structured set of protocols for moving and processing information in public networks. Standards are now in place for voice communications and for the exchange of data at low and moderate speeds. Thanks to recent work on broadband ISDN [1], a network architecture for communication of information from high-rate sources is emerging and keeping pace with advances in the technologies of transmission, switching, and signal processing.

Local area networks and metropolitan area networks, with more restricted coverage areas, are likewise progressing from research to standardization to implementation in an orderly manner. Evidence of this progress is the sequence of "IEEE 802" standards [2]–[4], covering existing and prospective networks including a short-range contention bus (IEEE 802.3), a token bus (IEEE 802.4), a token ring (IEEE 802.5), and a dual bus (IEEE 802.6). As with ISDN, research and development anticipate applications of local and metropolitan networks. As new requirements for integrated networking emerge, there is an abundance of technology to satisfy them.

The progress of wireless networking, on the other hand, stands in sharp contrast to that of fixed networks. Public demand for wireless

Paper approved by the Editor-in-Chief of the IEEE Communications Society. Manuscript received November 11, 1988; revised July 27, 1989. This paper was presented at the National Communications Forum, Chicago, IL, October 3, 1989.

D. J. Goodman is with the Department of Electrical and Computer Engineering, Rutgers University, Piscataway, NJ 08855-0909.

IEEE Log Number 9037484.

communications has consistently exceeded the capacity of available technology. Familiar examples are the congestion of citizen band radio, interference problems with cordless telephones, and the overloading of cellular networks in densely populated areas. Another marked departure from fixed networks is the state of wireless standards. With cellular telephony in the first decade of its existence, there are at least six different incompatible systems in use throughout the world. The AMPS system, of the United States and Canada, is still governed by a set of "interim standards." As industry works to adopt "full standards" for toady's cellular systems, work progresses at a rapid pace on second-generation technology. It is clear that, in common with the first-generation, second-generation services will be provided by at least five different incompatible systems, three for mobile telephony and two for cordless telephones.

The subject of this paper is third-generation wireless information networks, which will be required in the late 1990's. Our vision of the third-generation is a unified mode of wireless access, merging today's cellular and cordless applications in harmony with twenty-first century information services. As a step toward making this vision a reality, we are studying packetized information transfer in wireless networks. We believe that a network architecture based on packet transmission and switching will be capable of meeting a vastly expanded demand for wireless access to advanced information networks.

B. Wireless Networking Problems

Fig. 1 displays three network elements essential for wireless access to a fixed information network: wireless terminals, base stations, and switches. A base station is the fixed end of the radio channel that links wireless terminals to the remainder of the wireless information network. A switch connects the wireless and fixed networks. Switches also control communications within the wireless network infrastructure. This infrastructure is the set of base stations and switches, and the communications links that enable the system to coordinate the functions of its separate elements. Wireless information networks differ from conventional networks in several ways. For example,

- the communication channels linking base stations and terminals are unpredictable and highly variable with time;
- a wireless network is highly dynamic—its configuration changes many times each second as wireless terminals change location; and
- spectrum reuse is essential to high throughput because the radio bandwidth available for wireless networking is strictly limited.

1) Wireless Communication Channels: Wireless communication networks use ultrahigh frequencies to link terminals with the fixed infrastructure of base stations and switches. The propagation of radio signals between a wireless terminal and a base station is strongly influenced by

- the terminal-to-base distance;
- terrain, architectural features, and other environmental details;
- interference from other transmissions in the same system; and
- noise of many varieties.

All of these characteristics are unpredictable at the beginning of a communication session and are subject to substantial and rapid changes during the progress of the communication. Together they create a challenging transmission environment. Added to the chal-

Reprinted from *IEEE Transactions on Communications*, Vol. 38, No. 8, pp. 1272-1280, Aug. 1990.

453

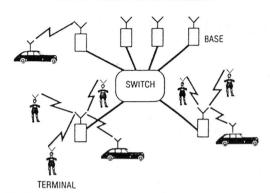

Fig. 1. Network elements for wireless access to fixed networks.

lenges of the channel impairments are those of limited bandwidth and the cost, size, power consumption, and physical robustness requirements of battery-powered consumer products. Reliable information transfer under these conditions requires sophisticated deployment of a wide range of techniques of advanced communications technology. New wireless access systems will employ low bit-rate speech coding [5], robust modulations [6], [7], novel error correcting codes matched to the speech codes [8], [9], frequency hopping [10], diversity reception [11], and nonlinear equalization [12], [13].

2) Network Rearrangements: In cellular systems, transmissions are confined to small service areas referred to as cells [14]. The most impressive properties of a cellular communication system are its ability to initiate communication with a user in any cell and to maintain the communication as the user moves from cell to cell. Thus, in contrast to other communication networks, wireless networks are subject to continual and rapid changes in configuration. Call set-up and handoff (the process of switching the communication from a channel served by one cell to another channel served by an adjacent cell) are currently performed by a mobile telephone switching office (MTSO) that controls all the calls in a large geographical area [15].

The rapidly growing demand for cellular telephone service strains the capacity of MTSO's to control all of the calls in their service areas. Future systems, with smaller cells and a higher volume of network control operations, will place considerably greater burdens on the processing capabilities of cellular controllers. These burdens will best be carried by distributed network controllers, rather than the centralized controllers of present systems.

3) Spectrum Reuse: The high capacity of a cellular system is due to the fact that any wireless channel can be used simultaneously for many communications in each service area. Cellular systems typically begin operation with a few large cells (on the order of 10 km rad) and, to meet increasing service demand, divide the original cells into smaller cells, each with the same number of assigned channels as the original large cell. Thus, the traffic density and spectrum efficiency increase with cell division. For various reasons, cellular operators find that, with present base station technology, there is a limit on minimum cell size of about 1 km rad.

On this scale, the maximum achievable traffic density in present cellular systems is less than 1 E/MHz/km^2. Although second-generation technology may be able to increase this density by nearly an order of magnitude, demand estimates for the end of this century are as high as 5000 E/km^2. With about 50 MHz bandwidth available for wireless networks, an increase of another factor of 10 will be required just to meet the demand for voice services. Even higher spectrum efficiency will be required to provide the variety of information services that will be required by third-generation network users.

II. TECHNOLOGY EVOLUTION

The most familiar means of wireless access to information networks are cellular telephones and residential cordless phones. Both are new products experiencing rapid growth in public acceptance. These products use first-generation technology, characterized by

analog (frequency modulation) voice transmission and limited flexibility.

In cordless telephony, the main limitation is vulnerability to interference. Most cordless phones have access to only one radio frequency channel. If two people in the same area have cordless telephones that operate over the same channel, they cannot use their cordless telephones at the same time.

The network control subsystem imposes a severe limitation on first-generation cellular systems. During the progress of a call, all communication between a mobile terminal and the network infrastructure is confined to a single voice channel. Network control messages—for handoff or power control, for example—are carried over the voice channel on a "blank and burst" basis. Because control messages interrupt speech transmission and produce audible clicks, the network control capacity of a cellular system is strictly limited.

A variety of second-generation systems are on the drawing boards in several countries. All of them will employ digital voice transmission. Second-generation cellular systems will have dedicated channels for the exchange of network control information between mobile terminals and the network infrastructure during a call. Cordless phones will have access to many channels, automatically selecting the best available channel at the beginning of a call and, in some cases, switching to a new channel as conditions change during the call.

The work reported here looks beyond second-generation technology to the end of this century. The aim is to provide a technological basis for third-generation wireless networking. In the third generation, the distinction between cordless phones and cellular mobile telecommunications will disappear. Millions of people will carry a lightweight, inexpensive terminal that will give them access to worldwide integrated information networks. Other more specialized terminals will provide wireless access to a wide range of broad-band ISDN information services. The challenges of third-generation wireless access are to create networking techniques that are capable of the following:

- carrying many types of information;
- serving a mass market in urban areas;
- operating efficiently in sparsely populated rural areas;
- operating indoors, outdoors, and within vehicles; and
- serving stationary terminals and terminals moving at high speed.

A. Second-Generation Wireless Access

Five specific systems, providing three types of digital wireless access, have received extensive attention in recent years. There are two cellular standards including a Pan-European mobile radio system, referred to as GSM [16], and an emerging North American mobile radio standard, referred to as TIA.[1] Likewise, there are two approaches to enhanced cordless telephony: CT2, a British standard [17], and DECT, digital European cordless telephone [18], [19], under consideration by the association of European operating telephone companies. It is highly likely that all four of these systems will be in commercial use in the early 1990's. In addition, Bell Communications Research has proposed Universal Digital Portable Communications [20] as a means of delivering basic services to telephone company subscribers.

B. Third-Generation Networks

Third-generation wireless information networks will provide access to advanced information services. They will accommodate a diverse set of information sources, operate in all transmission environments, and be capable of serving a very high user population. To make third-generation networking possible, technology must advance substantially beyond the second-generation state of the art. Second-generation systems will be optimized for voice transmission with limited capacity for other kinds of information. Each second-generation system is designed for a specific environment:

---

[1] A third digital cellular standard for Japan will be formulated in 1990.

mobile, indoor, or local telephone distribution, with undesirable characteristics in other environments. Moreover, it is expected that by the late 1990's demand for wireless services will overtake the capacity of second-generation technology.

To meet the challenges of the third generation, researchers have begun in the past few years to study microcells [21], [22] with terminal and base station power levels at least two orders of magnitude lower than in today's cellular systems. While first- and second-generation cellular systems are based on cell dimensions on the order of one to several kilometers, the minimum distance between base stations in third-generation systems will be on the order of 100 m. Thus, each base station will have a service area close to that of present and prospective cordless phones.

In contrast with cordless telephony, however, integrated wireless networks will have full call set-up and handoff capabilities. With small cell dimensions, the volume of handoffs will be high and the required network control will place heavy burdens on the infrastructure. To support these burdens, call control and call management functions will have to be decentralized and distributed over many processors.

C. Scope of this Paper

The remainder of this paper describes two new packet communication techniques devised for third-generation wireless networks. Among their advantages for this application are that

- they readily accommodate mixed information types;
- they simplify the routing of information in a network subject to continual reconfiguration; and
- they harmonize with the anticipated evolution of fixed networks.

The communication techniques are a wireless network infrastructure, referred to as a cellular packet switch, described in Sections III and IV, and a terminal-to-base transmission protocol, referred to as packet reservation multiple access, introduced in Section V. Sections VI and VII give design examples of networks that incorporate both a cellular packet switch and the PRMA transmission protocol. The emphasis throughout is on the ability of this new architecture to coordinate the operation wireless network elements in a voice communication application. With this architecture established, future work will focus on other important wireless networking issues including details of the transmission system, such as: coding, modulation, and equalization; efficient spectrum management; and movement of the diverse types of information associated with advanced information services.

III. CELLULAR PACKET SWITCH ARCHITECTURE

The cellular packet switch distributes network control among small processors, referred to as interface units, residing in all network elements. It uses the address fields of each packet to provide routing information corresponding to the changing locations of wireless terminals.

As indicated in Fig. 2, we view the infrastructure as a metropolitan area network (MAN) linking base stations, public switches, and a cellular controller. Information enters and leaves the MAN through cellular interface units including base station interface units (BIU); trunk interface units (TIU), each connected to a central office trunk of the public network; and a controller interface unit (CIU), connected to the cellular controller. The packet switching capability of a MAN works well with the PRMA protocol for information transfer between base stations and wireless terminals described in Section V. To marry PRMA to the cellular packet switch, we introduce a WIU (wireless terminal interface unit) to each terminal.

The WIU, BIU, TIU, and CIU organize information transfer among wireless terminals, base stations, central office trunks, and the cellular controller, respectively. Each packet contains a source address and a destination address. Sometimes an address is the permanent identifier of an interface unit. At other times, the address is a virtual circuit identifier associated with a particular communica-

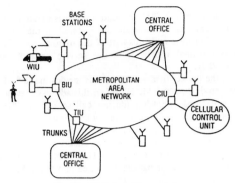

Fig. 2.   Cellular packet switch.

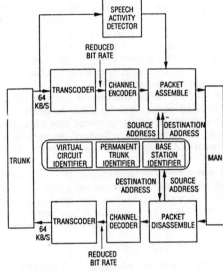

Fig. 3.   Cellular trunk interface unit.

tion session. The addressing procedures are discussed in Section IV in the context of specific communication and control functions. In this section, we describe the capabilities of the interface units.

A. Cellular Trunk Interface Unit

The TIU accepts and delivers information in the standard format of the public network. The speech format, for example, is 64 kbps companded pulse code modulation. As indicated in Fig. 3, the TIU converts this information to and from the format of the wireless access physical layer by means of transcoders and channel coders matched to the wireless access environment of the cellular packet switch. Each cellular packet switch can be customized for its own transmission environment (for example, urban mobile, indoor, or mobile satellite) by means of the transcoders and channel coders installed in the TIU's. An architecture that admits many terminal-base transmission technologies, each matched to a specific environment, is important to the successful operation of future wireless access systems [23].

In addition to transforming user information between the formats of the fixed network and the wireless access channels, the TIU contains a packet assembler and disassembler (PAD). A PAD combines user information or network control information with a packet header. The header contains flags, an error control field, a packet control field, and an address field. The address of the TIU, which is inserted in every outgoing packet, can be the permanent trunk identifier or it can be a virtual circuit identifier assigned by the cellular controller. A speech activity detector controls the generation of packets by the PAD. During silent intervals, no packets are produced, thus conserving transmission resources of the MAN and the wireless channels.

Packets sent from the TIU are routed to a base station by means of the permanent base station identifier stored in the simple TIU routing table. During a call, this part of the routing table changes as the wireless terminal moves from the service area of one base station to another.

In Fig. 3, the packet disassembler reads the destination address of all packets arriving on the MAN. When this address matches either the permanent trunk identifier (during call set-up), or the virtual circuit identifier (during a call), the packet disassembler processes the arriving packet. If the packet has arrived from a base station, the TIU records the source address of the packet in the base station identifier register of the routing table. This identifier then becomes the destination address for packets launched into the MAN from the TIU.

B. Wireless Terminal Interface Unit

In generating packets, the WIU is similar to the TIU. One difference is that the TIU receives user information from the public network, while the WIU is connected directly to the information source, through, for example, a 64 kbps analog-to-digital converter for speech signals. As indicated in Fig. 4, the packet assembler of the WIU delivers packets to the radio transmitter under the control of a PRMA protocol processor.

As in the TIU, the packet disassembler compares the destination address of received packets with the either the permanent terminal identifier (during call set-up) or the virtual circuit identifier (during a call). It extracts the information fields of speech packets destined for this terminal and converts them to a continuous 64 kbps signal stream.

In order to implement terminal-initiated handoff, the WIU refers to a channel quality monitor to determine a base station identifier. This monitor indicates the identity of the base station best able to serve the terminal in its current location. This base station becomes the destination of packets sent from the wireless terminal.

C. Base Station Interface Unit

This unit relays information between the TIU's and the wireless terminals. It also broadcasts, over its radio channel, the feedback packets called for by the PRMA protocol. The BIU multiplexes information packets that are sent to the radio transmitter. It also queues upstream packets for transmission over the MAN.

The BIU is always addressed by its permanent identifier. If an incoming packet arrives with a virtual circuit identifier in its address field, this identifier becomes the destination address when the packet is relayed, either to a TIU (upstream packet) or to a WIU (downstream packet). Certain network control packets arrive without virtual circuit identifiers. These packets are either relayed to the cellular controller or they are relayed to a WIU by means of the permanent identifier of the WIU. This identifier is extracted from the information field of the control packet. An example of this routing procedure appears in Section IV-C, Fig. 7.

D. Cellular Controller Interface Unit

The cellular controller receives, processes, and generates network control packets. It is always addressed by its permanent identifier. It assigns a virtual circuit identifier to each cellular call and sends this number to the TIU selected for the call and to the relevant WIU. To distribute the virtual circuit identifier to the TIU, the CIU uses the permanent identifier of the TIU. To send the virtual circuit identifier to the WIU, the CIU places the base station identifier in the destination address field of a control packet and the permanent terminal identifier in the information field of the control packet.

IV. CELLULAR PACKET SWITCH NETWORK CONTROL

By referring to three examples: conversational speech, handoff, and terminal-initiated call set-up, we show how a cellular packet switch organizes the flow of user information and system control information. The source and destination addresses of each packet control the routing of the packet to the correct interface unit. Prior

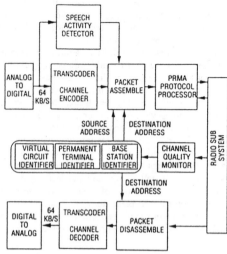

Fig. 4. Wireless terminal interface unit.

to call set-up, terminals and trunks are addressed by their permanent identifiers. In setting up a call, the cellular controller assigns to the call a virtual circuit identifier. This number then becomes the address of both the TIU and the WIU involved in the call. Base stations and the cellular controller are always addressed by their permanent identifiers.

A. Conversation (No Handoff)

While the wireless terminal remains in a single cell, packets move from terminal to base station to central office trunk (and in the opposite direction) in a straightforward manner. Fig. 5 uses the analogy of postcards to illustrate the bidirectional flow of speech and address information. Here we have a call in progress between wireless terminal W3011 and trunk T604. The terminal is in the cell served by base station B43. At the beginning of the call, the controller assigned virtual circuit identifier 4311 to the call and transmitted this number to the WIU at W3011 and the TIU at T604. Both interface units retain this virtual circuit identifier in their routing tables for the duration of the call. Each postcard (packet) in Fig. 5 contains a destination address (on the right side under the stamp), a return address (in the top left corner), and speech information (in quotes). The sequence of packet transfers is from top to bottom on the page. Under the control of speech activity detectors in the TIU and WIU, the packets generated during a conversation flow in bursts, referred to as talkspurts, each containing on average about 60 packets.

B. Handoff

The cellular packet switch hands a call from one base station to another when the wireless terminal determines that the call can best be handled by the new base station. As indicated in Fig. 6, the terminal initiates the handoff by sending a packet to the new base station, B19, instead of B43, which received earlier packets. Because the speech packet contains virtual circuit identifier 3011, it is intercepted by the TIU at T604. This TIU learns the identity of the new base station, writes B19 in its routing table, and sends new packets to the terminal through B19. Because the cellular controller plays no role in the handoff process, its workload is unaffected by the volume of handoffs, which can be very high in a microcellular system. In fact, the WIU can change base stations with every packet it sends without overloading the processing capability of the system because each handoff requires only that the TIU write a new base station identifier in its routing table.

If, during a conversation, there is a long silent gap in the speech entering the wireless terminal, the WIU will transmit special null packets in order to keep the TIU informed of the present location of the wireless terminal.

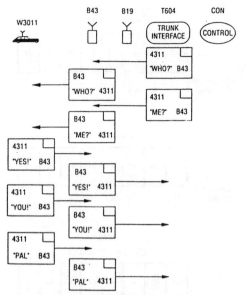

Fig. 5.   Conversation, no handoff. Each postcard represents one packet.

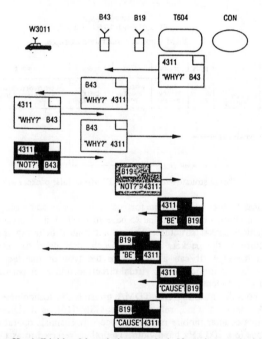

Fig. 6.   Handoff initiated by wireless terminal. No action required by the controller.

C. Terminal-Initiated Call Set-Up

Fig. 7 shows one possible scenario for a call set-up in a cellular packet switch. Here, we have a sequence of control packets leading to the transmission of the first speech packet ("Hello") from the public network to the wireless terminal.

First, the terminal sends an "off hook" message to the nearest base station. The base station relays the message, and the identity of the wireless terminal, to the cellular controller. The controller uses the information in this message to authenticate the calling party. If the caller is authorized to place a call, the controller returns a "dial tone" message to the base station. The BIU extracts the terminal identifier from the information field of the dial tone message and uses this identifier to relay the dial tone to the WIU. The response to this message is a packet containing the called party's number. This

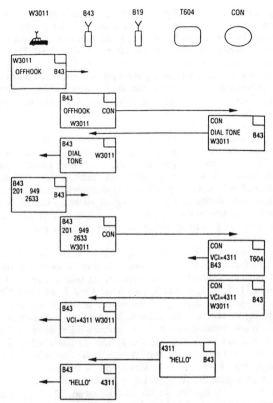

Fig. 7.   Terminal-initiated call set-up. Controller sends virtual circuit identifier (VCI) 4311 to TIU and WIU.

enables the controller to attempt to establish a connection, through the local central office, to the called party.

When the connection is established, the controller issues a virtual circuit identifier to the WIU that initiated the call and to the TIU assigned to this call by the controller. With the virtual circuit identifier recorded in both the WIU and the TIU, the conversation proceeds as in Fig. 5. Handoffs, as necessary, take place as shown in Fig. 6.

V. PACKET RESERVATION MULTIPLE ACCESS

Thus far, we have described the cellular packet switch, which moves user information packets and network control packets between base stations, public network trunks, and a cellular controller. In this section, we briefly describe packet reservation multiple access [24], [25] (PRMA), a mobile-to-base transmission protocol. In Section VI, we will show how a PRMA transmission system can be designed to provide an efficient interface between the wireless communication links and a specific metropolitan area network.

PRMA, a close relative of reservation ALOHA [26], [27], can be viewed as a combination of time division multiple access (TDMA) and slotted ALOHA [28]. The channel bit stream is first organized in *slots*, such that each slot can carry one packet from a terminal to the base station. The time slots, in turn, are grouped in frames. Within a frame, terminals recognize each slot as being either *available* or *reserved* on the basis of a feedback packet broadcast in the previous frame from the base station to all of the terminals. As in slotted ALOHA, terminals with new information to transmit contend for access to available slots. At the end of each slot, the base station broadcasts the feedback packet that reports the result of the transmission. A terminal that succeeds in sending a packet to the base station obtains a reservation for exclusive use of the corresponding time slot in subsequent frames.

The frame duration is chosen such that a speech terminal generates one packet per frame. Upon gaining a reservation, the terminal continues to use its reserved slot until it no longer has an information packet to transmit. The base station, on receiving no packet in the reserved slot, informs all terminals that the slot is once again available. In subsequent frames, terminals with new information to transmit contend for that slot and other available slots. The contention mechanism is based on a permission probability p, a design constant for all terminals in a given PRMA system. A contending terminal waits for an available slot and, with probability p, transmits a packet to the base in the available slot. In our studies to date, we have assumed that the transmission succeeds if and only if one terminal transmits in a slot.

Because each terminal with a reservation has exclusive use of one slot in each frame, the terminals with reservations share the channel as in TDMA. With terminals accessing the channel autonomously, PRMA lends itself well to operation with speech activity detectors, as indicated in Fig. 4. A speech activity detector classifies a speech signal at each instant as either *talking* or *silent* so that the voice terminal generates bursts of packets, corresponding to talkspurts. No packets are generated during the silent gaps that punctuate the talkspurts.

Fig. 8 illustrates the operation of PRMA. There are eight time slots per frame and the base station feedback packets for frame $k - 1$ have established that, in frame k, six slots are reserved by terminals 11, 5, 3, 1, 8, and 2. Two slots, 3 and 7, are available in frame k. At the beginning of the frame, terminals 6 and 4 are contending for access to the channel. Both of these terminals obtain permission to transmit in slot 3 and, because their packets collide, neither obtains a reservation. In slot 7, both terminals fail to obtain permission to transmit and, thus, remain in the contending state at the beginning of frame $k + 1$. Meanwhile, in frame $k - 1$, terminal 3 transmitted the final packet in its talkspurt. Therefore, in frame k (slot 4), it does not use its reservation. The base station feedback packet for slot 4 of frame k indicates that slot 4 is available in frame $k + 1$.

In frame $k + 1$, neither terminal 6 nor terminal 4 has permission to transmit in slot 3. In slot 4, terminal 4 has permission but terminal 6 does not. Thus, terminal 4 gains a reservation for slot 4. Terminal 6 obtains permission to transmit in slot 7 and reserves that slot in frame $k + 2$. In frame $k + 1$, terminal 8 gives up its reservation of slot 6, and a talkspurt begins at terminal 12 which enters the contending state. In frame $k + 2$, terminal 12 gains a reservation (slot 3) and terminal 1 releases its reservation (slot 5).

At any time, the number of available slots depends on the number of terminals in the talking state, a quantity that is subject to statistical fluctuation. When there are few available slots, terminals with new talkspurts are subject to relatively long delays in obtaining reservations. Unlike packet data systems, which respond to congestion by storing packets until they can be transmitted, packet speech systems must deliver packets promptly. In PRMA, any packet held beyond a certain specified time limit, D_{max}, is discarded by its terminal. Thus, a measure of PRMA performance is P_{drop}, the probability of packet dropping. To estimate system capacity, we set the limit, $P_{drop} \leq 0.01$, and find the maximum number of terminals that can operate within this limit. A mathematical analysis of PRMA [29] provides us with computing tools to assess the capacity of a cellular system that combines PRMA and a cellular packet switch operating with a specific MAN.

VI. COMBINED OPERATION OF CELLULAR PACKET SWITCH AND PRMA

As an example of a MAN to be incorporated in a cellular packet switch, we consider DQDB, distributed queue dual bus [30], [31], to be standardized as IEEE 802.6. In one proposed configuration of DQDB, there is a fixed packet size of 69 bytes, including a 5 byte header.[2] To unite PRMA with DQDB, we consider transmitting over the wireless channels an entire DQDB packet, which is received by a base station and inserted into the MAN. We postulate that, in addition to the DQDB header, PRMA will require an

---

[2] After this paper was written, the DQDB packet size was established at 53 bytes, the basis of revised PRMA designs.

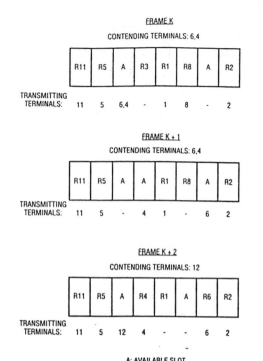

Fig. 8. PRMA protocol operation example.

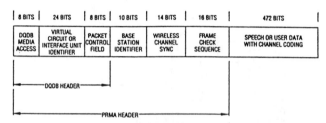

Fig. 9. Packet structure for combined PRMA cellular packet switch.

additional 5 bytes of overhead for such purposes as addressing and synchronization of the terminal-to-base receiver. Fig. 9 presents a hypothetical packet structure. The packet control field can specify the nature of the packet, such as speech, user data, or network control message. It can also indicate the type of interface unit address carried in the packet, virtual circuit identifier, or permanent interface unit identifier.

We consider adapting two TDMA transmission technologies for PRMA operation. One, referred to as GSM [16], is designed for wide area cellular mobile radio with one-way channels operating at 270 kbps in a 200 kHz bandwidth. The other, referred to as DECT [18], [19], is designed for indoor wireless communications with two-way channels operating at 1.344 Mb/s in a bandwidth of 2 MHz. To analyze PRMA operating with these transmission technologies, we specify two speech coding rates for each system: 8 kb/s and 16 kb/s for GSM, and 16 kb/s and 32 kb/s for DECT.

We thus have four PRMA configurations (two channel rates, each with two speech coding rates) with parameters presented in spreadsheet form in Table I. In all four, the packet size is 69 bytes (552 b) including 59 bytes (472 b) of speech information and 10 bytes of other information, including framing, synchronization, addresses, and error detection. In most applications, it is likely that the 59 bytes of speech information will include an error-correcting channel code, and that the error-detecting bits in a PRMA packet (most likely, a two-byte frame check sequence) will operate only on the header fields of the packet. Row 5 of Table I is the terminal-to-base channel transmission rate, which is 270 kb/s for the mobile radio configuration. For the indoor radio system it is 672 kb/s, half of the two-way time division duplex transmission rate.

TABLE I
SYSTEM DESIGN SPREADSHEET

		GSM Transmission Mobile Radio Applications		DECT Transmission Indoor Applications		Calculation (rx—>row x)
1	source rate, kb/s	8	16	16	32	
2	packet size, bits	552	552	552	552	
3	speech bits	472	472	472	472	
4	header bits	80	80	80	80	
5	one-way channel rate, kb/s	270	270	672	672	r3/r1
6	frame duration, ms	59	29.5	29.5	14.75	Equation (1)
7	slots per frame	28	14	35	17	Equation (1)
8	delay limit, ms	32	32	32	32	
9	contention interval, slots	15	15	37	36	Equation (2)
10	permission probability	0.35	0.5	0.25	0.4	
11	capacity, terminals	40	22	65	30	Figures 10 and 11
12	equivalent channels	33.75	16.875	42.0	21.0	r5/r1
13	terminals per channel	1.2	1.3	1.5	1.4	r11/r12
14	throughput (kb/s)	161	177	523	483	r1*r11*0.43*(r2/r3)
15	normalized throughput	0.60	0.66	0.78	0.72	r14/r5
16	PRMA channels per MAN	310	282	95	103	100,000/(2*r14)
17	minimum conversations/base	100	100	50	50	
18	PRMA channels/base	3	5	1	2	least integer≥(r17/r11)
19	base stations per MAN	103	56	95	51	int[r16/r18]
20	erlangs per base station	107.4	97.7	54.4	49.6	ref [32], page36-41
21	erlangs per MAN	11062	5471	5168	2530	r19*r20
22	base station grid, meters	300	300	50	50	
23	base station array	10x10	8x7	10x9	7x7	product≤r19
24	service area, km²	9	5	0.225	.123	(r22)²*r23*10⁻⁶
25	cellular reuse factor	25	25	25	25	
26	2-way bandwidth/ base, MHz	1.2	2.0	2.0	4.0	
27	total bandwidth, MHz	30	50	50	100	r25*r26
28	erlangs/km²/MHz	41	22	459	206	r21/(r24*r27)

If the speech rate is R_s b/s, a PRMA speech terminal generates 472 speech bits every $T = 472/R_s$ seconds, which is the PRMA frame duration indicated in row 6 of Table I. With the channel rate R_c b/s, the number of slots per PRMA frame, displayed in row 7, is

$$N = \text{int} \left[\frac{R_c T}{R_s T + H} \right] \qquad (1)$$

where int $[x]$ is the greatest integer $\leq x$, and $H = 80$ is the number of nonspeech bits per packet. For all four PRMA configurations, we have set a speech packet delay limit of 32 ms, which determines D, the number of slots over which a terminal can contend for a reservation before dropping the oldest speech packet in its memory:

$$D = \text{int} \left[\frac{0.032 N}{T} \right]. \qquad (2)$$

This contention interval is indicated in row 9 of Table I.

Given the parameters in Table I and a permission probability p, we can use the analysis of [29] to calculate the probability of packet dropping as a function of the number of speech terminals sharing the terminal-to-base channel. Based on measurements of recorded speech, we assume that the mean duration of a talkspurt is 1 s, and that the mean duration of a silent gap is 1.35 s. The speech activity factor is, therefore, $1.00/(1.00 + 1.35) = 0.43$, the fraction of time that a terminal generates packets during a conversation.

Packet dropping probabilities appear in Fig. 10 for the mobile radio transmission scheme and, in Fig. 11, for the indoor radio technique. Each curve pertains to the permission probability, shown in row 10 of Table I, that maximizes the number of terminals that can share the PRMA channel within the packet dropping constraint $P_{\text{drop}} \leq 0.01$. This number of terminals, an indication of PRMA system capacity, appears in row 11 of Table I and, in a normalized form, in row 13. Row 13 can be viewed as the number of PRMA terminals per baseband channel. The number of baseband channels, row 12, is defined as R_c/R_s, the number of speech signals carried in an idealized TDMA system, operating without speech activity detection and with no overheads at all.

The number of terminals per channel is greater than one because the speech activity detectors allow PRMA to operate as a statistical

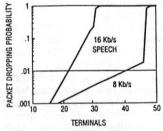

Fig. 10.   PRMA performance with GSM transmission system: 270 kbps wireless channel.

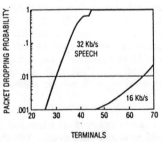

Fig. 11.   PRMA performance with DECT transmission system: 672 kbps wireless channel.

multiplexer in which the terminals in the silent state give up their channel resources (reserved time slots) to terminals in the talking state. The number of terminals per channel is bounded above by 2.35, the reciprocal of the 0.43 speech activity factor. The difference between 2.35 terminals per channel and an entry in row 13 is due to

- statistical fluctuations in the number of terminals with simultaneous talkspurts;
- collisions caused by terminals contending for the same time slot;
- an empty slot produced by PRMA at the end of each talkspurt; and
- nonspeech material that occupies $10/69 = 14\%$ of each packet.

VII. Network Topology Example

We consider a suggested configuration of the dual bus [31] in which the total bit rate is 156 Mb/s. Within this bit stream, we allocate 100 Mb/s for speech packets on the dual bus. The remainder of the bus traffic can be set aside for nonspeech information, including wireless network control messages and DQDB protocol control bits. This allocation ensures that the dual bus is never fully loaded and that the wireless channels, rather than the dual bus, limit the capacity and quality of communications. A cellular packet switch carrying 100 Mb/s of speech packets can handle the number of PRMA channels listed in row 16 of Table I. Row 16 is the ratio of 100 Mb/s to the two-way aggregate bit rate of the wireless channel. The one-way bit rate (row 14) is the product of the number of terminals (row 11), the source rate (row 1), the speech activity factor (0.43), and 69/59, the factor which accounts for the nonspeech content of each packet. The normalized PRMA throughput (row 15) is the ratio of row 14 to the channel rate (row 5).

To place these results in the context of an application environment, consider the GSM transmission system with 16 kb/s speech channels. Row 11 of Table I indicates that each PRMA channel can carry 22 simultaneous conversations. Let us postulate that the cost of installing a microcell base station is justified if the base station carries at least 100 simultaneous conversations. This implies that each base station should operate with five PRMA channels, with a total capacity of 110 conversations in a two-way bandwidth of 2 MHz. With a blocking probability of 0.02, the 110 voice channels can support 97.7 E of demand [32]. Furthermore, we find that the DQDB MAN, which carries the traffic of 282 PRMA channels (row 16 of Table I), can serve 56 base stations and carry 5471 E of teletraffic. If we place these base stations on a 300 meter grid in a city, as shown in Fig. 12, we have a cellular packet switch serving a rectangular area that is 2.4 by 2.1 km or 5 km^2 (2 square miles), which provides a traffic density of 1086 E/km^2. If this system operates conservatively with a channel reuse factor of 25, it will use a total bandwidth of 50 MHz, which is the allocation of second-generation (and some first-generation) cellular systems. In this case, the cellular efficiency is 1086/50 = 22 E/km^2/MHz, which exceeds present efficiencies by a factor of 20.

With 8 kb/s speech channels, similar reasoning (at least 100 channels per base station, reuse factor 25) leads to the conclusion that the cellular packet switch can provide 11062 E in a 9 km^2 service area. The total required bandwidth is only 30 MHz and the cellular efficiency is 41 E/km^2/MHz. If the systems in the two righthand columns of Table I are used indoors, it is reasonable to assume that they can operate economically with base stations serving on the order of 50 terminals each, and that the base stations are placed on a 50 m grid. This "pico-cell" arrangement leads to the very high service densities of the two last columns of row 28.

VIII. Conclusions

Cellular mobile radio and cordless telephones are technological infants, which will soon enter adolescence with the introduction of second-generation equipment. While more sophisticated than at present, the second-generation systems will retain many childhood characteristics. For example,

• they will function primarily for telephone communications with limited capacity for other types of information;
• their control structures will not support the highly dynamic networking that will be required at the end of this century;
• they will not use available radio spectrum with the efficiency needed to meet a mass-market demand for wireless access to advanced information networks.

This paper looks ahead to the third-generation, when wireless networks will assume the responsibilities of adulthood. They will conform to a common network architecture, so that the same terminals will be used for mobile communications in vehicles, by pedestrians in public areas, for residential applications, and for access to business information systems in offices, stores, factories,

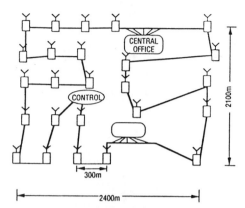

Fig. 12.   Example of cellular packet switch serving an area of 5 km^2.

etc. In common with plans for the evolving public switched telecommunications network, we are studying an architecture based on packet transmission and switching.

Our approach to switching, referred to as a cellular packet switch, places base stations and public network trunks on a high-speed metropolitan area network, such as a distributed queue dual bus. Functions currently performed by a single mobile telephone switching office will be distributed among thousands of interface units attached to base stations, public network trunks, and wireless terminals. This will allow future wireless information networks to grow gracefully and assume the call management burdens of the third-generation.

Matched to the cellular packet switch is a transmission technique referred to as packet reservation multiple access. A combination of slotted ALOHA and time division multiple access, PRMA makes efficient use of the radio channels that link base stations and wireless terminals. Four design examples, presented in an elaborate spreadsheet, show how two second-generation transmission techniques can be modified to operate with the PRMA protocol. each of these PRMA systems can be combined with a cellular packet switch based on the distributed queue dual bus metropolitan area network. The results are wireless information network configurations with many times the capacity of present systems.

Our work to date on third-generation wireless networks is confined to the ability of the message layer and the data link layer to support voice communications. Much remains to be done to understand how this architecture serves a wide variety of information sources, how it imposes requirements on the physical layer, and to what extent it can support the higher layers of an advanced network.

Acknowledgment

The author would like to thank M. Orsic of AT & T Bell Laboratories and S. Nanda of the Rutgers WINLAB.

References

[1] H. Armbuster and G. Arndt, "Broadband communication and its realization with broadband ISDN," *IEEE Commun. Mag.*, vol. 25, pp. 8–19, Nov. 1987.
[2] IEEE Project 802, Local Area Network Standards, IEEE, New York, Aug. 1984.
[3] G. E. Keiser, *Local Area Networks*. New York: McGraw-Hill, 1989, p. 54.
[4] J. F. Mollenauer, "Standards for metropolitan area networks," *IEEE Commun. Mag.*, vol. 26, no. 4, pp. 15–19, Apr. 1988.
[5] G. deBrito, "Low bit rate speech for Pan European mobile communication system," in *Proc. 38th IEEE Vehic. Technol. Conf.*, Philadelphia, PA, June 1988, pp. 147–152.
[6] J. Tarallo and G. I. Zysman, "Modulation techniques for digital cellular systems," in *Proc. 38th IEEE Vehic. Technol. Conf.*, Philadelphia, PA, June 1988, pp. 245–248.
[7] C.-E. Sundberg, "Continuous phase modulation," *IEEE Commun. Mag.*, vol. 24, no. 4, pp. 25–38, Apr. 1986.
[8] J. Hagenauer, C.-E. Sundberg, and N. Seshadri, "The performance

of rate-compatible punctured convolutional codes for future digital mobile radio," in *Proc. 38th IEEE Vehic. Technol. Conf.*, Philadelphia, PA, June 1988, pp. 22–29.

[9]   J. Hagenauer, N. Seshadri, and C.-E. Sundberg, "Variable rate sub-band speech coding and matched channel coding for mobile radio channels," in *Proc. 38th IEEE Vehic. Technol. Conf.*, Philadelphia, PA, June 1988, pp. 139–146.

[10] A. A. M. Saleh and L. J. Cimini, Jr., "Indoor radio communications using time-division multiple access with cyclical slow frequency hopping and coding," *IEEE J. Select. Areas Commun.*, vol. SAC-7, pp. 59–70, Jan. 1989.

[11] L. F. Chang and J. C.-I. Chuang, "Diversity selection using coding in a portable radio communications channel with frequency-selective fading," *IEEE J. Select. Areas Commun.*, vol. SAC-7, pp. 89–98, Jan. 1989.

[12] T. A. Sexton and K. Pahlavan, "Channel modeling and adaptive equalization in indoor radio channels," *IEEE J. Select. Areas Commun.*, vol. SAC-7, pp. 114–121, Jan. 1989.

[13] R. D'Avella, L. Moreno, and M. Sant'Agostino, "An adaptive MLSE reveiver for TDMA digital mobile radio," *IEEE J. Select. Areas Commun.*, vol. SAC-7, pp. 122–129, Jan. 1989.

[14] V. H. MacDonald, "The cellular concept," *Bell Syst. Tech. J.*, vol. 58, no. 1, part 3, pp. 15–42, Jan. 1979.

[15] K. J. S. Chadha, C. F. Hunnicutt, S. R. Peck, and J. Tebes, Jr., "Mobile telephone switching office," *Bell Syst. Tech. J.*, vol. 58, no. 1, part 3, pp. 71–96, Jan. 1979.

[16] B. J. T. Mallinder, "An overview of the GSM System," in *Third Nordic Semi. Digital Land Mobile Radio Commun.*, Copenhagen, Denmark, Sept. 1988, pp. 3.1.1–3.1.13.

[17] R. S. Swain and A. J. Motley, "Requirements of an advanced cordless telecommunication product," in *Proc. 1987 Int. Commun. Conf.*, Seattle, WA, June 1987, pp. 81–85.

[18] H. Ochsner, "DECT—Digital european cordless telecommunications," in *Proc. 39th IEEE Vehicular Technol. Conf.*, San Francisco, CA, May 1989, pp. 718–721.

[19] D. Akerberg, "Properties of a TDMA pico cellular office communication system," in *Proc. IEEE Global Commun. Conf. GLOBECOM '88*, Hollywood, FLA, Dec. 1988, pp. 1343–1349.

[20] D. C. Cox, "Universal digital portable radio communications," *Proc. IEEE*, vol. 75, pp. 436–477, Apr. 1987.

[21] R. Steele and V. K. Prabhu, "Mobile radio cell structure for high user density and large data rates," *Proc. IEE, Part F*, no. 5, pp. 396–404, Aug. 1985.

[22] R. Steele, "The cellular environment of lightweight handheld portables," *IEEE Commun. Mag.*, vol. 27, no. 7, pp. 20–29, June 1989.

[23] E. S. K. Chien, D. J. Goodman, and J. E. Russell, Sr., "Cellular access digital network. (CADN): Wireless access to networks of the future," *IEEE Commun. Mag.*, vol. 25, no. 6, pp. 22–31, June 1987.

[24] D. J. Goodman, R. A. Valenzuela, K. T. Gayliard, and B. Ramamurthi, "Packet reservation multiple access for local wireless communications," *IEEE Trans. Commun.*, vol. 37, pp. 885–890, Aug. 1989.

[25] D. J. Goodman and S. X. Wei, "Factors effecting the bandwidth efficiency of packet reservation multiple access," presented at *Proc. 39th IEEE Vehic. Technol. Conf.*, San Francisco, CA, May 1989, pp. 292–299.

[26] S. Tasaka, "Stability and performance of the R-ALOHA packet broadcast system," *IEEE Trans. Comp.*, vol. C-32, pp. 717–726, Aug. 1983.

[27] S. S. Lam, "Packet broadcast networks—A performance analysis of the *R*—ALOHA protocol," *IEEE Trans. Comp.*, vol. C-29, pp. 596–603, July 1980.

[28] A. S. Tanenbaum, *Computer Networks.* Englewood Cliffs, NJ: Prentice-Hall, 1981, pp. 249–273.

[29] D. J. Goodman, S. Nanda, and U. Timor, "Theory of packet reservation multiple access," submitted for publication.

[30] R. M. Newman, Z. L. Budrikis, and J. L. Hullett, "The QPSX MAN," *IEEE Commun. Mag.*, vol. 26, pp. 20–28, Apr. 1988.

[31] G. H. Clapp, M. Singh, and S. Karr, "Metropolitan area network architecture and services," in *Proc. IEEE Global Commun. Conf. GLOBECOM '88*, Hollywood, FLA, Dec. 1988, pp. 1246–1254.

[32] W. C. Y. Lee, *Mobile Cellular Telecommunications Systems.* New York: McGraw-Hill, 1989, pp. 34–41.

David J. Goodman (M'67–SM'86–F'88) was born in Brooklyn, NY, in 1939. He received the B.S. degree from Rensselaer Polytechnic Institute, Troy, NY, the M.S. degree from New York University, New York, NY, and the Ph.D. degree from Imperial College, University of London, London, U.K., all in electrical engineering.

Since September 1988, he has been Professor and Chairperson of the Deparment of Electrical and Computer Engineering at Rutgers, the State University of New Jersey, Piscataway, NJ. He is also Program Director of the Rutgers Wireless Information Network Laboratory and a Visiting Professor at Southampton University. Prior to joining Rutgers, he was with AT&T Bell Laboratories as a Department Head in the Communications Systems Research Laboratory. His research has spanned many areas of digital communications, including wireless information networks, digital signal processing, digital coding of speech signals, and speech quality assessment.

Dr. Goodman has held various positions in the IEEE Acoustics Speech and Signal Processing Society and the IEEE Communications Society. He is a member of the Board of Governors of the IEEE Vehicular Technology Society.

Chapter 9

CELLULAR SYSTEMS AND STANDARDS

Wireless Network Access for Personal Communications

Tetherless personal communications is a "paradigm shift" that
requires completely new economic and regulatory models.

Donald C. Cox

Reprinted from *IEEE Communications Magazine*, pp. 96-115, Dec. 1992.

xisting forms of wireless communications continue to experience rapid growth, and new applications and approaches are being spawned at an unprecedented rate. Increases in numbers and usage of cellular mobile telephones, cordless phones, and radio pagers remain large in spite of the difficult economic environment. Wireless data networks are also proliferating in numbers and kinds. Market studies continue to indicate large mass markets for new types of tetherless personal communications, perhaps exceeding 50 to 100 million subscribers in the United States alone.

The aim of this paper is to provide an explanation for the widely different technological approaches and technologies being proposed for different interpretations of wireless personal communications services (PCS).[1] The starting point is the different constraints and goals of the different interpretations. Technology compromises associated with the different constraints and goals are then explored. The issues are very complex, and are surrounded with considerable confusion. The issues themselves are subject to different interpretations, depending on one's vantage point. Such characteristics are typical of large paradigm shifts. The paper seeks to strip away the complexity and attempts to clarify the confusion.

The explosive interest and activity in tetherless personal communications in the United States and worldwide now permeates all forums for telecommunications, including regulatory and standards, as well as the technology and business communities. Papers, panels, and sessions on wireless and/or personal communications are numerous in related technical conferences and publications, e.g., in IEEE ICC '92, GLOBECOM '91, VTC '92, *Transactions on Communications, Transactions on Vehicular Technology*, and *Communications Magazine*, and are appearing at many technical conferences that are quite remote in subject matter from this topic, e.g., the 1992 Optical Fiber Conference (OFC '92) [1], the 1992 IEEE International Solid State Circuits Conference (ISSCC '92)[2], and the IEEE Conference on Computer Communications (INFOCOM '92) [3]. Many symposia, conferences, and special journal issues are being devoted solely to this topic [4-6].

On the regulatory/business front in the United States, 1) an FCC Notice of Inquiry (NOI) on PCS [7] in June 1990 received more than 100 comments from different entities, and again more than 100 reply comments to the original comments, 2) an FCC Policy Statement and Order [8] in October 1991 announced an en banc hearing on PCS that was held on December 5, 1991 with 20 witnesses who were selected to testify from about 80 applicants from many segments of the telecommunications industry, 3) an FCC Notice of Proposed Rule Making (NPRM) [9] to award Pioneer's Preference to proposals and effort aimed at novel new PCS received more than 70 petitions, 4) more than 100 experimental licenses have been awarded by the FCC to different business entities to do experiments as they pursue different interpretations of wireless PCS, 5) an FCC NPRM [10] considering spectrum between 1.85 and 2.2 GHz for new telecommunications technologies and applications, and 6) an FCC NPRM and Tentative Decision on amendment of rules to establish new PCSs [96]. State regulatory agencies also have become interested in this topic [11]. WARC '92 [12] designated spectrum around 2 GHz for future public land mobile telecommunications services (FPLMTS) that include personal communications applications as well as vehicular mobile applications.

In the United States standards arena, several bodies have started proceedings related to PCS, e.g., subcommittees T1E1 and T1P1 of committee T1 (Telecommunications) of the American National Standards Institute (ANSI), committee TR45 of the Telecommunications Industry Association (TIA), and Committee 802 of the IEEE. Some effort to coordinate the PCS/PCN activities of committee T1 and TIA has started. Bellcore also is interacting with manufacturers in its requirements process, having issued three rounds of technical advisories [13] (FAs and TAs) that are being revised with inputs from industry. Internationally there is wireless PCS related activity in the International Consultative Com-

DONALD C. COX is executive director of Radio Research at Bellcore, Red Bank, New Jersey.

[1] *In the United States and throughout this paper, the term PCS refers to personal communications services; however, in the Commission, Research on Advanced Communications in Europe (RACE), it refers to personal communications space, a concept quite different from, but related to, personal communications services.*

0163-6804/92/$03.00 1992 © IEEE

mittee on Radio (CCIR) in group TG 8/1 that continues several years of earlier activity in IWP 8/13 starting in 1985; and the International Consultative Committee on Telephone and Telegraph (CCITT), particularly as related to wireline personal communications in universal personal telecommunications (UPT).

Regional activities worldwide are equally intense and too numerous to cite here. In many cases, activities elsewhere have preceded those in the United States.

In the technology arena, manufacturers, service providers, and entrepreneurs worldwide have proposed many different technological approaches aimed at many different interpretations of wireless PCS. Much of the technical literature and many forums include the debates on access approaches, e.g., time division multiple access (TDMA), code division multiple access (CDMA) or spread spectrum (SS), and frequency division multiple access (FDMA).

Many different business entities also are participating in activities aimed at many different interpretations of wireless PCS. Participating business entities include equipment manufacturers, entrepreneurs, and service providers, including local exchange (telephone) companies, cellular mobile companies, cable TV companies (CATV), and interexchange (long distance) companies.

It is necessary to understand various wireless access technology alternatives and their relationships to communications networks in order to determine where they fit into the overall evolving telecommunications picture. Taken altogether, the wireless and PCS activities are arguably the fastest growing segment of the telecommunications industry. However, the result of all this activity by all these different participants is large and increasing confusion as to what personal communications [14] is.

Some of the interpretations of wireless PCS are:

1) Advanced, usually digital, cordless telephones, e.g., CT-2, DCT-900, DECT, and Handiphone. See [15], including some of the following: small, simple, low-power (a few milliwatts) radio base units at home connected to the telephone network and emulating a regular telephone to the network; small, simple, low-power (a few milliwatts) cordless handsets that can be used within several tens of meters of their own base units; phonepoints, i.e., public base units in airports, shopping centers, rail stations, etc., that can be accessed by low-power handsets within a few tens or up to 100 meters or so of a base unit; multiple, simple base units connected to and coordinated by a PBX or CENTREX switch, and accessible by low-power handsets throughout an area covered by the base units; and conventional radio pagers integrated with or used in conjunction with small handsets and phonepoints. More recently, entrepreneurs in the United States have been developing cordless telephones to operate in the 902 to 928 MHz or the 2400 to 2483.5 MHz industrial, scientific, and medical (ISM) bands under part 15 of FCC rules [16, 17]. These rules require spread spectrum access technology, permit up to 1 watt transmitter power for communications applications, and don't require licensing, but users must accept interference from any source and must not cause interference to any licensed radio service.

2) Cellular mobile radio, including some of the following: a) use of high-power (0.5 watt) pocket handsets in existing analog FM cellular mobile systems, b) use of such handsets, or lower power versions, in areas outside or in buildings covered by smaller lower-power cells (microcells) within an existing cellular system, c) use of high-power digital pocket handsets in next-generation digital cellular mobile systems[2] [15], whether they be European standard (GSM), Japanese digital cellular standard (JDC), or North American standard (IS-54), or United States nonstandard ETDMA, or CDMA, or d) use of lower-power handsets in next-generation digital cellular systems in microcell-equipped areas.

3) Personal communication networks (PCNs) with some of the following alternatives: a) overlay wire, fiber or fixed radio interconnected networks with some type of radio access (e.g., DCS-1800, next generation GSM digital cellular access modified for medium-power for pocket handsets at 1.8 GHz for PCN in the U.K. [18, 19] and standardized by the European Telecommunications Standards Institute (ETSI), the first PCN proposal to be allocated dedicated radio spectrum; low-to-medium-power spread spectrum access for handsets sharing spectrum with 1.8 GHz point-to-point microwave [20]; packet/TDMA radio access to metropolitan area networks for third generation wireless networks being researched at Rutgers WINLAB) [15], and b) low-power radio access to local exchange company fixed copper and fiber distribution networks for small, low-power pocket voice and data sets, with routing and radio system control provided by local exchange network intelligence [2, 13, 21-31]. This is sometimes viewed as a neighborhood, or in-building evolution [32] of existing use of radio in the telephone loop plant in rural areas [33-35], known as Basic Exchange Telecommunications Radio Service (BETRS), integrated with evolving exchange network intelligence.

4) Sparse-area coverage of large areas by satellite radio systems that can provide radio access over large regions of the earth for medium- to high-power portable handsets [36, 37].

5) Intelligent network features such as call forwarding and call transfer, and advanced intelligent network features (e.g., use of a personal number with wireline telephones including number translation and/or data base access like that employed for 800 services and credit card calls, such as the CCITT UPT concepts) [38, 39].

6) Wireless data, evolving rapidly in several specialized applications, and often separate from the more voice-oriented applications, such as: a) specialized low-data-rate, high-peak-power systems for wide-area coverage. Typical examples in the United States are the ARDIS network [40, 41] based on KDT technology and the RAM network [42] based on Mobitex technology. These systems operate within standard FM voice channels previously allocated to special mobile radio systems applications. b) Specialized lower peak-power wireless local area networks (WLANS) for various data rates from a few kb/s to several Mb/s [43, 44], and to the order of 10 Mb/s [44, 45]. These are all proprietary systems with little published technical detail. Some use spread spectrum and operate in the ISM bands under FCC part 15 while others operate with special authorization at other frequencies (e.g., ALTAIR [44, 45] at 18 GHz). c) Vehic-

Manufacturers, service providers, and entrepreneurs worldwide have proposed many technological approaches aimed at many different interpretations of wireless PCS.

[2] These different standards and technologies will be discussed later in the paper and most are described in more detail in [15].

ular cellular mobile channels used with voice band modems. These modems are designed to cope with the vagaries of the mobile radio multipath channels and with handoffs. d) A plan to share cellular mobile voice channels with quickly assigned data transmissions recently announced by data equipment manufacturers and United States cellular service providers [46].

It is the poorly defined PCS/PCN application that is surrounded with confusion. The confusion is created, at least partially, by a lack of focus on the type of tetherless service and the place or environment in which the service is to be provided. The confusion is characteristic of major paradigm shifts, and widespread tetherless personal communications appears to be such a paradigm shift in the way people communicate. There have been few communications technologies that have revolutionized the way the masses communicate (i.e.,

that have caused major shifts in a communications paradigm). Examples from the past include printing, that made books and papers available to everyone; the telephone, that made instantaneous voice communications available between most places; and radio and television broadcast, that made sound and picture news and entertainment available to almost everyone almost everywhere. Tetherless personal communications appears poised to take its place among these revolutionary technologies.

Under more careful scrutiny some of the current PCS/PCN debate that appears to be over technical issues (e.g., TDMA or CDMA, error correction or error detection, 32 kb/s or <10 kb/s speech coding, time division duplexing (TDD) or frequency division duplexing (FDD), etc.) sometimes becomes a disagreement in the fundamental applications, assumptions, and resulting requirements for the technology, more than on tech-

	Digital Cellular						Low Power Systems				
System	IS-54	GSM	CDMA	B-CDMA*	ISM-A†	DCS-1800	DECT	Handi-Phone	CT-2	CT-3 DCT-900	Bellcore UDPC
Multiple access	TDMA/ FDMA	TDMA/ FDMA	CDMA/ FDMA (DS)	CDMA (DS)	TDMA/ CDMA/ FDMA (DS)	TDMA/ FDMA	TDMA/ FDMA	TDMA/ FDMA	FDMA	TDMA/ FDMA	TDM/ TDMA/ FDMA
Freq. band Uplink (MHz) Downlink (MHz)	869-894 824-849 (USA)	935-960 890-915 (Eur.)	869-894 824-849 (USA)	Emerg. Tech.	902-928 (USA)	1710-1785 1805-1880 (UK)	1800-1900 (Eur)	1895-1907 (Japan)	864-868 (Eur. Asia)	862-866 (Sweden)	Emerg. Tech. (USA)
Duplexing	FDD	FDD	FDD	FDD	TDD	FDD	TDD	TDD	TDD	TDD	FDD
RF Ch. Spacing (KHz) Downlink (KHz) Uplink (KHz)	30 30	200 200	1250 1250	40 MHz 40 MHz	10 MHz	200 200	1728	300	100	1000	350 350
Modulation	π/4 DQPSK	GMSK	BPSK/ QPSK	M-PSK	2-BPSK	GMSK	GFSK	π/4 DQPSK	GFSK	GFSK	π/4 QPSK
Portable Txmit Power, Max/Avg	600 mW 200 mW	1W 125 mW	600 mW	600 mW	1W	1W 125 mW	250 mW 10 mW	80 mW 10 mW	10 mW 5 mW	80 mW 5 mW	200 mW 20 mW
Freq. assign.	Fixed	Dynamic			Dynamic	Dynamic	Dynamic	Dynamic	Dynamic	Dynamic	Autonomous automatic
Power control Handset Base	Y Y	Y Y	Y Y	Y N	Y Y	Y Y	N N		N N	N N	Y
Speech coding	VSELP	RPE-LTP	QCELP	ADM	ADPCM	RPE-LTP	ADPCM	ADPCM	ADPCM	ADPCM	ADPCM
Speech rate (kb/s)	7.95	13	8 (Var rate)	32	32	13	32	32	32	32	32
Speech Ch./RF Ch.	3	8			8	8	12	4	1	8	10
Ch. Bit Rate (kb/s) Uplink (kb/s) Downlink (kb/s)	48.6 48.6	270.833 270.833			1920	270.833 270.833	1152	96	72	640	500 500
Ch. coding	1/2 rate conv.	1/2 rate conv.	.5 rate fwd 1/3 rate rev; CRC		CRC	1/2 rate conv	CRC	CRC	No	CRC	CRC
Frame duration (ms)	40	4.615	20			4.615	10	5		16	
Chip rate (MHz)	N/A	N/A	1.2288	30		N/A	N/A	N/A	N/A	N/A	N/A

* Broadband CDMA proposed for sharing spectrum with 1.8 to 2.0 GHz point-to-point microwave; DS is direct sequence.
† One example of a technology proposed for use in ISM band.
~ Spectrum is 1.85 to 2.2 GHz being considered by FCC for emerging technologies.

■ Table 1. *Wireless technologies.*

nology issues themselves. Details of the technologies noted (e.g., CT-2, DCT-900, DECT, GSM, and IS-54) will not be repeated here because they have been described elsewhere [14, 15, 19]. Parameters for some of these wireless technologies are summarized in Table 1.

This paper begins with descriptions of different wireless communications applications, the environments in which they are applied, and issues surrounding the applications. These are followed by descriptions of the compromises that have been made in developing specific technologies optimized to serve the cordless telephone and the vehicular mobile radio applications. It is possible to use these specific technologies for some PCS. For example, pocket handsets are used in cellular mobile systems; however, they have limitations for these applications, largely related to power consumption and speech quality, that result from the technology compromises made for vehicular mobile applications. The limitations can be overcome by making different technology compromises that are more appropriate to the more general PCS applications characterized by pocket personal communicators in widespread pedestrian environments. The technology compromises needed for this more general pedestrian PCS application are then described along with the integration of this technology into the overall tetherless communications picture.

Characteristics are then outlined for a technology that is aimed at the widespread pedestrian PCS application and environment. The remainder of the paper makes comparisons among different specific wireless technologies that have been aimed at different applications and environments.

Interpretations and Technology Compromises

One way to look at interpretations of wireless personal communications is to focus on where people need, or are likely to use, such communications. A look at the constraints of the user needs and radio environments can then suggest technology compromises appropriate to these factors. This exercise can be helpful in sorting out some of the confusion surrounding the various interpretations of PCS and the technologies proposed, as noted in the previous section. The emphasis in what follows is on tetherless access to fixed communications networks, and not on radio-only communications systems or networks that stand alone, or are only loosely connected to other communications networks.

Places and Types of Tetherless Communications

Overall today's public appears to want various types[3] of tetherless communications in the following categories of places [47]:

1) In homes and neighborhoods, and when walking or sitting around their homes or with neighbors. These needs are partially served for voice in homes by today's analog cordless telephones, new digital cordless telephones (e.g., CT-2), and some spread-spectrum products being developed in the United States for use in ISM bands under part 15 of FCC regulations.

2) In buildings — voice. These needs are the target of several new wireless PBX/CENTREX technologies (e.g., DCT-900/CT-3/DECT, CT-2) and some ISM-band products being developed in the United States.

3) In homes and other buildings, outdoors for pedestrians in residential neighborhoods, and urban areas in places people congregate (e.g., airports, shopping malls, rail stations, and on campus), and any place where there are reasonable densities of people. These needs are being targeted by new proposals for low-power voice and data personal communications technologies for supporting pocket-sized personal communicators having long battery life when in use. These needs are only partially served by cordless telephones, telepoints, and wireless PBX/CENTREX using CT-2, DCT-900/CT-3/DECT, and ISM-band products.

4) In buildings — high speed data (or on campus). Several approaches to serving these needs exist, or are being developed. They include various ISM-band products with different data rates and protocols, and a product at 18 GHz. Technologies targeted at this specific need have been referred to as wireless local area networks (WLANs) or data PCS and often are aimed at data rates greater than a megabit per second.

5) Vehicular — voice-oriented. Voice-oriented needs of vehicular users traveling over wide areas, urban and suburban, and on highways between urban centers, are well served by high-power cellular mobile radio systems, existing analog systems now and next-generation digital systems in the near future. Some data needs also are being served using voiceband modems and will be served directly in next-generation digital systems.

6) Wide area — low-speed data. These needs are being served predominantly by two technologies designed specifically for this use (e.g., Motorola KDT and Ericsson Mobitex). New systems for providing more capacity by integrating special radio data technology with cellular systems are being proposed.

7) Large remote or sparsely populated areas. These needs are being partially met by special technologies used with geostationary orbit communications satellites (GEOSATs) that were designed for point-to-point or point-to-multipoint coverage of large regions (e.g., United States, Europe, or Japan). New specially designed low-earth-orbit satellite (LEOSAT) and GEOSAT systems have been proposed.

These seven types and categories do not necessarily require seven distinct technologies or seven distinct, service-dependent, spectrum allocations. They can be grouped in different ways to satisfy people's desires and to work within the capabilities and limitations of technology. It is unlikely, however, that all seven can be served by a single wireless access technology, or even by two technologies. Several different groupings appear technically possible. Perhaps services could be divided between indoor and outdoor, or on the basis of the degree of mobility needed.

Many individuals [48] involved in these issues have expressed the view that the public would like to access voice, and at least moderate-rate[4] data, services through a single, small, lightweight personal communicator that has a battery life of several hours while being used and that provides communications

It is unlikely that all seven types of tetherless communications mentioned here can be served by a single wireless access technology, or even by two technologies.

[3] In the future, multimedia wireless services may also become desirable, perhaps including video.

[4] Moderate rate data is interpreted here to be from a few kilobits per second to perhaps several hundred kilobits per second.

privacy equivalent to that of wireline telephones. This suggests serving with one optimized and standardized low-power technology, all of the PCS access needs of item three above, which also includes the first and second item and may partially overlap the sixth. The remaining needs of item five could perhaps be combined with those of seven in a different and more complex technology. The highly specialized needs of item four may need their own specific technology. The technological needs of these different combinations will be considered in more detail after considering other service issues.

Issues in Tetherless PCS

One figure of merit cannot adequately represent all of the complex issues surrounding the technology compromises needed for a particular application or interpretation of PCS [14, 19, 47]. Compromises must be made among several interrelated factors, some of which are:

- Circuit quality of radio channels provided in the environment. This includes speech distortion and transmission delay for both error-free channels and channels operating at a specified threshold error ratio.
- Percentage of area within a service region that can provide a specified service quality (threshold error ratio).
- Radio channel availability that depends on user traffic intensity and number of channels available in a coverage area.
- Economics of the fixed radio system and network that depend on the complexity of the fixed radio equipment and the network architecture, and the number of radio circuits per fixed radio transceiver and per radio frequency channel.
- Complexity of the personal communicator technology that affects cost, power consumption (battery size, weight, and operating time), and complexity of control functions required for network interaction.
- Radio spectrum utilization efficiency that affects the amount of spectrum needed and system economics.
- System capacity, e.g., the user density or the total number of users that can be supported in an area.
- Privacy, the degree of privacy and security provided by encryption of the radio link.
- Number of service providers, the division of traffic among them, and the degree of spectrum sharing or coordination among them.

Some forums have concentrated on one or two of these interrelated factors (e.g., radio spectrum utilization efficiency, or system capacity), while hardly acknowledging others. The next section discusses emphasis and compromises that are being made for some applications in some of the categories noted earlier.

Examples of Technology Compromises

Different compromises among these different issues are needed for the various applications and environments. These different compromises lead to diversified technologies and interpretations of PCS as noted in the introduction.

The need for these different compromises and resulting different technology solutions appears to be more difficult to grasp for wireless commu-

nications than it is for some other technologies, e.g., for airplanes. It appears obvious that different types of airplanes, i.e., having different size, fuel capacity, required takeoff distance, etc., are needed for transatlantic flights than are needed for short-distance commuter service. Transatlantic flights require large fuel capacity, can achieve economic benefit from large size (i.e., large payload capacity), and can tolerate long required takeoff distance since takeoffs are infrequent and are from a few major locations. In contrast, commuter flights require short takeoff distances to permit operation from numerous small airports, don't need as much fuel capacity because flights are short, and can experience an economic penalty from a size too large to be adequately filled for the short flights from the numerous relatively small locations. Thus, four-engine turbofan jumbo-jets for transatlantic flight represent quite a different technological solution from the small two-engine turboprop airplanes often used for short commuter airline flights. They are both airplanes, they both land and take off, and they both fly through the air, but they are quite different.

It seems much more difficult to see that technology and systems compromises for vehicular communications, for high-speed wireless LANs, or for pocket communicators will result in different radio solutions even though they all have transceivers (i.e., transmitters and receivers), they all interconnect transceivers to provide radio communications links, and they all use electromagnetic radiation. The existing confusion as to what constitutes PCS is a result of the different interpretations of applications noted previously, and resulting different compromises on issues that can result in the aforementioned different solutions. It is instructive to consider combinations of applications and compromises. The following groupings generally are representative of the particular applications noted; however, as in any attempt to categorize such complex entities, some entities that fit clearly in one category may have a few characteristics from a different category. Also, there are gray areas where an occasional entity may appear to have characteristics of several categories.

Cordless Telephone for Homes, Buildings, and Telepoints
— Cordless telephones are designed to provide limited coverage inside residences and buildings by eliminating the cord between the base unit and handset. These appliances compete directly with wireline telephones for people's voice communications needs. Therefore speech quality (distortion and transmission delay) of cordless phones must approach that of wireline phones, otherwise people often will choose the better quality of the tether over lower-quality limited tetherless access. This was demonstrated early in the United States during the evolution of simple cordless phones. People purchased inexpensive cordless phones that had poorer speech quality than they expected from wireline telephones, but soon discarded them. Having experienced the desirability of tetherless communications, many of them then purchased higher-quality, but more expensive cordless sets. Sales of the better-quality sets soon dominated to the point that vendors who continued to stress and improve quality now dominate the market. However, cost is still a significant factor;

few sets over $200 are being sold in the United States. Other characteristics that are required for cordless telephones as a result of the ready availability of wireline telephones in the home and building environment are low handset-weight and long usage times (a few hours) before requiring battery recharging or replacement. The characteristics in this environment drive design of analog cordless phones, and are driving design of digital cordless phones (CT-2, DCT-900/CT-3/DECT, and some ISM-band products) to emphasize the following when making technology compromises.

• Provide wireline telephone circuit quality for:
 –speech distortion,
 –transmission delay.
• Minimize transmitter power to:
 –minimize weight (and size of pocket sets),
 –minimize cost,
 –maximize usage time before battery attention.
• Minimize complexity to:
 –minimize dc power needed for signal processing in handsets,
 –minimize cost,
 –minimize weight (and size of pocket sets),
 –maximize usage time before battery attention.

In order to achieve the above, the following are accepted in design compromises:
• few user channels per MHz,
• few user channels per base unit on average,
• large numbers of base units per area,
• short transmission range,
• low complexity in network interaction.

One important radio-environment consequence that is a direct result of short transmission range and inside or low-height antennas is that multipath delay spread is generally less than 0.25 μsec, well over an order of magnitude less than for vehicular mobile radio environments. This is significant because no signal processing is needed for delay dispersion compensation for digital transmission rates up to one Mb/s or so. For slow pedestrian speeds and frequencies below 2 GHz, fading rates are slow, on the order of 10 Hz or less, so bit interleaving and error correction are ineffective. Also, significant signal decorrelation for diversity can be obtained in and around buildings from colocated antennas having different polarizations [49].

The results of these compromises in design for this environment are that most of the technology solutions for cordless telephones have a number of similarities, regardless of the access technology chosen. Some of these are listed below (D indicates "applies to digital implementations only"):
• (D) 32 kb/s adaptive differential pulse code modulation (ADPCM) for digital speech encoding:
 –low-complexity speech-encoding technique minimizes speech-processing delay and power consumption with distortion levels that produce wireline speech quality,
 –minimizes cost of digital speech-encoding function,
 –encoding rate obviously results in fewer users per MHz than would lower rates.
• Average transmitter power on the order of 10 milliwatts[5]:
 –minimizes weight, size and cost,
 –results in usage times of a few hours,
 –results in short transmission ranges up to several hundred meters or even less, depending on base unit height.
• (D) time division duplexing (TDD):

–requires only one band of frequencies,
–can provide diversity for active two-way radio links while using multiple antennas and signal processing only at the base unit to reduce complexity of handset,
–cordless telephone architecture with base units and handsets at about the same height in houses and buildings results in similar signal attenuation between base units and between handsets and corresponding base units. Thus, time synchronization is not required to minimize base-to-base interference [50].
• (D) low-complexity signal processing:
–no delay dispersion compensation (equalizers or multiple spread-spectrum correlators),[6]
–No high-complexity error-correction decoding.
• Low-transmission delay, less than 10 ms (single link round trip) for several, less than 50 ms for most:
–if less than 10 ms, echo cancellation may not be required at network interface.
• Simple frequency shift modulation, noncoherent detection:
–low complexity,
–reduced detection sensitivity and resistance to co-channel interference compared to coherently-detected phase modulation.
• Dynamic channel allocation:
–minimizes system planning and organization,
–cordless telephone architecture with few users per base unit realizes advantage in system capacity,
–can significantly slow call setup and transfer (handoff) processes in non-time-synchronized systems (cordless PBX in confined area can be readily time-synchronized; multi-PBX environment can result in nonsynchronized interference penalty).

Cordless telephone approaches tend to avoid involvement with the telephone network. That is, home base units do not look any different to the network than do wireline telephones, and a cordless PBX isolates its base units from the network. These limitations simplify control processes in the handsets, but do not permit the network to locate and route calls to tetherless users, who may be closer to some base unit other than their own, or than one connected to their own PBX. That is, no wireless user identity is available to the network for call routing any different from that for wireline telephones or PBXs. Some of the network limitations can be mitigated by a combination of wide-area (perhaps national) paging and cordless telephones/telepoints; however, the paging load could become large if very large numbers of users and sizes of regions were covered without any attempt to determine the area over which to page.

Vehicular Cellular Mobile Radio — Cellular mobile radio systems were designed to provide service to widely ranging vehicles outside in urban and suburban areas and along highways and in areas in between [81]. Although cellular pocket portables have become quite popular, existing analog systems and next-generation digital systems are not well-suited for serving them [22]. In order to cover large areas (e.g., adjacent to urban areas or along highways where user densities are low), mobile radio range needs to be large, up to 20 km, to collect enough traffic to be economical. These

Sales of the better-quality sets soon dominated to the point that vendors who continued to stress and improve quality now dominate the market.

[5] Some technologies based on TDMA have peak power levels of 100 milliwatts or so with duty cycles of 0.1 or less; however, it is the average power, i.e., (peak power duty cycle), that affects battery drain and range (range depends on energy per bit Eb, which is directly related to average power).

[6] United States ISM-band technologies must use spread-spectrum techniques; when applied to cordless telephones they are applied in simplified forms.

areas must be covered to make the service useful, because vehicular users do travel through them and do require service there. This requires high transmitter power, on the order of a watt, and complex high-power cell sites (base stations) with high antennas. These and other design constraints [51, 52] have resulted in expensive cell sites, on the order of $1 million. Of course, the availability of a large battery, fuel, an alternator, and an engine can readily supply the power needed by a relatively complex vehicular radio set. On the other hand, users in vehicles do not have any other convenient communications option. Stopping to use a public pay phone obviously requires suspending travel. Therefore, speech quality has been less of an imperative than with cordless telephones because the mobile users' limited choices are either to use what is there or to do without communications. The characteristics in this environment drive design of vehicular mobile radio systems to emphasize the following when making technology compromises:

• Minimize the number of cell sites:
–maximize user channels per cell site,
–maximize user channels per MHz,
–provide large transmission range, e.g., 20 km (12 mi.) in the environment,
–provide adequate transmitter power for range, e.g., on the order of one watt.

In order to achieve the above, the following are accepted in design compromise:
• High user set complexity:
–transmitter power dominates power needs for signal processing of considerable complexity.
• High transmitter and signal-processing power consumption relative to capacities of pocket-size batteries.
• Lower speech and circuit quality relative to wireline telephones.
• High network complexity.

A direct consequence of the long range and high (often 50 meters, i.e., 150 ft, or more) cell site antennas is the existence of large delay spreads (e.g., 5 or 10 μsec or more) on some paths. These mobile radio delay spreads can be between one and two orders of magnitude greater than the delay spreads that occur on shorter cordless telephone paths. The result is a need for considerable signal-processing complexity to compensate for significant delay dispersion. The sometimes high vehicular speeds and frequencies near 1 GHz can result in fading rates on the order of 100 Hz, significantly complicating delay dispersion compensation, but making bit interleaving and powerful error correction attractive for mitigating high-rate fading effects.

The technology solutions for analog and digital vehicular cellular mobile radio have a number of similarities that result from these compromises in design for this environment, again regardless of the access technology chosen. Some of these follow (D indicates "applies to digital implementations only"):
• (D) low-bit rate, ≤13 kb/s, digital speech encoding.
–encoding rate results in more users per MHz than does 32 kb/s encoding,
–high-complexity techniques increase power consumption and cost for this function,
–produces speech distortion resulting in less than wireline quality,

–usually incurs a few tens of ms of additional transmission delay.
• Average transmitter power on the order of one watt:
–transmission ranges up to 20 km (12 mi) when combined with high cell-site antennas,
–weight, size and cost penalty, particularly for pocket sets and their batteries,
–results in usage times of 45 minutes or less for pocket sets.
• Frequency division duplexing (FDD):
–cellular mobile radio architecture with high outside cell-site antennas results in less attenuation between cell sites than between mobiles and corresponding cell sites. With FDD, time synchronization is not required to maintain good frequency-reusing efficiency in the presence of strong cell site to cell site interference [50],
–requires two separated bands of frequencies,
–requires multiple antennas in mobile unit for antenna diversity.
• (D) high-complexity signal processing:
–delay dispersion compensation (complex equalizers or multiple spread-spectrum correlators),
–forward error correction decoding.
• (D) high transmission delay, on the order of 200 ms (single link round trip) for most system implementations.
• Fixed channel allocation:
–(D) architectures can provide more rapid mobile-assisted handoff in non-time-synchronized system (time synchronization complex to provide throughout widely dispersed cell sites),
–often requires considerable system planning and organization (this could be avoided by using quasi-static autonomous adaptive frequency-allocation techniques [53, 54]).
• (D) use of pauses in speech (i.e., inactivity) to increase capacity:
–proposed for several different multiple-access techniques,
–reduces perceived speech quality,
–Increases signal-processing complexity.
• Adaptive power control.

Cellular mobile systems are designed with co-channel interference and coverage requirements to provide good or better service to 90 percent of the streets and roads of a covered region [55], i.e., ten percent of users, or 1 in 10, by design, will receive service judged less than good; indoors, [22] the coverage is less than 90 percent. These systems have intense signaling interaction among cell sites, and network intelligence and control in the mobile-telephone switching office (MTSO). This requires complex protocols and messages among mobiles, cell sites, and MTSOs to permit calls to be routed to roving users. Thus, cellular mobile sets execute the complex control processes required for managing mobility, and cellular networks are becoming increasingly interconnected to facilitate mobility management over larger regions using more sophisticated network protocols (e.g., X.25 and SS7). Currently this sophisticated networking does not extend to interconnection with the wireline telephone networks, but trials and planning are underway to facilitate this interworking.

A Reality Check — Before we go on to consider other applications and compromises, perhaps it would be helpful to see if there is any indication that the previous discussion is valid. For this check, we should look at existing cordless telephones for telepoint use (i.e., pocketphones) and at pocket cellular telephones. Since digital cellular pocketphones are not in significant production, we can only consider existing analog units, and note expected effects of changes to digital.

Two products from one United States manufacturer are good for this comparison. One is a third-generation hand-portable analog FM cellular phone from this manufacturer that represents their second generation of pocketphones. The other is a first-generation digital cordless phone built to the United Kingdom CT-2 common air interface (CAI) standard. Both units are of flip phone style with the earpiece on the main handset body and the mouthpiece formed by, or on the flip-down part. Both operate near 900 MHz and have 1/4 wavelength pull-out antennas. Both are fully functional within their class of operation (i.e., full number of U.S. cellular channels, full number of CT-2 channels, automatic channel setup, etc.). Table 2 compares characteristics of these two wireless access pocketphones from the same manufacturer.

The most important items to note in the Table 2 comparison are:
- the talk time of the low-power pocketphone is four times that of the high-power pocketphone,
- the battery inside the low-power pocketphone is about one half the weight and size of the battery attached to the high-power pocketphone,
- the battery-usage ratio, talk time/weight of battery, is eight times greater, almost an order of magnitude, for the low-power pocketphone compared to the high-power pocketphone!

Additionally,
- the low-power (5 mw) digital cordless pocketphone is slightly smaller and lighter than the high-power (500 mw) analog FM cellular mobile pocketphone,
- even considering the vagaries of cost and price, the retail price difference between the two pocketphones is quite significant!

It should also be noted that:
1) The room for technology improvement of the CT-2 cordless phone may be greater since it is first generation and the cellular phone is second/third generation.
2) A digital cellular phone built to the IS-54, GSM, or JDC standard, or in the proposed United States CDMA technology, could be expected to either have less talk time or be heavier and larger than the analog FM phone, because: a) the low-bit rate digital speech coder is more complex and will consume more power than the analog speech processing circuits; b) the digital units will have complex digital signal-processing circuits for forward error correction — either for delay dispersion equalizing or for spread-spectrum processing — that will consume significant amounts of power and that have no equivalents in the analog FM unit; and c) power amplifiers for the shaped-pulse non-constant-envelope digital signals will be less efficient than the amplifiers for constant-envelope analog FM. Although it may be suggested that transmitter power control will reduce the weight and size of a CDMA handset and battery, if that handset is to be capable of operating at full power in fringe areas, it will have to have capabilities similar to other cellular sets. Similar power control applied to a CT-2-like low-maximum-power

Characteristic/parameter	CT-2	Cellular
Weight: flip-phone only	5.2 oz.	4.2 oz.
battery only	1.9 oz.	3.6 oz.
Total unit	7.1 oz.	7.8 oz.
Size (max. dimensions) flip-phone only	5.9 x 2.2 x 0.95 in. 8.5 cu. in.	5.5 x 2.4 x 0.9 in.
battery* only	1.9 x 1.3 x 0.5 in. internal	4.7 x 2.3 x 0.4 in. external
Total unit	5.9 x 2.2 x 0.95 in. 8.5 cu. in.	5.5 x 2.4 x 1.1 in. 11.6 cu. in.
Talk time: rechargeable battery non-rechargeable battery	180 min. (3 hr.) 600 min. (10 hr.)	45 min. N/A
Standby time rechargeable battery non-rechargeable battery	30 hr. 100 hr.	8 hr. N/A
Speech quality	32 kb/s telephone quality	30 kHz FM depends on channel quality
Transmit power avg.	0.005 watts	0.5 watts
Retail price‡	$375	$1500

* rechargeable battery
†ni-cad battery
‡U.S. mid 1992

■ **Table 2.** *Comparison of CT-2 and cellular pocket size flip-phones from the same manufacturer.*

set would also reduce its power consumption and thus also its weight and size.

The major difference in size, weight and talk time between the two pocketphones is directly attributable to the two orders of magnitude difference in average transmitter power. The generation of transmitter power dominates power consumption in the analog cellular phone. Power consumption in the digital CT-2 phone is more evenly divided between transmitter-power generation and digital signal-processing. Therefore, power consumption in complex digital signal-processing would have more impact on talk time in small low-power personal communicators than in cellular handsets where transmitter-power generation is so large. Other than reducing power consumption for both functions, the only alternative for increasing talk time and reducing battery weight is to invent new battery technology having greater energy density.

In contrast, lowering the transmitter power requirement, modestly applying digital signal-processing, and shifting some of the radio coverage burden to a higher density of small, low-power, low-complexity, low-cost fixed radio ports has the effect of shifting some of the talk time, weight and cost constraints from battery technology to solid-state electronics technology which continues to experience orders-of-magnitude improvements in the span of several years. However, digital signal-processing complexity cannot be permitted to overwhelm power consumption in low-power handsets; while small differences in complexity will not matter much, orders-of-magnitude differences in complexity will continue to be significant.

Thus, it can be seen from Table 2 that the size, weight, quality and cost arguments in the previous sections generally hold for these examples. It also is evident from the previous paragraphs that they will be even more notable when comparing digital cordless pocketphones with digital cellular pocketphones of the same development generations.

Compromises for Low-Power Radio Access for PCS — We will now take wireless PCS to be: 1) the provision of voice and moderate rate data, and perhaps in the future, multimedia including video, to small, lightweight, pocket-size personal communicators that can be used for tens of hours without attention to batteries, and 2) the provision of such communications over wide areas (i.e., in homes and other buildings, outdoors for pedestrians in neighborhoods and urban areas, in places where people congregate, and any place where there are reasonable densities of people). Thus, this is the cordless telephone compromise of previously mentioned material modified as needed to provide widespread outside coverage of pedestrians in neighborhoods, urban areas and congregating places. For this application, the wireless personal communicator will have to compete with wireline telephones having widespread availability, so the discussion about speech quality (distortion and transmission delay) in section II.C.1 applies here also, and the need for privacy over the radio link also is recognized. The emphasis in compromising on a technology for this definition of wireless PCS, then, is as follows:

• provide wireline telephone circuit quality for speech distortion, transmission delay, and privacy;
• minimize pocket set transmitter power to minimize weight (and size of pocket sets), minimize cost, and maximize usage time before battery attention;
• minimize complexity to minimize dc power needed for signal-processing power in pocket sets, minimize cost, minimize weight (and size of pocket sets), and maximize usage time before battery attention.

In order to achieve these things, the following design compromises are acceptable:
• moderate user channels per MHz,
• moderate number of fixed radio ports per area,
• moderate user channels per fixed radio port,
• moderate transmission range,
• moderate network interaction complexity,
• low (but not necessarily minimum) radio port cost.

An important radio environment consequence that results from a moderate (300 to 500M, i.e., 1000 to 1500 ft, or less) transmission range and inside or low-height (<10M, i.e., <30 ft.) antennas is that multipath delay spread is generally less than 0.5 μsec, again at least an order of magnitude less than vehicular mobile environments, but somewhat greater than even more restricted cordless phone environments. Therefore, no signal processing is needed for delay-dispersion compensation for transmission rates up to 1/2 MHz [27], i.e., rates somewhat less than for even shorter-range cordless phones, but significantly greater than for vehicular mobile radio. For slow pedestrian speeds and carrier frequencies of 2 GHz or so, fading rates are slow, on the order of 10 Hz or less, so bit interleaving and error correction are ineffective. Significant signal decorrelation for diversity can be obtained in and

around buildings from colocated antennas having different polarizations [49].

If fixed-port antennas are raised to street light or flagpole height, 20 or 30 feet, to provide better coverage in neighborhoods, suburban areas, or congregating places, attenuation between ports will be less than that between pocket sets and ports. This particular environmental situation becomes more like a vehicular environment than an indoor cordless phone environment. It requires either frequency-division duplexing or time synchronization of all port transmissions if time-division duplexing were used [50] in order to maintain reasonable spectrum utilization and thus system capacity using a moderate number of radio ports. For this type of PCS, the need to transfer active radio links from one port to another as users move, places constraints on radio link control that are not always taken into account in cordless telephone considerations. The compromises and environmental constraints for this definition of wireless PCS suggest the following system characteristics (only digital implementations have been considered):

• 32 kb/s ADPCM for digital speech encoding:
– low-complexity speech-encoding technique minimizes speech-processing delay and power consumption with distortion levels that produce wireline speech quality,
– minimizes cost of digital speech-encoding function,
– encoding rate obviously results in fewer users per MHz than would lower rates.
• Radio link privacy:
– digitally encrypt the radio link.
• Flexible radio-link architecture to accommodate different transmission bit rates:
– graceful evolution to lower-bit-rate speech when technology advancement permits,
– can support different user data needs.
• Average transmitter power on the order of 10 milliwatts.
– minimizes weight, size, and cost,
– results in usage times of a few hours,
– results in short transmission ranges up to a few hundred meters or even less, depending on base-unit height.
• Frequency-division duplexing (FDD):
– FDD avoids the need for time synchronization of ports to maintain good frequency reuse efficiency in the presence of strong port-to-port interference [50] where outside port antennas are deployed,
– permits more efficient use of radio transceivers; transmitters can transmit, and receivers receive for 100 percent of the time,
– requires two separated bands of frequencies,
– requires multiple antennas in pocket unit for antenna diversity (not difficult at 2 GHz since antennas are small).
• Low co-channel interference (less than one percent of users affected[7]):
– large frequency-reuse interval, i.e., less frequency reuse,
– large numbers of radio ports to cover any particular user density within a specified bandwidth.
• Large percentage coverage (99 percent) of a covered region.[8]
– large radio system margin, requiring closer port spacing,

[7] An order of magnitude fewer than cellular mobile requirements.

[8] An order of magnitude less uncovered area than cellular mobile requirements.

–moderate numbers of radio ports to cover a region.
• Low-complexity signal processing:
–no delay-dispersion compensation (equalizers or multiple spread spectrum correlators),
–no high-complexity error-correction decoding,
–simple error detection for blanking or speech extrapolation [79],
–minimizes power consumption.
• Low-transmission delay, less than 10 ms (single-link round trip).
• 4-level phase modulation and coherent detection:
–good detection sensitivity and resistance to co-channel interference for increased range and frequency-reusing efficiency,
–low-to-moderate-complexity signal processing.
• Autonomous adaptive quasi-fixed-frequency channel allocation:
–minimizes system planning and organization,
–architectures can provide more rapid portable-assisted automatic radio-link transfer in non-time-synchronized system,
–high occupancy, non-time-synchronized system incurs small decrease in capacity compared to dynamic channel allocation.
• Adaptive power control:
–minimizes battery drain,
–minimizes co-channel interference.

This set of characteristics contains many entries from the cordless-phone list that appear when cordless-phone compromises dominate the consideration. It also includes modifications of entries from the vehicular cellular mobile list where compromises become dominated by the need to provide pedestrian service from radio ports outside in neighborhoods and urban areas.

Proposals also have been made to arrive at a technology to satisfy this definition of PCS by using vehicular cellular-radio technology in microcells and power controlling cellular pocketphones for low-power operation. It is evident that this still leaves the mobile-optimized technology with unresolved complexity (power consumption and cost) and speech-quality penalties as illustrated in Table 2 and discussed previously. Also, it has been shown that the most satisfactory way to mix high-power transceivers (vehicular cellular technologies) and low-power transceivers (cordless phone or PCS technologies) in the same system is to divide the available frequency spectrum into separate high-power and low-power bands [56].[9]

The suggested performance-quality characteristics, such as co-channel interference and percentage coverage of a region, are an order of magnitude more stringent for wireless PCS than they are for cellular mobile radio [14, 19, 27]. This appears to affect system capacity, but in reality affects system economics, since capacity always can be made the same by adding low-cost radio ports [25]. This low-power PCS application will require a larger number of inexpensive fixed radio ports than a cellular radio system requires expensive cell sites, but it requires fewer ports than cordless telephone requires base units. The simple radio ports and network interfaces [25, 57] and use of public right-of-way can result in between two and three orders of magnitude lower cost for a radio port than for a complex high-power cell site.

Low-power fixed radio ports with separations of 2000 feet or so outside and 100 feet or so in-side will be quite numerous. Therefore, system economics will be strongly affected by physical, logical, and protocol interfaces with the fixed supporting network, whether it be the intelligent public switched telephone network (PSTN), the future integrated-services digital network (ISDN), or some other intelligent infrastructure network. Compatibility with standard digital transmission facilities [57], e.g., T1 lines, high-speed digital subscriber lines (HDSL), or ISDN lines, produces the most advantageous overall radio system/network economics. Minimizing the complexity and power consumption of radio ports also helps radio system reliability and maintenance, as well as economics. Since radio-port transceiver costs are only weakly dependent on radio transmission rate, multiplexed radio links that can support multiple users on one port transceiver also minimize radio system costs. These considerations suggest the use of simple radio ports that do not contain any per-user electronics. Simple radio ports that minimize complexity [57, 58] have the following attributes:

1) For transmitting: a) accept digitally multiplexed signals from a digital line, b) modulate them onto a radio frequency carrier, and c) amplify the modulated carrier, and

2) For receiving: a) amplify and downconvert the received signal, b) demodulate, synchronize, and regenerate the received digital signals, and c) transmit the digital signals down a digital line.

Integration of Low-Power Personal Communications with Other Tetherless Communications

Vehicular Cellular or Satellite Mobile Radio — There are approaches for integrating tetherless communications of personal pocket communicators with vehicular cellular mobile telephones [14]. The user's identity could be contained either in memory in the personal communicator set, or in a small smart card inserted into the set, as is a feature of the European GSM system. When entering an automobile, the small personal communicator or card could be inserted into a receptacle in a vehicular cellular or satellite mobile set installed in the automobile. The user's identity would then be transferred to the mobile set.[10] The mobile set could then initiate a data exchange with the cellular or satellite mobile system, indicating that the user could now receive calls at that mobile set. This information about the user's location would then be exchanged between the cellular or satellite mobile network and the exchange network intelligence so that calls to the user could be correctly routed.[11] Note that in this approach the radio sets are optimized for their specific environments, high-power vehicular or low-power pedestrian as discussed earlier, and the network access and call routing is coordinated by the interworking of network intelligence. This approach does not compromise the design of either radio set or radio system, and places the burden on network intelligence technology, a technology that benefits from the enormously large and rapid advances in computer technology.

The approach described above is consistent with what has actually happened in other applications of technology in significantly different environments. For example, consider the case of audio cassette tape players. Pedestrians often

The approach detailed here does not compromise the design of either radio set or radio system, and places the burden on network intelligence technology.

[9] A reviewer pointed out that a similar conclusion on spectrum partitioning was reached in RACE mobile project 1043.

[10] Inserting the small personal communicator in the vehicular set would also facilitate charging the personal communicator's battery.

[11] This is a feature proposed for FPLMTS in CCIR Rec. 687.

474

*Low-power
personal
portable
communi-
cations can
be provided
to occupants
of airplanes,
trains, and
buses by
installing
compatible
radio access
ports inside
these
vehicles.*

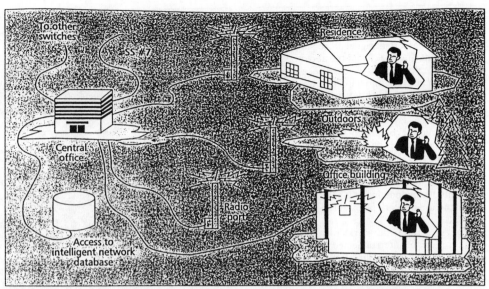

■ **Figure 1.** *Low-power exchange access digital radio integrated with network and intelligence.*

carry and listen to small portable tape players with lightweight headsets (e.g., a Walkman[12]). When one of these people enters an automobile, he or she often removes the tape from the Walkman and inserts it into a tape player installed in the automobile. The automobile player has speakers that fill the car with sound. The Walkman is optimized for a pedestrian, whereas the vehicular mounted player is optimized for an automobile. Both use the same tape, but they have separate tape heads, tape transports, audio preamps, etc. They do not attempt to share electronics. In this example, the tape cassette is the information-carrying entity similar to the user identification in the personal communications example discussed earlier. The main points are that the information is shared among different devices, but the devices are optimized for their environments and do not share electronics.

Similarly, a high-power vehicular cellular or satellite mobile set does not need to share oscillators, synthesizers, signal processing, or even frequency bands or protocols with a low-power pocket-size personal communicator. Only the information identifying the user and where he or she can be reached needs to be shared among the intelligence elements, e.g., routing logic, databases and common channel signaling [24-30, 59], of the infrastructure networks. This information exchange between network intelligence functions can be standardized and coordinated among infrastructure subnetworks owned and operated by different business entities (e.g., vehicular cellular mobile radio networks, satellite mobile networks, and intelligent local exchange networks). Such standardization and coordination are the same as are required to pass intelligence among local exchange networks and interexchange carrier networks.

Other Environments — Low-power personal portable communications can be provided to occupants of airplanes, trains, and buses by installing compatible radio access ports inside these vehicles.[13] The ports can be connected to high-power vehicular cellular mobile sets or to special air-ground or satellite-based mobile communication sets. Intelligence between the internal ports and mobile sets could interact with cellular mobile, air-ground,

or satellite networks in one direction using spectrum allocated for that purpose, and with personal communicators in the other direction to exchange user identification and route calls to and from users inside these large vehicles. Radio isolation between the low-power units inside the large metal vehicles and low-power systems outside the vehicles can be ensured by using windows that are opaque to the radio frequencies. Such an approach also has been considered for automobiles (i.e., a radio port for low-power personal communications connected to a cellular mobile set in a vehicle so that the personal communicator can access a cellular mobile network). This appears more difficult to implement since it would be more difficult to isolate the radio port environment inside an automobile from that of ports outside along streets. Opening a radio opaque window would couple the environments and complicate access issues.

Personal Communications Based on Wireless Network Access

A wireless technology has been synthesized and evolved with the compromises appropriate for low-power radio access for PCS as described earlier [13, 21, 27, 30, 53, 54, 57, 58]. An overall view of this wireless access is shown in Fig. 1. There are three major parts that are evident.
• The wireless access radio link:
– fixed radio port transceiver and antenna mounted either on a utility pole outside or within a large building,
– portable personal communicator transceiver including control and baseband circuitry for voice and moderate-rate data.
• The dense fixed distribution network:
– connects radio ports to centralized control and switching,
– wireline based, either copper or glass fiber.
• The network intelligence and circuit processing:
– radio port control and circuit multiplexing/demultiplexing.
– signaling,

[12] *Walkman is a registered trademark of the Sony Corporation.*

[13] *A reviewer noted that "CT-2 phones are being installed by at least one airline for use by passengers in flight."*

–databases for user information,
–switching,
–per-user or per-circuit signal processing.

In considering Fig. 1 and the above, it is evident that overall tetherless PCS contains a lot of fixed distribution network and a lot of network intelligence, but perhaps only a little wireless (i.e., only the last 1000 feet or so of the distribution network) [21-30, 57]. Thus, any PCS network will have an expensive infrastructure component that is separate from, but affected strongly by the wireless-access technology employed. This infrastructure economic fact has caused several revisions to the approach to separate personal communications networks in the United Kingdom, including the merger of the infrastructure network plans of two potential participants.

Sparse cellular networks with tall antenna structures can be interconnected with fixed point-to-point microwave or millimeter wave radio. Dense PCS networks with short antenna poles (<10 meters) are more difficult to so interconnect because of path obstruction by objects such as trees, buildings, terrain features, etc.

The local-exchange network is already ubiquitous, and is evolving rapidly to include significant intelligence. Its use to support PCS as the intelligent infrastructure for low-power digital radio used as the access technology for the last 1000 feet or so of loops can shorten PCS deployment time. It is important that the wireless access technology for such a PCS application be selected to be as compatible as possible with the existing infrastructure without incurring too much penalty to the radio system. The topic of fixed intelligent-network infrastructure is a large topic in itself, and is beyond the scope of this paper. However, wireless access technology impact on the fixed network is noted in several places throughout the paper. Network issues have also received some attention in [22-30, 57, 59].

Radio Link Characteristics and Parameters

The technical basis for selections and compromises for this wireless network access technology, sometimes referred to as Universal Digital Portable Communications (UDPC), are discussed in more detail in [22, 24, 27] and a preliminary requirements process has been started with industry [13]. The technology compromises have evolved [60] through this process to improve speech quality and reduce processing complexity by incorporating 32 kb/s ADPCM for startup and evolving to 16 kb/s speech encoding, and by reducing the inherent radio-link transmission delay from a 16 ms frame to a 2 ms frame. Reducing transmission delay can reduce processing complexity by removing the need for echo cancellers at a two-wire wireline interface.

The technology is TDMA-based with frequency division duplexing (FDD) as illustrated in Fig. 2. The radio link and technology, that are described in detail elsewhere [13, 22, 24, 27, 53, 54, 58, 60], have the following parameters and characteristics:
• Uplink TDMA/FDMA/FDD.
• Downlink TDM/FDMA/FDD.
• Link transmission rate 500 kb/s.
• User payload rate 320 kb/s.
• Rate allocated to real-time control 45 kb/s.
• 200 μs TDMA bursts.

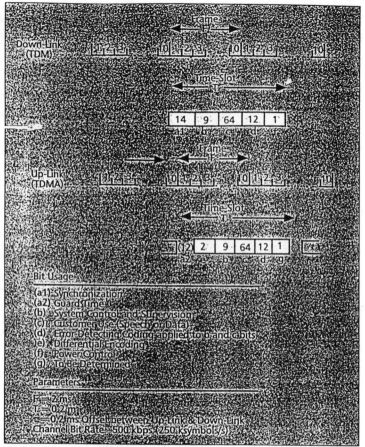

■ Figure 2. *Air interface frame and burst structure.*

• Low-complexity digital signal processing.
• Frequency range near 2 GHz:
–1/10 of the number of gates as digital cellular implementations.
• Integrated voice and data transmission.
• Data rates of 8 kb/s to 320 kb/s.
• 2 ms TDMA frame for 32 kb/s.
• 10 TDMA bursts/frame for 32 kb/s.
• Superframed for:
–4 ms frame for 16 kb/s,
–8 ms frame for 8 kb/s.
• One burst dedicated to alerting (paging) and call set up and control.
• 4-QAM modulation with Nyquist bandwidth shaping factor of a=0.5.
• 100 bits in each TDMA time slot:
–64 bits user payload.
–1-bit power control.
–9 bits system control.
–12 bits combined error-detection and burst synchronization.
–uplink 12-bit intervals (24 μsec) for guard time.
–uplink 2 bits for differential decoding.
–downlink 14 bits for burst synchronization.
• No forward error-correction coding/decoding.
• No bit interleaving.
• No overhead bits for symbol or bit timing.
• Combined error-detection and frame-synchronization coding/decoding [60, 61]:
–simple decoding,
–permits blanking or extrapolating errored speech frames to improve perception of link quality [79].
• Coherent detection.

Low-bit-rate speech encoding is not at all tolerant of the tandem speech encodings that will inevitably occur for PCS for many years.

- No bit overhead for carrier recovery.
- Guard time bits allotted for propagation delay, synchronization uncertainty, and transmitter turn on/off.
- Antenna diversity uplink and downlink:
  - uplink: 2-branch, 2-receiver selection plus error-controlled portable antenna switching (approximates 4-branch selection for slow fading);
  - downlink: 2-branch, 1-receiver preselection (equivalent to 2-branch selection for slow fading);
  - 2 antennas on portable and port (simple antennas at 2 GHz).
- No multipath delay dispersion equalization.
- Port transmitter power 400 mw continuous.
- Portable transmitter power 200 mw peak TDMA:
  - average portable transmitter power 20 mw per 32 kb/s, or 10 mw per 16 kb/s.
- Privacy encryption of radio link.

Radio system parameters and characteristics discussed elsewhere [13, 22, 24, 27] are:

- TDM and TDMA channel spacing 350 kHz.
- FDD spacing 80 to 120 MHz.
- Autonomous adaptive port-frequency channel assignment [53, 54].
- No time-synchronization of radio ports.
- No per-user signal processing at radio ports.
- Low co-channel interference environment:
  - 99 percent of locations or users with block error ratios less than two to three percent,
  - factor-of-ten more stringent than cellular design.
- High probability of coverage:
  - 99 percent of locations or users with block-error ratios less than two to three percent,
  - factor of ten more stringent than cellular design,
  - port spacing up to 2000 ft in a typical suburban residential area; or on order of 100 feet inside large buildings.
- Dynamic power control of uplink transmitters based on received-power and channel errors.
- Automatic link transfer of active channels among ports based on received-power, signal-quality, and channel errors.
- Automatic time slot transfer of active channels among TDMA time slots based on received-power and channel errors.

Comparisons with Other Technologies

Complexity/Coverage Area Comparisons—
Experimental research prototypes of radio ports and subscriber sets [62, 63] have been constructed to demonstrate the technical feasibility of the radio link requirements in [13]. These UDPC prototypes generally have the characteristics and parameters previously noted, with the exceptions that 1) the portable transmitter power is lower (10 mw average, 100 mw peak), 2) dynamic power control and automatic time slot transfer are not implemented, and 3) a rudimentary automatic link-transfer implementation is based only on received power. The experimental ports transmit near 2.17 GHz; the experimental subscriber sets transmit near 2.12 GHz. Both are operated under a Bellcore experimental license. The experimental prototypes incorporate application-specific, very large scale integrated circuits (ASIC) (VLSI) fabricated to demonstrate the feasibility of the low-complexity high-performance digital signal-processing techniques [63, 64] for symbol timing and coherent bit detection. These techniques permit the efficient short TDMA bursts having only 100 bits that are neces-

sary for low-delay TDMA implementations. Other digital signal-processing functions in the prototypes are implemented in programmable logic devices. All of the digital signal-processing functions combined require about 1/10 of the logic gates that are required for digital signal processing in vehicular digital cellular mobile implementations [64-66]; that is, this low-complexity PCS implementation having no delay-dispersion-compensating circuits and no forward error-correction decoding is about 1/10 as complex as the digital cellular implementations that include these functions.[14] The 32 kb/s ADPCM speech-encoding in the low-complexity PCS implementation is also about 1/10 as complex as the less-than-10 kb/s speech-encoding used in digital cellular implementations. This significantly lower complexity will continue to translate into lower power consumption and cost. It is particularly important for low-power pocket personal communicators with power control in which the DC power expended for radio frequency transmitting can be only tens of milliwatts for significant lengths of time.

The experimental radio links have been tested in the laboratory for detection sensitivity (BER vs. SNR) [62, 67, 68] and for performance against co-channel interference [69] and intersymbol interference caused by multipath delay spread [62]. These laboratory tests confirm the performance of the radio link techniques. In addition to laboratory tests, qualitative tests have been made in several PCS environments to compare these experimental prototypes with several United States CT-1 cordless telephones at 50 MHz, with CT-2 cordless telephones at 900 MHz, and with DCT-900 cordless telephones at 900 MHz. Some of these comparisons have been reported [70-73]. In general, depending on the criteria, e.g., either no degradation or limited degradation of circuit quality, these UDPC experimental prototypes covered areas inside buildings that ranged from 1.4 to 4 times the areas covered by the other technologies. The coverage areas for the experimental prototypes were always substantially limited in two or three directions by the outside walls of the buildings. These area factors could be expected to be even larger if the coverage were not limited by walls, i.e., once all of a building is covered in one direction, no more area can be covered no matter what the radio link margin. The earlier comparisons [71-73] were made with only two-branch uplink diversity before subscriber-set transmitting antenna switching was implemented, and with only one radio port before automatic radio-link transfer was implemented. The later tests [70] included these implementations. These reported comparisons agree with similar unreported comparisons made in our own laboratory building. Similar coverage comparison results have been noted for a 900 MHz ISM-band cordless telephone compared to the 2 GHz experimental prototype. The area coverage factors (e.g., x1.4 to x4) could be expected to be even greater if the cordless technologies had also been operated at 2 GHz since attenuation inside buildings between similar small antennas is about 7 dB greater at 2 GHz than at 900 MHz [74, 75] and the 900 MHz handsets transmitted only 3 dB less average power than the 2 GHz experimental prototypes. The greater area coverage demonstrated for this technology is expected because of the dif-

---

[14] *Some indication of VLSI complexity can be seen by the number of people required to design the circuits. For the low-complexity TDMA ASIC set, only one person part time plus a student part time were required; the complex CDMA ASIC has six authors on the paper alone.*

ferent compromises noted earlier; in particular:

1) Coherent detection of QAM provides more detection sensitivity than noncoherent detection of frequency-shift modulations [76].
2) Antenna diversity mitigates bursts of errors from multipath fading [62, 77, 78].
3) Error detection and blanking of TDMA bursts having errors significantly improves perceived speech quality [79]. (Undetected errors in the most significant bits cause sharp audio pops that seriously degrade perceived speech quality.)
4) Robust symbol timing, and burst and frame synchronization reduce the number of frames in error due to imperfect timing and synchronization [62].
5) Transmitting more power from the radio port compared to the subscriber set offsets the less sensitive subscriber set receiver compared to the port receiver that results from power and complexity compromises made in a portable set.

Of course, as expected, the low-power (10 mw) radio links cover less area than high-power (0.5 w) cellular mobile pocketphone radio links because of the 17 dB transmitter power difference resulting from the compromises discussed previously. In the case of vehicular mounted sets, even more radio-link advantage accrues to the mobile set because of the higher gain of vehicle-mounted antennas and higher transmitter power (3 w).

Speech Quality Issues

All of the PCS and cordless telephone technologies that use CCITT standardized 32 kb/s ADPCM speech encoding can provide similar error-free speech distortion quality. This quality often is rated on a five-point subjective mean opinion score (MOS) with 5 excellent, 4 good, 3 fair, 2 poor, and 1 very poor. The error-free MOS of 32 kb/s ADPCM is about 4.1 and degrades very slightly with tandem encodings. Tandem encodings could be expected in going from a digital-radio PCS access link, through a network using analog transmission or 64 kb/s PCM, and back to another digital-radio PCS access link on the other end of the circuit. In contrast, a low-bit-rate (<10 kb/s) vocoder proposed for a digital cellular system was recently reported [80] to yield an MOS of 3.4 on an error-free link without speech-activity detection. This score dropped to 3.2 when speech-activity detection was implemented to increase system capacity. This nearly one-full-point decrease on the five-point MOS score indicates significant degradation below accepted CCITT wireline speech distortion quality. Either almost half of the population must have rated it as poor, or most of the population must have rated it as only fair. In addition, the low-bit-rate speech encoding is not at all tolerant of the tandem speech encodings that will inevitably occur for PCS for many years. These comments in this paragraph are generally applicable to speech encoding at rates of 13 kb/sec or less.

In the arena of transmission delay, the short-frame (2 ms) FDMA/TDD and TDMA technologies (e.g., CT-2 and UDPC noted earlier) can readily provide single-radio-link round-trip delays of <10 ms, and perhaps even <5 ms. The longer frame (10 ms and greater) cordless-phone TDMA technologies, e.g., DCT-900/CT-3/DECT and some ISM-band implementations, inherently have a single-link round-trip delay of at least 20 ms and can range 30 ms to 40 ms or more in some implementations. As mentioned earlier, the digital vehicular-cellular technologies with low-bit-rate speech encoding, bit interleaving, forward error-correction decoding, and relatively long frame times (~16 to 20 ms) result in single link round-trip delays on the order of 200 ms, well over an order of magnitude greater than the short-frame technologies, and on the same order of magnitude as a single-direction synchronous satellite link. It should be noted that almost all United States domestic long-distance telephone circuits have been removed from such satellite links, and many international satellite links also are being replaced by undersea fiber links. These long-distance-circuit technology changes are made partially to reduce the perceptual impairment of long transmission delay.

Other Technology Issues

There is considerable debate in the cellular mobile industry over the use of TDMA or CDMA in digital systems, and in the cordless telephone arena over FDMA, TDMA, and CDMA. In many cases, issues discussed herein are scarcely considered and some are even forgotten in the debate over one or two issues. Since these debates and issues are spilling over into low-power radio access for PCS, it is useful to look at some of them here. Reference [15] gives technical details of several of the following technologies.

CT-2 — In the digital cordless arena, different targeted environments and goals have biased choices. For example, the choice of FDMA for CT-2 in the United Kingdom was dictated by the original goal of a simple, single-user, home cordless telephone with a simple technique for interference avoidance at call setup and no need for multi-channel multiplexing or handoff. FDMA/TDD meets the needs for simple single-user channelization and simple measurement of signal power for a frequency channel from both ends of a radio link. Dynamic channel allocation (DCA) by selection of a channel with a lowest combined interference index at base unit and handset is adequate for initial channel selection and access. In this environment a handset is expected to be closer to its base unit than to any other.

In a more widespread PCS application, this is not always true, and calls set up by a simple algorithm in a heavily loaded system can interfere with active calls causing a high premature forced call-termination rate. Handoff is cumbersome for simple non-time-synchronized FDMA so extension of these simple handset/base-unit cordless-telephone-oriented techniques to multi-user systems requires additional complexity that may produce limited benefit. For simple, low-power cordless telephones, a short (2 ms) TDD frame and 32 kb/s speech coding can provide wireline-quality circuits with low transmission delays when the handset and base unit are close to each other. These CT-2 technology choices were appropriate compromises for the cordless telephone applications and were extendable for the phonepoint application that emerged later, but are more difficult to extend further to widespread PCS applications.

DCT-900/DECT/CT-3 — The DCT-900/DECT/CT-3[15] choice of TDMA/TDD was dictated by the needs

There is considerable debate in the cellular mobile industry over the use of TDMA or CDMA in digital systems, and in the cordless telephone arena over FDMA, TDMA, and CDMA.

---

[15] DCT-900 and CT-3 are early 900 MHz implementations of the technology being defined by ETSI for use throughout Europe as DECT at frequencies closer to 2 GHz.

of multiple cordless telephones accessing multiple base units and connected to a PBX and by the shortage of paired frequency bands in Europe. In this case, multiplexing of multiple users at a base unit and handoff were readily implemented in single-frequency TDMA/TDD. As originally conceived with only one frequency, TDD permitted simple rapid monitoring of power in all channels from both ends of a radio link. Dynamic time slot allocation algorithms for DCA in conjunction with continuous transmission in at least one time slot as a "beacon" from all base units, provided a convenient mechanism for initial base unit and time-slot selection.

This combination was also effective for rapid handoff, either to another base unit when user motion caused its beacon slot to become sufficiently stronger, or to another time slot if the in-use slot experienced too many errors in leading or trailing bits protected by error detection coding. Delay spread inside buildings limited the link transmission rate to a little over 1 MHz if the complexity of delay-dispersion equalizers was to be avoided.

TDD permits two-way antenna diversity to be implemented from the base unit, and thus does not require two antennas at the handset for two-branch diversity for active calls.[16] However, it halves the capacity of a given radio link for any specified link transmission rate. DCA and TDD are readily workable in this environment because loops connecting base units to PBXs are very short, so adequate time synchronization of port transmission can be achieved with little effort. Additionally, the attenuation between base units inside buildings is not significantly less than that between base and portable units, so such synchronization is not even necessary for efficient TDD operation [50]. The rapid time-slot handoff possible in a single-frequency TDMA system could be expected to partially mitigate premature termination of active calls caused by new call attempts that interfere with them.

However, when the need for additional system capacity dictated adding frequency channels to make combined FDMA/TDMA, the picture became much more complex. Beacons are still needed so that portables can select base units, but blind slot problems can occur between frequency channels in the same time slots, and even in adjacent time slots unless either two frequency synthesizers are used or large enough guard times are allowed for synthesizer switching between TDMA time slots. It is not clear that all the technical issues have been resolved for DCA and FDMA/TDMA with many frequency channels, where a base unit can operate on any time slot at any frequency. This approach has the potential for very-high-capacity base units having simultaneous multi-frequency capability on all time slots, albeit with considerable additional complexity in control, with multiple high-switching-speed frequency synthesizers, and possibly with increased exposure to prematurely terminated calls when the system is heavily loaded with traffic.

In order to permit widespread use in outside environments where base units have less attenuation between themselves than between portable sets and base units, time synchronization of the base-unit transmissions is required to achieve good performance with TDD [50]. In any event, the efficient use of DCA in FDMA/TDMA systems requires time

synchronization of port transmissions, if handoffs are to be completed within reasonable time periods (10s of ms). The DCT-900/DECT/CT-3 technology choices were appropriate compromises for the cordless PBX application, but need modification for more widespread PCS applications, for which they also incur synchronization requirements and additional complexity.

U.S. ISM-Band Technology—Spread spectrum (SS) or CDMA cordless phones and wireless local area networks (WLANs) in the United States are based on FCC part-15 regulations that permit such operation with up to 1 watt transmitter power in ISM bands at 900 MHz, 2 GHz, and 5 GHz. The availability of spectrum without a license requirement has encouraged development of a number of ISM-band products [44]. There is no issue of spectrum utilization efficiency, as long as they meet part-15 spreading requirements. SS ISM-band developments run the gamut of radio-link techniques that include frequency hopping, direct sequence SS, TDMA, FDMA, TDD, and FDD. Some implementations use combinations of a few of these techniques. It is more difficult to follow these proprietary developments by entrepreneurs than more standards-oriented commercial developments because of the product secrecy that surrounds most of them. The aim in many of these developments appears to be to simplify the SS implementations while compromising on radio-link performance to reduce handset complexity and power consumption. Some have used power control to mitigate the near-far problem, but others use time or frequency to separate "in cell" users, and thus do not use SS processing gain for this in-cell separation. In these cases, SS is used only to gain access to the spectrum. It is not clear that SS would have been employed in these technologies if it were not a regulatory requirement for use of the spectrum [44]. Depending on the SS implementation, rapid handoff can be complicated or may not even be possible, and time synchronization of the transmissions from multiple base units often is required.

Digital Cellular Mobile Radio—Many of the vehicular mobile radio issues have been addressed in earlier sections; however, some remaining ones are noted here. While these technology choices include compromises appropriate for the vehicular application, they also include the limitations for widespread low-power personal communications noted previously.

European GSM — The European Global System for Mobile (GSM) from ETSI (formerly Group Special Mobile of CEPT) for pan-European vehicular digital cellular mobile radio was driven by the need for a common mobile standard throughout Europe, and the desire for digital transmission compatible with data and privacy. Analog cellular mobile service in Europe has been plagued from the beginning by different standards in various, often adjacent, countries. These include the Nordic NMT, the German Net C, the United Kingdom TACS and some AMPS systems, none of which will work together. Crossing a country boundary often results in the phone going dead. Spectrum was reallocated near 900 MHz throughout much of

[16] *The base-unit-only TDD diversity is not effective for handset receive-only functions, such as channel assessment for access and handoff, or paging.*

Europe and surrounding regions so that a completely new technology could be developed by GSM. The GSM effort in the early to mid 1980s considered several system implementations including TDMA, CDMA/SS, and FDMA technologies. Extensive system studies and laboratory implementations produced firm proposals that were evaluated by all concerned. A TDMA/FDMA/FDD technology was chosen with a radio link bit rate of 270 kb/s and thus requiring significant delay dispersion equalization [15]. The use of low-bit-rate (13 kb/s) speech encoding and Gaussian-minimum shift-keying (GMSK) modulation resulted in cell-site capacity within a given bandwidth similar to the existing analog FM systems. This technology has been modified for application to medium-power moderate-range PCN in the United Kingdom at frequencies near 2 GHz, for which it retains some of the vehicular-mobile-radio complexity to provide the range needed to cover sparsely populated areas [14, 19].

North American CTIA IS-54 — In contrast to the European situation, North America was not allocated different spectrum for digital cellular mobile radio. In the United States, the FCC opened the existing cellular bands to essentially any technology the service providers wanted to use, as long as they continued to serve the needs of mobile users. Also in contrast to Europe, analog FM AMPS [81] developed in the 1970s by Bell Labs was used exclusively throughout North America. The other major defining events in North America were that cellular-system growth was rapid, demand continued to outstrip system capacity in the largest cities, and cell sites and installation for cell splitting in such cities were very expensive.

The Cellular Telecommunications Industry Association (CTIA) members decided to abide by a common digital system standard, requested the TIA recommend one, and specified that it should retrofit into the existing AMPS systems. The high cost of new cell sites became the major driving force. Thus, the major factor in the new standard became maximizing the number of voice channels supportable at a cell site within the available cellular spectrum, a parameter that has become known as "The CAPACITY."

Although this CAPACITY factor enters into overall system capacity, an even stronger factor is the cell-site density (i.e., distance between cells). In fact, in the interim, the continuing rapid increase in cellular system demand has been met with AMPS technology by increasing the cell-site density, i.e., by splitting cells and by introducing microcells.

In any event, the rallying cry in the United States cellular industry became the CAPACITY by this definition. Several TDMA/FDMA and pure FDMA system proposals were considered before what became the IS-54 standard was selected. This standard fits 3 TDMA 8 kb/s encoded-speech channels into each 30 kHz AMPS channel.

Since immunity to co-channel interference for the chosen 4-level QAM modulation is expected to be similar to the 30 kHz analog FM, and some increase in trunking efficiency accrues from the larger numbers of available channels (larger trunk group), the CAPACITY of IS-54 is projected to be x3 to x4 that of the existing analog FM AMPS, where this increase is largely due to the low-bit-rate speech encoding. Although needed in only a few specific places for 48.6 kb/s transmission, a delay dispersion equalizer was required in the IS-54 standard.

U.S. CDMA Proposal — Toward the end of the IS-54 standards process, a new CDMA/SS/FDMA proposal was made to the CTIA [83]. This proposal is considerably more technically sophisticated than earlier SS proposals. It includes fast feedback control of mobile transmitter power, heavy forward error correction, speech detection and speech-encoding-rate adjustment to take advantage of speech inactivity, and multiple receiver correlators to latch onto and track resolvable multipath maxima [93]. The spreading sequence rate is 1.23 MHz.

The near-far problem is addressed directly and elegantly on the uplink by a combination of the fast-feedback power control, and a technique called "soft handoff" that permits the instantaneous selection of the best paths between a mobile and two cell sites. Path selection is done on a frame-by-frame basis when paths between a mobile and the two cell sites are within a specified average level (perhaps 6 dB) of each other. This soft handoff provides a form of macroscopic diversity [82] between pairs of cell sites when it is advantageous. Like the TDD and DCA discussed earlier, increasing capacity by soft handoff requires precise time synchronization (on the order of a μsec) among all cell sites in a system. An advantage of this proposal is that frequency coordination is not needed among cell sites since all sites can share a frequency channel. However, coordination of the absolute time delays of spreading sequences among cell sites is required, since these sequence delays are used to distinguish different cell sites for initial access and for soft handoff. Also, handoff from one frequency to another is complicated.

Initially, the projected CAPACITY (by the previous definition in the previous section) of this CDMA system, determined by mathematical analysis and computer simulation of simplified versions of the system, was x20 to x40 that of AMPS, with a coverage criterion of 99 percent of the covered area [83]. However, some other early estimates [85] suggested that the factors were more likely to be x6 to x8 of AMPS.

A limited experiment was run in San Diego, California, during the fourth quarter of 1991 under the observation of cellular equipment vendors and service providers. This experiment had 42 to 62 mobile units in fewer than that many vehicles,[17] and four or five cell sites, one with three sectors. Well over half of the mobiles needed to provide the interference environment for system capacity tests were simulated by hardware noise simulation by a method not yet revealed for technical assessment. Estimates of the CAPACITY from this CDMA experiment center around x10 that of AMPS [80, 86][18] with coverage criteria <99 percent, perhaps 90 percent to 95 percent, and with other CAPACITY estimates ranging between x8 and x15.

This experiment did not exercise several potential capacity-reducing factors, for example: 1) only four cells participated in capacity tests. The test mobiles were all located in a relatively limited area and had limited choices of cell sites with which to

In contrast to the European situation, North America was not allocated different spectrum for digital cellular mobile radio.

[17] Some vehicles contained more than one mobile unit.

[18] AT&T stated that the San Diego data supported a capacity improvement over analog cellular of at least x10.

communicate for soft handoffs. This excludes the effects of selecting a strong cell-site downlink for soft handoff that does not have the lowest uplink attenuation because of uncorrelated uplink and downlink multipath fading at slow vehicle speeds [87]; 2) the distribution of time-dispersed energy in hilly environments like San Diego usually is more concentrated around one or two delays than is the dispersed energy scattered about in heavily built-up urban areas like downtown Manhattan [88, 94] or Chicago. Energy concentrated at one or two delays is more fully captured by the limited number of receiver correlators than is energy more evenly dispersed in time; and 3) network delay in setting up soft-handoff channels can result in stronger paths to other cell sites than to the one controlling uplink transmitter power. This effect can be more pronounced when coming out of shadows of tall buildings at intersections in heavily built-up areas. The effect will not occur as frequently in a system with four or five cell sites as it will in a large, many-cell-site system. All of these effects and others [89, 95] will increase the interference in a large system, similarly to the increase in interference that results from additional mobiles, and thus will decrease the CAPACITY over that estimated in the San Diego trial. Factors like these have been shown to reduce the San Diego estimate of x10 to an expected CDMA CAPACITY of x5 or x6 of analog AMPS [87-89]. This further reduction in going from a limited experiment to a large-scale system in a large metropolitan area is consistent with the reduction already experienced in going from theoretical estimates to the limited experiment in a somewhat restricted environment [84].

The San Diego trial also indicated a higher rate of soft(er) handoffs [80] between antenna sectors at a single cell-site than expected for sectors well isolated by antenna patterns. This result suggests a lower realizable sectorization gain because of reflected energy than would be expected from more idealized antennas and locations. This could further reduce the estimated CAPACITY of a large-scale system.

Even considering the aforementioned factors, CAPACITY increases of x5 or x6 are significant. However, these estimates are consistent with the factor of x3 obtained from low-bit-rate (<10 kb/s) speech coding and the x2 to x2.5 obtained by taking advantage of speech pauses. These factors result in an expected increase of x6 to x7.5, with none of these speech-processing-related contributions being directly attributable to the spread-spectrum processing in CDMA. These results are consistent with the factor of x6 to x8 estimate made earlier [85], and are not far from the factor of x8 quoted recently [90].

U.S. E-TDMA Proposal — A proposal has been made to enhance the CAPACITY of the IS-54 TDMA radio links (i.e., enhanced TDMA or E-TDMA) by (1) incorporating statistical speech-multiplexing among time slots and frequency channels to take advantage of speech pauses[19] and (2) using even lower bit-rate (~4 kb/s) speech encoding. Combined, these have been projected to yield an additional factor of x5 for a total E-TDMA CAPACITY of x15 of analog AMPS. However, the lower speech encoding results in even further speech-quality degradation. The statistical multiplexing has the same

speech degradation noted previously where it is discussed as part of the CDMA proposal. Thus, E-TDMA would appear to yield similar CAPACITY to that of the CDMA proposal [90].

Japanese Digital Cellular (JDC) — The Japanese digital cellular air interface standard is closer to the United States IS-54 TDMA standard than to other digital cellular standards. It includes three-channel TDMA with interleaved 25 kHz channel spacing, uses $\pi/4$ shifted QDPSK modulation and an 11.2 kb/s combined speech and channel coding rate. The use of delay dispersion equalization is optional. The smaller channel spacing and interleaving is expected to yield a slightly larger increase in CAPACITY than IS-54. Comments on the IS-54 technology are generally applicable to this similar technology.

Percentage of Coverage of a Region — The percentage of a covered region that has a circuit quality better than some specified value has a very significant impact on spacing of base stations, on allowable frequency reuse, and on CAPACITY as defined earlier. Specified values for speech are typically either a block error ratio of several percent, or an average bit error ratio of several parts in 10^{-3}. As an example, consider a typical FDMA or TDMA/FDMA system with two-branch antenna diversity and many radio ports operating in an environment with Rayleigh distributed multipath fading, log normally distributed large-scale attenuation variation having a standard deviation of 10 dB, and average signal varying as inverse distance to the 4th power [24]. Such a system will need a base station spacing for 99 percent coverage that is about 70 percent of the spacing needed for 90 percent coverage [91]; that is, it will require about two times as many base stations to cover a region for a 99 percent criterion than for a 90 percent criterion.

For the same conditions, frequency reuse for 99 percent of a region meeting the specified co-channel interference objective will require about two times as many frequency-channel sets as would be required to meet a 90 percent objective [92]. This significant overall system-quality factor is frequently disregarded in systems comparisons where emphasis is placed on only one or two issues, e.g., CAPACITY for cellular systems. For example, vehicular cellular mobile systems are usually designed with a 90 percent coverage objective [81], while in contrast the wireless access for personal communications systems (UDPC) noted previously has a 99 percent coverage criterion. Thus, in order to be meaningful, any comparison of access or other technologies proposed for different applications must be compared using the same coverage criterion.

Network Time Synchronization — The need for synchronization of the timing of transmissions from all base stations in a network in order to achieve the desired performance has been mentioned for several technologies (e.g., TDMA/TDD with dynamic channel allocation (DCA) used in DCT-900/DECT/CT-3) and the United States Cellular CDMA proposal. This requirement for time synchronization among all base stations adds complexity to, and decreases robustness of large-scale networks. No existing or planned fixed-distribution network has the capability to provide such precise time synchronization [50, 54].

---

[19] *Such a technique has been used in undersea cable circuits and is known as time assignment speech interpolation (TASI) or for digital circuits, (DASI).*

The capacity of such radio networks is significantly degraded by loss of time synchronization. The UDPC technology for wireless network access for personal communications described previously has been selected to minimize its sensitivity to time-synchronization. Its CAPACITY could be increased by adding the requirement for time synchronization of radio ports and implementing capacity-increasing techniques like TDMA downlinks, and DCA over multiple frequency channels and time slots. Such changes would significantly increase the complexity of the radio ports and personal communicators while also affecting the complexity and fragility of the fixed distribution network.

Statistical Multiplexing, Speech Activity, CDMA, and TDMA — Factors of x2 to x2.5 have been projected for capacity increase possible by taking advantage of pauses in speech. It has been suggested that implementing statistical multiplexing is easier for CDMA systems because it is sometimes thought to be time consuming to negotiate channels for speech spurts for implementation in TDMA systems. However, the most negative quality-impacting factor in implementing statistical multiplexing for speech is not in obtaining a channel when needed, but is in the detection of the onset of speech, particularly in an acoustically noisy environment. The effect of clipping at the onset of speech is evident in the MOS scores noted for the speech-activity implementation in the United States cellular CDMA proposal discussed earlier (i.e., an MOS of 3.4 without statistical multiplexing and of 3.2 with it). The degradation in MOS can be expected to be even greater for encoding that starts with a higher MOS, e.g., 32 kb/s ADPCM.

It was noted earlier that the proposed cellular CDMA implementation was x10 as complex as the proposed UDPC wireless access for personal-communications TDMA implementation. From earlier discussion, the CDMA round-trip delay approaches 200 ms while the short 2-ms-frame TDMA delay is < 10 ms round trip. It should be noted that the TDMA architecture could permit negotiation for time slots when speech activity is detected. Since the TDMA frames already have capability for exchange of signaling data, added complexity for statistical multiplexing of voice could readily be added within less than 200 ms of delay and less than x10 in complexity. That TDMA implementation supports 10 circuits at 32 kb/s or 20 circuits at 16 kb/s for each frequency. These are enough circuits to gain benefit from statistical multiplexing. Even more gain could be obtained at radio ports that support two or three frequencies and thus have 20 to 60 circuits over which to multiplex.

A statistical multiplexing protocol for speech and data has been researched at Rutgers WINLAB [15]. The Rutgers packet reservation multiple access (PRMA) protocol has been used to demonstrate the feasibility of increasing CAPACITY on TDMA radio links. These PRMA TDMA radio links are equivalent to slotted ALOHA packet-data networks. Transmission delays of less than 50 ms are realizable. The CAPACITY increase achievable depends on the acceptable packet-dropping ratio. This increase is "soft" in that a small increase in users causes a small increase in packet-dropping ratio. This is analogous to the soft capacity claimed for CDMA.

Thus, for similar complexity and speech quality, there appears to be no inherent advantage of either CDMA or TDMA for the incorporation of statistical multiplexing. It is not included in the personal communications proposal, but is included in cellular proposals because of the different speech-quality/complexity design compromises discussed throughout this paper, not because of any inherent ease of incorporating it in any particular access technology.

Summary

*T*he need and demand for tetherless communications in several environments have been demonstrated well by the rapid growth of different wireless technologies that are optimized for particular applications and environments. Obvious examples are: (1) residential cordless telephones and their evolution to digital in CT-2 and to DCT-900/DECT/CT-3 for in-building PBX environments, (2) analog cellular radio for widespread vehicular service and its various digital evolutions to GSM, IS-54, JDC, U.S. CDMA, and E-TDMA, and (3) wireless data networks, both for low-rate wide-area coverage and higher-rate WLANs. However, in order to complete the paradigm shift to widespread tetherless personal communications for nearly everyone, what is needed is (1) standardized low-power technology that has the appropriate compromises for providing voice and moderate-rate data to small, lightweight, economical, pocket-size personal communicators that can be used for tens of hours without attention to batteries, and (2) providing such communications economically over wide areas, including in homes and other buildings, outdoors for pedestrians in neighborhoods and urban areas, in places where people congregate, and anywhere there are reasonable densities of people.

A combination of standards and regulation is needed [47] to encourage the development of such widespread low-power personal communications instead of replicating compromises and approaches aimed either at more restricted applications (e.g., cordless telephones), or at other equally important but different applications for different environments (e.g., high-data-rate WLANs or vehicular cellular mobile radio). A wireless technology described here has been configured with the compromises needed to meet this challenge, and experimental research prototypes have demonstrated its capability to meet economically the needs of this widespread personal communications application within our ability to see them today.

I may have left the reader with some remaining confusion with respect to applications, compromises and technologies. For this I apologize, but the entire complex topic of wireless PCS is not yet crystal clear. As is typical of such a massive paradigm shift potentially affecting every person and every existing communications paradigm, it is not possible to see clearly all the way through the shift at these early beginnings. As seen from any conventional viewpoint (e.g., cordless telephone, vehicular cellular mobile, or wireline telephone), it doesn't fit. Business and economic models, refined to assess such existing technologies, are not applicable, and may even yield wrong answers.

I have attempted to share one view of what is hap-

For similar complexity and speech quality, there appears to be no inherent advantage of either CDMA or TDMA for the incorporation of statistical multiplexing.

pening, a view developed over more than 10 years; however, we will have to play out the paradigm shift to see how it ends, and how it proceeds from now to arrive at that end. The only things that are clear are that it will continue to be a very complex process, and it will continue to be seen differently by many different groups in different positions and having different objectives.

However, if the increasingly complex electropolitical and regulatory issues can be successfully resolved, wireless access technology, distribution technology, and intelligence in large networks are poised to provide widespread, convenient, and economical tetherless personal communications to nearly everyone.

References

[1] D. J. Goodman, "Wireless Personal Communications Networks," *1992 Optical Soc. America and IEEE OFC '92,* Plenary Session paper TuA1, San Jose, CA, Feb. 4, 1992.
[2] G. H. Heilmeier, "Personal Communications: Quo Vadis," *IEEE 1992 Int. Solid-State Circuits Conf.,* Plenary Session, paper WA1.3, *Digest,* pp. 24-26, Feb. 19, 1992.
[3] Panel, "Challenges in Wireless Communications," *IEEE INFOCOM '92,* Florence, Italy, May 6, 1992.
[4] *IEEE/IEE Int. Sym. on Personal, Indoor and Mobile Radio Commun.,* London, UK, Sept. 23-25, 1991.
[5] *IEEE Int. Conf. on Universal Personal Commun.,* Dallas, TX, Sept. 9-Oct. 2, 1992.
[6] *IEEE Int. Sym. on Personal, Indoor and Mobile Radio Commun.,* Boston, MA, Oct. 19-21, 1992.
[7] Federal Communications Commission (FCC), Docket 90-314, Notice of Inquiry (NOI) on Personal Communications Services," released June 28, 1990.
[8] FCC, Docket No. 90-314, RM-7140, RM-7175, and RM-7618, "Policy Statement and Order," adopted Oct. 24, 1991.
[9] FCC, General Docket No. GEN90-217, "Establishment of the Procedures to Provide a Preference to Applicants Proposing an Allocation for New Services," released April 12, 1990, Report and Order, May 13, 1991.
[10] FCC, Docket No. 92-9, Notice of Proposed Rule Making (NPRM), "Redevelopment of Spectrum to Encourage Innovation in the Use of New Telecommunication Technologies," adopted: Jan. 16, 1992.
[11] "Personal Telecommunications Services-The Next Generation in Local Service Regulation," Session 9 at *6th Conf. on New Directions for State Telecommun. Reg.,* Feb. 13, 1991, Salt Lake City, UT.
[12] *ITU World Administrative Radio Conference (WARC '92),* Torremolinos, Spain, Feb. 3-March 3, 1992.
[13] Bellcore Technical Advisories, "Generic Framework Criteria for Universal Digital Personal Communications Systems (PCS)," FA-TSY-001013, Issue 1, March 1990 and FA-NWT-001013, Issue 2, Dec. 1990 and "Generic Criteria for Version 0.1 Wireless Access Communications Systems (WACS), TA-NWT-001313, Issue 1, July 1992.
[14] D. C. Cox, "Personal Communications — A Viewpoint," *IEEE Commun. Mag.,* pp. 8-20, 92, Nov. 1990.
[15] D. J. Goodman, "Trends in Cellular and Cordless Communications," *IEEE Commun. Mag.,* pp. 31-40, June 1991.
[16] T. Anderson, "Spread-Spectrum PCS Is Aim of Upcoming Research Project," *RCR,* p. 11, Nov. 5, 1990.
[17] "Wireless Office Phone System for ISM Bands Set for Late '91," *Adv. Wireless Commun.,* p. 8, Oct. 2, 1991.
[18] J. Loeber, "Keeping Two Feet on the Ground; Developing PCNs in the UK," *RCR,* pp. 32, 33, 48, Oct. 8, 1990.
[19] R. Steele, "Deploying Personal Communications Networks," *IEEE Commun. Mag.,* pp. 12-15, Sept. 1990.
[20] Petition to the FCC by PCN America, Inc., a Subsidiary of Millicom Incorporated, for the amendment of Section 2.106 of the Commission's Rules to Allocate Spectrum for a Personal Communications Network, received by the FCC Nov. 7, 1989.
[21] D. C. Cox, "Universal Portable Radio Communications," *IEEE Trans. on Veh. Tech.,* pp. 117-121, Aug. 1985.
[22] D. C. Cox, H. W. Arnold, and P. T. Porter, "Universal Digital Portable Communications-A System Perspective," *IEEE J. Sel. Areas in Commun.,* Vol. JSAC-5, pp. 764-773, June 1987.
[23] D. C. Cox, "Research Toward a Wireless Digital Loop," *Bellcore Exchange,* Vol. 2, pp. 2-7, Nov./Dec. 1986.
[24] D. C. Cox, "Universal Digital Portable Radio Communications," *Proc. IEEE,* Vol. 75, pp. 436-477, April 1987.
[25] D. C. Cox, "Portable Digital Radio Communications—An Approach to Tetherless Access," *IEEE Commun. Mag.,* pp. 30-40, July 1989.
[26] D. C. Cox, W. S. Gifford, and H. Sherry, "Low-Power Digital Radio as a Ubiquitous Subscriber Loop," *IEEE Commun. Mag.,* pp. 92-95, March 1991.
[27] D. C. Cox, "A Radio System Proposal for Widespread Low-Power Tetherless Communications," *IEEE Trans. on Commun.,* May 1991.
[28] D. C. Cox, "Approaches to PCN," *Commun. Int.,* pp. 45-48, Jan. 1991.
[29] R. R. Goldberg and G. G. Brush, "Getting Ready for PCS," *Telephony,* pp. 24-26, Feb. 3, 1992.
[30] G. G. Brush and C. H. Butler, "A More Personal Kind of Communications," *Bellcore Exchange,* Vol. 7, pp. 18-23, Sept./Oct. 1991.

[31] T. Hattori, *et al.,* "Personal Communications — Concept and Architecture," *IEEE ICC '90,* Atlanta, GA, pp. 1351-1357, April 16-19, 1990.
[32] D. Hochvert, "Wireline Carriers View," (NYNEX), *IEEE Personal Commun. Workshop,* Richardson, TX, April 16, 1991.
[33] Bellcore Special Report, SR-NPL-000676, "Comparison of the Economics of Radio and Conventional Distribution for Rural Areas," Issue 1, April 1987.
[34] FCC Report and Order, FCC Docket 86-495, "Basic Exchange Telecommunications Radio Service" RM 5442, adopted Dec. 10, 1987, released Jan. 19, 1988.
[35] S. H. Lin and R. S. Wolff, "A Radio Bridge to Remote Customers," *Bellcore Exchange.,* Vol. 5, pp. 32-36, Nov./Dec. 1989.
[36] J. L. Grubb, "The Traveler's Dream Come True," *IEEE Commun. Mag.,* Vol. 29, pp. 48-51, Nov. 1991.
[37] J. H. Lodge, "Mobile Satellite Communications Systems: Toward Global Personal Communications," *IEEE Commun. Mag.,* Vol. 29, pp. 24-30, Nov. 1991.
[38] CCITT Recommendations E.164, E.168, E.174, E.212.
[39] ANSI, Committee T1, subcommittee T1P1, documents T1P1.2/92-002R2 and T1P1.3/92-002R2, System Engineering Working Document for Personal Communications, April 1992.
[40] "Avis Jumps on ARDIS Bandwagon to Automate Airport Locations," *Industrial Commun.,* p.3, Aug. 30, 1991.
[41] L. Covens, "IBM and Motorola United in New Mobile Data Service," *RCR,* pp. 1,54, March 1991.
[42] "Sears Business Centers to Market RAM's Mobile Data Network," *Industr. Commun.,* p. 1, March 29, 1991.
[43] "NCR Introduces New Wireless Computer Networking Product" (WaveLan), *Indus. Commun.,* p. 3, Sept. 14, 1990.
[44] R. Schneiderman, "Spread Spectrum Gains Wireless Applications," *Microwaves and RF,* pp. 31-42, May 1992.
[45] "Motorola Announces Altair Wireless In-Building Network Ethernet Product," *Industr. Commun.,* p. 8, Feb. 22, 1991.
[46] A. Ramirez, "IBM and 9 Cellular Powers Team Up for Data Transfers," *New York Times,* April 22, 1992.
[47] D. C. Cox, "Response to Follow-Up Questions for Donald C. Cox from FCC Personal Communications Services en banc hearing," to FCC Chairman A. C. Sikes, Jan. 15, 1992.
[48] Testimony at FCC en banc hearing on Personal Communications Services, Washington, DC, Dec. 5, 1991.
[49] S. A. Bergmann and H. W. Arnold, "Polarization Diversity in the Portable Communications Environment," *Electron. Lett.,* Vol. 22, pp. 609-610, May 22, 1986.
[50] J. C-I Chuang, "Performance Limitations of TDD Wireless Personal Communications with Asynchronous Radio Ports," *Electron. Lett.,* Vol. 28, pp. 532-533, March 12, 1992.
[51] I. Dorros, "The New Future-Back to Technology," *IEEE Commun. Mag.,* p. 59, Jan. 1987.
[52] R. Stoffels, "Cellular Arrives at Frozen North," *TE&M,* p. 68, July 15, 1987.
[53] J. C-I Chuang, "Operation and Performance of a Self-Organizing Frequency Assignment Method for TDMA Portable Radio," *IEEE GLOBECOM '90,* San Diego, CA, pp. 1548-1552, Dec. 2-5, 1990.
[54] J. C-I Chuang, "Autonomous Frequency Assignment and Access for TDMA Personal Portable Radio Communications," *IEEE VTC '91,* St. Louis, MO, May 19-22, 1991.
[55] V. H. MacDonald, "The Cellular Concept," *Bell Sys. Tech. J.* (BSTJ), Vol. 58, pp. 15-42 (see p. 29), Jan. 1979.
[56] C.-L. I. L. J. Greenstein, and R. D. Gitlin, "A Microcell/Macrocell Cellular Architecture for Low- and High-Mobility Wireless Users," *IEEE GLOBECOM '91,* Phoenix, AZ, pp. 1006-1011 Dec. 2-5, 1991.
[57] H. W. Arnold, *et al.,* "Wireless Access Techniques and Fixed Facilities Architecture," *Wireless '92,* Calgary, Canada, July 8- 10, 1992.
[58] N. R. Sollenberger and A. Afrashteh, "A Remote Port Architecture for a Portable TDM/TDMA Radio Communications System," Fourth Nordic Seminar on Digital Mobile Radio Communications, Oslo, Norway, paper 12.3, June 26-28, 1990.
[59] M. J. Beller, "Call Delivery to Portable Telephones Away from Home Using the Local Exchange Network," *IEEE ICC '91,* Denver, CO, June 1991.
[60] V. K. Varma, *et al.,* "A Flexible Low- Delay TDMA Frame Structure," *IEEE ICC '91,* Denver, CO, June 23-26, 1991.
[61] L. F. Chang and N. R. Sollenberger, "Performance of a TDMA Portable Radio System Using a Block Code for Burst Synchronization and Error Detection," *IEEE GLOBECOM '84,* Dallas, TX, pp. 1371-1376, Nov. 27-30, 1989.
[62] N. R. Sollenberger, *et al.,* "Architecture and Implementation of an Efficient and Robust TDMA Frame Structure for Digital Portable Communications," *IEEE Trans. Veh. Tech.,* Vol. 40, pp. 250-260, Feb. 1991.
[63] N. R. Sollenberger and A. Afrashteh, "An Experimental Low-Delay TDMA Portable Radio Link," *Wireless '91,* Calgary, Canada, July 8-10, 1991.
[64] N. R. Sollenberger, "An Experimental TDMA Modulation/Demodulation CMOS VLSI Chip-Set," *IEEE CICC '91,* San Diego, CA, May 12-15, 1991.
[65] N. R. Sollenberger, "An Experimental VLSI Implementation of Low-Overhead Symbol Timing and Frequency Offset Estimation for TDMA Portable Radio Applications," *IEEE GLOBECOM '90,* San Diego, CA, pp. 1701-1711, Dec. 2-5, 1990.
[66] J. Hinderling, *et al.,* "CDMA Mobile Station Modem ASIC," *IEEE CICC '92,* Boston, MA, May 1992.
[67] N. R. Sollenberger and J. C-I Chuang, "Low Overhead Symbol Timing and Carrier Recovery for TDMA Portable Radio Systems," Third Nordic Sem. Digital Land Mobile Radio Commun., Copenhagen, Denmark, Sept. 13-15, 1988.
[68] J. C-I Chuang and N. R. Sollenberger, "Burst Coherent Detection with Robust Frequency and Timing Estimation for Portable Radio Communications," *IEEE GLOBECOM '88,* Hollywood, FL, Nov. 28-30, 1988.

[69] A. Afrashteh, N. R. Sollenberger, and D. D. Chukurov, "Signal to Interference Performance for a TDMA Portable Radio Link," *IEEE VTC '91*, St. Louis, MO, May 19-22, 1991.

[70] J. P. Tuthill, B. S. Granger, and J. L. Wurtz, Pacific Bell, "Request for a Pioneer's Preference" before the Federal Communications Commission, submission for FCC General Docket No. 90-314, RM-7140, and RM-7175, May 4, 1992.

[71] "Bellcore PCS Phone Shines in BellSouth Test; Others Fall Short," *Adv. Wireless Commun.*, pp. 2-3, Feb. 19, 1992.

[72] "Bellcore's Wireless Prototype, ...," *Microcell News*, pp. 5-6, Jan. 25, 1992.

[73] BellSouth Services, Inc., "Quarterly Progress Report Number 3 for Experimental Licenses KF2XFO and KF2XFN," to the FCC, Nov. 25, 1991.

[74] D. M. J. Devasirvatham, *et al.*, "Radio Propagation Measurements at 850 MHz, 1.7 GHz and 4 GHz Inside Two Dissimilar Office Buildings," *Electron. Lett.*, Vol. 26, No. 7, pp. 445-447, March 29, 1990.

[75] D. M. J. Devasirvatham, *et al.*, "Multi-Frequency Radiowave Propagation Measurements in the Portable Radio Environment," *IEEE ICC '90*, pp. 1334-1340, April 1990.

[76] J. C-I Chuang, "Comparison of Coherent and Differential Detection of BPSK and QPSK in a Quasistatic Fading Channel," *IEEE Trans. Commun.*, pp. 565-567, May 1990.

[77] L. F. Chang and J. C-I Chuang, "Outage Probability for a Frequency-Selective Fading Digital Portable Radio Channel with Selection Diversity Using Coding," *IEEE ICC '89*, Boston, MA, pp. 176-181, June 11-14, 1989.

[78] D. C. Cox, "Antenna Diversity Performance in Mitigating the Effects of Portable Radiotelephone Orientation and Multipath Propagation," *IEEE Trans. Commun.*, COM-31, pp. 620-628, May 1983.

[79] V. K. Varma, *et al.*, "Performance of Sub-Band and RPE Coders in the Portable Communication Environment," *Fourth Int. Conf. on Land Mobile Radio*, Coventry, UK, pp. 221-227, Dec. 14-17, 1987.

[80] Open Meeting on Status of CDMA Technology and Review of San Diego Experiment, Qualcomm Technology Forum, San Diego, CA, Jan. 16-17, 1992.

[81] *Bell Sys. Tech. J.* (BSTJ), Special Issue on Advanced Mobile Phone Service (AMPS), Vol. 58, Jan. 1979.

[82] R. C. Bernhardt, "Macroscopic Diversity in Frequency Reuse Radio Systems," *IEEE J. Sel. Areas in Commun.* (JSAC), SAC-5, pp. 862-870, June 1987.

[83] CTIA CDMA Digital Cellular Technology Open Forum, June 6, 1989.

[84] A. Viterbi, "When Not to Spread Spectrum-A Sequel," *IEEE Commun. Mag.*, p. 17, April 1985.

[85] "CDMA Capacity Seen as Less than Advertised," *Adv. Wireless Commun.*, pp. 1-2, Feb. 6, 1991.

[86] CTIA Meeting, Dec. 5-6, 1991, Washington, DC.

[87] S. Ariyavisitakul, "SIR-based Power Control in a CDMA System," *IEEE Globecom '92*, Orlando, FL, Paper 26.3, Dec. 6-9, 1992, to be publisahed in *IEEE Trans. Comm.*

[88] L. F. Chang, "Dispersive Fading Effects in CDMA Radio Systems," *Electron. Lett.*, Vol. 28, No. 19, pp. 1801-1802, Sept. 10, 1992.

[89] S. Ariyavisitakul, *et al.*, private communications.

[90] "TDMA Accelerates, but CDMA Could Still Be Second Standard," *Adv. Wireless Commun.*, p. 6, March 4, 1992.

[91] R. C. Bernhardt, "User Access in Portable Radio Systems in the ᵁᵁᵁ Limited Environment," *IEEE ICC '87*, Seattle, WA, pp. 97-104, June 8, 1987.

[92] R. C. Bernhardt, "An Improved Algorithm for Portable Radio Access," *IEEE VTC '88*, Philadelphia, PA, pp. 163-169, June 15-17, 1988.

[93] A. Salmasi and K. S. Gilhousen, "On the System Design Aspects of Code Division Multiple Access (CDMA) Applied to Digital Cellular and Personal Communications Networks," *IEEE VTC '91*, St. Louis, MO, pp. 57-62, May 1991.

[94] L. F. Chang and S. Ariyavisitakul, "Performance of a CDMA Radio Communications System with Feed-Back Power Control and Multipath Dispersion," *IEEE GLOBECOM '91*, Phoenix, AZ, pp. 1017-1021, Dec. 1991.

[95] S. Ariyavisitakul and L. F. Chang, "Signal and Interference Statistics of a CDMA System with Feedback Power Control," *IEEE GLOBECOM '91*, Phoenix, AZ, pp. 1490-1495, Dec. 1991.

[96] FCC, Docket No. 92-333, "Amendment of the Commission's Rules to Establish New Personal Communications Services," adopted: July 16, 1992.

Biography

DONALD C. COX [F] received B.S., M.S., and Honorary D.S. degrees from the University of Nebraska, Lincoln, and a Ph.D. degree from Stanford University. Dr. Cox is executive director of Radio Research at Bellcore. In 1983, prior to going to Bellcore, he had been at AT&T Bell Laboratories for 15 years where he was the department head of the Radio and Satellite Systems Research Department.

There are increasingly complex electropolitical and regulatory issues which must be resolved.

Reprinted from *Electrical Communications* 2nd quarter 1993, pp. 118-127.

What are GSM and DCS ?

C. Déchaux   Alcatel Mobile Communications Centre, Colombes, France
R. Scheller   Alcatel SEL, Stuttgart, Germany

This article is a brief outline of the major elements of the digital mobile radio system, introducing key terms and showing how the application of state-of-the-art system, hardware and software technologies can provide improved mobile services to more subscribers. It will serve as an introduction to the articles on specific GSM/DCS subsystems that follow in this issue. A list of all abbreviations used in these articles will be found on the last pages of the journal.

The Origin of the Cellular Concept

The potential for using radio waves to communicate between moving points was recognized early in the history of radio communications. The use of radio in police vehicles was reported in the 1920's, and in 1946 the idea of connecting a radio link to the fixed telephone network was implemented by Bell System in the USA to provide the so-called Public Correspondence Service. The interconnection with the mobile subscriber was made through a fixed channel, which was managed by an operator who was in effect providing a kind of secretarial service.

A first step in improvement of the efficiency of spectrum usage was made in 1964 by the introduction of automatic trunking (that is to say the allocation of a channel limited to the duration of each connection and chosen from a pool of channels).

The concept of Cellular which will be explained in this article was proposed by Bell System in 1971 following a request from the Federal Communications Commission. Real tests of the Advanced Mobile Phone System (AMPS) were carried out in Chicago where an operational system was installed in 1978.

The key objectives of such a Cellular System were stated as follows :
— Large subscriber capacity
— Efficient use of spectrum
— Nation-wide compatibility
— Widespread availability
— Adaptability to traffic density
— Service to vehicle and portable stations
— Regular telephone service and special services including dispatch
— Telephone quality of service
— Affordability.

Based on the Bell concept the effective introduction of commercial Cellular Services in Europe and in the USA started in the early 1980's and already cellular systems are serving more than 18 million subscribers around the world.

Today the key objectives are generally achieved by the main cellular system standards such as: AMPS in the 800 MHz band, its 900 MHz derivative TACS (total access cellular system), NMT (Nordic mobile telephone) and several other standards which can be considered as variations upon the same key concepts.

The Emergence of the GSM Standard

In 1982, at a time when the first commercial Cellular Services were emerging, the CEPT (Conférence Européenne des Postes et Télécommunications) took the initiative of launching a working group (called Groupe Spécial Mobile) in charge of specifying a common mobile communication system for Europe in a 900 MHz band which had been reserved in 1978 by the World Administrative Radio Conference.

Qualitatively speaking the 9 key objectives stated by Bell 15 years ago still apply; however a specific emphasis has been put on having access to ISDN related services and on the compatibility with CCITT recommendations. In terms of a quantitative targets, major improvements were efficiency of spectrum usage, and suitability for handheld terminal service.

Today the GSM standard has been successfully tested and more than 36 GSM networks will be in service in 1993 in 22 countries. At present more than 25 non-European countries have adopted or are considering adopting the GSM standard.

Mobile services can be associated with an individual subscriber rather than a piece of equipment or a line termination, thus providing what are known as Personal Communication Services. The development of Personal Communication Networks having radio access to the fixed telephone network is expected to affect a significant proportion of telephone subscribers within the next 10 years. From the various technologies which could be used to support such services, GSM technology adapted to the newly reserved frequency in the 1800 MHz frequency band (and renamed DCS 1800) has already been chosen by several operators. The GSM/DCS standard is expected to serve 20 to 40 million subscribers in the year 2000.

Basic Cellular Concept

Before presenting the main characteristics of the GSM/DCS system and services it is worthwhile to analyze the key aspects of the Cellular Concept.

Frequency reuse concept
Radio-electric waves appear to be the only effective means of providing communication with moving points. Unfortunately the radio spectrum is a limited resource, already widely used for fixed applications such as television and

radio broadcasting or microwave links. As frequency channels cannot be duplicated the idea is to reuse them, taking advantage of the limited distance of radio propagation at high frequencies. In this way separate areas, or cells, can be served by the same radio channel at the same time. The smaller the cell, the greater will be the number of channels which can be used at the same time in a given area consisting of many cells.

Figure 1 shows a typical reuse pattern in a 9-cell cluster where the same frequency channel is used in cells A and A1 . In this configuration three cells (A, A', A") are served by one base station of the infrastructure network.

Improvement of the frequency reuse factor
Although several practical constraints limit the reuse of radio channels, several technical solutions reducing the internal interference (co-channel interference) are used in GSM/DCS to overcome this limit. The transmitted radio signal power levels of the terminal and the base station are continuously adjusted to provide the requested quality of the link. This *power control mechanism* ensures a minimum of interference between the cells using the same radio channel.

Moreover, taking advantage of the discontinuous nature of speech signals in a normal dialogue (when

Figure 1 - Basic 9-cell cluster showing frequency reuse in cells A and A1

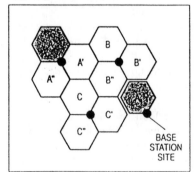

one is speaking the other one is generally listening, and every speaker's voice contains "holes" which occur between uttered syllables), the power of the transmitted signal can be reduced to zero during more than half of the time. This *discontinuous transmission mechanism* is activated by a voice activity detector.

Other ways of reducing the average interference are the use of *frequency hopping* (which will be explained later) and the choice of an appropriate modulation scheme.

Consequence of mobility in a cellular network
Two major consequences result from the mobility of subscribers in a cellular network :

— To ensure a connection the cell in which a mobile subscriber's terminal equipment is located must be known. Tracing the terminal within the network is done through location management
— The link between the terminal and the infrastructure must be maintained automatically whenever the subscriber's terminal moves from one cell to another (the process is called *handover*).

When the mobility of a subscriber is extended to different networks, in different countries for instance, a special process is initiated to "recognize" foreign subscribers and deal with their calls and associated charges. The facility of multi-network services is called *roaming*.

Organization of a cellular network
A cellular network consists of a set of base stations deployed in the geographical area to be serviced. The base stations which provide coverage of cells are linked through telephone lines to the *mobile*

switching centre in charge of managing the connection with the public system telephone network.

GSM/DCS Services and Facilities

All the services available in the integrated services digital network (ISDN) have been considered for implementation in GSM/DCS. However, due to limitations in the performance of the radio interface (in particular in terms of data transmission rate and error rate) some ISDN services have been implemented with restrictions.

Teleservices

Telephony is the most important teleservice in GSM/DCS. It allows calls between the general telephone network (PSTN/ISDN) and the mobile network. Emergency calls are also available on GSM/DCS to allow a direct and automatic connection with the nearest emergency service by dialling 112, using a unique and simple procedure. Group 3 Fax service is supported by the GSM/DCS network provided that interface adapters are installed at the terminal level and at the interface with the telephone network

Figure 2 - General scheme of GSM/DCS data transmission

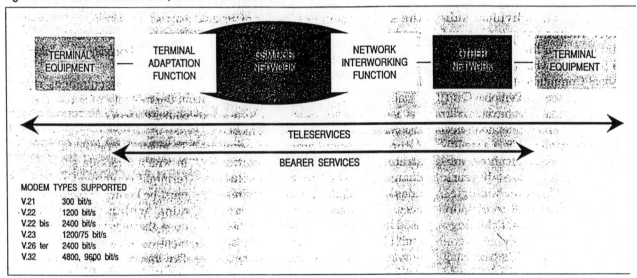

Figure 3 - GSM/DCS system architecture

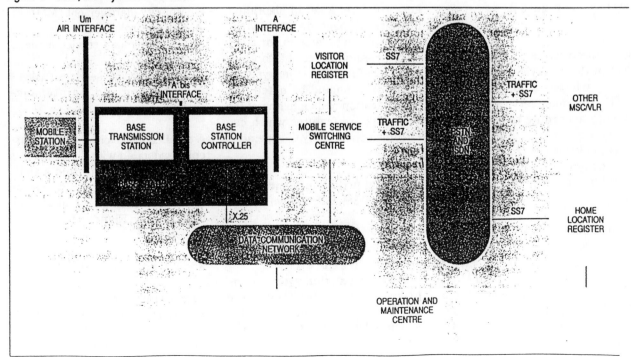

487

A kind of E-mail service is available in GSM/DCS. It is the *short message service* which can be considered as a bi-directional alphanumerical paging service. It is integrated in the GSM/DCS concept and can be economically implemented in a standard terminal. Any telecommunication server may be used to input a short message (140 bytes). Successful delivery of the message is acknowledged, which is a major advantage over paging. The short message service is available in both directions in a point to point mode but also in a cyclically broadcast mode in selected areas.

Bearer services

These data services, which are limited to layers 1, 2 and 3 of the OSI reference model, cover rates from 300 bit/s to 9.6 kbit/s. The mode of transmission can be either transparent or non-transparent. In the latter case an additional protection against errors is achieved with an acknowledgement procedure in the radio link protocol. The general scheme for bearer services and the type of modems supported by GSM/DCS are shown in **Figure 2**.

Supplementary services

Most of these services are equivalent to what is available in the PSTN. The main services are *call barring* according to a criterion such as out-going and international, *call forwarding* (if mobile busy or not reachable), and *connected line identification*.

Subscriber identity module

A GSM/DCS terminal cannot enter the network unless all subscriber specific data is present in the terminal. This data is embedded in a *smart card* (called SIM, subscriber identity module) which has to be plugged into the terminal. The SIM card, whose access is protected by a *personal identity number*, contains not only the subscriber data (ISDN number, personal key) but also personal information such as abbreviated dialling numbers, list of preferences of networks, and charging information. Short messages are also stored in the SIM card.

Security functions

The radio path of the link between the terminal and the infrastructure could be subject to various attempts at fraudulent use or unauthorized interception. In the GSM/DCS System, protection is achieved at 3 levels :
— Subscriber identity modules are authenticated by the system to prevent access by unregistered users
— The radio part of the link is ciphered to prevent unauthorized listening to calls (voice or data)
— Subscriber identity is protected.

The security is achieved by using ciphering algorithms which are resident in the terminals and in the infrastructure. A set of algorithms has already been defined by the CEPT administrations for their own use. Another set of algorithms will be defined for other types of operator. It should be noted that authentication and ciphering of the radio link are features which can be turned on or off by the operator.

Functional Architecture and Interface of a GSM/DCS System

The main functional entities and interfaces of a GSM/DCS system are shown in **Figure 3**.

The base station system comprises a controller (BSC) and transmit-receive equipment (BTS) which is deployed in the area to be covered.

The network subsystem part (NSS) includes switching equipment (MSC) dedicated to the mobile service and linking all system elements through leased lines, to PSTN and ISDN. The home and visitor location registers are databases used to store mobile subscriber data. Copies of the SIM subscribers' secret keys are stored in the authentication centre and the mobile equipment serial numbers are stored in the equipment identity register. All system elements are operated, controlled and maintained by the operation and maintenance centre.

The functions related to networking and switching are generally located "above" the A interface (in MSC, VLR, HLR) whereas those related to "radio aspects" are in the BSS. However, a number of functions are distributed and are performed by the co-operation of several functional entities.

Using the OSI model, the GSM system can be described by considering several functional "layers" seen across the main entities and interfaces :
— Transmission
— Radio resources management (RR)
— Mobility management (MM)
— Call control (CC)
— Supplementary service management (SS)
— Short message services (SMS)
— Operation and maintenance management.

A detailed description of the protocols associated with these layers is beyond the scope of this article. We will limit our analysis to some elements which are typical of the GSM system (**Figure 4**).

Radio Transmission - Layers 1 & 2

Basic transmission scheme

At the radio interface level (air interface) the access mode is a combination of time division

(TDMA) and frequency division (FDMA). So the quantum of signal transmitted is a burst of radio energy (in a repetitive pattern of 8 in GSM) transmitted in a frequency band (chosen from a set of radio carriers). An important option is frequency hopping which consists of changing at every time slot the frequency on which a burst is transmitted. The frequency change is made according to a pattern imposed by the base station.

The physical support (layer 1) consists of data transmitted at a rate of 270 kbit/s in 0.577 ms bursts on radio carriers separated by 200 kHz.

There are 124 radio channels in the 900 MHz band and 374 in the 1800 MHz band (mobile to fixed: 890 - 915 MHz and 1710 - 1785 MHz, fixed to mobile: 935 - 960 MHz and 1805 - 1880 MHz).

Besides the normal burst there are 3 other special types of burst used for access, synchronization and frequency correction (see **Figure 5**).

Logical channels

Bursts, which are able to carry different types of information, form logical channels which are organized into frames and higher structures as shown in **Figure 6**.

The "useful" end to end information exchanged through the network is voice or data. These are transmitted over a logical channel called the traffic channel (TCH). A full rate traffic channel allows the transmission of speech coded at 13 kbit/s or data at 3.6, 6 or 12.6 kbit/s. A half rate channel (achieved by using on average every second burst) allows speech transmission coded at about 6.5 kbit/s.

To manage and control the link between the mobile terminal and the infrastructure a set of control and signalling logical channels are

used. Each traffic channel is permanently associated with a slow associated control channel (SACCH) for non-urgent messages (for instance transmission of radio measurement performed by the terminal, or power control messages).

For urgent messages during the call or release process, a TCH is temporarily used for control and becomes a fast associated control channel (FACCH).

Mainly outside the traffic period, 8 other channels are used for the following main functions :
— Synchronization bursts are periodically broadcast by the base station to allow the terminal to fine tune the frequency and perform frame synchronization
— Base station identification and parameters (channel description, cell identity, location area identity, control channel description, options...) are broadcast periodically on the broadcast control channel (BCCH)

Figure 4 - GSM/DCS protocol architecture

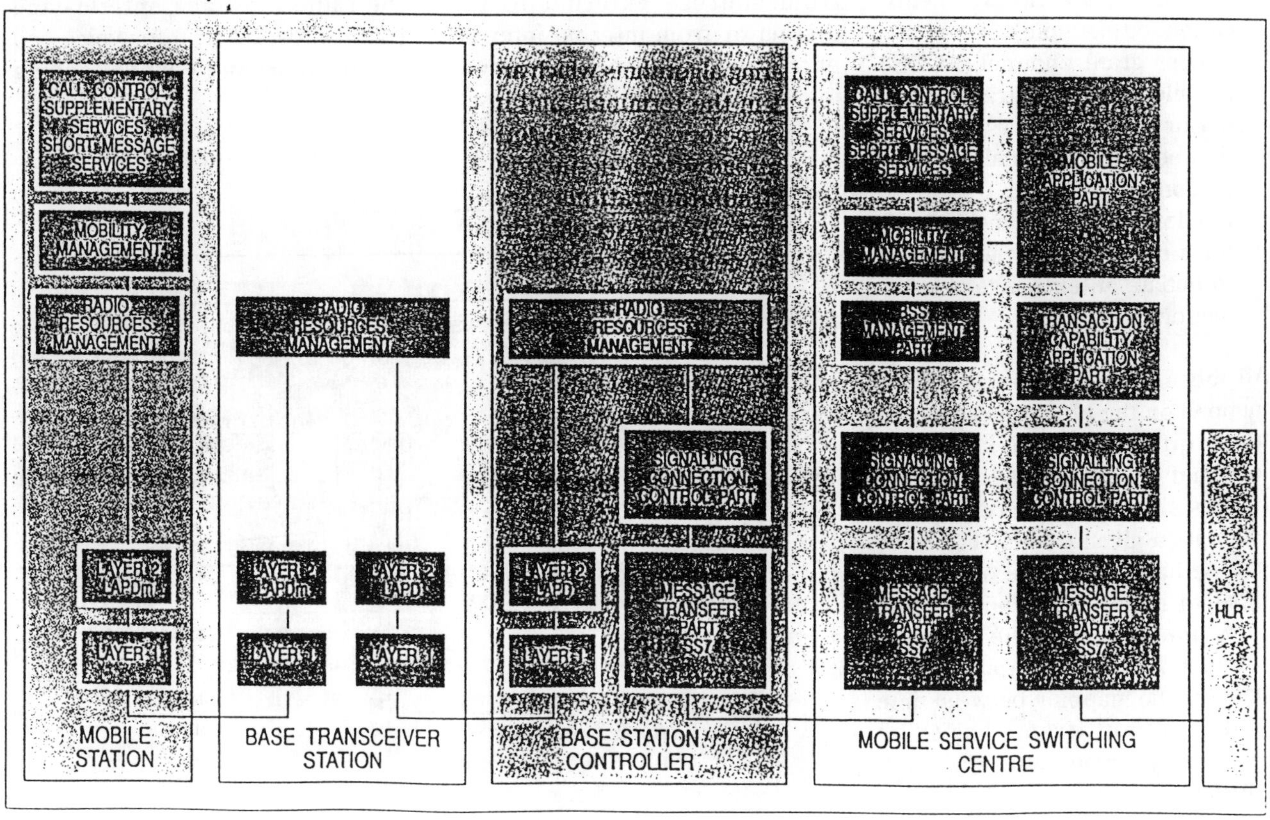

489

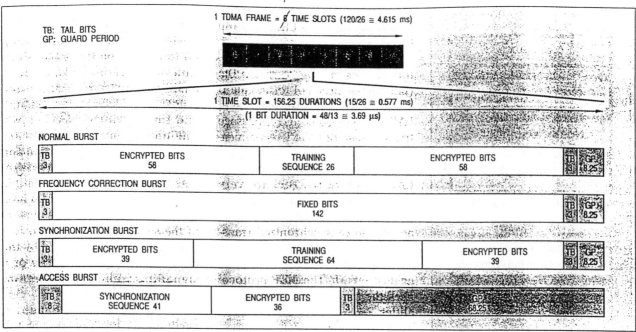

Figure 5 - Structure of GSM/DCS transmission bursts

— Paging channels (PCH) are used to call a terminal to initiate the call process

— Standalone dedicated control channels (SDCCH) are used for exchange of information before a traffic connection (authentication for example)

— Access grant channel (AGCH) for allocation of a traffic channel or a SDCCH for a connection between terminal and base station

— Random access channel (RACH) for a request from the terminal to initiate an exchange of information.

All these logical channels can be mapped on the physical channel in many different ways, hence the long and complex structure of the frames shown in Figure 6. As an example a given time slot (0 to 7) can be handled in one of the two types of mutiframe (26-multiframe or 51-multiframe). The 51-multiframe is more appropriate for signalling. The mapping between time slots and multiframe has to be decided by the operator when optimizing the network.

Channel coding

The radio transmission medium used for mobile communications is subject to a wide range of disturbances. The disturbing signals are mainly of burst-type, coming either from sources external to the system or from internal interference caused by frequency reuse

and multipath propagation. To achieve the requested level of acceptable errors (about 10^{-2} for voice for instance) many "error correction" techniques are implemented all along the data path. At the radio interface, interleaving (with a "depth" of 8 for voice and 19 for data) and channel coding

Figure 6 - Frame hierarchy

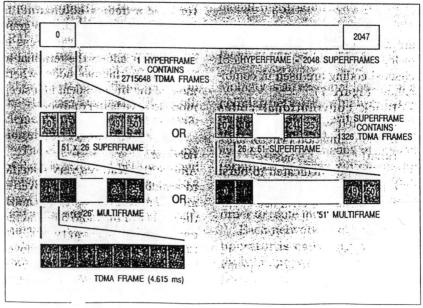

490

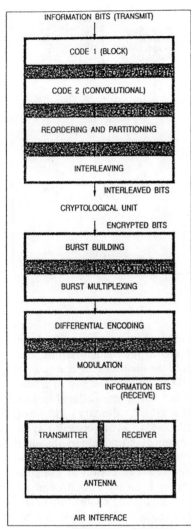

INFORMATION BITS (TRANSMIT)

CODE 1 (BLOCK)

CODE 2 (CONVOLUTIONAL)

REORDERING AND PARTITIONING

INTERLEAVING

INTERLEAVED BITS

CRYPTOLOGICAL UNIT

ENCRYPTED BITS

BURST BUILDING

BURST MULTIPLEXING

DIFFERENTIAL ENCODING

MODULATION

INFORMATION BITS (RECEIVE)

TRANSMITTER   RECEIVER

ANTENNA

AIR INTERFACE

Figure 7 - Sequence of coding and processing of bits in the transmission path

are used specifically on each type of logical channel. Three types of channel coding are used in combination (**Figure 7**).
— Block convolutional codes associated with maximum likelihood estimation (Viterbi algorithm)
— Cyclic code dedicated to burst error detection and correction
— Parity codes for error detection.

Transmission layer 2
The main purpose of layer 2 protocol is to provide link connections to exchange signalling between the

different entities: Mobile Station, BTS, BSC, MSC, VLR, HLR and CCITT No 7 Network.

In GSM, 3 types of layer 2 protocol are used (Figure 4): LAPDm (link access protocol for mobile D channel) on the air interface, LAPD on A-bis and MTP (message transfer part of CCITT recommendation).

The protocols LAPDm and LAPD used in the BSS are very similar to the ISDN protocol. However, the LAPDm takes advantage of synchronized transaction to avoid use of flags and increase speed and protection against errors. The MTP protocol uses the standard ISDN functionalities.

Radio Resource Management

This layer 3 set of protocols controls the links between the terminals and the infrastructure. When a terminal equipped with a SIM card is put into service it scans the radio channels to find the synchronization (logical) channel and become synchronized. It is then in an idle mode waiting either to be paged through the paging channel or for a request to access the network by sending a message on the random access channel. In the latter case a dedicated channel is allocated through the access grant channel. The paging process is such that it allows the terminal to be in discontinuous mode, where most of the time is spent in a sleep mode to save battery power.

Handover
Among the features very specific to a mobile network, handover is an essential process to optimize the quality of the radio link and the capacity of the network. In GSM/DCS the concept of handover has been extended to intra-cell, that is to say selection of different channels even in the same cell.

Before deciding to perform a handover, the base station controller accumulates information on traffic and on the actual situation of the radio links such as quality (error rate), transmitted power, level of received signal, and time advance (see below). The quality measurements are made both by the base station and by the terminals. Taking advantage of the time division multiple access structure of the radio link, the terminal measures parameters of the signal received from neighbouring cells (as requested by the base station) while a traffic or a signalling connection is established.

The algorithms for handover decision are implemented in the BSC, but are not specified by the GSM Recommendations. Thus each system supplier is free to design a proprietary algorithm. It is expected that suppliers will need to improve their algorithms to deal with high density areas where small cells are deployed.

Power Control
Both directions of the radio link between the terminal and the base station are subject to continuous (in fact every 60 ms) power adjustment over a range of about 26dB (for instance from 8 W to 20 mW). The power adjustment of the BTS and of the terminal are under the control of the BSC. This improves the spectrum efficiency by limiting the intra-system interference. It also increases, on average, the duration of terminal autonomy by saving battery power.

Time advance control
As all the terminals under control of a cell are at a variety of distances from the base station, the bursts received by the base station are spread due to different radio propagation delays. The time spread requires a large guard time

between bursts, resulting in a lower spectrum efficiency. In order to minimize this effect in the GSM DCS air interface, a time adjustment of the burst coming from the terminal is made in the terminal. The adjustment covers a range of 233 µs, which allows correction for cells of a maximum radius of 35 km. The time advance of each terminal is monitored by the BSC as a criterion for handover and also to help a terminal preset its time advance prior to synchronized handover.

Management of the radio channels

This function is under control of the BSC. Its freedom is limited by the basic physical and logical channel configurations which have been chosen by the operator when planning the network. Operation and maintenance commands are necessary to change the configuration.

The GSM Recommendations have a large flexibility to define a strategy of real-time allocation of common and dedicated channels. In particular, the set of traffic channels can be chosen dynamically using queuing, deferred assignment or priority for the "best quality" channels.

Restriction of access - priority subscribers

The access request from a terminal through the random access channel must be controlled in order to avoid blocking situations. Among the different means available, the most efficient is through the Access Class number. In each subscriber's SIM card a random number from 1 to 10 is stored. In the event of saturation the BSC can restrict, through the broadcast channel, the access to any class of subscribers. Special class numbers 11 to 15 are reserved for high priority subscribers (security services, emergency services, operator field staff).

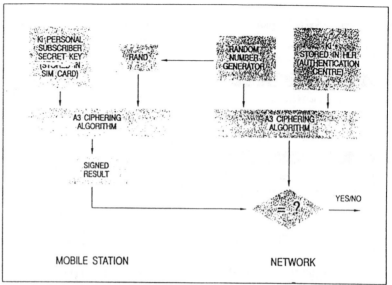

Figure 8 - Authentication process

Security Management

Authentication is performed by asking the terminal to give the result of a specific computation made on a random number (RAND) sent by the system and then comparing the result with what it should be.

This computation process is in fact dependent on a secret key (Ki) specific to each subscriber SIM card. The computation is done following a ciphering algorithm A3, whose property is that knowing the result (SRES) and one input (RAND), the other input (Ki) cannot be practically deduced. The secret key (Ki) and the A3 algorithm are stored with protection in the SIM card and in the HLR, both elements performing the calculation shown in **Figure 8**.

The ciphering of the radio burst of data is achieved with a second ciphering algorithm A5 applied to a key (Kc) chosen for each connection and a number changed at each burst.

The key Kc is computed in the terminal and in the HLR with a third algorithm A8 (see **Figure 9**) similar to A3. The A3 and A8 algo-

rithms are not specified in GSM/DCS recommendations, but left to the choice of the operator.

The personal number of a subscriber (IMSI, international mobile subscriber identity) is not ciphered, to avoid complexity in managing the preliminary dialogue between the terminal and the infrastructure. Protection of the subscriber identity is achieved by using a temporary substitute of the IMSI (called TMSI) allocated by the network at the first time of mobile registration in a given area.

Mobility and Call Management

Mobility is an essential feature of a cellular network, which requires basically that the terminal can be traced and located. The GSM/DCS system provides means of tracing terminals in areas which can be seen from a geographical and from an administrative point of view in order to route incoming calls.

Each network controlled by an operator is called a public land mobile network (PLMN). In Europe most of the geographical areas will be served by two or

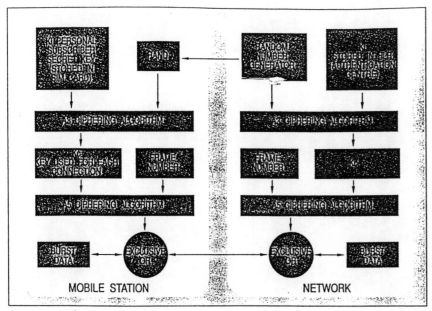

Figure 9 - Ciphering process

three operators and therefore two or three PLMNs.

In order to minimize the signalling load and the access time, every terminal identifies itself by its IMSI in location areas which correspond to a group of cells where it can be paged. The IMSI contains the personal subscriber number, the PLMN number and the country code. Location information is stored in the visitor location register and in the home location register of the subscriber's PLMN. When a call towards a mobile subscriber is initiated the PSTN looks for the relevant home location register. After finding in the HLR all the necessary information and in particular the location of the subscriber the call is routed to the relevant MSC.

A mobile application part (MAP) based on CCITT No 7 signalling has been specified so that calls can be routed to another network where a mobile station has roamed.

GSM/DCS Recommendations

The GSM/DCS Recommendations are a set of more than 130 specifications adding up to about 5500 pages

where the architecture, the functions and the interfaces of the system are described. The mobile station and base station conformity tests and associated measurement methods are detailed in the Recommendations.

A first level of specification called Phase 1 was approved by the European Telecommunications Standards Institute (ETSI) in 1990. A second and final level of the Recommendations (called Phase 2) is scheduled to be issued by the end of 1993. It will include a number of new features, in particular:
— Supplementary services:
  . Multiparty calls
  . Closed user group: a set of facilities similar to those found in a PBX to optimize or restrict the services to a limited number of subscribers
  . Advice of charge: charging information given in real-time at the terminal level
  . Line identification
— Support of half rate channels
— Support of terminals having different ciphering algorithms
— GSM extension band (+10 MHz)

It should be pointed out that the GSM/DCS Phase 2 network will

support Phase 1 terminals but not the other way round. Therefore a set of changes is being defined to upgrade the Phase 1 network so that it will be able to support Phase 2 terminals, although without servicing all the new features of Phase 2.

Conclusions

The GSM/DCS cellular standard represents a real breakthrough from the limitations of analog mobile techniques, both in terms of the range and quality of services that can be offered and in terms of technology. It has been designed with a large potential for improvement and extension for the benefit of both operator and subscriber. A special emphasis has been put on the service to hand-held personal equipment in a high density environment.

While the GSM/DCS technology is able to provide the so-called Cellular Service it can also be extended to provide a more widespread Personal Communications Service using an alternative radio access air interface.

The GSM/DCS standard provides not only a common air interface but also an open interface, allowing interconnection of different networks to provide multi-country services. The A-interface is an open internal interface of the system between the radio part and the network part, which allows multi-vendor networks.

Originally focused on the European market, the GSM/DCS standard has now achieved world-wide credibility. This is confirmed by the number of countries and operators that have invested in GSM/DCS networks, and by the number of companies who have committed themselves to developing the network equipment and terminals. This confidence in the system and in the market is a guarantee that competitively priced terminals and services will become widely available to subscribers.

References

1. Proceedings of the Digital Cellular Radio Conference. October 12th - 14th, 1988, Hagen, Westphalia, Germany.

2. Telecommutant - Les radiocommunications avec les mobiles - Numéro du 4 juin 1989.

3. The GSM System for Mobile Communications - Michel Mouly, Marie Bernadette Pautet, 1992.

Claude Déchaux was born in 1943. He obtained a degree from the Ecole Supérieure d'Electricité in France in 1967. He spent 17 years with Thomson-CSF where he worked for the telecommunications division, as Product Planning Manager of mobile telecommunications activities, and later as head of the Marketing and Product Planning Department for military tactical radiocommunications. He was appointed in 1982 as Technical Director of radiotelephone activities, which became part of Alcatel three years later. In 1987 he became Director of Alcatel Radiotelephone's Infrastructure Business and he was also appointed, within Alcatel Radio Space and Defence, Director of the GSM Radio Infrastructure Program. Claude Déchaux is at present Director of Mobile Communications Business Strategy at RSD.

Reinhard Scheller was born in Stuttgart. He finalized his studies in Electrical Engineering at the University of Stuttgart, with special emphasis on switching and teletraffic theory. He received his PhD in 1981. After joining Alcatel SEL in 1981 he worked in various areas of telecommunications, including System 12 public switching as well as the business system group. He was appointed Managing Director of an Alcatel SEL joint venture in Hungary in 1989. In 1991 Dr Scheller became Director of the Mobile Communication Product Group and is at present responsible for GSM and DCS 1800 infrastructure equipment within Alcatel SEL.

Reprinted from *Phillips Telecommunications Review*, Vol. 49, pp. 68-73, Sept. 1991.

DECT, a universal cordless access system

R.J.Mulder

DECT is particularly useful for wireless PABX environments, but it includes other areas as well, such as residential deployment, Telepoint services and local loop replacement. In addition, DECT positions itself as a natural complementary access system for Personal Communications Network (PCN) concepts in buildings. The article discusses the principles and characteristics of the DECT system, positions it vis-à-vis PCN and other cordless technology contestants, and gives some typical DECT applications.

1 Introduction

DECT, the Digital European Cordless Telephone system, is not just another cordless system. In comparison with competing systems it offers a superior quality and enhanced mobility. This makes DECT particularly useful for wireless PABX environments, but the DECT system pretensions include other areas as well, such as residential deployment, Telepoint services and local loop replacement. In addition, the access capabilities of DECT, combined with its mobility features, position it as a natural complementary access system for Personal Communications Network (PCN) concepts in buildings. The article discusses the principles and characteristics of the DECT system, positions it vis-à-vis PCN and other cordless technology contestants, and gives some typical DECT applications.

2 General principles of cellular systems

The DECT-specification brings cordless telephone technology into another class of applications by virtue of its capabilities to support a *micro-cellular concept*. It is this concept that has widened the scope of cordless technology to the somewhat "hyped" multi-purpose communication tool discussed today and therefore worth a closer look.

The cellular mobile radio systems we know today offer wide area mobility to moving users. They comprise, apart from the base stations and the mobiles/portables linked to them, an extensive fixed wide area infrastructure, overlaying the fixed telephone network, or partly deploying it.

All recent cellular mobile radio systems are based on the (theoretical) 7-cell cluster (**Fig.1**) to provide reuse of frequencies and hence proper communication channel capacity. The frequencies in these networks of cells are fixed in a rather painstaking process called *frequency planning*, based upon surveys of the areas to be covered and helped by extensive computerised tooling.

When users move from one cell to another (*roaming*), the fixed part of the network tracks them in order to know where they are when a call has to be

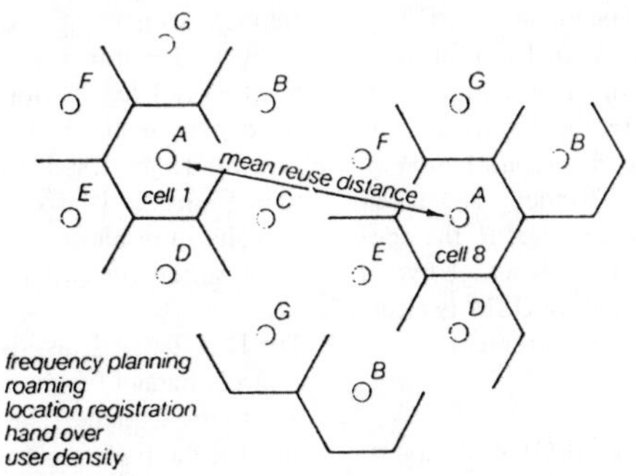

frequency planning
roaming
location registration
hand over
user density

Note: cells with the same letters use the same set of frequencies

Fig.1 Principles of cellular systems

passed on to them. That process is called *location registration*. When they move from one cell to another during a call the system executes a *handover* in order to allow the call to be continued.

Traffic capacity is obviously related to the user density. The main parameter to manage in this respect is the cell size. High user densities require more and smaller cells, associated obviously with lower transmitting power and a limited range of base stations and portables, to cater for the interference issues. The cell size foreseen for the new Pan-European digital Cellular Mobile Radio system GSM is reflecting this: a radius of 15 km should serve rural areas, while in

Table 1 Comparison DECT vs GSM

	DECT	GSM
net bit rate	32 kbit/s	13 kbit/s
frequency	1900 MHz/TDMA	900 MHz/TDMA
carriers/slots	10 x 12 = 120	44 x 8 = 352
channel allocation	dynamic	fixed
cell radius	25-100 m	2 km (urban)
		15 km (rural)
traffic density	10000 E/km	1000 E/km
pocketable	yes	restricted
location registration	> 6000 subscribers	yes

TDMA: Time Division Multiple Access

495

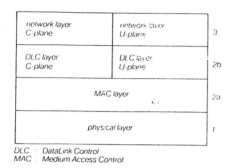

network layer C-plane	network layer U-plane	3
DLC layer C-plane	DLC layer U-plane	2b
MAC layer		2a
physical layer		1

DLC : DataLink Control
MAC : Medium Access Control

Fig.2 DECT layers

urban areas with hand-held coverage the radius might go down to 2 km.

A comparison of DECT with GSM (**Table 1**) makes clear what is meant with the statement that DECT serves a *micro-cellular* concept. Apart from bit rates, frequency spectrum and the number of channels per frequency or cell, the most paramount difference is the fact that DECT relies on a *dynamic allocation of channels* for communication. Channels are taken from the pool when needed, and returned to the pool after a call has been completed. The advantage over a fixed grid of channels as supported by GSM is clear: high traffic capacity, an obvious requirement for an office environment.

Another important difference is DECT's ability to page up to 6000 subscribers without the need to know in which cell they reside, obviating the need for a detailed and comprehensive location registration function in the majority of the office applications.

Finally, the handover principles applied in DECT are designed for in-building environments, where frequently changing conditions require an ample handover quality performance even within a single cell, which is not being met by GSM.
The micro-cellular nature of DECT, with typical cell sizes between 25 and 100 m, does not allow for a wide area coverage. DECT is, for that matter, there-

Fig.3 Structure of DECT frame and time slot

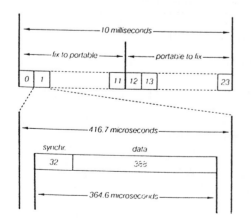

fore not a total system concept, like GSM, but restricts itself mainly to the provisions of local or metropolitan radio access, to be used in conjunction with a variety of wide-area networks, either of a fixed or mobile nature.

3 Technical description

The DECT system applies the OSI principles in a similar way as ISDN. A control-plane (C-plane) and a user-plane (U-plane) use the service offered by the lower layers, ie the physical layer and the Medium Access Control (MAC) layer (**Fig.2**).

The DECT physical layer uses an FDMA/TDMA/TDD radio transmission method, ie Frequency Division Multiple Access/Time Division Multiple Access/Time Division Duplex. Within a time slot a dynamic frequency selection of 1 out of 10 carrier frequencies is being used. The carrier frequencies are comprised within the 1880 to 1900 MHz frequency band. DECT has 24 time slots, ie 12 time slots for the fixed part to the portable and 12 time slots for the portable to the fixed part. The time slots fit within a time frame of 10 ms. Within a time slot a burst of 420 bits information is being transmitted, ie 32 synchronisation bits and 388 data bits (**Fig.3**).

The DECT MAC layer offers a paging channel and a control channel for the transfer of signalling information to the C-plane. The U-plane is served with channels for the transfer of user information. The normal bit rate of a user information channel is 32 kbit/s. DECT, however, also supports other bit rates, eg n x 32 kbit/s for ISDN and LAN-type applications. Apart from the evident MAC functions, such as the establishment of connections, the DECT MAC layer supports two specific functions: MAC handover and the provision of a so-called beacon.

The MAC handover, ie switching a call-in-progress to another channel, is initiated by the portable, when the transmission quality becomes poor. Handover can be either *intra-cell*, in case of interference, or *inter-cell*, when out of range. The DECT MAC layer provides a virtually seamless handover.

In the DECT system every fixed radio transmits on at least one channel. Thus every fixed radio can be considered as a beacon. This enables idle portables to find the best fixed radio and lock on it. As a consequence very fast connection set-up times can be achieved.

The datalink control (DLC) and network layer protocols in the C-plane of DECT are derived from ISDN and GSM protocols. In this way good interworking with both ISDN and GSM is ensured.

The DECT U-plane provides means to support ISDN basic services, such as 64 kbit/s unrestricted, but various packet mode services such as frame-relay and frame-switching can be included as well. The generic voice service is based on 32 kbit/s ADPCM techniques in accordance with CCITT Rec.G.721.

4 Mobility and DECT

As already indicated, DECT is primarily an *access system*. The different networks accessed with DECT are obviously not a part of the DECT Specification. The protocol supported via the "Common Interface" should however allow for mobility in conjunction with one or several networks. Mobility implies, amongst others, the following aspects:

- SUBSCRIPTION: determining the initial access rights and the user identities;
- ROAMING: checking access rights of calling and called users, paging and possibly location administration of called users;
- HANDOVER: changing of channels during a call;
- SECURITY: encryption and authentication.

The basis for the support of mobility in DECT networks is embodied by the DECT-identity structure. In order to discriminate a particular DECT-based cordless PBX network from another one, or from a Telepoint or residential system, DECT networks broadcast their identity and the associated access-right details. The portables or handsets carry identity and portable access right information as well. Actual communication between a handset and a DECT network is only allowed when there is a match between the fixed and portable access rights. As handsets may subscribe to more than one network, more than one match may be possible.

For encryption and authentication purposes, handsets accommodate an authentication key. This key is, together with a PIN-code, consulted at call establishment time to check the validity of the call attempt. The key also determines the DECT radio-link encryption used for the call.

The location registration principles applied in DECT are strongly based on the ones defined for GSM. Proper interworking with GSM is further ensured via

an external handover mechanism (as applied in the GSM architecture) in addition to the fast DECT handover within single-location areas.

5 DECT and its CT-contestants

As DECT is believed to feature above all in the office environment, as part of a PABX or even as a fully cordless PABX, it is useful to complete the facts-and-figures section in this article with a short comparison of the other candidate CT (Cordless Telephone) standards: CEPT's CT1 and CT2 (**Table 2**).

Both CT1 and CT2 support *dynamic channel allocation*, as that is a prerequisite in the office environment. But this is where the resemblance stops, as DECT appears to be superior when it comes to other important attributes of the specification.

It provides the highest traffic capacity in number of channels. Furthermore, the ability to support 12 channels per transceiver (thanks to TDMA) allows for more economic solutions when multi-handset arrangements have to be supported. Local call completion between handsets in such an arrangement can in fact be supported at a marginal increase of costs.

DECT supports, as already mentioned, location registration when it is used in multi-location environments or when the paging capacity of a single location is exceeded. This is, obviously, a very important characteristic when it comes to the implementation of user-mobility in private PABX networks.

Another important feature of DECT has to do with handover. As handover is initiated by the handset itself in the DECT system, and hence way before the actual communication is disrupted, a *seamless handover* mechanism results. Not only when moving from one cell to another, but also within a cell when changing conditions call for switching to another, better channel. In many office environments this is,

Table 2 Applicability to cordless PABXs

characteristic:	affects:	CT1 (CEPT)	CT2	DECT
analogue/digital	voice quality signalling capability	analogue	digital	digital
no. of channels	traffic density	40	40	120
no. of simultaneous calls per transceiver	installation mgmnt. cost effectiveness	1	1	12
location registration/ alerting for incoming calls	roaming	no/no	no/yes	yes/yes
channel allocation by	grade of service	initiator	initiator	portable
mux technique	grade of service spatial diversity	FDMA	FDMA/TDD	FDMA/ TDMA/TDD
encryption	fraud resistance	no	no	yes
handover	mobility during a call	no	yes (not seamless)	yes (seamless)

FDMA : Frequency Division Multiple Access
TDMA : Time Division Multiple Access
TDD : Time Division Duplex

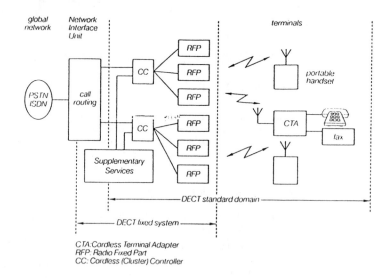

CTA: Cordless Terminal Adapter
RFP: Radio Fixed Part
CC: Cordless (Cluster) Controller

Fig.4 DECT functional concept

as experiments and experience with the early CT2 equipment learned, a valuable attribute.

This, however, does not make CT1 or CT2 useless. Not all applications require the provisions offered by DECT and CT1/CT2 solutions have of course one very important advantage over DECT: they are there. It is expected that the first DECT handsets will have to compete with 2nd or even 3rd generation CT1/CT2 handsets and this "learning curve" effect will position CT1/CT2 as serious contenders for many applications in the office.

As a result CT1/CT2 systems will serve the early applications in the office, due to their characteristics mainly in single-cell-few-handsets environments. When more cells are required they will probably be isolated, not requiring roaming or handover and serving few users.

DECT will initially (the learning curve issue!) be more expensive and therefore find its way to applications where it can excel and its costs are justifiable, ie the multi-cell/multi-handset environment or even the multi-location application. Therefore, DECT is typically suited for larger companies, allowing for the more complex arrangements required there, such

Fig.5 Cordless PABX intra-office

as flexible access rights to the network in order to serve "own" and "visiting" users in a proper way.

6 The functional concept of DECT

Irrespective of its application, a DECT system is always composed of a number of functional entities (**Fig.4**):

- The *portable handset* or terminal. When a more general access has to be provided a Cordless Terminal Adapter (CTA) may be applied as well.
- The *Radio Fixed Part* (RFP) supporting the physical layer of the DECT Common Interface within a cell.
- The *Cordless (Cluster) Controller* (CC) handling the MAC, DLC and Network layers for one or a cluster of RFPs. In a multi-cell system, the CC may take care of (seamless) handover between the cells served by the cluster.
- The *Network-specific Interface Unit*. In a multi-handset application, call completion may be supported by this entity as well.

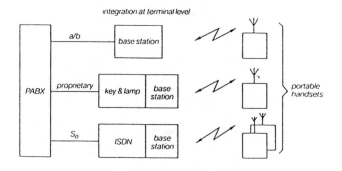

498

– The application-specific *Supplementary Services*. Examples are: centralised authentication and billing in a Telepoint system, or mobility management in a multi-location PABX-based network.

The combination of RFPs, CC, Network Interface and Supplementary Services is called the *DECT fixed system*. Depending on the scale and nature of the DECT application, functional entities are separately implemented or physically integrated.

7 The product scope

The DECT product scope comprises residential systems, small and large business systems, access to public services and to GSM/PCN, and the replacement of local loops. The next sections give more details on these aspects.

7.1 The residential system

This is obviously the archetype of all cordless applications. The single-cell DECT fixed system is integrated in one physical unit: the base station. In a multi-handset application local call completion may be included as well.

7.2 The small business system

This is basically a single-cell system, or a number of isolated cells each covering a logical part of an office. It differs from the residential system by the larger number of network connections and the support of enhanced facilities in the office.

As the small business environment is typically high-featured (key-systems, hybrid/ISDN switches), it is populated with a high penetration of system-dependent feature terminals. The best approach to cordless operation in this environment is by an integration of the DECT fixed system in these system terminals (Fig.5). The majority of the terminal features will remain in the fixed part. The DECT handset adds freedom of movement within the cell *(intra-office)* during the call and when calls are initiated or answered.

7.3 The large business system

In addition to the intra-office deployment, a large business will be a target for *inter-office* applications as well. The multi-cell or even multi-location concept allows ambulant users to be accessed or make calls whenever they are within reach of the DECT RFPs (Fig.6).

The architecture supporting this total mobility relies on dispersed RFPs and one or more Cordless Controllers (CCs) to take care of the mobility issues. The network interface to, for example the PABX, will support a private network signalling protocol (eg DPNSS1 or QSIG), enhanced with Location Registration provisions when multi-location mobility is supported. The latter results in a "Private (ie non-

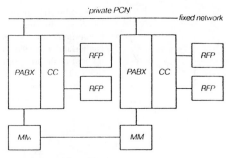

CC : Cordless (Cluster) Controller
RFP : Radio Fixed Part
MM : Mobility Manager

Fig.6 Cordless PABX inter-office

public) PCN" type of operation, allowing users to be addressed via a personal private-network number in the fixed and mobile parts of the network.

7.4 Public access service

Also known as Telepoint, the public access service provides "users-on-the-move" access to the public telephone network through base stations located in public areas. Initially for outgoing calls only, the service might evolve to two-way calling when the associated public network supports mobility management or wide-area paging. A Telepoint CC may take care of access management, billing and authentication for the region served.

7.5 Local loop replacement

Local loop replacement provides the final distribution to the public telephone network: the last cable

Fig.7 Local loop replacement

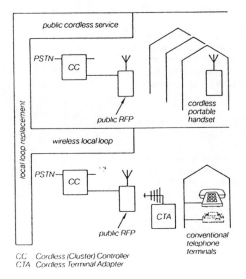

CC : Cordless (Cluster) Controller
CTA : Cordless Terminal Adapter

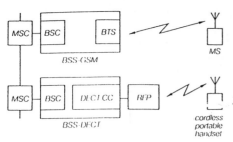

MSC: Mobile Service Switching Centre
BSS : Base Station Sub-system
BSC : Base Station Controller
BTS : Base Transceiver Station
MS : GSM Mobile Station
RFP : Radio Fixed Part

Fig.8 GSM/PCN access

section entering a house is replaced by a DECT radio link to a nearby distribution point (Fig.7).

7.6 GSM/PCN access

As discussed in the beginning of this paper, GSM-based PCNs (such as the DCS1800 system as defined within the UK and standardised in ETSI) are basically unfit to support in-building wireless communication. A DECT system for PCN access in buildings is an almost perfect and complementary solution. Where PCN access is required in buildings, a DECT-based Base Station Subsystem (BSS) replaces the GSM/PCN BSS used otherwise (Fig.8).

Mobility management, billing and authentication, and other functions reside in the PCN network. The DECT BSS provides a gateway role between the DECT handsets and the PCN network for this purpose. Specific applications of this type are the implementation of a Telepoint service in a mobile vehicle (bus, train or aircraft) or the provision of GSM access in cars from DECT handsets.

8 Conclusion

DECT is a flexible and versatile *access system* to a large variety of wide-area networks. Its main purpose is the addition of user mobility and freedom of movement to these networks, but it is also relevant as a physical loop replacement in rural public network applications, for example.

To support mobility, the DECT system is equipped with ample provisions to interwork with various network-based mobility management concepts. Its versatility puts it in a position to become a widely accepted *access "fabric"* to many networks. Philips, having recognised the importance of DECT, is actively supporting the DECT standardisation activity in ETSI. In contributing to the DECT standard, Philips is building up the expertise to allow a timely introduction of DECT technology in both the public and private communication domain.

Background of the author

Rob Mulder graduated from Delft Technical University in 1968 and joined the Philips Business Communication Systems activity in Hilversum (The Netherlands) in 1970. He was involved in the development of the low- to medium-sized members of the SOPHO-S family of digital PABXs and is currently heading the product development activities in BCS as Development Director.

500

The CDMA Standard

Code Division Multiple Access modulation has been standardized by TIA for the North American cellular telephone system. The author describes the system, its advantages and testing challenges.

David P. Whipple
Hewlett-Packard Company
Spokane, Washington

Reprinted from *Applied Microwave & Wireless*, pp. 24-39, Winter, 1994.

Code Division Multiple Access (CDMA) is a class of modulation that uses specialized codes to provide multiple communication channels in a designated segment of the electromagnetic spectrum. This article describes the implementation of CDMA that has been standardized by the Telecommunications Industry Association (the TIA) for the North American cellular telephone system.

The cellular telephone industry is faced with the problem of a customer base that is expanding while the amount of the electromagnetic spectrum allocated to cellular service is fixed. Capacity can be increased by installing additional cells (subdividing), but the degree of subdivision is limited because of the overhead needed to process handoffs between cells. In addition, property for cell sites is difficult to purchase in the areas where traffic is the highest.

The current analog system divides the available spectrum into 30-kHz-wide channels. This method of channelization (division of the spectrum into multiple channels) is commonly called FDMA, for Frequency Division Multiple Access (Figure 1). Alternate means of channelization are being developed to allow more users in the same region of the spectrum.

TDMA, or Time Division Multiple Access, uses the same 30-kHz channels, but adds a timesharing of three users on each frequency. All other factors being equal, this results in a threefold increase in capacity. CDMA, or Code Division Multiple Access, is a class of modulation that uses specialized codes[1,2] as the basis of channelization. These codes are shared by both the mobile station and the base station.

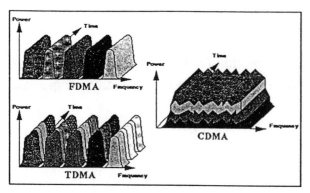

Figure 1. Cellular channelization methods. (a) Frequency Division Multiple Access (FDMA). (b) Time Division Multiple Access (TDMA). (c) Code Division Multiple Access (CDMA).

While CDMA is a class of modulation, this paper focuses on the implementation of CDMA for the North American cellular market, which was initially developed by QUALCOMM, Inc. and has been standardized by the Telecommunications Industry Association (the TIA).

Interference Effects

The analog system requires that the desired signal be at least 18 dB above any noise or interference on the same channel to provide acceptable call quality. The practical ramification of this is that only a portion of the available spectrum can be used in any given cell; not all of the channels can be used in every cell. A frequency reuse pattern of seven is commonly used to provide this attenuation (Figure 2). In other words, only one seventh of all of the cellular frequencies allocated to a carrier can be used in any one cell.

The use of omnidirectional cells does not allow for the required 18 dB attenuation. To overcome this, the cells are divided by sectored antennas (in the azimuthal plane). This reduces the frequencies available in any sector to only one out of 21.

In CDMA, signals can be and are received in the presence of high levels of interference. The practical limit depends on the channel conditions, but CDMA recep-

tion can take place in the presence of interference that is *18 dB larger than the signal*. Typically, the system operates with better conditions.

The frequencies are reused in every sector of every cell, and approximately half the interference on a given frequency is from outside cells. The other half is the user traffic from within the same cell on the same frequency.

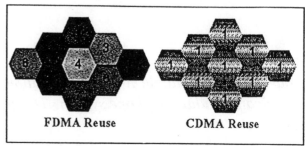

Figure 2. Cellular frequency reuse patterns. (a) FDMA reuse. (b) CDMA reuse.

A North American cellular CDMA system is described in Figure 3. CDMA starts with a basic data rate of 9600 bits/s. This is then spread[2] to a transmitted bit rate, or chip rate (the transmitted bits are called chips), of 1.2288 MHz. Spreading consists of applying digital codes to the data bits that increase the data rate while adding redundancy to the system. The chips are transmitted using a form of QPSK (quadrature phase shift keying) modulation that has been filtered to limit the bandwidth of the signal. This is added to the signal of all the other users in that cell.

When the signal is received, the coding is removed from the desired signal, returning it to a rate of 9600 bps. When the decoding is applied to the other users' codes, there is no despreading; the signals maintain the 1.2288-MHz bandwidth. The ratio of transmitted bits or chips to data bits is the coding gain. The coding gain for the North American CDMA system is 128, or 21 dB. Because of this coding gain of 21 dB, interference of up to 18 db above the signal level (3dB below the signal strength after coding gain) can be tolerated.

An analogy to CDMA is a crowded party. You can maintain a conversation with another person because your brain can track the sound of that person's voice and extract that voice from the interference of all other talkers. If the other talkers were to talk in different languages, discerning the desired speech would be even easier, because the crosscorrelation between the desired voice and the interference would be lower. The CDMA codes are designed to have very low cross correlation.

base link, which is called the reverse direction).

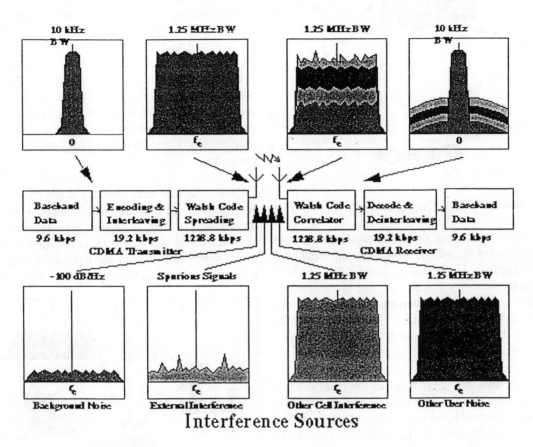

Figure 3. North American cellular CDMA system.

CDMA Features-

The data rate of 9600 bits/s can be thought of as a modem's transmission. Both the signaling overhead of the system as well as the useful data transmission payload must share this fundamental data rate. The system is designed so that multiple service options can use the modem. Currently, service option 1 is speech, service option 2 is a data loopback mode used for test purposes, and service option 3 is being defined as data services (both fax and asynchronous data, or "terminal" usage).

CDMA communication systems have many differences from analog systems:

1) Multiple users share one carrier frequency. In a fully loaded CDMA system, there are about 35 users on each carrier frequency. (There are actually two carrier frequencies per channel, 45 MHz away from each other. One is for the base-to-mobile link, which is called the forward direction, while the other is for the mobile-to-

2) The channel is defined by a code. There is a carrier frequency assignment, but the frequency band is 1.23 MHz wide.

3) The capacity limit is soft. Additional users add more interference to the system, which can cause a higher data error rate for all users, but this limit is not set by the number of physical channels.

CDMA makes use of multiple forms of diversity: spatial diversity, frequency diversity, and time diversity. The traditional form of spatial diversity, using multiple antennas, is used for the cell site receiver. Another form of spatial diversity is used during the process of handing off a call from one cell to the next. Called *soft handoff*, it is a make-before-break system in which two cell sites maintain a link with one mobile simultaneously (Figure 4).

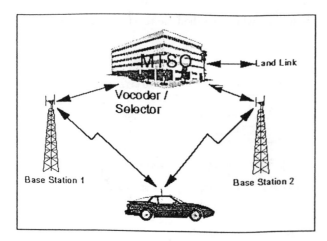

Figure 4. Spatial diversity during soft handoff.

The mobile station has multiple correlative receiver elements that are assigned to each incoming signal and can add these. There are at least four of these correlators—three that can be assigned to the link and one that searches for alternate paths. The cell sites send the received data, along with a quality index, to the MTSO (mobile telephone switch ing office) where a choice is made as to the better of the two signals.

Frequency diversity is provided in the bandwidth of the transmitted signal. A multipath environment will cause fading, which looks like a notch filter in the frequency domain (Figure 5). The width of the notch can vary, but typically will be less than 300 kHz. While this notch is sufficient to impair ten analog channels, it only removes about 25% of the CDMA signal.

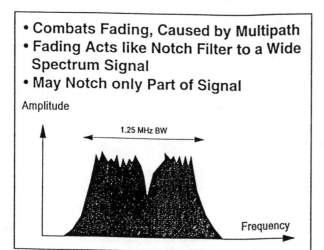

Figure 5. CDMA frequency diversity. The wide spectrum combats fading caused by multipath transmission. Fading has the effect of a notch filter to the wide-spectrum signal. Typically only a small part of the signal is lost.

Multipath signals are used to advantage, providing a form of time diversity. The multiple correlative receiver elements can be assigned to different, time delayed copies of the same signal. These can be combined in what is called a RAKE receiver[3] which has multiple elements called *fingers* (Figure 6). The term RAKE refers to the original block diagram of the receiver (Figure 6b), which includes a delay line with multiple taps. By weighting the signal at each tap in proportion to its strength, the time-diverse signals are combined in an optimal manner. The picture resembles a garden rake, hence the name.

Another form of time diversity is the use of forward error correcting codes followed by interleaving. Loss of transmitted bits tends to be grouped in time, while most error correction schemes work best when the bit errors are uniformly spread over time. Interleaving helps spread out errors and is common to most digital systems.

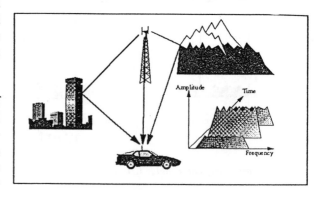

Figure 6. (a) The RAKE receiver takes advantage of multipath transmission to realize a form of time diversity.

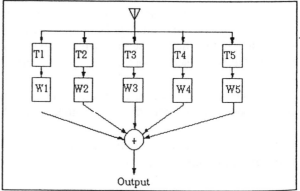

(b) RAKE receiver block diagram.

Mobile Station Power Control

Control of the mobile station power is essential for

CDMA to work. If one mobile station were to be received at the base station with too much power, it raises unnecessarily the amount of interference experienced by other users, and could jam them.

The goal is to have the signal of all mobile stations arrive at the base station with exactly the same and adequate power. Two forms of power control are used: *open-loop* and *closed-loop* .

Open-loop power control is based on the similarity of loss in the forward and reverse paths. The received power at the mobile station is used as a reference. If it is low, the mobile station is presumed to be far from the base station, therefore it transmits with high power. If it is high, the mobile station is assumed to be close and transmits with low power. The product of the two powers (or the sum of the two powers measured in dB) is a constant. This constant is -73 when the receive and transmit powers are measured in dBm. For example, when the received power is -85 dBm, the transmitted power is +12 dBm.

Closed-loop power control is used to force the power from the mobile station to deviate from the open-loop setting. This is done by an active feedback system from the base station to the mobile station. Power control bits are sent every 1.25 milliseconds to direct the mobile station either to increase or decrease its transmitted power in a 1 dB increment.

Because the CDMA mobile station transmits only enough power to maintain a link, the average transmitted power is much lower than that required for an analog system. An analog cellular phone always transmits enough power to overcome a fade, even though a fade does not exist most of the time. The CDMA's advantage in transmitting with lower power has the potential of longer battery life and smaller, lower-cost output amplifier design.

Speech Encoding

The speech is encoded before transmission. The purpose of encoding is to reduce the number of bits required to represent the speech. The CDMA voicecoder (*vocoder*, as it is called) has a data rate of 8550 bits per second. After additional bits are added for error detection up to the channel data rate of 9600 bits/s. This full channel capacity is not used, however, when the user is not speaking. The vocoder detects voice activity, and lowers the data rate during quiet periods. The lowest data rate is 1200 bits/s. Two intermediate rates of 2400 and 4800

bits/s are also used for special purposes.

The 2400 bits/s rate is used to transmit transients in the background noise, and the 4800 bits/s rate is used to mix vocoded speech and signaling data (signaling consists of link-management messages between the base station and the mobile station). In this last case, the channel data rate is 9600 bits/s, but half of the bits are assigned to voice and the other half to the message. This is called *dim and burst signaling*.

The mobile station pulses its output power during periods of lower-rate data. The power is turned on for 1/2, 1/4, or 1/8 of the time. The data rate is 9600 bits/s when the power is on, so the average data rate is 4800, 2400, or 1200 bits/s. This lowers the average power and the interference seen by other users.

The base station uses a different method to reduce power during quiet periods. It transmits with 100% duty cycle at 9600 bits/s, but uses only 1/2, 1/4, or 1/8 of full power and repeats the transmitted data 2, 4, or 8 times. The mobile station achieves the required signal-to-noise ratio by combining the multiple transmissions.

One important aspect of CDMA is the use of Walsh codes, or Hadamard codes[4]. These are based on the Walsh matrix, a square matrix with binary elements that always has a dimension that is a power of two. It is generated by seeding Walsh $(1) = W1 = 0$ and expanding as shown below and in Figure 7:

$$(1) \quad W_{2n} = \begin{matrix} W_n & W_n \\ W_n & \overline{W_n} \end{matrix}$$

Figure 7. Walsh matrices.

where n is the dimension of the matrix and the overscore denotes the logical NOT of the bits in the matrix.

The Walsh matrix has the property that every row is orthogonal to every other row and the logical NOT of every other row. Orthogonal means that the dot product of any two rows is zero. In simpler terms, it means that between any two rows exactly half the bits match and half the bits do not match. The CDMA system uses a 64-by-64-bit Walsh matrix.

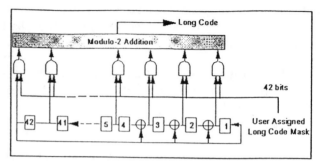

Figure 9. Long code privacy mask generation.

Forward Link Encoding

Walsh encoding is used in the forward link (base to mobile) as shown in Figure 8. The fundamental data rate of the channel is 9600 bits/s. The data is packetized into 20-ms blocks and has forward error correction applied by use of a convolutional encoder. This is done at half rate, which yields two bits out for every bit in. The data is then interleaved (a shuffling of the bits during the 20-ms period). This is done to better distribute bits lost during transmission. It has been shown that bit errors tend to come in groups rather than being spread out in time, while forward error correction works best when the errors are distributed uniformly over time. When the data is deinterleaved, the time-linked errors are more uniformly distributed.

The resulting data is then encoded using the Walsh matrix. One row of the Walsh matrix is assigned to a mobile station during call setup. If a 0 is presented to the Walsh cover, then the 64 bits of the assigned row of the Walsh matrix are sent. If a 1 is presented, then the NOT of the Walsh matrix row is sent. This has the effect of raising the data rate by a factor of 64, from 19.2 kbits /s to 1.2288 Mbits/s.

The last stage in coding is to convert from a binary signal to two binary channels in preparation for transmission using QPSK (quadrature phase shift keying) modulation. The data is split into I and Q (in-phase and quadrature) channels and the data in each channel is XORed with a unique PRBS short code. The short codes are spreading sequences that are generated much like the long code, with linear feedback shift registers. In the case of the short codes, there are two shift registers, each 15 bits long, with feedback taps that define specific sequences. These run at 1.2288 MHz. The short code sequences, each 32,768 bits long, are common to all CDMA radios, both mobile and base. They are used as a final level of spreading.

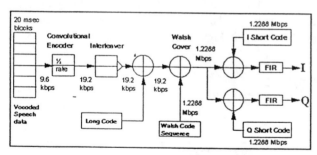

Figure 8. CDMA forward link physical layer.

Following the interleaver, the data is modified by the use of a long code, which serves only as a privacy mask. The long code is generated by a pseudorandom binary sequence (PRBS) that is generated by a 42-bit-long shift register (Figure 9). This register is also used as the master clock of the system, and is synchronized to the limit of propagation delays among all base stations and mobile stations. A mask is applied to the PRBS generator that selects a combination of the available bits. These are added modulo two by way of exclusive-OR gates to generate a single bit stream at 1.2288 MHz. For the forward link, a data rate of only 19.2 kbits/s is needed, so only 1 of 64 bits gets used. The long code generated in this way is XORed with the data from the interleaver.

After the data is XORed with the two short code sequences, the result is two channels of data at 1.2288 Mbits/s. Each channel is low-pass filtered digitally using an FIR (finite impulse response) filter. The filter cutoff frequency is approximately 615 kHz. A typical FIR filter implementation might output 9-bit-wide words at 4.9152 MHz. The resultant I and Q signals are converted to analog signals and are sent to a linear I/Q modulator. The final modulation is filtered QPSK.

Multiple channels in the base station are transmitted by combining the I and Q signals for each (Figure 10). Because all users share the composite signal from the cell, a reference signal called the pilot is transmitted. The pilot has all zero data and is assigned Walsh row number 0, which consists of all 0s. In other words, the pilot is made up of only the short spreading sequences. Typically 20% of the total energy of a cell is transmitted in the pilot signal.

The pilot signal forms a coherent phase reference for the mobile stations to use in demodulating the traffic data. It is also the timing reference for the code correlation. The short sequences allow the CDMA system to reuse all 64 Walsh codes in each adjacent cell. Each cell uses a different time offset on the short codes and is thereby uniquely identified while being able to reuse the 64 Walsh codes.

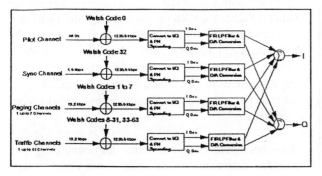

Figure 10. CDMA forward link channel format.

Reverse Link Encoding

The mobile station cannot afford the power of a pilot because it would then need to transmit two signals. This makes the demodulation job more difficult in the base stations. A different coding scheme is also used, as shown in Figure 11.

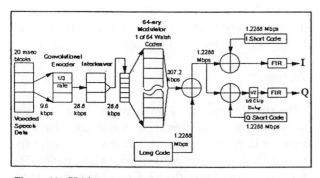

Figure 11. CDMA reverse link physical layer.

For speech, the same vocoder is used in both directions. Again, the data rate is 9600 bps. A 1/3-rate convolutional encoder is used, yielding an output rate of 28.8 kbits/s. The output of this is interleaved and then taken six bits at a time. A six-bit number can range from 0 to 63, and each group of six bits is used as a pointer to one row of the Walsh matrix. Every mobile station can transmit any row of the Walsh matrix as needed.

At this point, the data rate is 307.2 kbits/s, but there is no unique coding for channelization. The full-rate, long code is then applied, raising the rate to 1.2288 Mbits/s. This final data stream is split into I and Q channels and spread with the same short sequences as in the base station. There is one more difference. A time delay of 1/2 chip is applied to the Q channel before the FIR filter. This results in offset-QPSK modulation (Figure 12), and is used to avoid the amplitude transients inherent in QPSK. This makes the design of the output amplifier easier in the mobile station.

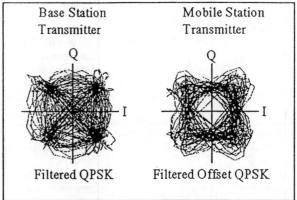

Figure 12. Constellation diagrams for CDMA modulation formats. (a) The base station transmitter uses filtered QPSK. (b) The mobile station transmitter uses filtered offset-QPSK.

The capacity is different in the forward and reverse links because of the differences in modulation. The forward link has the phase reference—the pilot signal—as well as orthogonal codes. The reverse link signal is not orthogonal because the long codes are applied after the use of the Walsh matrix. In this case the signals are uncorrelated but not orthogonal. The base station has the advantage of multiple receive antennas (diversity). All factors taken together, the reverse link sets system capacity.

The CDMA channelization functions are summarized in Table I.

Call Scenario

To better illustrate how the CDMA system operates, the system function will be described in terms of mobile station operation.

When the mobile station first turns on, it knows the assigned frequency for CDMA service in the local area. It will tune to that frequency and search for pilot signals. It is likely that multiple pilot signals will be found, each with a different time offset. This time offset is the

Parameter	Function	Notes
Frequency	Divides the spectrum into several 1.23-MHz frequency allocations.	Forward and reverse links are separated by 45 MHz.
Walsh Codes	Separates forward link users of the same cell.	Assigned by cell site. Walsh code 0 is always the pilot channel. Walsh code 32 is always the sync channel.
Long Code	Sparates reverse link users of the same cell.	Depends on time and user ID. The long code is composed of a 43-bit-long PRBS generator and a user specific mask.
Short Codes, also called the I and Q spreading sequences.	Separates cell sites or sectors of cells.	The I and Q codes are different but are based on 15-bit-long PRBS generators. Both codes repeat at 26.667 ms intervals. Base stations are differentiated by time offsets of the short sequences.

Table I. CDMA Channelization Functions.

means of distinguishing one base station from another. The mobile station will pick the strongest pilot, and establish a frequency reference and a time reference from that signal. It will then start demodulation of Walsh number 32, which is always assigned to the sync channel. The sync channel message contains the future contents of the 42-bit long code shift register. These are 320 ms early, so the mobile station has time to decode the message, load its register, and become synchronized with the base station's system time.

The mobile station may be required to register. This would be a power-on registration in which the mobile station tells the system that it is available for calls and also tells the system where it is. It is anticipated that a service area will be divided into zones, and if the mobile station crosses from one zone to another while no call is in progress, it will move its registration location by use of an idle state handoff. The design of the zones is left to the service provider and is chosen to minimize the support messages. Small zones result in efficient paging but a large number of idle state handoffs. Large zones minimize idle state handoffs, but require paging messages to be sent from a large number of cells in the zone.

At this point the user makes a call by entering the digits on the mobile station keypad and hitting the send button. The mobile station will attempt to contact the base station with an access probe. A long code mask is used that is based on cell site parameters. It is possible that multiple mobile stations may attempt a link on the access channel simultaneously, so collisions can occur. If the base station does not acknowledge (on the paging channel) the access attempt, the mobile station will wait a random time and try again. After making contact, the base station will assign a traffic channel with its Walsh number. At this point, the mobile station changes its long code mask to one based on its serial number, receives on the assigned Walsh number, and starts the conversation mode.

It is common for a mobile station communicating with one cell to detect another cell's pilot that is strong enough to be used. The mobile station will then request soft handoff. When this is set up, the mobile station will be assigned different Walsh numbers and pilot timing and use these in different correlative receiving elements. It is capable of combining the signals from both cells.

Eventually, the signal from the first cell will diminish and the mobile station will request from the second cell that soft handoff be terminated.

At the end of the call, the channels will be freed. When the mobile station is turned off, it will generate a power-down registration signal that tells the system that it is no longer available for incoming calls.

Testing

The complexity of the CDMA system raises substantial test issues. What needs to be tested, and what environment is needed for testing? To test the mobile station, the test equipment must emulate a base station. The tester needs to provide the pilot, sync, paging, and traffic channels. It must provide another signal that uses orthogonal Walsh symbols that represent the interference generated by other users of the same cell, and it must provide additive noise that simulates the combination of CDMA signals from other cells and background noise.

Bit error rate is not a meaningful measure, since substantial errors are expected at the chip rate and these are not available for test. The bits at the 9600-bit s/s rate are the only bits available for test, and these will either be all correct as a result of error correction or will have substantial errors. What is used instead is the frame error rate, a check of the received bits and the associated CRC (cyclic redundancy code) in each 20-ms block.

To test the transmitter, a new test has been defined: *waveform quality*. This is based on the crosscorrelation of the actual transmitted signal to the ideal signal transmitting the same data. This is important to the system because the CDMA receivers are correlators. In fact, they correlate the received signal with the ideal signal. If a signal deviates substantially from the ideal, the correlated portion of that signal will be used to make the link and the uncorrelated portion will act as additive interference. Closed-loop power control will maintain the correlated power at the needed level, and excess power will be transmitted. The specification is that the radios shall transmit with a waveform quality that limits the excess power to less than 0.25 dB. Other transmitter measurements include frequency and power control operation.

CDMA provides an advanced technology for cellular applications, providing high-quality service to a large number of users. It has been extensively tested and has been deployed in precommercial applications. Commercial service is scheduled to begin this year.

AMPS: Advanced Mobile Phone System.
This is the current analog FM system in North America. It uses 30-kHz channels and signaling is done superaudio, that is, at frequencies above the audio bandwidth for speech, which is 300 to 3000 Hz.

TACS: Total Access Communication System. This is the analog FM system used in the United Kingdom and Japan. It uses 25-kHz channels and signaling is super audio.

NMT: Nordic Mobile Telephone. Scandinavia led the world in cellular systems. The latest system uses 30-kHz channels, and signaling is done using 1200-Hz and 1800-Hz tones in much the same way as a modem.

J-TACS: This is a narrowband analog FM system in use in Japan. Channels are 12.5-kHz wide and signaling is subaudio, that is, at frequencies below the audio bandwidth for speech, which is 300 to 3000 Hz.

NAMPS: Narrow Analog Mobile Phone System. This is an analog FM system using 10-kHz-wide channels. Signaling is subaudio

GSM: Global System for Mobile Communications. This is the first digital cellula r system to be used commercially. It has

---

been adopted across Europe and in many countries of the Pacific rim. It uses 200-kHz channels with eight users per channel using TDMA, and has a vocoder rate of 13 kbits/s.

TDMA: Time Division Multiple Access. This is the first digital system standardi zed in North America. It uses 30-kHz channels, three users per channel using TDMA, and has a vocoder rate of 8 kbits/s.

E-TDMA: Extended TDMA. This system uses the same30-kHz channels as TDMA, but has six users per channel. The vocoder rate is cut to 4 kbits/s, and the channels are dynamically assigned based on voice activity detection. This is being propo sed as a follow-on to TDMA.

CDMA: Code Division Multiple Access. This system uses 1.23-MHz-wide channel sets, with a variable number of users on each carrier frequency. The full vocoder rate is 8.55 kbits/s, but voice activity detection and variable-rate coding can cut the data rate to 1200 bits/s. The effective data rate, determined empirically for simulated conversations, is 3700 bits/s. Access is by code.

This Article was edited from "North American Cellular CDMA," *Hewlett-Packard Journal*, Dec. 1993, pp. 90-97.

References:

1. R. Kerr, "CDMA Digital Cellular," *APPLIED MICROWAVE & WIRELESS*, Fall 1993, pp. 30-41.

2. J. White, "What is CDMA?," *APPLIED MICROWAVE & WIRELESS*, Fall 1993, pp. 5-7.

3. R. Price and P. E. Green, Jr., "A Communication Technique for Multipath Channels," *Proceedings of the IRE*, Vol. 46, March 1958, pp. 555-570.

4. J. G. Proakis, *Digital Communications, Second Edition*, McGraw-Hill Book Co., 1989

Dave Whipple is an R&D project manager for system architecture at Hewlett-Packard. He is working on CDMA test equipment and standards. With HP since 1973, he has worked at the Stanford Park Division and the Spokane Division. Initially he was a production engineer and then production engineering manager for signal generators. Later he was an R&D project manager for HP 8656B/57A signal generators, the HP 8920A communications test set, the HP 8922x GSM test sets, and the HP 8953DT TDMA test system.

He is the co-inventor of three patents, all dealing with FM in phase-locked loops.

Dave was born in Wilmington, Delaware and attended Purdue University, from which he received a BSEE degree in 1972 and an MSEE degree in 1973.

Overview of Cellular CDMA

William C. Y. Lee, *Fellow, IEEE*

Abstract—This paper is a general description of code division multiple access (CDMA). The analysis of power control schemes in CDMA is an original work. The wide-band wave propagation in the cellular environment presents an interesting result (the short-term fading reduction over the wide band-signal in cellular). Also less fading in urban areas than in suburban areas. The advantages of using CDMA listed in this paper have excited the cellular industry. Radio capacity is the key issue in selecting CDMA and is carefully described in this paper.

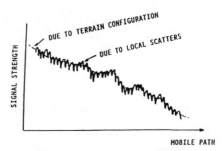

Fig. 1. Mobile radio environment.

I. INTRODUCTION

THE development of the code division multiple access (CDMA) scheme is mainly for capacity reasons. Ever since the analog cellular system started to face its capacity limitation in 1987, the promotion of developing digital cellular systems for increasing capacity has been carried out. In digital systems, there are three basic multiple access schemes, frequency division multiple access (FDMA), time division multiple access (TDMA), and code division multiple access (CDMA). In theory, it does not matter whether the spectrum is divided into frequencies, time slots, or codes, the capacity provided from these three multiple access schemes is the same. However, in the cellular system, we might find that one may be better than the another. Especially in the North American Cellular System, no additional spectrum will be allocated for digital cellular. Therefore, the analog and digital systems will co-exist in the same spectrum. Also, the problem of transition from analog to digital is another consideration. Although the CDMA has been used in satellite communications, the same CDMA system cannot be directly applied to the mobile cellular system. In order to design a cellular CDMA system, we first need to understand the mobile radio environment; then study whether the characteristics of CDMA are suitable for the mobile radio environment or not; and finally describe the natural beauty of applying CDMA in cellular systems.

II. MOBILE RADIO ENVIRONMENT

The propagation of a narrow-band carrier signal is a conventional means of communication. However, in a CDMA system, the propagation of a wide-band carrier signal is used. Therefore, we first describe the propagation of the narrow-band wave, then of the wide-band wave.

A. Narrow-Band (NB) Wave Propagation

A signal transmitted from the cell-site and received by either a mobile unit or a portable unit would propagate over a

Manuscript received August 1, 1990; revised October 1, 1990. This paper was presented at the 1990 IEEE GLOBECOM Conference, San Diego, CA.
The author is with PacTel Cellular, Irvine, CA 92714.
IEEE Log Number 9144471.

particular terrain configuration between two ends. Therefore, the effect of the terrain configuration generates a different long-term fading characteristic which follows a log-normal variation appearing on the envelope of the received signal, as shown in Fig. 1. Since the antenna height of a mobile or portable unit is close to the ground, three effects are observed [1]. First, the signal received is not only from the direct path but also from the strong reflected path due to the fact that the antennas of the mobile units are close to the ground. These two paths create an excessive path loss which is 40 dB/dec (fourth power law applied), i.e., doubling the path loss in decibels of the free-space path loss. Second, under the low antenna height condition at the mobile units, the human-made structures surrounding them would generate the multipath fading on the received signal called Rayleigh fading, as shown in Fig. 1. The multipath fading causes the burst error in digital transmission. The average duration of fades $\bar{t}$ as well as the level crossing rates $\bar{n}$ at 10 dB below the average power of a signal is a function of vehicle speed V and wavelength λ.

$$\bar{t} = 0.132\left(\frac{\lambda}{V}\right) \text{ s} \qquad (1)$$

$$\bar{n} = 0.75\left(\frac{V}{\lambda}\right) \text{ crossings/s.} \qquad (2)$$

For a frequency of 850 MHz and a speed of 15 m/h then $\bar{t} = 6$ ms and $\bar{n} = 16$ crossings/s. Third, a time delay spread phenomenon exists due to the time dispersive medium. In a mobile radio environment a single symbol, transmitted from one end and received at the other end, receives not only its own symbol but also many echoes of its symbol. The time delay spread intervals are measured from the first symbol to the last detectable echo, which are different in human-made environments. The average time delay spread due to the local scatterers in suburban areas is 0.5 μs and in urban areas is 3 μs. These local scatterers are in the near-end region as illustrated in Fig. 2, and the time delay spread corresponding to this region is illustrated in Fig. 3. There are other types of

Reprinted from *IEEE Transactions on Vehicular Technology*, Vol. 40, No. 2, pp. 291-302, May, 1991.

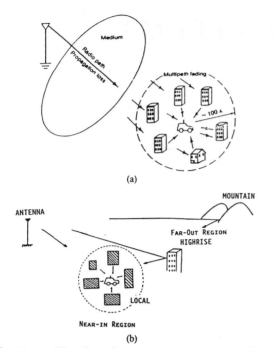

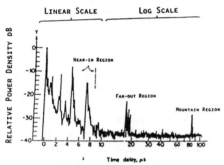

Fig. 2. (a) A mobile radio environment—two parts: propagation loss and multiple fading. (b) Time-delay spread scenario.

Fig. 3. An illustration on time-delay spread.

time delay spreads as illustrated in Fig. 2. One kind of delayed wave is due to the reflection of the high-rise buildings (far-out region), and one kind of delayed wave is due to the reflection from the mountains. Their corresponding time delays are illustrated in Fig. 3. In certain mountain areas, the time delay spread can be up to 100 μs. These time delay spreads would cause intersymbol interference (ISI) for data transmission [2]. In order to avoid the ISI, the transmission rate R_b should not exceed the inverse value of the delay spread Δ if the mobile unit is at a standstill (nonfading case),

$$R_b < 1/\Delta \qquad (3)$$

or R_b should not exceed the inverse value of $2\pi\Delta$ if the mobile unit is in motion (fading case)

$$R_b < 1/(2\pi\Delta). \qquad (4)$$

If the transmission rate R_b is higher than (3) or (4), both FDMA and TDMA need equalizers which are capable of reducing the ISI to a certain degree depending on the hardship of the time delay spread length and the wave arrival

distribution [3]–[5]. An FDMA system always requires less transmission rate than a TDMA system if both systems offer the same radio capacity. Usually an FDMA system can get away from using an equalizer as long as its transmission rate does not exceed too much above 10 kilosamples per second. The CDMA system does not need an equalizer but a simpler device called a correlator will be used. It will be described later.

B. Wide-Band Wave Propagation [6]

1) *Path Loss:* Suppose that a transmitted power P_t in watts is used to send a wide-band signal with a bandwidth B in hertz along a mobile radio path r. The power spectrum over the bandwidth B is $S_t(f)$, then the P_t can be expressed as

$$P_t = G_t \int_{f_0 - \frac{B}{2}}^{f_0 + \frac{B}{2}} S_t(f)\, df. \qquad (5)$$

The received power

$$P_r = \frac{P_t}{4\pi r^2} \times C(r.f) \times A_e(f) \qquad (6)$$

where

$$C(r, f) = \text{medium characteristic} = k/(r^2 f) \qquad (7)$$

$$A_e(f) = \text{effective aperture of the receiving antenna}$$

$$= \frac{c^2 G_r}{4\pi f^2} \qquad (8)$$

k is a constant factor, c is the speed of light, G_t and G_r are the gains of the transmitting and receiving antennas, respectively. Substituting (5), (7), and (8) into (6), we obtain

$$P_r = \frac{kc^2 G_R G_t}{(4\pi r^2)^2} \int_{f_0 - \frac{B}{2}}^{f_0 + \frac{B}{2}} S_t(f) \frac{1}{f^3}\, df. \qquad (9)$$

For simplicity but without losing much generality, let

$$S_t(f) = \text{constant}, \qquad (10)$$

$$\text{for } f_0 - B/2 \le f \le f_0 + B/2.$$

Then (9) becomes

$$P_r = \frac{kc^2 G_t G_R}{(4\pi r^2)^2} \frac{1}{f_0^3 \left[1 - \left(\frac{B}{2f_0} \right)^2 \right]^2}. \qquad (11)$$

Equation (11) is a general formula. For a narrow-band signal, $B \ll f_0$, then (11) becomes

$$P_r = \frac{kc^2 G_t G_R}{(4\pi r^2)^2 f_0^3} \quad \text{(narrow-band)}. \qquad (12)$$

From (11), we may find the B/f_0 ratio for the case of 1-dB difference in path loss between narrow-band and wide-band.

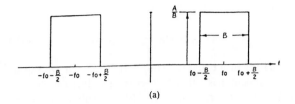

(a)

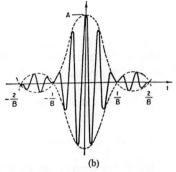

(b)

Fig. 4. Band-limited impulse. (a) Spectrum. (b) Waveshape.

That means by solving the denominator of (11) as follows:

$$10 \log \left[1 - \left(\frac{B}{2f_0} \right)^2 \right]^2 = -1 \text{ dB}$$

we obtain

$$B = 0.66 f_0.$$

In most wide-band applications, B will not be wider than $f_0/2$. Therefore, the narrow-band propagation path loss should be applied to the wide-band propagation path loss.

2) Multipath Fading Characteristic on Wide-Band: The wide-band pulse signaling $S_0(t)$ can be expressed as [7]

$$S_0(t) = A \frac{\sin (\pi B t)}{\pi t} \tag{13}$$

where A is the pulse amplitude shown in Fig. 4.

The received signal can be represented as

$$S(t) = (A/B) \sum_{m=-\infty}^{\infty} b_m(t) \frac{\sin \pi B \left(t - \frac{m}{B} \right)}{\pi \left(t - \frac{m}{B} \right)}. \tag{14}$$

The pulsewidth of $1/B$ is the time interval of the pulse occupied. Count all b_m that are not vanishing over a range of a finite number of m which is corresponding to a time delay spread Δ. Then the effective number of diversity branches, M, can be approximated by

$$M = \frac{\Delta + \frac{1}{B}}{\frac{1}{B}} = B \cdot \Delta + 1. \tag{15}$$

The effective number of diversity varies according to the human-made structures. The M is larger in the urban area than in the suburban area. Letting $\Delta = 0.5$ μs for suburban and $\Delta = 3$ μs for urban, and $B = 30$ kHz for narrow-band

and 1.25 MHz for wide-band, we find the effective number of diversity M in the following table.

Human-made environment	M diversity branches	
	$B = 30$ kHz	$B = 1.25$ MHz
$\Delta = 0.5$ μs Suburban	1.015	1.625
$\Delta = 3$ μs Urban	1.09	4.75

The wider the bandwidth, the less the fading. For $B = 1.25$ MHz, the fading of its received signal is reduced as if the diversity-branch receiver which equals $M = 1.625$ (between a single branch and two branches) is applied in suburban areas, and $M = 4.75$ (between four and five branches) is applied in urban areas. The wide-band signal would provide more diversity gain in urban areas than in suburban areas. For $B = 30$ kHz, no effective diversity gain is noticeable on its narrow-band received signal.

III. KEY ELEMENTS IN DESIGNING CELLULAR

The frequency reuse concept guides the cellular system design.

A. Cochannel Interference Reduction Factor (CIRF)

The minimum separation between two cochannel cells, D_s, is based on a cochannel interference reduction factor q which is expressed as

$$q = D_s/R \tag{16}$$

where R is the cell radius. The value of q is different for each system. For analog cellular systems, $q = 4.6$ is based on the channel bandwidth $B_c = 30$ kHz and the carrier-to-interference ratio (C/I) equals 18 dB.

B. Handoffs

The handoff is a unique feature in cellular. It switches the call to a new frequency channel in a new cell site without either interrupting the call or alerting the user. Reducing unnecessary handoffs and making necessary handoffs successfully are very important tasks for the cellular system operators in analog systems or in future FDMA or TDMA digital systems.

C. Frequency Management and Frequency Assignment

Based on the minimum distance D_s, the number of cells k, in a cell reuse pattern may be obtained,

$$K = (D_s/R)^2/3 = q^2/3. \tag{17}$$

The total allocated channels will be divided by K. There are K sets of frequencies; each cell operates its own set of frequencies managed by the system operator. This is the frequency management task. During a call process different frequencies are assigned to different calls. This is the frequency assignment task. Both tasks are critically impacted by interference and capacity.

D. Reverse-Link Power Control

The reverse-link power control is for reducing near-end to far-end interference. The interference occurs when a mobile unit close to the cell site can mask the received signal at the cell site so that the signal from a far-end mobile unit is unable to be received by the cell site at the same time. It is a unique type of interference occurring in the mobile radio environment.

E. Forward-Link Power Control

The forward-link power control is used to reduce the necessary interference outside its own cell boundary.

F. Capacity Enhancement

The capacity of cellular systems can be increased by handling q in two conditions.

1) Within standard cellular equipment—the value of q shown in (16) remains a constant. Reduce the cell radius R, thus D_s reduces. For a smaller D_s the same frequency can be used more often in the same geographical area: that is why we are trying to use small cells (sometimes called microcells or picocells) to increase capacity.

2) Chosen from different cellular systems—many different types of radio equipment can be chosen. Search for those cellular systems which can provide smaller values of q. When q shown in (16) is smaller, D_s can be less, even if the cell radius remains unchanged. We believe that q is smaller in properly designed digital cellular systems than q in analog systems. Choosing a smaller new q of a new system, we can increase the same amount of capacity without reducing the size of the cell based on the old q of an old system. That is why we are choosing a new digital system to replace the old analog system.

Reducing the size of cells in a system requires more cells. It is always costly. Therefore, the development of digital cellular systems properly is the right choice.

IV. SPREADING TECHNIQUES IN MODULATION

Spreading techniques in modulation are generally used in military systems for antijamming purposes. In general, there are two techniques: 1) spectrum spreading (spread spectrum) and 2) time spreading (time hopping) stated as follows:

A. Spread Spectrum (SS) Techniques

There are two general spread spectrum techniques, direct sequence (DS) and frequency hopping (FH).

1) Direct Sequence: In direct sequence, each information bit is symbolized by a large number of coded bits called chips. For example, if an information bit rate $R = 10$ kb/s is used and it needs an information bandwidth $B = 10$ kHz, and if each bit of 10 kb/s is coded by 100 chips, then the chip rate is 1 Mb/s which needs a DS bandwidth, $B_{ss} = 1$ MHz. The bandwidth is thus spreading from 10 kHz to 1 MHz. The spectrum spreading in DS is measured by the processing gain (PG) in decibels

$$PG = 10 \log \frac{B_{ss}}{B} \quad \text{(in dB)}. \quad (18a)$$

Then the PG of the above example is 20 dB. Or we say that this SS system has 20 dB processing gain. The first DS experiment was carried out in 1949 by DeRosa and Rogoff who established a link between New Jersey and California.

2) Frequency Hopping: An FH receiver would equip N frequency channels for an active call to hop over those N frequencies with a determined hopping pattern. If the information channel width is 10 kHz and there are 100 channels to hop, $N = 100$, the FH bandwidth $B_{ss} = 1$ MHz. The spectrum is spreading from 10 kHz (no hopping) to 1 MHz (frequency hopping). The spectrum spreading in FH is measured by the PG as

$$PG = 10 \log N \quad \text{(in dB)}. \quad (18b)$$

Then the PG of the above example is 20 dB. The total hopping frequency channels are called chips. There are two basic hopping patterns; one called fast hopping which makes two or more hops for each symbol. The other called slow hopping which makes two or more symbols for each hop. In general, the transmission data rate is the symbol rate. The symbol rate is equal to the bit rate at a binary transmission. Due to the limitation of today's technology, the FH is using a slow hopping pattern.

B. Time Hopping

A message transmitted with a data rate of R requiring a transmit time interval T is now allocated at a longer transmission time interval T_s. In time T_s the data are sent in bursts dictated by a hopping pattern. The time interval between bursts t_n also can be varied. The time spreading data rate R_s is always less than the information bit rate R. Assume that N bursts occurred in time T, then

$$R_s = \left(\frac{T_s}{T}\right) R = \left(1 - \frac{\sum_1^N t_n}{T}\right) R. \quad (19)$$

V. DESCRIPTION OF DS MODULATION

The spread spectra (DS and FH) are used for reducing intentional interference (enemy jamming), and now we are using it for increasing capacity instead of reducing the intentional interference. Immediately we realize that the FH with a slow hopping does not serve the purpose of increasing capacity. The slow hopping is to let good channels downgrade and bad channels upgrade. In order to have a system design for capacity, all the channels have to be deployed only marginally well. If bad channels do occur in this high capacity SS system, the system does not provide normal channels with excessive signal levels which can average with the poor signal levels of those bad channels to within an acceptable quality level. It just pulls down all the channels to an unacceptable level. The proper way should be either drop the

514

TECHNIQUE

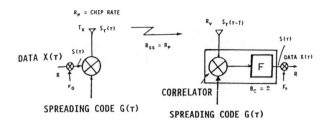

SPREADING PROCESS   DESPREADING PROCESS

Fig. 5. Basic spread—spectrum technique.

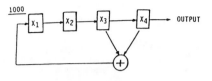

0 0 0 1 0 0 1 1 0 1 0 1 1 1 1   P = 2ᴺ - 1

N - NO; OF SHIFT REGISTERS
P - LENGTH OF SEQUENCE

Fig. 6. PN code (linear maximal length sequence) generator.

bad channels or correct the bad channels by other means. The fast hopping does help increase the capacity because of its advantage of applying diversity but the technology to have fast hopping at 800 MHz is not available.

1) Basic DS Technique: The basic DS technique is illustrated in Fig. 5. The data $x(t)$ transmitted with a data rate R is modulated by a carrier f_0 first, then by a spreading code $G(t)$ to form a DS signal $S_t(t)$ with a chip rate R_p which takes a DS bandwidth B_{ss}. The DS signal $S_t(t - T)$ after a propagation delay T is received and goes through a correlator using the same spreading code $G(t)$ prestored in it to despread the DS signal. Then the despread signal $S(t - T)$ is obtained. After demodulating it by f_0, $x(t)$ is recovered. Take a constant-envelop signal modulated on a carrier f_0 at transmitting end shown in Fig. 5. Let $x(t)$ be a data stream modulated by a binary phase shift keying (BPSK) that

$$x(t) = \pm 1 \qquad (20)$$

modulated by a binary shift keying

$$S(t) = x(t) \cos(2\pi f_0 t). \qquad (21)$$

At the transmitting end, the spreading sequence $G(t)$ modulation also uses BPSK

$$G(t) = \pm 1 \qquad (22)$$

then

$$S_t(t) = x(t) G(t) \cos(2\pi f_0 t). \qquad (23)$$

At the receiving end, the $S_t(t - T)$ is received after T seconds propagation delay. The despreading processing then takes place. The signal $S(t - T)$ coming out from the correlator is

$$S(t - T) = x(t - T)$$
$$\cdot G(t - T) G(t - \hat{T}) \cos(2\pi f_0(t - T)) \qquad (24)$$

where $\hat{T}$ is the estimated propagation delay generated in the receiver. Since $G(t) = \pm 1$,

$$G(t - T) G(t - \hat{T}) = 1 \qquad (25)$$

from a good correlator $T = \hat{T}$. Then

$$S(t - T) = x(t - T) \cos(2\pi f_0(t - T)). \qquad (26)$$

After it is demodulated by the carrier frequency f_0, the data $x(t - T)$ then are recovered as shown in Fig. 5.

2) Pseudonoise (PN) Code Generator: Pseudonoise code coming from a PN sequence is a deterministic signal [8]. For example, the sequence 00010011010111 is a PN sequence. It contains three properties.

a) *Balance property:* 7 zeros and 8 ones. The numbers of zeros and ones of a PN code are different only by one.

b) *Run property:* There are four "zero" runs (or "one" runs): runs = 4.
   1/2 of runs (i.e., 2) of length 1; i.e., two single "zeros (or ones)."
   1/4 of runs (i.e., 1) of length 2; i.e., one "2 consecutive zeros (or ones)."
   1/8 of runs (i.e., 0.5) of length 3; i.e., one "3 consecutive zeros (or ones)." In the above example, 1/8 of runs cannot be counted for too short a code.

c) *Correlation property:* Let D denote the "difference," and S denote the "same" by comparing two PN codes as follows:

$$\begin{array}{c} 0\ 0\ 0\ 1\ 0\ 0\ 1\ 1\ 0\ 1\ 0\ 1\ 1\ 1 \\ 1\ 0\ 0\ 0\ 1\ 0\ 0\ 1\ 1\ 0\ 1\ 0\ 1\ 1\ 1. \\ \hline D\ S\ S\ D\ D\ S\ D\ S\ D\ D\ D\ D\ S\ S\ S \end{array}$$

The value of the correlation of two N-bit sequences can be obtained by counting the number N_d of D's and the number N_s of S's and inserting them into the following equation:

$$P = \frac{1}{N}(N_s - N_d) = \frac{1}{15}(7 - 8) = -\frac{1}{15}. \qquad (27)$$

Then the correlation of a 15-b PN code is $-1/15$. The PN code generator of a four-shift register is shown in Fig. 6. The modulo 2 adder is summing the shift register X_3 and the shift register X_4. The summing signal then feeds back to the shift register X_1. Suppose that a 4-b sequence 1000 is fed into the shift register X_1. The output PN sequence from this PN code generator is 00010011010111. The code length L of any PN code generator is dependent upon the number of shift registers N:

$$L = 2^N - 1. \qquad (28)$$

The PN sequence generated in Fig. 6 is also called the linear maximal length sequence. For $N = 4$, L is 15.

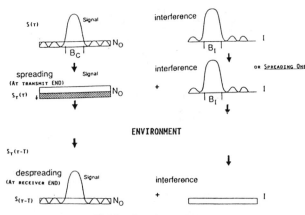

Fig. 7.   Spread spectrum.

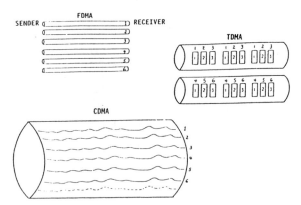

Fig. 8.   Illustration of different multiple access systems.

3) Reduction of Interference by a DS signal: The signal $S(t)$ of Fig. 5, before the spreading processing can be illustrated in both the frequency and time domains, is shown in Fig. 7. After spreading $S(t)$ with a given $G(t)$, the output $S_t(t)$ is transmitted out while the interference in the air could be a narrow-band signal or a DS signal with a different $G_I(t)$. When $S_t(t - T)$ is received after a propagation delay T, it is despreading with the same $G(t)$ and obtaining $S(t - T)$. The interference signal would spread to an SS signal by the $G(t)$ if it was a narrow-band signal, or stay as an SS signal because $G(t)$ and $G_I(t)$ do not agree with each other. Thus as a result, a low level of interference within the desired signal bandwidth B_c can be achieved.

VI. MULTIPLE ACCESS SCHEMES

The multiple access schemes are used to provide resources for establishing calls. There are five multiple access schemes. *FDMA* serves the calls with different frequency channels. *TDMA* serves the calls with different time slots. *CDMA* serves the calls with different code sequences. *PDMA* (polarization division multiple access) serves the calls with different polarization. PDMA is not applied to mobile radio [6]. *SDMA* (space division multiple access) serves the calls by spot beam antennas. The calls in different areas covered by the spot beams can be served by the same frequency—a frequency reuse concept. In the cellular system, the first three multiple access schemes can be applied. The illustration of the differences among three multiple access schemes are shown in Fig. 8. Assume that a set of six channels is assigned to a cell. In FDMA, six frequency channels serve six calls. In TDMA, the channel bandwidth is three times wider than that of FDMA channel bandwidth. Thus two TDMA channel bandwidths equal six FDMA channel bandwidths. Each TDMA channel provides three time slots. The total of six time slots serve six calls. In CDMA, one big channel has a bandwidth equal to six FDMA channels. The CDMA radio channel can provide six code sequences and serve six calls. Also, CDMA can squeeze additional code sequences in the same radio channel, but the other two multiple access schemes cannot. Adding additional code sequences, of course, degrades the voice quality.

A. Carrier-to-Interference Ratio (C/I)

In analog systems, only FDMA can be applied. The C/I received at the RF is closely related to the S/N at the baseband which is related to the voice quality. In digital systems, all three, FDMA, TDMA, and CDMA can be applied. The C/I received at the RF is closely related to the E_b/I_0 at the baseband.

$$C/I = (E_b/I_0)(R_b/B_c)$$
$$= (E_b/I_0)/(B_c/R_b) \qquad (29)$$

where E_b is the energy per bit; I_0 is the interference power per hertz, R_b is the bit per second, and B_c is the radio channel bandwidth in hertz. In digital FDMA or TDMA there are designated channels or time slots for calls. Thus R_b equals B_c and E_b/I_0 at the baseband is always greater than one, then C/I is also greater than one, i.e., a positive value in decibels. In CDMA, all the coded sequences say N, share one radio channel; thus B_c is much greater than R_b. The notation B_c is often replaced by B_{ss} which is the spread-spectrum channel. Within the radio channel, any one code sequence is interfered with N-1 of other code sequences. Therefore, the interference level is always higher than the signal level. C/I is less than one, i.e., a negative value in decibels.

B. Capacity of Cellular FDMA and TDMA [9]

In FDMA or TDMA, each frequency channel or each time slot is assigned to one call. During the call period, no other calls can share the same channel or slot. In this case, the cochannel interference would come from a distance of $D_s = qR$. Assume that the worst case of having six cochannel interferers (see Fig. 9) and the fourth power law pathloss are applied. The capacity of the cellular FDMA and cellular TDMA can be found by the radio capacity m expressed as

$$m = \frac{B_t/B_c}{K} = \frac{M}{\sqrt{\dfrac{2}{3}\left(\dfrac{C}{I}\right)_s}} \quad \text{number of channels/cell} \quad (30)$$

where

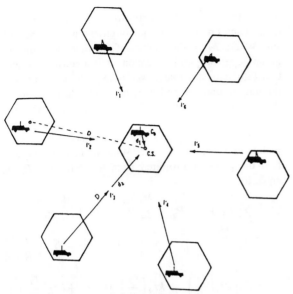

Fig. 9. Cochannel interference.

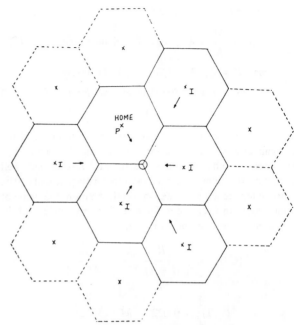

Fig. 10. CDMA system and its interference.

B_t total bandwidth (transmitted or received),

B_c channel bandwidth (transmitted or received) or equivalent channel bandwidth,

$M = B_t/B_c$ total number of channels or equivalent channels,

$(C/I)_s$ minimum required carrier-to-interference ratio per channel or per time slot.

Equation (30) can be directly applied to both analog FDMA and digital FDMA systems. In TDMA systems, B_c is an equivalent channel bandwidth. For example, a TDMA radio channel bandwidth of 30 kHz with three time slots can have an equivalent channel bandwidth of 10 kHz ($B_c = 10$ kHz). Therefore the minimum required $(C/I)_s$ of each time slot turns out to be the same as $(C/I)_s$ of the TDMA equivalent channel. The radio capacity is based on two parameters, B_c and $(C/I)_s$ as shown in (30). It has the same two parameters as appear in Shannon's channel capacity formula. The difference between (30) and Shannon's is that the two parameters are related in the former one and independent in the latter one. The $(C/I)_s$ of radio capacity can be found based on a standard voice quality as soon as the channel bandwidth B_c is given.

C. Radio Capacity of Cellular CDMA

Cellular CDMA is uniquely designed to work in cellular systems. The primary purpose of using this CDMA is for high capacity. In cellular CDMA, there are two CIRF values. One CIRF is called adjacent CIRF, $q_a = D_s/R = 2$. It means that the same radio channel can be reused in all neighboring cells. The other CIRF is called self-CIRF, $q_s = 1$. It means that different code sequences use the same radio channel to carry different traffic channels. The two CIRF's are shown in Fig. 10. With the smallest value of CIRF, the CDMA system is proven to be the most efficient frequency-reuse system we can find.

1) Required $(C/I)_s$ in Cellular CDMA: $(C/I)_s$ can be found from (29) depending on the value of E_b/I_0 which is

measured at the baseband determined by the voice quality. For example, the vocoder rate is $R_b = 8$ kb/s and the total wide-band channel bandwidth $B_t = 1.25$ MHz, then if E_b/I_0 is determined as follows:

$$E_b/I_0 = 7 \text{ dB, then } (C/I)_s = 0.032$$

$$E_b/I_0 = 4.5 \text{ dB, then } (C/I)_s = 0.01792.$$

The radio capacity of this system can be derived as follows. It can be calculated based on the forward link, and can also be further improved by the power control schemes.

2) Without Power Control Scheme: The radio capacity is calculated from the forward link C/I ratio. The $(C/I)_s$ received by a mobile unit at the boundary of a CDMA cell shown in Fig. 10 can be obtained based on nine interfering cells as follows:

$$(C/I)_s = \frac{\alpha \cdot R^{-4}}{\underbrace{\alpha(M-1)\cdot R^{-4}}_{\text{within the cell}} + \underbrace{\alpha \cdot 2M \cdot R^{-4}}_{\substack{\text{two closest} \\ \text{adjacent cells}}}}$$

$$+ \underbrace{\alpha \cdot 3M \cdot (2R)^{-4}}_{\substack{\text{three intermediate-} \\ \text{range cells}}} + \underbrace{\alpha \cdot 6M(2.633R)^{-4}}_{\text{six distant cells}}$$

$$= \frac{1}{3.3123M - 1} \tag{31}$$

where α is a constant factor, M is the number of traffic channels. $(C/I)_s$ can be determined based on E_b/I_0 and R_b/B_s as shown in (29). Then M can be found from (31)

$$(C/I)_s = 0.032 \quad M = 9.736$$

$$(C/I)_s = 0.01792 \quad M = 17.15.$$

The radio capacity defined in (30)

$$m = \frac{M}{K} \text{ number of traffic channels/cell.} \quad (32)$$

In this case $K = q_a^2/3 = 4/3 = 1.33$. Therefore,

$m = M/1.33 = 7.32$ traffic channels/cell for $E_b/I_0 = 7$ dB

$\quad = 12.9$ traffic channels/cell for $E_b/I_0 = 4.5$ dB.

3) With Power Control Scheme: We can increase the radio capacity by using a proper power control scheme. The power control scheme used at the forward link of each cell can reduce the interference to the other adjacent cells. The less the interference generated in a cell, the more the value of M increases. In (31), we notice that if we can neglect all the interference, then, as shown in Fig. 10

$$(C/I)_s = \frac{R^{-4}}{(M-1)R^{-4}} = \frac{1}{M-1} \quad (33)$$

for

$$(C/I)_s = 0.032 \quad M = 30.25$$
$$(C/I)_s = 0.01792 \quad M = 54.8.$$

Comparing (31) with (33), the total number of traffic channels M is drastically reduced due to the existence of interference. However, since interference is always existing in the adjacent cells, we can only reduce it by using a power control scheme. By using a power control scheme the total power after combining all traffic channels should be considered in two cases. a) The necessary power delivery to the close-in mobile unit and b) The total power reduced at the boundary.

a) The necessary power delivery to a close-in mobile unit. The transmitted power at the cell site for the jth mobile unit is P_j, which is proportional to r_j^n.

$$P_j \propto r_j^n \quad (34)$$

where r_j is the distance between the cell site and the jth mobile unit. n is a number. In examining the number n, we find that the power control scheme of using $n = 2$ in (34) can provide the optimum capacity and also meet the requirement that the forward link signal can still reach the near-end mobile unit at distance r_j from the cell site with a reduced power

$$P_j = P_R \left(\frac{r_j}{R}\right)^2 \quad (35)$$

where P_R is the power required to reach those mobile units at the cell boundary R. The M mobile units served by M traffic channels are assumed uniformly distributed in a cell. Then

$$p(M_l) = kr_l, \quad 0 \le r_l \le R \quad (36)$$

where $M = \sum_{l=1}^{L} M_l$. There are L groups of mobile units. Each one of L is equally circled around the cell site. Where M_e is the number of mobile units in the lth group depending on its location. k is a constant. Equation (36) indicates that fewer mobile units are closely circling around the cell site, more mobile units are at the outside ring of the cell site. Assume that the distance r_0 is from the cell site to a desired mobile unit, also assume that r_0 is a near-in distance between the mobile unit and the cell site. With the help of (34) and (35) the power transmitted from the cell site, P_t, is equal to

$$P_t = \sum^{M_1} P_1 + \sum^{M_2} P_2 + \sum^{M_3} P_3 + \cdots + \sum^{M_L} P_L$$
$$= P_R \left[\sum^{kr_1} \left(\frac{r_1}{R}\right)^2 + \sum^{kr_2} \left(\frac{r_2}{R}\right)^2 + \cdots + \sum^{kr_L} \left(\frac{r_L}{R}\right)^2 \right]$$
$$= P_R \left[kr_1 \left(\frac{r_1}{R}\right)^2 + kr_2 \left(\frac{r_2}{R}\right)^2 + \cdots + kr_L \left(\frac{r_L}{R}\right)^2 \right]. \quad (37)$$

Since r_L is the distance from the cell site to the cell boundary, $r_L = R$ then (37) becomes

$$P_t = P_R k \int_0^R \frac{r^3}{R^2} dr = P_R k \frac{R^2}{4}. \quad (38)$$

The total number of mobile units M can be obtained as

$$M = \sum_{l=1}^{L} M_l = k(r_1 + r_2 + \cdots + R)$$
$$= k \int_0^R r \, dr = k \frac{R^2}{2}. \quad (39)$$

Substituting (39) into (38):

$$P_t = P_R k \left[\frac{M}{2k}\right] = P_R \frac{M}{2}. \quad (40)$$

If the full power P_R is applied to every channel, then

$$P_t = MP_R. \quad (41)$$

Comparing (40) and (41), the total transmitted power reduces to one-half by using the power control scheme of (35). The $(C/I)_s$ of a mobile unit at a distance of r_0 which is close to the cell site is

$$(C/I)_{s1} = \frac{P_R(r_0/R)^2 \cdot r_0^{-4}}{P_R(M/2) \cdot r_0^{-4}} = \frac{(r_0/R)^2}{(M/2)}. \quad (42)$$

The interference from the adjacent cells can be neglected in (42) in this case.

b) The total power is reduced at the cell boundary. The $(C/I)_s$ of a mobile unit at a distance R which is at the cell boundary can be obtained similarly to (31).

$$(C/I)_{s2} = \frac{P_R}{P_R \left[\frac{M-1}{2} + 2\frac{M}{2} + 3\left(\frac{M}{2}\right) \cdot (2)^{-4} + 6\left(\frac{M}{2}\right)(2.633)^{-4} \right]} = \frac{1}{1.656M}. \quad (43)$$

The values of M and m can be found from (43) for the case of applying the power control scheme.

$$M = 18.87, \; m = 14.19, \; (C/I) = 0.032 \, (-15 \text{ dB})$$
$$M = 23.7, \; m = 28.33, \; (C/I) = 0.01792 \, (-17 \text{ dB}). \quad (44)$$

At this time, $(C/I)_s$ received by the mobile unit at the distance r_0 from (42) should be checked with (43) to see whether it is valid or not.

$$(C/I)_{s1} = \frac{(r_0/R)^2}{M/2} = \frac{3.3(r_0/R)^2}{3.3(M/2)} \geq \frac{1}{1.656M}. \quad (45)$$

In (44), the power reduction ratio $(r/R)^2$ has to be not less than 0.302 for those mobile units located less than the distance r_0 which is $0.55R$. If we set the lowest power to be $0.302P_R$ then the total power has to be changed.

$$P_t = P_R k \left[\frac{r_0^2}{R^2} r_1 + \frac{r_2^3}{R^2} + \frac{r_3^3}{R^2} + \cdots \right]$$
$$= P_R k \left[\left(\frac{r_0}{R}\right)^2 \int_0^{r_0} r \, dr + \int_{r_0}^{R} \frac{r^3}{R^2} \, dr \right]$$
$$= P_R k \frac{R^2}{4} \left[1 + \left(\frac{r_0}{R}\right)^4 \right]. \quad (46)$$

For $r_0/R = 0.55$, then $(r_0/R)^4 = 0.0913$.
The transmitted power P_t in (46) has to be adjusted as

$$P_t = P_R k (R^2/4) \times 1.0913 = P_R (M/2) \times 1.0913. \quad (47)$$

Equation (47) indicates that by setting the condition of the lowest power per traffic channel to be $0.302P_R$ at the cell site to serve the mobile units within and equal to the distance r_0, $r_0 = 0.55\,R$, the total power at the cell site is slightly increased by 1.0913 times as compared with (38). Under the adjusted transmitted power P_t as shown in (47), the actual values of M and m are reduced.

$$M = 18.87/1.0913 = 17.3,$$
$$m = 13 \quad \text{for } (C/I)_s = 0.032$$
$$M = 33.7/1.0913 = 30.9,$$
$$m = 25.96 \quad \text{for } (C/I)_s = 0.1792. \quad (48)$$

Comparing (48) with (44), we find no significant change of M and m when the adjusted transmitted power is applied.

D. Comparison of Different Cases in CDMA

Table I lists the performance of five different cases:

Case 1: No adjacent cell interference is considered (this is not a real case).
Case 2: No power control, adjacent cell interference is considered.
Case 3: Power control with $n = 1$, adjacent cell interference is considered.

Case 4: Power control with $n = 2$, adjacent cell interference is considered.
Case 5: Power control with $n = 3$, adjacent cell interference is considered.

In Table I, Case 1 is not a real case. In Case 2, without power control, the performance is poor. The power control schemes are used in Cases 3–5. In these cases, in order to provide the minimum transmitted power at the cell site for serving those mobile units within or equal to the distance of r_0, the total transmitted power at the cell site increases as indicated under the heading "after adjusting the transmitted power." Comparing the number of channels per cell m among Cases 3–5, we found that Case 4 has two channels more than Case 3 but one channel less than Case 5. However, Case 5 is harder to implement than Case 4. One channel gained in Case 5 over Case 4 can be washed out in the practical situation. When the power control schemes of $n > 3$ are used, no further improvement in radio capacity is found. Therefore, we conclude that $n = 2$ in Case 4 is a better choice.

VII. REDUCTION OF NEAR–FAR RATIO INTERFERENCE IN CDMA

In CDMA, all traffic channels are sharing one radio channel. Therefore a strong signal received from a near-in mobile unit will mask the weak signal from a far-end mobile unit at the cell site. To reduce this near-far ratio interference, a power control scheme should be applied on the reverse link. As a result, the signals received at the cell site from all the mobile units within a cell remain at the same level. The scheme is described as follows. The power transmitted from each mobile unit has to be adjusted based on its distance from the cell site, as

$$P_j = P_R \left(\frac{r_j}{R}\right)^4 \quad (49)$$

where P_R, r, and R are mentioned previously, and a fourth power rule is applied in (49). Neglecting the interfering signals from adjacent cell, the C/I received from a mobile unit J, at the cell site can be obtained as

$$C/I = \frac{P_R \left(\frac{r_J}{R}\right)^4 (r_J)^{-4}}{\sum_1^{M-1} P_R \left(\frac{r_j}{R}\right)^4 (r_j)^{-4}} = \frac{1}{M-1}. \quad (50)$$

The C/I of (48) has to be greater than or equal to the required $(C/I)_s$,

$$C/I \geq (C/I)_s. \quad (51)$$

Applying (51) in (50), we obtain

$$M = 30.25, \; m = 22.74, \quad \text{for } (C/I)_s = 0.032 \, (-15 \text{ dB})$$
$$M = 54.5, \quad m = 41.2, \quad \text{for } (C/I)_s = 0.01792 \, (-17 \text{ dB}).$$

The number of channels M obtained from the reverse link is much higher than that from the forward channel as shown in Table I. It indicates that the effort is to increase the number of channels on the forward link for more radio capacity.

TABLE I

| Performance in Different Cases | Adjacent Cell Interfering | | | | No Adjacent Cell Interference is Considered |
| | No Power Control | Power Control Schemes | | | |
	Case 2 $N = 0$	Case 3 $N = 1$	Case 4 $N = 2$	Case 5 $N = 3$	Case 1
Power Control due to the distance from the cell site	P_R	$P_R(r_j/R)$	$P_R(r_j/R)^2$	$P_R(r_j/R)^3$	P_R
R_0	N/A	$0.303R$	$0.55R$	$0.7R$	N/A
Before adjusting the TX power					
Total Transmitted Power at the Cell Site	MP_R	$P_R(2M/3)$	$P_R(M/2)$	$P_R(2M/5)$	MP_R
The $(C/I)_s$ Received at R_0	$\dfrac{1}{M-1}$	$(r_0/R)/(2M/3)$	$(r_0/R)^2/(M/2)$	$(r_0/R)3/(2M/5)$	$\dfrac{1}{M-1}$
At R (Cell Boundary)	$\dfrac{1}{3.3123M-1}$	$\dfrac{1}{2.2M}$	$\dfrac{1}{1.656M}$	$\dfrac{1}{1.32M}$	$\dfrac{1}{M-1}$
M at $(C/I)_s = 0.032$	9.736	14.2	18.87	23.67	30.25
$(C/I)_s = 0.0179$	17.15	25.36	33.7	42.27	54.8
M at $(C/I)_s = 0.032$	7.32	10.67	14.19	17.8	22.74
$= 0.0179$	12.9	19	28.33	31.78	41.2
After adjusting the TX power					
Total Transmitted Power at the Cell Site		$P_R(2M/3) \times 1.0139$	$P_R(M/2) \times 1.09$	$P_R(2M/5) \times 1.25$	
The $(C/I)_s$ Received at $R \le R_0$		$(r_0/R)/[(2M/3) \times 1.0139]$	$(r_0/R)^2/[(M/2) \times 1.09]$	$(r_0/R)^3/[2M/5 \times 1.25]$	
at $R > R_0$		$(r/R)/[(2M/3) \times 1.0139]$	$(r/R)^2/[(M/2) \times 1.09]$	$(r/R)^3/[(2M/5) \times 1.25]$	
at R		$\dfrac{1}{2.23M}$	$\dfrac{1}{1.8M}$	$\dfrac{1}{1.65M}$	
M at $(C/I)_s = 0.032$ (-15 dB)		14.2	17.3	19	
at $(C/I)_s = 0.010792$ (-17.4 dB)		25.36	31	33.8	
m at $(C/I)_s = -15$ dB		10.67	13	14	
at $(C/I)_s = -17.4$ dB		19	23.3	25.4	

VIII. NATURAL ATTRIBUTES OF CDMA [10]

There are many attributes of CDMA which are of great benefit to the cellular system.

1) Voice activity cycles: The real advantage of CDMA is the nature of human conversation. The human voice activity cycle is 35%. The rest of the time we are listening. In CDMA all the users are sharing one radio channel. When users assigned to the channel are not talking, all others on the channel benefit with less interference in a single CDMA radio channel. Thus the voice activity cycle reduces mutual interference by 65%, increasing the true channel capacity by three times. CDMA is the only technology that takes advantage of this phenomenon. Therefore, the radio capacity shown in (48) can be three times higher due to the voice activity cycle. It means that the radio capacity is about 40 channels per cell for C/I = -15 dB or $E_b/I_0 = 7$ dB.

2) No equalizer needed: When the transmission rate is much higher than 10 kb/s in both FDMA and TDMA, an equalizer is needed for reducing the intersymbol interference caused by time delay spread. However, in CDMA, only a correlator is needed instead of an equalizer at the receiver to despread the SS signal. The correlator is simpler than the equalizer.

3) One radio per site: Only one radio is needed at each site or at each sector. It saves equipment space and is easy to install.

4) No hard handoff: Since every cell uses the same CDMA radio, the only difference is the code sequences. Therefore, no handoff from one frequency to another frequency while moving from one cell to another cell. It is called a soft handoff.

5) No guard time in CDMA: The guard time is required in TDMA between time slots. The guard time does occupy the time period for certain bits. Those waste bits could be used to improve quality performance in TDMA. In CDMA, the guard time does not exist.

6) Sectorization for capacity: In FDMA and TDMA, the utilization of sectorization in each cell is for reducing the interference. The trunking efficiency of dividing channels in each sector also decreases. In CDMA, the sectorization is used to increase capacity by introducing three radios in three sectors and therefore, three times the capacity is obtained as compared with one radio in a cell in theory.

7) Less fading: Less fading is observed in the wide-band

signal while propagating in a mobile radio environment. More advantage of using a wide-band signal in urban areas than in suburban areas for fading reduction as described in Section II-B.

8) Easy transition: In a situation where two systems, analog and CDMA, have to share the same allocated spectrum, 10% of the bandwidth (1.25 MHz) will increase two times ($= 0.1 \times 20$) of the full bandwidth of FM radio capacity as shown below. Since only 5% (heavy users) of the total users take more than 30% of the total traffic, the system providers can let the heavy users exchange their analog units for the dual mode (analog/CDMA) units and convert 30% of capacity to CDMA on the first day of CDMA operations.

9) Capacity advantage: Given that

$B_t = 1.25$ MHz, the total bandwidth

$B_{ss} = 1.25$ MHz the CDMA radio channel

$B_c = 30$ kHz for FM

$B_c = 30$ kHz and three time slots for TDMA.

Capacity of FM 1.25/30 = 41.6

total numbers of channels $= \dfrac{1.25 \times 10^6}{30 \times 10^3}$

$= 41.67$ channels

the cell reuse pattern $K = 7$

the radio capacity $m_{FM} = \dfrac{41.67}{7} = 6$ channels/cell.

Capacity of TDMA

total number of channels $\dfrac{1.25 \times 10^6}{10 \times 10^3} = 125$ channels

the cell reuse pattern $K = 4$ (assumed)

the radio capacity $m_{TDMA} = \dfrac{125}{4} = 31.25$ channels/cell.

Capacity of CDMA

total number of channels / cell, $m = 13$

the cell reuse pattern $K = 1.33$

the radio capacity, take (48) at $E_b/I = 7$ dB

add voice activity cycle and sectorization,

$m_{CDMA} = 13 \times 3 \times 3 \approx 120$ channels/cell.

Therefore,

$$m_{CDMA} = 20 \times m_{FM}$$

$$= 4 \times m_{TDMA}.$$

10) No frequency management or assignment needed: In FDMA and TDMA, the frequency management is always a critical task to carry out. Since there is only one common radio channel in CDMA, no frequency management is needed. Also, the dynamic frequency would implement in TDMA and FDMA to reduce real-time interference, but needs a linear broad-band power amplifier which is hard to develop. CDMA does not need the dynamic frequency assignment.

11) Soft capacity: In CDMA, all the traffic channels share one CDMA radio channel. Therefore, we can add one

additional user so the voice quality is just slightly degraded as compared to that of the normal 40-channel cell. The difference in decibels is only $10 \log \dfrac{41}{40}$ which is 0.24 dB down in C/I ratio.

12) Coexistence: Both systems, analog and CDMA, can operate in two different spectras, and CDMA only needs 10 % of bandwidth to general 200 % of capacity. No interference would be considered between two systems.

13) For microcell and in-building systems: CDMA is a natural waveform suitable for microcell and in-building because of being susceptible to the noise and the interference.

IX. CONCLUSION

The overview of CDMA highlights the potential of increasing capacity in future cellular communications. This paper describes the mobile radio environment and its impact on narrow-band and wide-band propagation. The advantage of having CDMA in cellular systems is depicted. The concept of radio capacity in cellular is also introduced. The power control schemes in CDMA have been carefully analyzed. The natural attributes of CDMA provide the reader with the reasons that cellular is considering using it. This paper leads the reader to understand two CDMA papers [11], [12], which are analyzed in more depth in this issue, and to build interest in CDMA by reading other references [13]–[19].

REFERENCES

[1] W. C. Y. Lee, *Mobile Cellular Telecommunication System*. New York: McGraw-Hill, 1989, ch. 4.

[2] ——, *Mobile Communications Engineering*. New York: McGraw-Hill, 1982, pp. 340–399.

[3] J. G. Proakis, "Adaptive equalization for a TDMA digital mobile radio," *IEEE Trans. Veh. Technol.*, pp. 333–341, this issue.

[4] S. N. Crozier, D. D. Falconer, and S. Mahmond, "Short-block equalization techniques employing channel estimation for fading time-dispersive channels," in *Proc. IEEE Veh. Technol. Conf.*, San Francisco, CA, 1989, pp. 142–146.

[5] P. Monsen, "Theoretical and measured performance of a DEF modem on a fading multipath channel," *IEEE Trans. Commun.*, vol. COM-25, pp. 1144–1153, Oct. 1977.

[6] W. C. Y. Lee, *Mobile Communications Design Fundamentals*, New York: Howard W. Sams, 1986, p. 274.

[7] M. Schwartz, W. R. Bennett, and S. Stein, *Communications Systems and Techniques*, New York: McGraw-Hill, 1966, p. 561.

[8] B. Sklar, *Digital Communications, Fundamentals and Applications*. Englewood Cliffs, NJ: Prentice-Hall, 1988, p. 546.

[9] W. C. Y. Lee, "Spectrum efficiency in cellular," *IEEE Trans. Veh. Technol.*, vol. 38, pp. 69–75, May 1989.

[10] PacTel Cellular and Qualcomm, "CDMA cellular—The next generation," a pamphlet distributed at CDMA demonstration, Qualcomm, San Diego, CA, Oct. 20–Nov. 7, 1989.

[11] K. S. Gilhousen, I. M. Jacobs, R. Padovani, A. J. Viterbi, L. A. Weaver, and C. E. Wheatley, "On the capacity of a cellular CDMA system," *IEEE Trans. Veh. Technol.*, pp. 303–312, this issue.

[12] R. L. Pickholtz, L. B. Milstein, and D. L. Schilling, "Spread spectrum for mobile communications," *IEEE Trans. Veh. Technol.*, pp. 313–322, this issue.

[13] A. J. Viterbi, "When not to spread spectrum—A sequel," *IEEE Communications Mag.*, vol. 23, pp. 12–17, Apr. 1985.

[14] L. B. Milstein, R. L. Pickholtz, and D. L. Schilling, "Optimization of the processing gain of an FSK–FH system," *IEEE Trans. Commun.*, vol. COM-28, pp. 1062–1079, July 1980.

[15] G. K. Huth, "Optimization of coded spread spectrum system performance," *IEEE Trans. Commun.*, vol. COM-25, pp. 763–770, Aug. 1977.

[16] M. K. Simon, J. K. Omura, R. A. Scholtz, and B. K. Levitt, *Spread Spectrum Communications*, vol. 2. Rockville, MD: Computer Science Press, 1985.

[17] R. L. Pickholtz, D. L. Schilling, and L. B. Milstein, "Theory of spread-spectrum communications—A tutorial," *IEEE Trans. Commun.*, vol. COM-30, pp. 855–884, May 1982.

[18] R. A. Scholtz, "The origins of spread spectrum communications," *IEEE Trans. Commun.*, vol. COM-30, pp. 882–854, May 1982.

[19] A. J. Viterbi, "Spread spectrum communications—Myths and realities," *IEEE Communications Mag.*, pp. 11–18, May 1979.

William C. Y. Lee (M'64–SM'80–F'82) received the B.Sc. degree from the Chinese Naval Academy, Taiwan, and the M.S. and Ph.D. degrees from The Ohio State University, Columbus, in 1954, 1960, and 1963, respectively.

From 1959 to 1963 he was a Research Assistant at the Electroscience Laboratory, The Ohio State University. He was with AT&T Bell Laboratories from 1964 to 1979 where he was concerned with the study of wave propagation and systems, millimeter and optical waves propagation, switching systems, and satellite communications. He developed a UHF propagation model for use in planning the Bell System's new Advanced Mobile Phone Service and was a pioneer in mobile radio communication studies. He applied the field component diversity scheme over mobile radio communication links. While working in satellite communications, he discovered a method of calculating the rain rate statistics which would affect the signal attenuation at 10 GHz and above. He successfully designed a 4 × 4 element printed circuit antenna for tryout use. He studied and set a 3-mm wave link between the Empire State Building and Pan American Building in New York City, experimentally using the newly developed IMPATT diode. He also studied the scanning spot beam concept for satellite communication using the adaptive array scheme. From April 1979 until April 1985 he worked for ITT Defense Communications Division and was involved with advanced programs for wiring military communications system. He developed several simulation programs for the multipath fading medium and applied them to ground mobile communication systems. In 1982 he was Manager of the Advanced Development Department, responsible for the pursuit of new technologies for future communication systems. He developed an artificial intelligence application in the networking area at ITT and a patent based on his work was issued in March 1991. He joined PacTel Mobile Companies in 1985, where he is engaged in the improvement of system performance and capacity. He is currently the Vice President of Research and Technology at PacTel Cellular, Irvine, CA. He has written more than 100 technical papers and three textbooks, all in the mobile radio communications area. He was the founder and the first co-chairman of the Cellular Telecommunications Industry Association Subcommittee for Advanced Radio Techniques involving digital cellular standards. He has received the "Distinguished Alumni Award" from The Ohio State University, and the "Avant Garde" award from the IEEE Vehicular Technology Society in 1990.